MATRIX METHODS:
Theory, Algorithms and Applications

Dedicated to the Memory of ***Gene Golub***

MATRIX METHODS:
Theory, Algorithms and Applications

Dedicated to the Memory of ***Gene Golub***

Editors

Vadim Olshevsky
University of Connecticut, USA

Eugene Tyrtyshnikov
Russian Academy of Sciences, Russia

World Scientific

NEW JERSEY • LONDON • SINGAPORE • BEIJING • SHANGHAI • HONG KONG • TAIPEI • CHENNAI

Published by

World Scientific Publishing Co. Pte. Ltd.
5 Toh Tuck Link, Singapore 596224
USA office: 27 Warren Street, Suite 401-402, Hackensack, NJ 07601
UK office: 57 Shelton Street, Covent Garden, London WC2H 9HE

British Library Cataloguing-in-Publication Data
A catalogue record for this book is available from the British Library.

MATRIX METHODS: THEORY, ALGORITHMS AND APPLICATIONS
Dedicated to the Memory of Gene Golub

ISBN-13 978-981-283-601-4
ISBN-10 981-283-601-2

Printed in Singapore by World Scientific Printers.

To the memory of Gene Golub

PREFACE

This book is unique in covering the whole of a triptych consisting of algebraic theory, algorithmic problems, and numerical applications, all united by the essential use of matrix methods. This was the spirit of the 2nd International Conference on Matrix Methods and Operator Equations (23–27 July 2007, Moscow) hosted by the Institute of Numerical Mathematics of Russian Academy of Sciences and organized by Dario Bini, Gene Golub, Alexander Guterman, Vadim Olshevsky, Stefano Serra-Capizzano, Gilbert Strang, and Eugene Tyrtyshnikov.

Matrix methods provide the key to many problems in pure and applied mathematics. However, it often happens that linear algebra theory, numerical algorithms, and matrices in FEM/BEM applications live as if in three separate worlds. In this book, perhaps for the first time at all, they are put together as one entity, as it was in the Moscow meeting, where the algebraic part was represented by Hans Schneider, algorithms by Gene Golub, and applications by Guri Marchuk.

Among the authors of this book are several top-class experts in numerical mathematics, matrix analysis, and linear algebra applications including Dario Bini, Walter Gander, Alexander Guterman, Wolfgang Hackbusch, Khakim Ikramov, Valery Il'in, Igor Kaporin, Boris Khoromskij, Victor Pan, Stefano Serra-Capizzano, Reinhold Schneider, Vladimir Sergeichuk, Harald Wimmer and others.

The book assumes a good knowledge of basic linear algebra and general mathematical background. Besides professionals, it is intended as well for a wider audience, in academia and industry, of all those who consider using matrix methods in their work.

We are pleased to acknowledge that Alexander Guterman engaged in thorough editing "Algebra and Matrices" papers, Maxim Olshanskii and Yuri Vassilevski invested their time and expertise to the "Matrices and Applications" part, and Sergei Goreinov committed himself to producing a camera-ready manuscript.

Support for the Moscow meeting which created the foundation for this book, by the Russian Foundation for Basic Research, the Russian Academy of Sciences, International Foundation for Technology and Investments, Neurok Techsoft, and the University of Insubria (Como, Italy), is gratefully acknowledged.

The soul of the meeting was Gene Golub who rendered a charming "Golub's dimension" to the three main axes of the conference topics. This book is dedicated in gratitude to his memory.

Vadim Olshevsky

Eugene Tyrtyshnikov

Contents

Matrices and Algorithms

Matrices and Applications

ALGEBRA AND MATRICES

Operators Preserving Primitivity for Matrix Pairs

LeRoy B. Beasley[1] and Alexander E. Guterman[2,*]

[1] Department of Mathematics and Statistics, Utah State University, Logan, Utah 84322-4125, USA
lbeasley@math.usu.edu

[2] Department of Algebra, Faculty of Mathematics and Mechanics, Moscow State University, Moscow, 119991, GSP-1, Russia
guterman@list.ru

1 Introduction

A nonnegative matrix is called *primitive* if some power of it has only positive entries, or, equivalently, it is irreducible and its spectral radius is the only eigenvalue of maximal modulus, or, equivalently, the greatest common divisor of lengths of all circuits in the associate directed graph is equal to 1.

An alternative definition of primitivity arises in the asymptotic analysis of the homogeneous discrete time positive systems of the form

$$x(t+1) = Ax(t), \qquad t = 0, 1, \dots , \tag{1}$$

here a non-negative vector $x(0)$ represents the initial state. In this context the primitivity of A can be equivalently restated as the property that any positive initial condition $x(0)$ produces a state evolution which becomes strictly positive within a finite number of steps.

Such systems are described by the following equation, see [11],

$$x(h+1, k+1) = Ax(h, k+1) + Bx(h+1, k), \quad h, k \in \mathbb{Z}, \ \ h+k \geqslant 0, \tag{2}$$

where A and B are $n \times n$ nonnegative matrices and initial conditions $x(h, -h)$, $h \in \mathbb{Z}$, are nonnegative $n \times 1$ vectors. Positive discrete homogeneous 2D-dynamical systems are used to model diffusion processes, water pollution, etc., see [6, 7]. An entry of the vector $x(h, k)$ typically represents a quantity, such as pressure, concentration or density at a particular site along a stream. It can be seen that at each time-step the conditions of a site are determined by its previous conditions and the conditions of the site directly upstream from it, see [7, 11]. To investigate the systems of type 2, we need the following concept:

Definition 1. *Let* $A, B \in \mathcal{M}_n(\mathbb{Z})$*, and* h, k *be some non-negative integers. The* (h, k)-Hurwitz product*, which is denoted by* $(A, B)^{(h,k)}$*, is the sum of all matrices which are products of* h *copies of* A *and* k *copies of* B.

* The second author wishes to thank the grants RFBR 05-01-01048, NSh-5666.2006.1 and MK-2718.2007.1 for partial financial support.

Example 1. $(A,B)^{(1,0)} = A$ and

$$(A,B)^{(2,2)} = A^2B^2 + ABAB + AB^2A + BA^2B + BABA + B^2A^2.$$

In general the Hurwitz product satisfies the following recurrence relations:

$$\begin{aligned}(A,B)^{(h,0)} &= A^h,\\ (A,B)^{(0,k)} &= B^k,\\ (A,B)^{(h,k)} &= A(A,B)^{(h-1,k)} + B(A,B)^{(h,k-1)} \text{ for } h,k \geqslant 1.\end{aligned}$$

It can be directly checked, see [11], that the solution of (2) can be represented in the following way:

$$\begin{aligned}x(h,k) &= \textstyle\sum_{s=0}^{h+k}(A,B)^{(s,h+k-s)}x(h-s,s-h)\\ &= \textstyle\sum_{s=0}^{h+k}(A,B)^{(h+k-s,s)}x(s-k,k-s).\end{aligned}$$

Thus the Hurwitz products $(A,B)^{(h,k)}$ with $h+k=t$ and the initial condition determines the condition after t time-steps.

It is natural to ask for necessary and sufficient conditions on the matrix pair (A,B) in order that the solutions of (2) are eventually (i.e., for all (h,k) with $h+k$ sufficiently large) strictly positive for each appropriate sequence of initial values. As for the system (1), where the analogous question is answered in terms of primitivity, in this case primitivity for matrix pairs is needed which means the existence of integers h, k, $h+k > 0$, such that the Hurwitz product $(A,B)^{(h,k)}$ is a positive matrix.

Definition 2. *The* exponent *of the primitive pair* (A,B) *is the minimum value of* $h+k$ *taken over all pairs* (h,k) *such that* $(A,B)^{(h,k)}$ *is positive.*

An important issue in dealing with primitive matrices or matrix pairs is to find the complete list of matrix operators which map primitive matrices to primitive matrices or primitive matrix pairs to primitive matrix pairs. If such transformations exist then they allow us to simplify the system without losing its main property, namely, the primitivity. In this paper we deal with such transformations.

Following Frobenius, Schur and Dieudonné, many authors have studied the problems of determining the maps on the $n \times n$ matrix algebra $\mathcal{M}_n(\mathcal{F})$ over a field $\mathcal{F}$ that leave certain matrix relations, subsets, or properties invariant. For a survey of problems and results of this type see [9, 10].

The notion of primitivity is related to nonnegative matrices, i.e., matrices with the entries in the semiring of nonnegative real numbers. In the last decades much attention has been paid to Preserver Problems for matrices over various semirings, where completely different techniques are necessary to obtain classification of operators with certain preserving properties, see [10, Section 9.1] and references therein for more details. The notion of a semiring can be introduced as follows

Definition 3. *A* semiring $\mathcal{S}$ *consists of a set* $\mathcal{S}$ *and two binary operations, addition and multiplication, such that:*

- $\mathcal{S}$ *is an Abelian monoid under addition (identity denoted by 0);*
- $\mathcal{S}$ *is a semigroup under multiplication (identity, if any, denoted by 1);*
- *multiplication is distributive over addition on both sides;*
- $s0 = 0s = 0$ *for all* $s \in \mathcal{S}$.

In this paper we will always assume that there is a multiplicative identity 1 in $\mathcal{S}$ *which is different from 0.*

We need the following special class of semirings:

Definition 4. *A semiring is called* antinegative *if the zero element is the only element with an additive inverse.*

Standard examples of semirings, which are not rings, are antinegative, these include non-negative reals and integers, max-algebras, Boolean algebras.

Definition 5. *A* binary Boolean *semiring,* $\mathbb{B}$, *is a set* $\{0, 1\}$ *with the operations:*

$$\begin{array}{cc} 0+0=0 & 0\cdot 0=0 \\ 0+1=1+0=1 & 0\cdot 1=1\cdot 0=0 \\ 1+1=1 & 1\cdot 1=1. \end{array}$$

We will not use the term "binary" in the sequel.

Linear operators on certain antinegative semirings without zero divisors that strongly preserve primitivity were characterized by L. B. Beasley and N. J. Pullman in [3, 4].

Let us note that linear transformations $T : \mathcal{M}(\mathcal{S}) \to \mathcal{M}(\mathcal{S})$, preserving primitive matrix pairs, obviously preserve primitivity, so are classified in [3, 4]. To see this it is sufficient to consider primitive matrix pairs of the form $(A, 0)$. Thus their images are primitive matrix pairs of the form $(T(A), 0)$. Hence, $T(A)$ is primitive. However, if we consider operators on $\mathcal{M}^2(\mathbb{B}) = \mathcal{M}(\mathcal{S}) \times \mathcal{M}(\mathcal{S})$, then there is no easy way to reduce the problem of characterization of operators preserving primitive matrix pairs to the problem of characterization of certain transformations in each component.

In this paper we investigate the structure of surjective additive transformations on the Cartesian product $\mathcal{M}^2(\mathcal{S})$ preserving primitive matrix pairs. It turns out that for the characterization of these transformations we have to apply different and more involved techniques and ideas, such as primitive assignments, cycle matrices, etc. Our paper is organized as follows: in Section 2 we collect some basic facts, definitions and notations, in Section 3 we characterize surjective additive transformations $T : \mathcal{M}^2(\mathbb{B}) \to \mathcal{M}^2(\mathbb{B})$ preserving the set of primitive matrix pairs, in Section 4 we extend this result to matrices over arbitrary antinegative semiring without zero divisors.

Here $\mathcal{M}_{m,n}(\mathbb{B})$ denotes the set of $m \times n$ matrices with entries from the Boolean semiring $\mathbb{B}$.

2 Preliminaries

In this paper, unless otherwise is stated, $\mathcal{S}$ will denote any antinegative semiring without zero divisors and $\mathcal{M}_n(\mathcal{S})$ will denote the $n \times n$ matrices with entries from $\mathcal{S}$. Further, we denote by $\mathcal{M}_n^2(\mathcal{S})$ the Cartesian product of $\mathcal{M}_n(\mathcal{S})$ with itself, $\mathcal{M}_n(\mathcal{S}) \times \mathcal{M}_n(\mathcal{S})$.

The notions of primitivity and exponent for square matrices are classical.

Definition 6. *A matrix $A \in \mathcal{M}_n(\mathcal{S})$ is* primitive *if there is an integer $k > 0$ such that all entries of A^k are non-zero. In the case A is primitive, the* exponent *of A is the smallest such k.*

A classical example of primitive matrices is a so-called Wieland matrix.

Definition 7. *A Wieland matrix is*

$$W_n = \begin{bmatrix} 0 & 1 & & \\ & \ddots & \ddots & \\ 1 & & \ddots & 1 \\ 1 & & & 0 \end{bmatrix} \in \mathcal{M}_n(\mathcal{S}).$$

Also we consider the following primitive matrix

$$W_n' = \begin{bmatrix} 1 & 1 & & & \\ & 0 & \ddots & & \\ & & \ddots & \ddots & \\ 0 & & & \ddots & 1 \\ 1 & & & & 0 \end{bmatrix}.$$

These matrices are primitive and the Wieland matrix W_n is the matrix with the maximal possible exponent, see [8, Chapter 8.5].

Definition 8. *An operator $T : \mathcal{M}_{m,n}(\mathcal{S}) \to \mathcal{M}_{m,n}(\mathcal{S})$ is called* linear *if it is additive and $T(\alpha X) = \alpha T(X)$ for all $X \in \mathcal{M}_{m,n}(\mathcal{S})$, $\alpha \in \mathcal{S}$.*

Definition 9. *We say that an operator, $T : \mathcal{M}_n(\mathcal{S}) \to \mathcal{M}_n(\mathcal{S})$,* preserves (strongly preserves) *primitivity if for a primitive matrix A the matrix $T(A)$ is also primitive (A is primitive if and only if $T(A)$ is primitive).*

Definition 10. *A pair $(A, B) \in \mathcal{M}_n^2(\mathcal{S})$ is called* primitive *if there exist non-negative integers h, k such that the matrix $(A, B)^{(h,k)}$ is positive. In this case, we say that the* exponent *of (A, B) is (h, k) where $h+k$ is the smallest integer such that $(A, B)^{(h,k)}$ is positive, and if there is (a, b) such that $a + b = h + k$ and $(A, B)^{(a,b)}$ is positive then $h \geqslant a$.*

Example 2. The notion of primitive pairs generalizes the notion of primitivity. Indeed, pairs (A, B) with $k = 0$ and pairs (A, O) are primitive if and only if A is primitive. In particular, for any primitive matrix $A \in \mathcal{M}_n(\mathcal{S})$ the matrix pairs (A, O), (O, A), (A, A) are also primitive. For example, (W_n, O) and (O, W_n) are primitive. We note that there are primitive pairs (A, B) such that neither A nor B is primitive, for example $(A := E_{n1}, B := \begin{bmatrix} 1 & 1 & \dots & 1 \\ 0 & 1 & \dots & 1 \\ \vdots & \ddots & \ddots & \vdots \\ 0 & \dots & 0 & 1 \end{bmatrix})$.

We will use the notion of irreducible matrices and below we present the following two equivalent definitions of irreducibility, see [5] for details:

Definition 11. *A matrix* $A \in \mathcal{M}_n(\mathcal{S})$ *is called* irreducible *if* $n = 1$ *or the sum of the first* n *powers of* A *has no zero entries.* A *is* reducible *if it is not irreducible.*

Equivalently, A matrix A *is* reducible *if there is a permutation matrix* P *such that* $P^tAP = \begin{bmatrix} A_1 & O_{s,n-s} \\ A_2 & A_3 \end{bmatrix}$. *If* A *is not reducible it is* irreducible.

Definition 12. *An operator,* $T : \mathcal{M}_n^2(\mathcal{S}) \to \mathcal{M}_n^2(\mathcal{S})$, preserves *primitive pairs if for any primitive pair* (A_1, A_2) *we have that* $T(A_1, A_2)$ *is also primitive.*

In order to describe the final form of our operators we need the following notions.

Definition 13. *The matrix* $X \circ Y$ *denotes the* Hadamard *or* Schur product, *i.e., the* (i, j) *entry of* $X \circ Y$ *is* $x_{i,j}y_{i,j}$.

Definition 14. *An operator* $T : \mathcal{M}_{m,n}(\mathcal{S}) \to \mathcal{M}_{m,n}(\mathcal{S})$ *is called a* (U, V)-operator *if there exist invertible matrices* U *and* V *of appropriate orders such that* $T(X) = UXV$ *for all* $X \in \mathcal{M}_{m,n}(\mathcal{S})$, *or, if* $m = n$, $T(X) = UX^tV$ *for all* $X \in \mathcal{M}_{m,n}(\mathcal{S})$, *where* X^t *denotes the transpose of* X.

Definition 15. *An operator* T *is called a* (P, Q, B)-operator *if there exist permutation matrices* P *and* Q, *and a matrix* B *with no zero entries, such that* $T(X) = P(X \circ B)Q$ *for all* $X \in \mathcal{M}_{m,n}(\mathcal{S})$, *or, if* $m = n$, $T(X) = P(X \circ B)^tQ$ *for all* $X \in \mathcal{M}_{m,n}(\mathcal{F})$. *A* (P, Q, B)*-operator is called a* (P, Q)-operator *if* $B = J$, *the matrix of all ones.*

Definition 16. *A* line *of a matrix* A *is a row or a column of* A.

Definition 17. *We say that the matrix* A dominates *the matrix* B *if and only if* $b_{i,j} \neq 0$ *implies that* $a_{i,j} \neq 0$, *and we write* $A \geqslant B$ *or* $B \leqslant A$.

The matrix I_n is the $n \times n$ identity matrix, $J_{m,n}$ is the $m \times n$ matrix of all ones, $O_{m,n}$ is the $m \times n$ zero matrix. We omit the subscripts when the order is obvious from the context and we write I, J, and O, respectively. The matrix $E_{i,j}$, called a *cell*, denotes the matrix with exactly one nonzero entry, that being a one in the (i,j) entry. Let R_i denote the matrix whose i^{th} row is all ones and is zero elsewhere, and C_j denote the matrix whose j^{th} column is all ones and is zero elsewhere. We let $|A|$ denote the number of nonzero entries in the matrix A. We denote by $A[i,j|k,l]$ the 2×2-submatrix of A which lies on the intersection of the ith and j^{th} rows with the kth and l^{th} columns. A monomial matrix is a matrix which has exactly one non-zero entry in each row and each column.

3 Matrices over the binary Boolean semiring

The following lemma allows us to construct non-primitive matrix pairs:

Lemma 1. *Let $\mathcal{S}$ be an antinegative semiring without zero divisors, $(A,B) \in \mathcal{M}_n^2(\mathcal{S})$, and assume that at least one of the following two conditions is satisfied:*

1) $|A| + |B| < n+1$,

2) A and B together contain at most $n-1$ off-diagonal cells.

Then the pair (A,B) is not primitive.

Proof. 1. Let K be an irreducible matrix. We write $K = D + P$, where D is a certain diagonal matrix and P is a matrix with zero diagonal. Let $P_{i,j}$ denote the permutation matrix which corresponds to the transposition (i,j), i.e., $P_{i,j} = I - E_{i,i} - E_{j,j} + E_{i,j} + E_{j,i}$. If K has a row or column with no nonzero off-diagonal entry, say the ith row, then $P_{1,i}AP_{1,i} = \begin{bmatrix} \alpha & O_{1,n-1} \\ A_2 & A_3 \end{bmatrix}$ so that K is reducible. Thus, K must have a nonzero off diagonal entry in each row and each column. Hence $|P| \geqslant n$. Further, if K is irreducible and $|P| = n$ then P is a monomial matrix.

2. Note that the expansion of $(A+B)^{(h+k)}$ contains all the terms found in the (h,k)-Hurwitz product of (A,B). So, if (A,B) is a primitive pair in $\mathcal{M}_n^2(\mathcal{S})$ with exponent (h,k) then due to antinegativity of $\mathcal{S}$ we have that $(A+B)^{(h+k)}$ has all nonnegative entries, that is $A+B$ is primitive.

3. Assume to the contrary that (A,B) is a primitive pair. Then by Item 2 the matrix $A+B$ is primitive. Thus $A+B$ is irreducible. Hence by Item 1 the matrix $A+B$ has at least n nonzero off diagonal entries, and if $A+B$ has exactly n nonzero off diagonal entries then $(A+B) \circ (J \setminus I)$ is a monomial matrix. Since any power of a monomial matrix is a monomial matrix, we must have that $A+B$ has a nonzero diagonal entry. Since $|A| + |B| \geqslant |A+B|$ we have that $|A| + |B| \geqslant n+1$ and together, A and B have at least n nonzero off diagonal entries. This concludes the proof. □

Definition 18. *A graph is a full-cycle graph if it is a vertex permutation of the cycle*

$$1 \to 2 \to \cdots \to (n-1) \to n \to 1.$$

A $(0,1)$ full-cycle matrix *is the adjacency matrix of a full-cycle graph. If a matrix* A *with exactly* n *nonzero entries dominates a full-cycle* $(0,1)$*-matrix, we also say that* A *is a* full-cycle matrix.

Corollary 1. *Any primitive matrix* $A \in \mathcal{M}_n(\mathbb{B})$ *with exactly* $n+1$ *non-zero cells one of which is a diagonal cell dominates a full-cycle matrix.*

Proof. It follows from the proof of Lemma 1, item 1, that A dominates a permutation matrix P. Assume that P is not a full-cycle matrix. Since $|P| = n$, it follows that the graph of P is disconnected. Thus the graph of A is disconnected. Hence, A is not primitive. A contradiction. □

Lemma 2. *Let* $T : \mathcal{M}_{m,n}(\mathbb{B}) \to \mathcal{M}_{m,n}(\mathbb{B})$ *be a surjective additive operator. Then* $T(O) = O$ *and, hence,* T *is a bijective linear operator.*

Proof. By additivity we have $T(A) = T(A+O) = T(A) + T(O)$ for any A. By the definition of addition in $\mathbb{B}$ it follows that $T(O) \leqslant T(A)$ for any A.

Since T is surjective, for any i, $1 \leqslant i \leqslant m$, j, $1 \leqslant j \leqslant n$, there exists $A_{i,j} \in \mathcal{M}_{m,n}(\mathbb{B})$ such that $T(A_{i,j}) = E_{i,j}$. Thus for all i,j we have that $T(O) \leqslant T(A_{i,j}) = E_{i,j}$, i.e., $T(O) = O$.

Let us check the linearity of T now. Let $\lambda \in \mathbb{B}$, $X \in \mathcal{M}_{m,n}(\mathbb{B})$.

If $\lambda = 1$ then $T(\lambda X) = T(X) = \lambda T(X)$.

If $\lambda = 0$ then $T(\lambda X) = T(O) = O = \lambda T(X)$.

The bijectivity of T follows from the fact that any surjective operator on a finite set is injective, and $\mathcal{M}_{m,n}(\mathbb{B})$ is finite. □

Definition 19. *For matrices* $A = [a_{i,j}], B = [b_{i,j}] \in \mathcal{M}_n(\mathbb{B})$ *we denote by* $[A|B] \in \mathcal{M}_{n,2n}(\mathbb{B})$ *the* concatenation *of matrices* A *and* B, *i.e., the matrix whose* ith *row is* $(a_{i,1}, \ldots, a_{i,n}, b_{i,1}, \ldots, b_{i,n})$ *for all* $i, i = 1, \ldots, n$.

Definition 20. *Let* $T : \mathcal{M}_n^2(\mathbb{B}) \to \mathcal{M}_n^2(\mathbb{B})$ *be a surjective additive operator. Define the operator* $T^* : \mathcal{M}_{n,2n} \to \mathcal{M}_{n,2n}$ *by* $T^*([A|B]) = [C|D]$ *if* $T(A,B) = (C,D)$.

Lemma 3. *Let* $T : \mathcal{M}_n^2(\mathbb{B}) \to \mathcal{M}_n^2(\mathbb{B})$ *be a surjective additive operator, then the operator* T^* *is surjective and additive.*

Proof. Follows from the bijection between $\mathbb{B}$-semimodules $\mathcal{M}_n^2(\mathbb{B})$ and $\mathcal{M}_{n,2n}(\mathbb{B})$. □

Definition 21. *Let* $\mathcal{D} = \{D | D$ *is a diagonal matrix in* $\mathcal{M}_n(\mathbb{B})\}$. *Define the set* $\mathcal{D}^2 = \mathcal{D} \times \mathcal{D} = \{(A,B) | A, B \in \mathcal{D}\}$.

Definition 22. *Let* $\sigma : \{1, 2, \cdots, n\} \to \{1, 2, \cdots, n\}$ *be a bijection (permutation). We define the permutation matrix* P_σ *corresponding to* σ *by the formula* $P_\sigma = \sum_{i=1}^{n} E_{i,\sigma(i)}$.

We note that in this case $P_\sigma^t E_{i,j} P_\sigma = E_{\sigma(i),\sigma(j)}$ for all $i, j \in \{1, 2, \cdots, n\}$.

In the next lemma we show how to complete pairs of cells to a matrix which is similar to either W_n or W'_n by a permutation similarity matrix.

Lemma 4. *For any two pairs of distinct indices* (i,j), (k,l) *such that* $(i,j) \neq (l,k)$ *and either* $i \neq j$ *or* $k \neq l$ *or both, there exist a permutation matrix* P *and* $n-1$ *cells* $F_1, \ldots, F_{n-1}$ *such that* $E_{i,j} + E_{k,l} + F_1 + \ldots + F_{n-1} = PW_nP^t$ *or* PW'_nP^t.

Proof. Let i, j, k, l be four distinct integers in $\{1, 2, \cdots, n\}$. There are five cases to consider:

1. $(i,i), (i,l)$. Let σ be any permutation such that $\sigma(i) = n$, and $\sigma(l) = 1$ and $F_q = E_{\sigma^{-1}(q),\sigma^{-1}(q+1)}$, $q = 1, \cdots, n-1$. Then $P_\sigma^t(E_{i,i} + E_{i,l} + \sum_{q=1}^{n-1} F_q)P_\sigma = W'_n$.
2. $(i,i), (k,l)$. In this case, let σ be any permutation such that $\sigma(i) = 2, \sigma(k) = n$, and $\sigma(l) = 1$ and $F_q = E_{\sigma^{-1}(q),\sigma^{-1}(q+1)}$, $q = 1, \cdots, n-1$. Then $P_\sigma^t(E_{i,i} + E_{k,l} + \sum_{q=1}^{n-1} F_q)P_\sigma = W'_n$.
3. $(i,j), (i,l)$. In this case, let σ be any permutation such that $\sigma(i) = n-1, \sigma(j) = n$, and $\sigma(l) = 1$. Let $F_1 = E_{j,l}$, and $F_q = E_{\sigma^{-1}(q-1),\sigma^{-1}(q)}$ for $2 \leqslant q \leqslant n-1$, Then $P_\sigma^t(E_{i,j} + E_{k,l} + \sum_{q=1}^{n-1} F_q)P_\sigma = W_n$.
4. $(i,j), (k,j)$. In this case, let σ be any permutation such that $\sigma(i) = n-1, \sigma(k) = n$, and $\sigma(j) = 1$. Let $F_q = E_{\sigma^{-1}(q),\sigma^{-1}(q+1)}$ for $2 \leqslant q \leqslant n-1$, Then $P_\sigma^t(E_{i,j} + E_{k,l} + \sum_{q=1}^{n-1} F_q)P_\sigma = W_n$.
5. $(i,j), (k,l)$. In this case, let σ be any permutation such that $\sigma(i) = 1, \sigma(j) = 2, \sigma(k) = 3$, and $\sigma(l) = 4$. Let $F_1 = E_{j,k}$, $F_q = E_{\sigma^{-1}(q+2),\sigma^{-1}(q+3)}$ for $2 \leqslant q \leqslant n-3$, $F_{n-2} = E_{\sigma^{-1}(n),i}$, and $F_{n-1} = E_{\sigma^{-1}(n-1),i}$. Then $P_\sigma^t(E_{i,j} + E_{k,l} + \sum_{q=1}^{n-1} F_q)P_\sigma = W_n$.

□

Definition 23. *Let* $\mathcal{E} = \{E_{i,j} | 1 \leqslant i, j \leqslant n\}$, *the set of all cells. An* assignment *on* $\mathcal{E}$ *is a mapping* $\eta : \mathcal{E} \to \{0, 1\}$.

Definition 24. *We say that* η *is* nontrivial *if* η *is onto.*

Definition 25. *Let* $A \in \mathcal{M}_n(\mathbb{B})$, *we say that* η *is* A-nontrivial *if* $\eta|_{\mathcal{A}}$ *is onto where* $\mathcal{A} = \{E_{i,j} | A \geqslant E_{i,j}\}$.

That is, η is A-nontrivial if the restriction of η to the cells of A is onto.

Definition 26. *Further if* A *is primitive we say that* η *is* A-primitive *if*

$$\sum_{\{E_{i,j}\in\mathcal{A}|\eta(E_{i,j})=0\}}(E_{i,j},O)+\sum_{\{E_{i,j}\in\mathcal{A}|\eta(E_{i,j})=1\}}(O,E_{i,j})$$

is a primitive pair.

Definition 27. *If an assignment* η *is both* A*-nontrivial and* A*-primitive then we say that* η *is* A-nontrivial-primitive.

Remark 1. Assignment means the coloring of edges of the full graph in two colors. Assignment is non-trivial if both colors are used, it is A-nontrivial if both colors are used for the graph of the matrix A. Assignment is A-primitive if taking the sums of matrix units, corresponding to the edges of A of the different colors we get a primitive matrix pair.

Lemma 5. *Let* $(i,j,\alpha),(k,l,\beta)$ *be two triples such that* $1\leqslant i,j,k,l\leqslant n$, $k\neq l$, $\alpha,\beta\in\{0,1\}$, *and* $(i,j)\neq(k,l)$. *Let* $\mathcal{S}=\{\eta|\eta(E_{i,j})=\alpha,\eta(E_{k,l})=\beta\}$. *Then,* $\mathcal{S}$ *contains a* W_n*-nontrivial-primitive assignment and* $\mathcal{S}$ *contains a* W_n'*-nontrivial-primitive assignment.*

Proof. Since every primitive matrix has a primitive assignment [2, Theorem 2.1], the matrices W_n and W_n' have primitive assignments. Hence the lemma is trivial if $W_n\not\geqslant E_{i,j}+E_{k,l}$ and $W_n'\not\geqslant E_{i,j}+E_{k,l}$. Thus we assume that $W_n\geqslant E_{i,j}+E_{k,l}$ or $W_n'\geqslant E_{i,j}+E_{k,l}$. We shall define η to fulfill the requirements in each case.

Case 1. $W_n'\geqslant E_{i,j}+E_{k,l}$. Let us show that in this case there exists a W_n'-nontrivial-primitive assignment η such that $\eta(E_{i,j})=\alpha$ and $\eta(E_{k,l})=\beta$.

If $i=j=1$ and $l\equiv k+1 \bmod n$ and $\eta(E_{1,1})\neq\eta(E_{k,k+1})$, define $\eta(E_{p,q})=\eta(E_{1,1})$ for all $(p,q)\neq(k,k+1)$. If $\eta(E_{1,1})=\eta(E_{k,k+1})$, define $\eta(E_{k-1,k})=\eta(E_{1,1})$ and $\eta(E_{p,q})=\eta(E_{k,k+1})$ for all $(p,q)\neq(1,1),(k-1,k)$. This defines a W_n'-nontrivial-primitive assignment in $\mathcal{S}$. Note that here $W_n\not\geqslant E_{i,j}+E_{k,l}$, and hence there is a W_n-nontrivial-primitive assignment in $\mathcal{S}$.

If $i\neq j$ and $k\neq l$, then $j\equiv i+1 \bmod n$ and $l\equiv k+1 \bmod n$. If $\eta(E_{i,i+1})=\eta(E_{k,k+1})$ fix s, $1\leqslant s\leqslant n, s\neq i,k$, and let $\eta(E_{1,1})=\eta(E_{s,s+1})$ and $\eta(E_{p,q})=\eta(E_{i,i+1})$ for all $(p,q)\neq(1,1),(s,s+1)$. If $\eta(E_{i,i+1})\neq\eta(E_{k,k+1})$, let $\eta(E_{1,1})=\eta(E_{i,i+1})$ and $\eta(E_{p,q})=\eta(E_{k,k+1})$ for all $(p,q)\neq(1,1),(i,i+1)$.

In all cases, we have defined a W_n'-nontrivial-primitive assignment in $\mathcal{S}$.

Case 2 will deal with this case for a W_n-nontrivial-primitive assignment in $\mathcal{S}$.

Case 2. $W_n\geqslant E_{i,j}+E_{k,l}$. Let us show that in this case there exists a W_n-nontrivial-primitive assignment η such that $\eta(E_{i,j})=\alpha$ and $\eta(E_{k,l})=\beta$. We have the following subcases:

Subcase 1. $i,j,k,l\in\{1,n-1,n\}$. That is $(i,j)=(n,1)$ and $(k,l)=(n-1,1)$, or vice versa, or $(i,j)=(n,1)$ and $(k,l)=(n-1,n)$, or vice versa. If

$\eta(E_{i,j}) = \eta(E_{k,l})$ let $\eta(E_{1,2}) \neq \eta(E_{i,j})$ and $\eta(E_{p,q}) = \eta(E_{i,j})$ for all $(p,q) \neq (1,2),(k,l)$. If $\eta(E_{i,j}) \neq \eta(E_{k,l})$ then, since $E_{i,j}$ and $E_{k,l}$ are two of the cells $E_{n-1,n}, E_{n,1}, E_{n-1,1}$, let $E_{r,s}$ be the other of the three. If $(r,s) = (n-1,1)$, let $\eta(E_{r,s}) = \eta(E_{k,l})$ and $\eta(E_{p.q}) = \eta(E_{i,j})$ for all $(p,q) \neq (r,s),(k,l)$. If $(r,s) \neq (n-1,1)$ let $\eta(E_{r,s}) = \eta(E_{n-1,1})$ and $\eta(E_{p,p+1}) \neq \eta(E_{n-1,1})$ for all $p, 1 \leqslant p \leqslant n-2$.

Subcase 2. $i \in \{n, n-1\}, k \notin \{n-1, n\}$. (Equivalently, $k \in \{n, n-1\}, i \notin \{n-1, n\}$.) If $\eta(E_{i,j}) \neq \eta(E_{k,l})$, let $\eta(E_{p,q}) = \eta(E_{i,j})$ for all $(p,q) \neq (k,l)$. If $\eta(E_{i,j}) = \eta(E_{k,l})$, let $\eta(E_{s,s+1}) \neq \eta(E_{i,j})$ for some $s \neq k, s < n-1$, and let $\eta(E_{p,q}) = \eta(E_{i,j})$ for all $(p,q) \neq (s, s+1)$. Here, unless $n = 3$, the choice of s is always possible. The case $n = 3$ is an easy exercise.

Subcase 3. $i, k \notin \{n-1, n\}$. If $\eta(E_{i,j}) = \eta(E_{k,l})$ let $\eta(E_{n-1,n}) = \eta(E_{n-1,1}) \neq \eta(E_{i,j})$ and $\eta(E_{p,q}) = \eta(E_{i,j})$ for all other (p,q). If $\eta(E_{i,j}) \neq \eta(E_{k,l})$ let $\eta(E_{p,q}) = \eta(E_{i,j})$ for all $(p,q) \neq (k,l)$.

In all cases and subcases a W_n-nontrivial-primitive assignment in $\mathcal{S}$ has been defined. □

Lemma 6. *Let* $T : \mathcal{M}_n^2(\mathbb{B}) \to \mathcal{M}_n^2(\mathbb{B})$ *be a surjective additive operator which preserves primitive pairs. Then* $T(\mathcal{D}^2) = \mathcal{D}^2$.

Proof. Let us show that there is no element from $\mathcal{M}_n^2(\mathbb{B}) \setminus \mathcal{D}^2$ which are mapped by T to $\mathcal{D}^2$. Assume the converse, i.e., there is a matrix pair $(X,Y) \in \mathcal{M}_n^2(\mathbb{B}) \setminus \mathcal{D}^2$ such that $T(X,Y) \in \mathcal{D}^2$. Note that by Lemma 2 the operator T is bijective. Thus by [1, Theorem 1.2] the image of a cell must be a cell.

If $n = 1$ then all matrices are diagonal, so we can assume that $n \geqslant 2$ till the end of this proof.

Without loss of generality we may assume that X is non-diagonal. Thus there is $E_{i,j} \leqslant X$, $i \neq j$. By Lemma 2 the operator T is bijective and $T(O,O) = (O,O)$. Hence $T(E_{i,j}, O) \neq (O,O)$. Thus $T(E_{i,j}, O) \in \mathcal{D}^2$, since otherwise $T(X,O) \notin \mathcal{D}^2$ by antinegativity of $\mathbb{B}$. Since $n \geqslant 2$ we have that $|\mathcal{D}^2 \setminus \{(O,O)\}| \geqslant 15 > 2$. Thus by the surjectivity of T there is also some other pairs of matrices, whose image lies in $\mathcal{D}^2$, say $T(X',Y') \in \mathcal{D}^2$. Thus similar to the above we can say that there is a pair (r,s) such that either $T(E_{r,s}, O) \in \mathcal{D}^2$ (if $X' \neq O$) or $T(O, E_{r,s}) \in \mathcal{D}^2$ (if $Y' \neq O$), $(r,s) \neq (j,i)$ and $(r,s) \neq (i,j)$. We consider the first possibility now, i.e. there exists (r,s) such that $T(E_{r,s}, O) \in \mathcal{D}^2$.

Case 1. If $r = s$, by a permutational similarity of $\mathcal{M}_n^2(\mathbb{B})$ we can assume that $(r,r) = (1,1)$ and that $j \equiv (i+1) \bmod n$. By Lemma 5 there are $n-1$ cells $F_1, F_2, \cdots, F_{n-1}$ and a W_n'-nontrivial-primitive assignment η such that $W_n' \geqslant E_{i,j} + E_{r,r} + F_1 + \cdots + F_{n-1}$ and for $\mathcal{A} = \{E_{i,j}, E_{r,r}, F_1, \cdots, F_{n-1}\}$ the pair

$$(A,B) = \sum_{\{E_{k,l} \in \mathcal{A} \mid \eta(E_{k,l}) = 0\}} (E_{k,l}, O) + \sum_{\{E_{k,l} \in \mathcal{A} \mid \eta(E_{k,l}) = 1\}} (O, E_{k,l})$$

is a primitive pair. But $T(A,B)$ dominates two elements of $\mathcal{D}^2$ by the choice of i,j,r,s and hence cannot be primitive by Lemma 1, a contradiction.

Case 2. If $r \neq s$, by a permutational similarity of $\mathcal{M}_n^2(\mathbb{B})$ we can assume that $W_n \geqslant E_{i,j} + E_{r,s}$. By Lemma 5 there are $n-1$ cells $F_1, F_2, \cdots, F_{n-1}$ and a W_n-nontrivial-primitive assignment η such that $W_n \geqslant E_{i,j} + E_{r,s} + F_1 + \cdots + F_{n-1}$ and for $\mathcal{A} = \{E_{i,j}, E_{r,s}, F_1, \cdots, F_{n-1}\}$ the pair

$$(A, B) = \sum_{\{E_{k,l} \in \mathcal{A} | \eta(E_{k,l}) = 0\}} (E_{k,l}, O) + \sum_{\{E_{k,l} \in \mathcal{A} | \eta(E_{k,l}) = 1\}} (O, E_{k,l})$$

is primitive. But $T(A, B)$ dominates two elements of $\mathcal{D}^2$ and hence cannot be primitive by Lemma 1, a contradiction.

The cases $T(E_{i,j}, O), T(O, E_{r,s}) \in \mathcal{D}^2$ and X is diagonal, Y is non-diagonal can be considered in a similar way.

Thus, $T(\mathcal{M}_n^2(\mathbb{B}) \setminus \mathcal{D}^2) \subseteq \mathcal{M}_n^2(\mathbb{B}) \setminus \mathcal{D}^2$ since T is bijective by Lemma 2 and the set $\mathcal{M}_n^2(\mathbb{B})$ is finite it follows that $T(\mathcal{M}_n^2(\mathbb{B}) \setminus \mathcal{D}^2) = \mathcal{M}_n^2(\mathbb{B}) \setminus \mathcal{D}^2$ and thus we have that $T(\mathcal{D}^2) = \mathcal{D}^2$. □

Remark 2. Note that the sum of any three (or fewer) off-diagonal cells, no two of which are collinear, is dominated by a full-cycle permutation matrix unless one is the transpose of another. That is, if $i \neq p \neq r \neq i$ and $j \neq q \neq s \neq j$, and $(E_{i,j} + E_{p,q} + E_{r,s}) \circ (E_{i,j} + E_{p,q} + E_{r,s})^t = O$, then there is a full-cycle permutation matrix P such that $P \geqslant E_{i,j} + E_{p,q} + E_{r,s}$.

Let (A, B) be a matrix pair. For our purposes we will assume that if $a_{i,j} \neq 0$ then $b_{i,j} = 0$. Let G be the digraph whose adjacency matrix is A and let H be the digraph whose adjacency matrix is B. We color all the arcs in G color one and all the arcs in H color two, and then consider $G \cup H$, the two colored digraph with the same vertex set.

Definition 28. *We call this two colored digraph the digraph* associated *with the matrix pair* (A, B).

A useful tool in determining when a matrix pair is primitive is called the cycle matrix.

Definition 29. *If the digraph associated with the pair* (A, B) *has cycles* C_1, $C_2, \cdots, C_k$ *the cycle matrix* M *is a* $2 \times k$ *matrix of integers such that the* $(1, i)$ *entry is the number of arcs in cycle* C_i *that correspond to that part of the digraph associated with* A, *i.e., the arcs colored color 1, and the* $(2, i)$ *entry is the number of arcs in cycle* C_i *that correspond to that part of the digraph associated with* B, *the arcs colored color 2.*

The usefulness of this matrix is contained in the following result of Shader and Suwilo, see [11].

Theorem 1. [11] *Let* (A, B) *be a matrix pair with cycle matrix* M. *Then* (A, B) *is a primitive pair if and only if the greatest common divisor of all* 2×2 *minors of* M *is equal to 1.*

Lemma 7. *Let* (A, B) *be a matrix pair with* $A + B = W'_n$, $|A| + |B| = n + 1$ *and* $|A| \geqslant |B|$. *Then,* (A, B) *is a primitive pair if and only if* $B = O$ *or* B *is an off diagonal cell.*

Proof. Let M be the cycle matrix of the pair (A, B). If $B = O$ then $A = W'_n$ and (W'_n, O) is a primitive pair. If B is an off-diagonal cell then $M = \begin{bmatrix} 1 & n-1 \\ 0 & 1 \end{bmatrix}$ and $\det M = 1$ and hence (A, B) is a primitive pair by Theorem 1.

Now, assume that (A, B) is a primitive pair. We must show that $B = O$ or that B is an off-diagonal cell. If $B = O$ then we are done, so assume that $B \neq O$. By Lemma 1 either A or B or both contains a diagonal cell.

Case 1. Assume that $B \neq O$, and B dominates a diagonal cell. Then, since $A + B = W'_n$ and $|A| + |B| = n + 1$, it follows that non-zero cells of A and B are complementary. Thus $M = \begin{bmatrix} 0 & n-\alpha \\ 1 & \alpha \end{bmatrix}$ where α is the number of off-diagonal cells dominated by B. Since (A, B) is a primitive pair, we must have that $\det M = \pm 1$. Here, $\det M = n - \alpha$, so we have that $\alpha = n - 1$ and hence $|A| = 1$, a contradiction, since $|A| \geqslant |B|$ so that $|A| \geqslant \frac{n+1}{2} > 1$.

Case 2. Assume that $B \neq O$ and A has a nonzero diagonal entry. Here, the cycle matrix for (A, B) is $M = \begin{bmatrix} 1 & n-\alpha \\ 0 & \alpha \end{bmatrix}$ where α is the number of nonzero entries in B. Since, by Theorem 1, the determinant of M must be 1 or -1, we must have that $\alpha = 1$. That is B is an off-diagonal cell. □

Lemma 8. *Let* $n \geqslant 3$ *and* $T : \mathcal{M}^2_n(\mathbb{B}) \to \mathcal{M}^2_n(\mathbb{B})$ *be a surjective additive operator which preserves primitive pairs. Then, either* $T(\mathcal{D}, O) = (\mathcal{D}, O)$ *or* $T(\mathcal{D}, O) = (O, \mathcal{D})$.

Proof. By Lemma 2 T is a bijective linear operator. Suppose that $T(E_{i,i}, O) = (E_{k,k}, O)$ and $T(E_{j,j}, O) = (O, E_{l,l})$. Let $C = E_{1,2} + E_{2,3} + \cdots + E_{n-1,n} + E_{n,1}$, a full-cycle matrix and $T(C, O) = (X, Y)$. Then $(C + E_{i,i}, O)$ and $(C + E_{j,j}, O)$ are both primitive pairs, and hence their images must be a primitive pair. Since by Lemma 6 $T(\mathcal{D}^2) = \mathcal{D}^2$, we must have that $T(\mathcal{M}^2_n(\mathbb{B}) \setminus \mathcal{D}^2) = \mathcal{M}^2_n(\mathbb{B}) \setminus \mathcal{D}^2$, so that $T(C + E_{i,i}, O) = (X + E_{k,k}, Y)$ and $T(C + E_{j,j}, O) = (X, Y + E_{l,l})$ must both be primitive pairs. It was pointed out in the proof of Lemma 6 that T is bijective on the set of cells. Thus $T(C + E_{i,i}, O) = (X + E_{k,k}, Y)$ and $T(C + E_{j,j}, O) = (X, Y + E_{l,l})$ are primitive pairs which dominate exactly $n + 1$ cells. Since by Corollary 1, the only primitive matrices which dominate exactly $n + 1$ cells, one of which is a diagonal cell, dominate a full-cycle matrix, we must have that $X + Y + E_{k,k}$ and $X + Y + E_{l,l}$ dominate full-cycle matrices. It now follows that $X + Y$ is a full-cycle. Since $(X + E_{k,k}, Y)$ is a primitive pair we have by Lemma 7 that Y is an off-diagonal cell. Since $(X, Y + E_{l,l})$ is a primitive pair we have by Lemma 7 that X is an off diagonal cell. Since $X + Y$ is a full-cycle matrix, it follows that $n = 2$, a contradiction. Thus $T(\mathcal{D}, O) = (\mathcal{D}, O)$ or $T(\mathcal{D}, O) = (O, \mathcal{D})$ □

Henceforth, we let K denote the matrix with a zero main diagonal and ones everywhere else. That is K is the adjacency matrix of the complete loopless digraph.

Let us show that T acts on $\mathcal{M}_n^2(\mathbb{B})$ component wise.

Lemma 9. *Let $n \geqslant 3$ and $T : \mathcal{M}_n^2(\mathbb{B}) \to \mathcal{M}_n^2(\mathbb{B})$ be a surjective additive operator which preserves primitive pairs. Then, there is are bijective linear operators $L : \mathcal{M}_n(\mathbb{B}) \setminus \mathcal{D} \to \mathcal{M}_n(\mathbb{B}) \setminus \mathcal{D}$ and $S : \mathcal{D} \to \mathcal{D}$ such that either*
$T(X, O) = (L(X \circ K), O) + (S(X \circ I), O)$ *for all* X, *or*
$T(X, O) = (O, L(X \circ K)) + (O, S(X \circ I))$ *for all* X.

Proof. By Lemma 2, T is a bijective linear operator. Thus, by [1, Theorem 2.1] all cells in $\mathcal{M}_n^2(\mathbb{B})$ are mapped to cells. By virtue of Lemma 8 we may assume without loss of generality that for all l we have that $T(E_{l,l}, O) = (E_{l,l}, O)$ and $T(O, E_{l,l}) = (O, E_{\sigma(l),\sigma(l)})$ for some permutations σ.

Suppose that for some pairs $(p, q), (x, y)$ with $p \neq q$ we have $T(E_{p,q}, O) = (O, E_{x,y})$. Here, by Lemma 6, $x \neq y$. Let $F_1, F_2, \cdots, F_{n-1}$ be any cells such that $E_{p,q} + F_1 + F_2 + \cdots + F_{n-1}$ is a full-cycle. For an arbitrary k, let

$$(A, B) = (E_{k,k} + E_{p,q} + F_1 + F_2 + \cdots + F_{n-2}, F_{n-1}). \tag{3}$$

Then (A, B) is a primitive pair by Lemma 7. Thus the image must be a primitive pair.

As was pointed out, T maps cells to cells, thus $|T(A, B)| = |(A, B)| = n + 1$. Since $T(E_{k,k}, 0) = (E_{k,k}, 0) \in (\mathcal{D}, 0)$, it follows that the sum of two components of $T(A, B)$ is not a matrix which is similar to the Wieland matrix by a permutational transformation. Thus it is similar to W_n' and Lemma 7 can be applied. Therefore, $T(E_{k,k} + F_1 + F_2 + \cdots + F_{n-2}, F_{n-1})$ must be a pair of the form (C, O) since $T(E_{p,q}, O) = (O, E_{x,y})$ and the component of $T(A, B)$, which is without diagonal cells, can possess no more than one non-zero cell. By varying the choice of F_i's we get that if F is an off-diagonal cell not in row p or column q, then $T(F, O) \leqslant (J, O)$. That is, there are $n^2 - 3n + 3$ off diagonal cells F such that $T(F, O) \leqslant (J, O)$. Note however that in the expression $(A, B) = (E_{k,k} + E_{p,q} + F_1 + F_2 + \cdots + F_{n-2}, F_{n-1})$, see formula (3), the matrix F_{n-1} could be replaced by any of the other off-diagonal cells not in row p or column q. That is there are also $n^2 - 3n + 3$ off diagonal cells F such that $T(O, F) \leqslant (J, O)$. Further if $T(E_{p,r}, O) \leqslant (O, J)$ then as above $T(E_{i,q}, O) \leqslant (J, O)$ so that the number of off-diagonal cells F such that $T(F, O) \leqslant (J, O)$ is at least $n^2 - 3n + 4$. If $T(E_{p,r}, O) \leqslant (J, O)$ then again, the number of off-diagonal cells F such that $T(F, O) \leqslant (J, O)$ is at least $n^2 - 3n + 4$. It follows that $2[n^2 - 3n + 3] + 1 \leqslant n^2 - n$ since T is bijective by Lemma 2 and, therefore,

T is bijective on the set of cells. But that never happens. In this case we have arrived at a contradiction.

Define $L : \mathcal{M}_n(\mathbb{B}) \setminus \mathcal{D} \to \mathcal{M}_n(\mathbb{B}) \setminus \mathcal{D}$ by $T(X \circ K, O = (L(X \circ K, O)$ and $S : \mathcal{D} \to \mathcal{D}$ by $T(X \circ I, O) = S(X \circ I, O)$. The lemma now follows. □

Since the action of T is defined on $\mathcal{M}_n^2(\mathbb{B})$ independently in each component, the following definition is correct and makes sense.

Definition 30. *Let* $T : \mathcal{M}_n^2(\mathbb{B}) \to \mathcal{M}_n^2(\mathbb{B})$ *be a linear operator such that* $T(X, O) \in \mathcal{M}_n(\mathbb{B}) \times O$ *and* $T(O, X) \in O \times \mathcal{M}_n(\mathbb{B})$, *for all* $X \in \mathcal{M}_n(\mathbb{B})$. *Define the linear operators* T_1 *and* T_2 *on* $\mathcal{M}_n(\mathbb{B})$ *by* $T(X, Y) = (T_1(X), T_2(Y))$.

Corollary 2. *Let* $T : \mathcal{M}_n^2(\mathbb{B}) \to \mathcal{M}_n^2(\mathbb{B})$ *be a surjective additive operator which preserves primitive pairs. Then, there are bijective linear operators* $L_1 : \mathcal{M}_n(\mathbb{B}) \to \mathcal{M}_n(\mathbb{B})$ *and* $L_2 : \mathcal{M}_n(\mathbb{B}) \to \mathcal{M}_n(\mathbb{B})$ *which preserve primitivity such that* $T(X, Y) = (L_1(X), L_2(Y))$ *for all* $(X, Y) \in \mathcal{M}_n^2(\mathbb{B})$, *or* $T(X, Y) = (L_2(Y), L_1(X))$ *for all* $(X, Y) \in \mathcal{M}_n^2(\mathbb{B})$.

Proof. By Lemma 9 $T(X, O) = (L(X \circ K), O) + (S(X \circ I), O)$ for all X, or $T(X, O) = (O, L(X \circ K)) + (O, S(X \circ I))$ for all X. If $T(X, O) = (L(X \circ K), O) + (S(X \circ I), O)$, then by the bijectivity of T and Lemma 9, $T(O, X) = (O, L'(X \circ K)) + (O, S'(X \circ I))$ Here, define $L_1(X) = L(X \circ K) + S(X \circ I)$ and $L_2(X) = L'(X \circ K) + S'(X \circ I)$ so that $T(X, Y) = (L_1(X), L_2(Y))$. If $T(X, O) = (O, L(X \circ K)) + (O, S(X \circ I))$, then by the bijectivity of T and Lemma 9, $T(O, X) = (L'(X \circ K), O) + (S'(X \circ I), O)$. In this case $T(X, Y) = (L_2(Y), L_1(X))$. □

Lemma 10. *If* $L : \mathcal{M}_n(\mathbb{B}) \to \mathcal{M}_n(\mathbb{B})$ *is a bijective linear operator that preserves primitive matrices then* L *strongly preserves primitive matrices.*

Proof. Since the set $\mathcal{M}_n(\mathbb{B})$ is finite, the set of primitive matrices and non primitive matrices partition $\mathcal{M}_n(\mathbb{B})$. Since L is bijective, and the image of the set of primitive matrices is contained in the set of primitive matrices, the image of the set of primitive matrices must be equal to the set of primitive matrices and consequently the image of the set of nonprimitive matrices must be the set of nonprimitive matrices. That is, L strongly preserves primitive matrices. □

We now define a special operator that we need for Theorem 2 below.

Definition 31. *An operator* $D : \mathcal{M}_n(\mathbb{B}) \to \mathcal{M}_n(\mathbb{B})$ *is a* diagonal replacement operator *if* $D(E_{i,j}) = E_{i,j}$ *whenever* $i \neq j$, *and* $D(\mathcal{D}) \subseteq \mathcal{D}$. *It is* nonsingular *if* $D(E_{i,i}) \neq O$ *for all* i. *If* D *is bijective then there is a permutation* σ *of* $\{1, \cdots, n\}$ *such that* $D(E_{i,i}) = E_{\sigma(i),\sigma(i)}$ *for all* i. *In such a case we use the notation* D_σ *to denote the operator.*

Theorem 2. [4, Theorem 3.1] *The semigroup of linear operators on* $\mathcal{M}_n(\mathbb{B})$ *that strongly preserve primitive matrices is generated by transposition, the similarity operators and nonsingular diagonal replacement when* $n \neq 2$. *When* $n = 2$ *it is generated by those operators and the special operator defined by* $\begin{bmatrix} a & b \\ c & d \end{bmatrix} \to \begin{bmatrix} b & (a+d) \\ c & 0 \end{bmatrix}$ *for all* $a, b, c, d \in \mathcal{M}_n(\mathbb{B})$.

Let us now formulate our main theorem for matrix pairs.

Theorem 3. *Let* $n \geqslant 3$ *and* $T : \mathcal{M}_n^2(\mathbb{B}) \to \mathcal{M}_n^2(\mathbb{B})$ *be a surjective additive operator which preserves primitive pairs. Then there are permutation matrices* P, Q, *and* R *such that:*

$T(X,Y) = (P(X \circ K)P^t, P(Y \circ K)P^t) + (Q(X \circ I)Q^t, R(Y \circ I)R^t)$ *for all* $(X,Y) \in \mathcal{M}_n^2(\mathbb{B})$;

$T(X,Y) = (P(Y \circ K)P^t, P(X \circ K)P^t) + (Q(Y \circ I)Q^t, R(X \circ I)R^t)$ *for all* $(X,Y) \in \mathcal{M}_n^2(\mathbb{B})$;

$T(X,Y) = (P(X^t \circ K)P^t, P(Y^t \circ K)P^t) + (Q(X \circ I)Q^t, R(Y \circ I)R^t)$ *for all* $(X,Y) \in \mathcal{M}_n^2(\mathbb{B})$; *or*

$T(X,Y) = (P(Y^t \circ K)P^t, P(X^t \circ K)P^t) + (Q(Y \circ I)Q^t, R(X \circ I)R^t)$ *for all* $(X,Y) \in \mathcal{M}_n^2(\mathbb{B})$.

Proof. By Corollary 2 induced actions of T on $(\mathcal{M}_n(\mathbb{B}), O)$ and $(O, \mathcal{M}_n(\mathbb{B}))$ arise. According to the same corollary these actions are linear and defined correctly. By Lemma 10 these induced operators strongly preserve primitivity. Applying Theorem 2 now, we have that for some permutation matrices P and Q, and permutations σ and τ of $\{1, \cdots, n\}$, $T(X,Y) = (PD_\sigma(X)P^t, QD_\tau(Y)Q^t)$ for all $(X,Y) \in \mathcal{M}_n^2(\mathbb{B})$; or the similar transformations in the other three cases. Thus we only need show that $P = Q$ and it is impossible that there is a transposition in the first coordinate and no transposition in the second one.

We start with the transposition transformation. Without loss of generality assume that

$$T(X,O) = (PD_\sigma(X)P^t, O)$$

and

$$T(O,Y) = (O, QD_\tau(Y)Q^t).$$

Also without loss of generality we may assume that $P = I$ that is, $T(X,O) = (D_\sigma(X), O)$. Now, it is impossible that $T(O, E_{i,i+1}) = (O, E_{i,i+1})$ for all $i = 1, \ldots, n$, since there is no permutation matrix Q such that

$$Q(E_{1,2} + E_{2,3} + \cdots + E_{n-1,n} + E_{n,1})^t Q^t = E_{1,2} + E_{2,3} + \cdots + E_{n-1,n} + E_{n,1}.$$

Therefore, there is some i such that $T(O, E_{i,i+1}) \neq (O, E_{i,i+1})$ (subscripts taken modulo n). Say without loss of generality that $T(O, E_{n,1}) \neq (O, E_{n,1})$. Let

$$A_1 = E_{1,1} + E_{1,2} + E_{2,3} + \cdots + E_{n-1,n} \text{ and } A_2 = E_{n,1}.$$

Then (A_1, A_2) is primitive, whereas,

$$T(A_1, A_2) = (E_{i_1,i_1} + E_{1,2} + E_{2,3} + \cdots + E_{n-1,n}, E_{p,q}),$$

where $(p, q) \neq (n, 1)$. This matrix pair cannot be primitive since it has exactly n off diagonal entries and they do not form a full cycle, a contradiction. Thus, either X is transposed in both components or X is not transposed in both components.

Suppose that $P \neq Q$. Then there is some $E_{i,j}$ with $i \neq j$ such that $PE_{i,j}$ and $QE_{i,j}$ are cells in different rows. Let $k_1, k_2, \cdots k_{n-2}$ be distinct positive integers less than n such that $i, j \notin \{k_1, k_2, \cdots k_{n-1}\}$. Let $A = E_{1,1} + E_{j,k_1} + E_{k_1,k_2} + \cdots + E_{k_{n-3},k_{n-2}} + E_{k_{n-2},i}$. Then $(A, E_{i,j})$ is a primitive pair, but $T(A, E_{i,j}) = (X, Y)$ cannot be primitive as it has a row with no off diagonal entry in either X or Y, a contradiction. Thus $P = Q$. Now by splitting any matrix to its diagonal and off-diagonal parts, we obtain the form as in the statement of the theorem.

Note the special operator for $n = 2$ in Theorem 2 is not surjective. □

4 Matrices over antinegative semirings without zero divisors

Definition 32. *The* pattern, $\overline{A}$, *of a matrix* $A \in \mathcal{M}_n(\mathcal{S})$ *is the* $(0,1)$*-matrix whose* (i,j)*-th entry is 0 if* $a_{i,j} = 0$ *and 1 if* $a_{i,j} \neq 0$.

Remark 3. For a given matrix $A \in \mathcal{M}_n(\mathcal{S})$ we consider $\overline{A}$ as a matrix in $\mathcal{M}_n(\mathbb{B})$. If $\mathcal{S}$ is antinegative and without zero divisors then the mapping

$$\begin{array}{c} \mathcal{M}_n(\mathcal{S}) \rightarrow \mathcal{M}_n(\mathbb{B}) \\ A \rightarrow \overline{A} \end{array}$$

is a homomorphism of semirings.

Remark 4. Let $\mathcal{S}$ be antinegative and without zero divisors. Then direct computations show that $(A, B) \in \mathcal{M}_n(\mathcal{S})$ is primitive if and only if $(\overline{A}, \overline{B}) \in \mathcal{M}_n(\mathbb{B})$ is primitive.

Definition 33. *Let* T *be an additive operator on* $\mathcal{M}_n(\mathcal{S})$. *We say that its* pattern $\overline{T}$ *is an additive operator on* $\mathcal{M}_n(\mathbb{B})$ *defined by the rule* $\overline{T}(\overline{E_{i,j}}) = \overline{T(E_{i,j})}$ *and* $\overline{T}(\overline{O}) = \overline{T(O)}$.

Remark 5. It is easy to see that if $\mathcal{S}$ is antinegative and zero-divisor-free, then for any $A \in \mathcal{M}_n(\mathcal{S})$ we have that $\overline{T(A)} = \overline{T}(\overline{A})$.

Moreover, the following statement is true:

Lemma 11. *Let $\mathcal{S}$ be an antinegative semiring without zero divisors. Then the transformation which maps each additive operator T on $\mathcal{M}_n(\mathcal{S})$ to the operator $\overline{T}$ on $\mathcal{M}_n(\mathbb{B})$ is a homomorphism of semirings of additive operators on $\mathcal{M}_n(\mathcal{S})$ to additive operators on $\mathcal{M}_n(\mathbb{B})$.*

Proof. It is straightforward to see that if T is the zero operator, then $\overline{T}$ is the zero operator. The rest follows from [4, Lemma 2.1]. □

Let us apply the above lemma and Theorem 3 to obtain the characterization result over any antinegative semiring without zero divisors.

Corollary 3. *Let $T : \mathcal{M}_n^2(\mathcal{S}) \to \mathcal{M}_n^2(\mathcal{S})$ be a surjective additive operator which preserves primitive pairs. Then there is a permutation matrix $P \in \mathcal{M}_n(\mathcal{S})$, additive functions $\phi, \psi : \mathcal{S} \to \mathcal{S}$ with zero kernels, i.e., $\phi(x) = 0$ implies $x = 0$ and $\psi(y) = 0$ implies $y = 0$, and permutations σ and τ of $\{1, \cdots, n\}$ such that:*

$T(X,Y) = (PD_\sigma(X^\phi)P^t, PD_\tau(Y^\psi C)P^t)$ *for all* $(X,Y) \in \mathcal{M}_n^2(\mathbb{B})$, *where* X^ϕ *denotes the element-wise action of* ϕ *on the entries of* X;

$T(X,Y) = (PD_\tau(Y^\psi C)P^t, PD_\sigma(X^\phi B)P^t)$ *for all* $(X,Y) \in \mathcal{M}_n^2(\mathbb{B})$;

$T(X,Y) = (PD_\sigma((X^\phi)^t)P^t, PD_\tau((Y^\psi)^t)P^t)$ *for all* $(X,Y) \in \mathcal{M}_n^2(\mathbb{B})$; *or*

$T(X,Y) = (PD_\tau((Y^\psi)^t)P^t, PD_\sigma((X^\phi)^t)P^t)$ *for all* $(X,Y) \in \mathcal{M}_n^2(\mathbb{B})$.

References

1. L. B. Beasley, A. E. Guterman, Linear preservers of extremes of rank inequalities over semirings: Factor rank, *Journal of Mathematical Sciences* (New-York) **131**, no. 5, (2005) 5919–5938.
2. L. B. Beasley, S. J. Kirkland, A note on k-primitive directed graphs, *Linear Algebra and its Appl.*, **373** (2003) 67–74.
3. L. B. Beasley, N. J. Pullman, Linear operators that strongly preserve the index of imprimitivity, *Linear an Multilinear Algebra*, **31** (1992) 267–283.
4. L. B. Beasley, N. J. Pullman, Linear operators that strongly preserve primitivity, *Linear an Multilinear Algebra*, **25** (1989) 205–213.
5. R. Brualdi and H. Ryser, *Combinatorial Matrix Theory*, Cambridge University Press, New York, 1991.
6. E. Fornasini, A 2D systems approach to river pollution modelling, *Multidimensional System Signal Process* **2** (1991) 233–265.
7. E. Fornasini, M. Valcher, Primitivity of positive matrix pairs: algebraic characterization graph theoretic description and 2D systems interpretation, *SIAM J. Matrix Anal. Appl.* **19** (1998) 71–88.
8. R. A. Horn, C. R. Johnson, "Matrix Analysis", Cambridge University Press, New York.
9. C.-K. Li and N.-K. Tsing, Linear preserver problems: a brief introduction and some special techniques. Directions in matrix theory (Auburn, AL, 1990). *Linear Algebra Appl.* **162/164**, (1992), P. 217–235.

10. P. Pierce and others, A Survey of Linear Preserver Problems, *Linear and Multilinear Algebra*, **33** (1992) 1–119.
11. B. Shader, S. Suwilo, Exponents of non-negative matrix pairs, *Linear Algebra and its Appl.*, **363** (2003) 275–293.
12. H. Wielandt, Unzerlegbare, nicht negative Matrizen, Math. Z. **52** (1958) 642–645.

Decompositions of Quaternions and Their Matrix Equivalents

Drahoslava Janovská[1] and Gerhard Opfer[2]

[1] Institute of Chemical Technology, Prague, Department of Mathematics, Technická 5, 166 28 Prague 6, Czech Republic
janovskd@vscht.cz

[2] University of Hamburg, Faculty for Mathematics, Informatics, and Natural Sciences [MIN], Bundestraße 55, 20146 Hamburg, Germany
opfer@math.uni-hamburg.de

Dedicated to the memory of Gene Golub

Abstract. Since quaternions have isomorphic representations in matrix form we investigate various well known matrix decompositions for quaternions.

Keywords: decompositions of quaternions, Schur, polar, SVD, Jordan, QR, LU.

1 Introduction

We will study various decompositions of quaternions where we will employ the isomorphic matrix images of quaternions. The matrix decompositions allow in many cases analogue decompositions of the underlying quaternion.

Let us denote the skew field of quaternions by $\mathbb{H}$. It is well known that quaternions have an isomorphic representation either by certain complex (2×2)-matrices or by certain real (4×4)-matrices. Let $a := (a_1, a_2, a_3, a_4) \in \mathbb{H}$. Then the two isomorphisms $\jmath : \mathbb{H} \to \mathbb{C}^{2\times2}$, $\imath_1 : \mathbb{H} \to \mathbb{R}^{4\times4}$ are defined as follows:

$$\jmath(a) := \begin{pmatrix} \alpha & \beta \\ -\overline{\beta} & \overline{\alpha} \end{pmatrix} \in \mathbb{C}^{2\times2}, \quad \alpha := a_1 + a_2\mathbf{i},\ \beta := a_3 + a_4\mathbf{i}, \tag{1}$$

$$\imath_1(a) := \begin{pmatrix} a_1 & -a_2 & -a_3 & -a_4 \\ a_2 & a_1 & -a_4 & a_3 \\ a_3 & a_4 & a_1 & -a_2 \\ a_4 & -a_3 & a_2 & a_1 \end{pmatrix} \in \mathbb{R}^{4\times4}. \tag{2}$$

There is another very similar, but nevertheless different mapping, $\imath_2 : \mathbb{H} \to \mathbb{R}^{4\times4}$, the meaning of which will be explained immediately:

$$\imath_2(a) := \begin{pmatrix} a_1 & -a_2 & -a_3 & -a_4 \\ a_2 & a_1 & a_4 & -a_3 \\ a_3 & -a_4 & a_1 & a_2 \\ a_4 & a_3 & -a_2 & a_1 \end{pmatrix} \in \mathbb{R}^{4\times4}. \tag{3}$$

In the first equation (1) the overlined quantities $\overline{\alpha}, \overline{\beta}$ denote the complex conjugates of the non overlined quantities α, β, respectively. Let $b \in \mathbb{H}$ be another quaternion. Then, the isomorphisms imply $\jmath(ab) = \jmath(a)\jmath(b)$, $\imath_1(ab) = \imath_1(a)\imath_1(b)$. The third map, $\imath_2$, has the interesting property that it reverses the order of the multiplication:

$$\imath_2(ab) = \imath_2(b)\imath_2(a) \quad \forall a, b \in \mathbb{H}, \quad \imath_1(a)\imath_2(b) = \imath_2(b)\imath_1(a) \quad \forall a, b \in \mathbb{H}. \tag{4}$$

The mapping $\imath_2$ plays a central role in the investigations of linear maps $\mathbb{H} \to \mathbb{H}$. There is a formal similarity to the Kronecker product of two arbitrary matrices. See [16] for the mentioned linear maps and [11, Lemma 4.3.1] for the Kronecker product.

Definition 1. A complex (2×2)-matrix of the form introduced in (1) will be called a *complex q-matrix*. A real (4×4)-matrix of the form introduced in (2) will be called a *real q-matrix*. A real (4×4)-matrix of the form introduced in (3) will be called a *real pseudo q-matrix*. The set of all complex q-matrices will be denoted by $\mathbb{H}_\mathbb{C}$. The set of all real q-matrices will be denoted by $\mathbb{H}_\mathbb{R}$. The set of all real pseudo q-matrices will be denoted by $\mathbb{H}_\mathbb{P}$.

We introduce some common notation. Let $\mathbf{C}$ be a matrix of any size with real or complex entries. By $\mathbf{D} := \mathbf{C}^T$ we denote the *transposed matrix* of $\mathbf{C}$, where rows and columns are interchanged. By $\mathbf{E} := \overline{\mathbf{C}}$ we denote the *conjugate matrix* of $\mathbf{C}$ where all entries of $\mathbf{C}$ are changed to their complex conjugates. Finally, $\mathbf{C}^* := (\overline{\mathbf{C}})^T$. Let $a := (a_1, a_2, a_3, a_4) \in \mathbb{H}$. The first component, a_1, is called the *real part* of a, denoted by $\Re a$. The quaternion $a_v := (0, a_2, a_3, a_4)$ will be called *vector part* of a.

From the above representations it is clear how to recover a quaternion from the corresponding matrix. Thus, it is also possible to introduce inverse mappings

$$\jmath^{-1} : \mathbb{H}_\mathbb{C} \to \mathbb{H}, \quad \imath_1^{-1} : \mathbb{H}_\mathbb{R} \to \mathbb{H}, \quad \imath_2^{-1} : \mathbb{H}_\mathbb{P} \to \mathbb{H},$$

where $\jmath^{-1}$, $\imath_1^{-1}$ as well define isomorphisms. If we define a new algebra $\tilde{\mathbb{H}}$ where a new multiplication, denoted by $\star$ is introduced by $a \star b := ba$, then $\imath_2$ is also an isomorphism between $\tilde{\mathbb{H}}$ and $\mathbb{H}_\mathbb{P}$. This particularly implies that $\imath_2(ab) = \imath_2(b)\imath_2(a) \in \mathbb{H}_\mathbb{P}$ and $\imath_2(a^{-1}) = \imath_2(a)^{-1} = \imath_2(a)^T/|a|^2 \in \mathbb{H}_\mathbb{P}$ for all $a \in \mathbb{H}\backslash\{0\}$. Because of these isomorphisms it is possible to associate notions known from matrix theory with quaternions. Simple examples are:

$$\begin{aligned}
\det(a) &:= \det(\jmath(a)) = |a|^2, \quad \det(\imath_1(a)) = \det(\imath_2(a)) = |a|^4, && (5)\\
\operatorname{tr}(a) &:= \operatorname{tr}(\jmath(a)) = 2a_1, \quad \operatorname{tr}(\imath_1(a)) = \operatorname{tr}(\imath_2(a) = 4a_1, && (6)\\
\operatorname{eig}(a) &:= \operatorname{eig}(\jmath(a)) = [\sigma_+, \sigma_-], && (7)\\
\operatorname{eig}(\imath_1(a)) &= \operatorname{eig}(\imath_2(a)) = [\sigma_+, \sigma_+, \sigma_-, \sigma_-], \text{ where}\\
\sigma_+ &= a_1 + \sqrt{a_2^2 + a_3^2 + a_4^2}\,\mathbf{i} = a_1 + |a_v|\mathbf{i}, \quad \sigma_- = \overline{\sigma_+}, && (8)
\end{aligned}$$

$$\begin{aligned}
|a| &= \|\jmath(a)\|_2 = \|\imath_1(a)\|_2 = \|\imath_2(a)\|_2, && (9)\\
\operatorname{cond}(a) &:= \operatorname{cond}(\jmath(a)) = \operatorname{cond}(\imath_1(a)) = \operatorname{cond}(\imath_2(a)) = 1, && (10)\\
\jmath(a\overline{a}) &= \jmath(a)\jmath(a)^* = |a|^2\jmath(1) = |a|^2\mathbf{I}_2, && (11)\\
\imath_1(a\overline{a}) &= \imath_1(a)\imath_1(a)^{\mathrm{T}} = \imath_2(a\overline{a}) = \imath_2(a)^{\mathrm{T}}\imath_2(a) = |a|^2\mathbf{I}_4, && (12)
\end{aligned}$$

where det, tr, eig, cond refer to *determinant, trace, collection of eigenvalues, condition*, respectively. By $\mathbf{I}_2, \mathbf{I}_4$ we denote the identity matrices of order 2 and 4, respectively. We note that a general theory for determinants of quaternion valued matrices is not available. See [1]. We will review the classical matrix decompositions and investigate the applicability to quaternions. For the classical theory we usually refer to one of the books of HORN & JOHNSON, [10], [11].

In this connection it is useful to introduce another notion, namely that of equivalence between two quaternions. Such an equivalence may already be regarded as one of the important decompositions, namely the Schur decomposition, as we will see.

Definition 2. Two quaternions $a, b \in \mathbb{H}$ will be called *equivalent*, if there is an $h \in \mathbb{H}\backslash\{0\}$ such that

$$b = h^{-1}ah.$$

Equivalent quaternions a, b will be denoted by $a \sim b$. The set

$$[a] := \{s : s := h^{-1}ah,\ h \in \mathbb{H}\}$$

will be called *equivalence class of* a. It is the set of all quaternions which are equivalent to a.

Lemma 1. *The above defined notion of equivalence defines an equivalence relation. Two quaternions* a, b *are equivalent if and only if*

$$\Re a = \Re b, \quad |a| = |b|. \tag{13}$$

Furthermore, $a \in \mathbb{R} \Leftrightarrow \{a\} = [a]$. *Let* $a \in \mathbb{C}$. *Then* $\{a, \overline{a}\} \subset [a]$. *Let* $a = (a_1, a_2, a_3, a_4) \in \mathbb{H}$. *Then*

$$\sigma_+ := a_1 + \sqrt{a_2^2 + a_3^2 + a_4^2}\,\mathbf{i} \in [a].$$

Proof. See [13]. □

The complex number σ_+ occurring in the last lemma will be called *complex representative of* $[a]$. The equivalence $a \sim b$ can be expressed also in the form $ah - hb = 0$, with an $h \neq 0$. This is the homogeneous form of Sylvester's equation. This equation was investigated by JANOVSKÁ & OPFER [16]. It should be noted that algebraists refer to equivalent elements usually as *conjugate* elements. See [18, p. 35].

2 Decompositions of quaternions

A matrix decomposition of the form $\jmath(a) = \jmath(b)\jmath(c)$ or $\jmath(a) = \jmath(b)\jmath(c)\jmath(d)$ with $a, b, c, d \in \mathbb{H}$ and the same with $\imath_1$ also represents a direct decomposition of the involved quaternions, namely $a = bc$ or $a = bcd$ because of the isomorphy of the involved mappings $\jmath, \imath_1$. The same applies to $\imath_2$, only the multiplication order has to be reversed. We will study the possibility of decomposing quaternions with respect to various well known matrix decompositions. A survey paper on decompositions of quaternionic matrices was given by [19].

2.1 Schur decompositions

Let $\mathbf{U}$ be an arbitrary real or complex square matrix. If $\mathbf{U}\mathbf{U}^* = \mathbf{I}$ (identity matrix) then $\mathbf{U}$ will be called *unitary*. If $\mathbf{U}$ is real, then, $\mathbf{U}^* = \mathbf{U}^{\mathrm{T}}$. A real, unitary matrix will also be called *orthogonal*.

Theorem 1 (Schur 1). *Let* $\mathbf{A}$ *be an arbitrary real or complex square matrix. Then there exists a unitary matrix* $\mathbf{U}$ *of the same size as* $\mathbf{A}$ *such that*

$$\mathbf{D} := \mathbf{U}^*\mathbf{A}\mathbf{U} \tag{14}$$

is an upper triangular matrix and as such contains the eigenvalues of $\mathbf{A}$ *on its diagonal.*

Proof. See HORN & JOHNSON [10, p. 79]. □

Theorem 2 (Schur 2). *Let* $\mathbf{A}$ *be an arbitrary real square matrix of order* n. *Then there exists a real, orthogonal matrix* $\mathbf{V}$ *of order* n *such that*

$$\mathbf{H} := \mathbf{V}^{\mathrm{T}}\mathbf{A}\mathbf{V} \tag{15}$$

is an upper Hessenberg matrix with $k \leqslant n$ *block entries in the diagonal which are either real* (1×1) *matrices or real* (2×2) *matrices which have a pair of non real complex conjugate eigenvalues which are also eigenvalues of* $\mathbf{A}$.

Proof. See HORN & JOHNSON [10, p. 82]. □

The representation $\mathbf{A} = \mathbf{UDU}^*$ implied by (14) is usually referred to as *complex Schur decomposition of* $\mathbf{A}$, whereas $\mathbf{A} = \mathbf{VHV}^{\mathrm{T}}$ implied by (15) is usually referred to as *real Schur decomposition of* $\mathbf{A}$. Let a be a quaternion, then we might ask whether there is a Schur decomposition of the matrices $\jmath(a)$, $\imath_1(a)$, $\imath_2(a)$ in terms of quaternions. The (affirmative) answer was already given by Janovská & Opfer [15, 2007].

Theorem 3. *Let* $a \in \mathbb{H}\backslash\mathbb{R}$ *and* σ_+ *be the complex representative of* $[a]$. *There exists* $h \in \mathbb{H}$ *with* $|h| = 1$ *such that* $\sigma_+ = h^{-1}ah$ *and*

$$\begin{aligned} \jmath(a) &= \jmath(h)\jmath(\sigma_+)\jmath(h^{-1}), \qquad \imath_1(a) = \imath_1(h)\imath_1(\sigma_+)\imath_1(h^{-1}), \\ \imath_2(a) &= \imath_2(h^{-1})\imath_2(\sigma_+)\imath_2(h) \end{aligned} \tag{16}$$

are the Schur decompositions of $\jmath(a)$, $\imath_1(a)$, $\imath_2(a)$, *respectively, which includes that* $\jmath(h)$, $\imath_1(h)$, $\imath_2(h)$ *are unitary and* $\jmath(h^{-1}) = \jmath(h)^*$, $\imath_1(h^{-1}) = \imath_1(h)^{\mathrm{T}}$, $\imath_2(h^{-1}) = \imath_2(h)^{\mathrm{T}}$. *The first decomposition is complex, the other two are real.*

Proof. The first two decompositions given in (16) follow immediately from Lemma 1 and the fact that $\jmath, \imath_1$ are isomorphisms. See [15]. The last equation can be written as $\imath_2(h)\imath_2(a) = \imath_2(\sigma_+)\imath_2(h)$. Applying (4) one obtains $ah = h\sigma_+$ which coincides with the equation for σ_+ given in the beginning of the theorem. Matrix $\jmath(\sigma_+)$ is complex and diagonal: $\jmath(\sigma_+) = \mathrm{diag}(\sigma_+, \sigma_-)$. The other matrices $\imath_1(\sigma_+), \imath_2(\sigma_+)$ are upper Hessenberg with two real (2×2) blocks each:

$$\imath_1(\sigma_+) = \begin{pmatrix} a_1 & -|a_v| & 0 & 0 \\ |a_v| & a_1 & 0 & 0 \\ 0 & 0 & a_1 & -|a_v| \\ 0 & 0 & |a_v| & a_1 \end{pmatrix}, \quad \imath_2(\sigma_+) = \begin{pmatrix} a_1 & -|a_v| & 0 & 0 \\ |a_v| & a_1 & 0 & 0 \\ 0 & 0 & a_1 & |a_v| \\ 0 & 0 & -|a_v| & a_1 \end{pmatrix}.$$

□

If we have a look at the forms of $\imath_1$ and $\imath_2$, defined in (2), (3), respectively, we see that an upper (and lower) triangular matrix reduces immediately to a multiple of the identity matrix. This corresponds to the case where a is a real quaternion. Or in other words, it is not possible to find a complex Schur decomposition of $\imath_1(a), \imath_2(a)$ in $\mathbb{H}_{\mathbb{R}}, \mathbb{H}_{\mathbb{P}}$, respectively, if $a \notin \mathbb{R}$. In the mentioned paper [15, Section 8] we can also find, how to construct h which occurs in Theorem 3. One possibility is to put $h := \tilde{h}/|\tilde{h}|$, where

$$\tilde{h} := \begin{cases} (|a_v| + a_2, |a_v| + a_2, a_3 - a_4, a_3 + a_4) & \text{if } |a_3| + |a_4| > 0, \\ (1, 0, 0, 0) & \text{if } a_3 = a_4 = 0 \text{ and } a_2 > 0, \\ (0, 1, 0, 0) & \text{if } a_3 = a_4 = 0 \text{ and } a_2 < 0. \end{cases} \tag{17}$$

Let $\sigma_+ \sim a$ and multiply the defining equation $\sigma_+ = h^{-1}ah$ from the left by h, then $h\sigma_+ - ah = 0$ is the homogeneous form of Sylvester's equation and it was shown [16] that under the condition stated in (13) the homogeneous equation has a solution space (null space) which is a two dimensional subspace of $\mathbb{H}$ over $\mathbb{R}$.

2.2 The polar decomposition

The aim is to generalize the polar representation of a complex number. Let $z \in \mathbb{C}\backslash\{0\}$ be a complex number. Then, $z = |z|(z/|z|)$, and this representation of z is unique in the class of all two factor representations $z = pu$, where the first factor p is positive and the second, u, has modulus one. For matrices $\mathbf{A}$ one could correspondingly ask for a representation of the form $\mathbf{A} = \mathbf{PU}$, where the first factor $\mathbf{P}$ is positive semidefinite and the second, $\mathbf{U}$, is unitary. This is indeed possible, even for non square matrices $\mathbf{A} \in \mathbb{C}^{m\times n}, m \leqslant n$. Matrix $\mathbf{P}$ is always uniquely defined as $\mathbf{P} = (\mathbf{AA}^*)^{1/2}$ and $\mathbf{U}$ is uniquely defined if $\mathbf{A}$ has maximal rank m. If $\mathbf{A}$ is square and non singular, then $\mathbf{U} = \mathbf{P}^{-1}\mathbf{A}$. See Horn & Johnson [10, Theorem 7.3.2 and Corollary 7.3.3, pp. 412/413].

Let $a \in \mathbb{H}\backslash\{0\}$ be a non vanishing quaternion $a := (a_1, a_2, a_3, a_4)$. The quantity $a_v := (0, a_2, a_3, a_4)$ was called vector part of a as previously explained. The matrices $\jmath(a), \imath_1(a), \imath_2(a)$ are non singular square matrices where the columns are orthogonal to each other. See (11), (12) and its representation (in terms of quaternions) is obviously

$$a = |a|\frac{a}{|a|}. \tag{18}$$

The corresponding matrix representation in $\mathbb{H}_\mathbb{C}, \mathbb{H}_\mathbb{R}, \mathbb{H}_\mathbb{P}$, can be easily deduced by using (1) to (3) and the properties listed in (11), (12). We obtain

$$\jmath(a) = \operatorname{diag}(|a|, |a|)\,\jmath(\frac{a}{|a|}), \tag{19}$$

$$\imath_1(a) = \operatorname{diag}(|a|, |a|, |a|, |a|)\,\imath_1(\frac{a}{|a|}), \tag{20}$$

$$\imath_2(a) = \operatorname{diag}(|a|, |a|, |a|, |a|)\,\imath_2(\frac{a}{|a|}). \tag{21}$$

In all cases the first factor is positive definite and the second is unitary, orthogonal, respectively.

From a purely algebraic standpoint this representation of a is complete. However, already the name *polar* representation means more. In the complex case we have

$$\frac{z}{|z|} = \exp(\alpha\mathbf{i}), \quad z \neq 0$$

where $\alpha := \arg z$ is the angle between the x-axis and an arrow representing z emanating from the origin of the z-plane. As formula: $\alpha = \arctan(\Im z/\Re z)$. In the quaternionic case one finds (cf. [2, p. 11])

$$\frac{a}{|a|} = \exp(\alpha u), \quad a \neq 0,$$

with $u := a_v/|a_v|, \alpha := \arctan(|a_v|/a_1)$, and exp is defined by its Taylor series using $u^2 = -1$.

2.3 The singular value decomposition (SVD)

We start with the following well known theorem on a singular value decomposition of a given matrix $\mathbf{A}$. We restrict ourselves here to square matrices. The singular values of $\mathbf{A}$ are the square roots of the (non negative) eigenvalues of the positive semidefinite matrix $\mathbf{AA}^*$.

Theorem 4. *Let $\mathbf{A}$ be an arbitrary square matrix with real or complex entries. Then there are two unitary matrices $\mathbf{U}, \mathbf{V}$ of the same size as $\mathbf{A}$ such that*

$$\mathbf{D} := \mathbf{UAV}^*$$

is a diagonal matrix with the singular values of $\mathbf{A}$ in decreasing order on the diagonal. And the number of positive diagonal entries is the rank of $\mathbf{A}$.

Proof. See Horn & Johnson [10, 1991, p. 414]. □

Let a be a quaternion. The eigenvalues of $\jmath(a)$ are σ_+, σ_-, defined in (8) and

$$\jmath(a)\jmath(a)^* = \begin{pmatrix} |a|^2 & 0 \\ 0 & |a|^2 \end{pmatrix}.$$

Thus, the singular values of $\jmath(a)$ are $|a|, |a|$. The wanted decomposition must be of the form

$$\begin{pmatrix} |a| & 0 \\ 0 & |a| \end{pmatrix} = \mathbf{U} \begin{pmatrix} \alpha & \beta \\ -\overline{\beta} & \overline{\alpha} \end{pmatrix} \mathbf{V}^*$$

and the main question is whether $\mathbf{U}, \mathbf{V} \in \mathbb{H}_\mathbb{C}$. In order to solve this problem, we write it directly in terms of quaternions, namely

$$|a| = ua\overline{v}, \quad |u| = |v| = 1. \tag{22}$$

Theorem 5. *Let $a \in \mathbb{H}\backslash\mathbb{R}$. Choose $u \in \mathbb{H}$ with $|u| = 1$ and define $v := ua/|a|$ or, equivalently, choose v with $|v| = 1$ and define $u := v\overline{a}/|a|$. Then (22) defines a singular value decomposition of a and*

$$\jmath(|a|) = \jmath(u)\jmath(a)\jmath(v)^*$$

defines a corresponding SVD in $\mathbb{H}_\mathbb{C}$. A SVD with $u = v$ is impossible. The corresponding SVDs in $\mathbb{H}_\mathbb{R}$ and in $\mathbb{H}_\mathbb{P}$ are

$$\imath_1(|a|) = \imath_1(u)\imath_1(a)\imath_1(v)^{\mathrm{T}}, \quad \imath_2(|a|) = \imath_2(v)^{\mathrm{T}}\imath_2(a)\imath_2(u).$$

Proof. It is easy to see that (22) is valid if we choose u, v according to the given rules. If $u = v$ then $a = |a| \in \mathbb{R}$ follows, which was excluded. □

One very easy realization of (22) is to choose $u := 1$ and $v := a/|a|$ or to choose $v := 1$ and $u := \overline{a}/|a|$.

Example 1. Let $a := (1, 2, 2, 4)$. Then the three SVDs are:

$$\begin{pmatrix} 5 & 0 \\ 0 & 5 \end{pmatrix} = \begin{pmatrix} 1 & 0 \\ 0 & 1 \end{pmatrix} \begin{pmatrix} 1+2i & 2+4i \\ -2+4i & 1-2i \end{pmatrix} \begin{pmatrix} 1-2i & -2-4i \\ 2-4i & 1+2i \end{pmatrix} / 5.$$

$$\begin{pmatrix} 5 & 0 & 0 & 0 \\ 0 & 5 & 0 & 0 \\ 0 & 0 & 5 & 0 \\ 0 & 0 & 0 & 5 \end{pmatrix} = \begin{pmatrix} 1 & 0 & 0 & 0 \\ 0 & 1 & 0 & 0 \\ 0 & 0 & 1 & 0 \\ 0 & 0 & 0 & 1 \end{pmatrix} \begin{pmatrix} 1 & -2 & -2 & -4 \\ 2 & 1 & -4 & 2 \\ 2 & 4 & 1 & -2 \\ 4 & -2 & 2 & 1 \end{pmatrix} \begin{pmatrix} 1 & 2 & 2 & 4 \\ -2 & 1 & 4 & -2 \\ -2 & -4 & 1 & 2 \\ -4 & 2 & -2 & 1 \end{pmatrix} / 5.$$

$$\begin{pmatrix} 5 & 0 & 0 & 0 \\ 0 & 5 & 0 & 0 \\ 0 & 0 & 5 & 0 \\ 0 & 0 & 0 & 5 \end{pmatrix} = \begin{pmatrix} 1 & 2 & 2 & 4 \\ -2 & 1 & -4 & 2 \\ -2 & 4 & 1 & -2 \\ -4 & -2 & 2 & 1 \end{pmatrix} /5 \begin{pmatrix} 1 & -2 & -2 & -4 \\ 2 & 1 & 4 & -2 \\ 2 & -4 & 1 & 2 \\ 4 & 2 & -2 & 1 \end{pmatrix} \begin{pmatrix} 1 & 0 & 0 & 0 \\ 0 & 1 & 0 & 0 \\ 0 & 0 & 1 & 0 \\ 0 & 0 & 0 & 1 \end{pmatrix}.$$

2.4 The Jordan decomposition

Let $a := (a_1, a_2, a_3, a_4) \in \mathbb{H}\backslash\mathbb{R}$. Since the two eigenvalues $\sigma_\pm$ of $\jmath(a)$, defined in (8), are different there will be an $s \in \mathbb{H}\backslash\{0\}$ such that $a = s^{-1}\sigma_+ s$ which implies

$$\jmath(a) = \jmath(s^{-1})\jmath(\sigma_+)\jmath(s).$$

And this representation is the Jordan decomposition of $\jmath(a)$ and $\mathbf{J} := \jmath(\sigma_+) = \begin{pmatrix} \sigma_+ & 0 \\ 0 & \sigma_- \end{pmatrix}$ is the Jordan canonical form of $\jmath(a)$ [10, p. 126]. In this context this representation is almost the same as the Schur decomposition, only we do not require that $|s| = 1$. For the computation of s, we could use formula (17). In $\mathbb{H}_\mathbb{C}, \mathbb{H}_\mathbb{P}$ this decomposition reads

$$\imath_1(a) = \imath_1(s^{-1})\imath_1(\sigma_+)\imath_1(s), \quad \imath_2(a) = \imath_2(s)\imath_2(\sigma_+)\imath_2(s^{-1}),$$

where the explicit forms of $\imath_1(\sigma_+), \imath_2(\sigma_+)$ are given in the proof of Theorem 3.

2.5 The QR decomposition

Let $\mathbf{A}$ be an arbitrary complex square matrix. Then there is a unitary matrix $\mathbf{U}$ and an upper triangular matrix $\mathbf{R}$ of the same size as $\mathbf{A}$ such that

$$\mathbf{A} = \mathbf{UR}.$$

This well known theorem can be found in [10, p. 112]. And this decomposition is referred to as *QR-decomposition of* $\mathbf{A}$. All triangular matrices in $\mathbb{H}_\mathbb{C}$, in $\mathbb{H}_\mathbb{R}$, and in $\mathbb{H}_\mathbb{P}$ reduce to diagonal matrices. Therefore, the QR-decompositions of a quaternion $a \neq 0$ have the trivial form

$$a = \frac{a}{|a|}|a| \Leftrightarrow \jmath(a) = \jmath(\frac{a}{|a|})\,\jmath(|a|),\ \imath_1(a) = \imath_1(\frac{a}{|a|})\,\imath_1(|a|),\ \imath_2(a) = \imath_2(\frac{a}{|a|})\,\imath_2(|a|),$$

which is identical with the polar decomposition (18).

2.6 The LU decomposition

Let $\mathbf{A} \in \mathbb{C}^{n\times n}$ be given with entries a_{jk}, $j,k = 1,2,\dots,n$. Define the n submatrices $\mathbf{A}_\ell := (a_{jk})$, $j,k = 1,2,\dots,\ell, \ell = 1,2,\dots,n$. Then, following HORN & JOHNSON [10, p. 160] there is a lower triangular matrix $\mathbf{L}$ and an upper triangular matrix $\mathbf{U}$ such that

$$\mathbf{A} = \mathbf{LU}$$

if and only if all n submatrices $\mathbf{A}_\ell$, $\ell = 1,2,\dots,n$ are non singular. The above representation is called *LU-decomposition of* $\mathbf{A}$. Since triangular matrices in $\mathbb{H}_\mathbb{C}$, in $\mathbb{H}_\mathbb{R}$, and in $\mathbb{H}_\mathbb{P}$ reduce to diagonal matrices and since a product of two diagonal matrices is again diagonal an LU-decomposition of a quaternion a will in general not exist since $\jmath(a), \imath_1(a), \imath_2(a)$ are in general not diagonal. So we may ask for the ordinary LU-decomposion of $\jmath(a), \imath_1(a), \imath_2(a)$. In order that such a decomposition exist we must require that the mentioned submatrices are not singular. Let $a = (a_1, a_2, a_3, a_4)$. Then the two mentioned submatrices of $\jmath(a)$ are non singular if and only if the first (1×1) submatrix $\alpha := a_1 + a_2\mathbf{i} \neq 0$, since this implies that also the second (2×2) submatrix which is $\jmath(a)$ is non singular because its determinant is $|a|^2 = |\alpha|^2 + a_3^2 + a_4^2 > 0$.

Theorem 6. *Let* $a = (a_1, a_2, a_3, a_4) \in \mathbb{H}$. *Put* $\alpha := a_1 + a_2\mathbf{i}$ *and* $\beta := a_3 + a_4\mathbf{i}$. *An LU decomposition of* $\jmath(a)$ *exists if and only if* $\alpha \neq 0$. *If this condition is valid, then*

$$\jmath(a) = \begin{pmatrix} \alpha & \beta \\ -\overline{\beta} & \overline{\alpha} \end{pmatrix} = \begin{pmatrix} 1 & 0 \\ l_{21} & 1 \end{pmatrix}\begin{pmatrix} \alpha & \beta \\ 0 & u_{22} \end{pmatrix},$$

where

$$l_{21} = -\frac{\overline{\beta}}{\alpha}, \quad u_{22} = \frac{|\alpha|^2 + |\beta|^2}{\alpha} = \frac{|a|^2}{\alpha}.$$

Proof. The if and only part follows from the general theory. The above formula is easy to check. □

Theorem 7. *Let* $a = (a_1, a_2, a_3, a_4) \in \mathbb{H}$. *The four submatrices* $\mathbf{A}_l$ *of* $\imath_1(a)$ *and of* $\imath_2(a)$ *are non singular if and only if* $a_1 \neq 0$. *If this condition is valid, then*

$$\imath_1(a) := \begin{pmatrix} a_1 & -a_2 & -a_3 & -a_4 \\ a_2 & a_1 & -a_4 & a_3 \\ a_3 & a_4 & a_1 & -a_2 \\ a_4 & -a_3 & a_2 & a_1 \end{pmatrix} = \begin{pmatrix} 1 & 0 & 0 & 0 \\ l_{21} & 1 & 0 & 0 \\ l_{31} & l_{32} & 1 & 0 \\ l_{41} & l_{42} & l_{43} & 1 \end{pmatrix}\begin{pmatrix} a_1 & -a_2 & -a_3 & -a_4 \\ 0 & u_{22} & u_{23} & u_{24} \\ 0 & 0 & u_{33} & u_{34} \\ 0 & 0 & 0 & u_{44} \end{pmatrix},$$

where [results for $\imath_2(a)$ *are in parentheses]*

$$
\begin{aligned}
l_{j1} &:= a_j/a_1, j = 2,3,4, \quad (\textit{no change for } i_2(a)),\\
l_{32} &:= (a_1a_4 + a_2a_3)/(a_1^2 + a_2^2), (l_{32} := (-a_1a_4 + a_2a_3)/(a_1^2 + a_2^2) \textit{ for } \imath_2(a)),\\
l_{42} &:= (a_2a_4 - a_1a_3)/(a_1^2 + a_2^2), (l_{42} := (a_2a_4 + a_1a_3)/(a_1^2 + a_2^2) \textit{ for } \imath_2(a)),\\
u_{22} &:= (a_1^2 + a_2^2)/a_1, \quad (\textit{no change for } \imath_2(a)),\\
u_{23} &:= (-a_1a_4 + a_2a_3)/a_1, \quad (u_{23} := (a_1a_4 + a_2a_3)/a_1 \textit{ for } \imath_2(a)),\\
u_{24} &:= (a_1a_3 + a_2a_4)/a_1, \quad (u_{24} := (-a_1a_3 + a_2a_4)/a_1 \textit{ for } \imath_2(a)),\\
u_{33} &:= a_1 + l_{31}a_3 - l_{32}u_{23}, \quad (\textit{no change for } \imath_2(a)),\\
l_{43} &:= (a_2 + l_{41}a_3 - l_{42}u_{23})/u_{33}, (l_{43} := (-a_2 + l_{41}a_3 - l_{42}u_{23})/u_{33} \textit{ for } \imath_2(a)),\\
u_{34} &:= -a_2 + l_{31}a_4 - l_{32}u_{24}, \quad (u_{34} := a_2 + l_{31}a_4 - l_{32}u_{24} \textit{ for } \imath_2(a)),\\
u_{44} &:= a_1 + l_{41}a_4 - l_{42}u_{24} - l_{43}u_{34}, \quad (\textit{no change for } \imath_2(a)).
\end{aligned}
$$

A Cholesky decomposition cannot be achieved since all three matrices $\jmath(a)$, $\imath_1(a)$, $\imath_2(a)$ are missing symmetry.

Acknowledgment. The authors acknowledge with pleasure the support of the Grant Agency of the Czech Republic (grant No. 201/06/0356). The work is a part of the research project MSM 6046137306 financed by MSMT, Ministry of Education, Youth and Sports, Czech Republic.

References

1. J. FAN, *Determinants and multiplicative functionals on quaternionic matrices*, Linear Algebra Appl. **369** (2003), 193–201.
2. P. R. GIRARD, Quaternions, Clifford Algebras and Relativistic Physics, Birkhäuser, Basel, Boston, Berlin, 2007, 179 p.
3. R. A. HORN & C. R. JOHNSON, Matrix Analysis, Cambridge University Press, Cambridge, New York, 1992, 561 p.
4. R. A. HORN & C. R. JOHNSON, Topics in Matrix Analysis, Cambridge University Press, Cambridge, New York, 1991, 607 p.
5. D. JANOVSKÁ & G. OPFER, *Givens' transformation applied to quaternion valued vectors*, BIT **43** (2003), Suppl., 991–1002.
6. D. JANOVSKÁ & G. OPFER, *Fast Givens Transformation for Quaternionic Valued Matrices Applied to Hessenberg Reductions*, ETNA **20** (2005), 1–26.
7. D. JANOVSKÁ & G. OPFER, *Linear equations in quaternions*, In: Numerical Mathematics and Advanced Applications, Proceedings of ENUMATH 2005, A. B. de Castro, D. Gómez, P. Quintela, and P. Salgado, eds., Springer Verlag, New York, 2006, pp. 946–953.
8. D. JANOVSKÁ & G. OPFER, *Computing quaternionic roots by Newton's method*, ETNA **26** (2007), pp. 82–102.
9. D. JANOVSKÁ & G. OPFER, *On one linear equation in one quaternionic unknown*,

10. R. A. HORN & C. R. JOHNSON, Matrix Analysis, Cambridge University Press, Cambridge, New York, 1992, 561 p.
11. R. A. HORN & C. R. JOHNSON, Topics in Matrix Analysis, Cambridge University Press, Cambridge, New York, 1991, 607 p.
12. D. JANOVSKÁ & G. OPFER, *Givens' transformation applied to quaternion valued vectors*, BIT **43** (2003), Suppl., 991–1002.
13. D. JANOVSKÁ & G. OPFER, *Fast Givens Transformation for Quaternionic Valued Matrices Applied to Hessenberg Reductions*, ETNA **20** (2005), 1–26.
14. D. JANOVSKÁ & G. OPFER, *Linear equations in quaternions*, In: Numerical Mathematics and Advanced Applications, Proceedings of ENUMATH 2005, A. B. de Castro, D. Gómez, P. Quintela, and P. Salgado, eds., Springer Verlag, New York, 2006, pp. 946–953.
15. D. JANOVSKÁ & G. OPFER, *Computing quaternionic roots by Newton's method*, ETNA **26** (2007), pp. 82–102.
16. D. JANOVSKÁ & G. OPFER, *On one linear equation in one quaternionic unknown*, Hamburger Beiträge zur Angewandten Mathematik, Nr. 2007-14, September 2007, 34 p., dedicated to Bernd Fischer on the occasion of his 50th birthday.
17. D. JANOVSKÁ & G. OPFER, *Linear equations in quaternionic variables*, Mitt. Math. Ges. Hamburg **27** (2008), 223–234.
18. B. L. VAN DER WAERDEN, Algebra I, 5th ed., Springer, Berlin Göttingen, Heidelberg, 1960, 292 p.
19. F. ZHANG, *Quaternions and matrices of quaternions*, Linear Algebra Appl. **251** (1997), 21–57.

Sensitivity Analysis of Hamiltonian and Reversible Systems Prone to Dissipation-Induced Instabilities

Oleg N. Kirillov*

Institute of Mechanics, Moscow State Lomonosov University, Michurinskii pr. 1, 119192 Moscow, Russia
kirillov@imec.msu.ru;
Department of Mechanical Engineering, Technische Universität Darmstadt, Hochschulstr. 1, 64289 Darmstadt, Germany
kirillov@dyn.tu-darmstadt.de

Abstract. Stability of a linear autonomous non-conservative system in the presence of potential, gyroscopic, dissipative, and non-conservative positional forces is studied. The cases when the non-conservative system is close either to a gyroscopic system or to a circulatory one, are examined. It is known that marginal stability of gyroscopic and circulatory systems can be destroyed or improved up to asymptotic stability due to action of small non-conservative positional and velocity-dependent forces. We show that in both cases the boundary of the asymptotic stability domain of the perturbed system possesses singularities such as "Dihedral angle", "Break of an edge" and "Whitney's umbrella" that govern stabilization and destabilization as well as are responsible for the imperfect merging of modes. Sensitivity analysis of the critical parameters is performed with the use of the perturbation theory for eigenvalues and eigenvectors of non-self-adjoint operators. In case of two degrees of freedom, stability boundary is found in terms of the invariants of matrices of the system. Bifurcation of the stability domain due to change of the structure of the damping matrix is described. As a mechanical example, the Hauger gyropendulum is analyzed in detail; an instability mechanism in a general mechanical system with two degrees of freedom, which originates after discretization of models of a rotating disc in frictional contact and possesses the spectral mesh in the plane 'frequency' versus 'angular velocity', is analytically described and its role in the excitation of vibrations in the squealing disc brake and in the singing wine glass is discussed.

Keywords: matrix polynomial, Hamiltonian system, reversible system, Lyapunov stability, indefinite damping, perturbation, dissipation-induced instabilities, destabilization paradox, multiple eigenvalue, singularity.

* The work has been partly supported by the Alexander von Humboldt Foundation and by the German Research Foundation, Grant DFG HA 1060/43-1.

1 Introduction

Consider an autonomous non-conservative system

$$\ddot{\mathbf{x}} + (\Omega\mathbf{G} + \delta\mathbf{D})\dot{\mathbf{x}} + (\mathbf{K} + \nu\mathbf{N})\mathbf{x} = 0, \tag{1}$$

where dot stands for the time differentiation, $\mathbf{x} \in \mathbb{R}^m$, and real matrix $\mathbf{K} = \mathbf{K}^T$ corresponds to potential forces. Real matrices $\mathbf{D} = \mathbf{D}^T$, $\mathbf{G} = -\mathbf{G}^T$, and $\mathbf{N} = -\mathbf{N}^T$ are related to dissipative (damping), gyroscopic, and non-conservative positional (circulatory) forces with magnitudes controlled by scaling factors δ, Ω, and ν respectively. A *circulatory* system is obtained from (1) by neglecting velocity-dependent forces

$$\ddot{\mathbf{x}} + (\mathbf{K} + \nu\mathbf{N})\mathbf{x} = 0, \tag{2}$$

while a *gyroscopic* one has no damping and non-conservative positional forces

$$\ddot{\mathbf{x}} + \Omega\mathbf{G}\dot{\mathbf{x}} + \mathbf{K}\mathbf{x} = 0. \tag{3}$$

Circulatory and gyroscopic systems (2) and (3) possess fundamental symmetries that are evident after transformation of equation (1) to the form $\dot{\mathbf{y}} = \mathbf{A}\mathbf{y}$ with

$$\mathbf{A} = \begin{bmatrix} -\frac{1}{2}\Omega\mathbf{G} & \mathbf{I} \\ \frac{1}{2}\delta\Omega\mathbf{D}\mathbf{G} + \frac{1}{4}\Omega^2\mathbf{G}^2 - \mathbf{K} - \nu\mathbf{N} & \delta\mathbf{D} - \frac{1}{2}\Omega\mathbf{G} \end{bmatrix}, \quad \mathbf{y} = \begin{bmatrix} \mathbf{x} \\ \dot{\mathbf{x}} + \frac{1}{2}\Omega\mathbf{G}\mathbf{x} \end{bmatrix}, \tag{4}$$

where $\mathbf{I}$ is the identity matrix.

In the absence of damping and gyroscopic forces ($\delta = \Omega = 0$), $\mathbf{RAR} = -\mathbf{A}$ with

$$\mathbf{R} = \mathbf{R}^{-1} = \begin{bmatrix} \mathbf{I} & 0 \\ 0 & -\mathbf{I} \end{bmatrix}. \tag{5}$$

This means that the matrix $\mathbf{A}$ has a *reversible* symmetry, and equation (2) describes a reversible dynamical system [16, 19, 33]. Due to this property,

$$\det(\mathbf{A} - \lambda\mathbf{I}) = \det(\mathbf{R}(\mathbf{A} - \lambda\mathbf{I})\mathbf{R}) = \det(\mathbf{A} + \lambda\mathbf{I}), \tag{6}$$

and the eigenvalues of circulatory system (2) appear in pairs $(-\lambda, \lambda)$. Without damping and non-conservative positional forces ($\delta = \nu = 0$) the matrix $\mathbf{A}$ possesses the *Hamiltonian* symmetry $\mathbf{JAJ} = \mathbf{A}^T$, where $\mathbf{J}$ is a unit symplectic matrix [17, 23, 28]

$$\mathbf{J} = -\mathbf{J}^{-1} = \begin{bmatrix} 0 & \mathbf{I} \\ -\mathbf{I} & 0 \end{bmatrix}. \tag{7}$$

As a consequence,

$$\det(\mathbf{A} - \lambda\mathbf{I}) = \det(\mathbf{J}(\mathbf{A} - \lambda\mathbf{I})\mathbf{J}) = \det(\mathbf{A}^T + \lambda\mathbf{I}) = \det(\mathbf{A} + \lambda\mathbf{I}), \tag{8}$$

which implies that if λ is an eigenvalue of $\mathbf{A}$ then so is $-\lambda$, similarly to the reversible case. Therefore, an equilibrium of a circulatory or of a gyroscopic system is either unstable or all its eigenvalues lie on the imaginary axis of the complex plane implying marginal stability if they are semi-simple.

In the presence of all the four forces, the Hamiltonian and reversible symmetries are broken and the marginal stability is generally destroyed. Instead, system (1) can be asymptotically stable if its characteristic polynomial

$$P(\lambda) = \det(\mathbf{I}\lambda^2 + (\Omega\mathbf{G} + \delta\mathbf{D})\lambda + \mathbf{K} + \nu\mathbf{N}), \tag{9}$$

satisfies the criterion of Routh and Hurwitz. The most interesting for many applications, ranging from the rotor dynamics [3–5, 14, 25, 27, 30, 31, 48, 49, 59, 62] to physics of the atmosphere [9, 29, 62, 66] and from stability and optimization of structures [8, 10, 11, 15, 22, 26, 33, 39, 54, 55, 65, 69] to friction-induced instabilities and acoustics of friction [40, 42, 61, 67, 71–73, 75, 76], is the situation when system (1) is close either to circulatory system (2) with $\delta, \Omega \ll \nu$ (*near-reversible system*) or to gyroscopic system (3) with $\delta, \nu \ll \Omega$ (*near-Hamiltonian system*). The effect of small damping and gyroscopic forces on the stability of circulatory systems as well as the effect of small damping and non-conservative positional forces on the stability of gyroscopic systems are regarded as *paradoxical*, since the stability properties are extremely sensitive to the choice of the perturbation, and the balance of forces resulting in the asymptotic stability is not evident, as it happens in such phenomena as "tippe top inversion", "rising egg", and the onset of friction-induced oscillations in the squealing brake and in the singing wine glass [31, 48, 49, 59, 61, 62, 67, 71–73, 75–77].

Historically, Thomson and Tait in 1879 were the first who found that dissipation destroys the gyroscopic stabilization (*dissipation-induced instability*) [1, 28, 62, 66]. A similar effect of non-conservative positional forces on the stability of gyroscopic systems has been established almost a century later by Lakhadanov and Karapetyan [12, 13]. A more sophisticated manifestation of the dissipation-induced instabilities has been discovered by Ziegler on the example of a double pendulum loaded by a follower force with the damping, non-uniformly distributed among the natural modes [8]. Without dissipation, the Ziegler pendulum is a reversible system, which is marginally stable for the loads non-exceeding some critical value. Small dissipation of order $o(1)$ makes the pendulum either unstable or asymptotically stable with the critical load, which generically is lower than that of the undamped system by the quantity of order $O(1)$ (*the destabilization paradox*). Similar discontinuous change in the stability domain for the near-Hamiltonian systems has been observed by Holopainen [9, 66] in his study of the effect of dissipation on the stability of baroclinic waves in Earth's atmosphere, by Hoveijn and Ruijgrok on the example of a rotating shaft on an elastic foundation [30], and by Crandall, who investigated a gyroscopic pendulum with stationary and rotating damping [31]. Contrary to the Ziegler pendulum, the undamped gyropendulum is a gyroscopic system that is marginally stable when its spin exceeds a critical value. Despite the stationary damping, corresponding to a dissipative velocity-dependent force, destroys the gyroscopic stabilization [1], the Crandall gyropendulum with stationary and rotating damping, where the latter is related to a non-conservative positional force, can be asymptotically stable for the rotation rates exceeding considerably the critical spin of the undamped system. This is an example of the destabilization paradox in the Hamiltonian system.

As it was understood during the last decade, the reason underlying the destabilization paradox is that the multiparameter family of non-normal matrix operators of the system (1) generically possesses the multiple eigenvalues related to singularities of the boundary of the asymptotic stability domain, which were described and classified by Arnold already in 1970-s [17]. Hoveijn and Ruijgrok were, apparently, the first who associated the discontinuous change in the critical load in their example to the singularity Whitney umbrella, existing on the stability boundary [30]. The same singularity on the boundary of the asymptotic stability has been identified for the Ziegler pendulum [47], for the models of disc brakes [72, 76], of the rods loaded by follower force [54, 55], and of the gyropendulums and spinning tops [63, 70]. These examples reflect the general fact that the codimension-1 Hamiltonian (or reversible) Hopf bifurcation can be viewed as a singular limit of the codimension-3 dissipative resonant $1:1$ normal form and the essential singularity in which these two cases meet is topologically equivalent to Whitney's umbrella (Hamilton meets Hopf under Whitney's umbrella) [45, 66].

Despite the achieved qualitative understanding, the development of the sensitivity analysis for the critical parameters near the singularities, which is essential for controlling the stabilization and destabilization, is only beginning and is involving such modern disciplines as multiparameter perturbation theory of analytical matrix functions [7, 18, 20, 23, 24, 28, 29, 37, 41, 57, 58] and of non-self-adjoint boundary eigenvalue problems [51, 53–55], the theory of the structured pseudospectra of matrix polynomials [56, 73] and the theory of versal deformations of matrix families [30, 45, 47, 60]. The growing number of physical and mechanical applications demonstrating the destabilization paradox due to an interplay of non-conservative effects and the need for a justification for the use of Hamiltonian or reversible models to describe real-world systems that are in fact only near-Hamiltonian or near-reversible requires a unified treatment of this phenomenon.

The goal of the present paper is to find and to analyze the domain of asymptotic stability of system (1) in the space of the parameters δ, Ω, and ν with special attention to near-reversible and near-Hamiltonian cases. In the subsequent sections we will combine the study of the two-dimensional system, analyzing the Routh-Hurwitz stability conditions, with the perturbative approach to the case of arbitrary large m. Typical singularities of the stability boundary will be identified. Bifurcation of the domain of asymptotic stability due to change of the structure of the matrix $\mathbf{D}$ of dissipative forces will be thoroughly analyzed and the effect of gyroscopic stabilization of a dissipative system with indefinite damping and non-conservative positional forces will be described. The estimates of the critical parameters and explicit expressions, approximating the boundary of the asymptotic stability domain, will be extended to the case of $m > 2$ degrees of freedom with the use of the perturbation theory of multiple eigenvalues of non-self-adjoint operators. In the last section the general theory will be applied to the study of the onset of stabilization and destabilization in the models of gyropendulums and disc brakes.

2 A circulatory system with small velocity-dependent forces

We begin with the near-reversible case ($\delta, \Omega \ll \nu$), which covers Ziegler's and Nikolai's pendulums loaded by the follower force [8, 10, 11, 33, 47, 43, 44, 53, 66] (their continuous analogue is the viscoelastic Beck column [10, 39, 54, 55]), the Reut-Sugiyama pendulum [50], the low-dimensional models of disc brakes by North [67, 73], Popp [40], and Sinou and Jezequel [72], the model of a mass sliding over a conveyor belt by Hoffmann and Gaul [42], the models of rotors with internal and external damping by Kimball and Smith [3, 4] and Kapitsa [5, 66], and finds applications even in the modeling of the two-legged walking and of the dynamics of space tethers [32].

2.1 Stability of a circulatory system

Stability of system (1) is determined by its characteristic polynomial (8), which in case of two degrees of freedom has a convenient form provided by the Leverrier-Barnett algorithm [21]

$$P(\lambda, \delta, \nu, \Omega) = \lambda^4 + \delta \mathrm{tr}\mathbf{D}\,\lambda^3 + (\mathrm{tr}\mathbf{K} + \delta^2 \det \mathbf{D} + \Omega^2)\,\lambda^2 + \\ (\delta(\mathrm{tr}\mathbf{K}\mathrm{tr}\mathbf{D} - \mathrm{tr}\mathbf{K}\mathbf{D}) + 2\Omega\nu)\,\lambda + \det \mathbf{K} + \nu^2, \qquad (10)$$

where without loss of generality we assume that $\det \mathbf{G} = 1$ and $\det \mathbf{N} = 1$.

In the absence of damping and gyroscopic forces ($\delta = \Omega = 0$) the system (1) is circulatory, and the polynomial (10) has four roots $-\lambda_+$, $-\lambda_-$, λ_-, and λ_+, where

$$\lambda_\pm = \sqrt{-\frac{1}{2}\mathrm{tr}\mathbf{K} \pm \frac{1}{2}\sqrt{(\mathrm{tr}\mathbf{K})^2 - 4(\det \mathbf{K} + \nu^2)}}. \qquad (11)$$

The eigenvalues (11) can be real, complex or purely imaginary implying instability or marginal stability in accordance with the following statement.

Proposition 1. *If* $\mathrm{tr}\mathbf{K} > 0$ *and* $\det \mathbf{K} \leqslant 0$, *circulatory system (2) with two degrees of freedom is stable for* ${\nu_d}^2 < \nu^2 < {\nu_f}^2$, *unstable by divergence for*

$\nu^2 \leqslant {\nu_d}^2$, and unstable by flutter for $\nu^2 \geqslant {\nu_f}^2$, where the critical values ν_d and ν_f are

$$0 \leqslant \sqrt{-\det \mathbf{K}} =: \nu_d \leqslant \nu_f := \frac{1}{2}\sqrt{(\mathrm{tr}\mathbf{K})^2 - 4\det \mathbf{K}}. \tag{12}$$

If $\mathrm{tr}\mathbf{K} > 0$ and $\det \mathbf{K} > 0$, the circulatory system is stable for $\nu^2 < {\nu_f}^2$ and unstable by flutter for $\nu^2 \geqslant {\nu_f}^2$.
If $\mathrm{tr}\mathbf{K} \leqslant 0$, the system is unstable.

The proof is a consequence of formula (11), reversible symmetry, and the fact that time dependence of solutions of equation (2) is given by $\exp(\lambda t)$ for simple eigenvalues λ, with an additional—polynomial in t—prefactor (secular terms) in case of multiple eigenvalues with the Jordan block. The solutions monotonously grow for positive real λ implying static instability (divergence), oscillate with an increasing amplitude for complex λ with positive real part (flutter), and remain bounded when λ is semi-simple and purely imaginary (stability). For $\mathbf{K}$, having two equal eigenvalues, $\nu_f = 0$ and the circulatory system (2) is unstable in agreement with the Merkin theorem for circulatory systems with two degrees of freedom [34, 62].

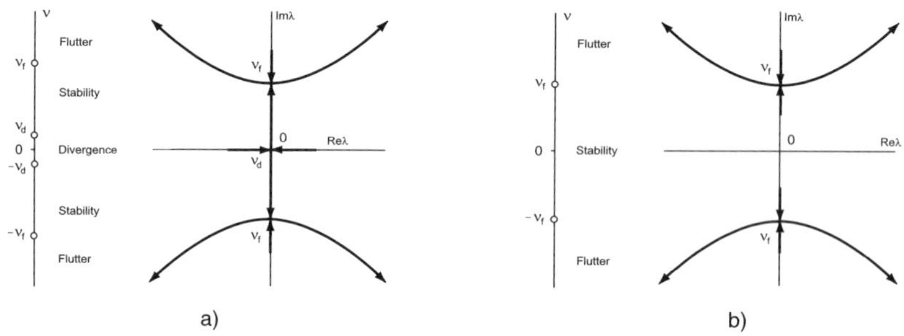

Fig. 1. Stability diagrams and trajectories of eigenvalues for the increasing parameter $\nu > 0$ for the circulatory system (2) with $\mathrm{tr}\mathbf{K} > 0$ and $\det \mathbf{K} < 0$ (a) and $\mathrm{tr}\mathbf{K} > 0$ and $\det \mathbf{K} > 0$ (b).

Stability diagrams and motion of eigenvalues in the complex plane for ν increasing from zero are presented in Fig. 1. When $\mathrm{tr}\mathbf{K} > 0$ and $\det \mathbf{K} < 0$ there are two real and two purely imaginary eigenvalues at $\nu = 0$, and the system is statically unstable, see Fig. 1(a). With the increase of ν both the imaginary and real eigenvalues are moving to the origin, until at $\nu = \nu_d$ the real pair merges and originates a double zero eigenvalue with the Jordan block. At $\nu = \nu_d$ the system is unstable due to linear time dependence of a solution corresponding to $\lambda = 0$. The further increase of ν yields splitting of the double zero eigenvalue into two purely imaginary ones. The imaginary eigenvalues of the same sign are then moving towards each other until at $\nu = \nu_f$ they originate a pair of double eigenvalues $\pm i\omega_f$ with the Jordan block, where

$$\omega_f = \sqrt{\frac{1}{2}\mathrm{tr}\mathbf{K}}. \tag{13}$$

At $\nu = \nu_f$ the system is unstable by flutter due to secular terms in its solutions. For $\nu > \nu_f$ the flutter instability is caused by two of the four complex eigenvalues lying on the branches of a hyperbolic curve

$$\mathrm{Im}\,\lambda^2 - \mathrm{Re}\,\lambda^2 = \omega_f^2. \tag{14}$$

The critical values ν_d and ν_f constitute the boundaries between the divergence and stability domains and between the stability and flutter domains respectively. For $\mathrm{tr}\mathbf{K} > 0$ and $\det \mathbf{K} = 0$ the divergence domain shrinks to a point $\nu_d = 0$ and for $\mathrm{tr}\mathbf{K} > 0$ and $\det \mathbf{K} > 0$ there exist only stability and flutter domains as shown in Fig. 1(b). For negative ν the boundaries of the divergence and flutter domains are $\nu = -\nu_d$ and $\nu = -\nu_f$.

In general, the Jordan chain for the eigenvalue $i\omega_f$ consists of an eigenvector $\mathbf{u}_0$ and an associated vector $\mathbf{u}_1$ that satisfy the equations [53]

$$(-\omega_f^2\mathbf{I} + \mathbf{K} + \nu_f\mathbf{N})\mathbf{u}_0 = 0, \quad (-\omega_f^2\mathbf{I} + \mathbf{K} + \nu_f\mathbf{N})\mathbf{u}_1 = -2i\omega_f\mathbf{u}_0. \tag{15}$$

Due to the non-self-adjointness of the matrix operator, the same eigenvalue possesses the left Jordan chain of generalized eigenvectors $\mathbf{v}_0$ and $\mathbf{v}_1$

$$\mathbf{v}_0^T(-\omega_f^2\mathbf{I} + \mathbf{K} + \nu_f\mathbf{N}) = 0, \quad \mathbf{v}_1^T(-\omega_f^2\mathbf{I} + \mathbf{K} + \nu_f\mathbf{N}) = -2i\omega_f\mathbf{v}_0^T. \tag{16}$$

The eigenvalues $\mathbf{u}_0$ and $\mathbf{v}_0$ are biorthogonal

$$\mathbf{v}_0^T\mathbf{u}_0 = 0. \tag{17}$$

In the neighborhood of $\nu = \nu_f$ the double eigenvalue and the corresponding eigenvectors vary according to the formulas [52, 53]

$$\begin{aligned}
\lambda(\nu) &= i\omega_f \pm \mu\sqrt{\nu - \nu_f} + o((\nu - \nu_f)^{\frac{1}{2}}),\\
\mathbf{u}(\nu) &= \mathbf{u}_0 \pm \mu\mathbf{u}_1\sqrt{\nu - \nu_f} + o((\nu - \nu_f)^{\frac{1}{2}}),\\
\mathbf{v}(\nu) &= \mathbf{v}_0 \pm \mu\mathbf{v}_1\sqrt{\nu - \nu_f} + o((\nu - \nu_f)^{\frac{1}{2}}),
\end{aligned} \tag{18}$$

where μ^2 is a real number given by

$$\mu^2 = -\frac{\mathbf{v}_0^T\mathbf{N}\mathbf{u}_0}{2i\omega_f\mathbf{v}_0^T\mathbf{u}_1}. \tag{19}$$

For $m = 2$ the generalized eigenvectors of the right and left Jordan chains at the eigenvalue $i\omega_f$, where the eigenfrequency is given by (13) and the critical value ν_f is defined by (12), are [52]

$$\mathbf{u}_0 = \begin{bmatrix} 2k_{12} + 2\nu_f \\ k_{22} - k_{11} \end{bmatrix}, \quad \mathbf{v}_0 = \begin{bmatrix} 2k_{12} - 2\nu_f \\ k_{22} - k_{11} \end{bmatrix}, \quad \mathbf{u}_1 = \mathbf{v}_1 = \begin{bmatrix} 0 \\ -4i\omega_f \end{bmatrix}. \tag{20}$$

Substituting (20) into equation (19) yields the expression

$$\mu^2 = -\frac{4\nu_f(k_{11} - k_{22})}{2i\omega_f \mathbf{v}_0^T \mathbf{u}_1} = \frac{\nu_f}{2\omega_f^2} > 0. \tag{21}$$

After plugging the real-valued coefficient μ into expansions (18) we obtain an approximation of order $|\nu - \nu_f|^{1/2}$ of the exact eigenvalues $\lambda = \lambda(\nu)$. This can be verified by the series expansions of (11) about $\nu = \nu_f$.

2.2 The influence of small damping and gyroscopic forces on the stability of a circulatory system

The one-dimensional domain of marginal stability of circulatory system (2) given by Proposition 1 blows up into a three-dimensional domain of asymptotic stability of system (1) in the space of the parameters δ, Ω, and ν, which is described by the Routh and Hurwitz criterion for the polynomial (10)

$$\delta \mathrm{tr}\mathbf{D} > 0, \;\; \mathrm{tr}\mathbf{K} + \delta^2 \det \mathbf{D} + \Omega^2 > 0, \;\; \det \mathbf{K} + \nu^2 > 0, \;\; Q(\delta, \Omega, \nu) > 0, \tag{22}$$

where

$$\begin{aligned} Q &:= -q^2 + \delta \mathrm{tr}\mathbf{D}(\mathrm{tr}\mathbf{K} + \delta^2 \det \mathbf{D} + \Omega^2)q - (\delta \mathrm{tr}\mathbf{D})^2(\det \mathbf{K} + \nu^2), \\ q &:= \delta(\mathrm{tr}\mathbf{K}\mathrm{tr}\mathbf{D} - \mathrm{tr}\mathbf{K}\mathbf{D}) + 2\Omega\nu. \end{aligned} \tag{23}$$

Considering the asymptotic stability domain (22) in the space of the parameters δ, ν and Ω we remind that the initial system (1) is equivalent to the first-order system with the real $2m \times 2m$ matrix $\mathbf{A}(\delta, \nu, \Omega)$ defined by expression (4). As it was established by Arnold [17], the boundary of the asymptotic stability domain of a multiparameter family of real matrices is not a smooth surface. Generically, it possesses singularities corresponding to multiple eigenvalues with zero real part. Applying the qualitative results of [17], we deduce that the parts of the ν-axis belonging to the stability domain of system (2) and corresponding to two different pairs of simple purely imaginary eigenvalues, form edges of the *dihedral angles* on the surfaces that bound the asymptotic stability domain of system (1), see Fig. 2(a). At the points $\pm\nu_f$ of the ν-axis, corresponding to the stability-flutter boundary of system (2) there exists a pair of double purely imaginary eigenvalues with the Jordan block. Qualitatively, the asymptotic stability domain of system (1) in the space (δ, ν, Ω) near the ν-axis

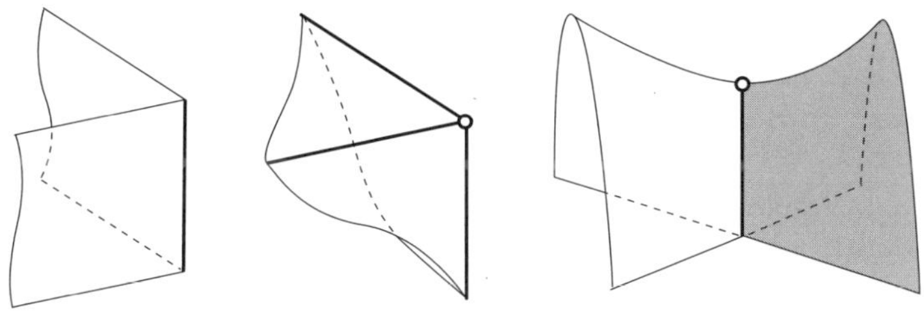

Fig. 2. Singularities *dihedral angle* (a), *trihedral angle* (b), and *deadlock of an edge* (or a half of the *Whitney umbrella* (c)) of the boundary of the asymptotic stability domain.

looks like a dihedral angle which becomes more acute while approaching the points $\pm\nu_f$. At these points the angle shrinks forming the *deadlock of an edge*, which is a half of the *Whitney umbrella* surface [17, 30, 45], see Fig. 2(c). In case when the stability domain of the circulatory system has a common boundary with the divergence domain, as shown in Fig. 1(a), the boundary of the asymptotic stability domain of the perturbed system (1) possesses the *trihedral angle* singularity at $\nu = \pm\nu_d$, see Fig. 2(b).

The first two of the conditions of asymptotic stability (22) restrict the region of variation of parameters δ and Ω either to a half-plane $\delta \mathrm{tr}\mathbf{D} > 0$, if $\det \mathbf{D} \geqslant 0$, or to a space between the line $\delta = 0$ and one of the branches of a hyperbola $|\det \mathbf{D}|\, \delta^2 - \Omega^2 = 2\omega_f^2$, if $\det \mathbf{D} < 0$. Provided that δ and Ω belong to the described domain, the asymptotic stability of system (1) is determined by the last two of the inequalities (22), which impose limits on the variation of ν. Solving the quadratic in ν equation $Q(\delta, \nu, \Omega) = 0$ we write the stability condition $Q > 0$ in the form

$$(\nu - \nu_{cr}^-)(\nu - \nu_{cr}^+) < 0, \tag{24}$$

with

$$\nu_{cr}^{\pm}(\delta, \Omega) = \frac{\Omega b \pm \sqrt{\Omega^2 b^2 + ac}}{a}\,\delta. \tag{25}$$

The coefficients a, b, and c are

$$\begin{aligned} a(\delta,\Omega) &= 4\Omega^2 + \delta^2(\mathrm{tr}\mathbf{D})^2, \quad b(\delta,\Omega) = 4\nu_f\beta_* + (\delta^2 \det \mathbf{D} + \Omega^2)\mathrm{tr}\mathbf{D}, \\ c(\delta,\Omega) &= \nu_f^2((\mathrm{tr}\mathbf{D})^2 - 4\beta_*^2) + (\omega_f^2 \mathrm{tr}\mathbf{D} - 2\nu_f\beta_*)(\delta^2 \det \mathbf{D} + \Omega^2)\mathrm{tr}\mathbf{D}, \end{aligned} \tag{26}$$

where

$$\beta_* := \frac{\mathrm{tr}(\mathbf{K} - \omega_f^2 \mathbf{I})\mathbf{D}}{2\nu_f}. \tag{27}$$

For $\det \mathbf{K} \leqslant 0$, the domain of asymptotic stability consists of two non-intersecting parts, bounded by the surfaces $\nu = \nu_{cr}^{\pm}(\delta, \Omega)$ and by the planes $\nu = \pm\nu_d$, separating it from the divergence domain. For $\det \mathbf{K} > 0$, inequality $\det \mathbf{K} + \nu^2 > 0$ is fulfilled, and in accordance with the condition (24) the asymptotic stability domain is contained between the surfaces $\nu = \nu_{cr}^{+}(\delta, \Omega)$ and $\nu = \nu_{cr}^{-}(\delta, \Omega)$.

The functions $\nu_{cr}^{\pm}(\delta, \Omega)$ defined by expressions (25) are singular at the origin due to vanishing denominator. Assuming $\Omega = \beta\delta$ and calculating a limit of these functions when δ tends to zero, we obtain

$$\nu_0^{\pm}(\beta) := \lim_{\delta \to 0} \nu_{cr}^{\pm} = \nu_f \frac{4\beta\beta_* \pm \mathrm{tr}\mathbf{D}\sqrt{(\mathrm{tr}\mathbf{D})^2 + 4(\beta^2 - \beta_*^2)}}{(\mathrm{tr}\mathbf{D})^2 + 4\beta^2}. \tag{28}$$

The functions $\nu_0^{\pm}(\beta)$ are real-valued if the radicand in (28) is non-negative.

Proposition 2. *Let $\lambda_1(\mathbf{D})$ and $\lambda_2(\mathbf{D})$ be eigenvalues of $\mathbf{D}$. Then,*

$$|\beta_*| \leqslant \frac{|\lambda_1(\mathbf{D}) - \lambda_2(\mathbf{D})|}{2}. \tag{29}$$

If $\mathbf{D}$ is semi-definite ($\det \mathbf{D} \geqslant 0$) or indefinite with

$$0 > \det \mathbf{D} \geqslant -\frac{(k_{12}(d_{22} - d_{11}) - d_{12}(k_{22} - k_{11}))^2}{4\nu_f^2}, \tag{30}$$

then

$$|\beta_*| \leqslant \frac{|\mathrm{tr}\mathbf{D}|}{2}, \tag{31}$$

and the limits $\nu_0^{\pm}(\beta)$ are continuous real-valued functions of β. Otherwise, there exists an interval of discontinuity $\beta^2 < \beta_*^2 - (\mathrm{tr}\mathbf{D})^2/4$.

Proof. With the use of the definition of β_*, (27), a series of transformations

$$\begin{aligned}
\beta_*^2 - \frac{(\mathrm{tr}\mathbf{D})^2}{4} &= \frac{1}{4\nu_f^2}\left(\frac{(k_{11} - k_{22})(d_{11} - d_{22})}{2} + 2k_{12}d_{12}\right)^2 \\
&\quad - \frac{(d_{11} + d_{22})^2}{4}\frac{((k_{11} - k_{22})^2 + 4k_{12}^2)}{4\nu_f^2} \\
&= -\det \mathbf{D} - \frac{(k_{12}(d_{22} - d_{11}) - d_{12}(k_{22} - k_{11}))^2}{4\nu_f^2}
\end{aligned} \tag{32}$$

yields the expression

$$\beta_*^2 = \frac{(\lambda_1(\mathbf{D}) - \lambda_2(\mathbf{D}))^2}{4} - \frac{(k_{12}(d_{22} - d_{11}) - d_{12}(k_{22} - k_{11}))^2}{4\nu_f^2}. \tag{33}$$

For real β_*, formula (32) implies inequality (30). The remaining part of the proposition follows from (33).

Inequality (30) subdivides the set of indefinite damping matrices into two classes.

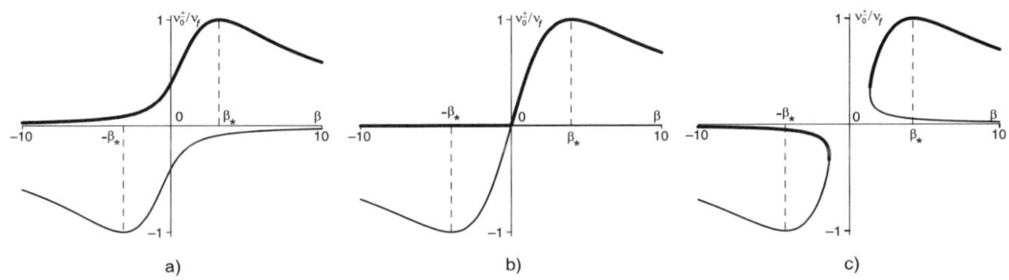

Fig. 3. The functions $\nu_0^+(\beta)$ (bold lines) and $\nu_0^-(\beta)$ (fine lines), and their bifurcation when $\mathbf{D}$ is changing from weakly to strongly indefinite.

Definition 1. *We call a* 2×2 *real symmetric matrix* $\mathbf{D}$ *with* $\det \mathbf{D} < 0$ *weakly indefinite, if* $4\beta_*^2 < (\mathrm{tr}\mathbf{D})^2$*, and strongly indefinite, if* $4\beta_*^2 > (\mathrm{tr}\mathbf{D})^2$.

As an illustration, we calculate and plot the functions $\nu_0^{\pm}(\beta)$, normalized by ν_f, for the matrix $\mathbf{K} > 0$ and indefinite matrices $\mathbf{D}_1$, $\mathbf{D}_2$, and $\mathbf{D}_3$

$$\mathbf{K}=\begin{bmatrix} 27 & 3 \\ 3 & 5 \end{bmatrix},\ \mathbf{D}_1=\begin{bmatrix} 6 & 3 \\ 3 & 1 \end{bmatrix},\ \mathbf{D}_2=\begin{bmatrix} 7 & \frac{4}{3}\sqrt{130}-11 \\ \frac{4}{3}\sqrt{130}-11 & 1 \end{bmatrix},\ \mathbf{D}_3=\begin{bmatrix} 7 & 5 \\ 5 & 1 \end{bmatrix}. \quad (34)$$

The graphs of the functions $\nu_0^{\pm}(\beta)$ bifurcate with a change of the damping matrix from the weakly indefinite to the strongly indefinite one. Indeed, since $\mathbf{D}_1$ satisfies the strict inequality (30), the limits are continuous functions with separated graphs, as shown in Fig. 3(a). Expression (30) is an equality for the matrix $\mathbf{D}_2$. Consequently, the functions $\nu_0^{\pm}(\beta)$ are continuous, with their graphs touching each other at the origin, Fig. 3(b). For the matrix $\mathbf{D}_3$, condition (30) is not fulfilled, and the functions are discontinuous. Their graphs, however, are joint together, forming continuous curves, see Fig. 3(c). The calculated $\nu_0^{\pm}(\beta)$ are bounded functions of β, non-exceeding the critical values $\pm\nu_f$ of the unperturbed circulatory system.

Proposition 3.

$$|\nu_0^{\pm}(\beta)| \leqslant |\nu_0^{\pm}(\pm\beta_*)| = \nu_f. \quad (35)$$

Proof. Let us observe that $\mu_0^{\pm} := \nu_0^{\pm}/\nu_f$ are roots of the quadratic equation

$$\nu_f^2 a_\beta \mu^2 - 2\delta\Omega b_0 \nu_f \mu - \delta^2 c_0 = 0, \quad (36)$$

with $\delta^2 a_\beta := a(\delta, \beta\delta)$, $b_0 := b(0,0)$, $c_0 := c(0,0)$. According to the Schur criterion [6] all the roots μ of equation (36) are inside the closed unit disk, if

$$\begin{aligned} \delta^2 c_0 + \nu_f^2 a_\beta &= (\mathrm{tr}\mathbf{D})^2 + 4(\beta^2 - \beta_*^2) + (\mathrm{tr}\mathbf{D})^2 \geqslant 0, \\ 2\delta\Omega\nu_f b_0 + \nu_f^2 a_\beta - \delta^2 c_0 &= (\beta + \beta_*)^2 \geqslant 0, \\ -2\delta\Omega\nu_f b_0 + \nu_f^2 a_\beta - \delta^2 c_0 &= (\beta - \beta_*)^2 \geqslant 0. \end{aligned} \quad (37)$$

The first of conditions (37) is satisfied for real $\nu_0^{\pm}$, implying $|\mu_0^{\pm}(\beta)| \leqslant 1$ with $|\mu_0^+(\beta_*)| = |\mu_0^-(-\beta_*)| = 1$.

The limits $\nu_0^{\pm}(\beta)$ of the critical values of the circulatory parameter $\nu_{cr}^{\pm}(\delta, \Omega)$, which are complicated functions of δ and Ω, effectively depend only on the ratio $\beta = \Omega/\delta$, defining the direction of approaching zero in the plane (δ, Ω). Along the directions $\beta = \beta_*$ and $\beta = -\beta_*$, the limits coincide with the critical flutter loads of the unperturbed circulatory system (2) in such a way that $\nu_0^+(\beta_*) = \nu_f$ and $\nu_0^-(-\beta_*) = -\nu_f$. According to Proposition 3, the limit of the non-conservative positional force at the onset of flutter for system (1) with dissipative and gyroscopic forces tending to zero does not exceed the critical flutter load of circulatory system (2), demonstrating a jump in the critical load which is characteristic of the destabilization paradox.

Power series expansions of the functions $\nu_0^{\pm}(\beta)$ around $\beta = \pm\beta_*$ (with the radius of convergence not exceeding $|\mathrm{tr}\mathbf{D}|/2$) yield simple estimates of the jumps in the critical load for the two-dimensional system (1)

$$\nu_f \mp \nu_0^{\pm}(\beta) = \nu_f \frac{2}{(\mathrm{tr}\mathbf{D})^2}(\beta \mp \beta_*)^2 + o((\beta \mp \beta_*)^2). \tag{38}$$

Leaving in expansions (38) only the second order terms and then substituting $\beta = \Omega/\delta$, we get equations of the form $Z = X^2/Y^2$, which is canonical for the Whitney umbrella surface [17, 30, 45]. These equations approximate the boundary of the asymptotic stability domain of system (1) in the vicinity of the points $(0, 0, \pm\nu_f)$ in the space of the parameters (δ, Ω, ν). An extension to the case when the system (1) has m degrees of freedom is given by the following statement.

Theorem 1. *Let the system (2) with m degrees of freedom be stable for $\nu < \nu_f$ and let at $\nu = \nu_f$ its spectrum contain a double eigenvalue $i\omega_f$ with the left and right Jordan chains of generalized eigenvectors $\mathbf{u}_0$, $\mathbf{u}_1$ and $\mathbf{v}_0$, $\mathbf{v}_1$, satisfying equations (15) and (16). Define the real quantities*

$$\begin{aligned} d_1 &= \mathrm{Re}(\mathbf{v}_0^T \mathbf{D} \mathbf{u}_0), \quad & d_2 &= \mathrm{Im}(\mathbf{v}_0^T \mathbf{D} \mathbf{u}_1 + \mathbf{v}_1^T \mathbf{D} \mathbf{u}_0), \\ g_1 &= \mathrm{Re}(\mathbf{v}_0^T \mathbf{G} \mathbf{u}_0), \quad & g_2 &= \mathrm{Im}(\mathbf{v}_0^T \mathbf{G} \mathbf{u}_1 + \mathbf{v}_1^T \mathbf{G} \mathbf{u}_0), \end{aligned} \tag{39}$$

and

$$\beta_* = -\frac{\mathbf{v}_0^T \mathbf{D} \mathbf{u}_0}{\mathbf{v}_0^T \mathbf{G} \mathbf{u}_0}. \tag{40}$$

Then, in the vicinity of $\beta := \Omega/\delta = \beta_*$ the limit of the critical flutter load ν_{cr}^+ of the near-reversible system with m degrees of freedom as $\delta \to 0$ is

$$\nu_0^+(\beta) = \nu_f - \frac{g_1^2(\beta - \beta_*)^2}{\mu^2(d_2 + \beta_* g_2)^2} + o((\beta - \beta_*)^2). \tag{41}$$

Proof. Perturbing a simple eigenvalue $i\omega(\nu)$ of the stable system (2) at a fixed $\nu < \nu_f$ by small dissipative and gyroscopic forces yields the increment

$$\lambda = i\omega - \frac{\mathbf{v}^T\mathbf{D}\mathbf{u}}{2\mathbf{v}^T\mathbf{u}}\delta - \frac{\mathbf{v}^T\mathbf{G}\mathbf{u}}{2\mathbf{v}^T\mathbf{u}}\Omega + o(\delta, \Omega). \tag{42}$$

Since the eigenvectors $\mathbf{u}(\nu)$ and $\mathbf{v}(\nu)$ can be chosen real, the first order increment is real-valued. Therefore, in the first approximation in δ and Ω, the simple eigenvalue $i\omega(\nu)$ remains on the imaginary axis if $\Omega = \beta(\nu)\delta$, where

$$\beta(\nu) = -\frac{\mathbf{v}^T(\nu)\mathbf{D}\mathbf{u}(\nu)}{\mathbf{v}^T(\nu)\mathbf{G}\mathbf{u}(\nu)}. \tag{43}$$

Substituting expansions (18) into formula (43), we obtain

$$\beta(\nu) = -\frac{d_1 \pm d_2\mu\sqrt{\nu_f - \nu} + o\left(\sqrt{\nu_f - \nu}\right)}{g_1 \pm g_2\mu\sqrt{\nu_f - \nu} + o\left(\sqrt{\nu_f - \nu}\right)}, \tag{44}$$

wherefrom expression (41) follows, if $|\beta - \beta_*| \ll 1$.

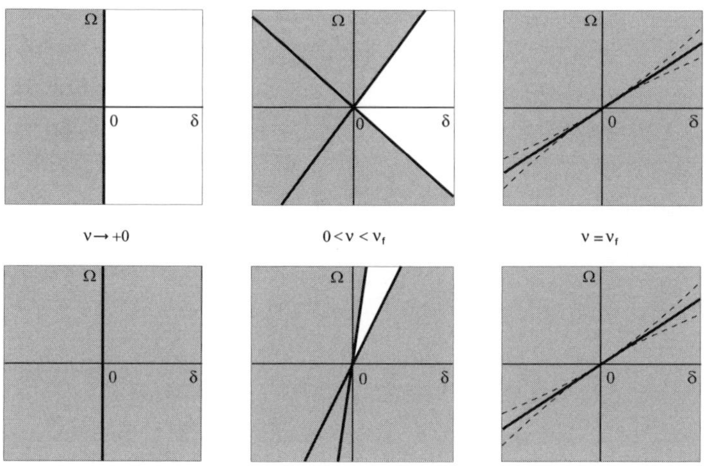

Fig. 4. For various ν, bold lines show linear approximations to the boundary of the asymptotic stability domain (white) of system (1) in the vicinity of the origin in the plane (δ, Ω), when $\mathrm{tr}\mathbf{K} > 0$ and $\det \mathbf{K} > 0$, and $4\beta_*^2 < (\mathrm{tr}\mathbf{D})^2$ (upper row) or $4\beta_*^2 > (\mathrm{tr}\mathbf{D})^2$(lower row).

After substituting $\beta = \Omega/\delta$ the formula (41) gives an approximation of the critical flutter load

$$\nu_{cr}^{\pm}(\delta, \Omega) = \nu_f - \frac{g_1^2(\Omega - \beta_*\delta)^2}{\mu^2(d_2 + \beta_* g_2)^2\delta^2}, \tag{45}$$

which has the canonical Whitney's umbrella form. The coefficients (21) and (39) calculated with the use of vectors (20) are

$$\begin{aligned} d_1 &= 2(k_{22}-k_{11})\mathrm{tr}(\mathbf{K}-\omega_f^2\mathbf{I})\mathbf{D}, \quad g_1 = 4(k_{11}-k_{22})\nu_f \\ d_2 &= -8\omega_f(2d_{12}k_{12}+d_{22}(k_{22}-k_{11})), \quad g_2 = 16\omega_f\nu_f. \end{aligned} \tag{46}$$

With (46) expression (41) is reduced to (38).

Using exact expressions for the functions $\omega(\nu)$, $\mathbf{u}(\nu)$, and $\mathbf{v}(\nu)$, we obtain better estimates in case when $m = 2$. Substituting the explicit expression for the eigenfrequency

$$\omega^2(\nu) = \omega_f^2 \pm \sqrt{\nu_f^2 - \nu^2}, \tag{47}$$

following from (11)–(13), into the equation (43), which now reads

$$\delta\left(2\nu_f\beta_* + \left(\omega^2(\nu)-\omega_f^2\right)\mathrm{tr}\mathbf{D}\right) - 2\Omega\nu = 0, \tag{48}$$

we obtain

$$\Omega = \frac{\nu_f}{\nu}\left[\beta_* \pm \frac{\mathrm{tr}\mathbf{D}}{2}\sqrt{1-\frac{\nu^2}{\nu_f^2}}\,\right]\delta. \tag{49}$$

Equation (49) is simply formula (28) inverted with respect to $\beta = \Omega/\delta$.

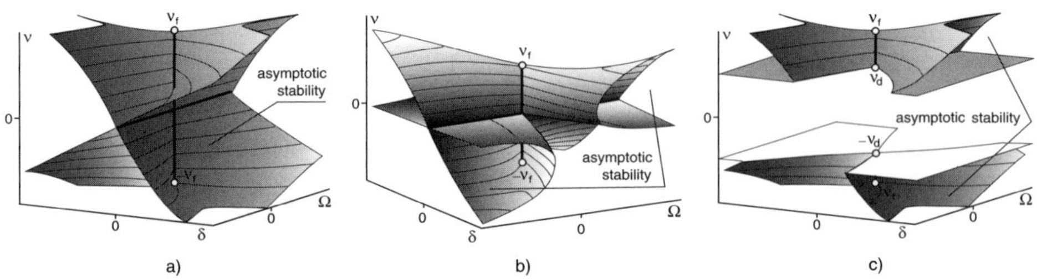

Fig. 5. The domain of asymptotic stability of system (1) with the singularities *Whitney umbrella*, *dihedral angle*, and *trihedral angle* when $\mathbf{K} > 0$ and $4\beta_*^2 < (\mathrm{tr}\mathbf{D})^2$ (a), $\mathbf{K} > 0$ and $4\beta_*^2 > (\mathrm{tr}\mathbf{D})^2$ (b), and when $\mathrm{tr}\mathbf{K} > 0$ and $\det\mathbf{K} < 0$ (c).

We use the linear approximation (49) to study the asymptotic behavior of the stability domain of the two-dimensional system (1) in the vicinity of the origin in the plane (δ, Ω) for various ν. It is enough to consider only the case when $\mathrm{tr}\mathbf{K} > 0$ and $\det\mathbf{K} > 0$, so that $-\nu_f < \nu < \nu_f$, because for $\det\mathbf{K} \leqslant 0$ the region $\nu^2 < \nu_d^2 \leqslant \nu_f^2$ is unstable and should be excluded.

For $\nu^2 < \nu_f^2$ the radicand in expression (49) is real and nonzero, so that in the first approximation the domain of asymptotic stability is contained between two lines intersecting at the origin, as depicted in Fig. 4 (central column). When

ν approaches the critical values $\pm\nu_f$, the angle becomes more acute until at $\nu = \nu_f$ or $\nu = -\nu_f$ it degenerates to a single line $\Omega = \delta\beta_*$ or $\Omega = -\delta\beta_*$ respectively. For $\beta_* \neq 0$ these lines are not parallel to each other, and due to inequality (31) they are never vertical, see Fig. 4 (right column). However, the degeneration can be lifted already in the second-order approximation in δ

$$\Omega = \pm\delta\beta_* \pm \frac{\omega_f \mathrm{tr}\mathbf{D}\sqrt{\det \mathbf{D} + \beta_*^2}}{2\nu_f}\delta^2 + O(\delta^3). \tag{50}$$

If the radicand is positive, equation (50) defines two curves touching each other at the origin, as shown in Fig. 4 by dashed lines. Inside the cusps $|\nu_{cr}^{\pm}(\delta,\Omega)| \geqslant \nu_f$.

The evolution of the domain of asymptotic stability in the plane (δ,Ω), when ν goes from $\pm\nu_f$ to zero, depends on the structure of the matrix $\mathbf{D}$ and is governed by the sign of the expression $4\beta_*^2 - (\mathrm{tr}\mathbf{D})^2$. For the negative sign the angle between the lines (49) is getting wider, tending to π as $\nu \to 0$, see Fig. 4 (upper left). Otherwise, the angle reaches a maximum for some $\nu^2 < \nu_f^2$ and then shrinks to a single line $\delta = 0$ at $\nu = 0$, Fig. 4 (lower left). At $\nu = 0$ the Ω-axis corresponds to a marginally stable gyroscopic system. Since the linear approximation to the asymptotic stability domain does not contain the Ω-axis at any $\nu \neq 0$, small gyroscopic forces cannot stabilize a circulatory system in the absence of damping forces ($\delta = 0$), which is in agreement with the theorems of Lakhadanov and Karapetyan [12, 13].

Reconstructing with the use of the obtained results the asymptotic stability domain of system (1), we find that it has three typical configurations in the vicinity of the ν-axis in the parameter space (δ,Ω,ν). In case of a positive-definite matrix $\mathbf{K}$ and of a semi-definite or a weakly-indefinite matrix $\mathbf{D}$ the addition of small damping and gyroscopic forces blows the stability interval of a circulatory system $\nu^2 < \nu_f^2$ up to a three-dimensional region bounded by the parts of a singular surface $\nu = \nu_{cr}^{\pm}(\delta,\Omega)$, which belong to the half-space $\delta\mathrm{tr}\mathbf{D} > 0$, Fig. 5(a). The stability interval of a circulatory system forms an edge of a dihedral angle. At $\nu = 0$ the angle of the intersection reaches its maximum (π), creating another edge along the Ω-axis. While approaching the points $\pm\nu_f$, the angle becomes more acute and ends up with the deadlock of an edge, Fig. 5(a).

When the matrix $\mathbf{D}$ approaches the threshold $4\beta_*^2 = (\mathrm{tr}\mathbf{D})^2$, two smooth parts of the stability boundary corresponding to negative and positive ν come towards each other until they touch, when $\mathbf{D}$ is at the threshold. After $\mathbf{D}$ becomes strongly indefinite this temporary glued configuration collapses into two pockets of asymptotic stability, as shown in Fig. 5(b). Each of the two pockets has a deadlock of an edge as well as two edges which meet at the origin and form a singularity known as the "break of an edge" [17].

The configuration of the asymptotic stability domain, shown in Fig. 5(c), corresponds to an indefinite matrix $\mathbf{K}$ with $\mathrm{tr}\mathbf{K} > 0$ and $\det\mathbf{K} < 0$. In this case the condition $\nu^2 > \nu_d^2$ divides the domain of asymptotic stability into two parts, corresponding to positive and negative ν. The intervals of ν-axis form edges of dihedral angles, which end up with the deadlocks at $\nu = \pm\nu_f$ and with the trihedral angles at $\nu = \pm\nu_d$, Fig. 5(c). Qualitatively, this configuration does not depend on the properties of the matrix $\mathbf{D}$.

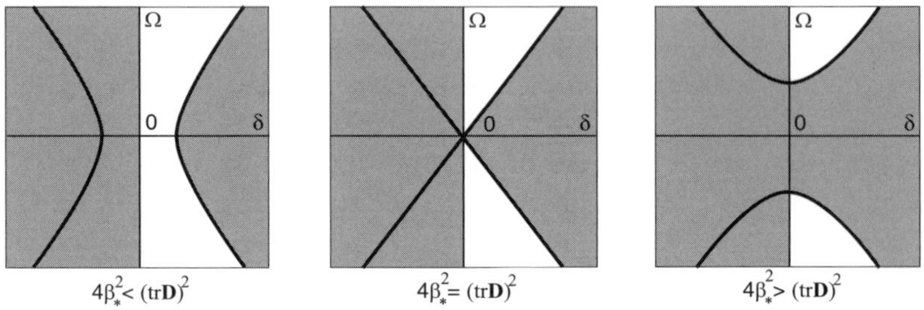

Fig. 6. Bifurcation of the domain of the asymptotic stability (white) in the plane (δ, Ω) at $\nu = 0$ due to the change of the structure of the matrix $\mathbf{D}$ according to the criterion (44).

We note that the parameter $4\beta_*^2 - (\mathrm{tr}\mathbf{D})^2$ governs not only the bifurcation of the stability domain near the ν-axis, but also the bifurcation of the whole stability domain in the space of the parameters δ, Ω, and ν. This is seen from the stability conditions (24)–(26). For example, for $\nu = 0$ the inequality $Q > 0$ is reduced to $c(\delta, \Omega) > 0$, where $c(\delta, \Omega)$ is given by (26). For positive semi-definite matrices $\mathbf{D}$ this condition is always satisfied. For indefinite matrices equation $c(\delta, \Omega) = 0$ defines either hyperbola or two intersecting lines. In case of weakly-indefinite $\mathbf{D}$ the stability domain is bounded by the ν-axis and one of the hyperbolic branches, see Figure 6 (left). At the threshold $4\beta_*^2 = (\mathrm{tr}\mathbf{D})^2$ the stability domain is separated to two half-conical parts, as shown in the center of Figure 6. Strongly-indefinite damping makes impossible stabilization by small gyroscopic forces, see Figure 6 (right). In this case the non-conservative forces are required for stabilization. Thus, we generalize the results of the works [35, 36], which were obtained for diagonal matrices $\mathbf{K}$ and $\mathbf{D}$. Moreover, the authors of the works [35, 36] did not take into account the non-conservative positional forces corresponding to the matrix $\mathbf{N}$ in equation (1) and missed the existence of the two classes of indefinite matrices, which lead to the bifurcation of the domain of asymptotic stability. We can also conclude that at least in two dimensions the requirement of definiteness of the matrix $\mathbf{D}$ established in [46] is not necessary for the stabilization of a circulatory system by gyroscopic and damping forces.

3 A gyroscopic system with weak damping and circulatory forces

A statically unstable potential system, which has been stabilized by gyroscopic forces can be destabilized by the introduction of small stationary damping, which is a velocity-dependent force [1]. However, many statically unstable gyropendulums enjoy robust stability at high speeds [31]. To explain this phenomenon a concept of rotating damping has been introduced, which is also proportional to the displacements by a non-conservative way and thus contributes not only to the matrix **D** in equation (1), but to the matrix **N** as well [3–5, 31]. This leads to a problem of perturbation of gyroscopic system (3) by weak dissipative and non-conservative positional forces [14, 27, 31, 32, 46, 48, 49, 59, 62, 63, 66, 74].

3.1 Stability of a gyroscopic system

In the absence of dissipative and circulatory forces ($\delta = \nu = 0$), the polynomial (10) has four roots $\pm\lambda_{\pm}$, where

$$\lambda_{\pm} = \sqrt{-\frac{1}{2}(\mathrm{tr}\mathbf{K} + \Omega^2) \pm \frac{1}{2}\sqrt{(\mathrm{tr}\mathbf{K} + \Omega^2)^2 - 4\det\mathbf{K}}}. \tag{51}$$

Analysis of these eigenvalues yields the following result, see e.g. [47].

Proposition 4. *If* $\det\mathbf{K} > 0$ *and* $\mathrm{tr}\mathbf{K} < 0$, *gyroscopic system (3) with two degrees of freedom is unstable by divergence for* $\Omega^2 < \Omega_0^{-2}$, *unstable by flutter for* $\Omega_0^{-2} \leqslant \Omega^2 \leqslant \Omega_0^{+2}$, *and stable for* $\Omega_0^{+2} < \Omega^2$, *where the critical values* Ω_0^- *and* Ω_0^+ *are*

$$0 \leqslant \sqrt{-\mathrm{tr}\mathbf{K} - 2\sqrt{\det\mathbf{K}}} =: \Omega_0^- \leqslant \Omega_0^+ := \sqrt{-\mathrm{tr}\mathbf{K} + 2\sqrt{\det\mathbf{K}}}. \tag{52}$$

If $\det\mathbf{K} > 0$ *and* $\mathrm{tr}\mathbf{K} > 0$, *the gyroscopic system is stable for any* Ω [2]. *If* $\det\mathbf{K} \leqslant 0$, *the system is unstable* [1].

Representing for $\det\mathbf{K} > 0$ the equation (51) in the form

$$\lambda_{\pm} = \sqrt{-\frac{1}{2}\left(\Omega^2 - \frac{1}{2}\left(\Omega_0^{-2} + \Omega_0^{+2}\right)\right) \pm \frac{1}{2}\sqrt{\left(\Omega^2 - \Omega_0^{-2}\right)\left(\Omega^2 - \Omega_0^{+2}\right)}}, \tag{53}$$

we find that at $\Omega = 0$ there are in general four real roots $\pm\lambda_{\pm} = \pm(\Omega_0^+ \pm \Omega_0^-)/2$ and system (3) is statically unstable. With the increase of Ω^2 the distance $\lambda_+ - \lambda_-$ between the two roots of the same sign is getting smaller. The roots are moving towards each other until they merge at $\Omega^2 = \Omega_0^{-2}$ with the origination of a pair of double real eigenvalues $\pm\omega_0$ with the Jordan blocks, where

$$\omega_0 = \frac{1}{2}\sqrt{\Omega_0^{+2} - \Omega_0^{-2}} = \sqrt[4]{\det\mathbf{K}} > 0. \tag{54}$$

Further increase of Ω^2 yields splitting of $\pm\omega_0$ to two couples of complex conjugate eigenvalues lying on the circle

$$\mathrm{Re}\lambda^2 + \mathrm{Im}\,\lambda^2 = \omega_0^2. \tag{55}$$

The complex eigenvalues move along the circle until at $\Omega^2 = {\Omega_0^+}^2$ they reach the imaginary axis and originate a complex-conjugate pair of double purely imaginary eigenvalues $\pm i\omega_0$. For $\Omega^2 > {\Omega_0^+}^2$ the double eigenvalues split into four simple purely imaginary eigenvalues which do not leave the imaginary axis, Fig. 7.

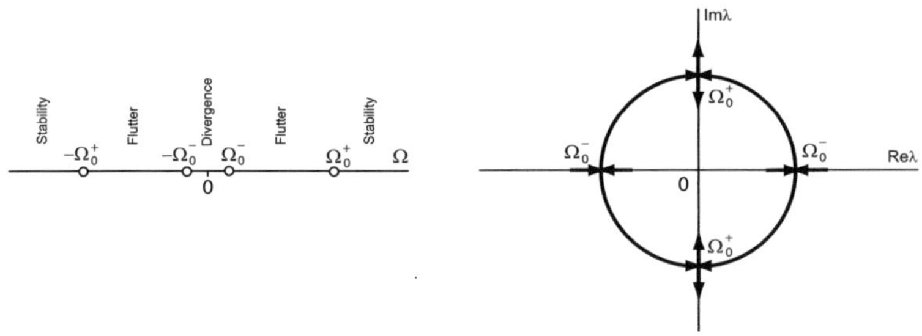

Fig. 7. Stability diagram for the gyroscopic system with $\mathbf{K} < 0$ (left) and the corresponding trajectories of the eigenvalues in the complex plane for the increasing parameter $\Omega > 0$ (right).

Thus, the system (3) with $\mathbf{K} < 0$ is statically unstable for $\Omega \in (-\Omega_0^-, \Omega_0^-)$, it is dynamically unstable for $\Omega \in [-\Omega_0^+, -\Omega_0^-] \cup [\Omega_0^-, \Omega_0^+]$, and it is stable (gyroscopic stabilization) for $\Omega \in (-\infty, -\Omega_0^+) \cup (\Omega_0^+, \infty)$, see Fig. 7. The values of the gyroscopic parameter $\pm\Omega_0^-$ define the boundary between the divergence and flutter domains while the values $\pm\Omega_0^+$ originate the flutter-stability boundary.

3.2 The influence of small damping and non-conservative positional forces on the stability of a gyroscopic system

Consider the asymptotic stability domain in the plane (δ, ν) in the vicinity of the origin, assuming that $\Omega \neq 0$ is fixed. Observing that the third of the inequalities (22) is fulfilled for $\det \mathbf{K} > 0$ and the first one simply restricts the region of variation of δ to the half-plane $\delta \mathrm{tr}\mathbf{D} > 0$, we focus our analysis on the remaining two of the conditions (22).

Taking into account the structure of coefficients (26) and leaving the linear terms with respect to δ in the Taylor expansions of the functions $\nu_{cr}^{\pm}(\delta, \Omega)$, we get the equations determining a linear approximation to the stability boundary

$$\nu = \frac{\mathrm{tr}\mathbf{KD} - \mathrm{tr}\mathbf{K}\mathrm{tr}\mathbf{D} - \mathrm{tr}\mathbf{D}\lambda_{\pm}^2(\Omega)}{2\Omega}\delta$$
$$= \frac{2\mathrm{tr}\mathbf{KD} + \mathrm{tr}\mathbf{D}(\Omega^2 - \mathrm{tr}\mathbf{K}) \pm \mathrm{tr}\mathbf{D}\sqrt{(\Omega^2+\mathrm{tr}\mathbf{K})^2 - 4\det\mathbf{K}}}{4\Omega}\delta, \qquad (56)$$

where the eigenvalues $\lambda_{\pm}(\Omega)$ are given by formula (51).

For $\det\mathbf{K} > 0$ and $\mathrm{tr}\mathbf{K} > 0$ the gyroscopic system is stable at any Ω. Consequently, the coefficients $\lambda_{\pm}^2(\Omega)$ are always real, and equations (56) define in general two lines intersecting at the origin, Fig. 8. Since $\mathrm{tr}\mathbf{K} > 0$, the second of the inequalities (22) is satisfied for $\det\mathbf{D} \geqslant 0$, and it gives an upper bound of δ^2 for $\det\mathbf{D} < 0$. Thus, a linear approximation to the domain of asymptotic stability near the origin in the plane (δ,ν), is an angle-shaped area between two lines (56), as shown in Fig. 8. With the change of Ω the size of the angle is varying and moreover, the stability domain rotates as a whole about the origin. As $\Omega \to \infty$, the size of the angle tends to $\pi/2$ in such a way that the stability domain fits one of the four quadrants of the parameter plane, as shown in Fig. 8 (right column). From (56) it follows that asymptotically as $\Omega \to 0$

$$\nu(\Omega) = \frac{\nu_f}{\Omega}\left(\beta_* \pm \frac{\mathrm{tr}\mathbf{D}}{2}\right) + o\left(\frac{1}{\Omega}\right). \qquad (57)$$

Consequently, the angle between the lines (56) tends to π for the matrices $\mathbf{D}$ satisfying the condition $4\beta_*^2 < (\mathrm{tr}\mathbf{D})^2$, see Fig. 8 (upper left). In this case in the linear approximation the domain of asymptotic stability spreads over two quadrants and contains the δ-axis. Otherwise, the angle tends to zero as $\Omega \to 0$, Fig. 8 (lower left). In the linear approximation the stability domain always belongs to one quadrant and does not contain δ-axis, so that in the absence of non-conservative positional forces gyroscopic system (3) with $\mathbf{K} > 0$ cannot be made asymptotically stable by damping forces with strongly-indefinite matrix $\mathbf{D}$, which is also visible in the three-dimensional picture of Fig. 5(b). The three-dimensional domain of asymptotic stability of near-Hamiltonian system (1) with $\mathbf{K} > 0$ and $\mathbf{D}$ semi-definite or weakly-idefinite is inside a dihedral angle with the Ω-axis as its edge, as shown in Fig. 5(a). With the increase in $|\Omega|$, the section of the domain by the plane $\Omega = \mathrm{const}$ is getting more narrow and is rotating about the origin so that the points of the parameter plane (δ,ν) that where stable at lower $|\Omega|$ can lose their stability for the higher absolute values of the gyroscopic parameter (*gyroscopic destabilization* of a statically stable potential system in the presence of damping and non-conservative positional forces).

To study the case when $\mathbf{K} < 0$ we write equation (56) in the form

$$\nu = \frac{\Omega_0^+}{\Omega}\left[\gamma_* + \frac{\mathrm{tr}\mathbf{D}}{4}\sqrt{\frac{\Omega^2}{{\Omega_0^+}^2} - 1}\left(\sqrt{\Omega^2 - {\Omega_0^+}^2} \pm \sqrt{\Omega^2 - {\Omega_0^-}^2}\right)\right]\delta, \qquad (58)$$

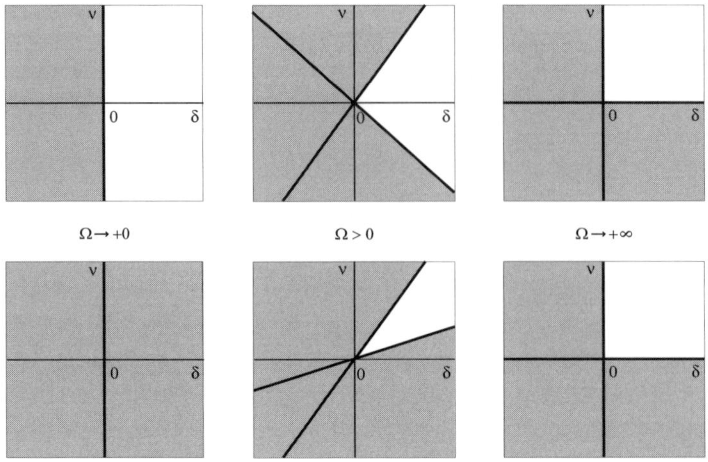

Fig. 8. For various Ω, bold lines show linear approximations to the boundary of the asymptotic stability domain (white) of system (1) in the vicinity of the origin in the plane (δ, ν), when $\mathrm{tr}\mathbf{K} > 0$ and $\det \mathbf{K} > 0$, and $4\beta_*^2 < (\mathrm{tr}\mathbf{D})^2$ (upper row) or $4\beta_*^2 > (\mathrm{tr}\mathbf{D})^2$ (lower row).

where

$$\gamma_* := \frac{\mathrm{tr}[\mathbf{K} + ({\Omega_0^+}^2 - \omega_0^2)\mathbf{I}]\mathbf{D}}{2\Omega_0^+}. \tag{59}$$

Proposition 5. *Let $\lambda_1(\mathbf{D})$ and $\lambda_2(\mathbf{D})$ be eigenvalues of $\mathbf{D}$. Then,*

$$|\gamma_*| \leqslant \Omega_0^+ \frac{|\lambda_1(\mathbf{D}) + \lambda_2(\mathbf{D})|}{4} + \Omega_0^- \frac{|\lambda_1(\mathbf{D}) - \lambda_2(\mathbf{D})|}{4}. \tag{60}$$

Proof. With the use of the Cauchy-Schwarz inequality we obtain

$$\begin{aligned} |\gamma_*| &\leqslant \Omega_0^+ \frac{|\mathrm{tr}\mathbf{D}|}{4} + \frac{\mathrm{tr}(\mathbf{K} - \frac{\mathrm{tr}\mathbf{K}}{2}\mathbf{I})(\mathbf{D} - \frac{\mathrm{tr}\mathbf{D}}{2}\mathbf{I})}{2\Omega_0^+} \\ &\leqslant \Omega_0^+ \frac{|\mathrm{tr}\mathbf{D}|}{4} + \frac{|\lambda_1(\mathbf{K}) - \lambda_2(\mathbf{K})||\lambda_1(\mathbf{D}) - \lambda_2(\mathbf{D})|}{4\Omega_0^+}. \end{aligned} \tag{61}$$

Taking into account that $|\lambda_1(\mathbf{K}) - \lambda_2(\mathbf{K})| = \Omega_0^- \Omega_0^+$, we get inequality (60).

Expression (58) is real-valued when $\Omega^2 \geqslant {\Omega_0^+}^2$ or $\Omega^2 \leqslant {\Omega_0^-}^2$. For sufficiently small $|\delta|$ the first inequality implies the second of the stability conditions (22), whereas the last inequality contradicts it. Consequently, the domain of asymptotic stability is determined by the inequalities $\delta \mathrm{tr}\mathbf{D} > 0$ and $Q(\delta, \nu, \Omega) > 0$, and its linear approximation in the vicinity of the origin in the (δ, ν)-plane has the form of an angle with the boundaries given by equations (58). For Ω tending to infinity the angle expands to $\pi/2$, whereas for $\Omega = \Omega_0^+$

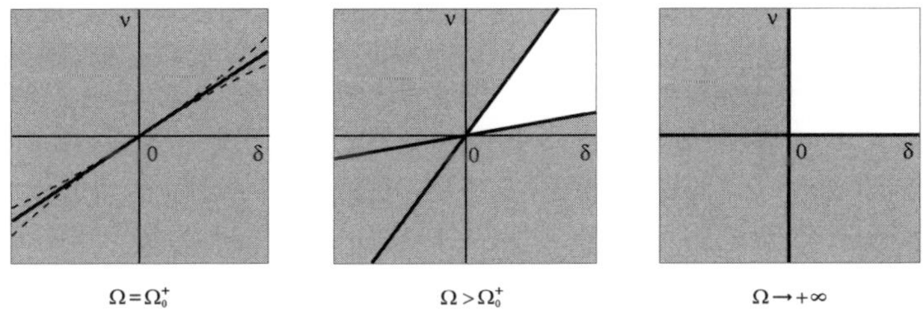

Fig. 9. For various Ω, bold lines show linear approximations to the boundary of the asymptotic stability domain (white) of system (1) in the vicinity of the origin in the plane (δ, ν), when $\mathbf{K} < 0$.

or $\Omega = -\Omega_0^+$ it degenerates to a single line $\nu = \delta\gamma_*$ or $\nu = -\delta\gamma_*$ respectively. For $\gamma_* \neq 0$ these lines are not parallel to each other, and due to inequality (60) they never stay vertical, see Fig. 9 (left). The degeneration can, however, be removed in the second-order approximation in δ

$$\nu = \pm\delta\gamma_* \pm \frac{\mathrm{tr}\mathbf{D}\sqrt{\omega_0^2 \det \mathbf{D} - \gamma_*^2}}{2\Omega_0^+}\delta^2 + O(\delta^3), \tag{62}$$

as shown by dashed lines in Fig. 9 (left). Therefore, gyroscopic stabilization of statically unstable conservative system with $\mathbf{K} < 0$ can be improved up to asymptotic stability by small damping and circulatory forces, if their magnitudes are in the narrow region with the boundaries depending on Ω. The lower the desirable absolute value of the critical gyroscopic parameter $\Omega_{cr}(\delta, \nu)$ the poorer choice of the appropriate combinations of damping and circulatory forces.

To estimate the new critical value of the gyroscopic parameter $\Omega_{cr}(\delta, \nu)$, which can deviate significantly from that of the conservative gyroscopic system, we consider the formula (58) in the vicinity of the points $(0, 0, \pm\Omega_0^+,)$ in the parameter space. Leaving only the terms, which are constant or proportional to $\sqrt{\Omega \pm \Omega_0^+}$ in both the numerator and denominator and assuming $\nu = \gamma\delta$, we find

$$\pm\Omega_{cr}^+(\gamma) = \pm\Omega_0^+ \pm \Omega_0^+ \frac{2}{(\omega_0 \mathrm{tr}\mathbf{D})^2}(\gamma \mp \gamma_*)^2 + o((\gamma - \gamma_*)^2), \tag{63}$$

After substitution $\gamma = \nu/\delta$ equations (63) take the form canonical for the Whitney umbrella. The domain of asymptotic stability consists of two pockets of two Whitney umbrellas, selected by the conditions $\delta\mathrm{tr}\mathbf{D} > 0$ and $Q(\delta, \nu, \Omega) > 0$. Equations (58) are a linear approximation to the stability boundary in the vicinity of the Ω-axis. Moreover, they describe in an implicit form a limit of the critical gyroscopic parameter $\Omega_{cr}(\delta, \gamma\delta)$ when δ tends to zero, as a function of the ratio $\gamma = \nu/\delta$, Fig. 10(b). Most of the directions γ give the limit value $|\Omega_{cr}^\pm(\gamma)| > \Omega_0^+$ with an exception for $\gamma = \gamma_*$ and $\gamma = -\gamma_*$, so that

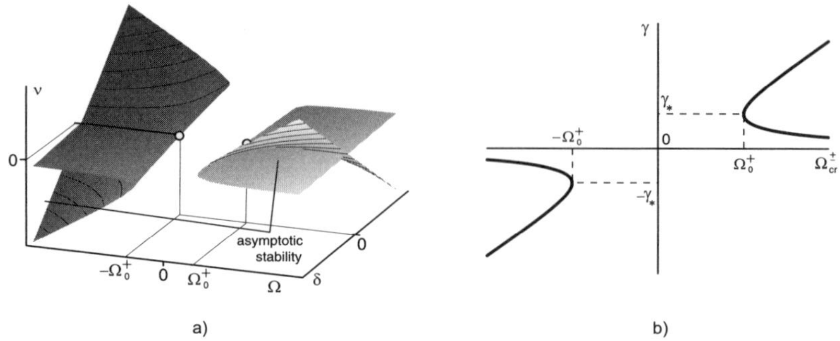

Fig. 10. Blowing the domain of gyroscopic stabilization of a statically unstable conservative system with $\mathbf{K} < 0$ up to the domain of asymptotic stability with the Whitney umbrella singularities (a). The limits of the critical gyroscopic parameter $\Omega_{cr}^{\pm}$ as functions of $\gamma = \nu/\delta$ (b).

$\Omega_{cr}^+(\gamma_*) = \Omega_0^+$ and $\Omega_{cr}^-(-\gamma_*) = -\Omega_0^+$. Estimates of the critical gyroscopic parameter (63) are extended to the case of arbitrary number of degrees of freedom by the following statement.

Theorem 2. *Let the system (3) with even number m of degrees of freedom be gyroscopically stabilized for $\Omega > \Omega_0^+$ and let at $\Omega = \Omega_0^+$ its spectrum contain a double eigenvalue $i\omega_0$ with the Jordan chain of generalized eigenvectors $\mathbf{u}_0$, $\mathbf{u}_1$, satisfying the equations*

$$
\begin{aligned}
(-\mathbf{I}\omega_0^2 + i\omega_0\Omega_0^+\mathbf{G} + \mathbf{K})\mathbf{u}_0 &= 0, \\
(-\mathbf{I}\omega_0^2 + i\omega_0\Omega_0^+\mathbf{G} + \mathbf{K})\mathbf{u}_1 &= -(2i\omega_0\mathbf{I} + \Omega_0^+\mathbf{G})\mathbf{u}_0.
\end{aligned} \tag{64}
$$

Define the real quantities d_1, d_2, n_1, n_2, and γ_* as

$$
\begin{aligned}
d_1 &= \mathrm{Re}(\overline{\mathbf{u}}_0^T\mathbf{D}\mathbf{u}_0), \quad & d_2 &= \mathrm{Im}(\overline{\mathbf{u}}_0^T\mathbf{D}\mathbf{u}_1 - \overline{\mathbf{u}}_1^T\mathbf{D}\mathbf{u}_0), \\
n_1 &= \mathrm{Im}(\overline{\mathbf{u}}_0^T\mathbf{N}\mathbf{u}_0), \quad & n_2 &= \mathrm{Re}(\overline{\mathbf{u}}_0^T\mathbf{N}\mathbf{u}_1 - \overline{\mathbf{u}}_1^T\mathbf{N}\mathbf{u}_0),
\end{aligned} \tag{65}
$$

$$
\gamma_* = -i\omega_0 \frac{\overline{\mathbf{u}}_0^T\mathbf{D}\mathbf{u}_0}{\overline{\mathbf{u}}_0^T\mathbf{N}\mathbf{u}_0}, \tag{66}
$$

where the bar over a symbol denotes complex conjugate.

Then, in the vicinity of $\gamma := \nu/\delta = \gamma_*$ the limit of the critical value of the gyroscopic parameter Ω_{cr}^+ of the near-Hamiltonian system as $\delta \to 0$ is

$$
\Omega_{cr}^+(\gamma) = \Omega_0^+ + \frac{n_1^2(\gamma - \gamma_*)^2}{\mu^2(\omega_0 d_2 - \gamma_* n_2 - d_1)^2}, \tag{67}
$$

which is valid for $|\gamma - \gamma_*| \ll 1$.

Proof. Perturbing the system (3), which is stabilized by the gyroscopic forces with $\Omega > \Omega_0^+$, by small damping and circulatory forces, yields an increment to a simple eigenvalue [53]

$$\lambda = i\omega - \frac{\omega^2 \overline{\mathbf{u}}^T \mathbf{D} \mathbf{u} \delta - i\omega \overline{\mathbf{u}}^T \mathbf{N} \mathbf{u} \nu}{\overline{\mathbf{u}}^T \mathbf{K} \mathbf{u} + \omega^2 \overline{\mathbf{u}}^T \mathbf{u}} + o(\delta, \nu). \tag{68}$$

Choose the eigenvalues and the corresponding eigenvectors that merge at $\Omega = \Omega_0^+$

$$\begin{aligned} i\omega(\Omega) &= i\omega_0 \pm i\mu\sqrt{\Omega - \Omega_0^+} + o(|\Omega - \Omega_0^+|^{\frac{1}{2}}), \\ \mathbf{u}(\Omega) &= \mathbf{u}_0 \pm i\mu \mathbf{u}_1 \sqrt{\Omega - \Omega_0^+} + o(|\Omega - \Omega_0^+|^{\frac{1}{2}}), \end{aligned} \tag{69}$$

where

$$\mu^2 = -\frac{2\omega_0^2 \overline{\mathbf{u}}_0^T \mathbf{u}_0}{\Omega_0^+ (\omega_0^2 \overline{\mathbf{u}}_1^T \mathbf{u}_1 - \overline{\mathbf{u}}_1^T \mathbf{K} \mathbf{u}_1 - i\omega_0 \Omega_0^+ \overline{\mathbf{u}}_1^T \mathbf{G} \mathbf{u}_1 - \overline{\mathbf{u}}_0^T \mathbf{u}_0)}. \tag{70}$$

Since $\mathbf{D}$ and $\mathbf{K}$ are real symmetric matrices and $\mathbf{N}$ is a real skew-symmetric one, the first-order increment to the eigenvalue $i\omega(\Omega)$ given by (68) is real-valued. Consequently, in the first approximation in δ and ν, simple eigenvalue $i\omega(\Omega)$ remains on the imaginary axis, if $\nu = \gamma(\Omega)\delta$, where

$$\gamma(\Omega) = -i\omega(\Omega) \frac{\overline{\mathbf{u}}^T(\Omega) \mathbf{D} \mathbf{u}(\Omega)}{\overline{\mathbf{u}}^T(\Omega) \mathbf{N} \mathbf{u}(\Omega)}. \tag{71}$$

Substitution of the expansions (69) into the formula (71) yields

$$\gamma(\Omega) = -(\omega_0 \pm \mu\sqrt{\Omega - \Omega_0^+}) \frac{d_1 \mp \mu d_2 \sqrt{\Omega - \Omega_0^+}}{n_1 \pm \mu n_2 \sqrt{\Omega - \Omega_0^+}}, \tag{72}$$

wherefrom the expression (67) follows, if $|\gamma - \gamma_*| \ll 1$.

Substituting $\gamma = \nu/\delta$ in expression (72) yields the estimate for the critical value of the gyroscopic parameter $\Omega_{cr}^+(\delta, \nu)$

$$\Omega_{cr}^+(\delta, \nu) = \Omega_0^+ + \frac{n_1^2 (\nu - \gamma_* \delta)^2}{\mu^2 (\omega_0 d_2 - \gamma_* n_2 - d_1)^2 \delta^2}. \tag{73}$$

We show now that for $m = 2$ expression (67) implies (63). At the critical value of the gyroscopic parameter Ω_0^+ defined by equation (52), the double eigenvalue $i\omega_0$ with ω_0 given by (54) has the Jordan chain

$$\mathbf{u}_0 = \begin{bmatrix} -i\omega_0 \Omega_0^+ - k_{12} \\ -\omega_0^2 + k_{11} \end{bmatrix}, \quad \mathbf{u}_1 = \frac{-1}{\omega_0^2 - k_{22}} \begin{bmatrix} 0 \\ i\omega_0 (k_{22} - k_{11}) - \Omega_0^+ k_{12} \end{bmatrix}. \tag{74}$$

With the vectors (74) equation (70) yields

$$\mu^2 = \frac{\Omega_0^+}{2}\frac{(\omega_0^2 - k_{11})(\omega_0^2 - k_{22})}{{\Omega_0^+}^2\omega_0^2 - k_{12}^2} = \frac{\Omega_0^+}{2} > 0, \tag{75}$$

whereas the formula (66) reproduces the coefficient γ_* given by (59). To show that (63) follows from (67) it remains to calculate the coefficients (65). We have

$$n_1 = -2\Omega_0^+\omega_0(\omega_0^2 - k_{11}), \quad \omega_0 d_2 - \gamma_* n_2 - d_1 = -2\omega_0^2(\omega_0^2 - k_{11})\mathbf{trD}. \tag{76}$$

Taking into account that $(\Omega_0^+)^2 = -\mathbf{tr}\mathbf{K} + 2\omega_0^2$, and using the relations (76) in (73) we exactly reproduce (63).

Therefore, in the presence of small damping and non-conservative positional forces, gyroscopic forces can both destabilize a statically stable conservative system (gyroscopic destabilization) and stabilize a statically unstable conservative system (gyroscopic stabilization). The first effect is essentially related with the dihedral angle singularity of the stability boundary, whereas the second one is governed by the Whitney umbrella singularity. In the remaining sections we demonstrate how these singularities appear in mechanical systems.

4 The modified Maxwell-Bloch equations with mechanical applications

The modified Maxwell-Bloch equations are the normal form for rotationally symmetric, planar dynamical systems [28, 48, 59]. They follow from equation (1) for $m = 2$, $\mathbf{D} = \mathbf{I}$, and $\mathbf{K} = \kappa\mathbf{I}$, and thus can be written as a single differential equation with the complex coefficients

$$\ddot{x} + i\Omega\dot{x} + \delta\dot{x} + i\nu x + \kappa x = 0, \quad x = x_1 - ix_2, \tag{77}$$

where κ corresponds to potential forces. Equations in this form appear in gyrodynamical problems such as the tippe top inversion, the rising egg, and the onset of oscillations in the squealing disc brake and the singing wine glass [14, 31, 48, 59, 62, 66, 68, 76].

According to stability conditions (22) the solution $x = 0$ of equation (77) is asymptotically stable if and only if

$$\delta > 0, \quad \Omega > \frac{\nu}{\delta} - \frac{\delta}{\nu}\kappa. \tag{78}$$

For $\kappa > 0$ the domain of asymptotic stability is a dihedral angle with the Ω-axis serving as its edge, Fig. 11(a). The sections of the domain by the planes $\Omega = \mathrm{const}$ are contained in the angle-shaped regions with the boundaries

$$\nu = \frac{\Omega \pm \sqrt{\Omega^2 + 4\kappa}}{2}\delta. \tag{79}$$

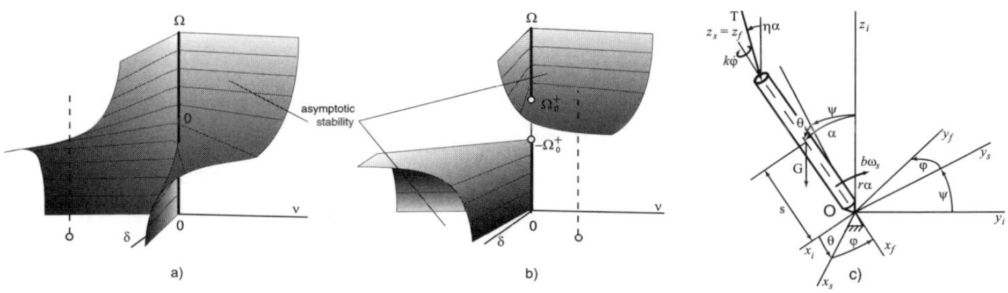

Fig. 11. Two configurations of the asymptotic stability domain of the modified Maxwell-Bloch equations for $\kappa > 0$ (a) and $\kappa < 0$ (b) corresponding to *gyroscopic destabilization* and *gyroscopic stabilization* respectively; Hauger's gyropendulum (c).

The domain shown in Fig. 11(a) is a particular case of that depicted in Fig. 5(a). For $\mathbf{K} = \kappa\mathbf{I}$ the interval $[-\nu_f, \nu_f]$ shown in Fig. 5(a) shrinks to a point so that at $\Omega = 0$ the angle is bounded by the lines $\nu = \pm\delta\sqrt{\kappa}$ and thus it is less than π. The domain of asymptotic stability is twisting around the Ω-axis in such a manner that it always remains in the half-space $\delta > 0$, Fig. 11(a). Consequently, the system stable at $\Omega = 0$ can become unstable at greater Ω, as shown in Fig. 11(a) by the dashed line. The larger magnitudes of circulatory forces, the lower $|\Omega|$ at the onset of instability.

As $\kappa > 0$ decreases, the hypersurfaces forming the dihedral angle approach each other so that, at $\kappa = 0$, they temporarily merge along the line $\nu = 0$ and a new configuration originates for $\kappa < 0$, Fig. 11(b). The new domain of asymptotic stability consists of two disjoint parts that are pockets of two Whitney umbrellas singled out by inequality $\delta > 0$. The absolute values of the gyroscopic parameter Ω in the stability domain are always not less than $\Omega_0^+ = 2\sqrt{-\kappa}$. As a consequence, the system unstable at $\Omega = 0$ can become asymptotically stable at greater Ω, as shown in Fig. 11(b) by the dashed line.

4.1 Stability of Hauger's gyropendulum

Hauger's gyropendulum [14] is an axisymmetric rigid body of mass m hinged at the point O on the axis of symmetry as shown in Figure (11)(c). The body's moment of inertia about the axis through the point O perpendicular to the axis of symmetry is denoted by I, the body's moment of inertia about the axis of symmetry is denoted by I_0, and the distance between the fastening point and the center of mass is s. The orientation of the pendulum, which is associated with the trihedron $Ox_fy_fz_f$, with respect to the fixed trihedron $Ox_iy_iz_i$ is specified by the angles ψ, θ, and ϕ. The pendulum experiences the force of gravity $G = mg$ and a follower torque T that lies in the plane of the z_i and z_f coordinate axes. The moment vector makes an angle of $\eta\alpha$ with the axis z_i, where η is a parameter ($\eta \neq 1$) and α is the angle between the z_i and z_f axes. Additionally, the pendulum experiences the restoring elastic moment $R = -r\alpha$ in the hinge and the dissipative moments $B = -b\omega_s$ and $K = -k\phi$, where ω_s is the angular velocity of an auxiliary coordinate system $Ox_sy_sz_s$ with respect to the inertial system and r, b, and k are the corresponding coefficients.

Linearization of the nonlinear equations of motion derived in [14] with the new variables $x_1 = \psi$ and $x_2 = \theta$ and the subsequent nondimensionalization yield the Maxwell-Bloch equations (77) where the dimensionless parameters are given by

$$\Omega = \frac{I_0}{I}, \quad \delta = \frac{b}{I\omega}, \quad \kappa = \frac{r - mgs}{I\omega^2}, \quad \nu = \frac{1-\eta}{I\omega^2}T, \quad \omega = -\frac{T}{k}. \tag{80}$$

The domain of asymptotic stability of the Hauger gyropendulum, given by (78), is shown in Fig. 11(a,b).

According to formulas (52) and (54), for the statically unstable gyropendulum ($\kappa < 0$) the singular points on the Ω-axis correspond to the critical values $\pm\Omega_0^+ = \pm 2\sqrt{-\kappa}$ and the critical frequency $\omega_0 = \sqrt{-\kappa}$. Noting that $\Omega_{cr}^+(\nu = \pm\sqrt{-\kappa}\delta, \delta) = \pm\Omega_0^+$ and substituting $\gamma = \nu/\delta$ into formula (78), we expand $\Omega_{cr}^+(\gamma)$ in a series in the neighborhood of $\gamma = \pm\sqrt{-\kappa}$

$$\Omega_{cr}^+(\gamma) = \pm 2\sqrt{-\kappa} \pm \frac{1}{\sqrt{-\kappa}}(\gamma \mp \sqrt{-\kappa})^2 + o\left((\gamma \mp \sqrt{-\kappa})^2\right). \tag{81}$$

Proceeding from γ to ν and δ in (81) yields approximations of the stability boundary near the singularities:

$$\Omega_{cr}^+(\nu, \delta) = \pm 2\sqrt{-\kappa} \pm \frac{1}{\sqrt{-\kappa}} \frac{(\nu \mp \delta\sqrt{-\kappa})^2}{\delta^2}. \tag{82}$$

They also follow from formula (63) after substituting $\omega_0 = \sqrt{-\kappa}$, and $\gamma_* = \sqrt{-\kappa}$, where the last value is given by (59). Thus, Hauger's gyropendulum, which is unstable at $\Omega = 0$, can become asymptotically stable for sufficiently large $|\Omega| \geqslant \Omega_0^+$ under a suitable combination of dissipative and nonconservative positional forces. Note that Hauger failed to find Whitney umbrella singularities on the boundary of the pendulum's gyroscopic stabilization domain.

4.2 Friction-induced instabilities in rotating elastic bodies of revolution

The modified Maxwell-Bloch equations (77) with $\Omega = 2\widetilde{\Omega}$, $\kappa = \rho^2 - \widetilde{\Omega}^2$, and $\nu = 0$ and $\delta = 0$, where $\rho > 0$ is the frequency of free vibrations of the potential system corresponding to $\delta = \widetilde{\Omega} = \nu = 0$, describe a two-mode approximation of the models of rotating elastic bodies of revolution after their linearization and discretization [67, 71, 76]. In the absence of dissipative and non-conservative positional forces the characteristic polynomial (10) corresponding to the operator $\mathbf{L}_0(\widetilde{\Omega}) = \mathbf{I}\lambda^2 + 2\lambda\widetilde{\Omega}\mathbf{G} + (\rho^2 - \widetilde{\Omega}^2)\mathbf{I}$, which belongs to the class of matrix polynomials considered, e.g., in [38], has four purely imaginary roots

$$\lambda_p^{\pm} = i\rho \pm i\widetilde{\Omega}, \quad \lambda_n^{\pm} = -i\rho \pm i\widetilde{\Omega}. \tag{83}$$

In the plane $(\widetilde{\Omega}, \mathrm{Im}\,\lambda)$ the eigenvalues (83) form a collection of straight lines intersecting with each other – *the spectral mesh* [64, 76]. Two nodes of the mesh at $\widetilde{\Omega} = 0$ correspond to the double semi-simple eigenvalues $\lambda = \pm i\rho$. The double semi-simple eigenvalue $i\rho$ at $\widetilde{\Omega} = \widetilde{\Omega}_0 = 0$ has two linearly-independent eigenvectors $\mathbf{u}_1$ and $\mathbf{u}_2$

$$\mathbf{u}_1 = \frac{1}{\sqrt{2\rho}}\begin{pmatrix} 0 \\ 1 \end{pmatrix}, \quad \mathbf{u}_2 = \frac{1}{\sqrt{2\rho}}\begin{pmatrix} 1 \\ 0 \end{pmatrix}. \tag{84}$$

The eigenvectors are orthogonal $\mathbf{u}_i^T\mathbf{u}_j = 0$, $i \neq j$, and satisfy the normalization condition $\mathbf{u}_i^T\mathbf{u}_i = (2\rho)^{-1}$. At the other two nodes at $\widetilde{\Omega} = \pm\widetilde{\Omega}_d$ there exist double semi-simple eigenvalues $\lambda = 0$. The range $|\widetilde{\Omega}| < \widetilde{\Omega}_d = \rho$ is called *subcritical* for the gyroscopic parameter $\widetilde{\Omega}$.

In the following, with the use of the perturbation theory of multiple eigenvalues, we describe the deformation of the mesh caused by dissipative ($\delta\mathbf{D}$) and non-conservative perturbations ($\nu\mathbf{N}$), originating, e.g. from the frictional contact, and clarify the key role of indefinite damping and non-conservative positional forces in the development of the subcritical flutter instability. This will give a clear mathematical description of the mechanism of excitation of particular modes of rotating structures in frictional contact, such as squealing disc brakes and singing wine glasses [67, 71, 76].

Under perturbation of the gyroscopic parameter $\widetilde{\Omega} = \widetilde{\Omega}_0 + \Delta\widetilde{\Omega}$, the double eigenvalue $i\rho$ into two simple ones bifurcates according to the asymptotic formula [58]

$$\lambda_p^{\pm} = i\rho + i\Delta\widetilde{\Omega}\frac{f_{11} + f_{22}}{2} \pm i\Delta\widetilde{\Omega}\sqrt{\frac{(f_{11} - f_{22})^2}{4} + f_{12}f_{21}} \tag{85}$$

where the quantities f_{ij} are

$$f_{ij} = \mathbf{u}_j^T \frac{\partial \mathbf{L}_0(\widetilde{\Omega})}{\partial \widetilde{\Omega}} \mathbf{u}_i \bigg|_{\widetilde{\Omega}=0, \lambda=i\rho} = 2i\rho\mathbf{u}_j^T\mathbf{G}\mathbf{u}_i. \tag{86}$$

The skew symmetry of $\mathbf{G}$ yields $f_{11} = f_{22} = 0$, $f_{12} = -f_{21} = i$, so that (86) gives the exact result (83).

4.2.1 Deformation of the spectral mesh.

Consider a perturbation of the gyroscopic system $\mathbf{L}_0(\widetilde{\Omega}) + \Delta\mathbf{L}(\widetilde{\Omega})$, assuming that the size of the perturbation $\Delta\mathbf{L}(\widetilde{\Omega}) = \delta\lambda\mathbf{D} + \nu\mathbf{N} \sim \varepsilon$ is small, where $\varepsilon = \|\Delta\mathbf{L}(0)\|$ is the Frobenius norm of the perturbation at $\widetilde{\Omega} = 0$. The behavior of the perturbed eigenvalue $i\rho$ for small $\widetilde{\Omega}$ and small ε is described by the asymptotic formula [58]

$$\lambda = i\beta + i\widetilde{\Omega}\frac{(f_{11}+f_{22})}{2} + i\frac{\epsilon_{11}+\epsilon_{22}}{2}$$
$$\pm i\sqrt{\frac{(\widetilde{\Omega}(f_{11}-f_{22})+\epsilon_{11}-\epsilon_{22})^2}{4} + (\widetilde{\Omega}f_{12}+\epsilon_{12})(\widetilde{\Omega}f_{21}+\epsilon_{21})}, \quad (87)$$

where f_{ij} are given by (86) and ϵ_{ij} are small complex numbers of order ε

$$\epsilon_{ij} = \mathbf{u}_j^T \Delta\mathbf{L}(0)\mathbf{u}_i = i\rho\delta\mathbf{u}_j^T\mathbf{D}\mathbf{u}_i + \nu\mathbf{u}_j^T\mathbf{N}\mathbf{u}_i. \quad (88)$$

With the use of the vectors (84) we obtain

$$\lambda = i\rho - \frac{\mu_1+\mu_2}{4}\delta \pm \sqrt{c}, \quad c = \left(\frac{\mu_1-\mu_2}{4}\right)^2\delta^2 + \left(i\widetilde{\Omega} + \frac{\nu}{2\rho}\right)^2, \quad (89)$$

where the eigenvalues μ_1, μ_2 of $\mathbf{D}$ satisfy the equation $\mu^2 - \mu\mathrm{tr}\mathbf{D} + \det\mathbf{D} = 0$.

Separation of real and imaginary parts in equation (89) yields

$$\mathrm{Re}\,\lambda = -\frac{\mu_1+\mu_2}{4}\delta \pm \sqrt{\frac{|c|+\mathrm{Re}\,c}{2}}, \quad \mathrm{Im}\,\lambda = \rho \pm \sqrt{\frac{|c|-\mathrm{Re}\,c}{2}}, \quad (90)$$

where

$$\mathrm{Re}\,c = \left(\frac{\mu_1-\mu_2}{4}\right)^2\delta^2 - \widetilde{\Omega}^2 + \frac{\nu^2}{4\rho^2}, \quad \mathrm{Im}\,c = \frac{\widetilde{\Omega}\nu}{\rho}. \quad (91)$$

The formulas (89)-(91) describe splitting of the double eigenvalues at the nodes of the spectral mesh due to variation of parameters.

Assuming $\nu = 0$ in formulas (90) we find that

$$\left(\mathrm{Re}\,\lambda + \frac{\mu_1+\mu_2}{4}\delta\right)^2 + \widetilde{\Omega}^2 = \frac{(\mu_1-\mu_2)^2}{16}\delta^2, \quad \mathrm{Im}\,\lambda = \rho \quad (92)$$

when

$$\widetilde{\Omega}^2 - \frac{(\mu_1-\mu_2)^2}{16}\delta^2 < 0, \quad (93)$$

and

$$\widetilde{\Omega}^2 - (\mathrm{Im}\,\lambda - \rho)^2 = \frac{(\mu_1-\mu_2)^2}{16}\delta^2, \quad \mathrm{Re}\,\lambda = -\frac{\mu_1+\mu_2}{4}\delta, \quad (94)$$

when the sign in inequality (93) is opposite. For a given δ equation (94) defines a hyperbola in the plane $(\widetilde{\Omega}, \mathrm{Im}\,\lambda)$, while (92) is the equation of a circle in the plane $(\widetilde{\Omega}, \mathrm{Re}\,\lambda)$, as shown in Fig. 12(a,c). For tracking the complex eigenvalues due to change of the gyroscopic parameter $\widetilde{\Omega}$, it is convenient to consider the eigenvalue branches in the three-dimensional space $(\widetilde{\Omega}, \mathrm{Im}\,\lambda, \mathrm{Re}\,\lambda)$. In this space the circle belongs to the plane $\mathrm{Im}\,\lambda = \rho$ and the hyperbola lies in the plane $\mathrm{Re}\,\lambda = -\delta(\mu_1+\mu_2)/4$, see Fig. 13(a,c).

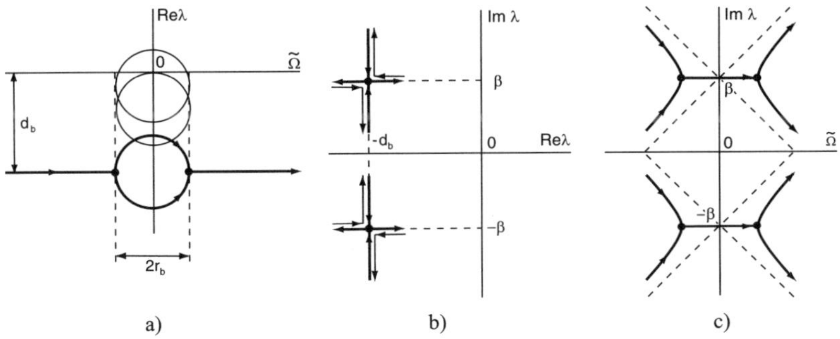

Fig. 12. Origination of a latent source of the subcritical flutter instability in presence of full dissipation: Submerged bubble of instability (a); coalescence of eigenvalues in the complex plane at two exceptional points (b); hyperbolic trajectories of imaginary parts (c).

The radius r_b of the circle of complex eigenvalues—*the bubble of instability*—and the distance d_b of its center from the plane $\mathrm{Re}\,\lambda = 0$ are expressed by means of the eigenvalues μ_1 and μ_2 of the matrix $\mathbf{D}$

$$r_b = |(\mu_1 - \mu_2)\delta|/4, \quad d_b = |(\mu_1 + \mu_2)\delta|/4. \tag{95}$$

Consequently, the bubble of instability is "submerged" under the surface $\mathrm{Re}\,\lambda = 0$ in the space $(\widetilde{\Omega}, \mathrm{Im}\,\lambda, \mathrm{Re}\,\lambda)$ and does not intersect the plane $\mathrm{Re}\,\lambda = 0$ under the condition $d_b \geqslant r_b$, which is equivalent to the positive-definiteness of the matrix $\delta\mathbf{D}$. Hence, the role of full dissipation or pervasive damping is to deform the spectral mesh in such a way that the double semi-simple eigenvalue is inflated to the bubble of complex eigenvalues (92) connected with the two branches of the hyperbola (94) at the points

$$\mathrm{Im}\,\lambda = \rho, \quad \mathrm{Re}\,\lambda = -\delta(\mu_1 + \mu_2)/4, \quad \widetilde{\Omega} = \pm\delta(\mu_1 - \mu_2)/4, \tag{96}$$

and to plunge all the eigenvalue curves into the region $\mathrm{Re}\,\lambda \leqslant 0$. The eigenvalues at the points (96) are double and have a Jordan chain of order 2. In the complex plane the eigenvalues move with the variation of $\widetilde{\Omega}$ along the lines $\mathrm{Re}\,\lambda = -d_b$ until they meet at the points (96) and then split in the orthogonal direction; however, they never cross the imaginary axis, see Fig. 12(b).

The radius of the bubble of instability is greater then the depth of its submersion under the surface $\mathrm{Re}\,\lambda = 0$ only if the eigenvalues μ_1 and μ_2 of the damping matrix have different signs, i.e. if *the damping is indefinite.* The damping with the indefinite matrix appears in the systems with frictional contact when the friction coefficient is decreasing with relative sliding velocity [35, 36, 40]. Indefinite damping leads to the emersion of the bubble of instability

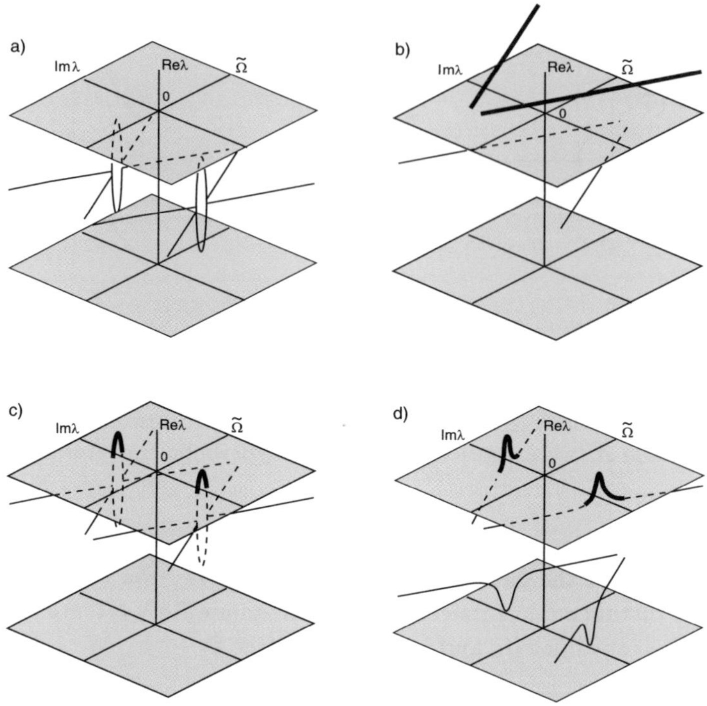

Fig. 13. The mechanism of subcritical flutter instability (bold lines): The ring (bubble) of complex eigenvalues submerged under the surface $\mathrm{Re}\,\lambda = 0$ due to action of dissipation with $\det \mathbf{D} \geqslant 0$ - a latent source of instability (a); repulsion of eigenvalue branches of the spectral mesh due to action of non-conservative positional forces (b); emersion of the bubble of instability due to indefinite damping with $\det \mathbf{D} < 0$ (c); collapse of the bubble of instability and immersion and emersion of its parts due to combined action of dissipative and non-conservative positional forces (d).

meaning that the eigenvalues of the bubble have positive real parts in the range $\widetilde{\Omega}^2 < \widetilde{\Omega}_{cr}^2$, where $\widetilde{\Omega}_{cr} = \frac{\delta}{2}\sqrt{-\det \mathbf{D}}$. Changing the damping matrix $\delta\mathbf{D}$ from positive definite to indefinite we trigger the state of the bubble of instability from latent ($\mathrm{Re}\,\lambda < 0$) to active ($\mathrm{Re}\,\lambda > 0$), see Fig. 13(a,c). Since for small δ we have $\widetilde{\Omega}_{cr} < \widetilde{\Omega}_d$, the flutter instability is subcritical and is localized in the neighborhood of the nodes of the spectral mesh at $\widetilde{\Omega} = 0$.

In the absence of dissipation, the non-conservative positional forces destroy the marginal stability of gyroscopic systems [12, 13]. Indeed, assuming $\delta = 0$ in the formula (89) we obtain

$$\lambda_p^{\pm} = i\rho \pm i\widetilde{\Omega} \pm \frac{\nu}{2\rho}, \quad \lambda_n^{\pm} = -i\rho \pm i\widetilde{\Omega} \mp \frac{\nu}{2\rho}. \tag{97}$$

According to (97), the eigenvalues of the branches $i\rho + i\widetilde{\Omega}$ and $-i\rho - i\widetilde{\Omega}$ of the spectral mesh get positive real parts due to perturbation by the non-conservative positional forces. The eigenvalues of the other two branches are shifted to the left from the imaginary axis, see Fig. 13(b).

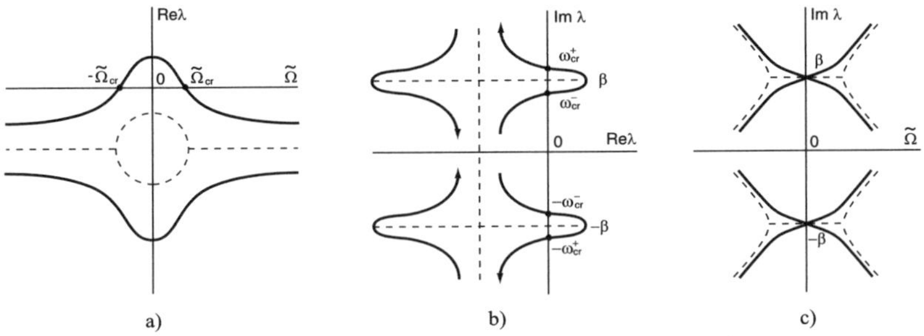

Fig. 14. Subcritical flutter instability due to combined action of dissipative and non-conservative positional forces: Collapse and emersion of the bubble of instability (a); excursions of eigenvalues to the right side of the complex plane when $\widetilde{\Omega}$ goes from negative values to positive (b); crossing of imaginary parts (c).

In contrast to the effect of indefinite damping the instability induced by the non-conservative forces only is not local. However, in combination with the dissipative forces, both definite and indefinite, the non-conservative forces can create subcritical flutter instability in the vicinity of diabolical points.

From equation (89) we find that in presence of dissipative and circulatory perturbations the trajectories of the eigenvalues in the complex plane are described by the formula

$$\left(\operatorname{Re}\lambda + \frac{\operatorname{tr}\mathbf{D}}{4}\delta\right)(\operatorname{Im}\lambda - \rho) = \frac{\widetilde{\Omega}\nu}{2\rho}. \tag{98}$$

Non-conservative positional forces with $\nu \neq 0$ destroy the merging of modes, shown in Fig. 12, so that the eigenvalues move along the separated trajectories. According to (98) the eigenvalues with $|\operatorname{Im}\lambda|$ increasing due to an increase in $|\widetilde{\Omega}|$ move closer to the imaginary axis then the others, as shown in Fig 14(b). In the space $(\widetilde{\Omega}, \operatorname{Im}\lambda, \operatorname{Re}\lambda)$ the action of the non-conservative positional forces separates the bubble of instability and the adjacent hyperbolic eigenvalue branches into two non-intersecting curves, see Fig 13(d). The form of each of the new eigenvalue curves carries the memory about the original bubble of instability, so that the real parts of the eigenvalues can be positive for the values of the gyroscopic parameter localized near $\widetilde{\Omega} = 0$ in the range $\widetilde{\Omega}^2 < \widetilde{\Omega}_{cr}^2$, where

$$\widetilde{\Omega}_{cr} = \delta\frac{\mathrm{tr}\mathbf{D}}{4}\sqrt{-\frac{\nu^2 - \delta^2\rho^2 \det \mathbf{D}}{\nu^2 - \delta^2\rho^2(\mathrm{tr}\mathbf{D}/2)^2}}. \tag{99}$$

follows from the equations (89)-(91).

The eigenfrequencies of the unstable modes from the interval $\widetilde{\Omega}^2 < \widetilde{\Omega}_{cr}^2$ are localized near the frequency of the double semi-simple eigenvalue at the node of the undeformed spectral mesh: $\omega_{cr}^- < \omega < \omega_{cr}^+$

$$\omega_{cr}^{\pm} = \rho \pm \frac{\nu}{2\rho}\sqrt{-\frac{\nu^2 - \delta^2\rho^2 \det \mathbf{D}}{\nu^2 - \delta^2\rho^2(\mathrm{tr}\mathbf{D}/2)^2}}. \tag{100}$$

When the radicand in formulas (99) and (100) is real, the eigenvalues make the excursion to right side of the complex plane, as shown in Fig. 14(b). In presence of non-conservative positional forces such excursions behind the stability boundary are possible, even when dissipation is full ($\det \mathbf{D} > 0$).

The equation (99) describes the surface in the space of the parameters δ, ν, and $\widetilde{\Omega}$, which is an approximation to the stability boundary. Extracting the parameter ν in (99) yields

$$\nu = \pm\delta\rho\mathrm{tr}\mathbf{D}\sqrt{\frac{\delta^2 \det \mathbf{D} + 4\widetilde{\Omega}^2}{\delta^2(\mathrm{tr}\mathbf{D})^2 + 16\widetilde{\Omega}^2}}. \tag{101}$$

If $\det \mathbf{D} \geqslant 0$ and $\widetilde{\Omega}$ is fixed, the formula (101) describes two independent curves in the plane (δ, ν) intersecting with each other at the origin along the straight lines given by the expression

$$\nu = \pm\frac{\rho\mathrm{tr}\mathbf{D}}{2}\delta. \tag{102}$$

However, in case when $\det \mathbf{D} < 0$, the radical in (101) is real only for $\delta^2 < -4\widetilde{\Omega}^2/\det \mathbf{D}$ meaning that (101) describes two branches of a closed loop in the plane of the parameters δ and ν. The loop is self-intersecting at the origin with the tangents given by the expression (102). Hence, the shape of the surface described by equation (101) is a cone with the "8"-shaped loop in a cross-section, see Fig. 15(a). The asymptotic stability domain is inside the two of the four pockets of the cone, selected by the inequality $\delta\mathrm{tr}\mathbf{D} > 0$, as shown in Fig. 15(a). The singularity of the stability domain at the origin is the degeneration of a more general configuration shown in Fig. 5(b).

The domain of asymptotic stability bifurcates when $\det \mathbf{D}$ changes from negative to positive values. This process is shown in Fig. 15. In case of indefinite damping there exists an instability gap due to the singularity at the origin. Starting in the flutter domain at $\widetilde{\Omega} = 0$ for any combination of the parameters

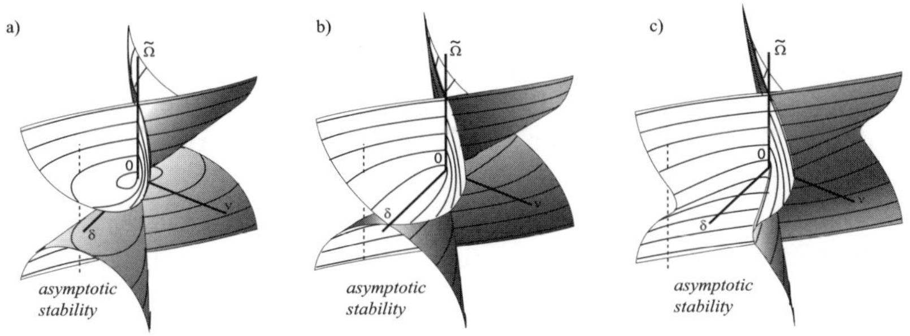

Fig. 15. Domains of asymptotic stability in the space $(\delta, \nu, \widetilde{\Omega})$ for different types of damping: Indefinite damping $\det \mathbf{D} < 0$ (a); semi-definite (pervasive) damping $\det \mathbf{D} = 0$ (b); full dissipation $\det \mathbf{D} > 0$ (c).

δ and ν one can reach the domain of asymptotic stability at higher values of $|\widetilde{\Omega}|$ (gyroscopic stabilization), as shown in Fig. 15(a) by the dashed line. The gap is responsible for the subcritical flutter instability localized in the vicinity of the node of the spectral mesh of the unperturbed gyroscopic system. When $\det \mathbf{D} = 0$, the gap vanishes in the direction $\nu = 0$. In case of full dissipation $(\det \mathbf{D} > 0)$ the singularity at the origin unfolds. However, the memory about it is preserved in the two instability gaps located in the folds of the stability boundary with the locally strong curvature, Fig. 15(c). At some values of δ and ν one can penetrate the fold of the stability boundary with the change of Ω, as shown in Fig. 15(c) by the dashed line. For such δ and ν the flutter instability is localized in the vicinity of $\widetilde{\Omega} = 0$.

The phenomenon of the local subcritical flutter instability is controlled by the eigenvalues of the matrix $\mathbf{D}$. When both of them are positive, the folds of the stability boundary are more pronounced if one of the eigenvalues is close to zero. If one of the eigenvalues is negative and the other is positive, the local subcritical flutter instability is possible for any combination of δ and ν including the case when the non-conservative positional forces are absent $(\nu = 0)$.

The instability mechanism behind the squealing disc brake or singing wine glass can be described as the emersion (or activation) due to indefinite damping and non-conservative positional forces of the bubbles of instability created by the full dissipation in the vicinity of the nodes of the spectral mesh.

Conclusions

Investigation of stability and sensitivity analysis of the critical parameters and critical frequencies of near-Hamiltonian and near-reversible systems is complicated by the singularities of the boundary of asymptotic stability domain, which are related to the multiple eigenvalues. In the paper we have developed the methods of approximation of the stability boundaries near the singularities and obtained estimates of the critical values of parameters in the case of arbitrary number of degrees of freedom using the perturbation theory of eigenvalues and eigenvectors of non-self-adjoint operators. In case of two degrees of freedom the domain of asymptotic stability of near-reversible and near-Hamiltonian systems is fully described and its typical configurations are found. Bifurcation of the stability domain due to change of the matrix of dissipative forces is discovered and described. Two classes of indefinite damping matrices are found and the explicit threshold, separating the weakly- and strongly indefinite matrices is derived. The role of dissipative and non-conservative forces in the paradoxical effects of gyroscopic stabilization of statically unstable potential systems as well as of destabilization of statically stable ones is clarified. Finally, the mechanism of subcritical flutter instability in rotating elastic bodies of revolution in frictional contact, exciting oscillations in the squealing disc brake and in the singing wine glass, is established.

Acknowledgments

The author is grateful to Professor P. Hagedorn for his interest to this work and useful discussions.

References

1. W. Thomson and P. G. Tait, *Treatise on Natural Philosophy*, Vol. 1, Part 1, New Edition, Cambridge Univ. Press, Cambridge, 1879.
2. E. J. Routh, *A treatise on the stability of a given state of motion*, Macmillan, London, 1892.
3. A. L. Kimball, *Internal friction theory of shaft whirling*, Phys. Rev., 21(6) (1923), pp. 703.
4. D. M. Smith, *The motion of a rotor carried by a flexible shaft in flexible bearings*, Proc. Roy. Soc. Lond. A., 142 (1933), pp. 92–118.
5. P. L. Kapitsa, *Stability and transition through the critical speed of fast rotating shafts with friction*, Zh. Tekh. Fiz., 9 (1939), pp. 124–147.
6. H. Bilharz, *Bemerkung zu einem Satze von Hurwitz*, Z. angew. Math. Mech., 24(2) (1944), pp. 77–82.
7. M. G. Krein, *A generalization of some investigations of linear differential equations with periodic coefficients*, Doklady Akad. Nauk SSSR N.S., 73 (1950), pp. 445-448.
8. H. Ziegler, *Die Stabilitätskriterien der Elastomechanik*, Ing.-Arch., 20 (1952), pp. 49–56.
9. E. O. Holopainen, *On the effect of friction in baroclinic waves*, Tellus, 13(3) (1961), pp. 363–367.

10. V. V. BOLOTIN, *Non-conservative Problems of the Theory of Elastic Stability*, Pergamon, Oxford, 1963.
11. G. HERRMANN AND I. C. JONG, *On the destabilizing effect of damping in non-conservative elastic systems*, ASME J. of Appl. Mechs., 32(3) (1965), pp. 592–597.
12. V. M. LAKHADANOV, *On stabilization of potential systems*, Prikl. Mat. Mekh., 39(1) (1975), pp. 53-58.
13. A. V. KARAPETYAN, *On the stability of nonconservative systems*, Vestn. MGU. Ser. Mat. Mekh., 4 (1975), pp. 109-113.
14. W. HAUGER, *Stability of a gyroscopic non-conservative system*, Trans. ASME, J. Appl. Mech., 42 (1975), pp. 739–740.
15. I. P. ANDREICHIKOV AND V. I. YUDOVICH, *The stability of visco-elastic rods*, Izv. Acad. Nauk SSSR. MTT, 1 (1975), pp. 150–154.
16. V. N. TKHAI, *On stability of mechanical systems under the action of position forces*, PMM U.S.S.R., 44 (1981), pp. 24–29.
17. V. I. ARNOLD, *Geometrical Methods in the Theory of Ordinary Differential Equations*, Springer, New York and Berlin, 1983.
18. A. S. DEIF, P. HAGEDORN, *Matrix polynomials subjected to small perturbations.* Z. angew. Math. Mech., 66 (1986), pp. 403–412.
19. M. B. SEVRYUK, *Reversible systems*, Lecture Notes in Mathematics 1211, Springer, Berlin, 1986.
20. N. V. BANICHUK, A. S. BRATUS, A. D. MYSHKIS, *Stabilizing and destabilizing effects in nonconservative systems*, PMM U.S.S.R., 53(2) (1989), pp. 158–164.
21. S. BARNETT, *Leverrier's algorithm: a new proof and extensions*, SIAM J. Matrix Anal. Appl., 10(4) (1989), pp. 551-556.
22. A. P. SEYRANIAN, *Destabilization paradox in stability problems of non-conservative systems*, Advances in Mechanics, 13(2) (1990), 89–124.
23. R. S. MACKAY, *Movement of eigenvalues of Hamiltonian equilibria under non-Hamiltonian perturbation*, Phys. Lett. A, 155 (1991), 266–268.
24. H. LANGER, B. NAJMAN, K. VESELIC, *Perturbation of the eigenvalues of quadratic matrix polynomials*, SIAM J. Matrix Anal. Appl. 13(2) (1992), pp. 474–489.
25. G. HALLER, *Gyroscopic stability and its loss in systems with two essential coordinates*, Intern. J. of Nonl. Mechs., 27 (1992), 113–127.
26. A. N. KOUNADIS, *On the paradox of the destabilizing effect of damping in non-conservative systems*, Intern. J. of Nonl. Mechs., 27 (1992), 597–609.
27. V. F. ZHURAVLEV, *Nutational vibrations of a free gyroscope*, Izv. Ross. Akad. Nauk, Mekh. Tverd. Tela, 6 (1992), 13–16.
28. A. M. BLOCH, P. S. KRISHNAPRASAD, J. E. MARSDEN, T. S. RATIU, *Dissipation-induced instabilities*, Annales de l'Institut Henri Poincaré, 11(1) (1994), pp. 37–90.
29. J. MADDOCKS AND M. L. OVERTON, *Stability theory for dissipatively perturbed Hamiltonian systems*, Comm. Pure and Applied Math., 48 (1995), pp. 583–610.
30. I. HOVEIJN AND M. RUIJGROK, *The stability of parametrically forced coupled oscillators in sum resonance*, Z. angew. Math. Phys., 46 (1995), pp. 384–392.
31. S. H. CRANDALL, *The effect of damping on the stability of gyroscopic pendulums*, Z. angew. Math. Phys., 46 (1995), pp. 761–780.
32. V. V. BELETSKY, *Some stability problems in applied mechanics*, Appl. Math. Comp., 70 (1995), pp. 117–141.

33. O. M. O'Reilly, N. K. Malhotra, N. S. Namachchivaya, *Some aspects of destabilization in reversible dynamical systems with application to follower forces*, Nonlin. Dyn. 10 (1996), pp. 63–87.
34. D. R. Merkin, *Introduction to the Theory of Stability*, Springer, Berlin, 1997.
35. P. Freitas, M. Grinfeld, P. A. Knight, *Stability of finite-dimensional systems with indefinite damping*, Adv. Math. Sci. Appl. 7(1) (1997), pp. 437–448.
36. W. Kliem and P. C. Müller, *Gyroscopic stabilization of indefinite damped systems*. Z. angew. Math. Mech. 77(1) (1997), pp. 163–164.
37. R. Hryniv and P. Lancaster, *On the perturbation of analytic matrix functions*, Integral Equations And Operator Theory, 34(3) (1999), pp. 325–338.
38. R. Hryniv and P. Lancaster, *Stabilization of gyroscopic systems*. Z. angew. Math. Mech. 81(10) (2001), pp. 675–681.
39. V. V. Bolotin, A. A. Grishko, M. Yu. Panov, *Effect of damping on the post-critical behavior of autonomous non-conservative systems*, Intern. J. of Nonl. Mechs. 37 (2002), pp. 1163–1179.
40. K. Popp, M. Rudolph, M. Kröger, M. Lindner, *Mechanisms to generate and to avoid friction induced vibrations*, VDI-Berichte 1736, VDI-Verlag, Düsseldorf, 2002.
41. P. Lancaster, A. S. Markus, F. Zhou, *Perturbation theory for analytic matrix functions: The semisimple case*, SIAM J. Matrix Anal. Appl. 25(3) (2003), pp. 606–626.
42. N. Hoffmann and L. Gaul, *Effects of damping on mode-coupling instability in friction induced oscillations*, Z. angew. Math. Mech. 83 (2003), pp. 524–534.
43. O. N. Kirillov, *How do small velocity-dependent forces (de)stabilize a non-conservative system?*, DCAMM Report 681, Lyngby, 2003.
44. A. P. Seiranyan and O. N. Kirillov, *Effect of small dissipative and gyroscopic forces on the stability of nonconservative systems*, Doklady Physics, 48(12) (2003), pp. 679–684.
45. W. F. Langford, *Hopf meets Hamilton under Whitney's umbrella*, in IUTAM symposium on nonlinear stochastic dynamics. Proceedings of the IUTAM symposium, Monticello, IL, USA, Augsut 26-30, 2002, Solid Mech. Appl. 110, S. N. Namachchivaya, et al., eds., Kluwer, Dordrecht, 2003, pp. 157–165.
46. A. P. Ivanov, *The stability of mechanical systems with positional non-conservative forces*, *J. Appl. Maths. Mechs.* 67(5) (2003), pp. 625–629.
47. A. P. Seyranian and A. A. Mailybaev, *Multiparameter stability theory with mechanical applications*, World Scientific, Singapore, 2003.
48. N. M. Bou-Rabee, J. E. Marsden, L. A. Romero, *Tippe Top inversion as a dissipation-induced instability*, SIAM J. Appl. Dyn. Sys. 3 (2004), pp. 352–377.
49. H. K. Moffatt, Y. Shimomura, M. Branicki, *Dynamics of an axisymmetric body spinning on a horizontal surface. I. Stability and the gyroscopic approximation*, Proc. Roy. Soc. Lond. A 460 (2004), pp. 3643–3672.
50. O. N. Kirillov, *Destabilization paradox*, Doklady Physics. 49(4) (2004), pp. 239–245.
51. O. N. Kirillov, A. P. Seyranian, *Collapse of the Keldysh chains and stability of continuous nonconservative systems*, SIAM J. Appl. Math. **64**(4) (2004), pp. 1383–1407.
52. O. N. Kirillov and A. P. Seyranian, *Stabilization and destabilization of a circulatory system by small velocity-dependent forces*, *J. Sound and Vibr.*, 283(3–5) (2005), pp. 781–800.

53. O. N. Kirillov, *A theory of the destabilization paradox in non-conservative systems*, Acta Mechanica, 174(3–4) (2005), pp. 145–166.
54. O. N. Kirillov and A. P. Seyranian, *Instability of distributed nonconservative systems caused by weak dissipation*, Doklady Mathematics, **71**(3) (2005), pp. 470–475.
55. O. N. Kirillov and A. O. Seyranian, *The effect of small internal and external damping on the stability of distributed non-conservative systems*, J. Appl. Math. Mech., 69(4) (2005), pp. 529–552.
56. P. Lancaster, P. Psarrakos, *On the Pseudospectra of Matrix Polynomials*, SIAM J. Matrix Anal. Appl., 27(1) (2005), pp. 115–120.
57. A. P. Seyranian, O. N. Kirillov, A. A. Mailybaev, *Coupling of eigenvalues of complex matrices at diabolic and exceptional points.* J. Phys. A: Math. Gen., 38(8) (2005), pp. 1723–1740.
58. O. N. Kirillov, A. A. Mailybaev, A. P. Seyranian, *Unfolding of eigenvalue surfaces near a diabolic point due to a complex perturbation*, J. Phys. A: Math. Gen., 38(24) (2005), pp. 5531-5546.
59. N. M. Bou-Rabee, J. E. Marsden, L. A. Romero, *A geometric treatment of Jellet's egg*, Z. angew. Math. Mech., 85(9) (2005), pp. 618–642.
60. A. A. Mailybaev, O. N. Kirillov, A. P. Seyranian, *Berry phase around degeneracies*, Dokl. Math., 73(1) (2006), pp. 129–133.
61. T. Butlin, J. Woodhouse, *Studies of the Sensitivity of Brake Squeal*, Appl. Mech. and Mater., 5-6 (2006), pp. 473–479.
62. R. Krechetnikov and J. E. Marsden, *On destabilizing effects of two fundamental non-conservative forces*, Physica D, 214 (2006), pp. 25–32.
63. O. N. Kirillov, *Gyroscopic stabilization of non-conservative systems*, Phys. Lett. A, 359(3) (2006), pp. 204–210.
64. U. Günther, O. N. Kirillov, *A Krein space related perturbation theory for MHD alpha-2 dynamos and resonant unfolding of diabolical points*, J. Phys. A: Math. Gen., 39 (2006), pp. 10057–10076
65. V. Kobelev, *Sensitivity analysis of the linear nonconservative systems with fractional damping*, Struct. Multidisc. Optim., 33 (2007), pp. 179-188.
66. R. Krechetnikov and J. E. Marsden, *Dissipation-induced instabilities in finite dimensions*, Rev. Mod. Phys., 79 (2007), pp. 519–553.
67. U. von Wagner, D. Hochlenert, P. Hagedorn, *Minimal models for disk brake squeal*, J. Sound Vibr., 302(3) (2007), pp. 527–539.
68. O. N. Kirillov, *On the stability of nonconservative systems with small dissipation*, J. Math. Sci., 145(5) (2007), pp. 5260–5270.
69. A. N. Kounadis, *Flutter instability and other singularity phenomena in symmetric systems via combination of mass distribution and weak damping*, Int. J. of Non-Lin. Mech., 42(1) (2007), pp. 24–35.
70. O. N. Kirillov, *Destabilization paradox due to breaking the Hamiltonian and reversible symmetry*, Int. J. of Non-Lin. Mech., 42(1) (2007), pp. 71–87.
71. G. Spelsberg-Korspeter, D. Hochlenert, O. N. Kirillov, P. Hagedorn, *In- and out-of-plane vibrations of a rotating plate with frictional contact: Investigations on squeal phenomena*, Trans. ASME, J. Appl. Mech., (2007) (submitted).
72. J.-J. Sinou and L. Jezequel, *Mode coupling instability in friction-induced vibrations and its dependency on system parameters including damping*, Eur. J. Mech. A., 26 (2007), 106–122.

73. P. Kessler, O. M. O'Reilly, A.-L. Raphael, M. Zworski, *On dissipation-induced destabilization and brake squeal: A perspective using structured pseudospectra*, J. Sound Vibr., 308 (2007), 1-11.
74. O. N. Kirillov *Gyroscopic stabilization in the presence of non-conservative forces*, Dokl. Math., 76(2) (2007), pp. 780–785.
75. G. Spelsberg-Korspeter, O. N. Kirillov, P. Hagedorn, *Modeling and stability analysis of an axially moving beam with frictional contact*, Trans. ASME, J. Appl. Mech. 75(3) (2008), 031001.
76. O. N. Kirillov, *Subcritical flutter in the acoustics of friction*, Proc. R. Soc. A, 464 (2008), pp.
77. J. Kang, C. M. Krousgrill, F. Sadeghi, *Dynamic instability of a thin circular plate with friction interface and its application to disc brake squeal*, J. Sound. Vibr. (2008).

Block Triangular Miniversal Deformations of Matrices and Matrix Pencils

Lena Klimenko[1] and Vladimir V. Sergeichuk[2]

[1] Information and Computer Centre of the Ministry of Labour and Social Policy of Ukraine, Esplanadnaya 8/10, Kiev, Ukraine
e.n.klimenko@gmail.com

[2] Institute of Mathematics, Tereshchenkivska 3, Kiev, Ukraine
sergeich@imath.kiev.ua

Abstract. For each square complex matrix, V. I. Arnold constructed a normal form with the minimal number of parameters to which a family of all matrices B that are close enough to this matrix can be reduced by similarity transformations that smoothly depend on the entries of B. Analogous normal forms were also constructed for families of complex matrix pencils by A. Edelman, E. Elmroth, and B. Kågström, and contragredient matrix pencils (i.e., of matrix pairs up to transformations $(A, B) \mapsto (S^{-1}AR, R^{-1}BS)$) by M. I. Garcia-Planas and V. V. Sergeichuk. In this paper we give other normal forms for families of matrices, matrix pencils, and contragredient matrix pencils; our normal forms are block triangular.

Keywords: canonical forms, matrix pencils, versal deformations, perturbation theory.

1 Introduction

The reduction of a matrix to its Jordan form is an unstable operation: both the Jordan form and the reduction transformations depend discontinuously on the entries of the original matrix. Therefore, if the entries of a matrix are known only approximately, then it is unwise to reduce it to Jordan form. Furthermore, when investigating a family of matrices smoothly depending on parameters, then although each individual matrix can be reduced to its Jordan form, it is unwise to do so since in such an operation the smoothness relative to the parameters is lost.

For these reasons, Arnold [1] constructed a *miniversal deformation* of any Jordan canonical matrix J; that is, a family of matrices in a neighborhood of J with the minimal number of parameters, to which all matrices M close to J can be reduced by similarity transformations that smoothly depend on the entries of M (see Definition 1).

Miniversal deformations were also constructed for:

(i) the Kronecker canonical form of complex matrix pencils by Edelman, Elmroth, and Kågström [9]; another miniversal deformation (which is *simple* in the sense of Definition 2) was constructed by Garcia-Planas and Sergeichuk [10];

(ii) the Dobrovol'skaya and Ponomarev canonical form of complex contragredient matrix pencils (i.e., of matrices of counter linear operators $U \rightleftarrows V$) in [10].

Belitskii [4] proved that each Jordan canonical matrix J is permutationally similar to some matrix $J^{\#}$, which is called a *Weyr canonical matrix* and possesses the property: all matrices that commute with $J^{\#}$ are block triangular. Due to this property, $J^{\#}$ plays a central role in Belitskii's algorithm for reducing the matrices of any system of linear mappings to canonical form, see [5, 11].

In this paper, we find another property of Weyr canonical matrices: they possess block triangular miniversal deformations (in the sense of Definition 2). Therefore, if we consider, up to smooth similarity transformations, a family of matrices that are close enough to a given square matrix, then we can take it in its Weyr canonical form $J^{\#}$ and the family in the form $J^{\#} + E$, in which E is block triangular.

We also give block triangular miniversal deformations of those canonical forms of pencils and contragredient pencils that are obtained from (i) and (ii) by replacing the Jordan canonical matrices with the Weyr canonical matrices.

All matrices that we consider are complex matrices.

2 Miniversal deformations of matrices

Definition 1 (see [1–3]). *A* deformation *of an n-by-n matrix A is a matrix function $\mathcal{A}(\alpha_1, \dots, \alpha_k)$ (its arguments $\alpha_1, \dots, \alpha_k$ are called* parameters*) on a neighborhood of $\vec{0} = (0, \dots, 0)$ that is holomorphic at $\vec{0}$ and equals A at $\vec{0}$. Two deformations of A are identified if they coincide on a neighborhood of $\vec{0}$.*

A deformation $\mathcal{A}(\alpha_1, \dots, \alpha_k)$ of A is versal *if all matrices $A+E$ in some neighborhood of A reduce to the form*

$$\mathcal{A}(h_1(E), \dots, h_k(E)) = \mathcal{S}(E)^{-1}(A + E)\mathcal{S}(E), \qquad \mathcal{S}(0) = I_n,$$

in which $\mathcal{S}(E)$ is a holomorphic at zero matrix function of the entries of E.

A versal deformation with the minimal number of parameters is called miniversal.

Definition 2. *Let a deformation $\mathcal{A}$ of A be represented in the form $A + \mathcal{B}(\alpha_1, \dots, \alpha_k)$.*

– *If* k *entries of* $\mathcal{B}(\alpha_1, \dots, \alpha_k)$ *are the independent parameters* $\alpha_1, \dots, \alpha_k$ *and the others are zero then the deformation* $\mathcal{A}$ *is called* simple[3].
– *A simple deformation is* block triangular *with respect to some partition of* A *into blocks if* $\mathcal{B}(\alpha_1, \dots, \alpha_k)$ *is block triangular with respect to the conformal partition and each of its blocks is either* 0 *or all of its entries are independent parameters.*

If $\mathcal{A}(\alpha_1, \dots, \alpha_k)$ is a miniversal deformation of A and $S^{-1}AS = B$ for some nonsingular S, then $S^{-1}\mathcal{A}(\alpha_1, \dots, \alpha_k)S$ is a miniversal deformation of B. Therefore, it suffices to construct miniversal deformations of canonical matrices for similarity.

Let

$$J(\lambda) := J_{n_1}(\lambda) \oplus \cdots \oplus J_{n_l}(\lambda), \qquad n_1 \geqslant n_2 \geqslant \cdots \geqslant n_l, \tag{1}$$

be a Jordan canonical matrix with a single eigenvalue equal to λ; the unites of Jordan blocks are written over the diagonal:

$$J_{n_i}(\lambda) := \begin{bmatrix} \lambda & 1 & & 0 \\ & \lambda & \ddots & \\ & & \ddots & 1 \\ 0 & & & \lambda \end{bmatrix} \qquad (n_i\text{-by-}n_i).$$

For each natural numbers p and q, define the $p \times q$ matrix

$$\mathcal{T}_{pq} := \begin{cases} \begin{bmatrix} * & 0 & \dots & 0 \\ \vdots & \vdots & & \vdots \\ * & 0 & \dots & 0 \end{bmatrix} & \text{if } p < q, \\ \begin{bmatrix} 0 & \dots & 0 \\ \vdots & & \vdots \\ 0 & \dots & 0 \\ * & \dots & * \end{bmatrix} & \text{if } p \geqslant q, \end{cases} \tag{2}$$

in which the stars denote independent parameters (alternatively, we may take $\mathcal{T}_{pq}$ with $p = q$ as in the case $p < q$).

Theorem 1 ([3, §30, Theorem 2]). (i) *Let* $J(\lambda)$ *be a Jordan canonical matrix of the form* (1) *with a single eigenvalue equal to* λ. *Let* $\mathcal{H} := [\mathcal{T}_{n_i,n_j}]$ *be the parameter block matrix partitioned conformally to* $J(\lambda)$ *with the blocks* $\mathcal{T}_{n_i,n_j}$ *defined in* (2). *Then*

$$J(\lambda) + \mathcal{H} \tag{3}$$

[3] Arnold's miniversal definitions presented in Theorem 1 are simple. Moreover, by [10, Corollary 2.1] the set of matrices of any quiver representation (i.e., of any finite system of linear mappings) over $\mathbb{C}$ or $\mathbb{R}$ possesses a simple miniversal deformation.

is a simple miniversal deformation of $J(\lambda)$.

(ii) *Let*

$$J := J(\lambda_1) \oplus \cdots \oplus J(\lambda_\tau), \qquad \lambda_i \neq \lambda_j \text{ if } i \neq j, \tag{4}$$

be a Jordan canonical matrix in which every $J(\lambda_i)$ *is of the form* (1), *and let* $J(\lambda_i) + \mathcal{H}_i$ *be its miniversal deformation* (3). *Then*

$$J + \mathcal{K} := (J(\lambda_1) + \mathcal{H}_1) \oplus \cdots \oplus (J(\lambda_\tau) + \mathcal{H}_\tau) \tag{5}$$

is a simple miniversal deformation of J.

Definition 3 ([13]). *The* Weyr canonical form $J^\#$ *of a Jordan canonical matrix* J (*and of any matrix that is similar to* J) *is defined as follows.*

(i) *If* J *has a single eigenvalue, then we write it in the form* (1). *Permute the first columns of* $J_{n_1}(\lambda)$, $J_{n_2}(\lambda)$, ..., *and* $J_{n_l}(\lambda)$ *into the first* l *columns, then permute the corresponding rows. Next permute the second columns of all blocks of size at least* 2×2 *into the next columns and permute the corresponding rows; and so on. The obtained matrix is the Weyr canonical form* $J(\lambda)^\#$ *of* $J(\lambda)$.

(ii) *If* J *has distinct eigenvalues, then we write it in the form* (4). *The Weyr canonical form of* J *is*

$$J^\# := J(\lambda_1)^\# \oplus \cdots \oplus J(\lambda_\tau)^\#. \tag{6}$$

Each direct summand of (6) has the form

$$J(\lambda)^\# = \begin{bmatrix} \lambda I_{s_1} & \begin{bmatrix} I_{s_2} \\ 0 \end{bmatrix} & & 0 \\ & \lambda I_{s_2} & \ddots & \\ & & \ddots & \begin{bmatrix} I_{s_k} \\ 0 \end{bmatrix} \\ 0 & & & \lambda I_{s_k} \end{bmatrix}, \tag{7}$$

in which s_i is the number of Jordan blocks $J_l(\lambda)$ of size $l \geqslant i$ in $J(\lambda)$. The sequence $(s_1, s_2, \ldots, s_k)$ is called the *Weyr characteristic* of J (and of any matrix that is similar to J) for the eigenvalue λ, see [12]. By [4] or [11, Theorem 1.2], all matrices commuting with $J^\#$ are block triangular.

In the next lemma we construct a miniversal deformation of $J^\#$ that is block triangular with respect to the most coarse partition of $J^\#$ for which all diagonal blocks have the form $\lambda_i I$ and each off-diagonal block is 0 or I. This means that the sizes of diagonal blocks of (7) with respect to this partition form the sequence obtained from

$$\begin{array}{l} s_k,\ s_{k-1} - s_k,\ \ldots,\ s_2 - s_3,\ s_1 - s_2, \\ s_k,\ s_{k-1} - s_k,\ \ldots,\ s_2 - s_3, \\ \ldots\ldots\ldots\ldots\ldots\ldots \\ s_k,\ s_{k-1} - s_k, \\ s_k \end{array}$$

by removing the zero members.

Theorem 2. (i) *Let* $J(\lambda)$ *be a Jordan canonical matrix of the form* (1) *with a single eigenvalue equal to* λ. *Let* $J(\lambda)+\mathcal{H}$ *be its miniversal deformation* (3). *Denote by*

$$J(\lambda)^{\#}+\mathcal{H}^{\#} \tag{8}$$

the parameter matrix obtained from $J(\lambda)+\mathcal{H}$ *by the permutations described in Definition* 3(i). *Then* $J(\lambda)^{\#}+\mathcal{H}^{\#}$ *is a miniversal deformation of* $J(\lambda)^{\#}$ *and its matrix* $\mathcal{H}^{\#}$ *is lower block triangular.*

(ii) *Let* J *be a Jordan canonical matrix represented in the form* (4) *and let* $J^{\#}$ *be its Weyr canonical form. Let us apply the permutations described in* (i) *to each of the direct summands of miniversal deformation* (5) *of* J. *Then the obtained matrix*

$$J^{\#}+\mathcal{K}^{\#}:=(J(\lambda_1)^{\#}+\mathcal{H}_1^{\#})\oplus\cdots\oplus(J(\lambda_\tau)^{\#}+\mathcal{H}_\tau^{\#}) \tag{9}$$

is a miniversal deformation of $J^{\#}$, *which is simple and block triangular* (*in the sense of Definition* 2).

Let us prove this theorem. The form of $J(\lambda)^{\#}+\mathcal{H}^{\#}$ and the block triangularity of $\mathcal{H}^{\#}$ become clearer if we carry out the permutations from Definition 3(i) in two steps.

First step. Let us write the sequence $n_1, n_2, \dots, n_l$ from (1) in the form

$$\underbrace{m_1,\dots,m_1}_{r_1 \text{ times}},\underbrace{m_2,\dots,m_2}_{r_2 \text{ times}},\dots,\underbrace{m_t,\dots,m_t}_{r_t \text{ times}},$$

where

$$m_1 > m_2 > \cdots > m_t. \tag{10}$$

Partition $J(\lambda)$ into t horizontal and t vertical strips of sizes

$$r_1m_1,\ r_2m_2,\ \dots,\ r_tm_t$$

(each of them contains Jordan blocks of the same size), produce the described permutations within each of these strips, and obtain

$$J(\lambda)^{+}:=J_{m_1}(\lambda I_{r_1})\oplus\cdots\oplus J_{m_t}(\lambda I_{r_t}), \tag{11}$$

in which

$$J_{m_i}(\lambda I_{r_i}):=\begin{bmatrix}\lambda I_{r_i} & I_{r_i} & & 0\\ & \lambda I_{r_i} & \ddots & \\ & & \ddots & I_{r_i}\\ 0 & & & \lambda I_{r_i}\end{bmatrix} \qquad (m_i \text{ diagonal blocks}).$$

By the same permutations of rows and columns of $J(\lambda) + \mathcal{H}$, reduce $\mathcal{H}$ to

$$\mathcal{H}^+ := [\tilde{\mathcal{T}}_{m_i,m_j}(r_i, r_j)],$$

in which every $\tilde{\mathcal{T}}_{m_i,m_j}(r_i, r_j)$ is obtained from the matrix $\mathcal{T}_{m_i,m_j}$ defined in (2) by replacing each entry 0 with the $r_i \times r_j$ zero block and each entry $*$ with the $r_i \times r_j$ block

$$\star := \begin{bmatrix} * & \dots & * \\ \vdots & & \vdots \\ * & \dots & * \end{bmatrix}. \tag{12}$$

For example, if

$$J(\lambda) = \underbrace{J_4(\lambda) \oplus \cdots \oplus J_4(\lambda)}_{p \text{ times}} \oplus \underbrace{J_2(\lambda) \oplus \cdots \oplus J_2(\lambda)}_{q \text{ times}} \tag{13}$$

then

$$J(\lambda)^+ = J_4(\lambda I_p) \oplus J_2(\lambda I_q) = \begin{array}{c} \\ (1,1) \\ (1,2) \\ (1,3) \\ (1,4) \\ (2,1) \\ (2,2) \end{array} \begin{array}{c} \begin{array}{cccccc} (1,1) & (1,2) & (1,3) & (1,4) & (2,1) & (2,2) \end{array} \\ \left[\begin{array}{cccc|cc} \lambda I_p & I_p & 0 & 0 & 0 & 0 \\ 0 & \lambda I_p & I_p & 0 & 0 & 0 \\ 0 & 0 & \lambda I_p & I_p & 0 & 0 \\ 0 & 0 & 0 & \lambda I_p & 0 & 0 \\ \hline 0 & 0 & 0 & 0 & \lambda I_q & I_q \\ 0 & 0 & 0 & 0 & 0 & \lambda I_q \end{array}\right] \end{array}. \tag{14}$$

A strip is indexed by (i, j) if it contains the j-th strip of $J_{m_i}(\lambda I_{r_i})$. Correspondingly,

$$\mathcal{H}^+ = \begin{array}{c} \\ (1,1) \\ (1,2) \\ (1,3) \\ (1,4) \\ (2,1) \\ (2,2) \end{array} \begin{array}{c} \begin{array}{cccccc} (1,1) & (1,2) & (1,3) & (1,4) & (2,1) & (2,2) \end{array} \\ \left[\begin{array}{cccc|cc} 0 & 0 & 0 & 0 & 0 & 0 \\ 0 & 0 & 0 & 0 & 0 & 0 \\ 0 & 0 & 0 & 0 & 0 & 0 \\ \star & \star & \star & \star & \star & \star \\ \hline \star & 0 & 0 & 0 & 0 & 0 \\ \star & 0 & 0 & 0 & \star & \star \end{array}\right] \end{array}. \tag{15}$$

Second step. We permute in $J(\lambda)^+$ the first vertical strips of

$$J_{m_1}(\lambda I_{r_1}),\ J_{m_2}(\lambda I_{r_2}),\ \dots,\ J_{m_t}(\lambda I_{r_t})$$

into the first t vertical strips and permute the corresponding horizontal strips, then permute the second vertical strips into the next vertical strips and permute the corresponding horizontal strips; continue the process until $J(\lambda)^\#$ is achieved. The same permutations transform $\mathcal{H}^+$ to $\mathcal{H}^\#$.

For example, applying there permutations to (14) and (15), we obtain

$$J(\lambda)^{\#} = \begin{array}{c|cc|cc|c|c|} & (1,1) & (2,1) & (1,2) & (2,2) & (1,3) & (1,4) \\ \hline (1,1) & \lambda I_p & 0 & I_p & 0 & 0 & 0 \\ (2,1) & 0 & \lambda I_q & 0 & I_q & 0 & 0 \\ \hline (1,2) & 0 & 0 & \lambda I_p & 0 & I_p & 0 \\ (2,2) & 0 & 0 & 0 & \lambda I_q & 0 & 0 \\ \hline (1,3) & 0 & 0 & 0 & 0 & \lambda I_p & I_p \\ \hline (1,4) & 0 & 0 & 0 & 0 & 0 & \lambda I_p \end{array} \tag{16}$$

and

$$\mathcal{H}^{\#} = \begin{array}{c|cc|cc|c|c|} & (1,1) & (2,1) & (1,2) & (2,2) & (1,3) & (1,4) \\ \hline (1,1) & 0 & 0 & 0 & 0 & 0 & 0 \\ (2,1) & \star & 0 & 0 & 0 & 0 & 0 \\ \hline (1,2) & 0 & 0 & 0 & 0 & 0 & 0 \\ (2,2) & \star & \star & 0 & \star & 0 & 0 \\ \hline (1,3) & 0 & 0 & 0 & 0 & 0 & 0 \\ \hline (1,4) & \star & \star & \star & \star & \star & \star \end{array}. \tag{17}$$

Proof of Theorem 2. (i) Following (14), we index the vertical (horizontal) strips of $J(\lambda)^+$ in (11) by the pairs of natural numbers as follows: a strip is indexed by (i, j) if it contains the j-th strip of $J_{m_i}(\lambda I_{r_i})$. The pairs that index the strips of $J(\lambda)^+$ form the sequence

$$\begin{array}{l} (1,1),\ (1,2),\ \ldots,\ (1,m_t),\ \ldots,\ (1,m_2),\ \ldots,\ (1,m_1), \\ (2,1),\ (2,2),\ \ldots,\ (2,m_t),\ \ldots,\ (2,m_2), \\ \cdots\cdots\cdots\cdots\cdots\cdots\cdots\cdots \\ (t,1),\ (t,2),\ \ldots,\ (t,m_t), \end{array} \tag{18}$$

which is is ordered lexicographically. Rearranging the pairs by the columns of (18):

$$(1,1),\ (2,1),\ \ldots,\ (t,1);\ \ldots;\ (1,m_t),\ (2,m_t),\ \ldots,\ (t,m_t);\ \ldots;\ (1,m_1) \tag{19}$$

(i.e., as in lexicographic ordering but starting from the second elements of the pairs) and making the same permutation of the corresponding strips in $J(\lambda)^+$ and $\mathcal{H}^+$, we obtain $J(\lambda)^{\#}$ and $\mathcal{H}^{\#}$; see examples (16) and (17).

The $((i,j),(i',j'))$-th entry of $\mathcal{H}^+$ is a star if and only if

$$\text{either } i \leqslant i' \text{ and } j = m_i, \text{ or } i > i' \text{ and } j' = 1. \tag{20}$$

By (10), in these cases $j \geqslant j'$ and if $j = j'$ then either $j = j' = m_i$ and $i = i'$, or $j = j' = 1$ and $i > i'$. Therefore, $\mathcal{H}^{\#}$ is lower block triangular.

(ii) This statement follows from (i) and Theorem 1(ii). □

Remark 1. Let $J(\lambda)$ be a Jordan matrix with a single eigenvalue, let $m_1 > m_2 > \cdots > m_t$ be the distinct sizes of its Jordan blocks, and let r_i be the number of Jordan blocks of size m_i. Then the deformation $J(\lambda)^\# + \mathcal{H}^\#$ from Theorem 2 can be formally constructed as follows:

- $J(\lambda)^\#$ and $\mathcal{H}^\#$ are matrices of the same size; they are conformally partitioned into horizontal and vertical strips, which are indexed by the pairs (19).
- The $((i,j),(i,j))$-th diagonal block of $J(\lambda)^\#$ is λI_{r_i}, its $((i,j),(i,j+1))$-th block is I_{r_i}, and its other blocks are zero.
- The $((i,j),(i',j'))$-th block of $\mathcal{H}^+$ has the form (12) if and only if (20) holds; its other blocks are zero.

3 Miniversal deformations of matrix pencils

By Kronecker's theorem on matrix pencils (see [6, Sect. XII, §4]), each pair of $m \times n$ matrices reduces by equivalence transformations

$$(A, B) \mapsto (S^{-1}AR, S^{-1}BR), \qquad S \text{ and } R \text{ are nonsingular},$$

to a *Kronecker canonical pair* (A_{kr}, B_{kr}) being a direct sum, uniquely determined up to permutation of summands, of pairs of the form

$$(I_r, J_r(\lambda)), \ (J_r(0), I_r), \ (F_r, G_r), \ (F_r^T, G_r^T),$$

in which $\lambda \in \mathbb{C}$ and

$$F_r := \begin{bmatrix} 1 & & 0 \\ 0 & \ddots & \\ & \ddots & 1 \\ 0 & & 0 \end{bmatrix}, \qquad G_r := \begin{bmatrix} 0 & & 0 \\ 1 & \ddots & \\ & \ddots & 0 \\ 0 & & 1 \end{bmatrix} \tag{21}$$

are matrices of size $r \times (r-1)$ with $r \geqslant 1$.

Definitions 1 and 2 are extended to matrix pairs in a natural way.

Miniversal deformations of (A_{kr}, B_{kr}) were obtained in [9, 10]. The deformation obtained in [10] is simple; in this section we reduce it to block triangular form by permutations of rows and columns. For this purpose, we replace in (A_{kr}, B_{kr})

- the direct sum (I, J) of all pairs of the form $(I_r, J_r(\lambda))$ by the pair $(I, J^\#)$, and
- the direct sum $(J(0), I)$ of all pairs of the form $(J_r(0), I_r)$ by the pair $(J(0)^\#, I)$,

in which $J^{\#}$ and $J(0)^{\#}$ are the Weyr matrices from Definition 3. We obtain a canonical matrix pair of the form

$$\bigoplus_{i=1}^{l}(F^T_{p_i}, G^T_{p_i}) \oplus (I, J^{\#}) \oplus (J(0)^{\#}, I) \oplus \bigoplus_{i=1}^{r}(F_{q_i}, G_{q_i}); \tag{22}$$

in which we suppose that

$$p_1 \leqslant \cdots \leqslant p_l, \qquad q_1 \geqslant \cdots \geqslant q_r. \tag{23}$$

(This special ordering of direct summands of (22) admits to construct its miniversal deformation that is block triangular.)

Denote by

$$0^{\uparrow} := \begin{bmatrix} * & \cdots & * \\ & 0 & \end{bmatrix}, \quad 0^{\downarrow} := \begin{bmatrix} & 0 & \\ * & \cdots & * \end{bmatrix}, \quad 0^{\leftarrow} := \begin{bmatrix} * & \\ \vdots & 0 \\ * & \end{bmatrix}, \quad 0^{\rightarrow} := \begin{bmatrix} & * \\ 0 & \vdots \\ & * \end{bmatrix}$$

the matrices, in which the entries of the first row, the last row, the first column, and the last column, respectively, are stars and the other entries are zero, and write

$$\mathcal{Z} := \begin{bmatrix} * & \cdots & * & 0 & \cdots & 0 \\ & & & \vdots & \ddots & \vdots \\ & 0 & & 0 & \cdots & 0 \end{bmatrix}$$

(the number of zeros in the first row of $\mathcal{Z}$ is equal to the number of rows). The stars denote independent parameters.

In the following theorem we give a simple miniversal deformation of (22) that is block triangular with respect to the partition of (22) in which $J^{\#}$ and $J(0)^{\#}$ are partitioned as in Theorem 2 and all blocks of $(F^T_{p_i}, G^T_{p_i})$ and (F_{q_i}, G_{q_i}) are 1-by-1.

Theorem 3. *Let* (A, B) *be a canonical matrix pair of the form* (22) *satisfying* (23). *One of the block triangular simple miniversal deformations of* (A, B) *has the form* $(\mathcal{A}, \mathcal{B})$, *in which*

$$\mathcal{A} := \left[\begin{array}{cccc|c|c|cccc}
F^T_{p_1} & & & & & & & & & \\
& F^T_{p_2} & & & & & & & 0 & \\
& & \ddots & & & & & & & \\
0 & & & F^T_{p_l} & & & & & & \\ \hline
& & 0 & & I & & & & & \\ \hline
0^{\rightarrow} & 0^{\rightarrow} & \ldots & 0^{\rightarrow} & 0 & J(0)^{\#} + \mathcal{H}^{\#} & & & & \\ \hline
& & & & & 0^{\downarrow} & F_{q_1} & & & \\
0^{\rightarrow} & 0^{\rightarrow} & \ldots & 0^{\rightarrow} & 0 & 0^{\downarrow} & & F_{q_2} & & \\
& & & & & \vdots & & & \ddots & \\
& & & & & 0^{\downarrow} & 0 & & & F_{q_r}
\end{array}\right] \tag{24}$$

and

$$
\mathcal{B} := \left[\begin{array}{cccc|c|c|cccc}
G^T_{p_1} & & & & & & & & & \\
\mathcal{Z}^T & G^T_{p_2} & & & & & & & 0 & \\
\vdots & \ddots & \ddots & & & & & & & \\
\mathcal{Z}^T & \dots & \mathcal{Z}^T & G^T_{p_l} & & & & & & \\ \hline
0^{\leftarrow} & 0^{\leftarrow} & \dots & 0^{\leftarrow} & J^{\#} + \mathcal{K}^{\#} & & & & & \\ \hline
 & & 0 & & 0 & I & & & & \\ \hline
 & & 0^{\uparrow} & & 0^{\uparrow} & & G_{q_1} & & & \\
 & & 0^{\uparrow} & & 0^{\uparrow} & & \mathcal{Z} & G_{q_2} & & \\
 & & \vdots & & \vdots & 0 & \vdots & \ddots & \ddots & \\
 & & 0^{\uparrow} & & 0^{\uparrow} & & \mathcal{Z} & \dots & \mathcal{Z} & G_{q_r}
\end{array}\right], \tag{25}
$$

where $J(0)^{\#} + \mathcal{H}^{\#}$ *and* $J^{\#} + \mathcal{K}^{\#}$ *are the block triangular miniversal deformations* (8) *and* (9).

Proof. The following miniversal deformation of matrix pairs was obtained in [10]. The matrix pair (22) is equivalent to its Kronecker canonical form

$$
(A_{\mathrm{kr}}, B_{\mathrm{kr}}) := \bigoplus_{i=1}^{r} (F_{q_i}, G_{q_i}) \oplus (I, J) \oplus (J(0), I) \oplus \bigoplus_{i=1}^{l} (F^T_{p_i}, G^T_{p_i}).
$$

By [10, Theorem 4.1], one of the simple miniversal deformations of $(A_{\mathrm{kr}}, B_{\mathrm{kr}})$ has the form $(\mathcal{A}_{\mathrm{kr}}, \mathcal{B}_{\mathrm{kr}})$, in which

$$
\mathcal{A}_{\mathrm{kr}} := \left[\begin{array}{cccc|c|c|cccc}
F_{q_r} & & & 0 & & 0^{\downarrow} & & & & \\
 & F_{q_{r-1}} & & & & 0^{\downarrow} & & & & \\
 & & \ddots & & 0 & \vdots & 0^{\rightarrow} & 0^{\rightarrow} & \dots & 0^{\rightarrow} \\
0 & & & F_1 & & 0^{\downarrow} & & & & \\ \hline
 & & & & I & 0 & & & 0 & \\ \hline
 & & & & & J(0) + \mathcal{H} & 0^{\rightarrow} & 0^{\rightarrow} & \dots & 0^{\rightarrow} \\ \hline
 & & & & & & F^T_{p_l} & & & 0 \\
 & & 0 & & & & & F^T_{p_{l-1}} & & \\
 & & & & & & & & \ddots & \\
 & & & & & & 0 & & & F^T_{p_1}
\end{array}\right]
$$

and

$$
\mathcal{B}_{kr} := \left[\begin{array}{cccc|c|c|cccc}
G_{q_r} & \mathcal{Z} & \dots & \mathcal{Z} & 0^{\uparrow} & & & 0^{\uparrow} & & \\
 & G_{q_{r-1}} & \ddots & \vdots & 0^{\uparrow} & & & 0^{\uparrow} & & \\
 & & \ddots & \mathcal{Z} & \vdots & 0 & & \vdots & & \\
0 & & & G_{q_1} & 0^{\uparrow} & & & 0^{\uparrow} & & \\
\hline
 & & & & J+\mathcal{K} & 0 & 0^{\leftarrow} & 0^{\leftarrow} & \dots & 0^{\leftarrow} \\
\cline{5-10}
 & & & & & I & & 0 & & \\
\cline{6-10}
 & & & & & & G^T_{p_l} & \mathcal{Z}^T & \dots & \mathcal{Z}^T \\
 & & 0 & & & & & G^T_{p_{l-1}} & \ddots & \vdots \\
 & & & & & & & & \ddots & \mathcal{Z}^T \\
 & & & & & & 0 & & & G^T_{p_1}
\end{array}\right].
$$

In view of Theorem 2, the deformation $(\mathcal{A}_{kr}, \mathcal{B}_{kr})$ is permutationally equivalent to the deformation $(\mathcal{A}, \mathcal{B})$ from Theorem 3. (The blocks $\mathcal{H}$ and $\mathcal{K}$ in $(\mathcal{A}_{kr}, \mathcal{B}_{kr})$ are lower block triangular; because of this we reduce $(\mathcal{A}_{kr}, \mathcal{B}_{kr})$ to $(\mathcal{A}, \mathcal{B})$, which is lower block triangular.) □

Remark 2. Constructing $J(\lambda)^\#$, we for each r join all r-by-r Jordan blocks $J_r(\lambda)$ of $J(\lambda)$ in $J_r(\lambda I)$; see (11). We can join analogously pairs of equal sizes in (22) and obtain a pair of the form

$$
\bigoplus_{i=1}^{l'} (\hat{F}^T_{p'_i}, \hat{G}^T_{p'_i}) \oplus (I, J^\#) \oplus (J(0)^\#, I) \oplus \bigoplus_{i=1}^{r'} (\hat{F}_{q'_i}, \hat{G}_{q'_i}), \tag{26}
$$

in which $p'_1 < \dots < p'_{l'}$ and $q'_1 > \dots > q'_{r'}$. This pair is permutationally equivalent to (22). Producing the same permutations of rows and columns in (24) and (25), we join all F^T_p, G^T_p, F_q, G_q in $\hat{F}^T_p, \hat{G}^T_p, \hat{F}_q, \hat{G}_q$, and $0, 0^{\uparrow}, 0^{\downarrow}, 0^{\leftarrow}, 0^{\rightarrow}, \mathcal{Z}$ in $\hat{0}, \hat{0}^{\uparrow}, \hat{0}^{\downarrow}, \hat{0}^{\leftarrow}, \hat{0}^{\rightarrow}, \hat{\mathcal{Z}}$ which consist of blocks 0 and $\star$ defined in (12); the obtaining pair is a block triangular miniversal deformation of (26).

4 Miniversal deformations of contragredient matrix pencils

Each pair of $m \times n$ and $n \times m$ matrices reduces by transformations of contragredient equivalence

$$
(A, B) \mapsto (S^{-1}AR, R^{-1}BS), \qquad S \text{ and } R \text{ are nonsingular,}
$$

to the *Dobrovol'skaya and Ponomarev canonical form* [7] (see also [8]) being a direct sum, uniquely determined up to permutation of summands, of pairs of the form

$$
(I_r, J_r(\lambda)), \ (J_r(0), I_r), \ (F_r, G^T_r), \ (F^T_r, G_r), \tag{27}
$$

in which $\lambda \in \mathbb{C}$ and the matrices F_r and G_r are defined in (21).

For each matrix M, define the matrices

$$M_\triangle := \begin{bmatrix} 0 \dots 0 \\ M \end{bmatrix}, \qquad M_\triangleright := \begin{bmatrix} M & \begin{matrix} 0 \\ \vdots \\ 0 \end{matrix} \end{bmatrix}$$

that are obtained by adding the zero row to the top and the zero column to the right, respectively. Each block matrix whose blocks have the form $\mathcal{T}_\triangle$ (in which $\mathcal{T}$ is defined in (2)) is denoted by $\mathcal{H}_\triangle$. Each block matrix whose blocks have the form $\mathcal{T}_\triangleright$ is denoted by $\mathcal{H}_\triangleright$.

Theorem 4. *Let*

$$(I, J) \oplus (A, B) \tag{28}$$

be a canonical matrix pair for contragredient equivalence, in which J *is a nonsingular Jordan canonical matrix,*

$$(A, B) := \bigoplus_{i=1}^{l} (F_{p_i}, G_{p_i}^T) \oplus (I, J(0)) \oplus (J'(0), I) \oplus \bigoplus_{i=1}^{r} (F_{q_i}^T, G_{q_i}),$$

$J(0)$ *and* $J'(0)$ *are Jordan matrices with the single eigenvalue* 0*, and*

$$p_1 \geqslant p_2 \geqslant \cdots \geqslant p_l, \qquad q_1 \leqslant q_2 \leqslant \cdots \leqslant q_r.$$

Then one of the simple miniversal deformations of (28) *has the form*

$$(I, J + \mathcal{K}) \oplus (\mathcal{A}, \mathcal{B}), \tag{29}$$

in which $J + \mathcal{K}$ *is the deformation* (5) *of* J *and* $(\mathcal{A}, \mathcal{B})$ *is the following deformation of* (A, B)*:*

$$\mathcal{A} := \left[\begin{array}{cccc|c|c|cccc}
F_{p_1} & \mathcal{T} & \dots & \mathcal{T} & & & & & & \\
 & F_{p_2} & \ddots & \vdots & \mathcal{H}_\triangle & \mathcal{H} & & \mathcal{H} & & \\
 & & \ddots & \mathcal{T} & & & & & & \\
 & & & F_{p_l} & & & & & & \\
\hline
 & & & & I & \mathcal{H} & & \mathcal{H}_\triangleright & & \\
\hline
 & & & & & J'(0) + \mathcal{H} & & \mathcal{H} & & \\
\hline
 & & & & & & G_{q_1}^T & \mathcal{T} & \dots & \mathcal{T} \\
 & 0 & & & & & & G_{q_2}^T & \ddots & \vdots \\
 & & & & & & & & \ddots & \mathcal{T} \\
 & & & & & & & & & G_{q_r}^T
\end{array}\right]$$

and

$$
\mathcal{B} := \left[\begin{array}{c|c|c|c}
\begin{matrix} G^T_{p_1}+\mathcal{T} & & \\ \mathcal{T} & \ddots & \\ \vdots\ \ddots & & \ddots \\ \mathcal{T}\dots & \mathcal{T} & G^T_{p_l}+\mathcal{T} \end{matrix} & & & 0 \\
\hline
\mathcal{H} & J(0)+\mathcal{H} & & \\
\hline
\mathcal{H}_{\triangleright} & \mathcal{H} & I & \\
\hline
\mathcal{H} & \mathcal{H} & \mathcal{H}_{\triangle} & \begin{matrix} F_{q_1}+\mathcal{T} & & \\ \mathcal{T} & \ddots & \\ \vdots\ \ddots & & \ddots \\ \mathcal{T}\dots & \mathcal{T} & F_{q_r}+\mathcal{T} \end{matrix}
\end{array}\right].
$$

Proof. The following simple miniversal deformation of (28) was obtained in [10, Theorem 5.1]: up to obvious permutations of strips, it has the form

$$
(I, J+\mathcal{K}) \oplus (\mathcal{A}', \mathcal{B}'), \tag{30}
$$

in which $J+\mathcal{K}$ is (5),

$$
\mathcal{A}' := \left[\begin{array}{c|c|c|c}
\begin{matrix} F_{p_1}+\mathcal{T} & \mathcal{T}\ \dots & \mathcal{T} \\ & F_{p_2}+\mathcal{T}\ \ddots & \vdots \\ & \ddots & \mathcal{T} \\ 0 & & F_{p_l}+\mathcal{T} \end{matrix} & 0 & \mathcal{H} & \mathcal{H} \\
\hline
0 & I & 0 & 0 \\
\hline
\mathcal{H} & 0 & J'(0)+\mathcal{H} & \mathcal{H} \\
\hline
0 & 0 & \mathcal{H} & \begin{matrix} G^T_{q_1} & \mathcal{T} & \dots & \mathcal{T} \\ & G^T_{q_2} & \ddots & \vdots \\ & & \ddots & \mathcal{T} \\ 0 & & & G^T_{q_r} \end{matrix}
\end{array}\right],
$$

and

$$
\mathcal{B}' := \left[\begin{array}{c|c|c|c}
\begin{matrix} G^T_{p_1} & & 0 \\ \mathcal{T} & G^T_{p_2} & \\ \vdots & \ddots & \ddots \\ \mathcal{T} & \dots\ \mathcal{T} & G^T_{p_l} \end{matrix} & \mathcal{H} & 0 & 0 \\
\hline
\mathcal{H} & J(0)+\mathcal{H} & \mathcal{H} & \mathcal{H} \\
\hline
0 & \mathcal{H} & I & 0 \\
\hline
\mathcal{H} & \mathcal{H} & 0 & \begin{matrix} F_{q_1}+\mathcal{T} & & 0 \\ \mathcal{T} & F_{q_2}+\mathcal{T} & \\ \vdots\ \ddots & & \ddots \\ \mathcal{T}\dots & \mathcal{T} & F_{q_r}+\mathcal{T} \end{matrix}
\end{array}\right];
$$

Let (C, D) be the canonical pair (28), and let $(\mathcal{P}, \mathcal{Q})$ be any matrix pair of the same size in which each entry is 0 or $*$. By [10, Theorem 2.1], see also the beginning of the proof of Theorem 5.1 in [10], $(C+\mathcal{P}, D+\mathcal{Q})$ is a versal (respectively, miniversal) deformation of (C, D) if and only if for every pair (M, N) of size of (C, D) there exist square matrices S and R and a pair (respectively, a unique pair) (P, Q) obtained from $(\mathcal{P}, \mathcal{Q})$ by replacing its stars with complex numbers such that

$$(M, N) + (CR - SC,\ DS - RD) = (P, Q). \tag{31}$$

The matrices of (C, D) are block diagonal:

$$C = C_1 \oplus C_2 \oplus \cdots \oplus C_t, \qquad D = D_1 \oplus D_2 \oplus \cdots \oplus D_t,$$

in which $(C_i,\ D_i)$ are of the form (27). Partitioning conformally the matrices of (M, N) and $(\mathcal{P}, \mathcal{Q})$ and equating the corresponding blocks in (31), we find that $(C+\mathcal{P}, D+\mathcal{Q})$ is a versal deformation of (C, D) if and only if

> for each pair of indices (i, j) and every pair (M_{ij}, N_{ij}) of the size of $(\mathcal{P}_{ij}, \mathcal{Q}_{ij})$ there exist matrices S_{ij} and R_{ij} and a pair (P_{ij}, Q_{ij}) obtained from $(\mathcal{P}_{ij}, \mathcal{Q}_{ij})$ by replacing its stars with complex numbers such that
>
> $$(M_{ij}, N_{ij}) + (C_i R_{ij} - S_{ij} C_j,\ D_i S_{ij} - R_{ij} D_j) = (P_{ij}, Q_{ij}).$$

Let $(C+\mathcal{P}', D+\mathcal{Q}')$ be the deformation (30) of (C, D). Since it is versal,

> for each pair of indices (i, j) and every pair (M_{ij}, N_{ij}) of the size of $(\mathcal{P}'_{ij}, \mathcal{Q}'_{ij})$ there exist matrices S_{ij} and R_{ij} and a pair (P'_{ij}, Q'_{ij}) obtained from $(\mathcal{P}'_{ij}, \mathcal{Q}'_{ij})$ by replacing its stars with complex numbers such that
>
> $$(M_{ij}, N_{ij}) + (C_i R_{ij} - S_{ij} C_j,\ D_i S_{ij} - R_{ij} D_j) = (P'_{ij}, Q'_{ij}).$$

Let $(C+\mathcal{P}, D+\mathcal{Q})$ be the deformation (29). In order to prove that it is versal, let us verify the condition (32). If $(\mathcal{P}_{ij}, \mathcal{Q}_{ij}) = (\mathcal{P}'_{ij}, \mathcal{Q}'_{ij})$ then (32) holds by (33).

Let $(\mathcal{P}_{ij}, \mathcal{Q}_{ij}) \neq (\mathcal{P}'_{ij}, \mathcal{Q}'_{ij})$ for some (i, j). Since the condition (33) holds, it suffices to verify that for each (P'_{ij}, Q'_{ij}) obtained from $(\mathcal{P}'_{ij}, \mathcal{Q}'_{ij})$ by replacing its stars with complex numbers there exist matrices S and R and a pair (P_{ij}, Q_{ij}) obtained from $(\mathcal{P}_{ij}, \mathcal{Q}_{ij})$ by replacing its stars with complex numbers such that

$$(P'_{ij}, Q'_{ij}) + (C_i R - S C_j,\ D_i S - R D_j) = (P_{ij}, Q_{ij}). \tag{34}$$

The following 5 cases are possible.

Case 1: $(C_i, D_i) = (F_p, G_p^T)$ *and* $i = j$. Then

$$(P'_{ii}, Q'_{ii}) = (T, 0) = \left(\begin{bmatrix} & 0 & \\ \alpha_1 & \cdots & \alpha_{p-1} \end{bmatrix}, 0\right)$$

(we denote by T any matrix obtained from $\mathcal{T}$ by replacing its stars with complex numbers). Taking

$$S := \begin{bmatrix} 0 & & & & 0 \\ \alpha_{p-1} & \ddots & & & \\ \ddots & \ddots & \ddots & & \\ \alpha_2 & \ddots & \ddots & \ddots & \\ \alpha_1 & \alpha_2 & \ddots & \alpha_{p-1} & 0 \end{bmatrix}, \qquad R := \begin{bmatrix} 0 & & & & 0 \\ \alpha_{p-1} & \ddots & & & \\ \ddots & \ddots & \ddots & & \\ \alpha_3 & \ddots & \ddots & \ddots & \\ \alpha_2 & \alpha_3 & \ddots & \alpha_{p-1} & 0 \end{bmatrix}$$

in (34), we obtain

$$(P_{ii}, Q_{ii}) = \left(0, \begin{bmatrix} \alpha_{p-1} & \\ \vdots & 0 \\ \alpha_1 & \end{bmatrix}\right) = (0, T).$$

Case 2: $(C_i, D_i) = (F_p, G_p^T)$ *and* $(C_j, D_j) = (I_m, J_m(0))$. Then $(P'_{ij}, Q'_{ij}) = (0, T)$. Taking $S := -T_\Delta$ and $R := 0$ in (34), we obtain $(P_{ij}, Q_{ij}) = (T_\Delta, 0)$.

Case 3: $(C_i, D_i) = (I_m, J_m(0))$ *and* $(C_j, D_j) = (J_n(0), I_n)$. Then $(P'_{ij}, Q'_{ij}) = (0, T)$. Taking $S := 0$ and $R := T$ in (34), we obtain $(P_{ij}, Q_{ij}) = (T, 0)$.

Case 4: $(C_i, D_i) = (I_m, J_m(0))$ *and* $(G_j, D_j) = (G_q^T, F_q)$. Then $(P'_{ij}, Q'_{ij}) = (0, T)$. Taking $S := 0$ and $R := T_\triangleright$ in (34), we obtain $(P_{ij}, Q_{ij}) = (T_\triangleright, 0)$.

Case 5: $(C_i, D_i) = (J_n(0), I_n)$ *and* $(G_j, D_j) = (F_p, G_p^T,)$. Then $(P'_{ij}, Q'_{ij}) = (T, 0)$. Taking $S := T_\triangleright$ and $R := 0$ in (34), we obtain $(P_{ij}, Q_{ij}) = (0, T_\triangleright)$.

We have proved that the deformation (29) is versal. It is miniversal since it has the same number of parameters as the miniversal deformation (30). □

Remark 3. The deformation $(I, J + \mathcal{K}) \oplus (\mathcal{A}, \mathcal{B})$ from Theorem 4 can be made block triangular by the following permutations of its rows and columns, which are transformations of contragredient equivalence:

- First, we reduce $(I, J + \mathcal{K})$ to the form $(I, J^\# + \mathcal{K}^\#)$, in which $J^\# + \mathcal{K}^\#$ is defined in (9).
- Second, we reduce the diagonal block $J(0) + \mathcal{H}$ in $\mathcal{B}$ to the form $J(0)^\# + \mathcal{H}^\#$ (defined in (8)) by the permutations of rows and columns of $\mathcal{B}$ described in Definition 3. Then we make the contragredient permutations of rows and columns of $\mathcal{A}$.

– Finally, we reduce the diagonal block $J'(0)+\mathcal{H}$ in $\mathcal{A}$ to the form $J'(0)^{\#}+\mathcal{H}^{\#}$ (defined in (8)) by the permutations of rows and columns of $\mathcal{A}$ described in Definition 3, and make the contragredient permutations of rows and columns of $\mathcal{B}$. The obtained deformation $J'(0)^{\#}+\mathcal{H}^{\#}$ is lower block triangular, we make it upper block triangular by transformations

$$P(J'(0)^{\#}+\mathcal{H}^{\#})P, \qquad P := \begin{bmatrix} 0 & & 1 \\ & \cdot^{\cdot^{\cdot}} & \\ 1 & & 0 \end{bmatrix}$$

(i.e., we rearrange in the inverse order the rows and columns of $\mathcal{A}$ that cross $J'(0)^{\#}+\mathcal{H}^{\#}$ and make the contragredient permutations of rows and columns of $\mathcal{B}$).

References

1. V. I. ARNOLD, *On matrices depending on parameters*, Russian Math. Surveys, 26 (no. 2) (1971), pp. 29–43.
2. V. I. ARNOLD, *Lectures on bifurcations in versal families*, Russian Math. Surveys, 27 (no. 5) (1972), pp. 54–123.
3. V. I. ARNOLD, *Geometrical Methods in the Theory of Ordinary Differential Equations*, Springer-Verlag, New York, 1988.
4. G. R. BELITSKII, *Normal forms in a space of matrices*, in Analysis in Infinite-Dimensional Spaces and Operator Theory, V. A. Marchenko, ed., Naukova Dumka, Kiev, 1983, pp. 3-15 (in Russian).
5. G. R. BELITSKII, *Normal forms in matrix spaces*, Integral Equations Operator Theory, 38 (2000), pp. 251–283.
6. F. R. Gantmacher, *Matrix Theory*, Vol. 2, AMS Chelsea Publishing, Providence, RI, 2000.
7. N. M. DOBROVOL'SKAYA AND V. A. PONOMAREV, *A pair of counter operators*, Uspehi Mat. Nauk, 20 (no. 6) (1965), pp. 80–86.
8. R. A. HORN AND D. I. MERINO, *Contragredient equivalence: a canonical form and some applications*, Linear Algebra Appl., 214 (1995), pp. 43–92.
9. A. EDELMAN, E. ELMROTH, AND B. KÅGSTRÖM, *A geometric approach to perturbation theory of matrices and matrix pencils. Part I: Versal deformations*, SIAM J. Matrix Anal. Appl., 18 (1997), pp. 653–692.
10. M. I. GARCIA-PLANAS AND V. V. SERGEICHUK, *Simplest miniversal deformations of matrices, matrix pencils, and contragredient matrix pencils*, Linear Algebra Appl., 302–303 (1999), pp. 45–61 (some misprints of this paper were corrected in its preprint arXiv:0710.0946).
11. V. V. SERGEICHUK, *Canonical matrices for linear matrix problems*, Linear Algebra Appl., 317 (2000), pp. 53–102.
12. H. SHAPIRO, *The Weyr characteristic*, Amer. Math. Monthly, 106 (1999), pp. 919–929.
13. E. WEYR, *Répartition des matrices en espèces et formation de toutes les espèces*, C. R. Acad. Sci. Paris, 100 (1885), pp. 966–969.

Determining the Schein Rank of Boolean Matrices

Evgeny E. Marenich*

Murmansk State Pedagogic University, Russia
marenich1@yandex.ru

Abstract. In this paper we present some results of Schein rank of Boolean matrices. A notion of the intersection number of a bipartite graph is defined and its applications to Schein rank of Boolean matrices are derived. We discuss minimal and maximal matrices of given Schein rank, the number of $m \times n$ Boolean matrices with given Schein rank. The Schein ranks of some $m \times n$ Boolean matrices are determined. In the last section, we give some further result concerning the Schein rank of Boolean matrices.

Keywords: Boolean matrix, Schein rank, coding functions for bipartite graphs.

1 Introduction

The following are described in Sections 2 and 3:

1. the set of all $m \times n$ minimal Boolean matrices of Schein rank k;
2. the set of all $m \times n$ maximal Boolean matrices of Schein rank 2,3;
3. some maximal $m \times n$ Boolean matrices of Schein rank k.

In Section 4 we define the intersection number of a bipartite graph Γ and prove that the intersection number is equal to the minimum number of maximal complete bipartite subgraphs whose union includes all edges of Γ.

In Section 5 we define a k-canonical family $CS(k)$ of bipartite graphs, obtain the family $CS(2)$ and some graphs in the family $CS(3)$.

In Section 6, we apply the intersection number and canonical families to determining the Schein rank of Boolean matrices. In particular, formulas for the number of all $m \times n$ Boolean matrices of Schein rank k are obtained.

In Section 7, coding of bipartite graphs is studied. In Section 8, we define the bipartite intersection graphs and investigate the Schein rank of associated matrices.

In Section 9, we give some further result concerning the Schein rank of Boolean matrices.

* This research is conducted in accordance with the Thematic Plan of Russian Federal Educational Agency, theme №1.03.07.

2 The Schein rank of Boolean matrices

Our notation and terminology are similar to those of [1], [4]. We collect in this section a number of result and definitions required latter. Where possible we state simple corollaries of these results without proof. We collect in this section a number of results and definitions required latter. A detailed treatment may be found in [3], [4].

Let $\mathfrak{U}$ be a finite set, $2^{\mathfrak{U}}$ be the collection of all subsets of $\mathfrak{U}$. The number of elements in $\mathfrak{U}$ is denoted by $|\mathfrak{U}|$.

Let $\mathrm{Bul}(\mathfrak{U}) = (2^{\mathfrak{U}}, \subseteq)$ be the Boolean algebra (or poset) of all subsets of a finite set $\mathfrak{U}$ partially ordered by inclusion. Let $\mathrm{Bul}(k)$ be the Boolean algebra of all subsets of a finite set of k elements.

Let $P = \{\widetilde{0}, \widetilde{1}\}$ be a two-element Boolean lattice with the greatest element $\widetilde{1}$ and the least element $\widetilde{0}$. The lattice operations meet $\wedge$ and join $\vee$ are defined as follows:

$$\begin{array}{c|cc} \wedge & \widetilde{0} & \widetilde{1} \\ \hline \widetilde{0} & \widetilde{0} & \widetilde{0} \\ \widetilde{1} & \widetilde{0} & \widetilde{1} \end{array} \quad \begin{array}{c|cc} \vee & \widetilde{0} & \widetilde{1} \\ \hline \widetilde{0} & \widetilde{0} & \widetilde{1} \\ \widetilde{1} & \widetilde{1} & \widetilde{1} \end{array}.$$

Following [4], we recall some definitions.

Let $P^{m\times n}$ denote the set of all $m \times n$ (Boolean) matrices with entries in P. Matrices with all entries in P will be denoted by ROMAN capitals $A = \|a_{ij}\|_{m\times n}$, $B = \|b_{ij}\|_{m\times n}$, $C = \|c_{ij}\|_{m\times n}$, $X = \|x_{ij}\|_{m\times n}$, ...

Then the usual definitions for addition and multiplication of matrices over field are applied to Boolean matrices as well. The $n\times n$ identity matrix $E = E_{n\times n}$ is the matrix such that

$$e_{ij} = \begin{cases} \widetilde{1}, & \text{if } i = j, \\ \widetilde{0}, & \text{if } i \neq j. \end{cases}$$

Denote by $\overline{E}_{n\times n}$ the $n \times n$ matrix with $\widetilde{0}$' entries on the main diagonal and $\widetilde{1}$ elsewhere.

The $m \times n$ zero matrix $0_{m\times n}$ is the matrix all of whose entries are $\widetilde{0}$. The $m \times n$ universal matrix $J_{m\times n}$ is the matrix all of whose entries are $\widetilde{1}$.

The transpose of A will be denoted by $A^{(t)}$.

Define a partial ordering $\leqslant$ on $P^{m\times n}$ by $A \leqslant B$ iff $a_{ij} \leqslant b_{ij}$ for all i, j.

Let $A^{(r)}$ ($A_{(r)}$) denote the rth column (row) of A.

A subspace of $P^{m\times 1}$ ($P^{1\times n}$) is a subset of $P^{m\times 1}$ ($P^{1\times n}$) containing the zero vector and closed under addition. A column space $\mathrm{Column}(A)$ of a matrix A is the span of the set of all columns of A. Likewise one has a row space $\mathrm{Row}(A)$ of A.

Definitions of the column rank $\mathrm{rank}_c(A)$ (row rank $\mathrm{rank}_r(A)$) of A is due to Kim, [4].

Theorem 1 (Kim, Roush, [3]). *Let* $A \in P^{m\times n}$, $A \neq 0_{m\times n}$. *Then the following conditions are equivalent:*

(i) $\mathrm{rank}_s(A) = k$.
(ii) k *is the least integer such that* A *is a product of an* $m \times k$ *matrix and an* $k \times n$ *matrix.*
(iii) k *is the smallest dimension of a subspace* W *such that* W *contains the column space* $\mathrm{Column}(A)$ *(row space* $\mathrm{Row}(A)$*).*

Example. We have $\mathrm{Column}(E_{n\times n}) = P^{n\times 1}$. From Theorem 1 (iii), it follows that

$$\mathrm{rank}_s(E_{n\times n}) = n.$$

The following theorem is due to Kim [4].

Theorem 2. *Let* $A \in P^{m\times n}$. *Then:*

(i) $\mathrm{rank}_s(A) = \mathrm{rank}_s(A^{(t)})$.
(ii) $\mathrm{rank}_s(A) \leqslant \min\{\mathrm{rank}_c(A), \mathrm{rank}_r(A)\}$.
(iii) $\mathrm{rank}_s(A) \leqslant \min\{m, n\}$.
(iv) *If* $\mathrm{Column}(A) \leqslant \mathrm{Column}(B)$ *then* $\mathrm{rank}_s(A) \leqslant \mathrm{rank}_s(B)$.

Corollary 1. *Let* $A \in P^{m\times n}$. *If* B *is a submatrix of* A *then*

$$\mathrm{rank}_s(B) \leqslant \mathrm{rank}_s(A).$$

Corollary 2. *Let* $A_1, \ldots, A_k \in P^{m\times n}$. *Then*

$$\mathrm{rank}_s(A_1 + A_2 + \ldots + A_k) \leqslant \mathrm{rank}_s(A_1) + \mathrm{rank}_s(A_2) + \ldots + \mathrm{rank}_s(A_k).$$

Corollary 3. *Let* $A_1, \ldots, A_k$ *be Boolean matrices. If the product* $A_1A_2\ldots A_k$ *is defined, then*

$$\mathrm{rank}_s(A_1A_2\ldots A_k) \leqslant \mathrm{rank}_s(A_i),\ i = 1, \ldots, k,$$

$$\mathrm{rank}_s(A_1A_2\ldots A_k) \leqslant \min\{\mathrm{rank}_s(A_1), \mathrm{rank}_s(A_2), \ldots, \mathrm{rank}_s(A_k)\}.$$

Example. If $A \in P^{n\times n}$ is invertible, then $\mathrm{rank}_s(A) = n$.

A square matrix is called a permutation matrix if every row and every column contains only one $\widetilde{1}$.

Corollary 4. *Let* $A \in P^{m\times n}$ *and* $\pi \in P^{m\times m}$, $\sigma \in P^{n\times n}$ *be permutation matrices. Then*

$$\mathrm{rank}_s(\pi A) = \mathrm{rank}_s(A\sigma) = \mathrm{rank}_s(A).$$

Corollary 5. *Let* $A \in P^{n\times n}$. *Then*

$$\mathrm{rank}_s(A) \geqslant \mathrm{rank}_s(A^2) \geqslant \mathrm{rank}_s(A^3) \geqslant \ldots.$$

3 Matrices of Schein rank 2, 3

Let $A \in P^{m\times n}$. By $\rho(A)$ denote the number of $\widetilde{1}$'s in A.

By $Chr_k(m,n)$ we denote the set of all matrix $A \in P^{m\times n}$ such that $rank_s(A) = k$, where $\min\{m,n\} \geqslant k$.

The term $\widetilde{1}$-rank of a matrix $A \in P^{m\times n}$ is the maximum number of $\widetilde{1}$'s entries of A no two of which share a row or column of A. We denote the term $\widetilde{1}$-rank of A by $\rho_t(A)$. By Konig theorem [1], it follows that the $\widetilde{1}$-term rank of A is the minimum number of rows and columns of A containing all $\widetilde{1}$'s entries of A.

An element a of a poset $(Q, \leqslant)$ is maximal if whenever $a \leqslant x$, then $a = x$. We dually define minimal elements.

The set of all minimal matrices in $(Chr_k(m,n), \leqslant)$ is described in the following theorem.

Theorem 3. *Let* $m,n \geqslant k$. *A matrix* A *is minimal in* $(Chr_k(m,n), \leqslant)$ *iff* $\rho(A) = k$ *and* A *has a* $k\times k$ *permutation submatrix.*

Proof. If $\rho(A) = k$ and A has a $k\times k$ permutation submatrix, then A is minimal in $(Chr_k(m,n), \leqslant)$.

Let $C \in Chr_k(m,n)$. We first show that $\rho_t(C) \geqslant k$. Suppose, to the contrary, that $\rho_t(C) < k$. By Konig theorem [1], it follows that $\rho_t(C)$ rows and columns of C containing all $\widetilde{1}$'s entries of C.

We see that $rank_s(C) \leqslant \rho_t(C) < k$. This is a contradiction since $\rho_t(C) \geqslant k$. Therefore exists a matrix $A \in P^{m\times n}$ such that $A \leqslant C$, $\rho(A) = k$ and A has a $k \times k$ permutation submatrix. □

The number of all minimal matrices in $(Chr_k(m,n), \leqslant)$ is

$$n(n-1)\dots(n-k+1)m(m-1)\dots(m-k+1).$$

Let $\Delta_k \in P^{k\times k}$ have the following form:

$$\Delta_k = \begin{pmatrix} \widetilde{1} & \widetilde{1} & \widetilde{1} & \dots \widetilde{1} \\ \widetilde{0} & \widetilde{1} & \widetilde{1} & \dots \widetilde{1} \\ \widetilde{0} & \widetilde{0} & \widetilde{1} & \dots \widetilde{1} \\ & \dots & & \dots \\ \widetilde{0} & \widetilde{0} & \widetilde{0} & \dots \widetilde{1} \end{pmatrix}.$$

From [5] it follows that $rank_s(\Delta_k) = k$, $k \geqslant 1$.

Let $\sim$ be the equivalence relation on $P^{m\times n}$ defined by $B \sim C$ iff $C = \pi B \sigma$ for some permutation matrices $\pi \in P^{m\times m}$, $\sigma \in P^{n\times n}$.

Now we obtain some maximal matrices in $(Chr_k(m,n), \leqslant)$.

Theorem 4. *Let* $A \in P^{m\times n}$. *If there exists a submatrix* B *of* A *such that* $B \sim \Delta_k$ *and* B *contains all* $\widetilde{0}$*'s entries of* A, *then* A *is maximal in* $(Chr_k(m,n), \leqslant)$.

Proof. We have $rank_s(A) = rank_s(\Delta_k) = k$. It suffices to show that Δ_k is maximal in $(Chr_k(k,k), \leqslant)$.

Let B be obtained from Δ_k by replacing a selection of the $\widetilde{0}$'s by $\widetilde{1}$'s. Let r be the least integer such that $b_{ir} \neq (\Delta_k)_{ir}$ for some i. Then $B_{(r)}$ is a span of some rows $B_{(i)}$, $i \neq r$. Therefore $rank_s(B) < k$. □

The set of all maximal matrices of $(Chr_2(m,n), \leqslant)$ is described in the following theorem.

Theorem 5. *Let* $A \in P^{m\times n}$ *and* $m, n \geqslant 2$. *Then a matrix* A *is maximal in* $(Chr_2(m,n), \leqslant)$ *iff only one entry of* A *is* $\widetilde{0}$.

The number of all maximal elements in the poset $(Chr_2(m,n), \leqslant)$ is nm.

The set of all maximal matrices in the poset $(Chr_3(m,n), \leqslant)$ is described in the following theorem.

Theorem 6. *Let* $A \in P^{m\times n}$ *and* $m, n \geqslant 3$. *A matrix* A *is maximal in* $(Chr_3(m,n), \leqslant)$ *iff there exists a submatrix* B *of* A *such that* $B \sim \Delta_3$ *or* $B \sim \overline{E}_{3\times 3}$, *and* B *contains all* $\widetilde{0}$*'s entries of* A.

Proof. Let $C \in Chr_3(m,n)$. By Konig theorem, it follows that the $\widetilde{0}$-term rank of C is the minimum number of rows and columns of C containing all $\widetilde{0}$'s entries of C. By Konig theorem the proof is now divided into following cases.

Case 1: there exist three $\widetilde{0}$'s entrees such that no two $\widetilde{0}$ entries share a row or column of A. The matrix A obtained from C by replacing other $\widetilde{0}$ entrees by $\widetilde{1}$ is maximal in $(Chr_3(m,n), \leqslant)$.

Case 2: there exist two rows and columns of C containing all $\widetilde{0}$ entries of C. If there exists a row (a column) of C containing all $\widetilde{0}$'s en-tries of C, then $rank_s(C) \leqslant 2$, which is a contradiction. Therefore there exist a row and a column of C containing all $\widetilde{0}$'s entries of C.

Case 2.1: there exist two columns of C containing all $\widetilde{0}$'s entries of C. Then there exists a submatrix B of C such that $rank_s(B) = rank_s(C) = 3$ and each row of B is a row of the matrix

$$D = \begin{pmatrix} \widetilde{0} & \widetilde{0} & \widetilde{1} \\ \widetilde{1} & \widetilde{0} & \widetilde{1} \\ \widetilde{0} & \widetilde{1} & \widetilde{1} \\ \widetilde{1} & \widetilde{1} & \widetilde{1} \end{pmatrix}.$$

By considering all matrices B such that $rank_s(B) = 3$, we conclude the proof in this case.

Case 2.2: there exist a row and a column of C containing all $\widetilde{0}$ entries of C. It is easy to see that $rank_s(C) = rank_s(A)$ for some matrix A such that $\rho(A) = n-3$ and A has a submatrix B such that $B \sim \Delta_3$. □

Remark. The matrix $\overline{E}_{k\times k}$ is not maximal in $(Chr_k(k,k), \leqslant)$ for $k \geqslant 5$.

4 On coding of bipartite graphs by sets

Let $\Gamma = \Gamma(V_1 \cup V_2, E)$ be a bipartite graph with bipartition $V_1 = \{1,2,3,\ldots\}$, $V_2 = \{1',2',3',\ldots\}$ and U a finite set.

A function $f : V_1 \cup V_2 \to 2^U$ is called U- coding function for Γ if for any vertices v_1, v_2 conditions $\{v_1, v_2\} \in E$ and $f(v_1) \cap f(v_2) \neq \varnothing$ are equivalent. We call $f(v)$ the code of $v \in V_1 \cup V_2$.

Note that there exist coding functions for any bipartite graph Γ.

The intersection number $nint_{bp}(\Gamma)$ of a bipartite graph $\Gamma = \Gamma(V_1 \cup V_2, E)$ is the least number $|U|$ such that there exists a U-coding function for Γ.

Note that every maximal complete bipartite subgraph has at least one edge.

The following example clarifies the above definitions.

Example. Let Γ_1 be the following graph:

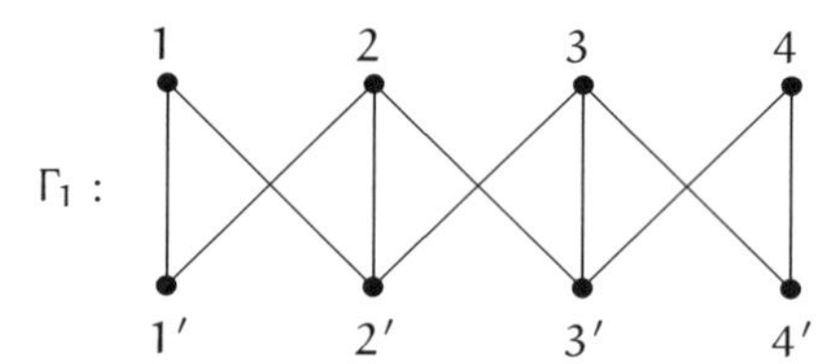

Then some maximal complete bipartite subgraphs of Γ_1 are:

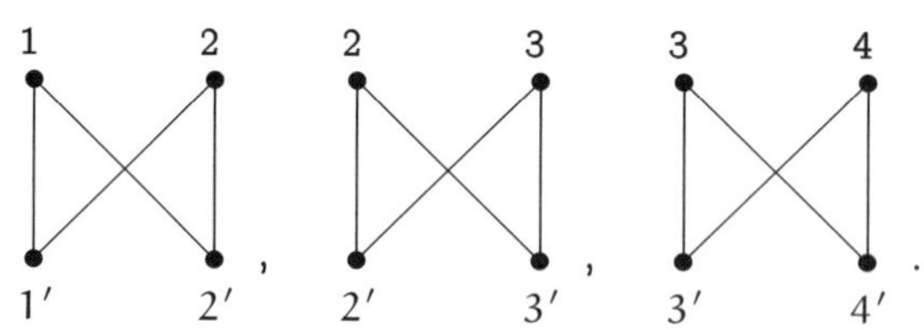

In the following theorem we show that the intersection number $nint_{bp}(\Gamma)$ of a bipartite graph is closely connected to the set of all complete bipartite subgraphs of Γ.

Theorem 7. *Let $\Gamma = \Gamma(V_1 \cup V_2, E)$ be a bipartite graph. The intersection number $nint_{bp}(\Gamma)$ is equal to the minimum number of maximal complete bipartite subgraphs whose union includes all edges of Γ.*

Proof. Let $\mathrm{nint}_{bp}(\Gamma) = k$ and f be a U- coding function for Γ, where $U = \{1,\ldots,k\}$. Define sets

$$V_r = \{v \mid v \in V_1 \cup V_2, r \in f(v)\}, \quad r = 1,\ldots,k.$$

Note that $V_r \neq \varnothing$, $r = 1,\ldots,k$. Let Γ_r be a subgraph such that V_r is the set of vertices of Γ_r. Then Γ_r is a complete bipartite subgraph of Γ. The union of subgraphs $\Gamma_1,\ldots,\Gamma_k$ includes all edges of Γ. Any subgraph $\Gamma_1,\ldots,\Gamma_k$ is contained in some maximal complete bipartite subgraph. Therefore the minimum number of maximal complete bipartite subgraphs whose union includes all edges of Γ is less than or equal to $k = \mathrm{nint}_{bp}(\Gamma)$.

Let the minimum number of complete bipartite subgraphs whose union includes all edges of Γ is equal to k. Let the union of $\Gamma_1,\ldots,\Gamma_k$ includes all edges of Γ. For any $v \in V_1 \cup V_2$ define the set $f(v)$: $r \in f(v)$ iff v is a vertex of Γ_r.

We now prove that $f : V_1 \cup V_2 \to 2^U$ is a U-coding function for Γ.

Let $v_1 \in V_1$, $v_2 \in V_2$ and $\{v_1, v_2\} \in E$. Then $\{v_1, v_2\}$ is an edge of some Γ_r. Therefore $r \in f(v_1), f(v_2)$ and $f(v_1) \cap f(v_2) \neq \varnothing$.

Let $v_1 \in V_1$, $v_2 \in V_2$, and $f(v_1) \cap f(v_2) \neq \varnothing$. Then there exists $r \in f(v_1), f(v_r)$. Therefore $\{v_1, v_2\}$ is an edge of Γ_r. Thus $\{v_1, v_2\} \in E$.

We have proved that $f : V_1 \cup V_2 \to 2^U$ is a U-coding function for Γ. Then $\mathrm{nint}_{bp}(\Gamma)$ is less than or equal to the minimum number of maximal complete bipartite subgraphs whose union includes all edges of Γ.

Thus $\mathrm{nint}_{bp}(\Gamma)$ equals the minimum number of maximal complete bipartite subgraphs whose union includes all edges of Γ. □

Example. The minimum number of maximal complete bipartite subgraphs whose union includes all edges of Γ_1 is equal to 3. Therefore $\mathrm{nint}_{bp}(\Gamma_1) = 3$.

5 On canonical bipartite graphs

Let $\Gamma = \Gamma(V_1 \cup V_2, E)$ be a bipartite graph with bipartition $V_1 = \{1,2,3,\ldots\}$, $V_2 = \{1',2',3',\ldots\}$ and U a finite set.

Denote by $\hat{V}_1$ the set of all nonisolated vertices of V_1. In the same way, we define $\hat{V}_2$. Define the following sets

$$E(v) = \{z \mid \{v,z\} \in E\}, \quad v \in V_1 \cup V_2.$$

Let $\sim$ be the equivalence relation on $\hat{V}_1 \cup \hat{V}_2$ defined by:

$$u \sim v \text{ whenever } E(u) = E(v).$$

Let $\Gamma_c = \Gamma(V_1' \cup V_2', E')$ be a bipartite graph with bipartition V_1',V_2', where

$$V_1' = \hat{V}_1/\sim, \quad V_2' = \hat{V}_2/\sim$$

are quotient sets and E' is defined by:

$$\{\hat{\imath}, \hat{\jmath}'\} \in E' \text{ iff } \{i, j\} \in E.$$

We call Γ_c a canonical representation of Γ.

Example. Consider the graph Γ and its canonical representation Γ_c,

Γ : 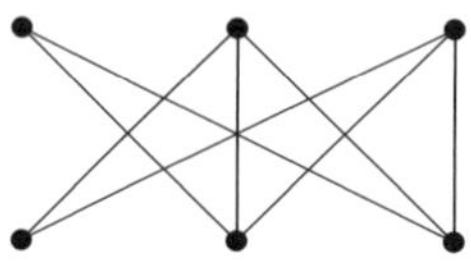 , Γ_c : 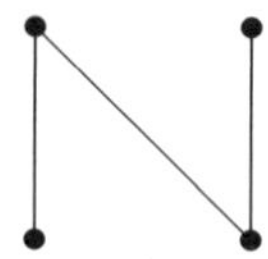 .

Lemma 1. *For any bipartite graph Γ, the following statements are valid.*

(i) $\mathrm{nint}_{bp}(\Gamma) = \mathrm{nint}_{bp}(\Gamma_c)$.
(ii) *If* $\mathrm{nint}_{bp}(\Gamma) = k$, *then* $k \leqslant V_1', V_2' \leqslant 2^k - 1$.

Let $CS(k)$ be the set of all nonisomorphic canonical representations for bipartite graphs of intersection number k. We call $CS(k)$ a k-canonical family. Any canonical representation of a bipartite graph is called a canonical graph.

Example.

1. The canonical family $CS(1)$ contains the unique graph

2. The canonical family $CS(2)$ contains four graphs

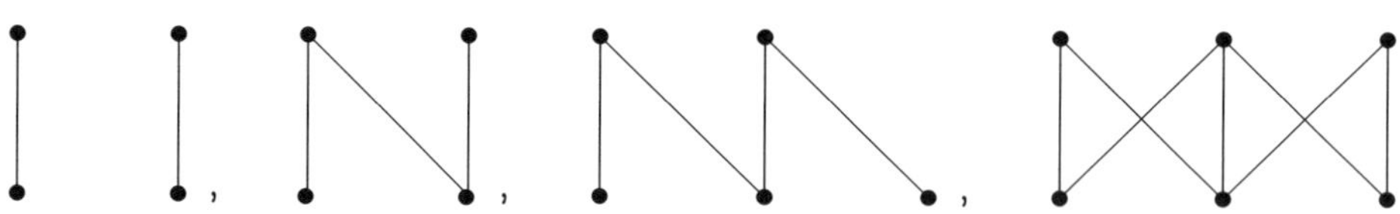

3. In $CS(3)$ we consider all graphs with three vertices in bipartition V_1', V_2':

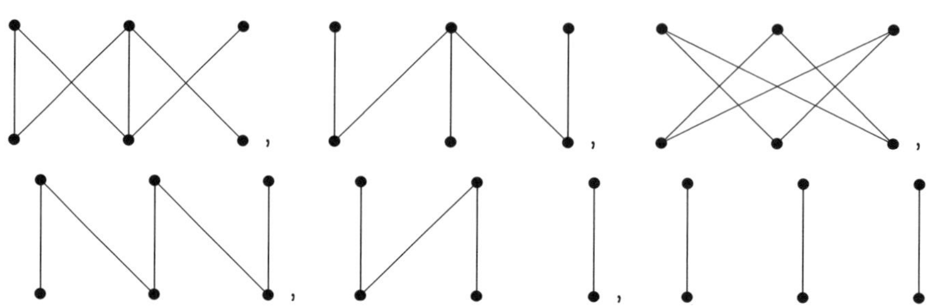

The canonical family $CS(k)$ give us all bipartite graphs $\Gamma = \Gamma(V_1 \cup V_2, E)$ such that $\text{nint}_{bp}(\Gamma) = k$.

Let $F_k(m, n)$ is the number of all bipartite graphs $\Gamma = \Gamma(V_1 \cup V_2, E)$ such that $V_1 = \{1, \dots, m\}$, $V_2 = \{1', 2', \dots, n'\}$, $\text{nint}_{bp}(\Gamma) = k$.

We have

$$F_1(m, n) = (2^m - 1)(2^n - 1), \quad F_1(n, n) = (2^n - 1)^2. \tag{1}$$

For the canonical family $CS(2)$, we obtain the following theorem.

Theorem 8. *For all* $m, n \geqslant 1$

$$\begin{aligned} F_2(m, n) = {} & \tfrac{3}{2}(3^m - 2 \cdot 2^m + 1)(3^n - 2 \cdot 2^n + 1) \\ & + \tfrac{1}{2}(3^m - 2 \cdot 2^m + 1)(4^n - 3 \cdot 3^n + 3 \cdot 2^n - 1) \\ & + \tfrac{1}{2}(3^n - 2 \cdot 2^n + 1)(4^m - 3 \cdot 3^m + 3 \cdot 2^m - 1) \\ & + \tfrac{1}{2}(4^m - 3 \cdot 3^m + 3 \cdot 2^m - 1)(4^n - 3 \cdot 3^n + 3 \cdot 2^n - 1). \end{aligned} \tag{2}$$

In particular, for all $n \geqslant 1$

$$\begin{aligned} F_2(n, n) = {} & \tfrac{3}{2}(3^n - 2 \cdot 2^n + 1)^2 \\ & + (3^n - 2 \cdot 2^n + 1)(4^n - 3 \cdot 3^n + 3 \cdot 2^n - 1) + \tfrac{1}{2}(4^n - 3 \cdot 3^n + 3 \cdot 2^n - 1)^2. \end{aligned} \tag{3}$$

6 On the Schein rank of Boolean matrices and the intersection number of associated graphs

Let $A \in P^{m \times n}$, U be a finite set.

To a matrix $A \in P^{m \times n}$ associate a bipartite graph $\Gamma(A) = \Gamma(V_1 \cup V_2, E)$ with bipartition $V_1 = \{1, \dots, m\}$, $V_2 = \{1', 2', \dots, n'\}$ by taking $a_{ij} = \widetilde{1}$ if and only if there is an edge between i and j'.

To a matrix $A \in P^{m \times n}$ associate a bipartite graph $\Gamma(A) = \Gamma(V_1 \cup V_2, E)$ by taking bipartition $V_1 = \{1, \dots, m\}$, $V_2 = \{1', 2', \dots, n'\}$ and a set of edges E such that $\{i, j'\} \in E$ if and only if $a_{ij} = \widetilde{1}$.

The following theorem reduces the Schein rank problem for any matrix A to determining the intersection number of $\Gamma(A)$.

Theorem 9. *The Schein rank of* A *equals the intersection number of* $\Gamma(A)$.

Proof. We first prove that $\text{nint}_{bp}(\Gamma) \leqslant \text{rank}_s(A)$. Let $\text{rank}_s(A) = k$. Then

$$A = C_1 D_1 + C_2 D_2 + \ldots + C_k D_k$$

for some $C_1, C_2, \dots, C_k \in P^{m \times 1}$, $D_1, D_2, \dots, D_k \in P^{1 \times n}$. Define sets:

$$f(i) = \{j \mid (C_j)_{(i)} = \widetilde{1},\ j = 1, \dots, k\}, \quad i = 1, \dots, m,$$
$$f(j') = \{i \mid (D_i)^{(j)} = \widetilde{1},\ i = 1, \dots, k\}, \quad j = 1, \dots, n.$$

Let $f : V_1 \cup V_2 \to 2^U$ and $U = \{1, \ldots, k\}$ be a function and a set. We now prove that f is a U-coding function of $\Gamma(A)$. The following statements are equivalent:

- $a_{ij} = \tilde{1}$;
- $\tilde{1} = (C_r D_r)_{ij} = (C_r)_{(i)} \wedge (D_r)^{(j)}$ for some r;
- there exists r such that $r \in f(i)$, $r \in f(j')$;
- $f(i) \cap f(j') \neq \varnothing$.

We have proved that $a_{ij} = \tilde{1}$ iff $f(i) \cap f(j') \neq \varnothing$. Therefore f is a U-coding function of $\Gamma(A)$. Thus $\mathrm{nint}_{bp}(\Gamma) \leqslant k = \mathit{rank}_s(A)$.

We now prove that $\mathrm{nint}_{bp}(\Gamma) \leqslant \mathit{rank}_s(A)$. Let $\mathrm{nint}_{bp}(\Gamma) = k$ and f is a U-coding function for $\Gamma(A)$. We have $f : V_1 \cup V_2 \to 2^U$, where $U = \{1, \ldots, k\}$. Define column vectors $C_1, C_2, \ldots, C_k \in P^{m \times 1}$ by setting:

$$(C_r)_{(i)} = \tilde{1} \text{ iff } r \in f(i),\ i = 1, \ldots, m,\ r = 1, \ldots, k.$$

Similarly, define row vectors $D_1, D_2, \ldots, D_k \in P^{1 \times n}$ by setting:

$$(D_r)^{(j)} = \tilde{1} \text{ iff } r \in f(j'),\ j = 1, \ldots, n,\ r = 1, \ldots, k.$$

We claim that $A = C_1 D_1 + C_2 D_2 + \ldots + C_k D_k$. Indeed, the following statements are equivalent:

- $a_{ij} = \tilde{1}$;
- $f(i) \cap f(j') \neq \varnothing$;
- there exists r such that $r \in f(i)$, $r \in f(j')$;
- there exists r such that $(C_r)_{(i)} = \tilde{1}$, $(D_r)^{(j)} = \tilde{1}$;
- $(C_r)_{(i)} \wedge (D_r)^{(j)} = (C_r D_r)_{ij} = \tilde{1}$ for some r;
- $(C_1 D_1 + C_2 D_2 + \ldots + C_k D_k)_{ij} = \tilde{1}$.

Therefore

$$A = C_1 D_1 + C_2 D_2 + \ldots + C_k D_k, \quad \mathit{rank}_s(A) \leqslant k = \mathrm{nint}_{bp}(\Gamma).$$

We have proved that $\mathit{rank}_s(A) = \mathrm{nint}_{bp}(\Gamma)$. □

From Theorem 9 and [2], [Remark 6.7], we obtain the following corollary.

Corollary 6 ([5]). *Let $A \in P^{m \times n}$. The Schein rank of A is equal to the minimum number of complete bipartite subgraphs whose union includes all edges of $\Gamma(A)$.*

Example. Let $A \in P^{n \times n}$ have the following form:

$$A = \begin{pmatrix} \tilde{1}\,\tilde{1}\,\tilde{0} \ldots \tilde{0}\,\tilde{0} \\ \tilde{0}\,\tilde{1}\,\tilde{1} \ldots \tilde{0}\,\tilde{0} \\ \cdots \quad \cdots \quad \cdots \\ \tilde{0}\,\tilde{0}\,\tilde{0} \ldots \tilde{1}\,\tilde{1} \\ \tilde{1}\,\tilde{0}\,\tilde{0} \ldots \tilde{0}\,\tilde{1} \end{pmatrix}.$$

Then $\Gamma(A)$ have the following form:

$\Gamma(A)$:

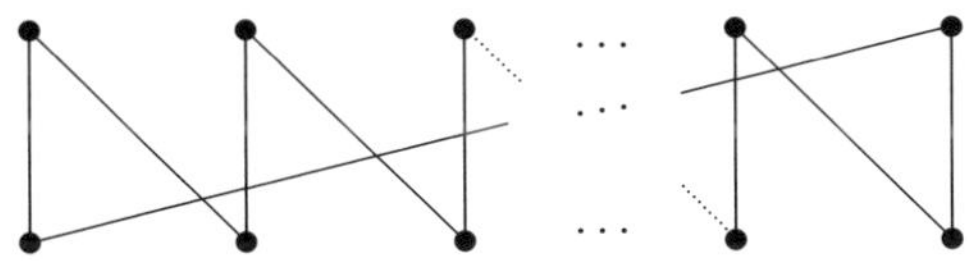

Note that $\Gamma(A)$ has $2n$ edges and any maximal complete bipartite subgraph contains two edges. Therefore the minimum number of maximal complete bipartite subgraphs whose union includes all edges of Γ is n. Thus $rank_s(A) = n$.

The canonical family $CS(k)$ give us all matrices $A \in P^{m\times n}$ such that $rank_s(A) = k$.

Theorem 10. *Let* $m, n \geqslant 1$ *and* $\min\{m, n\} \geqslant k$. *Then* $|Chr_k(m, n)| = F_k(m, n)$.

Proof. The number of all matrices $A \in P^{m\times n}$ such that $rank_s(A) = k$ is equal to the number of all bipartite graphs $\Gamma = \Gamma(V_1 \cup V_2, E)$ such that $V_1 = \{1, \ldots, m\}$, $V_2 = \{1', 2', \ldots, n'\}$, $nint_{bp}(\Gamma) = k$. □

The results of section 5 give us the formulas for $|Chr_1(m, n)|$ and $|Chr_2(m, n)|$.

Example. 1. The number of all matrices $A \in P^{2\times 2}$ such that $rank_s(A) = k$ is equal to $F_k(2, 2)$. Using canonical families, we get:

$$F_0(2,2) = 1,\ F_1(2,2) = 9,\ F_2(2,2) = 6.$$

2. The number of all matrices $A \in P^{3\times 3}$ such that $rank_s(A) = k$ is equal to $F_k(3, 3)$. Using canonical families, we get:

$$F_0(3,3) = 1,\ F_1(3,3) = 49,\ F_2(3,3) = 306,\ F_3(3,3) = 156.$$

From the proof of Theorem 9 we obtain the following statements.

If $f : V_1 \cup V_2 \to 2^U$ is a U-coding function for $\Gamma(A)$ and $U = \{1, \ldots, k\}$ is a set, then $A = XY$ where $X \in P^{m\times k}$, $Y \in P^{k\times n}$ are given by:

$$x_{ij} = \widetilde{1} \quad \text{iff} \quad j \in f(i),\ i \in V_1,\ j \in U; \tag{4}$$

$$y_{ij} = \widetilde{1} \quad \text{iff} \quad i \in f(j'),\ i \in U,\ j' \in V_2. \tag{5}$$

Thus $X_{(i)}$ associates to the set $f(i)$ and $Y^{(j)}$ associates to the set $f(j')$.

If $A = XY$, where $X \in P^{m\times k}$, then $f : V_1 \cup V_2 \to 2^U$ given by (4), (5) is a U-coding function for $\Gamma(A)$.

7 On coding of bipartite graphs by antichains

Let $A \in P^{m\times n}$ be a matrix, $\Gamma(A) = \Gamma(V_1 \cup V_2, E)$ a bipartite graph associated to A, $f : V_1 \cup V_2 \to 2^U$ a U-coding function for $\Gamma(A)$.

For given real number x, denote by $\lfloor x \rfloor$ the greatest integer that is less than or equal to x. Similarly, $\lceil x \rceil$ is the least interger that $\geqslant x$.

Denote by $l = N(k)$ the least number such that $k \leqslant \binom{l}{\lfloor l/2 \rfloor}$, where $k \in N$. We have:

$N(1) = 1$, $N(2) = 2$, $N(3) = 3$, $N(4) = N(5) = N(6) = 4$,
$N(7) = \ldots = N(10) = 5$, $N(11) = \ldots = N(20) = 6$,
$N(21) = \ldots = N(35) = 7$, $N(36) = \ldots = N(70) = 8$,
$N(71) = \ldots = N(126) = 9$, $N(127) = \ldots = N(252) = 10$,
$N(253) = \ldots = N(462) = 11$, $N(463) = \ldots = N(924) = 12$,
$N(925) = \ldots = N(1716) = 13$.

Consider the following properties of $N(k)$.

Lemma 2. *Let* $q, t, k \in N$, $1 \leqslant q \leqslant k$. *Then:*

(i) $k \geqslant N(\binom{k}{q})$;
(ii) $k = N(\binom{k}{q})$ *for any given* $t \geqslant 1$ *and sufficiently large* $k = 2q - t$;
(iii) $k = N(\binom{k}{q})$ *for any given* $t \geqslant 1$ *and sufficiently large* $k = 2q + t$.

Proof. (i) The equality $k = N(\binom{k}{q})$ is equivalent to

$$\binom{k-1}{\lfloor (k-1)/2 \rfloor} < \binom{k}{q}. \tag{6}$$

Let t be even, $t = 2a$. Then (6) is equivalent to

$$q(q-1)\ldots(q-a+1) < 2(q-a)(q-a-1)\ldots(q-2a+1). \tag{7}$$

Both sides of (7) are polynomials in one variable q. These polynomials have degree a. We compare their highest coefficients and see that (6) holds for any sufficiently large q.

Let t be odd. Similar reasoning gives (6). □

For $k = 2q \pm t$, we can get more precise result.

Corollary 7. *Let* $q, k \in N$, $1 \leqslant q \leqslant k$. *The equality* $k = N(\binom{k}{q})$ *holds if:*
$k = 2q$ *for all* q;
$k = 2q - 1$ *for all* $q \geqslant 2$;
$k = 2q - 2$ *for all* $q \geqslant 3$;
$k = 2q - 3$ *for all* $q \geqslant 3$;
$k = 2q - 4$ *for all* $q \geqslant 8$.

Corollary 8. *Let* $q, k \in N$, $1 \leqslant q \leqslant k$. *The equality* $k = N(\binom{k}{q})$ *holds if:*
$k = 2q + 1$ *for all* $q \geqslant 1$;
$k = 2q + 2$ *for all* $q \geqslant 1$;
$k = 2q + 3$ *for all* $q \geqslant 4$.

In particular,

$$N\left(\binom{k}{\lfloor k/2\rfloor}\right) = N\left(\binom{k}{\lceil k/2\rceil}\right) = k, \quad k \geqslant 1.$$

A subset B of a poset $(Q, \leqslant)$ is an $\leqslant$-antichain if for any pair of distinct elements x and y of B, both $x \nleqslant y$ and $y \nleqslant x$.

The following lemma is useful for calculation of the Schein rank of Boolean matrices.

Lemma 3. *Let* $A \in P^{m\times n}$. *Then:*

(i) *If the family of all rows of* A *is a* $\leqslant$*-antichain, then* $\mathrm{rank}_s(A) \geqslant N(m)$.
(ii) *If the family of all columns of* A *is a* $\leqslant$*-antichain, then* $\mathrm{rank}_s(A) \geqslant N(n)$.

Proof. (i) Let $f : V_1 \cup V_2 \to 2^U$ be a U-coding function for $\Gamma(A)$, where $|U| = \mathrm{rank}_s(A)$. We now prove that $\{f(i) \mid i \in V_1\}$ is a $\leqslant$-antichain.

Suppose $f(i_1) \subseteq f(i_2)$ for some $i_1,\ i_2 \in V_1$. Then $A = XY$, where $X \in P^{m\times k}$, are given by (4) and (5). According the definition of X,

$$\text{if } x_{i_1 j} = \tilde{1} \text{ then } j \in f(i_1) \subseteq f(i_2) \text{ and } x_{i_2 j} = \tilde{1} \text{ for any } j \in V_2.$$

Therefore

$$X_{(i_1)} \leqslant X_{(i_2)}, \quad A_{(i_1)} = X_{(i_1)}Y \leqslant X_{(i_2)}Y = A_{(i_2)}.$$

Since the family of all rows of A is a $\leqslant$-antichain, we see that $i_1 = i_2$ and $\{f(i) \mid i \in V_1\}$ is a $\subseteq$-antichain. The family of sets $\{f(i) \mid i \in V_1\}$ is a $\leqslant$-antichain of a poset $\mathrm{Bul}(U)$. By Sperner's theorem, [1], we have

$$m \leqslant \binom{|U|}{\lfloor |U|/2\rfloor}. \square$$

We say that $A \in P^{n\times n}$ is an (n, k, λ) design if each column and each row of A has exactly k $\tilde{1}$'s, and each two rows of A has exactly λ $\tilde{1}$'s in common.

Example. Let $A \in P^{n\times n}$ be an (n, k, λ) design, where $\lambda < k < n$. Then

$$n \geqslant \mathrm{rank}_s(A) \geqslant \max\{\min\{n, \frac{nk}{\lambda^2}\}, N(n)\}. \tag{8}$$

Since $\lambda < k$, the family of all rows of A is a $\leqslant$-antichain. Therefore

$$\mathrm{rank}_s(A) \geqslant N(n).$$

Combining this with

$$\mathrm{rank}_s(A) \geqslant \min\{n, \frac{nk}{\lambda^2}\}, \tag{9}$$

obtained in [5], we get (8).

Note that the inequality (8) is exact (the inequality (9) is not exact) for $\overline{E}_{n\times n}$.

8 Bipartite intersection graphs $\Gamma_{k,p,q}$

Let $k, p, q \in N$ and $U = \{1, \ldots, k\}$. We renumerate l-element subsets of U in lexicographical ordering

$$W_l(U) = \{w_{k,l,1}, \ldots, w_{k,l,b(k,l)}\}, \quad \text{where} \quad b(k,l) = \binom{k}{l}.$$

Define the bipartite graph $\Gamma_{k,p,q} = \Gamma(V_1 \cup_2, E)$ by setting:

$$V_1 = W_p(U), \quad V_2 = W_q(U);$$

$$\{w_{k,p,i}, w_{k,q,j}\} \in E \text{ iff } w_{k,p,i} \cap w_{k,q,j} \neq \varnothing.$$

We have $|V_1| = \binom{k}{p}$ and $|V_2| = \binom{k}{q}$. Note that $\Gamma_{k,p,q}$ is a regular graph,

$$\deg(v) = \binom{k-p}{q}, \quad v \in V_1,$$

$$\deg(v) = \binom{k-q}{p} \quad v \in V_2.$$

The graph $\Gamma(A) = \Gamma(V_1 \cup V_2, E)$ is associated to the matrix

$$A(k,p,q) = (a(k,p,q)_{ij}) \in P^{bin(k,p) \times bin(k,q)},$$

where

$$a(k,p,q)_{ij} = \tilde{1} \quad \text{iff} \quad w_{k,p,i} \cap w_{k,q,j} \neq \varnothing.$$

If $p + q \leqslant k$ then the sets of all rows and all columns of $A(k,p,q)$ are $\leqslant$-antichains.

The rows of $A(k,p,1)$ associate to p-element subsets of U that is in lexicographical ordering.

Let $C(k,p) \in P^{k \times k}$ be the circulant matrix, obtained by cycling the row whose first p entries are $\tilde{1}$ and whose last $k-p$ entries are $\tilde{0}$.

Theorem 11. *Let* $p, k \in N$, $1 \leqslant p \leqslant k$. *Then:*

(i) $k \geqslant rank_s(A(k,p,1)) \geqslant N(\binom{k}{p})$.
(ii) *If* $k = N(\binom{k}{p})$, *then* $rank_s(A(k,p,1)) = k$.
(iii) *If* $k \geqslant 2p-1$, *then* $rank_s(A(k,p,1)) = k$.

Proof. (i) The set of all rows of $A(k,p,1)$ is a $\leqslant$-antichain. (iii)The circulant $C(k,p)$ is a submatrix in $A(k,p,1)$. Therefore $k \geqslant rank_s(A(k,p,1)) \geqslant rank_s(C(k,p))$. From [5], if $k \geqslant 2p-1$, then $rank_s(C(k,p)) = k$. □

Example. Consider the following matrix

$$A(4,2,1) = \begin{pmatrix} \tilde{1}\tilde{1}\tilde{0}\tilde{0} \\ \tilde{1}\tilde{0}\tilde{1}\tilde{0} \\ \tilde{1}\tilde{0}\tilde{0}\tilde{1} \\ \tilde{0}\tilde{1}\tilde{1}\tilde{0} \\ \tilde{0}\tilde{1}\tilde{0}\tilde{1} \\ \tilde{0}\tilde{0}\tilde{1}\tilde{1} \end{pmatrix}.$$

Since $4 = N(\binom{4}{2})$, we see that $rank_s(A(4,2,1)) = 4$.

It is easy to prove that

$$\begin{aligned} A(k,p,q) &= A(k,p,1) \cdot A(k,1,q), \\ A(k,p,1) &= (A(k,1,p))^{(t)}. \end{aligned} \tag{10}$$

The matrix $A(k,p,1)$ is a block matrix. Indeed,

$$A(k,p,1) = \begin{pmatrix} J_{bin(k-1,p-1)\times 1} & A(k-1,p-1,1) \\ O_{bin(k-1,p)\times 1} & A(k-1,p,1) \end{pmatrix}.$$

Combining this with (10), we get that $A(k,p,q)$ is the following block matrix:

$$A(k,p,q) = \begin{pmatrix} J_{bin(k-1,p-1)\times bin(k-1,q-1)} & A(k-1,p-1,q) \\ A(k-1,p,q-1) & A(k-1,p,q) \end{pmatrix}.$$

Example. The graph $\Gamma_{5,2,2}$ is

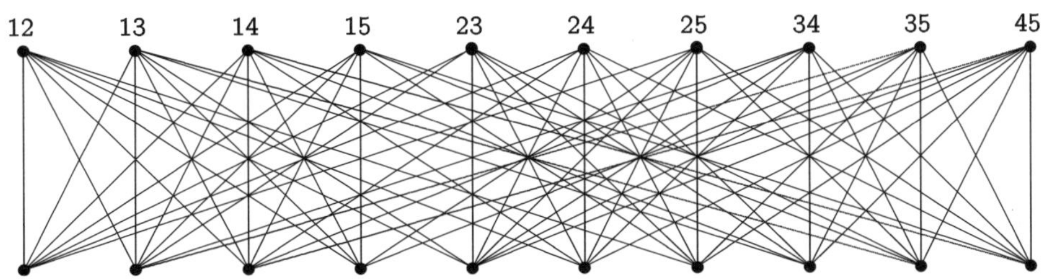

Therefore

$$A(5,2,2) = \begin{pmatrix} \tilde{1}\tilde{1}\tilde{1}\tilde{1}\tilde{1}\tilde{1}\tilde{1}\tilde{0}\tilde{0}\tilde{0} \\ \tilde{1}\tilde{1}\tilde{1}\tilde{1}\tilde{1}\tilde{0}\tilde{0}\tilde{1}\tilde{1}\tilde{0} \\ \tilde{1}\tilde{1}\tilde{1}\tilde{1}\tilde{0}\tilde{1}\tilde{0}\tilde{1}\tilde{0}\tilde{1} \\ \tilde{1}\tilde{1}\tilde{1}\tilde{1}\tilde{0}\tilde{0}\tilde{1}\tilde{0}\tilde{1}\tilde{1} \\ \tilde{1}\tilde{1}\tilde{0}\tilde{0}\tilde{1}\tilde{1}\tilde{1}\tilde{1}\tilde{1}\tilde{0} \\ \tilde{1}\tilde{0}\tilde{1}\tilde{0}\tilde{1}\tilde{1}\tilde{1}\tilde{1}\tilde{0}\tilde{1} \\ \tilde{1}\tilde{0}\tilde{0}\tilde{1}\tilde{1}\tilde{1}\tilde{1}\tilde{0}\tilde{1}\tilde{1} \\ \tilde{0}\tilde{1}\tilde{1}\tilde{0}\tilde{1}\tilde{1}\tilde{0}\tilde{1}\tilde{1}\tilde{1} \\ \tilde{0}\tilde{1}\tilde{0}\tilde{1}\tilde{1}\tilde{0}\tilde{1}\tilde{1}\tilde{1}\tilde{1} \\ \tilde{0}\tilde{0}\tilde{1}\tilde{1}\tilde{0}\tilde{1}\tilde{1}\tilde{1}\tilde{1}\tilde{1} \end{pmatrix} = \begin{pmatrix} \tilde{1}\tilde{1}\tilde{0}\tilde{0}\tilde{0} \\ \tilde{1}\tilde{0}\tilde{1}\tilde{0}\tilde{0} \\ \tilde{1}\tilde{0}\tilde{0}\tilde{1}\tilde{0} \\ \tilde{1}\tilde{0}\tilde{0}\tilde{0}\tilde{1} \\ \tilde{0}\tilde{1}\tilde{1}\tilde{0}\tilde{0} \\ \tilde{0}\tilde{1}\tilde{0}\tilde{1}\tilde{0} \\ \tilde{0}\tilde{1}\tilde{0}\tilde{0}\tilde{1} \\ \tilde{0}\tilde{0}\tilde{1}\tilde{1}\tilde{0} \\ \tilde{0}\tilde{0}\tilde{1}\tilde{0}\tilde{1} \\ \tilde{0}\tilde{0}\tilde{0}\tilde{1}\tilde{1} \end{pmatrix} \begin{pmatrix} \tilde{1}\tilde{1}\tilde{1}\tilde{1}\tilde{0}\tilde{0}\tilde{0}\tilde{0}\tilde{0}\tilde{0} \\ \tilde{1}\tilde{0}\tilde{0}\tilde{0}\tilde{1}\tilde{1}\tilde{1}\tilde{0}\tilde{0}\tilde{0} \\ \tilde{0}\tilde{1}\tilde{0}\tilde{0}\tilde{1}\tilde{0}\tilde{0}\tilde{1}\tilde{1}\tilde{0} \\ \tilde{0}\tilde{0}\tilde{1}\tilde{0}\tilde{0}\tilde{1}\tilde{0}\tilde{1}\tilde{0}\tilde{1} \\ \tilde{0}\tilde{0}\tilde{0}\tilde{1}\tilde{0}\tilde{0}\tilde{1}\tilde{0}\tilde{1}\tilde{1} \end{pmatrix}.$$

Now we obtain the following properties of the Schein rank for $A(k,p,q)$.

Theorem 12. *Let* $k,p,q \in N$, $1 \leqslant p,q \leqslant k$.
Then:

(i) $rank_s(A(k,p,q)) \leqslant \min\{\, rank_s(A(k,1,q)),\ rank_s(A(k,p,1))\,\} \leqslant k$.
(ii) *If* $p+q \leqslant k$, *then* $rank_s(A(k,p,q)) \geqslant \max\left\{N\left(\binom{k}{p}\right),\ N\left(\binom{k}{q}\right)\right\}$.

Proof. The inequality (i) follows from (10). (ii) The families of all rows and all columns of $A(k,p,q)$ are $\leqslant$-antichains. This completes the proof. □

The following is an immediate consequence of Theorem 11 and Corollary 8.

Corollary 9. *If* $k,p \in N$, $1 \leqslant p \leqslant k$. *Then:*

(i) *If* $p \leqslant k/2$ *and* $k = N(\binom{k}{p})$, *then* $rank_s(A(k,p,p)) = k$.
(ii) $rank_s(A(2p,p,p)) = 2p$.
(iii) $rank_s(A(2p+1,p,p+1)) = 2p+1$.

Example. If $k = 4,5,6,7$, then $k = N(\binom{k}{2})$. Therefore

$$rank_s(A(k,2,2)) = k,\ k = 4,5,6,7.$$

If $k = 6,7,8,9$, then $k = N(\binom{k}{3})$. Therefore

$$rank_s(A(k,3,3)) = k,\ k = 6,7,8,9.$$

The following corollary is an application of Theorem 11.

Corollary 10. *Let* $k,p \in N$, $1 \leqslant p < k$. *Then*

$$rank_s(\begin{pmatrix} A(k,p,p) & \overline{E}_{bin(k,p)\times bin(k,p)} \\ \overline{E}_{bin(k,p)\times bin(k,p)} & A(k,k-p,k-p) \end{pmatrix}) = k.$$

Proof. Consider the product of block matrices:

$$\begin{pmatrix} A(k,p,1) \\ A(k,k-p,1) \end{pmatrix} \begin{pmatrix} A(k,1,p) & A(k,1,k-p) \end{pmatrix}$$

$$= \begin{pmatrix} A(k,p,p) & \overline{E}_{bin(k,p)\times bin(k,p)} \\ \overline{E}_{bin(k,p)\times bin(k,p)} & A(k,k-p,k-p) \end{pmatrix}. \tag{11}$$

Taking into account (10), we obtain

$$k \geqslant rank_s(\begin{pmatrix} A(k,p,p) & \overline{E}_{bin(k,p)\times bin(k,p)} \\ \overline{E}_{bin(k,p)\times bin(k,p)} & A(k,k-p,k-p) \end{pmatrix}) \geqslant rank_s(A(k,p,p)) = k. \square$$

In particular, for $p = 1$ we have

$$rank_s \begin{pmatrix} E_{k\times k} & \overline{E}_{k\times k} \\ \overline{E}_{k\times k} & J_{k\times k} \end{pmatrix} = k.$$

9 The Schein rank of $\overline{E}_{n\times n}$

The following exercise is due to Kim [4], [p. 63, Exercise 24].

Exercise. Prove that the Schein rank of the matrix $\overline{E}_{n\times n}$ is k if $n = \binom{k}{[k/2]}$.

The ranks of all square matrices with $\widetilde{0}$ on the main diagonal and $\widetilde{1}$ elsewhere are determined in [8]. From Theorem 9 and Sperner's theorem, we get the following result.

Theorem 13. *The Schein rank of* $\overline{E}_{n\times n}$ *is equal to* $N(n)$.

Proof. The matrix $\overline{E} = \overline{E}_{n\times n}$ is associated to a bipartite graph

$$\Gamma(\overline{E}) = \Gamma(V_1 \cup V_2, E).$$

We have:

$$V_1 = \{1,\dots,n\},\ V_2 = \{1',2',\dots,n'\};$$
$$\text{and}$$
$$\{i,j'\} \in E \text{ is an edge of } \Gamma(\overline{E}) \text{ whenever } i \neq j.$$

We now calculate $\text{nint}_{bp}(\Gamma(\overline{E}))$. Let $\text{nint}_{\varnothing}(\Gamma(\overline{E})) = m$ and f be a U-coding function for $\Gamma(\overline{E})$, where $|U| = m$. Denote:

$$f(i) = a_i,\quad f(i') = b_i,\quad i = 1,\dots,n.$$

Consider the following sets:

$$g(i) = a_i,\quad g(i') = \overline{a}_i = U - a_i,\quad i = 1,\dots,n.$$

It is easy to prove that $g : V_1 \cup V_2 \to 2^U$ is a U-coding function for $\Gamma(\overline{E})$. In particular,

$$a_i \cap \overline{a}_j \neq \varnothing \quad \text{for all} \quad i \neq j.$$

If $a_i \subseteq a_j$ for some $i \neq j$, then

$$\overline{a}_j \subseteq \overline{a}_i,\quad a_i \cap \overline{a}_j \subseteq a_i \cap \overline{a}_i = \varnothing,\qquad a_i \cap \overline{a}_j = \varnothing.$$

This is a contradiction. Therefore the family $\{a_1, a_2, \dots, a_n\}$ is an $\subseteq$-antichain in $\text{Bul}(U)$. According Sperner's theorem, $n \leqslant \binom{m}{\lfloor m/2 \rfloor}$.

We now prove that there exists a U-coding function for $\Gamma(\overline{E})$, where $|U| = k$. By Sperner's theorem the size of a maximal $\subseteq$-antichain in $\text{Bul}(k)$ equals $\binom{k}{\lfloor k/2 \rfloor}$. Let $\{a_1, a_2, \dots, a_n\}$ be some n-element $\subseteq$-antichain in $\text{Bul}(U)$ such that $\{a_1, a_2, \dots, a_n\}$ is contained in a maximal $\subseteq$-antichain and $|a_i| = \lfloor k/2 \rfloor$ for all $i = 1,\dots,n$. Then

$$|\overline{a}_i| = \lceil k/2 \rceil \quad \text{for all} \quad i = 1,\dots,n.$$

Denote:

$$f(i) = a_i, \quad f(i') = \overline{a}_i = U - a_i, \quad i = 1, \ldots, n.$$

Suppose $a_i \cap \overline{a}_j = \varnothing$ for some i, j. We have

$$|a_i| + |\overline{a}_j| = \lfloor k/2 \rfloor + \lceil k/2 \rceil = k, \quad a_i \cup \overline{a}_j = U.$$

Therefore $\overline{a}_i = \overline{a}_j$, $a_i = a_j$, $i = j$. Thus the equality $f(i) \cap f(j') = a_i \cap \overline{a}_j = \varnothing$ is equivalent to $i = j$. We have proved that $f : V_1 \cup V_2 \to 2^U$ is a U-coding function for $\Gamma(\overline{E})$. □

Corollary 11. *Let* $n = \binom{k}{\lfloor k/2 \rfloor}$. *The following statements are valid.*

(i) $\mathrm{rank}_s(\overline{E}_{n\times n}) = k$.

(ii) *If* $\overline{E}_{n\times n} = XY$, *where* $X \in P^{n\times N(n)}$, $Y \in P^{N(n)\times n}$, *then*

$$X = \pi A(k, \lfloor k/2 \rfloor, 1), \quad Y = A(k, 1, \lceil k/2 \rceil)\, \pi^{(t)}, \quad \text{or} \tag{12}$$

$$X = \pi A(k, \lceil k/2 \rceil, 1), \quad Y = A(k, 1, \lfloor k/2 \rfloor)\, \pi^{(t)}, \tag{13}$$

where $\pi \in P^{\mathrm{bin}(k,\lfloor k/2\rfloor)\times \mathrm{bin}(k,\lfloor k/2\rceil}$ *is a permutation matrix.*

Proof. Using Theorem 13 and properties of numbers $N(n)$, we get

$$\mathrm{rank}_s(\overline{E}_{n\times n}) = N\left(\binom{k}{\lfloor k/2 \rfloor}\right) = k.$$

By Sperner's theorem, see [1], there exist only two $\subseteq$-antichains of maximal length in $\mathrm{Bul}(k)$, which are the $\lfloor k/2 \rfloor$-element set and the $\lceil k/2 \rceil$-element set. From the proof of Theorem 13 we get (ii). □

If k is even, then (12) coincides with (13). If k is odd, then (12) does not coincide with (13).

Example. The matrix $\overline{E}_{6\times 6}$ is the product of two matrices

$$\overline{E}_{6\times 6} = \begin{pmatrix} \tilde{0} & \tilde{1} & \tilde{1} & \tilde{1} & \tilde{1} & \tilde{1} \\ \tilde{1} & \tilde{0} & \tilde{1} & \tilde{1} & \tilde{1} & \tilde{1} \\ \tilde{1} & \tilde{1} & \tilde{0} & \tilde{1} & \tilde{1} & \tilde{1} \\ \tilde{1} & \tilde{1} & \tilde{1} & \tilde{0} & \tilde{1} & \tilde{1} \\ \tilde{1} & \tilde{1} & \tilde{1} & \tilde{1} & \tilde{0} & \tilde{1} \\ \tilde{1} & \tilde{1} & \tilde{1} & \tilde{1} & \tilde{1} & \tilde{0} \end{pmatrix} = \begin{pmatrix} \tilde{1} & \tilde{1} & \tilde{0} & \tilde{0} \\ \tilde{1} & \tilde{0} & \tilde{1} & \tilde{0} \\ \tilde{1} & \tilde{0} & \tilde{0} & \tilde{1} \\ \tilde{0} & \tilde{1} & \tilde{1} & \tilde{0} \\ \tilde{0} & \tilde{1} & \tilde{0} & \tilde{1} \\ \tilde{0} & \tilde{0} & \tilde{1} & \tilde{1} \end{pmatrix} \begin{pmatrix} \tilde{0} & \tilde{0} & \tilde{0} & \tilde{1} & \tilde{1} & \tilde{1} \\ \tilde{0} & \tilde{1} & \tilde{1} & \tilde{0} & \tilde{0} & \tilde{1} \\ \tilde{1} & \tilde{0} & \tilde{1} & \tilde{0} & \tilde{1} & \tilde{0} \\ \tilde{1} & \tilde{1} & \tilde{0} & \tilde{1} & \tilde{0} & \tilde{0} \end{pmatrix}.$$

Example. Let $B = B(n)$ be the $n \times n$ matrix with $\tilde{0}$ on the main and back diagonals and $\tilde{1}$ elsewhere.

In particular, consider

$$B(5) = \begin{pmatrix} \tilde{0} & \tilde{1} & \tilde{1} & \tilde{1} & \tilde{0} \\ \tilde{1} & \tilde{0} & \tilde{1} & \tilde{0} & \tilde{1} \\ \tilde{1} & \tilde{1} & \tilde{0} & \tilde{1} & \tilde{1} \\ \tilde{1} & \tilde{0} & \tilde{1} & \tilde{0} & \tilde{1} \\ \tilde{0} & \tilde{1} & \tilde{1} & \tilde{1} & \tilde{0} \end{pmatrix}, \quad B(6) = \begin{pmatrix} \tilde{0} & \tilde{1} & \tilde{1} & \tilde{1} & \tilde{1} & \tilde{0} \\ \tilde{1} & \tilde{0} & \tilde{1} & \tilde{1} & \tilde{0} & \tilde{1} \\ \tilde{1} & \tilde{1} & \tilde{0} & \tilde{0} & \tilde{1} & \tilde{1} \\ \tilde{1} & \tilde{1} & \tilde{0} & \tilde{0} & \tilde{1} & \tilde{1} \\ \tilde{1} & \tilde{0} & \tilde{1} & \tilde{1} & \tilde{0} & \tilde{1} \\ \tilde{0} & \tilde{1} & \tilde{1} & \tilde{1} & \tilde{1} & \tilde{0} \end{pmatrix}$$

We have

$$B_{(r)} = B_{(n-r+1)}, \quad B^{(r)} = B^{(n-r+1)}, \quad r = 1, \ldots, n. \tag{14}$$

By removing rows of numbers $\lceil n/2 \rceil + 1$, $\lceil n/2 \rceil + 2$, …, n from $B(n)$, we get the matrix X.

From (14), we have $rank_s(B(n)) = rank_s(X)$. By removing columns of numbers $\lceil n/2 \rceil + 1$, $\lceil n/2 \rceil + 2$, …, n from X, we get $\overline{E}_{k\times k}$, where $k = \lceil n/2 \rceil$.

From (14), we have $rank_s(X) = rank_s(\overline{E}_{k\times k})$. Therefore

$$rank_s(B(n)) = N(\lceil n/2 \rceil).$$

Example. Let $C(n)$ be the $n \times n$ matrix with $\widetilde{1}$ on the main and back diagonals and $\widetilde{0}$ elsewhere. We have $rank_s(C(n)) = \lceil n/2 \rceil$.

Acknowledgments

This research is conducted in accordance with the Thematic plan of Russian Federal Educational Agency, theme №1.03.07.

References

1. M. Aigner, *Combinatorial Theory*, Grundlehren Math. Wiss. 234, Springer-Verlag, Berlin, 1979.
2. J. Orlin, *Contentment in graph theory: Covering graphs with cliques*, K. Nederlandse Ak. van Wetenschappen Proc. Ser. A, 80 (1977), pp. 406–424.
3. Kim, Ki Hang and F.W. Roush, *Generalized fuzzy matrices*, Fuzzy Sets Systems, 4 (1980), pp. 293–315.
4. Kim, Ki Hang, *Boolean matrix theory and applications*, Marcel Dekker, New York and Basel, 1982.
5. D.A. Gregory, N.J. Pullman, *Semiring rank: Boolean rank and nonnegative rank factorization*, Journal of Combinatorics, Information & System Sciences, v. 8, No.3 (1983), pp. 223–233.
6. Di Nola, S. Sessa, *On the Schein rank of matrices over linear lattice*, Linear Algebra Appl., 118 (1989), pp. 155–158.
7. Di Nola, S. Sessa *Determining the Schein rank of matrices over linear lattices and finite relational equations*, The Journal of Fuzzy Mathematics, vol. 1, No. 1 (1993), pp. 33–38.
8. D. de Caen, D.A. Gregory and N.J. Pullman, *The Boolean rank of zero-one matrices.*, Proc. Third Caribbean Conference on Combinatorics, Graph Theory, and Computing, Barbados, 1981,pp. 169–173 SIAM J. Numer. Anal., 19 (1982), pp. 400–408.

Lattices of Matrix Rows and Matrix Columns. Lattices of Invariant Column Eigenvectors

Valentina Marenich*

Murmansk State Pedagogic University, Russia
VMarenich@yandex.ru

Abstract. We consider matrices over a Brouwerian lattice. The linear span of columns of a matrix A form a semilattice. We call it a column semilattice for A. The questions are: when column semilattice is a lattice, when column semilattice is a distributive lattice, and what formulas can be obtained for the meet and the join operations? We prove that for any lattice matrix A, the column semilattice is a lattice. We also obtain formulas for the meet and the join operations. If A is an idempotent or A is a regular matrix, then the column semilattice is a distributive lattice.

We also consider invariant eigenvectors of a square matrix A over a Brouwerian lattice. It is proved that all A-invariant eigenvectors form a distributive lattice and the simple formulas for the meet and the join operations are obtained.

Keywords: lattice matrix, lattices of columns, invariant eigenvectors of lattice matrices.

1 Introduction

In Section 2 we recall some definitions: lattice matrices, column vectors over a lattice, operations over lattice matrices, Brouwerian and Boolean lattices, systems of linear equations over a lattice. Also we recall the solvability criterion for a system of linear equations over a Brouwerian lattice and some its corollaries, which are needed for the sequal (for more details see [1]).

In Section 3, we define a column semilattice $(\mathrm{Column}(A), \leqslant)$ that is the linear span of columns of a matrix A. Similarly, a row semilattice can be defined.

The questions are: when $(\mathrm{Column}(A), \leqslant)$ is a lattice, when $(\mathrm{Column}(A), \leqslant)$ is a distributive lattice, and what formulas can be obtained for the meet and the join operations? In 1962, K.A. Zarezky proved that for a square matrix A over the two-element Boolean lattice, the column semilattice is a lattice whenever A is a regular matrix.

We consider some cases when the column semilattice is a lattice and get formulas for the meet $\widetilde{\wedge}$ and the join $\widetilde{\vee}$ operations. Note that the similar results

* This research is conducted in accordance with the Thematic Plan of Russian Federal Educational Agency, theme №1.03.07.

can be obtained for a row semilattice. The main result of this section is the following. For a regular matrix A over a Brouwerian lattice,

1. the formula for the meet operation $\widetilde{\wedge}$ in the lattice $(Column(A), \leqslant)$ is
$$u \widetilde{\wedge} v = C(u \wedge v), \text{ for all } u, v \in Column(A),$$
where C is an idempotent such that $Column(A) = Column(C)$.
2. $(Column(A), \leqslant)$ is a distributive lattice.

In section 4, we recall the definition of invariant column eigenvectors that is due to L.A. Skornyakov, see [6]. The set of all invariant column eigenvectors form a subspace. We prove that for any $m \times m$ matrix A over a distributive lattice:

1. the subspace of all invariant column eigenvectors coincides with
$$Column\ ((A + A^2)^k), \text{ where } k \geqslant m,$$
2. matrix $(A + A^2)^k$ is an idempotent.

In section 5, we consider a square matrix A and A-invariant eigenvectors over a Brouwerian lattice. From previous results it follows that all A-invariant eigenvectors form a distributive lattice. Also the simple formulas for the meet and the join operations are obtained.

2 Preliminaries

The following notations will be used throughout. Denote by $(P, \wedge, \vee, \leqslant)$ a lattice.

2.1 Lattice matrices and column vectors

Let $P^{m \times n}$ be a set of all $m \times n$ matrices over P and $A = \|a_{ij}\| \in P^{m \times n}$. We define the following matrix operations:

– for any matrices $A, B \in P^{m \times n}$

$$A + B = \|a_{ij} \vee b_{ij}\|;$$

– for any matrices $A \in P^{m \times n}, B \in P^{n \times k}$

$$AB = \| \bigvee_{r=1}^{n} (a_{ir} \wedge b_{rj}) \|_{m \times k}.$$

A square lattice matrix $A \in P^{m \times m}$ is an *idempotent* if $A^2 = A$. The *transpose* of A is defined by analogy with linear algebra and is denoted by $A^{(t)}$.

Any element $(p_1, \dots, p_m)^t$ in $P^{m\times 1}$ is called a *column vector*. We define a partial order on $P^{m\times 1}$:

$$(p_1, \dots, p_m)^t \leqslant (p'_1, \dots, p'_m)^t \Leftrightarrow p_1 \leqslant p'_1, \dots, p_m \leqslant p'_m$$

and the folowing operations. For any $(p_1, \dots, p_m)^t, (p'_1, \dots, p'_m)^t \in P^{m\times 1}$, and $\lambda \in P$,

$$(p_1, \dots, p_m)^t + (p'_1, \dots, p'_m)^t = (p_1 \vee p'_1, \dots, p_m \vee p'_m)^t; \tag{1}$$

$$\lambda(p_1, \dots, p_m)^t = (\lambda \wedge p_1, \dots, \lambda \wedge p_m)^t. \tag{2}$$

With these notations define a *linear span* of column vectors (by analogy with linear algebra).

Any set $S \subseteq P^{m\times 1}$ closed under the operations (1) and (2) is called a *subspace*.

The partially ordered set $(P^{m\times 1}, \leqslant)$ is a lattice with meet $\wedge$ and join $\vee$ operations defined as follows. For any $(p_1, \dots, p_m)^t, (p'_1, \dots, p'_m)^t \in P^{m\times 1}$,

$$(p_1, \dots, p_m)^t \vee (p'_1, \dots, p'_m)^t = (p_1 \vee p'_1, \dots, p_m \vee p'_m)^t,$$
$$(p_1, \dots, p_m)^t \wedge (p'_1, \dots, p'_m)^t = (p_1 \wedge p'_1, \dots, p_m \wedge p'_m)^t.$$

Recall that any partially ordered set is called, more simply, a *poset*.

2.2 Brouwerian lattices

Let us recall the definition of Brouwerian lattices. If for given elements $a, b \in P$ the greatest solution of the inequality $a \wedge x \leqslant b$ exists then it is denoted by $\left\langle \frac{b}{a} \right\rangle$ and is called the *relative pseudocomplement* of a in b. If $\left\langle \frac{b}{a} \right\rangle$ exists for all $a, b \in P$ then $(P, \wedge, \vee, \leqslant)$ is called a *Brouwerian lattice*. Note that:

- any Brouwerian lattice has the greatest element denoted by $\hat{1}$;
- any Brouwerian lattice is a distributive lattice;
- any finite distributive lattice is a Brouwerian lattice.

Let $(P, \wedge, \vee, \leqslant)$ be a Brouwerian lattice, $A \in P^{m\times n}$ a matrix, $c = (c_1, \dots, c_m)^t \in P^{m\times 1}$ a column vector. Define a vector

$$\left\langle \frac{c}{A} \right\rangle = {}^t\left(\bigwedge_{i=1}^{m} \left\langle \frac{c_i}{a_{i1}} \right\rangle, \dots, \bigwedge_{i=1}^{m} \left\langle \frac{c_i}{a_{in}} \right\rangle \right) \in P^{n\times 1}.$$

2.3 Boolean lattices

Let $(P, \wedge, \vee, \leqslant)$ be a distributive lattice with the least element $\widehat{0}$ and the greatest element $\widehat{1}$. If for any $a \in P$ there exists $\overline{a} \in P$ such that $a \vee \overline{a} = \widehat{1}$ and $a \wedge \overline{a} = \widehat{0}$, then $(P, \wedge, \vee, \leqslant)$ is called a *Boolean lattice*.

Any Boolean lattice is a Brouwerian lattice, where $\left\langle {b \atop a} \right\rangle = \overline{a} \vee b$.

Denote by $\overline{A}$ a matrix $\overline{A} = \|\overline{a}_{ij}\|$.

Let U be a finite set, 2^U be the collection of all subsets of U. Denote by $\mathrm{Bul}(U) = (2^U, \subseteq)$ the poset of all subsets of U partially ordered by inclusion (we call it a *Boolean algebra*). Let $\mathrm{Bul}(k)$ be the Boolean algebra of all subsets of a finite k-element set. It is obvious that $\mathrm{Bul}(U)$ $(\mathrm{Bul}(k))$ is a Boolean lattice.

2.4 Systems of linear equations over Brouwerian lattices

Befor continuing, we require the following results, which are known from [1].

Let $A \in P^{m \times n}$, $c \in P^{m \times 1}$. Define a system of linear equations

$$Ax = c, \tag{3}$$

and a system of linear inequations

$$Ax \leqslant c. \tag{4}$$

Theorem 1. *Let $(P, \wedge, \vee, \leqslant)$ be a Brouwerian lattice. Then*

(i) $x = \left\langle {c \atop A} \right\rangle$ *is the greatest solution for the system of inequations (4).*
(ii) *System (3) is solvable whenever $x = \left\langle {c \atop A} \right\rangle$ is the solution of (3). If System (3) is solvable, then $x = \left\langle {c \atop A} \right\rangle$ is the greatest solution of it.*

Corollary 1. *Let $(P, \wedge, \vee, \leqslant)$ be a Brouwerian lattice. Then $x = \left\langle {c \atop A} \right\rangle$ is the greatest vector in the set $\{Ax \mid Ax \leqslant c,\ x \in P^{n \times 1}\}$.*

Theorem 2. *Let $(P, \wedge, \vee, \leqslant)$ be a Boolean lattice. Then the following conditions are equivalent:*

(i) *System (3) is solvable.*
(ii) *The greatest solution of System (3) is*

$$x = \left\langle {c \atop A} \right\rangle = \overline{A^t \cdot \overline{c}}.$$

The solvability for systems of linear equations over Boolean lattices was studied in details by Rudeany in [2].

Corollary 2. *Let $(P, \wedge, \vee, \leqslant)$ be a a Boolean lattice. Then*

$$x = A\left\langle {c \atop A} \right\rangle = A \cdot \overline{A^t \cdot \overline{c}}$$

is the greatest vector in the set $\{Ax \mid Ax \leqslant c,\ x \in P^{n \times 1}\}$.

3 Semilattices and lattices of matrix columns and matrix rows

Let $A = \|a_{ij}\| \in P^{m\times n}$ be a matrix, $A^{(j)} = (a_{1j}, \dots, a_{mj})^t$ the j-th column.

The linear span of columns we denote by $\mathrm{Column}(A)$. Let $u \in \mathrm{Column}(A)$, then $u = Ax$ for some column vector $x \in P^{n\times 1}$.

Define a poset $(\mathrm{Column}(A), \leqslant)$ with respect to the partial order $\leqslant$ induced by the lattice $(P^{m\times 1}, \leqslant)$.

Let $(P, \wedge, \vee, \leqslant)$ be a distributive lattice and $A \in P^{m\times n}$. Then $(\mathrm{Column}(A), \leqslant)$ is an upper semilattice with the join operation $\widetilde{\vee}$ given by

$$\begin{aligned}(p_1, \dots, p_m)^t \widetilde{\vee} (p_1', \dots, p_m')^t &= (p_1, \dots, p_m)^t + (p_1', \dots, p_m')^t \\ &= (p_1 \vee p_1', \dots, p_m \vee p_m')^t,\end{aligned}$$

for any $(p_1, \dots, p_m)^t, (p_1', \dots, p_m')^t \in \mathrm{Column}(A)$. We call $(\mathrm{Column}(A), \leqslant)$ a *column semilattice.*

Similarly, a *row semilattice* $(\mathrm{Row}(A), \leqslant)$ can be defined.

The questions are: *when* $(\mathrm{Column}(A), \leqslant)$ *is a lattice, when* $(\mathrm{Column}(A), \leqslant)$ *is a distributive lattice, and what formulas can be obtained for the meet and the join operations?* In 1962, K.A. Zarezky obtained the following result.

Theorem 3 (Zarezky's criterion). *Let* $P = \{\hat{0}, \hat{1}\}$ *be a two-element Boolean lattice and* $A \in P^{m\times m}$ *a square matrix. Then* $(\mathrm{Column}(A), \leqslant)$ *is a distributive lattice whenever* A *is a regular matrix.*

Recall that a square matrix $A \in P^{m\times m}$ is called a regular matrix if there exists $B \in P^{m\times m}$ such that

$$ABA = A.$$

It is known that A is a regular matrix whenever there exists an idempotent C such that

$$\mathrm{Column}(A) = \mathrm{Column}(C), \tag{5}$$

see [3].

In the following theorem we consider some cases when the column semilattice is a lattice and obtain formulas for the meet $\widetilde{\wedge}$ and join $\widetilde{\vee}$ operations.

Theorem 4. *Let* $(P, \wedge, \vee, \leqslant)$ *be a lattice and* $A \in P^{m\times n}$.

(i) *If* $(P, \wedge, \vee, \leqslant)$ *is a Brouwerian lattice, then* $(\mathrm{Column}(A), \leqslant)$ *is a lattice, where the formulas for the meet* $\widetilde{\wedge}$ *and join* $\widetilde{\vee}$ *operations are*

$$u \widetilde{\vee} v = u + v,$$

$$u \tilde{\wedge} v = A\left\langle \frac{u \wedge v}{A} \right\rangle = A\left(\left\langle \frac{u}{A}\right\rangle \wedge \left\langle \frac{v}{A}\right\rangle\right),$$

for all $u, v \in \mathrm{Column}(A)$.

(ii) *If* $(P, \wedge, \vee, \leqslant)$ *is a Boolean lattice, then* $(\mathrm{Column}(A), \leqslant)$ *is a lattice, where formulas for the operations meet* $\widetilde{\wedge}$ *and join* $\widetilde{\vee}$ *are*

$$u \widetilde{\vee} v = u + v = u \vee v,$$
$$u \widetilde{\wedge} v = A \cdot \overline{A^t \cdot (\overline{u} + \overline{v})},$$

for all $u, v \in \mathrm{Column}(A)$.

Proof. (i) Let $\widetilde{w} = u \widetilde{\wedge} v$. Then $\widetilde{w}$ is the greatest vector in the set

$$\{w = Ax,\ x \in P^{n \times 1} | w \leqslant u,\ w \leqslant v\} = \{w = Ax,\ x \in P^{n \times 1} | w \leqslant u \wedge v\}.$$

Accoding Corollary 1,

$$\widetilde{w} = A \left\langle {u \wedge v \atop A} \right\rangle.$$

(ii) Accoding i) and Corollary 2,

$$u \widetilde{\wedge} v = A \left\langle {u \wedge v \atop A} \right\rangle = A \cdot \overline{A^t \cdot (\overline{u \wedge v})} = A \cdot \overline{A^t \cdot (\overline{u} + \overline{v})}. \quad \square$$

Corollary 3. *Let* $(P, \wedge, \vee, \leqslant)$ *be a finite distributive lattice. Then the column semilattice* $(\mathrm{Column}(A), \leqslant)$ *is a lattice, in which the meet and join operations can be calculated by formulas from Theorem 4 (i).*

For some column lattices, we can express the meet operation $\widetilde{\wedge}$ more simply.

Theorem 5. *Let* $(P, \wedge, \vee, \leqslant)$ *be a Brouwerian lattice and matrix* $A \in P^{m \times m}$ *an idempotent. Then:*

(i) *the formulas for meet and join operations* $\widetilde{\wedge}$, $\widetilde{\vee}$ *in the lattice* $(\mathrm{Column}(A), \leqslant)$ *are*

$$u \widetilde{\vee} v = u + v, \quad u \widetilde{\wedge} v = A\,(u \wedge v)\,,$$

for all $u, v \in \mathrm{Column}(A)$.
(ii) $(\mathrm{Column}(A), \leqslant)$ *is a distributive lattice.*

Proof. (i) According Theorem 4,

$$u \widetilde{\wedge} v = A \left(\left\langle {u \atop A} \right\rangle \wedge \left\langle {v \atop A} \right\rangle \right).$$

Note that u is the solution of $Ax = u$. According Theorem 1,

$$u \leqslant \left\langle {u \atop A} \right\rangle.$$

Since $u \leqslant \langle {u \atop A} \rangle$ and $v \leqslant \langle {v \atop A} \rangle$, we get

$$A(u \wedge v) \leqslant A \left(\left\langle {u \atop A} \right\rangle \wedge \left\langle {v \atop A} \right\rangle \right) = u \widetilde{\wedge} v.$$

Note that $w = Aw$ for any idempotent matrix $A \in P^{m\times m}$ and any vector $w \in \mathrm{Column}(A)$. We have

$$A(u \wedge v) \leqslant u\widetilde{\wedge}v = A(u\widetilde{\wedge}v) \leqslant A(u \wedge v).$$

Therefore

$$u\widetilde{\wedge}v = A(u \wedge v).$$

(ii) For any $u, v, w \in \mathrm{Column}(A)$,

$$\begin{aligned} w\widetilde{\vee}(u\widetilde{\wedge}v) &= w + A(u \wedge v) = Aw + A(u \wedge v) = A(w \vee (u \wedge v)) \\ &= A((w\widetilde{\vee}u) \wedge (w\widetilde{\vee}v)) = (w\widetilde{\vee}u)\widetilde{\wedge}(w\widetilde{\vee}v), \end{aligned}$$

This completes the proof.

Note that the results similar to Theorem 4, Corrolary 3 and Theorem 5 can be obtained for a row semilattice $(\mathrm{Row}(A), \leqslant)$.

The following statement is an analog of Zarezky's theorem.

Theorem 6. *Let* $(P, \wedge, \vee, \leqslant)$ *be a Brouwerian lattice,* $A \in P^{m\times m}$ *a regular matrix, and* $C \in P^{m\times m}$ *an idempotent such that* $\mathrm{Column}(A) = \mathrm{Column}(C)$. *Then:*

(i) *the formular for the meet operation* $\widetilde{\wedge}$ *in the lattice* $(\mathrm{Column}(A), \leqslant)$ *is*

$$u\widetilde{\wedge}v = C(u \wedge v),$$

for all $u, v \in \mathrm{Column}(A)$.
(ii) $(\mathrm{Column}(A), \leqslant)$ *is a distributive lattice.*

Proof. Recall that for any regular matrix A there always exists an idempotent C such that $\mathrm{Column}(A) = \mathrm{Column}(C)$, see (5). The proof of (i) and (ii) follows from Theorem 5.

The result similar to the statement (ii) was proved by K.A. Zarezky for semigroups of binary relations, see [4]. Kim and Roush obtained the similar result for the fuzzy lattice, see [5].

4 Subspaces of A-invariant column eigenvectors

Let $(P, \wedge, \vee, \leqslant)$ be a distributive lattice and $A \in P^{m\times m}$ a square matrix. The following definition of invariant column eigenvectors over a lattice is due to L.A. Skornyakov, see [6]. (We say "*A-invariant column vectors*" instead of "*invariant column eigenvectors*".)

A column vector $u \in P^{m\times 1}$ is called an *A-invariant* if $Au = u$. If $u, v \in P^{m\times 1}$ are invariant column vectors and $p \in P$, then $u+v$ and pv are A-invariant column vectors. Therefore the set all A-invariant column vectors form a subspace.

Our purpose is to describe the subspace of all A-invariant column vectors.

The following two lemmas are needed for the sequel.

Lemma 1. *Let* $(P, \wedge, \vee, \leqslant)$ *be a distributive lattice and* $A \in P^{m\times m}$ *a square matrix. Then*

$$A^m \leqslant \sum_{r=m+1}^{2m} A^r \leqslant \sum_{r>m} A^r \leqslant \sum_{1\leqslant r\leqslant m} A^r. \tag{6}$$

These inequalities was prooved by K. Chechlarova in [7].

Lemma 2. *Let* $(P, \wedge, \vee, \leqslant)$ *be a distributive lattice,* $A \in P^{m\times m}$ *a square matrix and* $k \geqslant m$. *Then*

$$A^k \leqslant A^{k+1} + A^{k+2} + \ldots + A^{k+m}, \tag{7}$$

$$A^{2k+1} \leqslant A^k + A^{k+1} + \ldots + A^{2k}. \tag{8}$$

Proof. Since (6), we get

$$A^m \leqslant A^{m+1} + A^{m+2} + \ldots + A^{2m}.$$

Multiplying both sides by A^{k-m}, we obtain (7).

Now let us prove (8). First we prove (8) for $\{\widetilde{0}, \widetilde{1}\}$ -Boolean matrix A. Let $A \in \{0, 1\}^{m\times m}$; then for any column vector $\xi \in \{\widetilde{0}; \widetilde{1}\}^{m\times 1}$

$$A^k\xi \leqslant (A^k + A^{k+1})\xi \leqslant \ldots \leqslant (A^k + A^{k+1} + \ldots + A^{2k})\xi.$$

If all inequalities are strict, then there exists a $<$-chain of length $k+1 > m$ in the lattice $(\{\widetilde{0}, \widetilde{1}\}^m, \leqslant)$. This is a contradiction, because the length of Boolean algebra $(\{\widetilde{0}, \widetilde{1}\}^m, \leqslant) \cong \mathrm{Bul}(m)$ is equal to m. Suppose

$$(A^k + A^{k+1} + \ldots + A^{k+s})\xi = (A^k + A^{k+1} + \ldots + A^{k+s+1})\xi$$

for some s, where $1 \leqslant s \leqslant k-1$. Then

$$A^{k+s+1}\xi \leqslant (A^k + A^{k+1} + \ldots + A^{k+s})\xi.$$

We prove by induction on $r \geqslant s+1$ the following inequalities

$$A^{k+r}\xi \leqslant (A^k + A^{k+1} + \ldots + A^{k+s})\xi. \tag{9}$$

For $r = s+1$ the inequality is already proved. We assume that the inequality holds for r and prove it for $r+1$. Indeed,

$$A^{k+r+1}\xi \leqslant (A^{k+1} + \ldots + A^{k+s+1})\xi \leqslant (A^k + A^{k+1} + \ldots + A^{k+s+1})\xi$$
$$= (A^k + A^{k+1} + \ldots + A^{k+s})\xi + A^{k+s+1}\xi = (A^k + A^{k+1} + \ldots + A^{k+s})\xi.$$

Thus inequalities (9) are valid.

By setting $r = k + 1 \geqslant s + 1$ in (9), we obtain

$$A^{2k+1}\xi \leqslant (A^k + A^{k+1} + \ldots + A^{2k})\xi.$$

Since ξ is an arbitrary column vector, we see that the inequality (8) is valid for any Boolean matrix $A \in \{0, 1\}^{m \times m}$.

Now we suppose that A is a lattice matrix, $A \in P^{m \times m}$. Using the inequality (8) for Boolean $\{\widetilde{0}, \widetilde{1}\}$-matrices and the decomposition of A into the linear span of sections (constituents), we see that (8) is valid over the lattice $(P, \wedge, \vee, \leqslant)$. (The linear span of sections are defined in [5].)

Lemma 2 gives us the following.

Theorem 7. *Let* $(P, \wedge, \vee, \leqslant)$ *be a distributive lattice,* $k \geqslant m$ *and* $A \in P^{m \times m}$. *Then:*

(i) $(A + A^2)^k = (A + A^2)^k \cdot A$,
(ii) $(A + A^2)^k$ *is an idempotenet matrix.*

Proof. (i) It follows from (7) that

$$A^k \leqslant A^{k+1} + \ldots + A^{2k}.$$

Combining this inequality and (8), we get

$$\begin{aligned}(A + A^2)^k = A^k + A^{k+1} + \ldots + A^{2k} &= A^k + A^{k+1} + \ldots + A^{2k} + A^{2k+1} \\ &= A^{k+1} + \ldots + A^{2k} + A^{2k+1} = (A + A^2)^k \cdot A.\end{aligned}$$

(ii) Using (i), we see that $(A+A^2)^k = (A+A^2)^k A^s$ for any $s \geqslant 0$. Therefore

$$\begin{aligned}\left((A + A^2)^k\right)^2 = (A + A^2)^k (A + A^2)^k &= (A + A^2)^k \left(A^k + A^{k+1} + \ldots + A^{2k}\right) \\ &= \sum_{s=0}^{2k} (A + A^2)^k A^s = \sum_{s=0}^{2k} (A + A^2)^k = (A + A^2)^k,\end{aligned}$$

and $(A + A^2)^k$ is an idempotenet matrix.

The result similar to Theorem 7 was proved by K. H. Kim for the two-element Boolean lattice $P = \{\widetilde{0}, \widetilde{1}\}$, see [3].

In the following lemma we describe invariant vectors of idempotent matrices.

Lemma 3. *Let* $(P, \wedge, \vee, \leqslant)$ *be a distributive lattice,* $B \in P^{m \times m}$ *an idempotent and* $\xi \in P^{m \times 1}$ *a column vector. Then* ξ *is a* B*-invariant vector whenever* $\xi \in \mathrm{Column}(B)$.

Proof. Suppose $B\xi = \xi$; then $\xi \in \mathrm{Column}(B)$.

Suppose $\xi \in \mathrm{Column}(B)$. Since B is an idempotent, we get $B = B^2$ and $B^{(j)} = B \cdot B^{(j)}$ for any column B^j, $j = 1, \ldots, n$. By definition of $\mathrm{Column}(B)$,

$$\xi = \beta_1 B^{(1)} + \ldots + \beta_n B^{(m)},$$

for some $\beta_1, \ldots, \beta_m \in P$. Therefore

$$B\xi = B(\beta_1 B^{(1)} + \ldots + \beta_m B^{(m)}) = \beta_1 B \cdot B^{(1)} + \ldots + \beta_m B \cdot B^{(m)} = \xi. \quad \square$$

Now we can describe all invariant eigenvectors of $m \times m$-matrices over a lattice.

Theorem 8. *Let $(P, \wedge, \vee, \leqslant)$ be a distributive lattice, $A \in P^{m\times m}$ and $k \geqslant m$. Then the subspace of all A-invariant column vectors coincides with $\mathrm{Column}((A + A^2)^k)$ and matrix $(A + A^2)^k$ is an idempotent.*

Proof. Accoding Theorem 7, $(A + A^2)^k$ is an idempotent and $(A + A^2)^k = (A + A^2)^k \cdot A$.

First we shall prove that conditions $A\xi = \xi$ and $(A + A^2)^k\xi = \xi$ are equavalent for any $\xi \in P^{m\times 1}$. Suppose $A\xi = \xi$, then obviously $(A + A^2)^k\xi = \xi$. Suppose $(A + A^2)^k\xi = \xi$, then $A\xi = A(A + A^2)^k\xi = (A + A^2)^k\xi = \xi$.

Since $(A+A^2)^k$ is an idempotent, using Lemma 3, we see that $(A+A^2)^k\xi = \xi$ is aquavalent to $\xi \in \mathrm{Column}((A + A^2)^k)$.

For the two-element Boolean lattice $P = \{\widetilde{0}, \widetilde{1}\}$, Theorem 8 is a corollary of results obtained by T.S. Blyth, see [10].

5 Lattices of invariant column vectors

Let $(P, \wedge, \vee, \leqslant)$ be a Brouwerian lattice, $A \in P^{m\times m}$ a square matrix. From previous results it follows that all A-invariant vectors form a distributive lattice. Also the simple formulas for the meet and the join operations are obtained.

Theorem 9. *Let $(P, \wedge, \vee, \leqslant)$ be a Brouwerian lattice, $A \in P^{m\times m}$ a square matrix, $J_{m\times 1} = (\hat{1}, \ldots, \hat{1})^t \in P^{m\times n}$ a universal column vector, $k \geqslant m$. Then:*

(i) *all A-invariant vectors form a lattice, which coincides with*

$$\left(\mathrm{Column}((A + A^2)^k), \widetilde{\wedge}, \widetilde{\vee}, \leqslant\right)$$

(ii) *the formulas for the meet $\widetilde{\wedge}$ and the join $\widetilde{\vee}$ operations in the lattice of all A-invariant vectors are*

$$u\widetilde{\vee}v = u + v = u \vee v, \ u\widetilde{\wedge}v = (A + A^2)^k(u \wedge v),$$

for all A-invariant column vectors $u, v \in P^{m\times 1}$;

(iii) *the lattice of all A-invariant vectors is distributive, with the greatest element* $A^m J_{m\times 1}$. *If* $(P, \wedge, \vee, \leqslant)$ *is a Brouwerian lattice with* $\hat{0}$, *then the lattice of all A-invariant vectors has the least element* $\widetilde{0} = (\hat{0}, \ldots, \hat{0})^t \in P^{m\times 1}$.

Proof. Statements (i) and (ii) are immediate consequences of Theorems 8, 4, 5. Let us prove (iii). First we shall prove that $A^m \cdot J_{m\times 1}$ is an A-invariant vector. From the obvious inequality $AJ_{m\times 1} \leqslant J_{m\times 1}$, it folows that

$$A^{r+1} J_{m\times 1} \leqslant A^r J_{m\times 1}, \quad r = 1, 2, \ldots. \tag{10}$$

According to (6),

$$A^m J_{m\times 1} \leqslant \sum_{r>m} A^r J_{m\times 1}.$$

Consider the right part of this inequality. Since (10), we see that $A^{m+1} J_{m\times 1}$ is the greatest summand, therefore

$$\sum_{r>m} A^r J_{m\times 1} = A^{m+1} J_{m\times 1}.$$

Applying (10), we get

$$A^m J_{m\times 1} \leqslant \sum_{r>m} A^r J_{m\times 1} = A^{m+1} J_{m\times 1} \leqslant A^m J_{m\times 1}.$$

To conclude the proof, it remains to note that $A^m \cdot J_{m\times 1}$ is the greatest A-invariant vector. Indeed, if ξ is A-invariant vector, then

$$\xi = A\xi = \ldots = A^m \xi \leqslant A^m J_{m\times 1}. \quad \square$$

Acknowledgments

This research is conducted in accordance with the Thematic plan of Russian Federal Educational Agency, theme №1.03.07.

References

1. E.E. Marenich, V.G. Kumarov *Inversion of matrices over a pseudocomplemented lattice*, Journal of Mathematical Sciences, Volume 144, Number 2, Springer New York (2007), pp. 3968–3979
2. S. Rudeanu, *Lattice Functions and Equations*, Springer - Verlag, London, 2001.
3. Kim, Ki Hang, *Boolean matrix theory and applications*, Marcel Dekker, New York and Basel, 1982.
4. K.A. Zarezky, *Regular elements in semigroups of binary relations*, Uspeki mat. nauk, 17-3 (1962), pp. 105–108.

5. K.X. KIM, F.W. ROUSH, *Generalized fuzzy matrices*, Fuzzy Sets Systems, 4 (1980), pp. 293–315.
6. L.A. SKORNYAKOV, *Eigenvectors of a matrix over a distributive lattice*, Vestnik Kievskogo Universiteta, 27 (1986), pp. 96–97.
7. K. CHECHLAROVA, *Powers of matrices over distributive lattices - a review*, Fuzzy Sets and Systems, 138 (2003), pp. 627–641.
8. S. KIRKLAND, N.J. PULLMAN *Boolean spectral theory*, Linear Algebra Appl., 175 (1992), pp. 177–190.
9. Y.-J. TAN , *On the powers of matrices over distributive lattices*, Linear Algebra Appl., 336 (2001), pp. 1–14.
10. T.S. BLYTH, *On eigenvectors of Boolean matrices*, Proc. Royal Soc., Edinburgh Sect A 67 (1966), pp. 196–204.

Matrix Algebras and Their Length

Olga V. Markova

Moscow State University, Mathematics and Mechanics Dept., Russia
ov_markova@mail.ru

Abstract. Let $\mathbb{F}$ be a field and let $\mathcal{A}$ be a finite-dimensional $\mathbb{F}$-algebra. We define the length of a finite generating set of this algebra as the smallest number k such that words of the length not greater than k generate $\mathcal{A}$ as a vector space, and the length of the algebra is the maximum of lengths of its generating sets. In this paper we study the connection between the length of an algebra and the lengths of its subalgebras. It turns out that the length of an algebra can be smaller than the length of its subalgebra. To investigate, how different the length of an algebra and the length of its subalgebra can be, we evaluate the difference and the ratio of the lengths of an algebra and its subalgebra for several representative families of algebras. Also we give examples of length computation of two and three block upper triangular matrix algebras.

Keywords: length, finite-dimensional associative algebras, matrix subalgebras, upper triangular matrices, block matrices.

1 Main definitions and notation

Let $\mathbb{F}$ be an arbitrary field and let $\mathcal{A}$ be a finite-dimensional associative algebra over $\mathbb{F}$. Each finite-dimensional algebra is certainly finitely generated. Let $\mathcal{S} = \{a_1, \ldots, a_k\}$ be a finite generating set for $\mathcal{A}$.

Notation 1. Let $\langle S \rangle$ denote the linear span, i.e. the set of all finite linear combinations with coefficients from $\mathbb{F}$, of the set S.

Definition 1. *A* length *of a word* $a_{i_1} \cdots a_{i_t}$, $a_{i_j} \in \mathcal{S}$, $a_{i_j} \neq 1$, *is* t. *If* $\mathcal{A}$ *is an algebra with* 1, *then it is said that* 1 *is a word of elements from* $\mathcal{S}$ *of* length 0.

Notation 2. Let $\mathcal{S}^i$ denote the set of all words in the alphabet $a_1, \ldots, a_k$ of a length less than or equal to i, $i \geqslant 0$.

Notation 3. Let $\mathcal{L}_i(\mathcal{S}) = \langle \mathcal{S}^i \rangle$ and let $\mathcal{L}(\mathcal{S}) = \bigcup_{i=0}^{\infty} \mathcal{L}_i(\mathcal{S})$ be the linear span of all words in the alphabet $a_1, \ldots, a_k$. Note that $\mathcal{L}_0(\mathcal{S}) = \mathbb{F}$, if $\mathcal{A}$ is unitary, and $\mathcal{L}_0(\mathcal{S}) = 0$, otherwise.

Since $\mathcal{S}$ is a generating set for $\mathcal{A}$, any element of $\mathcal{A}$ can be written as a finite linear combination of words in $a_1, \dots, a_k$, i.e., $\mathcal{A} = \mathcal{L}(\mathcal{S})$. The definition of $\mathcal{S}^i$ implies that $\mathcal{L}_{i+j}(\mathcal{S}) = \langle \mathcal{L}_i(\mathcal{S})\mathcal{L}_j(\mathcal{S})\rangle$ and $\mathcal{L}_0(\mathcal{S}) \subseteq \mathcal{L}_1(\mathcal{S}) \subseteq \dots \subseteq \mathcal{L}_h(\mathcal{S}) \subseteq \dots \subseteq \mathcal{L}(\mathcal{S}) = \mathcal{A}$. Since $\mathcal{A}$ is finite dimensional, there exists an integer $h \geqslant 0$ such that $\mathcal{L}_h(\mathcal{A}) = \mathcal{L}_{h+1}(\mathcal{A})$.

Definition 2. *A number* $l(\mathcal{S})$ *is called a* length of a finite generating set $\mathcal{S}$ *provided it equals the smallest number* h*, such that* $\mathcal{L}_h(\mathcal{S}) = \mathcal{L}_{h+1}(\mathcal{S})$.

Note that if for some $h \geqslant 0$ it holds that $\mathcal{L}_h(\mathcal{S}) = \mathcal{L}_{h+1}(\mathcal{S})$, then

$$\mathcal{L}_{h+2}(\mathcal{S}) = \langle \mathcal{L}_1(\mathcal{S})\mathcal{L}_{h+1}(\mathcal{S})\rangle = \langle \mathcal{L}_1(\mathcal{S})\mathcal{L}_h(\mathcal{S})\rangle = \mathcal{L}_{h+1}(\mathcal{S})$$

and similarly $\mathcal{L}_i(\mathcal{S}) = \mathcal{L}_h(\mathcal{S})$ for all $i \geqslant h$. Thus $l(\mathcal{S})$ is defined correctly. Since $\mathcal{S}$ is a generating set for $\mathcal{A}$, it follows that $\mathcal{L}_h(\mathcal{S}) = \mathcal{L}(\mathcal{S}) = \mathcal{A}$.

The following definition is crucial for this paper.

Definition 3. *The* length of the algebra $\mathcal{A}$*, denoted by* $l(\mathcal{A})$*, is the maximum of lengths of all its generating sets.*

Definition 4. *The word* $v \in \mathcal{L}_j(\mathcal{S})$ *is called* reducible over $\mathcal{S}$ *if there exists* $i < j$*, such that* $v \in \mathcal{L}_i(\mathcal{S})$ *and* $\mathcal{L}_i(\mathcal{S}) \neq \mathcal{L}_j(\mathcal{S})$.

Notation 4. Let $M_n(\mathbb{F})$ be the full matrix algebra of order n over $\mathbb{F}$, $T_n(\mathbb{F})$ be the algebra of $n \times n$ upper triangular matrices over matrices over $\mathbb{F}$, $D_n(\mathbb{F})$ be the algebra of $n \times n$ diagonal matrices over matrices over $\mathbb{F}$, and $N_n(\mathbb{F})$ be the subalgebra of nilpotent matrices in $T_n(\mathbb{F})$.

Notation 5. We denote by E the identity matrix, by $E_{i,j}$ the matrix unit, i.e. the matrix with 1 on (i,j)-position and 0 elsewhere.

2 Introduction

The problem of evaluating the length of the full matrix algebra in terms of its order was posed in 1984 by A. Paz in [4] and has not been solved yet. The case of 3×3 matrices was studied by Spencer ad Rivlin [5], [6] in connection with possible applications in mechanics. Some known upper bounds for the length of the matrix algebra are not linear.

Theorem 6. [4, Theorem 1, Remark 1] *Let* $\mathbb{F}$ *be an arbitrary field. Then* $l(M_n(\mathbb{F})) \leqslant \lceil (n^2+2)/3 \rceil$.

Theorem 7. [3, Corollary 3.2] *Let* $\mathbb{F}$ *be an arbitrary field. Then* $l(M_n(\mathbb{F})) < n\sqrt{2n^2/(n-1)+1/4}+n/2-2$.

In [4] Paz also suggested a linear bound:

Conjecture 1. [4, Conjecture 6.4] Let $\mathbb{F}$ be an arbitrary field. Then $l(M_n(\mathbb{F})) = 2n-2$.

Some generating sets of lengths not greater than $2n-2$, has been considered by Longstaff in [2]. An example of a generating set of length $2n-2$ in the case of an algebraically closed field of characteristic 0 is constructed in [1, Section 4].

In order to develop a general theory of length function we consider its algebraic properties.

Sharp lower and upper bounds for the lengths of direct sums of algebras were obtained in [7]. Namely, for further discussions we need the following result:

Theorem 8. [7, Theorem 2] *Let $\mathcal{A}$ and $\mathcal{B}$ be finite-dimensional algebras over a field $\mathbb{F}$ of lengths $l_{\mathcal{A}}$ and $l_{\mathcal{B}}$, correspondingly. Then the following inequalities hold:*

$$\max\{l_{\mathcal{A}}, l_{\mathcal{B}}\} \leqslant l(\mathcal{A} \oplus \mathcal{B}) \leqslant l_{\mathcal{A}} + l_{\mathcal{B}} + 1. \tag{1}$$

This paper is devoted to the connection between the length of an algebra and the lengths of its subalgebras. A positive answer has been obtained to the question whether the length of an algebra can be smaller than the length of its subalgebra.

Namely in this paper we construct several families of algebras, which contain subalgebras with the length exceeding the lengths of the algebras.

Then the natural question is to understand, how different the length of an algebra and the length of its subalgebra can be? In order to answer this question we consider the difference and the ratio of the lengths of an algebra and its subalgebra. Let $(\mathcal{A}, \mathcal{A}')$ be a pair, where $\mathcal{A}$ is an algebra over an arbitrary field $\mathbb{F}$ and $\mathcal{A}' \subseteq \mathcal{A}$ be its subalgebra. We show that there exist families of such pairs, such that $l(\mathcal{A}') \geqslant l(\mathcal{A})$ and for any natural number k the difference of the lengths is $l(\mathcal{A}') - l(\mathcal{A}) = k$ (Theorem 9).

Also we investigate the ratio between $l(\mathcal{A}')$ and $l(\mathcal{A})$. The question on the possible values of length ratio remains open yet in general. But in the Sections 3.1 and 3.2 we give some examples of length computation of two and three block upper triangular matrix algebras. Apart from their intrinsic interest, these examples give the following result: for any rational number $r \in [1,2]$ there exist such $\mathbb{F}$-algebra $\mathcal{A}$ and its subalgebra $\mathcal{A}'$, that $l(\mathcal{A}')/l(\mathcal{A}) = r$ (Corollary 2).

We note that there are still very few examples of algebras with exactly evaluated length. In this papers we give some new series of such examples: algebras $\mathcal{A}_{n,m}$, cf. Theorem 11, and $\mathcal{A}_{n_1,n_2,n_3}$, cf. Theorem 14.

In addition in Section 3.3 we give some examples of algebras $\mathcal{A}$ satisfying the inequality $l(\mathcal{A}) \geqslant l(\mathcal{A}')$ for any subalgebra $\mathcal{A}' \subseteq \mathcal{A}$.

3 On the lengths of algebra and its subalgebras

Notice that generally speaking the length function unlike the dimension function can increase when passing from an algebra to its subalgebras.

We first consider two types of transformations preserving the length of a generating set.

Proposition 1. *Let $\mathbb{F}$ be an arbitrary field and let $\mathcal{A}$ be a finite-dimensional associative $\mathbb{F}$-algebra. If $\mathcal{S} = \{a_1, \ldots, a_k\}$ is a generating set for $\mathcal{A}$ and $C = \{c_{ij}\} \in M_k(\mathbb{F})$ is non-singular, then the set of coordinates of the vector*

$$C\begin{pmatrix} a_1 \\ \vdots \\ a_k \end{pmatrix} = \begin{pmatrix} c_{11}a_1 + c_{12}a_2 + \ldots + c_{1k}a_k \\ \vdots \\ c_{k1}a_1 + c_{k2}a_2 + \ldots + c_{kk}a_k \end{pmatrix}, \tag{2}$$

i.e. the set

$$\mathcal{S}_c = \{c_{11}a_1 + c_{12}a_2 + \ldots + c_{1k}a_k, \ldots, c_{k1}a_1 + c_{k2}a_2 + \ldots + c_{kk}a_k\}$$

is also a generating set for $\mathcal{A}$ and $l(\mathcal{S}_c) = l(\mathcal{S})$.

Proof. Let us prove using the induction on n that $\mathcal{L}_n(\mathcal{S}) = \mathcal{L}_n(\mathcal{S}_c)$ holds for every n. Since any linear combination $\gamma_1 a_1 + \ldots + \gamma_k a_k \in \mathcal{L}_1(\mathcal{S})$, then $\mathcal{L}_1(\mathcal{S}_c) \subseteq \mathcal{L}_1(\mathcal{S})$. The non-singularity of C provides that $a_i \in \mathcal{L}_1(\mathcal{S}_c)$, $i = 1, \ldots, k$, i.e. $\mathcal{L}_1(\mathcal{S}) \subseteq \mathcal{L}_1(\mathcal{S}_c)$. Hence $\mathcal{L}_1(\mathcal{S}_c) = \mathcal{L}_1(\mathcal{S})$. Let us take $n > 1$ and suppose that for $n-1$ the equality holds. Then

$$\mathcal{L}_n(\mathcal{S}) = \langle \mathcal{L}_1(\mathcal{S})\mathcal{L}_{n-1}(\mathcal{S}) \rangle = \langle \mathcal{L}_1(\mathcal{S}_c)\mathcal{L}_{n-1}(\mathcal{S}_c) \rangle = \mathcal{L}_n(\mathcal{S}_c).$$

Proposition 2. *Let $\mathbb{F}$ be an arbitrary field and let $\mathcal{A}$ be a finite-dimensional associative unitary $\mathbb{F}$-algebra. Let $\mathcal{S} = \{a_1, \ldots, a_k\}$ be a generating set for $\mathcal{A}$ such that $1_{\mathcal{A}} \notin \langle a_1, \ldots, a_k \rangle$. Then $\mathcal{S}_1 = \{a_1 + \gamma_1 1_{\mathcal{A}}, \ldots, a_k + \gamma_k 1_{\mathcal{A}}\}$ is also a generating set for $\mathcal{A}$ and $l(\mathcal{S}_1) = l(\mathcal{S})$.*

Proof. The proof is analogous to that of Proposition 1, but simpler.

For further considerations we need the following class of matrices:

Definition 5. *Let $\mathbb{F}$ be an arbitrary field. A matrix $C \in M_n(\mathbb{F})$ is called* nonderogatory *provided* $\dim_{\mathbb{F}}(\langle E, C, C^2, \ldots, C^{n-1} \rangle) = n$.

Lemma 1. [8, Lemma 7.7] *Let $\mathbb{F}$ be an arbitrary field and let $\mathcal{A}$ be a commutative subalgebra of $M_n(\mathbb{F})$. If there exists a nonderogatory matrix $A \in \mathcal{A}$ then $\mathcal{A}$ is a subalgebra generated by A, and $l(\mathcal{A}) = n - 1$.*

Proposition 3. *Let $\mathbb{F}$ be an arbitrary field and let $\mathcal{A}_4 \subset T_4(\mathbb{F})$ be an algebra generated by matrices E, $E_{4,4}$, $E_{1,2}$, $E_{1,3}$ and $E_{2,3}$. Then $l(\mathcal{A}_4) = 2$.*

Proof. The dimension of any subalgebra of $M_4(\mathbb{F})$ generated by a single matrix does not exceed 4, but $\dim_{\mathbb{F}} \mathcal{A}_4 = 5$. Hence for any generating set $\mathcal{S} = \{A_1, \ldots, A_k\}$ for $\mathcal{A}_4$ it holds that $k \geqslant 2$ and if $k = 2$, then $E \notin \langle A_1, A_2 \rangle$. If the generating set $\mathcal{S}$ contains 3 matrices A_1, A_2, A_3 such that $E \notin \langle A_1, A_2, A_3 \rangle$, then $\dim_{\mathbb{F}} \mathcal{L}_1(\mathcal{S}) \geqslant 4$ and in this case $\dim_{\mathbb{F}} \mathcal{L}_2(\mathcal{S}) = 5$, that is $l(\mathcal{S}) \leqslant 2$. Let us consider the case when $\mathcal{S} = \{A, B\}$, $E \notin \langle A, B \rangle$. It follows from Proposition 2 that matrices A and B can be taken in the following form

$$A = \begin{pmatrix} 0 & a_{12} & a_{13} & 0 \\ 0 & 0 & a_{23} & 0 \\ 0 & 0 & 0 & 0 \\ 0 & 0 & 0 & a_{44} \end{pmatrix}, \quad B = \begin{pmatrix} 0 & b_{12} & b_{13} & 0 \\ 0 & 0 & b_{23} & 0 \\ 0 & 0 & 0 & 0 \\ 0 & 0 & 0 & b_{44} \end{pmatrix}.$$

Since $\mathcal{S}$ is a generating set, then $a_{44} \neq 0$ or $b_{44} \neq 0$. Without loss of generality we will assume that $a_{44} \neq 0$. Then by Proposition 1 we can take $b_{44} = 0$. Then

$$A^2 = a_{12}a_{23}E_{1,3} + a_{44}^2 E_{4,4}, \; AB = a_{12}b_{23}E_{1,3}, \; BA = a_{23}b_{12}E_{1,3},$$
$$B^2 = b_{12}b_{23}E_{1,3}, \; A^3 = a_{44}^3 E_{4,4}.$$

Other products in A and B of length greater than or equal to 3 are equal to zero. Hence, we obtain that for a generating set $\mathcal{S}$ the vectors (a_{12}, a_{23}) and (b_{12}, b_{23}) are always linearly independent. But in this case $AB \neq 0$ or $BA \neq 0$, that is $E_{1,3} \in \mathcal{L}_2(\mathcal{S})$. Hence $E_{4,4} = a_{44}^{-2}(A^2 - a_{12}a_{23}E_{1,3}) \in \mathcal{L}_2(\mathcal{S})$, $E_{1,2}, E_{1,3} \in \langle A, B, E_{1,3}, E_{4,4} \rangle \subseteq \mathcal{L}_2(\mathcal{S})$. Consequently, $\mathcal{L}_2(\mathcal{S}) = \mathcal{A}_4$ and $l(\mathcal{S}) = 2$. That is $l(\mathcal{A}_4) = 2$.

Example 1. Let $\mathbb{F}$ be an arbitrary field and let $\mathcal{A}_4 \subset T_4(\mathbb{F})$, generated by matrices E, $E_{4,4}$, $E_{1,2}$, $E_{1,3}$ and $E_{2,3}$. There exists a subalgebra $\mathcal{A}'$ of $\mathcal{A}_4$, generated by a nonderogatory matrix $A = E_{1,2} + E_{2,3} + E_{4,4}$, $l(\mathcal{A}') = 3 > 2 = l(\mathcal{A}_4)$.

Corollary 1. *For all $n \geqslant 4$ and for any field $\mathbb{F}$ with the number of elements greater than $n-4$ there exist such subalgebras $\mathcal{A}'_n \subset \mathcal{A}_n \subset M_n(\mathbb{F})$ that $l(\mathcal{A}'_n) = n - 1 > l(\mathcal{A}_n)$.*

Proof. Let $f_4, \ldots, f_n \in \mathbb{F}$ be distinct nonzero elements. Consider the following subalgebras:

$$\mathcal{A}_n = \langle \{E_{1,1} + E_{2,2} + E_{3,3}, E_{1,2}, E_{1,3}, E_{2,3}, E_{4,4}, \ldots, E_{n,n}\} \rangle,$$
$$\mathcal{A}'_n = \langle E_{1,2} + E_{1,3} + \sum_{i=4}^{n} f_i E_{i,i} \rangle.$$

Then $l(\mathcal{A}'_n) = n-1$, since the matrix $E_{1,2} + E_{1,3} + \sum_{i=4}^{n} f_i E_{i,i}$ is nonderogatory. It follows from [8, Theorem 4.5] that $l(D_{n-4}(\mathbb{F})) = n-5$. Consequently, $l(\mathcal{A}_n) = l(\mathcal{A}_4 \oplus D_{n-4}(\mathbb{F})) \leqslant 2 + (n-5) + 1 = n-2$ by consequent application of Example 1 and Theorem 8.

Thus Proposition 3 and Example 1 provide a positive answer to the question whether the length of an algebra can be smaller than the length of its subalgebra. Consequently, the next natural question is: *what values can be taken on by the difference and the ratio of the lengths of an algebra and of its subalgebra.*

Let $(\mathcal{A}, \mathcal{A}')$ be a pair, where $\mathcal{A}$ is an $\mathbb{F}$-algebra over and arbitrary field $\mathbb{F}$ and $\mathcal{A}' \subseteq \mathcal{A}$ be its subalgebra. The next theorem shows that there exist families of such pairs, such that $l(\mathcal{A}') \geqslant l(\mathcal{A})$ and the difference can be arbitrary large, and thus answers the first question. The second question is considered in the next two sections.

Theorem 9. *For any natural number* k *there exist a number* n *and such algebras* $\mathcal{A}' \subset \mathcal{A} \subset M_n(\mathbb{F})$, *that* $l(\mathcal{A}') - l(\mathcal{A}) = k$.

Proof. Example 2 and Proposition 4 below give an explicit construction of a pair $(\mathcal{A}, \mathcal{A}')$ of $\mathbb{F}$-algebras such that $\mathcal{A}' \subseteq \mathcal{A}$ and $l(\mathcal{A}') - l(\mathcal{A}) = k$. This construction is based on Example 1.

Example 2. Let $\mathbb{F}$ be a sufficiently large field, let k be a fixed positive number, $n = 4k$. Let $\mathcal{A} = \underbrace{\mathcal{A}_4 \oplus \ldots \oplus \mathcal{A}_4}_{k \text{ times}} \subset M_n(\mathbb{F})$, $A_i = E_{4i-3,4i-3} + E_{4i-3,4i-2} + E_{4i-2,4i-2} + E_{4i-2,4i-1} + E_{4i-1,4i-1}$, $i = 1, \ldots, k$, let us assign

$$\mathcal{A}' = \langle \sum_{i=1}^{k} (a_i A_i + b_i E_{4i,4i}) \rangle \subset \mathcal{A},$$

here a_i, b_i, $i = 1, \ldots, k$, are distinct nonzero elements from $\mathbb{F}$.

$l(\mathcal{A}) = 3k - 1$ as shown below, while $l(\mathcal{A}') = n - 1 = 4k - 1$ by Lemma 1.

Proposition 4. *Let* k *be a fixed natural number, let* $\mathbb{F}$ *be a field with* $|\mathbb{F}| \geqslant 2k + 1$. *Let* $\mathcal{A} = \underbrace{\mathcal{A}_4 \oplus \ldots \oplus \mathcal{A}_4}_{k \text{ times}} \subset M_n(\mathbb{F})$. *Then* $l(\mathcal{A}) = 3k - 1$.

Proof. It follows from Theorem 8 and Example 1 that $l(\mathcal{A}) \leqslant 2k + k - 1 = 3k - 1$. Consider a generating set

$$\mathcal{S}_{\mathcal{A}} = \{A = \sum_{i=1}^{k} (\alpha_i (A_i - E_{4i-2,4i-1}) + \beta_i E_{4i,4i}),\ E_{4j-2,4j-1},\ j = 1, \ldots,\ k\},$$

where α_i, β_i, $i = 1, \ldots, k$, are distinct nonzero elements from $\mathbb{F}$, $A_i = E_{4i-3,4i-3} + E_{4i-3,4i-2} + E_{4i-2,4i-2} + E_{4i-2,4i-1} + E_{4i-1,4i-1}$, $i = 1, \ldots, k$. Since $AE_{4j-2,4j-1} = \alpha_j (E_{4j-3,4j-1} + E_{4j-2,4j-1})$, $E_{4j-2,4j-1} A = \alpha_j E_{4j-2,4j-1}$, and the degree of the minimal polynomial of A is $3k$, then $l(\mathcal{S}_{\mathcal{A}}) = 3k - 1 = l(\mathcal{A})$.

3.1 Two block subalgebras in upper triangular matrix algebra

We note that in Example 2 the value $m = l(\mathcal{A}') - l(\mathcal{A})$ is an arbitrary number, however the ratio $r = (l(\mathcal{A}')+1) : (l(\mathcal{A})+1) = 4:3$ is a constant. The main aim of this and the next sections is to show that for any rational number $r \in [1,2]$ there exist $\mathbb{F}$-algebra $\mathcal{A}$ and its subalgebra $\mathcal{A}'$, such that $l(\mathcal{A}')/l(\mathcal{A}) = r$. In this section we consider the following 2-parametric family of algebras $\mathcal{A}_{n,m}$,

$$\mathcal{A}_{n,m} = \left\langle E,\ \sum_{i=1}^{n} E_{ii},\ E_{i,j},\ \middle|\ \begin{array}{l} 1 \leqslant i < j \leqslant n, \\ \text{or} \\ n+1 \leqslant i < j \leqslant n+m \end{array} \right\rangle \subset T_{m+n}(\mathbb{F}),$$

where $n \geqslant m$ are natural numbers, over an arbitrary field $\mathbb{F}$. We compute their lengths explicitly and found the subalgebra $\mathcal{A}'_{n,m}$ with $l(\mathcal{A}'_{n,m}) > l(\mathcal{A}_{n,m})$ in each algebra of this family, then choosing appropriate values of parameters n and m we obtain the required behavior of the ratio $l(\mathcal{A}'_{n,m})/l(\mathcal{A}_{n,m})$, see Corollary 2.

Remark 1. The aforementioned constructions generalize Example 1, namely, we obtain a series of algebras $\mathcal{A}(n) = \mathcal{A}_{n,m}$ and their subalgebras $\mathcal{A}'(n)$ with the fixed length difference m, for which the length ratio $r = r(n)$ is a non-constant linear-fractional function.

Remark 2. Algebra $\mathcal{A}_4$ described in Example 1 coincides with $\mathcal{A}_{3,1}$.

Notation 10. Any $A \in \mathcal{A}_{n,m}$ is of the following form $A = \begin{pmatrix} A' & 0 \\ 0 & A'' \end{pmatrix}$, where $A' \in T_n(\mathbb{F})$, $A'' \in T_m(\mathbb{F})$. From now on we will use the following notation $A = A' \oplus A''$.

In the following two lemmas we mark special elements in generating sets which are significant for computation of the length of $\mathcal{A}_{n,m}$.

Lemma 2. *Let $n \in \mathbb{N}$, $n \geqslant 3$ and let $\mathcal{S}$ be a generating set for $\mathcal{A}_{n,m}$. Then there exists a generating set $\widetilde{\mathcal{S}}$ for $\mathcal{A}_{n,m}$ such that the following conditions hold:*

1. $\dim \mathcal{L}_1(\widetilde{\mathcal{S}}) = |\widetilde{\mathcal{S}}| + 1$;

2. there exist a matrix $A_0 = A_0' \oplus A_0'' \in \widetilde{\mathcal{S}}$ *such that* $A_0' = \sum\limits_{1 \leqslant i<j \leqslant n} a_{i,j} E_{i,j}$ *and* $A_0'' = \sum\limits_{1 \leqslant i<j \leqslant m} a_{i+n,j+n} E_{i+n,j+n} + E$;

3. for all $S \in \widetilde{\mathcal{S}}$, $S \neq A_0$, *it holds that* $(S)_{i,i} = 0$, $i = 1, \ldots, n+m$;

4. there exist $B_1, \ldots, B_{n-1} \in \widetilde{\mathcal{S}}$ *such that either*

(i) for all $r = 1, \ldots, n-1$

$$B_r' = E_{r,r+1} + \sum_{i=1}^{n} \sum_{j=i+2}^{n} b_{i,j;r} E_{i,j},\ B_r'' \in N_m(\mathbb{F}),$$

or

(ii) there exists $k \in \{1, \dots, n-1\}$ *such that*

$$B'_r = E_{r,r+1} + b_r E_{k,k+1} + \sum_{i=1}^{n} \sum_{j=i+2}^{n} b_{i,j;r} E_{i,j},$$
$$B''_r \in N_m(\mathbb{F}), \quad r = 1, \dots, n-1, r \neq k,$$
$$B'_k = A'_0 = a_k E_{k,k+1} + \sum_{i=1}^{n} \sum_{j=i+2}^{n} \hat{a}_{i,j} E_{i,j}, \quad a_k \neq 0,$$
$$B''_k = A''_0 = \sum_{1 \leqslant i < j \leqslant m} \hat{a}_{i+n,j+n} E_{i+n,j+n} + E;$$

5. $l(\widetilde{S}) = l(S)$.

Proof. Let us consequently transform the set S into a generating set satisfying the conditions 1–4.

1. This condition is equivalent to the fact that all elements from S are linearly independent and $E \notin \langle S \rangle$. Otherwise we remove all redundant elements from S. Notice that it does not change the length of S.

2. In order to prove the existence of A_0 it is sufficient to show that S contains a matrix with two distinct eigenvalues. But if all matrices in S had only one eigenvalue, then S would not be a generating set for $\mathcal{A}_{n,m}$.

3. Proposition 2 allows us to transform the given generating set into a generating set $S' = \{S - (S)_{1,1} E, \ S \in S\}$ preserving its length. Then by Proposition 1 the transformation of S' into a generating set $S'' = \{A_0, \ S - (S)_{n+1,n+1} A_0, \ S \in S', \ S \neq A_0\}$ also does not change its length. For the sake of simplicity of the subsequent text let us redenote S'' by S.

4. Since $E_{i,i+1} \in \mathcal{A}_{n,m}$, but for any $t \geqslant 2$ and $S \in S^t \setminus S$ the coefficient $(S)_{i,i+1} = 0$, $i = 1, \dots, n-1$, then there exist $B_1, \dots, B_{n-1} \in S$ such that vectors

$$u_i = ((B_i)_{1,2}, (B_i)_{2,3}, \dots, (B_i)_{n-1,n}),$$

$i = 1, \dots, n-1$ are linearly independent.

Let us next do the following transformation $\mathcal{F}$ of the set S (by Proposition 1 $\mathcal{F}$ preserves the length of S), which is identical for all elements $S \in S$, $S \neq B_i, i = 1, \dots, n-1$, i.e. $\mathcal{F}(S) = S$, and its action on the set of matrices B_j, $j = 1, \dots, n-1$ depends on belonging of A_0 to this set as follows:

(i) Assume that $A_0 \notin \{B_1, \dots, B_{n-1}\}$. There exists a non-singular linear transformation $F = \{f_{i,j}\} \in M_{n-1}(\mathbb{F})$ that maps the set $\{u_i\}$ into the set

$$\{e_1 = (1, 0, \dots, 0), e_2 = (0, 1, \dots, 0), \dots, e_{n-1} = (0, 0, \dots, 1)\} \subset \mathbb{F}^{n-1},$$

i.e. $e_i = \sum_{j=1}^{n-1} f_{i,j} u_j$. Then let us assign $\mathcal{F}(B_r) = \sum_{j=1}^{n-1} f_{r,j} B_j$.

Then $\mathcal{F}(B_r)' = E_{r,r+1} + \sum_{i=1}^{n} \sum_{j=i+2}^{n} b_{i,j;r} E_{i,j}$.

(ii) Assume that $A_0 \in \{B_1, \dots, B_{n-1}\}$, i.e. $A_0 = B_p$ for some $p \in \{1, \dots, n-1\}$. Since any matrix in $M_{n-1,n-2}(\mathbb{F})$ of rank $n-2$ contains a non-singular submatrix of order $n-2$, then there exists a number $k \in \{1, \dots, n-1\}$ such that the vectors

$$v_i = ((B_i)_{1,2}, \dots, (B_i)_{k-1,k}, (B_i)_{k+1,k+2}, \dots, (B_i)_{n-1,n}),$$

$i = 1, \dots, n-1$, $i \neq p$ are linearly independent. Since the matrices B_j were numbered arbitrarily, we would assume that $p = k$. There exists a non-singular linear transformation $G = \{g_{i,j}\} \in M_{n-2}(\mathbb{F})$ that maps the set $\{v_i\}$ into the set

$$\{e_1 = (1, 0, \dots, 0), e_2 = (0, 1, \dots, 0), \dots, e_{n-2} = (0, 0, \dots, 1)\} \subset \mathbb{F}^{n-2},$$

i.e. $e_i = \sum_{j=1}^{k-1} g_{i,j} v_j + \sum_{j=k}^{n-2} g_{i,j} v_{j+1}$. Then let us assign

$$\mathcal{F}(B_r) = \sum_{j=1}^{k-1} g_{r,j} B_j + \sum_{j=k}^{n-2} g_{r,j} B_{j+1}, r \neq k, \ \mathcal{F}(A_0) = A_0 - \sum_{\substack{i=1 \\ i \neq k}}^{n-1} (A_0)_{i,i+1} \mathcal{F}(B_i).$$

Then

$$\mathcal{F}(B_r)' = E_{r,r+1} + b_r E_{k,k+1} + \sum_{i=1}^{n} \sum_{j=i+2}^{n} b_{i,j;r} E_{i,j} \text{ if } r \neq k,$$

$$\mathcal{F}(A_0)' = a_k E_{k,k+1} + \sum_{i=1}^{n} \sum_{j=i+2}^{n} \hat{a}_{i,j} E_{i,j}, \quad a_k \neq 0,$$

$$\mathcal{F}(A_0)'' = \sum_{1 \leqslant i < j \leqslant m} \hat{a}_{i+n,j+n} E_{i+n,j+n} + E.$$

For the sake of simplicity of the subsequent text let us redenote $\mathcal{F}(A_0)$ and $\mathcal{F}(B_r)$ by A_0 and B_r, correspondingly. Then the generating set $\widetilde{\mathcal{S}} = \mathcal{F}(\mathcal{S})$ is of required type.

5. The transformations applied to the set $\mathcal{S}$ in paragraphs 1–4 did not change its length, consequently, the length of the new generating set is equal to the length of the initial one.

Lemma 3. *Let $n \geqslant 3$, $m \geqslant 2$ and let $\mathcal{S}$ be a generating set for $\mathcal{A}_{n,m}$, satisfying the conditions 1–4 of Lemma 2. Then one of the following conditions holds:*

(i) there exist such $C, C_1, \dots, C_{m-1} \in \langle \mathcal{S} \setminus A_0 \rangle$ that

$$C_r'' = E_{r+n,r+n+1} + \sum_{i=1}^{m} \sum_{j=i+2}^{m} c_{i+n,j+n;r} E_{i+n,j+n}, \ r = 1, \dots, m-1,$$

and if $\tilde{A}_0 = E - A_0 + C$ then $\tilde{A}_0'' = \sum_{i=1}^{m} \sum_{j=i+2}^{m} \tilde{a}_{i+n,j+n} E_{i+n,j+n}$,

or

(ii) there exist such $C, C_r \in \langle \mathcal{S} \setminus A_0 \rangle$, $r = 1, \dots, m-1$, $r \neq s$ that

$$C_r'' = E_{r+n,r+n+1} + c_r E_{s+n,s+n+1} + \sum_{i=1}^{m} \sum_{j=i+2}^{m} c_{i+n,j+n;r} E_{i+n,j+n},$$

and if $\tilde{A}_0 = E - A_0 + C$ *then*

$$\tilde{A}_0'' = \tilde{a}_s E_{s+n,s+n+1} + \sum_{i=1}^{m} \sum_{j=i+2}^{m} \tilde{a}_{i+n,j+n} E_{i+n,j+n}, \quad \tilde{a}_s \neq 0,$$

where $s \in \{1, \ldots, m-1\}$ *(in this case let us assign* $C_s = \tilde{A}_0$*).*

Proof. Since all $S_1, S_2 \in \mathcal{S}$ and all $i = n+1, \ldots, n+m-1$ satisfy $(S_1 S_2)_{i,i+1} = 0$ if $S_1 \neq A_0, S_2 \neq A_0$ and

$(S_1 A_0)_{i,i+1} = (S_1)_{i,i+1}, \quad (A_0 S_2)_{i,i+1} = (S_2)_{i,i+1}$ and $(A_0^2)_{i,i+1} = 2(A_0)_{i,i+1}$,

then there exist $\tilde{C}_1, \ldots, \tilde{C}_{m-1} \in \mathcal{S}$ such that the vectors

$$w_i = ((\tilde{C}_i)_{n+1,n+2}, (\tilde{C}_i)_{n+2,n+3}, \ldots, (\tilde{C}_i)_{n+m-1,n+m}),$$

$i = 1, \ldots, m-1$ are linearly independent.

Matrices C_i can be obtained from $\tilde{C}_i$, $i = 1, \ldots, m-1$, if we apply transformations similar to those in paragraph 4 of Lemma 2. Assign $C = \sum_{i=1}^{m-1} \hat{a}_{i+n,i+n+1} C_i$. That is $(C)_{i+n,i+n+1} = (A_0)_{i+n,i+n+1}$ for all $i = 1, \ldots, m-1$.

Further length computation of $\mathcal{A}_{n,m}$ is carried out separately for different values of n and m.

Lemma 4. *Let* $\mathbb{F}$ *be an arbitrary field. Then* $l(\mathcal{A}_{1,1}) = 1$ *and* $l(\mathcal{A}_{2,2}) = 3$.

Proof. **1.** Algebra $\mathcal{A}_{1,1} = D_2(\mathbb{F})$. Consequently, it follows from [8, Theorem 4.5] that $l(\mathcal{A}_{1,1}) = 1$.

2. Algebra $\mathcal{A}_{2,2}$ is generated by a nonderogatory matrix $E_{1,1} + E_{1,2} + E_{2,2} + E_{3,4}$. Consequently, by Lemma 1 we have $l(\mathcal{A}_{2,2}) = 3$.

Lemma 5. *Let* $\mathbb{F}$ *be an arbitrary field. Then* $l(\mathcal{A}_{2,1}) = 2$ *and* $l(\mathcal{A}_{3,2}) = 3$.

Proof. **1.** Algebra $\mathcal{A}_{2,1}$ is generated by a nonderogatory matrix $E_{1,1}+E_{1,2}+E_{2,2}$. Consequently, by Lemma 1 we have $l(\mathcal{A}_{2,1}) = 2$.

2. Let $\mathcal{S}$ be a generating set for $\mathcal{A}_{3,2}$. Without loss of generality we assume $\mathcal{S}$ to satisfy the conditions 1–4 of Lemma 2 and therefore one of the conditions of Lemma 3. Let us show that $\mathcal{L}_3(\mathcal{S}) = \mathcal{A}_{3,2}$. We have $B_1 B_2 (E - A_0) = a E_{1,3}$, $a \neq 0$, $B_1(E-A_0)^2 = b_{11} E_{1,2} + b_{12} E_{2,3} + b_{13} E_{1,3}$, $B_2(E-A_0)^2 = b_{21} E_{1,2} + b_{22} E_{2,3} + b_{23} E_{1,3}$, with linearly independent vectors (b_{11}, b_{12}) and (b_{21}, b_{22}). $C_1 A_0^2 = b E_{4,5} + c E_{1,3}$, $b \neq 0$. Then $E_{4,4} + E_{5,5} = (A_0 - (A_0)_{1,2} E_{1,2} - (A_0)_{1,3} E_{1,3} - (A_0)_{2,3} E_{2,3} - (A_0)_{4,5} E_{4,5}) \in \mathcal{L}_3(\mathcal{S})$.

Consider $\mathcal{S} = \{A = E_{1,2},\ B = E_{2,3} + E_{4,4} + E_{4,5} + E_{5,5},\ E\}$. It follows from the following equations $A^2 = 0$, $AB = E_{1,3}$, $BA = 0$, $B^2 = E_{4,4} + 2E_{4,5} + E_{5,5}$ and $B^3 - B^2 = E_{4,5}$ that $l(\mathcal{S}) = 3 = l(\mathcal{A}_{3,2})$.

Lemma 6. *Let $\mathbb{F}$ be an arbitrary field, let $n, m \in \mathbb{N}$ and $n - m \geqslant 2$. Then* $l(\mathcal{A}_{n,m}) = n - 1$.

Proof. Let us first prove the upper bound $l(\mathcal{A}_{n,m}) \leqslant n-1$. Consider a generating set $\mathcal{S}$ for $\mathcal{A}_{n,m}$. Without loss of generality we assume $\mathcal{S}$ satisfying the conditions 1–4 of Lemma 2.

1. We use induction on $p = n - (j - i)$ to prove that $E_{i,j} \in \mathcal{L}_{n-1}(\mathcal{S})$ for $1 \leqslant i < j \leqslant n$, $j - i \geqslant 2$.

If $p = 1$ then $B_1 B_2 \dots B_{n-1} = (a_k)^t E_{1,n} \in \mathcal{L}_{n-1}(\mathcal{S})$, $t \in \{0, 1\}$, since $B_1'' B_2'' \dots B_{n-1}'' = 0$ as a product of $n-1$ nilpotent matrices of order $m \leqslant n-2$ when $t = 0$, and as a product of $n-2$ nilpotent and one unitriangular matrices of order $m \leqslant n-2$ when $t = 1$.

Consider the following matrix products

$B_{j,\ j+n-p-1} = B_j B_{j+1} \dots B_{j+n-p-1}(E - A_0)^{p-1} \in \mathcal{L}_{n-1}(\mathcal{S}), j = 1, \dots, p.$

We have $B'_{j,\ j+n-p-1} = (a_k)^t E_{j,j+n-p} + \sum\limits_{h=1}^{n} \sum\limits_{i=h+n-p+1}^{n} d_{h,i;j,p} E_{h,i}$, $t \in \{0,1\}$, and $B''_{j,\ j+n-p-1} = 0$ as a product of $n-1$ nilpotent matrices of order $m \leqslant n-2$ when $t = 0$, and as a product of $n-2$ nilpotent and one unitriangular matrices of order $m \leqslant n-2$ when $t = 1$.

Applying the induction hypothesis we obtain that $E_{i,i+n-q-1} \in \mathcal{L}_{n-1}(\mathcal{S})$ for all $q = 1, \dots, p-1$, $i = 1, \dots, q$. Hence

$$B_{j,\ j+n-p-1} - (a_k)^t E_{j,j+n-p} \in \mathcal{L}_{n-1}(\mathcal{S}).$$

Since by definition it holds that $B_{j,\ j+n-p-1} \in \mathcal{L}_{n-1}(\mathcal{S})$, then

$$E_{j,\ j+n-p-1} = (a_k)^{-t}(B_{j,\ j+n-p-1} - (B_{j,\ j+n-p-1} - (a_k)^t E_{j,j+n-p})) \in \mathcal{L}_{n-1}(\mathcal{S}),$$

$j = 1, \dots, p.$

2. Let us now consider $B_{j,\ j} = B_j(E - A_0)^{n-2} \in \mathcal{L}_{n-1}(\mathcal{S})$, $j = 1, \dots, n-1$. It follows immediately that

$$(B_{j,\ j})_{r,r+1} = (B_j)_{r,r+1},\ j, r = 1, \dots, n-1,$$

that is

$$B'_{j,\ j} = E_{j,j+1} + \gamma_j E_{k,k+1} + \sum_{h=1}^{n} \sum_{i=h+2}^{n} d_{h,i;j} E_{h,i},\ j \neq k,$$

$$B'_{k,\ k} = a_k^t E_{k,k+1} + \sum_{h=1}^{n} \sum_{i=h+2}^{n} d_{h,i;k} E_{h,i}, t \in \{0,1\},$$

and $B''_{r,\ r} = 0$ as a product of $n-1$ nilpotent matrices of order $m \leqslant n-2$ when $r \neq k$, and as a product of $n-2$ nilpotent and one unitriangular matrices of order $m \leqslant n-2$ when $r = k$.

It follows from paragraph 1 that $\sum_{h=1}^{n}\sum_{i=h+2}^{n} d_{h,i;j}E_{h,i} \in \mathcal{L}_{n-1}(\mathcal{S})$, $j = 1,\ldots,n$. Therefore $E_{k,k+1} = a_k^{-t}(B_{k,k} - \sum_{h=1}^{n}\sum_{i=h+2}^{n} d_{h,i;k}E_{h,i}) \in \mathcal{L}_{n-1}(\mathcal{S})$. Then $E_{j,j+1} = B_{j,j} - \sum_{h=1}^{n}\sum_{i=h+2}^{n} d_{h,i;j}E_{h,i} - \gamma_j E_{k,k+1} \in \mathcal{L}_{n-1}(\mathcal{S})$.

Consequently, $E_{i,j} \in \mathcal{L}_{n-1}(\mathcal{S})$, $1 \leqslant i < j \leqslant n$. Hence for any $N \in N_n(\mathbb{F})$ it holds that $N \oplus 0 \in \mathcal{L}_{n-1}(\mathcal{S})$.

3. Let $S_1,\ldots,S_n \in \mathcal{S}$ and assume there exists some $S_i \neq A_0$. It follows from [7, Equation (1)] that there exists $V \in \mathcal{L}_{n-1}(\mathcal{S})$ such that $S_1\cdots S_n + V = S' \oplus 0$, $S' \in N_n(\mathbb{F})$, but it follows from paragraphs 1 and 2 that $S' \oplus 0 \in \mathcal{L}_{n-1}(\mathcal{S})$. Therefore $S_1\cdots S_n$ is reducible. By Caley–Hamilton Theorem it holds that $(A_0'')^{m+1} \in \langle A_0'', (A_0'')^2, \ldots, (A_0'')^m\rangle$. Consequently, there exists $V_A \in \mathcal{L}_{n-1}(\mathcal{S})$ such that $A_0^n + V_A = A' \oplus 0$, $A' \in N_n(\mathbb{F})$, but it follows from paragraphs 1 and 2 that $A' \oplus 0 \in \mathcal{L}_{n-1}(\mathcal{S})$. Therefore A_0^n is also reducible.

So any word of length n in elements of $\mathcal{S}$ is reducible, therefore $\mathcal{L}_n(\mathcal{S}) = \mathcal{L}_{n-1}(\mathcal{S})$ and $l(\mathcal{S}) \leqslant n-1$.

By Theorem 8 we obtain that $l(\mathcal{A}_{n,m}) \geqslant n-1$. Consequently, $l(\mathcal{A}_{n,m}) = n-1$.

Lemma 7. *Let $\mathbb{F}$ be an arbitrary field, $n \in \mathbb{N}$ and $n > 3$. Then $l(\mathcal{A}_{n,n-1}) = n-1$.*

Proof. Let us first prove the upper bound $l(\mathcal{A}_{n,n-1}) \leqslant n-1$. Let $\mathcal{S}$ be a generating set for $\mathcal{A}_{n,n-1}$. Without loss of generality we assume $\mathcal{S}$ to satisfy the conditions 1–4 of Lemma 2 and therefore one of the conditions of Lemma 3.

1. We use induction on $p = n-(j-i)$ to prove that $E_{i,j} \in \mathcal{L}_{n-1}(\mathcal{S})$ for $1 \leqslant i < j \leqslant n$, $j-i \geqslant 2$.

Consider the case when $p = 1$.

(i) Assume that there is no such number k that $A_0 = B_k$. Then we obtain $B_1B_2\cdots B_{n-1} = E_{1,n} \in \mathcal{L}_{n-1}(\mathcal{S})$, since $B_1''B_2''\cdots B_{n-1}'' = 0$ as a product of $n-1$ nilpotent matrices of order $n-1$. Also it holds that $C_1\cdots C_{n-2}A_0 = aE_{n+1,2n-1} + bE_{1,n}$, $a \neq 0$, that is $E_{n+1,2n-1} \in \mathcal{L}_{n-1}(\mathcal{S})$.

(ii) Assume now that there exists a number k such that $A_0 = B_k$. Then we obtain $B_1B_2\cdots B_{n-1} = a_kE_{1,n} + \alpha E_{n+1,2n-1}$. Assume that $\alpha = 0$. It follows from the equalities $(C_1\cdots C_{n-2})'' = aE_{1,n-1}$, $a \neq 0$ and $(C_1\cdots C_{n-2})' = \beta_1E_{1,n-2} + \beta_2E_{1,n-1} + \beta_3E_{1,n} + \beta_4E_{2,n-1} + \beta_5E_{2,n} + \beta_6E_{3,n} \in N_n(\mathbb{F})$ for $n > 3$ that if $k = n-1$ then $A_0C_1\cdots C_{n-2} = aE_{n+1,2n-1}$, and if $k \neq n-1$ then $C_1\cdots C_{n-2}A_0 = aE_{n+1,2n-1}$, consequently, $E_{1,n}$, $E_{n+1,2n-1} \in \mathcal{L}_{n-1}(\mathcal{S})$. Assume now that $\alpha \neq 0$. Therefore $(B_1\cdots B_{k-1}B_{k+1}\cdots B_{n-1})'' = \alpha E_{1,n-1}$ and $(B_1\cdots B_{k-1}B_{k+1}\cdots B_{n-1})' = \beta_1E_{1,n-1} + \beta_2E_{1,n} + \beta_3E_{2,n}$. Since $n > 3$, then $k \neq 1$ or $k \neq n-1$. If $k \neq 1$ we obtain that $A_0B_1\cdots B_{k-1}B_{k+1}\cdots B_{n-1} =$

$\alpha E_{n+1,2n-1}$, and if $k \neq n-1$ we obtain that $B_1 \cdots B_{k-1} B_{k+1} \cdots B_{n-1} A_0 = \alpha E_{n+1,2n-1}$, consequently, $E_{1,n}$, $E_{n+1,2n-1} \in \mathcal{L}_{n-1}(\mathcal{S})$.

Therefore in all cases it holds that $E_{1,n}, E_{n+1,2n-1} \in \mathcal{L}_{n-1}(\mathcal{S})$.

Consider matrices $B_{j,j+n-p-1} \in \mathcal{L}_{n-1}(\mathcal{S}), j = 1, \ldots, p$ defined in Lemma 6. $B''_{j,j+n-p-1} = b(j,p)E_{n+1,2n-1}$, $b(j,p) \in \mathbb{F}$, as a product of $n-1$ nilpotent matrices of order $n-1$ when $t = 0$, and as a product of $n-2$ nilpotent and one unitriangular matrices of order $n-1$ when $t = 1$.

Hence using induction hypothesis and arguments similar to those of paragraph 1 of Lemma 6 we obtain that $E_{j,\ j+n-p-1} \in \mathcal{L}_{n-1}(\mathcal{S})$, $j = 1, \ldots, p$.

2. Consider $B_{j,\ j} \in \mathcal{L}_{n-1}(\mathcal{S})$, $j = 1, \ldots, n-1$ defined in Lemma 6. It follows immediately that $B''_{r,\ r} = b(r)E_{n+1,2n-1}$, $b(r) \in \mathbb{F}$, as a product of $n-1$ nilpotent matrices of order $n-1$ when $r \neq k$, and as a product of $n-2$ nilpotent and one unitriangular matrices of order $n-1$ when $r = k$.

Hence using arguments similar to those of paragraph 2 of Lemma 6 we obtain that $E_{j,j+1} \in \mathcal{L}_{n-1}(\mathcal{S})$.

Consequently, $E_{i,j} \in \mathcal{L}_{n-1}(\mathcal{S})$, $1 \leqslant i < j \leqslant n$. Hence for any $N \in N_n(\mathbb{F})$ it holds that $N \oplus 0 \in \mathcal{L}_{n-1}(\mathcal{S})$. Therefore, as it was shown in paragraph 3 of Lemma 6 any word of length n in elements of $\mathcal{S}$ is reducible, thus $\mathcal{L}_n(\mathcal{S}) = \mathcal{L}_{n-1}(\mathcal{S})$ and $l(\mathcal{S}) \leqslant n-1$. Then $l(\mathcal{A}_{n,n-1}) \leqslant n-1$.

By Theorem 8 we obtain that $l(\mathcal{A}_{n,n-1}) \geqslant n-1$. Consequently, $l(\mathcal{A}_{n,n-1}) = n-1$.

Lemma 8. *Let $\mathbb{F}$ be a field, $n \in \mathbb{N}$ and $n > 2$. Then $l(\mathcal{A}_{n,n}) = n$.*

Proof. **I.** Let us first prove the upper bound $l(\mathcal{A}_{n,n}) \leqslant n$. Let $\mathcal{S}$ be a generating set for $\mathcal{A}_{n,n}$. Without loss of generality we assume $\mathcal{S}$ to satisfy the conditions 1–4 of Lemma 2 and therefore one of the conditions of Lemma 3.

1. We use induction on $p = n-(j-i)$ to prove that $E_{i,j}$, $E_{i+n,j+n} \in \mathcal{L}_{n-1}(\mathcal{S})$ for $1 \leqslant i < j \leqslant n$, $j - i \geqslant 2$.

Consider the case when $p = 1$. Assume that there does not exist such number k that $A_0 = B_k$. Then we obtain $B_1 B_2 \cdots B_{n-1}(E - A_0) = E_{1,n} \in \mathcal{L}_n(\mathcal{S})$, since $B''_1 B''_2 \cdots B''_{n-1}(E - A_0)'' = 0$ as a product of n nilpotent matrices of order n. Assume now that there exist a number k such that $A_0 = B_k$. Then we obtain $B_1 B_2 \cdots B_{n-1} = a_k E_{1,n} + \alpha_1 E_{n+1,2n-1} + \alpha_2 E_{n+1,2n} + \alpha_3 E_{n+2,2n}$. Notice that since $n = m > 2$, then if condition (ii) of Lemma 3 holds, then the number s introduced there satisfies one of the inequalities $s \neq 1$ or $s \neq n-1$. And if condition (i) of Lemma 3 holds, both inequalities hold true. If $s \neq 1$ then $\tilde{A}_0 B_1 B_2 \cdots B_{n-1} = E_{1,n}$, and if $s \neq n-1$ then $B_1 B_2 \cdots B_{n-1} \tilde{A}_0 = E_{1,n}$, that is $E_{1,n} \in \mathcal{L}_n(\mathcal{S})$. Also it holds that $C_1 \cdots C_{n-1} A_0 = aE_{1,n} + bE_{n+1,2n} \in \mathcal{L}_n(\mathcal{S})$, $b \neq 0$, therefore, $E_{n+1,2n} = b^{-1}(C_1 \cdots C_{n-1} A_0 - aE_{1,n}) \in \mathcal{L}_n(\mathcal{S})$.

Consider the following matrix products

$$B_{j,j+n-p-1} = B_j B_{j+1} \cdots B_{j+n-p-1}(E - A_0)^p \in \mathcal{L}_n(\mathcal{S}), j = 1, \ldots, p.$$

$$B'_{j,\ j+n-p-1} = a_k^t E_{j,j+n-p} + \sum_{h=1}^{n} \sum_{i=h+n-p+1}^{n} b_{h,i;j,p} E_{h,i},\ t \in \{0,1\},$$

and $B''_{j,j+n-p-1} = b(j,p)E_{n+1,2n}$, $b(j,p) \in \mathbb{F}$, as a product of n nilpotent matrices of order n when $t = 0$, and as a product of $n-1$ nilpotent and one unitriangular matrices of order n when $t = 1$.

Consider $C_{j,j+n-p-1} = C_j C_{j+1} \cdots C_{j+n-p-1} A_0^p \in \mathcal{L}_n(\mathcal{S}), j = 1,\ldots,p$.

$$C''_{j,\ j+n-p-1} = (\tilde{a}_s)^t E_{j+n,j+2n-p} + \sum_{h=1}^{n} \sum_{i=h+n-p+1}^{n} c_{h,i;j,p} E_{h+n,i+n},\ t \in \{0,1\},$$

and $C'_{j,\ j+n-p-1} = c(j,p)E_{1,n}$ as a product of n nilpotent matrices of order n when $t = 0$, and as a product of $n-1$ nilpotent and one unitriangular matrices of order n when $t = 1$.

Applying the induction hypothesis we obtain that

$$E_{i,i+n-q-1}, E_{i+n,i+2n-q-1} \in \mathcal{L}_n(\mathcal{S}) \text{ for all } q = 2,\ldots,p-1,\ i = 1,\ldots,q,$$

and $E_{1,n}$, $E_{n+1,2n} \in \mathcal{L}_n(\mathcal{S})$ as was shown above. Therefore,

$$B_{j,\ j+n-p-1} - (a_k)^t E_{j,j+n-p}, C_{j,\ j+n-p-1} - (a_s)^t E_{j+n,j+2n-p} \in \mathcal{L}_n(\mathcal{S}).$$

Since by definition it holds that $B_{j,\ j+n-p-1}, C_{j,\ j+n-p-1} \in \mathcal{L}_n(\mathcal{S})$ then

$$E_{j,\ j+n-p-1} = (a_k)^{-t}(B_{j,\ j+n-p-1} - (B_{j,\ j+n-p-1} - (a_k)^t E_{j,j+n-p})) \in \mathcal{L}_n(\mathcal{S}),$$

$E_{j+n,\ j+2n-p-1} = (a_s)^{-t}(C_{j,\ j+n-p-1} - (C_{j,\ j+n-p-1} - (a_s)^t E_{j+n,j+2n-p})) \in \mathcal{L}_n(\mathcal{S}),\ j = 1,\ldots,p.$

2. Consider next $B_{j,j} = B_j(E-A_0)^{n-1} \in \mathcal{L}_n(\mathcal{S})$ and $C_{j,j} = C_j A_0^{n-1} \in \mathcal{L}_n(\mathcal{S})$, $j = 1,\ldots,n-1$. It follows immediately that

$$B'_{j,j} = E_{j,j+1} + \beta_j E_{k,k+1} + \sum_{h=1}^{n} \sum_{i=h+2}^{n} b_{h,i;j,n-1} E_{h,i},\ j \neq k,$$

$$B'_{k,k} = (a_k)^t E_{k,k+1} + \sum_{h=1}^{n} \sum_{i=h+2}^{n} b_{h,i;k,n-1} E_{h,i},\ t \in \{0,1\},$$

$$C''_{j,j} = E_{j+n,j+n+1} + \gamma_j E_{s+n,s+n+1} + \sum_{h=1}^{n} \sum_{i=h+2}^{n} c_{h,i;j,n-1} E_{h+n,i+n},\ j \neq s,$$

$$C''_{s,s} = (\tilde{a}_s)^t E_{s+n,s+n+1} + \sum_{h=1}^{n} \sum_{i=h+2}^{n} c_{h,i;s,n-1} E_{h,i},\ t \in \{0,1\},$$

$B''_{r,r} = b(j)E_{1,n}$ as a product of n nilpotent matrices of order n when $t = 0$, and as a product of $n-1$ nilpotent and one unitriangular matrices of order n when

$t=1$, and $C'_{r,r}=c(r)E_{1,n}$ as a product of $n-1$ nilpotent and one unitriangular matrices of order n.

It follows from paragraph 1 that

$$B_{k,k}-(a_k)^t E_{k,k+1}, C_{s,s}-(a_s)^t E_{s+n,s+n+1} \in \mathcal{L}_n(\mathcal{S}).$$

Therefore $E_{k,k+1}$, $E_{s+n,s+n+1} \in \mathcal{L}_n(\mathcal{S})$. Then

$$E_{j,j+1}=B_{j,j}-\sum_{h=1}^{n}\sum_{i=h+2}^{n} b_{h,i;j,n-1}E_{h,i}-\beta_j E_{k,k+1}-b(j)E_{n+1,2n}\in \mathcal{L}_{n-1}(\mathcal{S}),$$

$$E_{j+n,j+n+1}=C_{j,j}-\sum_{h=1}^{n}\sum_{i=h+2}^{n} c_{h,i;j,n-1}E_{h,i}-\gamma_j E_{s+n,s+n+1}-c(j)E_{1,n}\in \mathcal{L}_{n-1}(\mathcal{S}).$$

Consequently, $E_{j,j+n-p}$, $E_{j+n,j+2n-p} \in \mathcal{L}_n(\mathcal{S})$, $j=1,\dots,p$.

3. Then it holds that

$$0\oplus E''=(A_0-a_kE_{k,k+1}-\sum_{h=1}^{n}\sum_{i=h+n-p+1}^{n}\hat{a}_{ij}E_{i,j}-\sum_{1\leqslant i<j\leqslant n}\hat{a}_{i+n\ j+n}E_{i+n,j+n}),$$

that is $0\oplus E \in \mathcal{L}_n(\mathcal{S})$.

Hence any generating set $\mathcal{S}$ satisfies $\mathcal{L}_n(\mathcal{S})=\mathcal{A}_{n,n}$, therefore $l(\mathcal{A}_{n,n})\leqslant n$.

II. Let us construct a generating set for $\mathcal{A}_{n,n}$ of length n. Let

$$\mathcal{S}_n=\{A_i=E_{i,i+1}+E_{n+i,n+i+1},\ i=1,\dots,n-1,\ E,\ E_n=\sum_{j=1}^{n}E_{j,j}\}.$$

Since $A_iA_j=0$ for $j\neq i+1$ and $E_nA_i=A_iE_n=E_{i,i+1}$, then $E_{1,n}\notin \mathcal{L}_{n-1}(\mathcal{S}_n)$, where

$$\mathcal{L}_{n-1}(\mathcal{S}_n)=\left\langle E,\ E_n,\ E_{i,j},\ E_{i+n,j+n},\ E_{1,n}+E_{n+1,2n}\ \middle|\ \begin{matrix}1\leqslant i<j\leqslant n,\\ j-i\leqslant n-2\end{matrix}\right\rangle,$$

but $E_{1,n}=A_1\cdots A_{n-1}E_n\in\mathcal{L}_n(\mathcal{S})$ and therefore, $\mathcal{L}_n(\mathcal{S})=\mathcal{A}_{n,n}$. Consequently, $l(\mathcal{A}_{n,n})=n$.

The combination of Lemmas 4–8 implies

Theorem 11. *Let $\mathbb{F}$ be an arbitrary field, let $n\geqslant m$ be natural numbers and let*

$$\mathcal{A}_{n,m}=\left\langle E,\ \sum_{i=1}^{n}E_{ii},\ E_{i,j},\ \middle|\ \begin{matrix}1\leqslant i<j\leqslant n,\\ \text{or}\\ n+1\leqslant i<j\leqslant n+m\end{matrix}\right\rangle\subset T_{m+n}(\mathbb{F}).$$

Then

$$l(\mathcal{A}_{n,m})=\begin{cases}n-1, & \text{for } n-m\geqslant 2,\\ n-1 & \text{for } n=m+1,\ n>3,\\ n+1 & \text{for } n=m=2,\\ n, & \text{for } n=m\neq 2,\\ n & \text{for } n=m+1,\ m=1,2.\end{cases}$$

The following Corollary shows in particular that the length ratio for a two block algebra and its subalgebra can take on any rational value in $[1,2]$.

Corollary 2. *Let $\mathbb{F}$ be an arbitrary field, let $n \geqslant m$ be fixed natural numbers. Let*

$$C_{n,m} = \sum_{i=1}^{n-1} E_{i,i+1} + \sum_{j=1}^{m-1} (E_{j+n,j+n} + E_{j+n,j+n+1}) + E_{n+m,n+m} \in \mathcal{A}_{n,m}$$

be a nonderogatory matrix, and let

$$\mathcal{A}'_{n,m} = \langle C^j_{n,m}, |\ 0 \leqslant j \leqslant n+m-1 \rangle \subseteq \mathcal{A}_{n,\ m}.$$

Then

1. $l(\mathcal{A}'_{n,m}) = n+m-1$;
2. *for* $n = m = 1,\ 2$ *or* $n = 2,\ m = 1$ $\mathcal{A}_{n,\ m} = \mathcal{A}'_{n,m}$;
3. $l(\mathcal{A}'_{n,\ m}) - l(\mathcal{A}_{n,\ m}) = \begin{cases} m, & \text{for } n-m \geqslant 2, \text{or } n = m+1,\ n > 3, \\ m-1, & \text{for } n = m \neq 2,\ \text{ or } n = 3,\ m = 2, \end{cases}$ *and*

$$\frac{l(\mathcal{A}'_{n,\ m})+1}{l(\mathcal{A}_{n,\ m})+1} = \begin{cases} 1+\dfrac{m}{n}, & \text{for } n-m \geqslant 2, \text{or } n = m+1,\ n > 3, \\ 1+\dfrac{m-1}{n+1}, & \text{for } n = m \neq 2,\ \text{ or } n = 3,\ m = 2. \end{cases}$$

3.2 Three block subalgebras in upper triangular matrix algebra

In this section we consider the 3-parametric family of algebras $\mathcal{A}_{n_1,n_2,n_3} \subset T_{n_1}(\mathbb{F}) \oplus T_{n_2}(\mathbb{F}) \oplus T_{n_3}(\mathbb{F})$,

$$\mathcal{A}_{n_1,n_2,n_3} = \left\langle E,\ \sum_{i=1}^{n_1} E_{i,i},\ \sum_{i=n_1+1}^{n_1+n_2} E_{i,i},\ E_{i,j},\ \middle|\ \begin{array}{l} 1 \leqslant i < j \leqslant n_1, \\ \text{or} \\ n_1+1 \leqslant i < j \leqslant n_1+n_2, \\ \text{or} \\ n_1+n_2+1 \leqslant i < j \leqslant n_1+n_2+n_3 \end{array} \right\rangle$$

over an arbitrary field $\mathbb{F}$, compute their lengths explicitly and found the subalgebras $\mathcal{A}'_{n_1,n_2,n_3} \subset \mathcal{A}_{n_1,n_2,n_3}$ with $l(\mathcal{A}'_{n_1,n_2,n_3}) > l(\mathcal{A}_{n_1,n_2,n_3})$, then choosing appropriate values of parameters n_1, n_2, n_3 we obtain arbitrary rational ratios $\dfrac{l(\mathcal{A}'_{n_1,n_2,n_3})}{l(\mathcal{A}_{n_1,n_2,n_3})} \in [1,2)$, see Corollary 3.

Notation 12. Any $A \in \mathcal{A}_{n_1,\ n_2,\ n_3}$ is of the following form

$$A = \begin{pmatrix} A' & 0 & 0 \\ 0 & A'' & 0 \\ 0 & 0 & A''' \end{pmatrix}, \text{where } A' \in T_{n_1}(\mathbb{F}),\ A'' \in T_{n_2}(\mathbb{F}),\ A''' \in T_{n_3}(\mathbb{F}).$$

From now on we will use the following notation $A = A' \oplus A'' \oplus A'''$.

In the following three lemmas we mark special elements in generating sets which are significant for computation of the length of $\mathcal{A}_{n_1,n_2,n_3}$.

Lemma 9. *Let $\mathcal{S}$ be a generating set for $\mathcal{A}_{n_1,n_2,n_3}$. Then there exists a generating set $\widetilde{\mathcal{S}}$ for $\mathcal{A}_{n_1,n_2,n_3}$ such that the following conditions hold:*
1. $\dim \mathcal{L}_1(\widetilde{\mathcal{S}}) = |\widetilde{\mathcal{S}}| + 1$;
2. any $S \in \widetilde{\mathcal{S}}$ *satisfies* $(S)_{ii} = 0,\ i = 1, \ldots, n_1$;
3. either

(i) there exist matrices $A_1 = (a_{i,j;1}),\ A_2 = (a_{i,j;2}) \in \widetilde{\mathcal{S}}$ *such that*

$$A_1'' = E + N_1,\ N_1 \in N_{n_2}(\mathbb{F}),\ A_1''' \in N_{n_3}(\mathbb{F}),$$
$$A_2'' \in N_{n_2}(\mathbb{F}),\ A_2''' = E + N_2,\ N_2 \in N_{n_3}(\mathbb{F}),$$

and all $S \in \widetilde{\mathcal{S}},\ S \neq A_1, A_2$, *satisfy* $(S)_{i,i} = 0,\ i = 1, \ldots, n_1 + n_2 + n_3$;

or

(ii) there exists a matrix $A_0 = (a_{i,j;0})$ *such that*

$$A_0'' = E + N,\ N \in N_{n_2}(\mathbb{F}),\ A_0''' = aE + M,\ M \in N_{n_3}(\mathbb{F}),\ a \notin \{0, 1\},$$

and all $S \in \mathcal{S},\ S \neq A_0$ *satisfy* $(S)_{i,i} = 0,\ i = 1, \ldots, n_1 + n_2 + n_3$;

4. $l(\widetilde{\mathcal{S}}) = l(\mathcal{S})$.

Proof. Let us consequently transform the set $\mathcal{S}$ into a generating set satisfying the conditions 1–3.

1. We use the same arguments as in point 1 of Lemma 2.

2. Proposition 2 allows us to transform the given generating set into a generating set $\mathcal{S}_1 = \{S - (S)_{1,1}E,\ S \in \mathcal{S}\}$ preserving its length.

3. (i) Assume there exist $C_1,\ C_2 \in \mathcal{S}_1$ such that vectors

$$c_1 = ((C_1)_{n_1+1,n_1+1},\ (C_1)_{n_1+n_2+1,n_1+n_2+1}),$$
$$c_2 = ((C_2)_{n_1+1,n_1+1},\ (C_2)_{n_1+n_2+1,n_1+n_2+1})$$

are linearly independent. Thus there exists a non-singular matrix $F = (f_{i,j}) \in M_2(\mathbb{F}_2)$ such that $(1,0) = f_{1,1}c_1 + f_{1,2}c_2$, $(0,1) = f_{2,1}c_1 + f_{2,2}c_2$. Let us assign $A_i = f_{i,1}C_1 + f_{i,2}C_2,\ i = 1, 2$. Then Proposition 1 allows us to transform the given generating set into a generating set

$$\mathcal{S}_2 = \{A_1, A_2, S |\ S \in \mathcal{S},\ S \neq C_1, C_2\}$$

preserving its length. And by Proposition 1 the transformation of $\mathcal{S}_2$ into a generating set $\mathcal{S}_3 = \{A_1, A_2, S - (S)_{n_1+1,n_1+1}A_1 - (S)_{n_1+n_2+1,n_1+n_2+1}A_2 |\ S \in \mathcal{S}_1,\ S \neq A_1, A_2\}$ also does not change its length. In this case we assign $\widetilde{\mathcal{S}} = \mathcal{S}_3$.

(ii) Otherwise there exists such a matrix A in $\mathcal{S}$ that vectors

$$((A)_{n_1+1\ n_1+1},\ (A)_{n_1+n_2+1\ n_1+n_2+1}),\ ((A^2)_{n_1+1\ n_1+1},\ (A^2)_{n_1+n_2+1\ n_1+n_2+1})$$

are linearly independent. Thus matrix A has two distinct non-zero eigenvalues. Then we can replace matrix A in s $\mathcal{S}_1$ with the matrix $A_0 = (A)^{-1}_{n_1+1,n_1+1}A$.

Then Proposition 1 allows us to transform the given generating set into a generating set $\mathcal{S}_2 = \{A_0,\ S - (S)_{n_1+1,n_1+1}A_0,\ S \in \mathcal{S},\ S \neq A_0\}$. Let us assign $\widetilde{\mathcal{S}} = \mathcal{S}_2$.

Lemma 10. *Let $\mathcal{S}$ be a generating set for $\mathcal{A}_{n_1,n_2,n_3}$ satisfying the conditions 1, 2 and 3(i) of Lemma 9. The there exist a generating set $\widetilde{\mathcal{S}}$ for $\mathcal{A}_{n_1,n_2,n_3}$ satisfying $l(\widetilde{\mathcal{S}}) = l(\mathcal{S})$, such matrices $B_1, \dots, B_{n_1-1} \in \widetilde{\mathcal{S}}$ and k_1, $k_2 \in \{1, \dots, n_1 - 1\}$ that one of the following conditions holds:*

1. $$B'_r = E_{r,r+1} + \sum_{i=1}^{n_1}\sum_{j=i+2}^{n_1} b_{i,j;r}E_{i,j},\ B''_r \in N_{n_2}(\mathbb{F}),$$
$$B'''_r \in N_{n_3}(\mathbb{F}),\ r = 1, \dots, n-1;$$

2. *there exists* $j \in \{1, 2\}$ *such that*
$$B'_r = E_{r,r+1} + b_{rj}E_{k_j,k_j+1} + \sum_{h=1}^{n_1}\sum_{i=h+2}^{n_1} b_{h,i;r}E_{h,i},$$
$$B''_r \in N_{n_2}(\mathbb{F}),\ B'''_r \in N_{n_3}(\mathbb{F}),\ r = 1, \dots, n_1 - 1,\ r \neq k_j,$$
$$A'_j = B'_{k_j} = a(k_j, j)E_{k_j,k_j+1} + \sum_{h=1}^{n_1}\sum_{i=h+2}^{n_1} a_{h,i;j}E_{h,i},\ a(k_j, j) \neq 0,$$
$$B''_{k_j} = A''_j,\ B'''_{k_j} = A'''_j;$$

3. $$B'_r = E_{r,r+1} + b_{r1}E_{k_1,k_1+1} + b_{r2}E_{k_2,k_2+1} + \sum_{h=1}^{n_1}\sum_{i=h+2}^{n_1} b_{h,i;r}E_{h,i},$$
$$B''_r \in N_{n_2}(\mathbb{F}),\ B'''_r \in N_{n_3}(\mathbb{F}),\ r = 1, \dots, n_1 - 1,\ r \neq k_1, k_2,$$
$$A'_j = B'_{k_j} = a(k_1, j)E_{k_1,k_1+1} + a(k_2, j)E_{k_2,k_2+1} + \sum_{h=1}^{n_1}\sum_{i=h+2}^{n_1} a_{h,i;j}E_{h,i},$$
$$a(k_j, j) \neq 0,\ a(k_1, 1)a(k_2, 2) - a(k_2, 1)a(k_1, 2) \neq 0,$$
$$B''_{k_j} = A''_j,\ B'''_{k_j} = A'''_j, j = 1, 2.$$

Proof. Since $E_{i,i+1} \in \mathcal{A}_{n_1,n_2,n_3}$, but for any $t \geqslant 2$ and $S \in \mathcal{S}^t \setminus \mathcal{S}$ the coefficient $(S)_{i,i+1} = 0$, $i = 1, \dots, n_1 - 1$, then there exist $B_1, \dots, B_{n_1-1} \in \mathcal{S}$ such that the vectors $((B_i)_{1,2}, (B_i)_{2,3}, \dots, (B_i)_{n_1-1,n_1})$, $i = 1, \dots, n_1 - 1$ are linearly independent.

Consider next the following transformation $\mathcal{F}$ of the set $\mathcal{S}$ (by Proposition 1 $\mathcal{F}$ preserves the length of $\mathcal{S}$), which is identical for all elements $S \in \mathcal{S}$, $S \neq B_i, i = 1, \dots, n_1 - 1$, i.e. $\mathcal{F}(S) = S$, and its action on the set of matrices B_j, $j = 1, \dots, n_1 - 1$ depends on belonging of A_1 and A_2 to this set as follows:

If $|\{B_1, \dots, B_{n_1-1}\} \bigcap \{A_1, A_2\}| \leqslant 1$, then $\mathcal{F}$ is constructed similarly to the transformation described in point 4 of Lemma 2.

Assume that both $A_1, A_2 \in \{B_1, \dots, B_{n_1-1}\}$, i.e. $A_1 = B_p$, $A_2 = B_q$ for some distinct $p, q \in \{1, \dots, n_1 - 1\}$. Since any matrix in $M_{n_1-1,n_1-3}(\mathbb{F})$ of rank $n_1 - 3$ contains a non-singular submatrix of order $n_1 - 3$, then there exist numbers k_1, $k_2 \in \{1, \dots, n_1 - 1\}$, $k_1 < k_2$ such that the vectors

$$v_i = ((B_i)_{1,2}, \dots, (B_i)_{k_1-1,k_1}, (B_i)_{k_1+1,k_1+2}, \dots,$$
$$(B_i)_{k_2-1,k_2}, (B_i)_{k_2+1,k_2+2}(B_i)_{n_1-1,n_1}),$$

$i = 1, \ldots, n_1 - 1$, $i \neq p$, q are linearly independent.

Since the matrices B_j were numbered arbitrarily, we would assume that $p = k_1$, $q = k_2$.

There exists a non-singular linear transformation $G = \{g_{i,j}\} \in M_{n_1-3}(\mathbb{F})$ that maps the set $\{v_i\}$ into the set

$$\{e_1 = (1, 0, \ldots, 0),\ e_2 = (0, 1, \ldots, 0), \ldots, e_{n_1-3} = (0, 0, \ldots, 1)\} \subset \mathbb{F}^{n_1-3}.$$

i.e. $e_i = \sum\limits_{j=1}^{n_1-3} g_{i,j} v_j$. Then let us assign

$$\mathcal{F}(B_r) = \sum_{j=1}^{k_1-1} g_{r,j} B_j + \sum_{j=k_1}^{k_2-1} g_{r,j} B_{j+1} + \sum_{j=k_2}^{n_1-1} g_{r,j} B_{j+2}, r \neq k_1, k_2,$$

$$\mathcal{F}(A_s) = A_s - \sum_{\substack{i=1 \\ i \neq k_1, k_2}}^{n_1-1} (A_s)_{i,i+1} \mathcal{F}(B_i),\ s = 1, 2.$$

For the sake of simplicity of the subsequent text let us redenote $\mathcal{F}(A_1)$, $\mathcal{F}(A_2)$ and $\mathcal{F}(B_r)$ by A_1, A_2 and B_r, correspondingly.

Lemma 11. *Let $\mathcal{S}$ be a generating set for $\mathcal{A}_{n_1,n_2,n_3}$ satisfying conditions 1, 2 and 3(ii) of Lemma 9. Then there exist a generating set $\widetilde{\mathcal{S}}$ for $\mathcal{A}_{n_1,n_2,n_3}$ satisfying $l(\widetilde{\mathcal{S}}) = l(\mathcal{S})$, such matrices $B_1, \ldots, B_{n_1-1} \in \widetilde{\mathcal{S}}$ and $k_0 \in \{1, \ldots, n_1 - 1\}$ that one of the following conditions holds:*

1.
$$B_r' = E_{r,r+1} + \sum_{i=1}^{n_1} \sum_{j=i+2}^{n_1} b_{i,j;r} E_{i,j},\ B_r'' \in N_{n_2}(\mathbb{F}),$$
$$B_r''' \in N_{n_3}(\mathbb{F}),\ r = 1, \ldots, n-1;$$

2.
$$B_r' = E_{r,r+1} + b_{r0} E_{k_0,k_0+1} + \sum_{h=1}^{n_1} \sum_{i=h+2}^{n_1} b_{h,i;r} E_{h,i},$$
$$B_r'' \in N_{n_2}(\mathbb{F}),\ B_r''' \in N_{n_3}(\mathbb{F}),\ r = 1, \ldots, n_1 - 1,\ r \neq k,$$
$$A_0' = B_{k_0}' = a(k_0, 0) E_{k_0,k_0+1} + \sum_{h=1}^{n_1} \sum_{i=h+2}^{n_1} a_{h,i;j} E_{h,i},\ a(k_0, 0) \neq 0,$$
$$B_{k_0}'' = A_0'',\ B_{k_0}''' = A_0'''.$$

Proof. The proof is analogous to the proof of point 4 of Lemma 2.

Theorem 13. *Let $\mathbb{F} = \mathbb{F}_2$, let n_1, n_2, $n_3 \in \mathbb{N}$, $n_1 \geqslant n_2 + 2$, $n_2 \geqslant n_3$, $(n_2, n_3) \neq (1, 1)$, $(2, 1)$, $(2, 2)$. Then $l(\mathcal{A}_{n_1, n_2, n_3}) = n_1 - 1$.*

Proof. Let us first prove the upper bound $l(\mathcal{A}_{n_1,n_2,n_3}) \leqslant n_1 - 1$. Let $\mathcal{S}$ be a generating set for $\mathcal{A}_{n_1,n_2,n_3}$. Without loss of generality we assume $\mathcal{S}$ to satisfy the conditions 1–2 of Lemma 9. Since by [8, Theorem 6.1] $l(D_3(\mathbb{F}_2) = 1$, then the only possibility for $\mathcal{S}$ is to satisfy condition 3(i) of Lemma 9, and consequently, we assume $\mathcal{S}$ to satisfy one of the conditions of Lemma 10.

1. We use induction on $p = n_1 - (j-i)$ to prove that $E_{i,j} \in \mathcal{L}_{n_1-1}(\mathcal{S})$ for $1 \leqslant i < j \leqslant n_1$, $j - i \geqslant 2$.

Notice that $B_1B_2\cdots B_{n_1-1} = E_{1,n_1} \in \mathcal{L}_{n_1-1}(\mathcal{S})$, since $(B_1B_2\cdots B_{n_1-1})'' = 0$ and $(B_1B_2\cdots B_{n_1-1})''' = 0$ as products of $n_1 - 1$ nilpotent matrices of order $n_{s+1} \leqslant n_1 - 2$, if $\mathcal{S}$ satisfies condition 1 of Lemma 10 and as products of $n_1 - 2$ nilpotent and one unitriangular matrices of order $n_{s+1} \leqslant n_1 - 2$, if $\mathcal{S}$ satisfies condition 2 or 3 of Lemma 10.

Consider the following matrix products

$$B_{j,j+n_1-p-1} = B_jB_{j+1}\cdots B_{j+n_1-p-1}(E - A_1 - A_2)^{p-1} \in \mathcal{L}_{n_1-1}(\mathcal{S}), j = 1,\ldots,p.$$

We have $B'_{j,j+n_1-p-1} = E_{j,j+n_1-p} + \sum\limits_{h=1}^{n_1}\sum\limits_{i=h+n_1-p+1}^{n_1} c_{h,i;j}E_{h,i}$, and $B''_{j,j+n_1-p-1} = 0$ and $B'''_{j,j+n_1-p-1} = 0$ as products of $n_1 - 1$ nilpotent matrices of order $n_{s+1} \leqslant n_1 - 2$, if $\mathcal{S}$ satisfies condition 1 of Lemma 10 or for k_s defined in points 2 and 3 of Lemma 10 it holds that $k_s \notin \{j,\ldots,j+n_1-p-1\}$, $s = 1,2$, and as products of $n_1 - 2$ nilpotent and one unitriangular matrices of order $n_{s+1} \leqslant n_1 - 2$ otherwise. Applying the induction hypothesis we obtain that $E_{i,i+n_1-q-1} \in \mathcal{L}_{n_1-1}(\mathcal{S})$ for all $q = 1,\ldots,p-1$, $i = 1,\ldots,q$. Therefore $B_{j,\ j+n_1-p-1} - E_{j,j+n-p} \in \mathcal{L}_{n_1-1}(\mathcal{S})$. Hence we obtain that $E_{j,\ j+n_1-p-1} = (B_{j,\ j+n_1-p-1} - (B_{j,\ j+n_1-p-1} - E_{j,j+n_1-p})) \in \mathcal{L}_{n_1-1}(\mathcal{S})$, $j = 1,\ldots,p$.

2. Consider next $B_{j,j} = B_j(E - A_1 - A_2)^{n_1-2} \in \mathcal{L}_{n_1-1}(\mathcal{S}), j = 1,\ldots,n_1-1$.

$$B'_{j,j} = E_{j,j+1} + \gamma_{j,1}E_{k_1,k_1+1} + \gamma_{j,2}E_{k_2,k_2+1} + \sum_{h=1}^{n_1}\sum_{i=h+2}^{n_1} c_{h,i;j}E_{h,i},\ j \neq k_1,\ k_2$$

$$B'_{k_r,k_r} = a_{k_1,r}E_{k_1,k_1+1} + a_{k_2,r}E_{k_2,k_2+1} + \sum_{h=1}^{n_1}\sum_{i=h+2}^{n_1} c_{h,i;r}E_{h,i},\ r = 1,2,$$

$B''_{j,j} = 0$ and $B'''_{j,j} = 0$ as products of $n_1 - 1$ nilpotent matrices of order $n_{s+1} \leqslant n_1 - 2$, if $k_s \neq j$ or does not exist, and as products of $n_1 - 2$ nilpotent and one unitriangular matrices of order $n_{s+1} \leqslant n_1 - 2$, if $k_s = j$, $s = 1,2$. It follows from paragraph 1 that $E_{j,j+1} + \gamma_{j,1}E_{k_1,k_1+1} + \gamma_{j,2}E_{k_2,k_2+1}$, $a_{k_1,r}E_{k_1,k_1+1} + a_{k_2,r}E_{k_2,k_2+1} \in \mathcal{L}_{n_1-1}(\mathcal{S}), j \neq k_1,\ k_2, r = 1,2$, and hence $E_{j,j+1} \in \mathcal{L}_{n_1-1}(\mathcal{S})$, $j = 1,\ldots,n_1-1$.

Consequently, $E_{i,j} \in \mathcal{L}_{n_1-1}(\mathcal{S})$, $1 \leqslant i < j \leqslant n_1$.

3. From paragraphs 1 and 2 we obtain that $(E - A_1 - A_2)^{n_2} = \sum\limits_{i=1}^{n_1} E_{i,i} + \sum\limits_{1\leqslant h<i\leqslant n_1} \lambda_{h,i}E_{h,i} \in \mathcal{L}_{n_2}(\mathcal{S})$, and $\sum\limits_{i=1}^{n_1} E_{i,i} \in \mathcal{L}_{n_1-1}(\mathcal{S})$.

4. Let $S_1,\ldots,S_{n_1} \in \mathcal{S}$. The set $\mathcal{S}_{2,3} = \{S'' \oplus S'',\ S \in \mathcal{S}\}$ is a generating set for $\mathcal{A}_{n_2,\ n_3}$. By Theorem 11 $\mathcal{A}_{n_2,n_3} = \mathcal{L}_{n_2}(\mathcal{S}')$. Consequently, there exists $S_0 \in \mathcal{L}_{n_2}(\mathcal{S}_{2,3})$ such that $(S_1\cdots S_{n_1})'' \oplus (S_1\cdots S_{n_1})''' - S_0 = 0$. Let $T_0 \in \mathcal{L}_{n_2}(\mathcal{S})$

be the counter image of S_0. Then $S_1 \cdots S_{n_1} - T_0 = S' \oplus 0 \oplus 0 \in \mathcal{L}_{n_1-1}(\mathcal{S})$. That is any word of length n_1 in elements of $\mathcal{S}$ is reducible, $\mathcal{L}_{n_1}(\mathcal{S}) = \mathcal{L}_{n_1-1}(\mathcal{S})$ and $l(\mathcal{S}) \leqslant n_1 - 1$.

It follows from Theorem 8 that $l(\mathcal{A}_{n_1,n_2,n_3}) \geqslant n_1 - 1$. Consequently, $l(\mathcal{A}_{n_1,n_2,n_3}) = n_1 - 1$.

Theorem 14. *Let* $\mathbb{F}$ *be an arbitrary field,* $|\mathbb{F}| \geqslant 3$*, and let* $n_1,\ n_2,\ n_3 \in \mathbb{N}$, $n_1 \geqslant n_2 + n_3 + 2$, $n_2 \geqslant n_3$, $(n_2,\ n_3) \neq (1,\ 1),\ (2,\ 1),\ (2,\ 2)$. *Then* $l(\mathcal{A}_{n_1,\ n_2,\ n_3}) = n_1 - 1$.

Proof. **I.** Let us first prove the upper bound $l(\mathcal{A}_{n_1,n_2,n_3}) \leqslant n_1 - 1$. Let $\mathcal{S}$ be a generating set for $\mathcal{A}_{n_1,n_2,n_3}$. Without loss of generality we assume $\mathcal{S}$ to satisfy the conditions 1–2 of Lemma 9. If $\mathcal{S}$ satisfies condition 3(i) of Lemma 9, then the proof is analogous to the proof of Theorem 13. Consequently, we assume $\mathcal{S}$ to satisfy condition 3(ii) of Lemma 9, and therefore one of the conditions of Lemma 11.

1. We use induction on $p = n_1 - (j - i)$ to prove that $E_{i,j} \in \mathcal{L}_{n_1-1}(\mathcal{S})$ for $1 \leqslant i < j \leqslant n_1$, $j - i \geqslant 2$. Let us denote $m = n_1 + n_2 + 1$.

If $p = 1$, then $B_1 \cdots B_{n_1-1} = bE_{1,n_1} \in \mathcal{L}_{n_1-1}(\mathcal{S})$, $b \neq 0$, since $(B_1 \cdots B_{n_1-1})'' = 0$ and $(B_1 \cdots B_{n_1-1})''' = 0$ as products of $n_1 - 2$ nilpotent and one unitriangular matrices or $n_1 - 1$ nilpotent matrices of orders n_2 and n_3, correspondingly.

If $p \leqslant n_1 - n_3 - 2$ and $j = 1, \ldots, p$ consider

$$B_{j,j+n_1-p-1} = B_j \cdots B_{j+n_1-p-1}(E - A_0)^{p-1} \in \mathcal{L}_{n_1-1}(\mathcal{S}),$$

$$B'_{j,j+n_1-p-1} = a(k_0, 0)^t E_{j,j+n_1-p} + \sum_{h=1}^{n_1} \sum_{i=h+n_1-p+1}^{n_1} c_{h,i;j} E_{h,i},\ t \in \{0, 1\},$$

and $B''_{j,j+n_1-p-1} = 0$, $B''_{j,j+n_1-p-1} = 0$, as products of nilpotent matrices of lengths smaller than orders of factors.

If $n_1 - n_3 - 1 \leqslant p < n_1 - 1$ and $j = 1, \ldots, p$ consider

$$B_{j,j+n_1-p-1} = B_j \cdots B_{j+n_1-p-1}(E - a^{-1}A_0)^{n_3-n_1+p}(E - A_0)^{n_1-n_3-1},$$

$$B_{j,j+n_1-p-1} \in \mathcal{L}_{n_1-1}(\mathcal{S}),$$

$$B'_{j,j+n_1-p-1} = a(k_0, 0)^t E_{j,j+n_1-p} + \sum_{h=1}^{n_1} \sum_{i=h+n_1-p+1}^{n_1} c_{h,i;j} E_{h,i},\ t \in \{0, 1\},$$

$B''_{j,j+n_1-p-1} = 0$ and $B''_{j,j+n_1-p-1} = 0$, as products of nilpotent matrices of lengths smaller than orders of factors.

Applying the induction hypothesis we obtain that $E_{i,i+n_1-q-1} \in \mathcal{L}_{n_1-1}(\mathcal{S})$ for all $q = 2, \ldots, p - 1$, $i = 1, \ldots, q$, and $E_{1,n_1} \in \mathcal{L}_{n_1-1}(\mathcal{S})$ as shown above.

Therefore, $B_{j,\ j+n_1-p-1} - a(k_0,0)^t E_{j,j+n_1-p} \in \mathcal{L}_{n_1-1}(\mathcal{S})$. Hence we obtain that $E_{j,\ j+n_1-p-1} = (a(k_0,0))^{-t}(B_{j,j+n_1-p-1} - (B_{j,j+n_1-p-1} - (a(k_0,0))^t E_{j,j+n_1-p})) \in \mathcal{L}_{n_1-1}(\mathcal{S})$, $j = 1,\dots,p$.

2. For $j = 1,\dots,n_1-1$ consider products $B_{j,j} = B_j(E - a^{-1}A_0)^{n_3-1}(E - A_0)^{n_1-n_3-1}$, $j \neq k_0$, and $B_{k_0,k_0} = B_{k_0}(E - a^{-1}A_0)^{n_3}(E - A_0)^{p-n_3-1}$, $B_{j,j} \in \mathcal{L}_{n_1-1}(\mathcal{S})$. We have

$$B'_{j,j} = E_{j,j+1} + \gamma_j E_{k_0,k_0+1} + \sum_{h=1}^{n_1}\sum_{i=h+2}^{n_1} c_{h,i;j} E_{h,i},\ j \neq k_0$$

$$B'_{k_0,k_0} = a(k_0,0)E_{k_0,k_0+1} + \sum_{h=1}^{n_1}\sum_{i=h+2}^{n_1} c_{h,i;k_0} E_{h,i},$$

$B''_{r,r} = 0$ and $B'''_{r,r} = 0$ as products of nilpotent matrices of lengths smaller than orders of factors. With paragraph 1 it gives $E_{j,j+n_1-p} \in \mathcal{L}_{n_1-1}(\mathcal{S})$, $j = 1,\dots,p$.

3. We have $(E - A_0)^{n_2}(E - a^{-1}A_0)^{n_3} = \sum_{i=1}^{n_1} E_{i,i} + \sum_{1\leqslant h<i\leqslant n_1} \lambda_{h,i}E_{h,i} \in \mathcal{L}_{n_2+n_3}(\mathcal{S})$, and $\sum_{i=1}^{n_1} E_{i,i} \in \mathcal{L}_{n_1-1}(\mathcal{S})$.

4. Therefore, as it was shown in paragraph 4 of Theorem 13 any word of length n_1 in elements of $\mathcal{S}$ is reducible, $\mathcal{L}_{n_1}(\mathcal{S}) = \mathcal{L}_{n_1-1}(\mathcal{S})$ and $l(\mathcal{S}) \leqslant n_1 - 1$.

II. It follows from Theorem 8 that $l(\mathcal{A}_{n_1,n_2,n_3}) \geqslant n_1 - 1$.
Consequently, $l(\mathcal{A}_{n_1,n_2,n_3}) = n_1 - 1$.

The following Corollary shows in particular that the length ratio for a three block algebra and its subalgebra also can take on many different values, namely any rational value in $[1,2)$.

Corollary 3. *Let $\mathbb{F}$ be an arbitrary field, $|\mathbb{F}| \geqslant 3$, and let $n_1,\ n_2,\ n_3 \in \mathbb{N}$, $n_1 \geqslant n_2 + n_3 + 2$, $n_2 \geqslant n_3 \geqslant 3$. Let $a \in \mathbb{F}, a \neq 0,1$ and let*

$$C_{n_1,n_2,n_3} = \sum_{i=1}^{n_1-1} E_{i,i+1} + \sum_{j=n_1+1}^{n_1+n_2-1} (E_{j,j} + E_{j,j+1}) + E_{n_1+n_2,n_1+n_2} + \sum_{k=n_1+n_2+1}^{n_1+n_2+n_3-1} (aE_{k,k} + E_{k,k+1}) + aE_{n_1+n_2+n_3,n_1+n_2+n_3} \in \mathcal{A}_{n_1,n_2,n_3}$$

be a nonderogatory matrix, let

$$\mathcal{A}'_{n_1,n_2,n_3} = \langle C^j_{n_1,n_2,n_3},|\ 0 \leqslant j \leqslant n_1 + n_2 + n_3 - 1\rangle \subseteq \mathcal{A}_{n_1,n_2,n_3}.$$

Then
1. $l(\mathcal{A}'_{n_1,n_2,n_3}) = n_1 + n_2 + n_3 - 1$;

2. $l(\mathcal{A}'_{n_1,n_2,n_3}) - l(\mathcal{A}_{n_1,n_2,n_3}) = n_2 + n_3$;

3. $\dfrac{l(\mathcal{A}'_{n_1,n_2,n_3}) + 1}{l(\mathcal{A}_{n_1,n_2,n_3}) + 1} = 1 + \dfrac{n_2 + n_3}{n_1} < 2.$

Remark 3. Let us denote

$$\mathcal{A}_{n_1} = \langle E^{(n_1)},\ E_{i,j},\ 1 \leqslant i < j \leqslant n_1 \rangle \subset T_{n_1}(\mathbb{F}).$$

Notice that $\mathcal{A}_{n_1,n_2,n_3} = \mathcal{A}_{n_1} \oplus \mathcal{A}_{n_2,n_3}$ and $l(\mathcal{A}_{n_1,n_2,n_3}) = l(\mathcal{A}_{n_1}) = \max(l(\mathcal{A}_{n_1}),\ l(\mathcal{A}_{n_2,n_3}))$. That is we obtain another example providing sharpness of the lower bound in (1).

3.3 Examples

We now give the examples of algebras with length bounding the lengths of subalgebras.

Corollary 4. *Let $\mathbb{F}$ be an arbitrary field, $n,\ m \in \mathbb{N}$, $n \geqslant m-2$, and let $\mathcal{A}_{n,\ m}$ be the algebra introduced in Theorem 11. Let also*

$$\mathcal{B} = \langle E_{i,j},\ 1 \leqslant i < j \leqslant n,\ E,\ \sum_{i=1}^{n} E_{i,i},\ N_1, \ldots, N_p \in 0 \oplus N_m(\mathbb{F}) \rangle \subseteq \mathcal{A}_{n,\ m}.$$

Then $l(\mathcal{B}) = n - 1 = l(\mathcal{A}_{n,\ m})$.

Example 3. Let $\mathbb{F}$ be an arbitrary field, let $\mathcal{A} \subseteq T_n(\mathbb{F}))$ be a subalgebra of upper triangular matrix algebra. Then $l(\mathcal{A}) \leqslant l(T_n(\mathbb{F}))$.

Proposition 5. *Let $\mathbb{F}$ be an arbitrary field, let $\mathcal{A}$ be a finite-dimensional $\mathbb{F}$-algebra, and $\mathcal{B} \subseteq \mathcal{A}$ such that there exist $a_1, \ldots, a_n \in \mathcal{A}$ satisfying $\langle \mathcal{B}, a_1, \ldots, a_n \rangle = \mathcal{A}$ and $a_i b,\ b a_i \in \langle a_1, \ldots, a_n \rangle$ for all $b \in \mathcal{B}$. Then $l(\mathcal{B}) \leqslant l(\mathcal{A})$.*

Proof. Let $\mathcal{S}_{\mathcal{B}}$ be a generating set for $\mathcal{B}$. Then $\mathcal{S}_{\mathcal{A}} = \mathcal{S}_{\mathcal{B}} \bigcup \{a_1, \ldots, a_n\}$ is a generating set for $\mathcal{A}$ of the length $l(\mathcal{S}_{\mathcal{B}})$. Then $l(\mathcal{A}) \geqslant l(\mathcal{S}_{\mathcal{A}}) = l(\mathcal{S}_{\mathcal{B}})$ and therefore $l(\mathcal{A}) \geqslant \max\limits_{\mathcal{S}_{\mathcal{B}}} l(\mathcal{S}_{\mathcal{B}}) = l(\mathcal{B})$.

Let us give some examples of algebras satisfying the condition of Proposition 5.

Example 4. Let $\mathbb{F}$ be an arbitrary field, let $\mathcal{A}$ be a subalgebra of $T_n(\mathbb{F})$ and let $\mathcal{B} = \mathcal{A} \bigcap D_n(\mathbb{F})$. Then $l(\mathcal{B}) \leqslant l(\mathcal{A})$.

Example 5. Let $\mathbb{F}$ be an arbitrary field and let $\mathcal{A}$, $\mathcal{B}$ be finite-dimensional $\mathbb{F}$-algebras. Then $\mathcal{A} \subset \mathcal{A} \oplus \mathcal{B}$ and $l(\mathcal{A}) \leqslant l(\mathcal{A} \oplus \mathcal{B})$.

The author is greatly indebted to her supervisor Dr. A. E. Guterman for the attention given to the work and for useful discussions.

References

1. T. J. Laffey, Simultaneous Reduction of Sets of Matrices under Similarity, Linear Algebra and its Applications, **84**(1986), 123–138
2. W. E. Longstaff, Burnside's theorem: irreducible pairs of transformations, Linear Algebra and its Applications, **382**(2004), 247–269
3. C. J. Pappacena, An Upper Bound for the Length of a Finite-Dimensional Algebra, Journal of Algebra, **197**(1997), 535–545
4. A. Paz, An Application of the Cayley–Hamilton Theorem to Matrix Polynomials in Several Variables, Linear and Multilinear Algebra, **15**(1984), 161–170
5. A. J. M. Spencer, R. S. Rivlin, The Theory of Matrix Polynomials and its Applications to the Mechanics of Isotropic Continua, Archive for Rational Mechanics and Analysis, **2**(1959), 309–336
6. A. J. M. Spencer, R. S. Rivlin, Further Results in the Theory of Matrix Polynomials, Archive for Rational Mechanics and Analysis, **4**(1960), 214–230
7. O. V. Markova, On the length of upper-triangular matrix algebra, Uspekhi Matem. Nauk, 60(2005), no. 5, 177-178, [in Russian]; English translation: Russian Mathematical Surveys, **60**(2005), no. 5, 984–985.
8. O. V. Markova, Length computation of matrix subalgebras of special type, Fundamental and Applied Mathematics, **13** (2007), Issue 4, 165–197.

On a New Class of Singular Nonsymmetric Matrices with Nonnegative Integer Spectra

Tatjana Nahtman[1,*] and Dietrich von Rosen[2]

[1] Institute of Mathematical Statistics, University of Tartu, Estonia
tatjana.nahtman@ut.ee;
Department of Statistics, University of Stockholm, Sweden
tatjana.nahtman@statistics.su.se
[2] Department of Biometry and Engineering, Swedish University of Agricultural Sciences, Sweden
dietrich.von.rosen@bt.slu.se

Abstract. The objective of this paper is to consider a class of singular nonsymmetric matrices with integer spectrum. The class comprises generalized triangular matrices with diagonal elements obtained by summing the elements of the corresponding column. If the size of a matrix belonging to the class equals $n \times n$, the spectrum of the matrix is given by the sequence of distinct nonnegative integers up to $n-1$, irrespective of the elements of the matrix. Right and left eigenvectors are obtained. Moreover, several interesting relations are presented, including factorizations via triangular matrices.

Keywords: eigenvectors, generalized triangular matrix, integer spectrum, nonsymmetric matrix, triangular factorization, Vandermonde matrix.

1 Introduction

In this paper we consider a new class of singular matrices with remarkable algebraic properties. For example, the spectrum of a matrix belonging to this class depends only on the size of the matrix and not on the specific elements of this matrix. Moreover, the spectrum entirely consists of successive non-negative integer values $0, 1, \ldots, n-1$. A special case of this class of matrices originates from statistical sampling theory (Bondesson & Traat, 2005, 2007). In their papers, via sampling theory (the Poisson sampling design) as well as some analytic proofs, eigenvalues and eigenvectors were presented. Their proofs remind on the use of Lagrangian polynomials which for example are used when finding the inverse of a Vandermonde matrix (e.g. see Macon & Spitzbart, 1958; El-Mikkawy, 2003). We have not found any other work related to the matrix class which we are going to consider.

* The work of T. Nahtman was supported by the grant GMTMS6702 of Estonian Research Foundation.

The main purpose of this paper is to introduce the class, show some basic algebraic properties, show how to factor the class and demonstrate how to find eigenvalues and eigenvectors of matrices belonging to the class. The paper focuses more on presentation of results than giving complete proofs of the most general versions of the theorems.

Definition 1. *A square nonsymmetric matrix* $B = (b_{ij})$ *of order* n *belongs to the* $\mathbb{B}_n$*-class if its elements satisfy the following conditions:*

$$b_{ii} = \sum_{\substack{j=1,\\ j\neq i}}^{n} b_{ji}, \quad i = 1, \ldots, n, \tag{1}$$

$$b_{ij} + b_{ji} = 1, \quad j \neq i,\ i, j = 1, \ldots, n, \tag{2}$$

$$b_{ij} - b_{ik} = \frac{b_{ij} b_{ki}}{b_{kj}},\ b_{kj} \neq 0, \quad j \neq k,\ i \neq k, j;\ i, j, k = 1, \ldots, n. \tag{3}$$

Instead of (3) one may use $b_{kj} = \frac{b_{ij} b_{ki}}{b_{ij} - b_{ik}}$ or $b_{ij} b_{kj} = b_{ik} b_{kj} + b_{ij} b_{ki}$. Relation (2) defines a generalized triangular structure and it can be shown that (3) is a necessary and sufficient condition for the class to have the non-negative integer spectra consisting of the distinct integers $\{0, 1, \ldots, n-1\}$. Due to (1), the sum of the elements in each row equals $n-1$.

Another matrix with integer eigenvalues and row element sum equal to $n-1$, with many applications in various fields, is the well known tridiagonal Kac-matrix (Clement matrix); see Taussky & Todd (1991). Moreover, for any $B \in \mathbb{B}_n$ with positive elements we may consider $(n-1)^{-1} B$ as a transition matrix with interesting symmetric properties reflected by the equidistant integer spectra.

When $B \in \mathbb{B}_3$,

$$B = \begin{pmatrix} b_{21} + b_{31} & b_{12} & b_{13} \\ b_{21} & b_{12} + b_{32} & b_{23} \\ b_{31} & b_{32} & b_{13} + b_{23} \end{pmatrix}$$

$$= \begin{pmatrix} b_{21} + b_{31} & 1 - b_{21} & 1 - b_{31} \\ b_{21} & 1 - b_{21} + b_{32} & 1 - b_{32} \\ b_{31} & b_{32} & 2 - b_{31} - b_{32} \end{pmatrix}. \tag{4}$$

It is worth observing that any $B \in \mathbb{B}_n$ is a sum of three matrices: an upper triangular matrix, a diagonal matrix and a skew-symmetric matrix. For (4) we have

$$B = \begin{pmatrix} 0 & 1 & 1 \\ 0 & 1 & 1 \\ 0 & 0 & 2 \end{pmatrix} + \begin{pmatrix} b_{21} + b_{31} & 0 & 0 \\ 0 & -b_{21} + b_{32} & 0 \\ 0 & 0 & -b_{31} - b_{32} \end{pmatrix} + \begin{pmatrix} 0 & -b_{21} & -b_{31} \\ b_{21} & 0 & -b_{32} \\ b_{31} & b_{32} & 0 \end{pmatrix}.$$

Note that the eigenvalues $\{0, 1, 2\}$ of B are found on the diagonal of the upper triangular matrix, irrespective of the values of (b_{ij}) as long as they satisfy (1)–(3).

In the Conditional Poisson sampling design (e.g., see Aires, 1999)

$$b_{ij} = \frac{p_i(1-p_j)}{p_i - p_j}$$

are used to calculate conditional inclusion probabilities, where the p_i's are inclusion probabilities under the Poisson design. Bondesson & Traat (2005, 2007) generalized this expression somewhat and considered

$$b_{ij} = \frac{r_i}{r_i - r_j}, \tag{5}$$

where the r_i's are arbitrary distinct values. In this paper, instead of (5), we assume (3) to hold. Note that any b_{ij} satisfying (5) also satisfies (3). For the matrix defined via the elements in (5) Bondesson & Traat (2005, 2007) presented eigenvalues, and right and left eigenvectors. They expressed their results as functions of r_i in (5) whereas in this paper we will express the results in terms of b_{ij}, i.e. the elements of $B \in \mathbb{B}_n$. Moreover, the proofs of all results in this paper are pure algebraic whereas Bondesson & Traat (2005, 2007) indicated proofs based on series expansions and identification of coefficients. It is however not clear how to apply their results to the $\mathbb{B}_n$-class of matrices, given in Definition 1. Moreover, the algebraic approach of this paper opens up a world of interesting relations. In particular, the triangular factorization of matrices in the $\mathbb{B}_n$-class, presented in Section 4.

As noted before it follows from (3) that

$$b_{kj} = \frac{b_{ij}b_{ki}}{b_{ij} - b_{ik}} = \frac{b_{ij}(1-b_{ik})}{b_{ij} - b_{ik}}. \tag{6}$$

Hence, any row in B, $B \in \mathbb{B}_n$, generates all other elements and thus, there are at most $n-1$ functionally independent elements in B. For example, we may use b_{1j}, $j = 2, 3, \ldots, n$, to generate all other elements in B. Furthermore, if we choose for r_j in (5), without loss of generality, $r_1 = 1$ and

$$r_j = -\frac{b_{j1}}{b_{1j}}, \quad j = 2, 3, \ldots, n,$$

it follows that

$$b_{1j} = \frac{1}{1 - r_j}$$

and

$$b_{ij} = \frac{\frac{1}{1-r_j}\left(1 - \frac{1}{1-r_i}\right)}{\frac{1}{1-r_j} - \frac{1}{1-r_i}} = \frac{r_i}{r_i - r_j}.$$

Thus, all b_{ij}'s can be generated by the above choice of r_j. This means that a matrix defined by (5), as considered in Bondesson & Traat (2005, 2007), is a canonical version of any matrix defined through (3), assuming that (1) and (2) hold.

The class $\mathbb{B}_n$ can be generalized in a natural way.

Definition 2. *The matrix* $B_{n,k}$: $(n-k+2) \times (n-k+2)$, $k = 2, \dots, n$, *is obtained from the matrix B,* $B \in \mathbb{B}_n$, *by elimination of* $k-2$ *consecutive rows and columns starting from the second row and column, with corresponding adjustments in the main diagonal.*

The paper consists of five sections. In Section 2 some basic and fundamental relations for any $B \in \mathbb{B}_n$ are given which will be used in the subsequent. Section 3 consists of a straightforward proof concerning the spectrum of any $B \in \mathbb{B}_n$. In Section 4 we consider a factorization of $B \in \mathbb{B}_n$ into a product of three triangular matrices. Finally, in Section 5 expressions of left and right eigenvectors are presented. Several proofs of theorems are omitted due to lengthy calculations. However, for further details it is referred to the technical report Nahtman & von Rosen (2007). All proofs of this paper could easily have been presented for, say $n < 7$, but for a general n we rely on induction which is more difficult to look through. Their is certainly space for improving the proofs and this is a another reason for omitting them.

In the present paper only real-valued matrices are considered, although the generalization to matrices with complex-valued entries could be performed fairly easy.

2 Preparations

This section shows some relations among the elements in $B \in \mathbb{B}_n$ which are of utmost importance for the subsequent presentation.

Theorem 1. *Let* $B \in \mathbb{B}_n$. *For all* $n > 1$

(i) *The sum of the products of the off-diagonal row elements equals 1:*

$$\sum_{i=1}^{n} \prod_{\substack{j=1 \\ j \neq i}}^{n} b_{ij} = 1.$$

(ii) *The sum of the products of the off-diagonal column elements equals 1:*

$$\sum_{j=1}^{n} \prod_{\substack{i=1 \\ i \neq j}}^{n} b_{ij} = 1.$$

Proof. Because of symmetry only (i) is proven. For $n = 2$ the trivial relation $b_{12} + b_{21} = 1$ is obtained. Moreover, for $n = 3$

$$\sum_{i=1}^{3}\prod_{\substack{j=1\\j\neq i}}^{3} b_{ij} = b_{12}b_{13} + b_{21}b_{23} + b_{31}b_{32} = b_{12} - b_{12}b_{31} + b_{21} - b_{21}b_{32} + b_{31}b_{32}$$

$$= 1 - (b_{12} - b_{13})b_{32} - b_{21}b_{32} + b_{31}b_{32}$$
$$= 1 - (b_{12} + b_{21})b_{32} + (b_{13} + b_{31})b_{32} = 1,$$

where in the second equality (3) is utilized. Now it is assumed that the theorem is true for $n - 1$, i.e.

$$\sum_{i=1}^{n-1}\prod_{\substack{j=1\\j\neq i}}^{n-1} b_{ij} = 1, \tag{7}$$

which by symmetry yields

$$\sum_{\substack{i=1\\i\neq k}}^{n}\prod_{\substack{j=1\\j\neq i\\j\neq k}}^{n} b_{ij} = 1, \qquad k = 1, 2, \ldots, n. \tag{8}$$

From here on a chain of calculations is started:

$$\sum_{i=1}^{n}\prod_{\substack{j=1\\j\neq i}}^{n} b_{ij} = \sum_{i=1}^{n-1}\prod_{\substack{j=1\\j\neq i}}^{n-1} b_{ij}b_{in} + \prod_{j=1}^{n-1} b_{nj}$$
$$= \sum_{i=1}^{n-2}\prod_{\substack{j=1\\j\neq i}}^{n-2} b_{ij}b_{in-1}b_{in} + \prod_{j=1}^{n-2} b_{n-1j}b_{n-1n} + \prod_{j=1}^{n-2} b_{nj}b_{nn-1}$$
$$= \sum_{i=1}^{n-2}\prod_{\substack{j=1\\j\neq i}}^{n-2} b_{ij}b_{in-1}(1 - b_{ni}) + \prod_{j=1}^{n-2} b_{n-1j}(1 - b_{nn-1}) + \prod_{j=1}^{n-2} b_{nj}b_{nn-1}. \tag{9}$$

Since by the induction assumption

$$\sum_{i=2}^{n-2}\prod_{\substack{j=1\\j\neq i}}^{n-2} b_{ij}b_{in-1} + \prod_{j=1}^{n-2} b_{n-1j} = 1$$

the last expression in (9) equals

$$1 - \sum_{i=1}^{n-2}\prod_{\substack{j=1\\j\neq i}}^{n-2} b_{ij}(b_{in-1} - b_{in})b_{nn-1} - \prod_{j=1}^{n-2} b_{n-1j}b_{nn-1} + \prod_{j=1}^{n-2} b_{nj}b_{nn-1}, \tag{10}$$

where (3) has been used: $b_{in-1}b_{ni} = (b_{in-1} - b_{in})b_{nn-1}$. Reshaping (10) we obtain

$$1 - \sum_{i=1}^{n-1} \prod_{\substack{j=1 \\ j\neq i}}^{n-1} b_{ij} b_{nn-1} + \sum_{\substack{i=1 \\ i\neq n-1}}^{n} \prod_{\substack{j=1 \\ j\neq i \\ j\neq n-1}}^{n} b_{ij} b_{nn-1}, \tag{11}$$

and using the induction assumption, i.e. (7) as well as (8), we see that (11) is indeed equal to

$$1 - b_{nn-1} + b_{nn-1} = 1,$$

and the theorem is proved. □

Corollary 1. *Let $B \in \mathbb{B}_n$. For all $n > 1$,*

$$\sum_{i=1}^{n-1} \prod_{\substack{j=1 \\ j\neq i}}^{n} b_{ij} = 1 - \prod_{j=1}^{n-1} b_{nj}.$$

Corollary 2. *Let $B \in \mathbb{B}_n$. For every integer a such that $a < n$,*

$$\sum_{i=a}^{n} \prod_{\substack{j=a \\ j\neq i}}^{n} b_{ij} = 1.$$

Theorem 2. *Let $B \in \mathbb{B}_n$ and put $c_{ij} = b_{ij}^{-1} b_{ji}$. Then,*

(i) $c_{ij}^{-1} = c_{ji}, \quad i \neq j,$

(ii) $c_{ki}c_{jk} = -c_{ji}, \quad k \neq i,\ j \neq k,\ i \neq j,$ *(cancellation)*

(iii) $c_{ki}c_{lj} = c_{kj}c_{li}, \quad k \neq i, j;\ l \neq i, j.$ *(exchangeability)*

Proof. (i) follows immediately from the definition of c_{ij}. For (ii) it is observed that (see (3))

$$\frac{b_{ij}b_{ki}}{b_{kj}} = -\frac{b_{ji}b_{ik}}{b_{jk}}$$

and hence,

$$c_{ki}c_{jk} = b_{ki}^{-1}b_{ik}b_{jk}^{-1}b_{kj} = b_{ki}^{-1}b_{ik}b_{jk}^{-1}b_{kj}b_{ij}b_{ij}^{-1} = -b_{ki}^{-1}b_{ki}b_{jk}^{-1}b_{jk}b_{ij}b_{ij}^{-1} = -c_{ji}.$$

Concerning (iii) it is noted that

$$c_{ki}c_{lj} = c_{ki}c_{lj}c_{il}c_{li} = -c_{ki}c_{ij}c_{li} = c_{kj}c_{li}. \square$$

Throughout the paper the following abbreviations for two types of multiple sums will be used. Both will frequently be applied in the subsequent:

$$\sum_{i_1 \leqslant \cdots \leqslant i_k}^{[m,n]} = \sum\nolimits_{i_1=m}^{n} \sum\nolimits_{i_2=i_1}^{n} \cdots \sum\nolimits_{i_k=i_{k-1}}^{n}, \qquad i_0 = m, \tag{12}$$

$$\sum_{i_1 < \cdots < i_k}^{[m,n]} = \sum\nolimits_{i_1=m}^{n-k+1} \sum\nolimits_{i_2=i_1+1}^{n-k+2} \cdots \sum\nolimits_{i_k=i_{k-1}+1}^{n}, \qquad i_0 = m-1. \tag{13}$$

The next theorem is an ancillary result which because of lengthy and tedious calculations will not be proven.

Theorem 3. *Let* $B \in \mathbb{B}_n$. *Then,*

$$b_{i+11} + \sum_{k=i+1}^{n-1} \sum_{i_1 \leqslant \cdots \leqslant i_{k-1}}^{[k+1,n]} \prod_{m=i+1}^{k} b_{i_{m-i}m}^{-1} b_{mi_{m-i}} b_{1m} b_{k+11} = \prod_{m=i+1}^{n} b_{m1}, \tag{14}$$

$$i = 1, 2, \ldots, n-2.$$

3 Preliminaries

The aim of this section is to derive, in a very elementary algebraic way, the eigenvalues of the matrix $B \in \mathbb{B}_n$. Moreover, the section serves as an introduction to Section 4, i.e. a similar technique for proving the next theorem will also be used in Section 4.

Throughout the paper we denote by e_i, (d_i), $i = 1, \ldots, n$, standard unit vectors of size n, i.e. the ith column of the identity matrix I_n. Define the following lower triangular matrix

$$U = e_1 e_1' + b_{12}(e_2 e_1' - e_2 e_2') + \sum_{i=3}^{n} b_{i2}(e_i e_i' - e_i e_2') \tag{15}$$

$$= \begin{pmatrix} 1 & 0 & 0 & 0 & 0 & \ldots & 0 \\ b_{12} & -b_{12} & 0 & 0 & 0 & \ldots & 0 \\ 0 & -b_{32} & b_{32} & 0 & 0 & \ldots & 0 \\ 0 & -b_{42} & 0 & b_{42} & 0 & \ldots & 0 \\ 0 & -b_{52} & 0 & 0 & b_{52} & \ldots & 0 \\ \vdots & \vdots & \vdots & \vdots & \vdots & \ddots & 0 \\ 0 & -b_{n2} & 0 & 0 & 0 & \ldots & b_{n2} \end{pmatrix}$$

with inverse

$$U^{-1} = \sum_{i=1}^{n} e_i e_1' - b_{12}^{-1} \sum_{i=2}^{n} e_i e_2' + \sum_{i=3}^{n} b_{i2}^{-1} e_i e_i' \tag{16}$$

$$= \begin{pmatrix} 1 & 0 & 0 & 0 & 0 & \dots & 0 \\ 1 & -b_{12}^{-1} & 0 & 0 & 0 & \dots & 0 \\ 1 & -b_{12}^{-1} & b_{32}^{-1} & 0 & 0 & \dots & 0 \\ 1 & -b_{12}^{-1} & 0 & b_{42}^{-1} & 0 & \dots & 0 \\ 1 & -b_{12}^{-1} & 0 & 0 & b_{52}^{-1} & \dots & 0 \\ \vdots & \vdots & \vdots & \vdots & \vdots & \ddots & 0 \\ 1 & -b_{12}^{-1} & 0 & 0 & 0 & \dots & b_{n2}^{-1} \end{pmatrix}.$$

Theorem 4. *Let $B = B_{n,2} \in \mathbb{B}_n$. Then there exists a nonsingular lower triangular matrix U, given in* (15), *such that $UB_{n,2}U^{-1}$ has the following structure*

$$B_{n,2} = \left(\begin{array}{c|c} n-1 & -\sum_{j=1}^{n} b_{1j} b_{12}^{-1} \quad \sum_{k=3}^{n} b_{1k} b_{k2}^{-1} d_k' \\ \hline 0 & B_{n,3} \end{array} \right),$$

where d_k is a standard unit vector of size $n-2$ and $B_{n,3}$: $(n-1) \times (n-1)$ is defined in Definition 2.

Proof. First notice that $B_{n,2}$ can be written

$$B_{n,2} = \sum_{j=1}^{n} \sum_{\substack{i=1 \\ i \neq j}}^{n} b_{ji} e_i e_i' + \sum_{i=1}^{n} \sum_{\substack{j=1 \\ i \neq j}}^{n} b_{ij} e_i e_j'.$$

Thus

$$\begin{aligned} B_{n,2}U^{-1} = & \sum_{j=1}^{n} \sum_{\substack{i=1 \\ i \neq j}}^{n} b_{ji} e_i e_1' + \sum_{i=1}^{n} \sum_{\substack{j=1 \\ i \neq j}}^{n} b_{ij} e_i e_1' - \sum_{j=1}^{n} \sum_{\substack{k=2 \\ k \neq j}}^{n} b_{jk} b_{12}^{-1} e_k e_2' - \\ & \sum_{k=1}^{n} \sum_{\substack{j=2 \\ j \neq k}}^{n} b_{kj} b_{12}^{-1} e_k e_2' + \sum_{j=1}^{n} \sum_{\substack{k=3 \\ k \neq j}}^{n} b_{jk} b_{k2}^{-1} e_k e_k' + \sum_{j=1}^{n} \sum_{k=3}^{n} b_{jk} b_{k2}^{-1} e_j e_k'. \end{aligned} \tag{17}$$

Premultiplying (17) by $e_1 e_1'$ yields

$$(n-1) e_1 e_1' - \sum_{j=2}^{n} b_{1j} b_{12}^{-1} e_1 e_2' + \sum_{k=3}^{n} b_{1k} b_{k2}^{-1} e_1 e_k' \tag{18}$$

and premultiplying (17) by $b_{12}(e_2 e_1' - e_2 e_2')$ results in

$$\begin{aligned} & \sum_{\substack{j=1 \\ j \neq 2}}^{n} b_{j2} e_2 e_2' - \textstyle\sum_{j=2}^{n} b_{1j} e_2 e_2' + \sum_{j=3}^{n} b_{2j} e_2 e_2' + \sum_{k=3}^{n} b_{1k} b_{k2}^{-1} b_{12} e_2 e_k' - \\ & \qquad \textstyle\sum_{k=3}^{n} b_{2k} b_{k2}^{-1} b_{12} e_2 e_k' = \sum_{j=3}^{n} b_{j1} e_2 e_2' + \sum_{k=3}^{n} b_{1k} e_2 e_k', \end{aligned}$$

where in the last equality properties (2)–(3) have been used. It remains to premultiply (17) with $\sum_{i=3}^{n} b_{i2}(e_i e_i' - e_i e_2')$ which gives

$$\sum_{i=3}^{n} b_{i1} e_i e_2' + \sum_{i=3}^{n} \sum_{\substack{j=3 \\ j \neq i}}^{n} b_{ji} e_i e_i' + \sum_{i=3}^{n} \sum_{\substack{k=3 \\ k \neq i}}^{n} b_{ik} e_i e_k'.$$

Hence, by summarizing the above calculations we obtain

$$UB_{n,2}U^{-1} = \left(\begin{array}{c|cc} n-1 & -\sum_{j=1}^{n} b_{1j} b_{12}^{-1} & \sum_{k=3}^{n} b_{1k} b_{k2}^{-1} d_k' \\ \hline 0 & & B_{n,3} \end{array}\right)$$

whixh is the statement of the theorem. □

Thus, a useful recursive relation between $B_{n,2}$ and $B_{n,3}$ has been established. An important consequence of Theorem 4 is the following result which indeed was our starting point of interest for studying $B \in \mathbb{B}_n$.

Theorem 5. *Let* $B \in \mathbb{B}_n$. *Then B has the nonnegative integer eigenvalues* $\{0, 1, \ldots, n-1\}$.

Let us now introduce the following auxiliary matrices: a lower unit triangular matrix L_n: $n \times n$, with all non-zero elements equal to one, and a lower bidiagonal matrix K_n with 1 in the main diagonal and -1 in the the first subdiagonal which actually is the inverse of L_n, i.e. $K_n = L_n^{-1}$.

Theorem 6. *The lower triangular matrix* U *given in* (15) *may be decomposed as*

$$U = \mathrm{Diag}(1, b_{12}, b_{32}, \ldots, b_{n2})\mathrm{Diag}(I_2, L_{n-2})K_n$$

$$= \begin{pmatrix} 1 & 0 & 0 & 0 & \cdots & 0 \\ 0 & b_{12} & 0 & 0 & \cdots & 0 \\ 0 & 0 & b_{32} & 0 & \cdots & 0 \\ 0 & 0 & 0 & b_{42} & \cdots & 0 \\ \vdots & \vdots & \vdots & \vdots & \ddots & \vdots \\ 0 & 0 & 0 & 0 & \cdots & b_{n2} \end{pmatrix} \cdot \left(\begin{array}{cc|cccc} 1 & 0 & 0 & 0 & \cdots & 0 \\ 0 & 1 & 0 & 0 & \cdots & 0 \\ \hline 0 & 0 & 1 & 0 & \cdots & 0 \\ 0 & 0 & 1 & 1 & \cdots & 0 \\ \vdots & \vdots & \vdots & \vdots & \ddots & \vdots \\ 0 & 0 & 1 & 1 & \cdots & 1 \end{array}\right) \cdot \begin{pmatrix} 1 & 0 & 0 & 0 & \cdots & 0 \\ -1 & 1 & 0 & 0 & \cdots & 0 \\ 0 & -1 & 1 & 0 & \cdots & 0 \\ 0 & 0 & -1 & 1 & \cdots & 0 \\ \vdots & \vdots & \vdots & \ddots & \ddots & \vdots \\ 0 & 0 & 0 & 0 & -1 & 1 \end{pmatrix}.$$

Its inverse equals

$$\begin{aligned} U^{-1} &= K_n^{-1}\mathrm{Diag}(I_2, L_{n-2}^{-1})\mathrm{Diag}(1, b_{12}^{-1}, b_{32}^{-1}, \ldots, b_{n2}^{-1}) \\ &= L_n \mathrm{Diag}(I_2, K_{n-2})\mathrm{Diag}(1, b_{12}^{-1}, b_{32}^{-1}, \ldots, b_{n2}^{-1}) \end{aligned}$$

$$= \begin{pmatrix} 1 & 0 & 0 & 0 & 0 & 0 \\ 1 & 1 & 0 & 0 & 0 & 0 \\ 1 & 1 & 1 & 0 & 0 & 0 \\ 1 & 1 & 1 & 1 & 0 & 0 \\ \vdots & \vdots & \vdots & \vdots & \ddots & \vdots \\ 1 & 1 & 1 & 1 & 1 & 1 \end{pmatrix} \cdot \left(\begin{array}{cc|cccc} 1 & 0 & 0 & 0 & 0 & 0 \\ 0 & 1 & 0 & 0 & 0 & 0 \\ \hline 0 & 0 & 1 & 0 & 0 & 0 \\ 0 & 0 & -1 & 1 & 0 & 0 \\ \vdots & \vdots & \vdots & \vdots & \ddots & \vdots \\ 0 & 0 & 0 & 0 & -1 & 1 \end{array}\right) \cdot \begin{pmatrix} 1 & 0 & 0 & 0 & 0 & 0 \\ 0 & b_{12}^{-1} & 0 & 0 & 0 & 0 \\ 0 & 0 & b_{32}^{-1} & 0 & 0 & 0 \\ 0 & 0 & 0 & b_{42}^{-1} & 0 & 0 \\ 0 & 0 & 0 & 0 & \ddots & 0 \\ 0 & 0 & 0 & 0 & 0 & b_{n2}^{-1} \end{pmatrix}.$$

Next the matrix $U_{n,k} : (n-k+2) \times (n-k+2)$, $k = 2, \dots, n$, is defined as

$$U_{n,k} = e_1 e_1' + b_{1k}(e_2 e_1' - e_2 e_2') + \sum_{i=3}^{n} b_{i+k-2,k}(e_i e_i' - e_i e_2'). \tag{19}$$

The inverse of $U_{n,k}$ is given by

$$U_{n,k}^{-1} = \sum_{i=1}^{n} e_i e_1' - b_{1k}^{-1} \sum_{i=2}^{n} e_i e_2' + \sum_{i=3}^{n} b_{i+k-2,k}^{-1} e_i e_i'. \tag{20}$$

Note, that $U_{n,2} = U$, and thus $U_{n,k}$ is a direct generalization of U. For example,

$$U_{5,3} = \begin{pmatrix} 1 & 0 & 0 & 0 \\ b_{13} & -b_{13} & 0 & 0 \\ 0 & -b_{43} & b_{43} & 0 \\ 0 & -b_{53} & 0 & b_{53} \end{pmatrix}, \quad U_{5,4} = \begin{pmatrix} 1 & 0 & 0 \\ b_{14} & -b_{14} & 0 \\ 0 & -b_{54} & b_{54} \end{pmatrix},$$

$$U_{5,3}^{-1} = \begin{pmatrix} 1 & 0 & 0 & 0 \\ 1 & -b_{13}^{-1} & 0 & 0 \\ 1 & -b_{13}^{-1} & b_{43}^{-1} & 0 \\ 1 & -b_{13}^{-1} & 0 & b_{53}^{-1} \end{pmatrix}, \quad U_{5,4}^{-1} = \begin{pmatrix} 1 & 0 & 0 \\ 1 & -b_{14}^{-1} & 0 \\ 1 & -b_{14}^{-1} & b_{54}^{-1} \end{pmatrix}.$$

It has been shown that in order to triangularize $B = B_{n,2}$ the matrices U and U^{-1} could be used. Furthermore, the matrix $U_{n,k}$ and its inverse $U_{n,k}^{-1}$, given in (19) and (20), respectively, triangularize $B_{n,k}$. Hence, we may state the following interesting result.

Theorem 7. *Let* $D_{l,m} = \mathrm{Diag}(I_{l-2+m}, U_{n,l+m})$, *and* $U_{n,k}$ *is defined in (19). For* $B_{n,k}$, *given in Definition 2, the following relations hold (* $*$ *indicates an unspecified element):*

(i) $$U_{n,k} B_{n,k} U_{n,k}^{-1} = \left(\begin{array}{c|c} n-k+1 & * \\ \hline 0 & B_{n,k+1} \end{array}\right),$$

(ii) $$\prod_{i=0}^{h} D_{k,h-i} B_{n,k} \prod_{i=0}^{h} D_{k,i}^{-1} = \left(\begin{array}{c|c|c|c} n-k+1 & * & * & * \\ \hline 0 & \ddots & * & * \\ \hline 0 & 0 & n-h-1 & * \\ \hline 0 & 0 & 0 & B_{n,k+h+1} \end{array}\right),$$

(iii) $$\prod_{i=0}^{n-2} D_{2,n-2-i} B_{n,2} \prod_{i=0}^{n-2} D_{2,i}^{-1} = \left(\begin{array}{c|c|c|c} n-1 & * & * & * \\ \hline 0 & \ddots & * & * \\ \hline 0 & 0 & 1 & -1 \\ \hline 0 & 0 & 0 & 0 \end{array}\right).$$

In particular, Theorem 7 is important since it shows a way of how to triangularize any $B \in \mathbb{B}_n$. This will be further exploited in the next section.

Theorem 8. *Let* $B_{n,k}$ *be given in Definition 2. Then* $B_{n,k}$ *has the nonnegative integer eigenvalues* $\{0, 1, \dots, n-k+1\}$.

4 Triangular factorization

In this section Theorem 7 is explored and $B \in \mathbb{B}_n$ is facorized into a product of three triangular matrices, which will be called a VTU-decomposition. For notational convenience it is assumed that $\prod_{i=h}^{n} a_i = 1$, $\sum_{i=h}^{n} a_i = 0$, if $n < h$, and $\prod_{\substack{i=2 \\ i\neq 2}}^{2} a_i = 1$. This convention will be applied throughout the rest of the paper. Indeed, it implies that Theorem 1 also is true for $n = 1$. The next matrix will later appear as one of the matrices in the triangular VTU-decomposition of $B \in \mathbb{B}_n$:

$$U_n = e_1e_1' - b_{12}e_2e_2' + \sum_{i=2}^{n}\prod_{m=2}^{i} b_{1m}e_ie_1' - \sum_{\substack{j=2 \\ j\neq 2}}^{n}\sum_{i=j}^{n} b_{1j}\prod_{\substack{m=2 \\ m\neq j}}^{i} b_{jm}e_ie_j' \quad (21)$$
$$= \sum_{i=1}^{n}\prod_{m=2}^{n} b_{1m}e_ie_1' - \sum_{j=2}^{n}\sum_{i=j}^{n} b_{1j}\prod_{\substack{m=2 \\ m\neq j}}^{i} b_{jm}e_ie_j',$$

where as before e_i is a standard unit vector of size n. Its inverse is presented without a proof but it is interesting to observe that a strict proof heavily depends on Theorem 2.

Theorem 9. *Let* U_n *be given by* (21), *and let* $V_n = U_n^{-1}$. *Then,*

$$V_n = \sum_{k=1}^{n} e_ke_1' - b_{12}^{-1}e_2e_2' + \sum_{k=3}^{n} b_{2k}b_{1k}^{-1}b_{k2}^{-1}e_ke_2' \quad (22)$$
$$- \sum_{k=3}^{n} b_{1k}^{-1}\prod_{m=2}^{k-1} b_{km}^{-1}e_ke_k' - \sum_{k=4}^{n}\sum_{l=3}^{k-1} b_{1l}b_{1l}^{-1}\prod_{m=1}^{l-1} b_{km}^{-1}e_ke_l'$$
$$= \sum_{k=1}^{n} e_ke_1' - \sum_{k=2}^{n}\sum_{l=2}^{k} b_{1l}b_{1l}^{-1}\prod_{m=1}^{l-1} b_{km}^{-1}e_ke_l'.$$

Let

$$I_{\{j>k\}} = \begin{cases} 1, & \text{if } j > k, \\ 0, & \text{otherwise.} \end{cases}$$

In the next U_n and V_n from the previous theorem are presented elementwise.

Theorem 10. *Let* $U_n = (u_{ij})$ *and* $V_n = (v_{ij})$ *be given by* (21) *and* (22), *respectively. Then,*

$$u_{ij} = (-b_{1j})^{I_{\{j>1\}}}\prod_{\substack{k=2 \\ k\neq j}}^{i} b_{jk},\ i \geqslant j \quad (23)$$

and

$$v_{ij} = \left(-\frac{b_{j1}}{b_{1j}}\right)^{I_{\{j>1\}}} \prod_{k=1}^{j-1} b_{ik}^{-1}, \ i \geqslant j. \tag{24}$$

Example 1. For $n = 4$ the matrices U_4 and V_4 are given by

$$U_4 = \begin{pmatrix} 1 & 0 & 0 & 0 \\ b_{12} & -b_{12} & 0 & 0 \\ b_{12}b_{13} & -b_{12}b_{23} & -b_{13}b_{32} & 0 \\ b_{12}b_{13}b_{14} & -b_{12}b_{23}b_{24} & -b_{13}b_{32}b_{34} & -b_{14}b_{42}b_{43} \end{pmatrix},$$

$$V_4 = \begin{pmatrix} 1 & 0 & 0 & 0 \\ 1 & -b_{12}^{-1} & 0 & 0 \\ 1 & -b_{21}/(b_{12}b_{31}) & -1/(b_{13}b_{32}) & 0 \\ 1 & -b_{21}/(b_{12}b_{41}) & -b_{31}/(b_{13}b_{41}b_{42}) & -1/(b_{14}b_{42}b_{43}) \end{pmatrix}.$$

The matrices U_n and V_n may also be related to Theorem 7.

Theorem 11. *Let U_n and V_n be given by (4.1) and (4.2), respectively. Then,*

$$U_n = \prod_{i=0}^{n-2} \mathrm{Diag}(I_{n-i-2}, U_{n,n-i}), \tag{25}$$

$$V_n = \prod_{i=0}^{n-2} \mathrm{Diag}(I_i, U_{n,2+i}), \tag{26}$$

where $U_{n,k}$ is defined in (19).

Before considering the VTU-decomposition, i.e. the factorization $U_n B V_n = T_n$ which is one of the main theorems of the paper, where T_n is a triangular matrix specified below in Theorem 12, a technical lemma stating another basic property of $B \in \mathbb{B}_n$ is presented. Once again the proof is omitted.

Lemma 1. *Let $B \in \mathbb{B}_n$ and let $(U_n^{21} : U_n^{22})$ be the last row in U_n, given in (21). Then,*

$$(U_n^{21} : U_n^{22})B = 0.$$

Theorem 12. *(VTU-decomposition) Let $B \in \mathbb{B}_n$, U_n and $V_n = U_n^{-1}$ be the triangular matrices given by (21) and (22), respectively. Then $U_n B V_n = T_n$, where the upper triangular T_n equals*

$$T_n = \sum_{k=1}^{n} (n-k)e_k e_k' + \sum_{r=3}^{n} \sum_{k=1}^{r-2} \sum_{l=k+1}^{r-1} \prod_{m=k+1}^{l} b_{rm}^{-1} b_{lr} e_k e_l' - \sum_{r=3}^{n} \sum_{k=1}^{r-2} \prod_{m=k+1}^{r-1} b_{rm}^{-1} e_k e_r' - \sum_{k=1}^{n-1} e_k e_{k+1}'.$$

Proof. After the proof we show some details for $n = 3$. Suppose that $U_{n-1}B_{n-1}V_{n-1} = T_{n-1}$ holds, where $B_{n-1} \in \mathbb{B}_{n-1}$. Using the notation of Theorem 9

$$U_nBV_n = \begin{pmatrix} U_{n-1} & 0 \\ U_n^{21} & U_n^{22} \end{pmatrix} B \begin{pmatrix} V_{n-1} & 0 \\ V_n^{21} & V_n^{22} \end{pmatrix}$$

and let B be partitioned as

$$B = \begin{pmatrix} B^{11} & B^{12} \\ B^{21} & B^{22} \end{pmatrix}, \qquad \begin{pmatrix} n-1 \times n-1 & n-1 \times 1 \\ 1 \times n-1 & 1 \times 1 \end{pmatrix}.$$

From Lemma 1 it follows that $(U_n^{21} : U_n^{22})B = 0$ and thus

$$U_nBV_n = \begin{pmatrix} U_{n-1}B^{11}V_{n-1} + U_{n-1}B^{12}V_n^{21} & U_{n-1}B^{12}V_n^{22} \\ 0 & 0 \end{pmatrix}. \tag{27}$$

The blocks of the non-zero elements should be studied in some detail. Thus, one has to show that $U_{n-1}B^{12}V_n^{22}$ equals the first $n-1$ elements in the nth column of T_n. Let $T = (t_{ij})$, where $t_{ij} = 0$, if $i > j$. For example, for the second element in $U_{n-1}B^{12}V_n^{22}$:

$$-(-b_{12}b_{2n} + b_{12}b_{1n})b_{1n}^{-1}\prod_{m=2}^{n-1} b_{nm}^{-1} = -b_{n2}b_{1n}b_{1n}^{-1}\prod_{m=2}^{n-1} b_{nm}^{-1} = -\prod_{m=3}^{n-1} b_{nm}^{-1},$$

which equals t_{2n}.

For $U_{n-1}B^{11}V_{n-1} + U_{n-1}B^{12}V_n^{21}$, given in (27), it is noted that this expression equals

$$U_{n-1}B_{n-1}V_{n-1} + I - \sum_{i=1}^{n-1} b_{in}U_{n-1}d_id_i'V_{n-1} + \sum_{i=1}^{n-1} U_{n-1}b_{in}d_iV_n^{21} \tag{28}$$

and then the two last terms in (28) should be exploited. After some calculations this will give a useful recursive relation between U_nBV_n and $U_{n-1}B_{n-1}V_{n-1}$:

$$\begin{aligned} U_nB_nV_n =& (I_{n-1} : 0)'U_{n-1}B_{n-1}V_{n-1}(I_{n-1} : 0) - \sum_{k=1}^{n-2}\prod_{m=k+1}^{n-1} b_{nm}^{-1}e_ke_n' - \\ & e_{n-1}e_n' + \sum_{k=1}^{n-1} e_ke_k' + \sum_{k=1}^{n-2}\sum_{l=k+1}^{n-1}\prod_{m=k+1}^{l} b_{nm}^{-1}b_{ln}e_ke_l'. \end{aligned}$$

By utilizing this expression together with the induction assumption about $U_{n-1}B_{n-1}V_{n-1} = T_{n-1}$ leads to the T_n of the theorem. □

Corollary 3. *Let* $T_n = (t_{ij})$ *be the upper triangular matrix defined in Theorem 12. Then the elements of* T_n *are given by*

$$t_{ij} = \sum_{k=j+1}^{n} \prod_{l=i+1}^{j} b_{kl}^{-1} - \sum_{k=i}^{j-1} t_{ik} = \sum_{k=j+1}^{n} \prod_{l=i+1}^{j} b_{kl}^{-1} - I_{\{j>i\}} \sum_{k=j}^{n} \prod_{l=i+1}^{j-1} b_{kl}^{-1}.$$

Observe that the expression implies that $t_{ii} = n - i$. Moreover, $T_n 1 = 0$. The structure of the matrix T_n is the following

$$T_n' = \begin{pmatrix} n-1 & 0 & \dots & 0 & 0\,0 \\ \sum_{i'=3}^{n} b_{i'2}^{-1} - (n-1) & n-2 & \dots & 0 & 0\,0 \\ \sum_{i'=4}^{n} \prod_{j=2}^{3} b_{i'j}^{-1} - \sum_{i'=3}^{n} b_{i'2}^{-1} & \sum_{i'=4}^{n} b_{i'3}^{-1} - (n-2) & \ddots & 0 & 0\,0 \\ \vdots & \sum_{i'=5}^{n} \prod_{j=3}^{4} b_{i'j}^{-1} - \sum_{i'=4}^{n} b_{i'3}^{-1} & \dots & 2 & 0\,0 \\ \vdots & \vdots & \dots & b_{nn-1}^{-1} - 2 & 1\,0 \\ -\prod_{j'=2}^{n-1} b_{nj'}^{-1} & -\prod_{j=3}^{n} b_{i'j}^{-1} & \dots & -b_{nn-1}^{-1} & -1\,0 \end{pmatrix}.$$

This section is ended by showing some detailed calculations for $n = 3$.

Example 2. For $n = 3$

$$T_3 = \begin{pmatrix} 2 & b_{32}^{-1} - 2 & -b_{32}^{-1} \\ 0 & 1 & -1 \\ 0 & 0 & 0 \end{pmatrix}.$$

From (23) and (24) in Theorem 10 we have

$$U_3 = \begin{pmatrix} 1 & 0 & 0 \\ b_{12} & -b_{12} & 0 \\ b_{12}b_{13} & -b_{12}b_{23} & -b_{13}b_{32} \end{pmatrix}, \qquad V_3 = \begin{pmatrix} 1 & 0 & 0 \\ 1 & -b_{12}^{-1} & 0 \\ 1 & b_{23}b_{13}^{-1}b_{32}^{-1} & -b_{13}^{-1}b_{32}^{-1} \end{pmatrix}.$$

We are going to show that $V_3T_3U_3 = B \in \mathbb{B}_3$. Now

$$T_3U_3 = \begin{pmatrix} 2 + b_{32}^{-1}b_{12} - 2b_{12} - b_{32}^{-1}b_{12}b_{13} & b_{12} - b_{12}b_{13} & 0 \\ +2b_{12} + b_{32}^{-1}b_{23}b_{12} - b_{12}b_{32}^{-1} & -b_{12} + b_{12}b_{23} & 0 \\ b_{13} & b_{13}b_{32} & 0 \end{pmatrix}'$$

$$= \begin{pmatrix} 2 - 2b_{12} + (b_{12} - b_{13}) & b_{12}b_{32}^{-1}(1 - b_{23}) + 2b_{12} & b_{13} \\ b_{12}b_{31} & -b_{12}b_{32} & b_{13}b_{32} \\ 0 & 0 & 0 \end{pmatrix}$$

$$= \begin{pmatrix} b_{21} + b_{31} & b_{12} & b_{13} \\ b_{12}b_{31} & -b_{12}b_{32} & b_{13}b_{32} \\ 0 & 0 & 0 \end{pmatrix}$$

and

$$
\begin{aligned}
&V_3T_3U_3\\
&=\begin{pmatrix} b_{21}+b_{31} & b_{21} & b_{21}+b_{31}+b_{23}b_{13}^{-1}b_{32}^{-1}b_{12}b_{31} \\ b_{12} & b_{12}+b_{32} & b_{12}-b_{12}b_{32}b_{23}b_{13}^{-1}b_{32}^{-1} \\ b_{13} & -b_{13}b_{12}^{-1}b_{32}+b_{13} & b_{13}+b_{13}b_{32}b_{23}b_{13}^{-1}b_{32}^{-1} \end{pmatrix}'\\
&=\begin{pmatrix} b_{21}+b_{31} & b_{12} & b_{13} \\ b_{21} & b_{12}+b_{32} & -(b_{32}-b_{31})+b_{13} \\ b_{31} & b_{12}-(b_{23}-b_{21}) & b_{13}+b_{23} \end{pmatrix}\\
&=\begin{pmatrix} b_{21}+b_{31} & b_{12} & b_{13} \\ b_{21} & b_{12}+b_{32} & b_{23} \\ b_{31} & b_{32} & b_{13}+b_{23} \end{pmatrix}=B,
\end{aligned}
$$

where in the above calculations we have used (3) and Theorem 2 (ii).

5 Eigenvectors of the matrix B

It is already known from Theorem 5 that the matrix $B \in \mathbb{B}_n$ has eigenvalues $\{0, 1, \ldots, n-1\}$. This can also be seen from the structure of the matrix T_n given in Corollary 3 and the fact that the matrices B and T_n are similar, i.e. $U_n B U_n^{-1} = T_n$. The right eigenvectors of the matrix B are of special interest in sampling theory when B is a function of the inclusion probabilities, outlined in the Introduction. We are going to present the eigenvectors of the matrix $B \in \mathbb{B}_n$ in a general form. From Section 2 we know that $U_n B U_n^{-1} = T_n$, where the matrix T_n is an upper-triangular matrix given by Theorem 12. Since B and T_n are similar, they have the same eigenvalues and then the eigenvectors of B are rather easy to obtain using the eigenvectors of T_n. In the next theorem we shall obtain explicit expressions for the eigenvectors of the matrix T_n.

Theorem 13. *Let T_n be given by Theorem 12. Then there exist upper triangular matrices V_T and U_T such that*

$$
\begin{aligned}
T_n &= U_T \Lambda V_T,\\
\Lambda &= \mathrm{diag}(n-1, n-2, \ldots, 1, 0),\\
U_T &= V_T^{-1}.
\end{aligned}
\tag{29}
$$

The matrix $U_T = (u_{ij})$ is given by

$$
u_{ij} = 1 + \sum_{g=1}^{j-i} (-1)^g \sum_{i_1<\cdots<i_g}^{[j+1,n]} \sum_{j_1<\cdots<j_g}^{[i+1,j]} \prod_{k=1}^{g} \frac{b_{j_k i_k}}{b_{i_k j_k}}, \qquad i \leqslant j,
$$

where $\sum_{i_1<\cdots<i_g}^{[j+1,n]}$ is defined in (13).

The matrix $V_T = (v_{ij})$ *satisfies*

$$v_{ii} = 1,\ i = 1, \ldots, n, \tag{30}$$

$$v_{ii+1} = -1 + \sum_{i_1=i+2}^{n} \frac{b_{i+1 i_1}}{b_{i_1 i+1}}, \tag{31}$$

$$v_{ij} = \sum_{h=j-i-1}^{j-i} (-1)^{j-i-h} \sum_{i_1 \leqslant \cdots \leqslant i_h}^{[i+h+1,n]} \prod_{k=i+1}^{i+h} \frac{b_{k i_{k-i}}}{b_{i_{k-i} k}}, \quad j-i > 1, \tag{32}$$

where $\sum_{i_1 \leqslant \cdots \leqslant i_h}^{[i+h+1,n]}$ *is defined in (12).*

Proof. By presenting the main ideas we are going to indicate that $V_T = U_T^{-1}$. Let as in Theorem 2 $c_{ij} = b_{ij}^{-1} b_{ji}$.

The basic idea of the proof is to note that

$$U_T = R + Z^u, \tag{33}$$

$$V_T = R^{-1} + Z^v, \tag{34}$$

where R is a unit upper triangular matrix, i.e. an upper triangular of ones, $Z^u = (z_{ij}^u)$ is a strictly upper triangular matrix given by

$$z_{ij}^u = \sum_{g=1}^{j-i} (-1)^g \sum_{i_1 < \cdots < i_g}^{[j+1,n]} \sum_{j_1 < \cdots < j_g}^{[i+1,j]} \prod_{k=1}^{g} c_{i_k j_k}, \tag{35}$$

and z_{ij}^v in the strictly upper triangular matrix $Z^v = (z_{ij}^v)$ are given by

$$z_{ii+1}^v = \sum_{i_1=i+2}^{n} c_{i_1 i+1}, \tag{36}$$

$$z_{ij}^v = \sum_{h=j-i-1}^{j-i} (-1)^{j-i-h} \sum_{i_1 \leqslant \cdots \leqslant i_h}^{[i+h+1,n]} \prod_{k=i+1}^{i+h} c_{i_{k-i} k},\ j-i > 1. \tag{37}$$

The aim is to show that $V_T U_T = I$. It follows immediately that

$$V_T U_T = (R^{-1} + Z^v)(R + Z^u) = I + R^{-1} Z^u + Z^v R + Z^v Z^u.$$

Therefore

$$Z^v Z^u = -R^{-1} Z^u - Z^v R$$

is considered which is equivalent to the interesting equation

$$Z^v R R^{-1} Z^u = -R^{-1} Z^u - Z^v R. \tag{38}$$

Some calculations yield

$$R^{-1}Z^{u} = \sum_{i=1}^{n-1}\sum_{j=i+1}^{n}(z^{u}_{ij} - z^{u}_{i+1j})e_i e_j'$$
$$= \sum_{i=1}^{n-2}\sum_{j>i}^{n-1}\sum_{g=1}^{j-i}\sum_{i_1<\cdots<i_g}^{[j+1,n]}\sum_{j_2<\cdots<i_k}^{[i+2,j]}\prod_{k=2}^{g} c_{i_k j_k} c_{i_1 i+1} e_i e_j' \tag{39}$$

and

$$Z^{v}R = \sum_{i=1}^{n-2}\sum_{j=i+1}^{n-1}\sum_{i_1\leqslant\cdots\leqslant i_{j-i}}^{[j+1,n]}\prod_{k=1}^{j-i} c_{i_k i+k} e_i e_j'. \tag{40}$$

Relation (38) can be converted into

$$R^{-1}Z^{u} = (I + Z^{v}R)^{-1} - I. \tag{41}$$

Since $Z^{v}R$ is strictly upper triangular the right hand side of (41) is easy to interpret. The next lemma provides a useful result (e.g. see Harville, 1997, for related material).

Lemma 2. *Let* $T = (t_{ij}) : n \times n$ *be any strictly upper triangular matrix. Then*

$$\{(I+T)^{-1}\}_{ii} = 1, \quad i = 1, 2, \ldots, n,$$
$$\{(I+T)^{-1}\}_{ij} = -t_{ij} + \sum_{g=2}^{j-i}(-1)^{g}\sum_{i_1<\cdots<i_{g-1}}^{[i+1,j-1]} t_{ii_1}\prod_{k=2}^{g-1} t_{i_{k-1}i_k} t_{i_{g-1}j}.$$

Observe that the elements in $\{(I+T)^{-1}\}_{ij}$ and $\{(I+T)^{-1}\}_{i+1j+1}$ are of the same form. For example the second expression is obtained from the first if reindexing: $i \to i+1, i_1 \to i_1+1, i_2 \to i_2+1, \ldots, j \to j+1$. Therefore it follows that the most difficult part in verifying (41) is to verify that the last elements in the first row in both sides are equal, i.e.

$$-\sum_{k=3}^{n-2}(R^{-1}Z^{u})_{1k}(Z^{v}R)_{kn-1} - (Z^{v}R)_{1n-1} + (Z^{v}R)_{12}(Z^{v}R)_{2n-1}$$
$$= (R^{-1}Z^{u})_{1n-1}. \tag{42}$$

To verify the equality of the other elements in (41) follows by symmetry. However, to prove (42) is a fairly straightforward exercise where the key trick is to use Theorem 2 and therefore is omitted.

In the next it will be shown that $T_n = U_T\Lambda V_T$ which is the same as verifying

$$(I + Z^{v}R)R^{-1}TR = \Lambda(I + Z^{v}R). \tag{43}$$

Direct calculations via Corollary 3 give

$$(R^{-1}TR)_{il} = (n-i)I_{\{i=l\}} + \sum_{m=l+1}^{n} c_{mi+1} \prod_{s=i+2}^{l} b_{ms}^{-1} I_{\{i<l\}} \tag{44}$$

and from (40) it follows that

$$(I + Z^V R)_{ki} = I_{\{k=i\}} + \sum_{i_1 \leqslant \cdots \leqslant i_{i-k}}^{[i+1,n]} \prod_{r=1}^{i-k} c_{i_r k+r} I_{\{k<i, k\leqslant n-2, i\leqslant n-1\}}. \tag{45}$$

Thus (43) is true if

$$\sum_{i=1}^{n-1} (I + Z^V R)_{ki} (R^{-1}TR)_{il} = (n-k)(I + Z^V R)_{kl}$$

which by (44) and (45) can be written

$$\begin{aligned}
&\sum_{i=k}^{l} \{I_{\{k=i\}} + \sum_{i_1 \leqslant \cdots \leqslant i_{i-k}}^{[i+1,n]} \prod_{r=1}^{i-k} c_{i_r k+r}^{-1} I_{\{k<i, k\leqslant n-2, i\leqslant n-1\}}\} \\
&\qquad \times \{(n-i)I_{\{i=l\}} + \sum_{m=l+1}^{n} c_{mi+1} \prod_{s=i+2}^{l} b_{ms}^{-1} I_{\{i<l\}}\} \\
&\qquad = (n-k)(I + Z^V R)_{kl}.
\end{aligned} \tag{46}$$

Note that (46) immediately holds if $k = l$. For the off-diagonal elements in (46) one has to perform rather lengthy calculations and frequently use Theorem 2 (ii) and (iii). □

Now we are ready to present eigenvectors of the matrix B.

Theorem 14. *Let* $B \in \mathbb{B}_n$ *and the matrices* U_n, V_n *and* T_n *be given by* (21), (22) *and Theorem* 12, *respectively. Then the matrix of left eigenvectors* $W_L = (w_{ij}^L)$ *for* B *is given by*

$$w_{i1}^L = \prod_{k=2}^{i} b_{1k} \prod_{l=i+1}^{n} b_{l1}, \quad i = 1, \ldots, n, \tag{47}$$

$$w_{ii}^L = (-b_{1i})^{I_{\{i>1\}}} \prod_{k=2}^{i-1} b_{ik} \prod_{l=i+1}^{n} b_{li}, \tag{48}$$

$$w_{ij}^L = (-1)^{I_{\{i>j\}}} b_{1j} \prod_{\substack{k=2 \\ k\neq j}}^{i} b_{jk} \prod_{\substack{l=i+1 \\ l\neq j}}^{n} b_{lj}, \qquad j > 1. \tag{49}$$

Proof. Let W_L^n be W_L where n indicates the size of the matrix. Thus, from (47)–(49), it follows that

$$\begin{aligned} W_L^n = &\sum_{i=1}^{n}\prod_{k=2}^{i} b_{1k}\prod_{l=i+1}^{n} b_{l1}e_ie_1' + \sum_{i=1}^{n}(-b_{1i})^{I_{\{i>1\}}}\prod_{k=2}^{i-1} b_{ik}\prod_{l=i+1}^{n} b_{li}e_ie_i' \\ &-\sum_{i=3}^{n}\sum_{j=2}^{i-1} b_{1j}\prod_{\substack{k=2\\k\neq j}}^{i} b_{jk}\prod_{\substack{l=i+1\\l\neq j}}^{n} b_{lj}e_ie_j' + \sum_{j=2}^{n}\sum_{i=1}^{j} b_{1j}\prod_{\substack{k=2\\k\neq j}}^{i} b_{jk}\prod_{\substack{l=i+1\\l\neq j}}^{n} b_{lj}e_ie_j'. \end{aligned} \tag{50}$$

From Theorem 12 it follows that

$$B = V_nT_nU_n,$$

and from Theorem 13

$$T_n = U_T\Lambda V_T.$$

Hence, V_TU_n satisfies

$$V_TU_nB = \Lambda V_TU_n,$$

which means that V_TU_n is a matrix consisting of left eigenvectors. We are going to show that

$$W_L^n = V_TU_n.$$

From (50) it follows that

$$W_L^n = \begin{pmatrix} W_L^{n-1}D_n & 0 \\ 0 & 0 \end{pmatrix} + \begin{pmatrix} 0 & H_{\bullet n} \\ H_{n\bullet} & h_{nn} \end{pmatrix}, \tag{51}$$

where

$$\begin{aligned} D_n &= \sum_{i=1}^{n-1} b_{ni}d_id_i',\quad d_i : (n-1)\times 1, \\ H_{\bullet n} &= (h_{in}) = \sum_{i=1}^{n-1} b_{1n}\prod_{k=2}^{i} b_{nk}\prod_{l=i+1}^{n-1} b_{ln}d_i, \\ H_{n\bullet} &= (h_{ni}) = \prod_{k=2}^{n} b_{1k}d_1' - \sum_{j=2}^{n-1} b_{1j}\prod_{\substack{k=2\\k\neq j}}^{n} b_{jk}d_j, \\ h_{nn} &= -b_{1n}\prod_{k=2}^{n-1} b_{nk}. \end{aligned} \tag{52}$$

For small n, say $n \leqslant 6$, it is fairly easy to show that $W_L^n = V_T U_n$. Because of (51), we will indicate a proof for a general n, which is based on induction. Suppose that

$$W_L^{n-1} = (R_{n-1}^{-1} + Z_{n-1}^V)U_{n-1},$$

where it has been used that $V_T = R_{n-1}^{-1} + Z_{n-1}^V$ (see 34), and it will be shown that

$$W_L^n = (R_n^{-1} + Z_n^V)U_n,$$

which for small n directly can be verified. First observe that

$$(H_{n\bullet} : h_{nn}) = e_n' U_n = W_L^{n\bullet},$$

where $e_n : n \times 1$ and $W_L^{n\bullet}$ is the last row in W_L^n. Let $W_L^n = (w_{ij}^n)$ and $U_{\bullet n}$ stands for the last column in U_n. Then

$$w_{n-1n}^n = (0, 0, \ldots, 0, -1)U_{\bullet n} = h_{n-1n}. \tag{53}$$

It remains to show that for the first $n-2$ elements of $H_{\bullet n}$

$$h_{in} = ((R_n^{-1} + Z_n^V)U_n)_{in}, \quad i = 1, 2, \ldots, n-2, \tag{54}$$

and

$$((R_{n-1}^{-1} + Z_{n-1}^V)U_{n-1}D_{n-1})_{ij} = ((R_n^{-1} + Z_n^V)U_n)_{ij}, \ i, j = 1, 2, \ldots, n-1. \tag{55}$$

First consider (55). As in the proof of Theorem 13 we will use the products $Z_n^V R_n$ and $R_n^{-1} U_n$, and thus work with

$$(I + Z_n^V R_n) R_n^{-1} U_n, \tag{56}$$

which is simpler than $(R_n^{-1} + Z_n^V)U_n$. The expression for $Z_n^V R_n$ was given in (40), and thus

$$I + Z_n^V R_n = I + \sum_{i=1}^{n-2} \sum_{j=i+1}^{n-1} \sum_{i_1 \leqslant \cdots \leqslant i_{j-1}}^{[j+1,n]} \prod_{m=1}^{j-i} b_{i_m i+m}^{-1} b_{i+m i_m} e_i e_j'. \tag{57}$$

Moreover, (21) gives that

$$\begin{aligned} R_n^{-1} U_n = &\sum_{i=1}^{n} \prod_{m=2}^{i} b_{1m} b_{i+11}^{I_{\{i<n\}}} e_i e_1' + \sum_{j=2}^{n} b_{1j} \prod_{m=2}^{j-1} b_{jm} e_{j-1} e_j' - \\ &\sum_{j=2}^{n} \sum_{i=j}^{n} b_{1j} \prod_{\substack{m=2 \\ m \neq j}}^{i} b_{jm} b_{i+1j}^{I_{\{i<n\}}} e_i e_j'. \end{aligned} \tag{58}$$

It will be shown that the first column of the left hand side of (55) equals the first column of the right hand side. It is, besides lengthy calculations, fairly easy to verify (55) for the other columns too.

From (57) and (58) it follows that the first column in (55) can be written

$$\{\sum_{i=1}^{n-3}\sum_{j=i+1}^{n-2}\sum_{i_1\leqslant\cdots\leqslant i_{j-i}}^{[j+1,n-1]}\prod_{m=1}^{j-i} b^{-1}_{i_m i+m} b_{i+m i_m} e_i e_j' + e_i e_i'\}\{\sum_{k=1}^{n-1}\prod_{\underline{m}=2}^{k} b_{1\underline{m}} b_{k+11} b_{n1} e_k\} =$$
$$\{\sum_{i=1}^{n-3}\sum_{j=i+1}^{n-1}\sum_{i_1\leqslant\cdots\leqslant i_{j-i}}^{[j+1,n]}\prod_{m=1}^{j-i} b^{-1}_{i_m i+m} b_{i+m i_m} e_i e_j' + e_1 e_1'\}\{\sum_{k=1}^{n}\prod_{\underline{m}=2}^{k} b_{1\underline{m}} b_{k+11} e_k\}. \tag{59}$$

Furthermore, (59) is equivalent to

$$\prod_{\underline{m}=2}^{i} b_{1\underline{m}} b_{i+11} b_{n1} + \sum_{j=i+1}^{n-2}\sum_{i_1\leqslant\cdots\leqslant i_{j-i}}^{[j+1,n-1]}\prod_{m=1}^{j-i} b^{-1}_{i_m m+i} b_{m+i i_m}\prod_{\underline{m}=2}^{j} b_{1\underline{m}} b_{j+11} b_{n1} =$$
$$\prod_{\underline{m}=2}^{i} b_{1\underline{m}} b_{i+11} + \sum_{j=i+1}^{n-1}\sum_{i_1\leqslant\cdots\leqslant i_{j-i}}^{[j+1,n]}\prod_{m=1}^{j-i} b^{-1}_{i_m m+i} b_{m+i i_m}\prod_{\underline{m}=2}^{j} b_{1\underline{m}} b_{j+11}. \tag{60}$$

The validity of (60) will be exploited and first the left hand side is considered. Some manipulations give that it equals

$$\prod_{\underline{m}=2}^{i} b_{1\underline{m}}(b_{i+11} + \sum_{j=i+1}^{n-2}\sum_{i_1\leqslant\cdots\leqslant i_{j-i}}^{[j+1,n-1]}\prod_{m=i+1}^{j} b^{-1}_{i_{m-i} m} b_{m i_{m-i}} b_{1m} b_{j+11}) b_{n1}$$

which by applying Theorem 3 can be written

$$\prod_{\underline{m}=2}^{i} b_{1\underline{m}}\prod_{i+1}^{n-1} b_{m1} b_{n1} = \prod_{\underline{m}=2}^{i} b_{1\underline{m}}\prod_{i+1}^{n} b_{m1}.$$

When performing the same calculations to the right hand side of (60) we see that (60) is true.

Now, briefly consider (54). From (57) and (58) it follows that

$$\{(R_n^{-1} + Z_n^V)U_n\}_{in} = \{(I + Z_n^V R_n)R_n^{-1}U_n\}_{in} = \{Z_n^V R_n R_n^{-1} U_n\}_{in} =$$
$$\prod_{m=1}^{n-1-i} b^{-1}_{ni+m} b_{i+mn} b_{1n}\prod_{\underline{m}=2}^{n-1} b_{n\underline{m}} =$$
$$b_{1n}\prod_{m=i+1}^{n-1} b^{-1}_{nm} b_{mn} b_{1n}\prod_{\underline{m}=i+1}^{n-1} b_{n\underline{m}}\prod_{\underline{m}=2}^{i} b_{n\underline{m}} = b_{1n}\prod_{m=i+1}^{n-1} b_{mn}\prod_{\underline{m}=2}^{i} b_{n\underline{m}},$$

which implies that (52) as well as (54) hold. □

Theorem 15. *Let* $B \in \mathbb{B}_n$ *and* $c_{ij} = b_{ji}^{-1} b_{ij}$, $i \neq j$. *Then the matrix of left eigenvectors,* W_L, *defined in Theorem* 14 *can be decomposed as* $W_L = D_L V D_R$, *where* D_L *and* D_R *are diagonal matrices, and* V *is a Vandermonde matrix given by*

$$D_L = \sum_{i=1}^{n} \prod_{k=2}^{i} c_{k1} e_i e_i' \tag{61}$$

$$D_R = \prod_{l=2}^{n} b_{l1} e_1 e_1' + \sum_{i=2}^{n} b_{1i} \prod_{\substack{l=2 \\ l \neq i}}^{n} b_{li} e_i e_i' = \sum_{i=1}^{n} b_{1i}^{I_{\{i>1\}}} \prod_{\substack{l=2 \\ l \neq i}}^{n} b_{li} e_i e_i' \tag{62}$$

$$V = \sum_{i=1}^{n} e_i e_1' + \sum_{i=1}^{n} \sum_{j=2}^{n} (-1)^{i-1} c_{1j}^{i-1} e_i e_j' = \sum_{i=1}^{n} \sum_{j=1}^{n} ((-c_{1j})^{i-1})^{I_{\{j>1\}}} e_i e_j'. \tag{63}$$

Proof. The matrix W_L, defined in Theorem 14, can be written

$$W_L = D_L D_L^{-1} \{ \sum_{i=1}^{n} \prod_{k=2}^{i} c_{k1} e_i e_1' - \sum_{i=2}^{n} \prod_{k=2}^{i-1} c_{ki} e_i e_i' + \sum_{i=1}^{n} (-1)^{I_{\{i>j\}}} \prod_{\substack{k=2 \\ k \neq j}}^{n} c_{kj} e_i e_j' \} D_R.$$

Thus, it has to be shown that

$$D_L^{-1} \{ \sum_{i=1}^{n} \prod_{k=2}^{i} c_{k1} e_i e_1' - \sum_{i=2}^{n} \prod_{k=2}^{i-1} c_{ki} e_i e_i' + \sum_{i=1}^{n} (-1)^{I_{\{i>j\}}} \prod_{\substack{k=2 \\ k \neq j}}^{n} c_{kj} e_i e_j' \} = V,$$

where

$$D_L^{-1} = \sum_{i=1}^{n} \prod_{k=2}^{i} c_{1k} e_i e_i'.$$

However, from Theorem 2 (ii) it follows that

$$\prod_{k=2}^{i} c_{1k} \prod_{l=2}^{i-1} c_{li} = (-1)^{i-2} c_{1i}^{i-1}$$

which verifies the theorem. □

When obtaining the right eigenvectors for $B \in \mathbb{B}_n$ Theorem 15 will be utilized. Therefore a general expression for the inverse of a Vandermonde matrix is required, which is well-known (e.g. see El-Mikkaway, 2003). In the next lemma we present, without a proof, the result for the inverse Vandermonde matrix.

Lemma 3. *Let* V *be given in Theorem 15. Then* $W = (w_{ij})$, *where* $W = V^{-1}$, *is given by*

$$w_{ij} = \sum_{\substack{i_1<\cdots<i_{n-j}\\ i_1\neq i,\ldots,i_{n-j}\neq i}}^{[1,n]} (-c_{1i_1})^{I_{\{i_1>1\}}} \prod_{m=2}^{n-j}(-c_{1i_m})(-1)^{n-j} \times$$

$$\prod_{\substack{k=2\\k\neq i}}^{n}((c_{1k}+(-c_{1i})^{I_{\{i>1\}}})(-c_{1i}-1)^{I_{\{i>1\}}})^{-1}, \qquad j \leqslant n-1$$

$$w_{in} = \prod_{\substack{k=2\\k\neq i}}^{n}((c_{1k}+(-c_{1i})^{I_{\{i>1\}}})(-c_{1i}-1)^{I_{\{i>1\}}})^{-1}.$$

Remark 1. If $Q = (q_{ij})$ is a Vandermomde matrix, where $q_{ij} = (c_j)^{I_{\{i>1\}}}$, $c_k \neq c_l$, its inverse may immediately be obtained by noting that

$$\sum_{j=1}^{n} \sum_{\substack{i_1<\cdots<i_{n-j}\\ i_1\neq i,\ldots,i_{n-j}\neq i}}^{[1,n]} \prod_{m=1}^{n-j} c_{i_m} c_i^{j-1}(-1)^{n-j} = \prod_{\substack{j=1\\j\neq i}}^{n}(c_i - c_j),$$

$$\sum_{j=1}^{n} \sum_{\substack{i_1<\cdots<i_{n-j}\\ i_1\neq i,\ldots,i_{n-j}\neq i}}^{[1,n]} \prod_{m=1}^{n-j} c_{i_m} c_l^{j-1}(-1)^{n-j} = 0, \qquad l \neq i.$$

Before stating the last theorem it is noted that by applying Theorem 2

$$\prod_{\substack{k=2\\k\neq i}}^{n}((c_{1k}+(-c_{1i})^{I_{\{1>i\}}})(-c_{1i}-1)^{I_{\{1>i\}}})^{-1} = ((-c_{i1})^{n-2}(-b_{1i}))^{I_{\{i>1\}}} \prod_{\substack{k=2\\k\neq i}}^{n} b_{ik}. \quad (64)$$

The left hand side of this expression appears in Lemma 3. Using (64) a theorem concerning the right eigenvectors will be formulated and proved.

Theorem 16. *Let* $B \in \mathbb{B}_n$. *Then the matrix of right eigenvectors* $W_R = (w^R_{ij})$ *for* B *is the following:*

$$w^R_{1j} = \prod_{l=j+1}^{n} c_{l1} \sum_{i_1<\cdots<i_{n-j}}^{[2,n]} \prod_{m=1}^{n-j} c_{1i_m}, \qquad j = 1,2,\ldots,n-1,$$

$$w^R_{1n} = 1, \qquad w^R_{in} = -c_{1i}, \qquad i = 2,3,\ldots,n,$$

$$w^R_{ij} = \prod_{\substack{l=j+1\\l\neq i}}^{n} c_{l1} c_{1i}^{I_{\{i\leqslant j\}}} \sum_{\substack{i_1<\cdots<i_{n-j}\\ i_1\neq i,\ldots,i_{n-j}\neq i}}^{[1,n]} (-c_{1i_1})^{I_{\{i_1>1\}}} \prod_{m=2}^{n-j} c_{1i_m},$$

$$i = 2,3,\ldots,n, \quad j = 1,2,\ldots,n-1.$$

Proof. First the inverse of D_R, given by (62), will be multiplied with (64) which yields

$$(-1)^{I_{\{i>1\}}} \sum_{i=1}^{n} \prod_{\substack{l=2 \\ l \neq i}}^{n} c_{l1} e_i e_i'. \tag{65}$$

Moreover, multiplying

$$\prod_{\substack{k=2 \\ k \neq i}}^{n} (c_{1k} + (-c_{1i})^{I_{\{1>i\}}})(-c_{1i} - 1)^{I_{\{1>i\}}} V,$$

where V is the Vandermonde matrix presented in Theorem 15, with (65) from the left and with the inverse of D_L, given by (61), from the right we obtain after some manipulations

$$\sum_{i=1}^{n} (-c_{1i})^{I_{\{i>1\}}} e_i e_n' + \sum_{j=1}^{n-1} \prod_{l=j+1}^{n} c_{l1} \sum_{i_1 < \cdots < i_{n-j}}^{[2,n]} \prod_{m=1}^{n-j} c_{1i_m} e_1 e_j'$$
$$+ \sum_{i=2}^{n} \sum_{j=1}^{n-1} \prod_{\substack{l=j+1 \\ l \neq i}}^{n} c_{l1} c_{1i}^{I_{\{i \leqslant j\}}} \sum_{\substack{i_1 < \cdots < i_{n-j} \\ i_1 \neq i, \ldots, i_{n-j} \neq i}}^{[1,n]} (-c_{1i_1})^{I_{\{i_1>1\}}} \prod_{m=2}^{n-j} c_{1i_m} e_i e_j'.$$

This relation establishes the statement of the theorem. □

The theorem can be further simplified but this will be omitted because there is no really use of it and it burdens the presentation. However, in the next example one can see how the matrix given below in (68) reduces to the matrix in (67).

Example 3. In this example Theorem 2 will frequently be utilized. According to Theorem 14 the matrix of left eigenvectors, in the case $n = 4$, equals

$$\begin{pmatrix} b_{21}b_{31}b_{41} & b_{32}b_{42}b_{12} & b_{13}b_{23}b_{43} & b_{14}b_{24}b_{34} \\ b_{12}b_{31}b_{41} & -b_{12}b_{32}b_{42} & b_{13}b_{32}b_{43} & b_{14}b_{42}b_{34} \\ b_{12}b_{13}b_{41} & -b_{12}b_{23}b_{42} & -b_{13}b_{32}b_{43} & b_{14}b_{42}b_{43} \\ b_{12}b_{13}b_{14} & -b_{12}b_{23}b_{24} & -b_{13}b_{32}b_{34} & -b_{14}b_{42}b_{43} \end{pmatrix},$$

which can be decomposed as

$$\begin{pmatrix} 1 & 0 & 0 & 0 \\ 0 & c_{21} & 0 & 0 \\ 0 & 0 & c_{21}c_{31} & 0 \\ 0 & 0 & 0 & c_{21}c_{31}c_{41} \end{pmatrix} \begin{pmatrix} 1 & 1 & 1 & 1 \\ 1 & -c_{12} & -c_{13} & -c_{14} \\ 1 & c_{12}^2 & c_{13}^2 & c_{14}^2 \\ 1 & -c_{12}^3 & -c_{13}^3 & -c_{14}^3 \end{pmatrix} \times$$
$$\begin{pmatrix} b_{21}b_{31}b_{41} & 0 & 0 & 0 \\ 0 & b_{12}b_{32}b_{42} & 0 & 0 \\ 0 & 0 & b_{13}b_{23}b_{43} & 0 \\ 0 & 0 & 0 & b_{14}b_{24}b_{34} \end{pmatrix}. \tag{66}$$

The inverse of the Vandermonde matrix in this expression equals

$$\begin{pmatrix} b_{12}b_{13}b_{14} & 0 & 0 & 0 \\ 0 & -c_{21}^2 b_{12}b_{23}b_{24} & 0 & 0 \\ 0 & 0 & -c_{31}^2 b_{13}b_{32}b_{34} & 0 \\ 0 & 0 & 0 & -c_{21}^2 b_{14}b_{42}b_{43} \end{pmatrix} \times$$

$$\begin{pmatrix} c_{12}c_{13}c_{14} & c_{12}c_{13}+c_{12}c_{14}+c_{13}c_{14} & c_{12}+c_{13}+c_{14} & 1 \\ -c_{13}c_{14} & -c_{13}-c_{14}+c_{13}c_{14} & c_{13}+c_{14}-1 & 1 \\ -c_{12}c_{14} & -c_{12}-c_{14}+c_{12}c_{14} & c_{12}+c_{14}-1 & 1 \\ -c_{12}c_{13} & -c_{12}-c_{13}+c_{12}c_{13} & c_{12}+c_{13}-1 & 1 \end{pmatrix}.$$

Therefore, when taking the inverse of (66), the right eigenvectors are given by

$$\begin{pmatrix} c_{21}c_{31}c_{41} & 0 & 0 & 0 \\ 0 & -c_{21}^2 c_{32}c_{42} & 0 & 0 \\ 0 & 0 & -c_{31}^2 c_{23}c_{43} & 0 \\ 0 & 0 & 0 & -c_{41}^2 c_{24}c_{34} \end{pmatrix} \times$$

$$\begin{pmatrix} c_{12}c_{13}c_{14} & c_{12}c_{13}+c_{12}c_{14}+c_{13}c_{14} & c_{12}+c_{13}+c_{14} & 1 \\ -c_{13}c_{14} & -c_{13}-c_{14}+c_{13}c_{14} & c_{13}+c_{14}-1 & 1 \\ -c_{12}c_{14} & -c_{12}-c_{14}+c_{12}c_{14} & c_{12}+c_{14}-1 & 1 \\ -c_{12}c_{13} & -c_{12}-c_{13}+c_{12}c_{13} & c_{12}+c_{13}-1 & 1 \end{pmatrix} \times$$

$$\begin{pmatrix} 1 & 0 & 0 & 0 \\ 0 & c_{12} & 0 & 0 \\ 0 & 0 & c_{12}c_{13} & 0 \\ 0 & 0 & 0 & c_{12}c_{13}c_{14} \end{pmatrix} = \begin{pmatrix} c_{21}c_{31}c_{41} & 0 & 0 & 0 \\ 0 & -c_{31}c_{41} & 0 & 0 \\ 0 & 0 & -c_{21}c_{31} & 0 \\ 0 & 0 & 0 & -c_{21}c_{31} \end{pmatrix} \times$$

$$\begin{pmatrix} c_{12}c_{13}c_{14} & c_{12}^2c_{13}+c_{12}c_{14}+c_{13}c_{14} & c_{12}^2c_{13}+c_{12}c_{13}^2+c_{12}c_{13}c_{14} & c_{12}c_{13}c_{14} \\ -c_{13}c_{14} & -c_{13}c_{12}-c_{14}c_{12}+c_{12}c_{13}c_{14} & c_{12}c_{13}^2+c_{12}c_{13}c_{14}-c_{12}c_{13} & c_{12}c_{13}c_{14} \\ -c_{12}c_{14} & -c_{12}^2-c_{12}c_{14}+c_{12}^2c_{14} & c_{12}^2c_{13}+c_{12}c_{13}c_{14}-c_{12}c_{13} & c_{12}c_{13}c_{14} \\ -c_{12}c_{13} & -c_{12}^2-c_{12}c_{13}+c_{12}^2c_{13} & c_{12}^2c_{13}+c_{12}c_{13}^2-c_{12}c_{13} & c_{12}c_{13}c_{14} \end{pmatrix} =$$

$$\begin{pmatrix} 1 & -c_{42}-c_{32}+1 & 1-c_{43}-c_{42} & 1 \\ 1 & -c_{42}-c_{32}-c_{12} & c_{12}c_{43}-c_{12}-c_{42} & -c_{12} \\ 1 & -c_{42}+1-c_{12} & c_{12}c_{43}-c_{13}-c_{43} & -c_{13} \\ 1 & -c_{32}+1-c_{12} & -c_{12}-c_{13}+1 & -c_{14} \end{pmatrix}. \tag{67}$$

The last relation will now be compared to what is obtained from Theorem 16. It follows from the theorem that

$$W_R = \begin{pmatrix} 1 & c_{31}c_{41}(c_{12}c_{13}+c_{12}c_{14}+c_{13}c_{14}) & c_{41}(c_{12}+c_{13}+c_{14}) & 1 \\ 1 & c_{31}c_{41}c_{12}(c_{13}+c_{14}-c_{13}c_{14}) & c_{41}c_{12}(1-c_{13}-c_{14}) & -c_{12} \\ 1 & c_{41}(c_{12}+c_{14}-c_{12}c_{14}) & c_{41}c_{13}(-c_{12}-c_{14}+1) & -c_{13} \\ 1 & c_{31}(c_{12}+c_{13}-c_{12}c_{13}) & 1-c_{12}-c_{13} & -c_{14} \end{pmatrix} \tag{68}$$

which by some manipulations can be shown to be identical to (67). For example, by using Theorem 2 the element $w_{12}^R = c_{31}c_{41}(c_{12}c_{13}+c_{12}c_{14}+c_{13}c_{14}) = -c_{42}-c_{32}+1$.

References

1. N. Aires, *Algorithms to find exact inclusion probabilities for conditional Poisson sampling and Pareto* πps *sampling designs*, Methodol. Comput. Appl. Probab., (1999), pp. 457-469.
2. L. Bondesson, I. Traat, *On a matrix with integer eigenvalues and its relation to conditional Poisson sampling*, Res. Lett. Inf. Math. Sci., 2005, Vol.8, pp. 155–163.
3. L. Bondesson, I. Traat, *A nonsymmetric matrix with integer eigenvalues.* Linear Multilinear Algebra, 55, (2007), pp. 239–247.
4. D.A. Harville, *Matrix algebra from a statistician's perspective.* Springer-Verlag, New York, (1997).
5. T. Nahtman, D. von Rosen, *On a class of singular nonsymmetric matrices with nonnegative integer apectra.* Research Report 2007:10, Centre of Biostochastics, SLU.
6. N. Macon, A. Spitzbart, *Inverses of Vandermonde matrices.* Amer. Math. Monthly, 65, (1958), pp. 95–100.
7. M.E.A. El-Mikkawy, *Explicit inverse of a generalized Vandermonde matrix.* Appl. Math. Comput., 146, (2003), pp. 643–651.
8. O. Taussky, J. Todd, *Another look at a matrix of Mark Kac.* Proceedings of the First Conference of the International Linear Algebra Society (Provo, UT, 1989). Linear Algebra Appl. 150, (1991), pp. 341–360.

Reduction of a Set of Matrices over a Principal Ideal Domain to the Smith Normal Forms by Means of the Same One-Sided Transformations

V. M. Prokip

Pidstryhach Institute for Applied Problems of Mechanics and Mathematics
NAS of Ukraine,
3b Naukova Str., L'viv, 79060, Ukraine
vprokip@mail.ru

Abstract. We show that for $n \times n$ nonsingular matrices $A_1, A_2, \ldots, A_k$ $(k \geqslant 2)$ over a commutative principal ideal domain R with relatively coprime determinants there exist invertible $n \times n$ matrices $U, V_1, \ldots, V_k$ over R such that $UA_iV_i = S_{A_i}$ are the Smith normal forms of the matrices A_i.

Keywords: matrix, principal ideal domain, Smith normal form, equivalence of matrices.

Let R be a commutative principal ideal domain with an identity $e \neq 0$ (see [1], [4]) and let R_n denote the ring of $n \times n$ matrices over R. We denote by $UT(n, R)$ and by $LT(n, R)$ the multiplicative groups of upper and lower triangular matrices of order n over R respectively. By I_n we denote the identity matrix of order n and by 0_m^k we denote the zero $(m \times k)$-matrix.

Let $A \in R_n$ and rank$A = r$. For the matrix A there exist matrices $U, V \in GL(n, R)$ such that

$$UAB = S_A = \text{diag}(a_1,\ a_2, \ldots, a_r,\ 0, \ldots, 0)$$

is the diagonal matrix, where a_j for all integer $j = 1, 2, \ldots, r$ are nonzero elements of R such that $a_i | a_{i+1}$ (divides), $1 \leqslant i \leqslant r-1$. The diagonal matrix S_A is known as the Smith normal form of the matrix A (see [1], [4]). We denote by d_A^k the greatest common divisor (g.c.d.) of the minors of order k of the matrix A and by (a, b) the g.c.d. of nonzero elements $a, b \in R$.

V. Dlab and C. M. Ringel [2] have established canonical forms of the pairs of complex matrices A_1, A_2 by transformation $(A_1, A_2) \to (QA_1P_1, QA_2P_2)$, where Q is an invertible complex matrix and P_1, P_2 are invertible real matrices. In [5]–[7] authors studied when a pair of matrices (A, B) over a commutative principal ideal domain is reducible to a pair of diagonal matrices by transformations $(A, B) \to (UAV, UBW)$ with invertible matrices U, V, W. In this paper we establish conditions under which for matrices $A_1, A_2, \ldots, A_k \in R_n$, $(k \geqslant 2)$, there exist matrices $U, V_1, V_2, \dots, V_k \in GL(n, R)$ such that $UA_iV_i = S_{A_i}$ for all $i = 1, 2, \dots, k$.

To prove the main result, we need the following proposition.

Proposition 1. *Let* $A, B \in R_n$ *such that* $\operatorname{rank} A = \operatorname{rank} B = \operatorname{rank}\begin{bmatrix} A & B \end{bmatrix} = r$. *If* $(d_A^r, d_B^r) = e$ *then, for the matrices* A *and* B, *there exist matrices* $U, V_1, V_2 \in GL(n, R)$ *such that*

$$UAV_1 = S_A \quad \text{and} \quad UBV_2 = S_B.$$

Proof. Suppose that $A, B \in R_n$ are nonsingular matrices. It is obvious that

$$\operatorname{rank} A = \operatorname{rank} B = \operatorname{rank}\begin{bmatrix} A & B \end{bmatrix} = n.$$

Let $S_A = \operatorname{diag}(a_1, a_2, \dots, a_{n-1}, a_n)$ and $S_B = \operatorname{diag}(b_1, b_2, \dots, b_{n-1}, b_n)$ be the Smith normal form of the matrices A and B respectively.

Bearing in mind Corollary 1 in [8], for the matrices A and B there exist matrices $U_0, V_{01}, V_{02} \in GL(n, R)$ such that

$$U_0AV_{01} = S_A = \operatorname{diag}(a_1, a_2, \dots, a_n)$$

and

$$U_0BV_{02} = TS_B = \begin{bmatrix} b_1 & 0 & \dots & \dots & \dots & 0 \\ b_1t_{21} & b_2 & 0 & \dots & \dots & 0 \\ b_1t_{31} & b_2t_{32} & b_3 & 0 & \dots & 0 \\ \dots & \dots & \dots & \dots & \dots & \dots \\ b_1t_{n1} & b_2t_{n2} & b_3t_{n3} & \dots & b_{n-1}t_{n,n-1} & b_n \end{bmatrix},$$

where $T \in LT(n, R)$.

It is obvious that $a_k = a_1f_{k1}$, where $f_{k1} \in R$ for all $k = 2, 3, \dots, n$. Let $x_{21}, x_{31}, \dots, x_{n1}$ be arbitrary elements in R. Consider invertible matrices

$$U_1(\bar{x}_1) = \begin{bmatrix} e & 0 & \dots\dots\dots & 0 \\ f_{21}x_{21} & e & 0 \ \dots\dots & 0 \\ f_{31}x_{31} & 0 & e \quad 0 \ \dots & 0 \\ \dots & \dots\dots\dots & \dots\dots\dots & \dots \\ f_{n1}x_{n1} & 0 & 0 \ \dots \ 0 & e \end{bmatrix} \in LT(n, R)$$

and

$$V_1(\bar{x}_1) = \begin{bmatrix} e & 0 & \dots\dots\dots & 0 \\ -x_{21} & e & 0 \ \dots\dots & 0 \\ -x_{31} & 0 & e \quad 0 \ \dots & 0 \\ \dots & \dots\dots\dots & \dots\dots\dots & \dots \\ -x_{n1} & 0 & 0 \ \dots \ 0 & e \end{bmatrix} \in LT(n, R).$$

It is easy to make sure that for the matrices $U_1(\bar{x}_1)$, $V_1(\bar{x}_1)$ and S_A the following equality holds

$$U_1(\bar{x}_1)S_AV_1(\bar{x}_1) = S_A.$$

Now prove that for the matrix TS_B there exist $n \times n$ invertible matrices $U_1(\bar{\alpha}_1)$, $V_1(\bar{\alpha}_1)$ and $V_2(\bar{\beta}_1)$ of $LT(n, R)$ such that

$$U_1(\bar{\alpha}_1) S_A V_1(\bar{\alpha}_1) = S_A,$$

and

$$U_1(\bar{\alpha}_1) TS_B V_2(\bar{\beta}_1) = \begin{bmatrix} b_1 & 0 & \dots & \dots & \dots & 0 \\ 0 & b_2 & 0 & \dots & \dots & 0 \\ 0 & b_2 t_{3,2} & b_3 & 0 & \dots & 0 \\ \dots & \dots & \dots & \dots & \dots & \dots \\ 0 & b_2 t_{n,2} & b_3 t_{n,3} & \dots & b_{n-1} t_{n,n-1} & b_n \end{bmatrix}.$$

Consider the equation

$$U_1(\bar{y}_1) TS_B \begin{bmatrix} e \\ z_{21} \\ z_{31} \\ \vdots \\ z_{n1} \end{bmatrix} = \begin{bmatrix} b_1 \\ 0 \\ 0 \\ \vdots \\ 0 \end{bmatrix}. \tag{1}$$

Write equation (1) in the form

$$\begin{bmatrix} b_1 & 0 & \dots & \dots & & \dots & 0 \\ b_1(f_{21}y_{21} + t_{21}) & b_2 & 0 & \dots & & \dots & 0 \\ b_1(f_{31}y_{31} + t_{31}) & b_2 t_{32} & b_3 & 0 & \dots & \dots & 0 \\ b_1(f_{41}y_{41} + t_{41}) & b_2 t_{42} & b_3 t_{43} & b_4 & 0 & \dots & 0 \\ \dots & \dots & \dots & \dots & \dots & \dots & \dots \\ b_1(f_{n1}y_{n1} + t_{n1}) & b_2 t_{n2} & b_3 t_{n3} & b_4 t_{n4} & \dots & b_{n-1} t_{n,n-1} & b_n \end{bmatrix} \cdot \begin{bmatrix} e \\ z_{21} \\ z_{31} \\ z_{41} \\ \vdots \\ z_{n1} \end{bmatrix} = \begin{bmatrix} b_1 \\ 0 \\ \vdots \\ \vdots \\ \vdots \\ 0 \end{bmatrix}, \tag{2}$$

where $y_{21}, y_{31}, \dots, y_{n1}, z_{21}, z_{31}, \dots, z_{n1}$ are unknowns. It is evident that $b_i = b_j g_{ij}$, where $g_{ij} \in R$ for all $1 \leqslant j < i \leqslant n$. From equation (2) it follows

$$\begin{cases} f_{21}y_{21} + g_{21}z_{21} = -t_{21}, \\ f_{31}y_{31} + g_{21}t_{32}z_{21} + g_{31}z_{31} = -t_{31}, \\ f_{41}y_{41} + g_{21}t_{42}z_{21} + g_{31}t_{43}z_{31} + g_{41}z_{41} = -t_{31}, \\ \dots\dots\dots\dots\dots\dots\dots\dots\dots\dots \\ f_{n1}y_{n1} + g_{21}t_{n2}z_{21} + g_{31}t_{n3}z_{31} + \dots + g_{n1}z_1 = -t_{n1}. \end{cases} \tag{3}$$

Consider the first equation in (3)

$$f_{21}y_{21} + g_{21}z_{21} = -t_{21}.$$

Since $(f_{21}, g_{21}) = e$, then this equation is solvable. Let $y_{21} = \alpha_{21} \in R$ and $z_{21} = \beta_{21} \in R$ be the solution of this equation. The second equation in (3) can be rewritten in the form:

$$f_{31}y_{31} + g_{31}z_{31} = -(g_{21}t_{32}\beta_{21} + t_{31}).$$

Since $(f_{31}, g_{31}) = e$, we have that the last equality is solvable. Let $y_{31} = \alpha_{31} \in R$ and $z_{31} = \beta_{21} \in R$ be the solution of the last equation. We represent the third equation in (3) in the form

$$f_{41}y_{41} + g_{41}z_{41} = -(g_{21}t_{42}\beta_{21} + g_{31}t_{43}\beta_{31} + t_{31}). \tag{4}$$

Since $(f_{41}, g_{11}) = e$, then the equation (4) is also solvable.

After a finite number of steps we obtain that the system of linear equation (3) is solvable. Let

$$y_{j1} = \alpha_{j1} \in R \quad \text{and} \quad z_{j1} = \beta_{j1} \in R, \quad j = 2, 3, \dots, n,$$

be the solution of system linear equations (3). Thus, for the matrices S_A and TS_B there exist the invertible matrices

$$U_1(\bar{\alpha}_1) = \begin{bmatrix} e & 0 & \dots & \dots & \dots & 0 \\ f_{21}\alpha_{21} & e & 0 & \dots & \dots & 0 \\ f_{31}\alpha_{31} & 0 & e & 0 & \dots & 0 \\ \dots & \dots & \dots & \dots & \dots & \dots \\ f_{n1}\alpha_{n1} & 0 & 0 & \dots & 0 & e \end{bmatrix} \in LT(n, R),$$

$$V_1(\bar{\alpha}_1) = \begin{bmatrix} e & 0 & \dots & \dots & \dots & 0 \\ -\alpha_{21} & e & 0 & \dots & \dots & 0 \\ -\alpha_{31} & 0 & e & 0 & \dots & 0 \\ \dots & \dots & \dots & \dots & \dots & \dots \\ -\alpha_{n1} & 0 & 0 & \dots & 0 & e \end{bmatrix} \in LT(n, R)$$

and

$$V_2(\bar{\beta}_1) = \begin{bmatrix} e & 0 & \dots & \dots & \dots & 0 \\ \beta_{21} & e & 0 & \dots & \dots & 0 \\ \beta_{31} & 0 & e & 0 & \dots & 0 \\ \dots & \dots & \dots & \dots & \dots & \dots \\ \beta_{n1} & 0 & 0 & \dots & 0 & e \end{bmatrix} \in LT(n, R)$$

such that:

$$U_1(\bar{\alpha}_1)S_A V_1(\bar{\alpha}_1) = S_A$$

and

$$U_1(\bar{\alpha}_1)TS_B V_2(\bar{\beta}_1) = \begin{bmatrix} b_1 & 0 & \dots & \dots & \dots & 0 \\ 0 & b_2 & 0 & \dots & \dots & 0 \\ 0 & b_2t_{3,2} & b_3 & 0 & \dots & 0 \\ \dots & \dots & \dots & \dots & \dots & \dots \\ 0 & b_2t_{n,2} & b_2t_{n,3} & \dots & b_{n-1}t_{n,n-1} & b_n \end{bmatrix} = \operatorname{diag}(e, T_1)S_B,$$

where $T_1 \in LT(n-1, R)$.

We now carry on the similar reasoning for the matrices S_A and $\mathrm{diag}(e, T_1)S_B$. Evidently, that $a_k = a_2 f_{k2}$ for all $k = 3, 4, \dots, n$, where $f_{k2} \in R$. Let x_{32}, x_{42}, $\dots$, x_{n2} be arbitrary elements from R. For invertible matrices

$$U_2(\bar{x}_2) = \begin{bmatrix} e & 0 & \dots\dots & & \dots & 0 \\ 0 & e & 0 & \dots\dots & \dots & 0 \\ 0 & f_{32}x_{32} & e & 0 & \dots\dots & 0 \\ 0 & f_{42}x_{42} & 0 & e & 0\ \dots & 0 \\ \dots & \dots & \dots & \dots & \dots & \dots \\ 0 & f_{n2}x_{n2} & 0 & 0 & \dots\ 0 & e \end{bmatrix} \in LT(n, R)$$

and

$$V_2(\bar{x}_1) = \begin{bmatrix} e & 0 & \dots\dots & & \dots & 0 \\ 0 & e & 0 & \dots\dots & \dots & 0 \\ 0 & -x_{32} & e & 0 & \dots\dots & 0 \\ 0 & -x_{42} & 0 & e & 0\ \dots & 0 \\ \dots & \dots & \dots & \dots & \dots & \dots \\ 0 & -x_{n2} & 0 & 0 & \dots\ 0 & e \end{bmatrix} \in LT(n, R)$$

we have

$$U_1(\bar{x}_1)S_A V_1(\bar{x}_1) = S_A.$$

It is easy to make sure that for the matrices S_A and $\mathrm{diag}(e, T_1)S_B$ there exist the invertible matrices $U_2(\bar{\alpha}_2)$, $V_2(\bar{\alpha}_2) \in LT(n, R)$ and

$$V_2(\bar{\beta}_2) = \begin{bmatrix} e & 0 & \dots\dots & & \dots & 0 \\ 0 & e & 0 & \dots\dots & \dots & 0 \\ 0 & \beta_{32} & e & 0 & \dots\dots & 0 \\ 0 & \beta_{42} & 0 & e & 0\ \dots & 0 \\ \dots & \dots & \dots & \dots & \dots & \dots \\ 0 & \beta_{n2} & 0 & 0 & \dots\ 0 & e \end{bmatrix} \in LT(n, R)$$

such that:

$$U_1(\bar{\alpha}_2)S_A V_1(\bar{\alpha}_2) = S_A$$

and

$$U_1(\bar{\alpha}_2)\mathrm{diag}(e,T_1)S_B V_2(\bar{\beta}_2) =$$

$$\begin{bmatrix} b_1 & 0 & \dots & \dots & \dots & \dots & 0 \\ 0 & b_2 & 0 & \dots & \dots & \dots & 0 \\ 0 & 0 & b_3 & 0 & \dots & \dots & 0 \\ 0 & 0 & b_3 t_{43} & b_3 & 0 & \dots & 0 \\ \dots\dots & & \dots & \dots & \dots & \dots & \dots \\ 0 & 0 & b_3 t_{n3} & b_4 t_{n4} & \dots & b_{n-1}t_{n,n-1} & b_n \end{bmatrix} = \mathrm{diag}(I_2, T_2)S_B,$$

where $T_2 \in LT(n-2, R)$.

After a finite number of steps we reduce the nonsingular matrices A and B to their Smith normal forms S_A and S_B respectively.

Now consider the case $\operatorname{rank}A = \operatorname{rank}B = \operatorname{rank}\begin{bmatrix} A & B \end{bmatrix} = r < n$. It is obvious, that for the matrix $\begin{bmatrix} A & B \end{bmatrix}$ there exists a matrix $U_0 \in GL(n, R)$ such that

$$U_0 \begin{bmatrix} A & B \end{bmatrix} = \begin{bmatrix} A_0 & B_0 \\ 0^n_{n-r} & 0^n_{n-r} \end{bmatrix},$$

where A_0 and B_0 are $r \times n$ matrices over R. For the matrices A_0 and B_0 there exist matrices $V_{01}, V_{02} \in GL(n, R)$ such that $A_0 V_{01} = \begin{bmatrix} A_1 & 0^{n-r}_r \end{bmatrix}$ and $B_0 V_{02} = \begin{bmatrix} B_1 & 0^{n-r}_r \end{bmatrix}$, where $A_1, B_1 \in R_r$. Thus, for the matrices A and B there exist matrices $U_0, V_{01}, V_{02} \in GL(n, R)$ such that

$$U_0 A V_{01} = \begin{bmatrix} A_1 & 0^{n-r}_r \\ 0^r_{n-r} & 0^{n-r}_{n-r} \end{bmatrix} \quad \text{and} \quad U_0 B V_{02} = \begin{bmatrix} B_1 & 0^{n-r}_r \\ 0^r_{n-r} & 0^{n-r}_{n-r} \end{bmatrix}.$$

Since $(d^r_A, d^r_B) = e$ and $A_1, B_1 \in R_r$, we have $(\det A_1, \det B_1) = e$. Thus, for the matrices A_1 and B_1 there exist matrices $U_1, V_{11}, V_{12} \in GL(r, R)$ such that $U_1 A_1 V_{11} = S_{A_1}$ and $U_1 B_1 V_{12} = S_{B_1}$. In this particular case it is not difficult to check that for the invertible matrices

$$U = \begin{bmatrix} U_1 & 0^{n-r}_r \\ 0^r_{n-r} & I_{n-r} \end{bmatrix} U_0, \quad V_1 = V_{01} \begin{bmatrix} V_{11} & 0^{n-r}_r \\ 0^r_{n-r} & I_{n-r} \end{bmatrix} \quad \text{and} \quad V_2 = V_{02} \begin{bmatrix} V_{11} & 0^{n-r}_r \\ 0^r_{n-r} & I_{n-r} \end{bmatrix}$$

the following relations hold

$$UAV_1 = \begin{bmatrix} S_{A_1} & 0^{n-r}_r \\ 0^r_{n-r} & 0^{n-r}_{n-r} \end{bmatrix} = S_A \quad \text{and} \quad UAV_2 = \begin{bmatrix} S_{B_1} & 0^{n-r}_r \\ 0^r_{n-r} & 0^{n-r}_{n-r} \end{bmatrix} = S_B.$$

This completes the proof of Proposition 1. □

Theorem 1. *Let* $A_1, A_2, \ldots, A_m \in R_n$ *be nonsingular matrices. If* $(\det A_i, \det A_j) = e$ *for all* $i \neq j$; $i, j = 1, 2, \ldots, m$; *then there exist matrices* $U, V_1, V_2, \ldots, V_m \in GL(n, R)$ *such that*

$$UA_jV_j = S_{A_j}, \quad j = 1, 2, \ldots, m.$$

Proof. We prove the theorem by induction on m. For $m = 2$ the proof follows from the proposition. Suppose that the theorem holds for $m = k$. Let $m = k + 1$ and $A_1, A_2, \ldots, A_k, A_{k+1} \in R_n$ be nonsingular matrices with relatively coprime determinants, i.e. $(\det A_i, \det A_j) = e$ for all $i \neq j$. By the induction hypothesis, the theorem is true for the matrices $A_1, A_2, \ldots, A_k \in R_n$. Let $W_0, W_1, W_2, \ldots, W_k \in GL(n, R)$ such that

$$W_0 A_l W_l = S_{A_l} = \operatorname{diag}\Big(s_1(A_l), s_2(A_l), \ldots, s_n(A_l)\Big),$$

be the Smith normal forms of the matrices A_l , $l = 1, 2, \ldots, k$.

We denote by $B = W_0 A_{k+1}$ and by $S_B = \mathrm{diag}(b_1, b_2, \ldots, b_{n-1}, b_n)$. It is obvious that $S_B = S_{Ak+1}$. In accordance with Remark 1 in [8] for the matrix B there exist invertible matrices

$$Q = \begin{bmatrix} e & q_{12} & q_{13} & \ldots & q_{1n} \\ 0 & e & q_{23} & \ldots & q_{2n} \\ \ldots & \ldots & \ldots & \ldots & \ldots \\ 0 & 0 & \ldots & 0 & e \end{bmatrix} \in UT(n, R) \quad \text{and} \quad W_{k+1} \in GL(n, R)$$

such that

$$QBW_{m+1} = TS_B,$$

where $T \in LT(n, R)$. It is easily verified that for the matrices Q and S_{A_l} there exist matrices $Q_l \in UT(n, R)$ such that

$$QS_{A_l} = S_{A_l} Q_l, \quad \text{for all} \quad l = 1, 2, \ldots, k.$$

So, we have the invertible $n \times n$ matrices QW_0, $W_1 Q_1^{-1}$, $W_2 Q_2^{-1}$, ..., $W_k Q_k^{-1}$, W_{k+1} such that

$$QW_0 A_1 W_1 Q_1^{-1} = S_{A_1}, \quad QW_0 A_2 W_2 Q_2^{-1} = S_{A_2}, \quad \ldots, \quad QW_0 A_k W_k Q_k^{-1} = S_{A_k},$$

and

$$QW_0 A_{k+1} W_{k+1} = TS_{A_{k+1}} = TS_B, \quad \text{where} \quad T \in LT(n, R).$$

Let us define

$$f_i = s_i(A_1) s_i(A_2) \cdots s_i(A_k), \quad i = 1, 2, \ldots, n;$$

$$\begin{aligned} h_i^1 &= s_i(A_2) s_i(A_3) \cdots s_i(A_k), \\ h_i^l &= s_i(A_1) \cdots s_i(A_{l-1}) s_i(A_{l+1}) \cdots s_i(A_k), \quad l = 2, 3, \ldots, k-1, \\ h_i^k &= s_i(A_1) s_i(A_2) \cdots s_i(A_{k-1}). \end{aligned}$$

It is easy to check that

$$f_i = s_i(A_l) h_i^l, \quad i = 1, 2, \ldots, n;$$

$$f_i = f_{ij} f_j \quad \text{and} \quad h_i^l = h_{ij}^l h_j^l, \quad 1 \leqslant j < i \leqslant n,$$

for all $l = 1, 2, \ldots, k$.

Let $x_{21}, x_{31}, \ldots, x_{n1}$ be arbitrary elements from R. Consider invertible matrices

$$U_1(\bar{x}_1) = \begin{bmatrix} e & 0 & \ldots\ldots\ldots & 0 \\ f_{21} x_{21} & e & 0 \ \ldots\ldots & 0 \\ f_{31} x_{31} & 0 & e \ \ 0 \ \ldots & 0 \\ \ldots & \ldots\ldots & \ldots\ldots & \ldots \\ f_{n1} x_{n1} & 0 & 0 \ \ldots \ 0 & e \end{bmatrix} \in LT(n, R)$$

and

$$V_{1l}(\bar{x}_1) = \begin{bmatrix} e & 0 & \dots & \dots & \dots & 0 \\ -h^l_{21}x_{21} & e & 0 & \dots & \dots & 0 \\ -h^l_{31}x_{31} & 0 & e & 0 & \dots & 0 \\ \dots & \dots & \dots & \dots & \dots & \dots \\ -h^l_{n1}x_{n1} & 0 & 0 & \dots & 0 & e \end{bmatrix} \in LT(n, R), \quad l = 1, 2, \dots, k.$$

It is easy to see that for the matrices $U_1(\bar{x}_1)$, $V_{11}(\bar{x}_1)$, $V_{12}(\bar{x}_1)$, ..., $V_{1k}(\bar{x}_1)$ and S_{A_l} the following equalities hold:

$$U_1(\bar{x}_1)S_{A_l}V_{1l}(\bar{x}_1) = S_{A_l} \quad \text{for all} \quad l = 1, 2, \dots, k.$$

Since $(\det A_i, \det A_j) = e$ for all $i \neq j$, we have $(f_{k1}, b_m) = e$ for all $2 \leqslant k \leqslant n$ and $1 \leqslant m \leqslant n$. Now we prove that for matrix TS_B there exist matrices $U_1(\bar{\alpha}_1)$ and $V_B(\bar{\beta}_1)$ such that

$$U_1(\bar{\alpha}_1)TS_BV_B(\bar{\beta}_1) = \begin{bmatrix} b_1 & 0 & \dots & \dots & \dots & 0 \\ 0 & b_2 & 0 & \dots & \dots & 0 \\ 0 & b_2t_{32} & b_3 & 0 & \dots & 0 \\ \dots & \dots & \dots & \dots & \dots & \dots \\ 0 & b_2t_{n2} & b_3t_{n3} & \dots & b_{n-1}t_{n,n-1} & b_n \end{bmatrix}. \tag{5}$$

Consider the equation (see also the proof of Proposition 1)

$$U_1(\bar{y}_1)TS_B \begin{bmatrix} e \\ z_{21} \\ z_{31} \\ \vdots \\ z_{n1} \end{bmatrix} = \begin{bmatrix} b_1 \\ 0 \\ 0 \\ \vdots \\ 0 \end{bmatrix}. \tag{6}$$

Applying arguments analogous to those used in the proof of Proposition 1 to equation (6), we have that there exist invertible matrices $U_1(\bar{\alpha}_1)$ and $V_B(\bar{\beta}_1)$ such that equality (5) is true. Further, proceeding by analogy, we get the proof of Theorem 1. □

Note. The theorem presented above is true for a commutative adequate ring, [3]. Furthermore, the obtained Theorem 1 can be easily extended to the commutative elementary divisor domains R, where for any three elements $a \neq 0$, $b \neq 0$ and c from R, there exists an element $t \in R$ such that the greatest common divisor of (a, b, c) coincides with the greatest common divisor of $(a + tc, tb)$.

Acknowledgments

The author is grateful to the referee for his careful reading of the manuscript, helpful comments and suggestions.

References

1. C.W. Brown. Matrices over commutative rings. Monographs and Textbooks in Pure and Applied Math., Vol.160. Marcel Dekker, New York, 1993.
2. V.Dlab and C. M.F.Ringel. Canonical forms of pairs of complex matrices. Linear Algebra Appl., **147**(1991), pp. 387-410.
3. O. Helmer. The elementary divisors for certain rings without chain condition. Bull. Amer. Math. Soc., **49**(1943), pp. 225–236.
4. M. Newman. Integral Matrices. – Academic Press: New York, 1972.
5. V.M.Petrychkovych. A generalized equivalence of a pair of matrices. Mat. Stud., **8**(1997), pp. 147-152.
6. V.M.Prokip. On the problem of multiplicativity of canonical diagonal forms of matrices. Ukrain. Math. Journal., **47**(1995), pp. 1581–1585.
7. V.M.Prokip. On the structure of divisors of matrices over the principal ideal domain, Dopov. Nats. Akad. Nauk Ukr. 2002, no. 6, 27–32.
8. V.M.Prokip. Structure of matrices and their divisors over the domain of principal ideals. Ukrain. Math. Journal., **54**(2002), pp. 1378–1385.

MATRICES AND ALGORITHMS

Nonsymmetric Algebraic Riccati Equations Associated with an M-Matrix: Recent Advances and Algorithms*

Dario A. Bini[1], Bruno Iannazzo[2], Beatrice Meini[1], and Federico Poloni[3]

[1] Dipartimento di Matematica, Università di Pisa,
bini@dm.unipi.it,
meini@dm.unipi.it

[2] Dipartimento di Fisica e Matematica, Università dell'Insubria, Via Valleggio, 11, 22100 Como, Italy
iannazzo@mail.dm.unipi.it

[3] Scuola Normale Superiore, Pisa
f.poloni@sns.it

Abstract. We survey theoretical properties and algorithms concerning the problem of solving a nonsymmetric algebraic Riccati equation, and we report on some known methods and new algorithmic advances. In particular, some results on the number of positive solutions are proved and a careful convergence analysis of Newton's iteration is carried out in the cases of interest where some singularity conditions are encountered. From this analysis we determine initial approximations which still guarantee the quadratic convergence.

Keywords: matrix equation, M-matrix, nonsymmetric algebraic Riccati equation, Newton's method, shift technique.

1 Introduction

Nonsymmetric Algebraic Riccati equations (NARE) are quadratic matrix equations of the kind

$$XCX - AX - XD + B = 0, \tag{1}$$

where the unknown X is an $m \times n$ matrix, and the coefficients A, B, C and D have sizes $m \times m$, $m \times n$, $n \times m$ and $n \times n$, respectively.

The term *nonsymmetric* distinguishes this case from the widely studied continuous-time algebraic Riccati equations, defined by the quadratic matrix equation $XCX - AX - XA^T + B = 0$, where B and C are symmetric. We refer the reader to the books [38, 44] for a comprehensive analysis of continuous-time algebraic Riccati equations.

The matrix coefficients of the NARE (1) define the $(m+n) \times (m+n)$ matrix

$$M = \begin{bmatrix} D & -C \\ -B & A \end{bmatrix}, \tag{2}$$

* This work was supported by MIUR grant number 2004015437.

which, throughout the paper, we assume to be an M-matrix. This assumption is motivated by the increasing applicative interest of this kind of NAREs, and by the recent theoretical and algorithmic advances that have been achieved.

We recall that M is an M-matrix if it can be written as $M = \alpha I - N$ where N has nonnegative entries, $\alpha \geqslant \rho(N)$, and $\rho(N)$ is the spectral radius of N. We say that equation (1) is associated with an M-matrix if its coefficients, arranged as in (2), form an M-matrix.

There are two important applications where nonsymmetric algebraic Riccati equations associated with M-matrices play a very important role: the study of fluid queues models [47, 46, 48], and the analysis of transport equations [35, 34]. In both cases the solution of interest is the matrix S with nonnegative entries, which among all the nonnegative solutions is the one with componentwise minimal entries. We call any solution S sharing this property *minimal nonnegative solution*. These applications will be outlined in Sections 1.1 and 1.2.

The research activity concerning the analysis of NAREs associated with M-matrices and the design of numerical algorithms for their solution has had a strong acceleration in the last decade. Important progress has been made concerning theoretical properties of this class of matrix equations and new effective algorithms relying on the properties of M-matrices have been designed and analyzed [6, 13, 11, 10, 15, 18, 20, 21, 23–26, 32, 35, 41–43].

In this paper we provide a survey of the most important results and of the most effective algorithms concerning the analysis and the numerical treatment of the NAREs associated with M-matrices together with some new results. We also provide a unifying framework where different techniques and properties can be described in a simpler form and where more insights into the properties of matrix equations are given.

In particular, we report on results concerning the existence of a minimal nonnegative solution S and prove some new results on the number of nonnegative solutions of the NARE (1). We analyze the spectral properties of the matrix

$$H = \begin{bmatrix} D & -C \\ B & -A \end{bmatrix},$$

relate them to the eigenvalues of S and use this relation when H is singular to classify the problem into three classes according to the sign of a scalar parameter referred as the *drift* associated with the equation.

After reporting on perturbation results of the solution, we present, analyze and compare different algorithms for computing the minimal nonnegative solution S. Besides a "direct" method based on the Schur decomposition we consider functional iterations having linear convergence and then the Newton iteration which has a generally quadratic rate of convergence. The class of doubling algorithms is discussed: this class includes cyclic reduction and the Structure-preserving Doubling Algorithm (SDA). Here we report very recent results relating these two algorithms, in particular the proof that SDA turns out to be cyclic reduction applied to a specific problem.

We give a separate treatment on the case of interest where the associated matrix M is singular. Here we prove that a particular choice of the initial point in Newton's iteration can restore the quadratic speed of convergence which otherwise would be linear. In fact, we provide a simple but general result concerning the "structured" convergence of Newton's iteration.

Furthermore, we discuss the possibility of replacing the original equation with a different one, having the same solution S, but where the singularity is removed. The advantage that we get with this technique is twofold: on one hand we can accelerate the speed of iterative methods by switching from the linear to the quadratic convergence; on the other hand we may guarantee the full machine accuracy ε in the solution which otherwise would be $O(\sqrt{\varepsilon})$.

Numerical experiments which validate our theoretical analysis conclude the paper.

The paper is structured as follows: in Sections 1.1 and 1.2 we describe the applications of NAREs; in Section 2 we deal with theoretical properties and in Section 3 with algorithms; the case where M is singular is discussed in Section 4 while Section 5 reports the results of numerical experiments and the concluding remarks.

1.1 Application to fluid queues

In the analysis of two dimensional continuous-time Markov processes, called fluid queues, a crucial step is to compute the element-wise minimal nonnegative solution S of the NARE (1). In [3, 46, 47, 17, 1, 6], the fluid flow models are described in terms of a two-dimensional continuous-time Markov process denoted by $\{(X(t), \varphi(t)), \quad t \geqslant 0\}$ where $X(t)$ represents the level, while $\varphi(t)$ represents the phase. The phase process $\{\varphi(t) : \quad t \geqslant 0\}$ is an irreducible Markov chain with space state $\mathcal{S}_1 \cup \mathcal{S}_2$, $\mathcal{S}_1 = \{1, 2, \dots, m\}$, $\mathcal{S}_2 = \{m+1, m+2, \dots, m+n\}$, and infinitesimal generator the opposite of (2). The minimal nonnegative solution $S = (s_{i,j})$ of (1) is such that $s_{i,j}$ is the probability that, starting from level x in phase $i \in \mathcal{S}_2$, the process $(X(t), \varphi(t))$ first returns to level x in finite time and does so in phase $j \in \mathcal{S}_1$, while avoiding levels below x. A detailed description of this kind of models can be found in [46].

1.2 Application to transport equation

Riccati equations associated with M-matrices also appear in a problem in neutron transport theory, a variation of the one-group neutron transport equation, described in [35] where the mathematical model consists in solving an integrodifferential equation. After discretization of this integrodifferential equation, the problem can be expressed as the following equation for an unknown matrix $X \in \mathbb{R}^{n\times n}$

$$\Delta X + X\widehat{\Delta} = (Xq + e)(e^T + q^T X), \tag{3}$$

with

$$\Delta = \mathrm{diag}(\delta_1, \ldots, \delta_n), \qquad \widehat{\Delta} = \mathrm{diag}(\hat{\delta}_1, \ldots, \hat{\delta}_n),$$
$$\delta_i = \frac{1}{cx_i(1-\alpha)}, \qquad \hat{\delta}_i = \frac{1}{cx_i(1+\alpha)}, \quad i = 1, \ldots, n,$$
$$e = [1\ 1\ \cdots\ 1]^T, \qquad q_i = \frac{w_i}{2x_i}, \quad i = 1, \ldots, n.$$

The matrices and vectors above depend on the two parameters $0 < c \leqslant 1$, $0 \leqslant \alpha < 1$, and on the sequences $(x_i)_{i=1}^n$ and $(w_i)_{i=1}^n$, which are the nodes and weights of a Gaussian quadrature on $[0, 1]$, ordered such that (x_i) is decreasing. The solution of physical interest is the minimal nonnegative one, whose existence can be proved thanks to Theorem 7 that we report in Section 2.3.

Equation (3) coincides with the NARE (1) with

$$A = \widehat{\Delta} - eq^T, \qquad B = ee^T,$$
$$C = qq^T, \qquad D = \Delta - qe^T.$$

Under these hypotheses it is easy to prove that M is a diagonal-plus-rank-1 M-matrix. Due to this additional structure, ad-hoc algorithms can be developed, such as the ones described in [43, 42, 11]. Moreover, the iterates appearing when implementing most of the traditional algorithms are structured and can be completely described with $O(n)$ parameters. Therefore, structured versions of these algorithms can be developed, resulting in quadratically convergent iterations for (3) that require only $O(n^2)$ operations per step, as shown in [11] for Newton's method.

2 Theoretical properties

Before analyzing the numerical methods for the effective solution of equation (1), it is worth giving some theoretical properties of the NARE.

A large amount of properties concerning equation (1) have been stated in [18, 20, 21, 25, 35]; we summarize some of them. These results concern algebraic Riccati equations associated with nonsingular or singular irreducible M-matrices. The case in which M is singular and reducible is of minor interest.

A nonzero matrix $A = (a_{ij})$ is said *nonnegative* (*nonpositive*) if $a_{ij} \geqslant 0$ ($a_{ij} \leqslant 0$). In this case one writes $A \geqslant 0$ ($A \leqslant 0$). A matrix $A = (a_{ij})$ is said *positive* (*negative*) if $a_{ij} > 0$ ($a_{ij} < 0$). In this case one writes $A > 0$ ($A < 0$).

A matrix B is called a Z-matrix, if there exists a nonnegative matrix A such that $B = sI - A$, for a suitable scalar s. In other words a Z-matrix is a matrix all whose off-diagonal elements are nonpositive.

A matrix B is called an M-matrix, if it can be written in the form $B = sI - A$, where A is nonnegative, $s > 0$ and $s \geqslant \rho(A)$. If $s = \rho(A)$ the M-matrix is singular. Observe that an M-matrix is a Z-matrix.

We denote the set of the eigenvalues of A by $\sigma(A)$. Throughout the paper, e will denote the vector with components equal to 1, whose length is specified by the context.

2.1 Some useful facts about nonnegative matrices

A nonnegative matrix maps the cone of nonnegative vectors into itself; this cone contains an eigenvector as stated by the following celebrated result [8].

Theorem 1 (Perron–Frobenius theorem). *Any nonnegative matrix A has a real eigenvalue $\lambda \geqslant 0$ such that $|\mu| \leqslant \lambda$ for each $\mu \in \sigma(A)$. Moreover, there exists a vector $v \geqslant 0$ such that $Av = \lambda v$.*

Any irreducible nonnegative matrix A has a real eigenvalue $\lambda > 0$ such that $|\mu| \leqslant \lambda$ for each $\mu \in \sigma(A)$. Moreover, λ is simple and there exists a vector $v > 0$ such that $Av = \lambda v$.

If A is positive, then $|\mu| < \lambda$ for each $\mu \in \sigma(A)\backslash\{\lambda\}$.

We state a useful corollary of the Perron–Frobenius theorem.

Corollary 2. *Let A be an irreducible nonnegative matrix and let $v_1, \ldots, v_n$ be a set of Jordan chains of A. Then there exists only one positive or negative vector among the v_i's and it is a scalar multiple of v, where v is the positive vector of Theorem 1.*

From the Perron–Frobenius theorem many interesting properties of Z- and M-matrices follow. For instance, a Z-matrix has a "leftmost" (in the complex plane) real eigenvalue corresponding to a nonnegative eigenvector, for an M-matrix this eigenvalue is nonnegative. In particular, one deduces that the eigenvalues of an M-matrix have nonnegative real part.

A very common problem is to check if a given Z-matrix is an M-matrix. The following result states some equivalent conditions for a Z-matrix to be a nonsingular M-matrix. The proofs can be found in [8].

Theorem 3. *For a Z-matrix A, the following conditions are equivalent:*

(a) *A is a nonsingular M-matrix;*
(b) $A^{-1} \geqslant 0$;
(c) $Au > 0$ *for some vector* $u > 0$;
(d) *All the eigenvalues of A have positive real parts.*

Theorem 4. *For a Z-matrix* A *it holds that:* A *is an M-matrix if and only if there exists a nonzero vector* $v \geqslant 0$ *such that* $Av \geqslant 0$ *or a nonzero vector* $w \geqslant 0$ *such that* $w^T A \geqslant 0$.

The equivalence of (a) and (c) in Theorem 3 implies the next result.

Lemma 5. *Let* A *be a nonsingular M-matrix. If* $B \geqslant A$ *is a Z-matrix, then* B *is also a nonsingular M-matrix.*

The following well-known result concerns the properties of Schur complements of M-matrices.

Lemma 6. *Let* M *be a nonsingular M-matrix or an irreducible singular M-matrix. Partition* M *as*

$$M = \begin{bmatrix} M_{11} & M_{12} \\ M_{21} & M_{22} \end{bmatrix},$$

where M_{11} *and* M_{22} *are square matrices. Then* M_{11} *and* M_{22} *are nonsingular M-matrices. The Schur complement of* M_{11} *(or* M_{22}*) in* M *is also an M-matrix (singular or nonsingular according to* M*). Moreover, the Schur complement is irreducible if* M *is irreducible.*

2.2 The dual equation

Reverting the coefficients of equation (1) yields the *dual equation*

$$YBY - YA - DY + C = 0, \tag{4}$$

which is still a NARE, associated with the matrix

$$N = \begin{bmatrix} A & -B \\ -C & D \end{bmatrix}$$

that is a nonsingular M-matrix or an irreducible singular M-matrix if and only if the matrix M is so. In fact N is clearly a Z-matrix and $N = \Pi M \Pi$, where $\Pi = \Pi^{-1}$ is the matrix which permutes the blocks of M. So, if $Mv \geqslant 0$, for $v \geqslant 0$, then $N\Pi v \geqslant 0$ and by Theorem 4, N is an M-matrix.

2.3 Existence of nonnegative solutions

The special structure of the matrix M of (2) allows one to prove the existence of a minimal nonnegative solution S of (1), i.e., a solution $S \geqslant 0$ such that $X - S \geqslant 0$ for any solution $X \geqslant 0$ to (1). See [20] and [21] for more details.

Theorem 7. *Let* M *in* (2) *be an M-matrix. Then the NARE* (1) *has a minimal nonnegative solution* S. *If* M *is irreducible, then* $S > 0$ *and* $A - SC$ *and* $D - CS$ *are irreducible M-matrices. If* M *is nonsingular, then* $A - SC$ *and* $D - CS$ *are nonsingular M-matrices.*

Observe that the above theorem holds for the dual equation (4) and guarantees the existence of a minimal nonnegative solution of (4) which is denoted by T.

2.4 The eigenvalue problem associated with the matrix equation

A useful technique frequently encountered in the theory of matrix equations consists in relating the solutions to some invariant subspaces of a matrix polynomial.

In particular, the solutions of (1) can be described in terms of the invariant subspaces of the matrix

$$H = \begin{bmatrix} D & -C \\ B & -A \end{bmatrix}, \tag{5}$$

which is obtained premultiplying the matrix M by $J = \begin{bmatrix} I_n & 0 \\ 0 & -I_m \end{bmatrix}$.

In fact, if X is a solution of equation (1), then, by direct inspection,

$$H \begin{bmatrix} I_n \\ X \end{bmatrix} = \begin{bmatrix} I_n \\ X \end{bmatrix} R, \tag{6}$$

where $R = D - CX$. Moreover, the eigenvalues of the matrix R are a subset of the eigenvalues of H. Conversely, if the columns of the $(n+m) \times n$ matrix $\begin{bmatrix} Y \\ Z \end{bmatrix}$ span an invariant subspace of H, and Y is a nonsingular $n \times n$ matrix, then ZY^{-1} is a solution of the Riccati equation, in fact

$$H \begin{bmatrix} Z \\ T \end{bmatrix} = \begin{bmatrix} Z \\ T \end{bmatrix} V,$$

for some V, from which post-multiplying by Z^{-1} one obtains

$$H \begin{bmatrix} I \\ TZ^{-1} \end{bmatrix} = \begin{bmatrix} Z \\ T \end{bmatrix} VZ^{-1} = \begin{bmatrix} I \\ TZ^{-1} \end{bmatrix} ZVZ^{-1};$$

setting $X = TZ^{-1}$ one has $D - CX = ZVZ^{-1}$ and $B - AX = XD - XCX$.

Similarly, for the solutions of the dual equation it holds that

$$H \begin{bmatrix} Y \\ I_m \end{bmatrix} = \begin{bmatrix} Y \\ I_m \end{bmatrix} U,$$

where $U = BY - A$. The eigenvalues of the matrix U are a subset of the eigenvalues of H.

2.5 The eigenvalues of H

We say that a set $\mathcal{A}$ of k complex numbers has a (k_1, k_2) splitting with respect to the unit circle if $k = k_1 + k_2$, and $\mathcal{A} = \mathcal{A}_1 \cup \mathcal{A}_2$, where $\mathcal{A}_1$ is formed by k_1 elements of modulus at most 1 and $\mathcal{A}_2$ is formed by k_2 elements of modulus at least 1. Similarly, we say that $\mathcal{A}$ has a (k_1, k_2) splitting with respect to the imaginary axis if $k = k_1 + k_2$, and $A = \mathcal{A}_1 \cup \mathcal{A}_2$, where $\mathcal{A}_1$ is formed by k_1 elements with nonpositive real part and $\mathcal{A}_2$ is formed by k_2 elements with nonnegative real part. We say that the splitting is *complete* if at lest one set $\mathcal{A}_1$ or $\mathcal{A}_2$ has no eigenvalues in its boundary.

Since the eigenvalues of an M-matrix have nonnegative real part, it follows that the eigenvalues of H have an (m, n) splitting with respect to the imaginary axis. This property is proved in the next

Theorem 8. *Let* M *be an irreducible M-matrix. Then the eigenvalues of* $H = JM$ *have an* (m, n) *splitting with respect to the imaginary axis. Moreover, the only eigenvalue that can lie on the imaginary axis is* 0.

Proof. Let $v > 0$ be the only positive eigenvector of M, and let $\lambda \geqslant 0$ be the associate eigenvalue; define $D_v = \mathrm{diag}(v)$. The matrix $\overline{M} = D_v^{-1} M D_v$ has the same eigenvalues as M; moreover, it is an M-matrix such that $\overline{M}e = \lambda e$. Due to the sign structure of M-matrices, this means that $\overline{M}$ is diagonal dominant (strictly in the nonsingular case). Notice that $\overline{H} = D_v^{-1} H D_v = J\overline{M}$, thus $\overline{H}$ is diagonal dominant as well, with m negative and n positive diagonal entries. We apply Gershgorin's theorem [30, Sec. 14] to $\overline{H}$; due to the diagonal dominance, the Gershgorin circles never cross the imaginary axis (in the singular case, they are tangent in 0). Thus, by using a continuity argument we can say that m eigenvalues of $\overline{H}$ lie in the negative half-plane and n in the positive one, and the only eigenvalues on the imaginary axis are the zero ones. But since H and $\overline{H}$ are similar, they have the same eigenvalues. □

We can give a more precise result on the location of the eigenvalues of H, after defining the *drift* of the Riccati equation. Indeed, when M is a singular irreducible M-matrix, by the Perron–Frobenius theorem, the eigenvalue 0 is simple, there are positive vectors u and v such that

$$u^T M = 0, \quad Mv = 0, \tag{7}$$

and both the vectors u and v are unique up to a scalar factor.

Writing $u = \begin{bmatrix} u_1 \\ u_2 \end{bmatrix}$ and $v = \begin{bmatrix} v_1 \\ v_2 \end{bmatrix}$, with $u_1, v_1 \in \mathbb{R}^n$ and $u_2, v_2 \in \mathbb{R}^m$, one can define

$$\mu = u_2^T v_2 - u_1^T v_1 = -u^T J v. \tag{8}$$

The number μ determines some properties of the Riccati equation. Depending on the sign of μ and following a Markov chain terminology, one can call μ the *drift* as in [6], and can classify the Riccati equations associated with a singular irreducible M-matrix in three categories:

(a) *positive recurrent* if $\mu < 0$;
(b) *null recurrent* if $\mu = 0$;
(c) *transient* if $\mu > 0$.

In fluid queues problems, v coincides with the vector of ones. In general v and u can be computed by performing the LU factorization of the matrix M, say $M = LU$, and solving the two triangular linear systems $u^T L = [0, \ldots, 0, 1]$ and $Uv = 0$ (see [30, Sec. 54]).

The location of the eigenvalues of H is made precise in the following [20, 23]:

Theorem 9. *Let M be a nonsingular or a singular irreducible M-matrix, and let $\lambda_1, \ldots, \lambda_{m+n}$ be the eigenvalues of $H = JM$ ordered by nonincreasing real part. Then λ_n and λ_{n+1} are real and*

$$\mathrm{Re}\lambda_{n+m} \leqslant \cdots \leqslant \mathrm{Re}\lambda_{n+2} < \lambda_{n+1} \leqslant 0 \leqslant \lambda_n < \mathrm{Re}\lambda_{n-1} \leqslant \cdots \leqslant \mathrm{Re}\lambda_1.$$

The minimal nonnegative solutions S and T of the equation (1) and of the dual equation (4), respectively, are such that $\sigma(D - CS) = \{\lambda_1, \ldots, \lambda_n\}$ and $\sigma(A - SC) = \sigma(A - BT) = \{-\lambda_{n+1}, \ldots, -\lambda_{n+m}\}$.

If M is nonsingular then $\lambda_{n+1} < 0 < \lambda_n$. If M is singular and irreducible then:

1. *if $\mu < 0$ then $\lambda_n = 0$ and $\lambda_{n+1} < 0$;*
2. *if $\mu = 0$ then $\lambda_n = \lambda_{n+1} = 0$ and there exists only one eigenvector, up to a scalar constant, for the eigenvalue 0;*
3. *if $\mu > 0$ then $\lambda_n > 0$ and $\lambda_{n+1} = 0$.*

We call λ_n and λ_{n+1} the *central eigenvalues* of H. If H (and thus M) is nonsingular, then the central eigenvalues lie on two different half planes so the splitting is complete. In the singular case the splitting is complete if and only if $\mu \neq 0$.

The *close to null recurrent* case, i.e., the case $\mu \approx 0$, deserves particular attention, since it corresponds to an ill-conditioned null eigenvalue for the matrix H. In fact, if u and v are normalized such that $\|u\|_2 = \|v\|_2 = 1$, then $1/|\mu|$ is the condition number of the null eigenvalue for the matrix H (see [19]).

When M is singular irreducible, for the Perron–Frobenius theorem the eigenvalue 0 is simple, therefore $H = JM$ has a one dimensional kernel and $u^T J$ and v are the unique (up to a scalar constant) left and right eigenvectors, respectively, corresponding to the eigenvalue 0. However the algebraic multiplicity of 0 as an eigenvalue of H can be 2; in that case, the Jordan form of H has a 2×2 Jordan block corresponding to the 0 eigenvalue and it holds $u^T J v = 0$ [31].

The next result, presented in [25], shows the reduction from the case $\mu < 0$ to the case $\mu > 0$ and conversely, when M is singular irreducible. This property enable us to restrict our interest only to the case $\mu \leqslant 0$.

Lemma 10. *The matrix S is the minimal nonnegative solution of* (1) *if and only if $Z = S^T$ is the minimal nonnegative solution of the equation*

$$XC^TX - XA^T - D^TX + B^T = 0. \tag{9}$$

Therefore, if M is singular and irreducible, the equation (1) *is transient if and only if the equation* (9) *is positive recurrent.*

Proof. The first part is easily shown by taking transpose on both sides of the equation (1). The M-matrix corresponding to (9) is

$$M_t = \begin{bmatrix} A^T & -C^T \\ -B^T & D^T \end{bmatrix}.$$

Since

$$[v_2^T \; v_1^T]\, M_t = 0, \quad M_t \begin{bmatrix} u_2 \\ u_1 \end{bmatrix} = 0,$$

the second part readily follows. □

2.6 The differential of the Riccati operator

The matrix equation (1) defines a Riccati operator

$$\mathcal{R}(X) = XCX - AX - XD + B,$$

whose differential $d\mathcal{R}_X$ at a point X is

$$d\mathcal{R}_X[H] = HCX + XCH - AH - HD. \tag{10}$$

The differential $H \to d\mathcal{R}_X[H]$ is a linear operator which can be represented by the matrix

$$\Delta_X = (CX - D)^T \otimes I_m + I_n \otimes (XC - A), \tag{11}$$

where $\otimes$ denotes the Kronecker product (see [30, Sec. 10]).

We say that a solution X of the matrix equation (1) is critical if the matrix Δ_X is singular.

From the properties of Kronecker product [30, Sec. 10], it follows that the eigenvalues of Δ_X are the sums of those of $CX - D$ and $XC - A$. If $X = S$, where S is the minimal nonnegative solution, then $D - CX$ and $A - XC$ are M-matrices (compare Theorem 7), and thus all the eigenvalues of Δ_S have nonpositive real parts. Moreover, since $D - CS$ and $A - SC$ are M-matrices then $-\Delta_S$ is an M-matrix. The minimal nonnegative solution S is critical if and only if both M-matrices $D - CS$ and $A - SC$ are singular, thus, in view of Theorem 9, the minimal solution is critical if and only if M is irreducible singular and $\mu = 0$.

Moreover, if $0 \leqslant X \leqslant S$ then $D - CX \geqslant D - CS$ and $A - XC \geqslant A - SC$ are nonsingular M-matrices by lemma 5, thus $-\Delta_X$ is a nonsingular M-matrix.

2.7 The number of positive solutions

If the matrix M is irreducible, Theorem 7 states that there exists a minimal positive solution S of the NARE. In the study of nonsymmetric Riccati differential equations associated with an M-matrix [18, 34] one is interested in all the positive solutions.

In [18] it is shown that if M is nonsingular or singular irreducible with $\mu \neq 0$, then there exists a second solution S_+ such that $S_+ > S$ and S_+ is obtained by a rank one correction of the matrix S. More precisely, the following result holds [18].

Theorem 11. *If M is irreducible nonsingular or irreducible singular with $\mu \neq 0$, then there exists a second positive solution S_+ of (1) given by*

$$S_+ = S + kab^T,$$

where $k = (\lambda_n - \lambda_{n+1})/b^T Ca$, a is such that $(A - SC)a = -\lambda_{n+1}a$ and b is such that $b^T(D - CS) = \lambda_n b^T$.

We prove that there are exactly two nonnegative solutions in the noncritical case and only one in the critical case. In order to prove this result it is useful to study the form of the Jordan chains of an invariant subspace of H corresponding to a positive solution.

Lemma 12. *Let M be irreducible and let Σ be any positive solution of (1). Denote by $\eta_1, \ldots, \eta_n$ the eigenvalues of $D - C\Sigma$ ordered by nondecreasing real part. Then η_1 is real, and there exists a positive eigenvector v of H associated with η_1. Moreover, any other vector independent of v, belonging to Jordan chains of H corresponding to $\eta_1, \ldots, \eta_n$ cannot be positive or negative.*

Proof. Since Σ is a solution of (1), then from (6) one has

$$H \begin{bmatrix} I \\ \Sigma \end{bmatrix} = \begin{bmatrix} I \\ \Sigma \end{bmatrix} (D - C\Sigma).$$

Since $D - CS$ is an irreducible M-matrix for Theorem 7, and $\Sigma \geqslant S$ (S is the minimal positive solution), then $D - C\Sigma$ is an irreducible Z-matrix and thus can be written as $sI - N$ with N nonnegative and irreducible. Then by Theorem 1 and Corollary 2 η_1 is a simple real eigenvalue of $D - C\Sigma$, the corresponding eigenvector can be chosen positive and there are no other positive or negative eigenvectors or Jordan chains corresponding to any of the eigenvalues. Let $P^{-1}(D - C\Sigma)P = K$ be the Jordan canonical form of $D - C\Sigma$, where the first column of P is the positive eigenvector corresponding to η_1. Then we have

$$H \begin{bmatrix} P \\ \Sigma P \end{bmatrix} = \begin{bmatrix} P \\ \Sigma P \end{bmatrix} K.$$

Thus, the columns of $\begin{bmatrix} P \\ \Sigma P \end{bmatrix}$ are the Jordan chains of H corresponding to $\eta_1, \ldots, \eta_n$, and there are no positive or negative columns, except for the first one. □

Theorem 13. *If M is an irreducible nonsingular M-matrix or an irreducible singular M-matrix with $\mu \neq 0$, then (1) has exactly two positive solutions. If M is irreducible singular with $\mu = 0$, then (1) has a unique positive solution.*

Proof. From Lemma 12 applied to S it follows that H has a positive eigenvector corresponding to λ_n, and no other positive or negative eigenvectors or Jordan chains corresponding to $\lambda_1, \ldots, \lambda_n$. Let T be the minimal nonnegative solution of the dual equation (4). Then

$$H \begin{bmatrix} T \\ I \end{bmatrix} = \begin{bmatrix} T \\ I \end{bmatrix} (-(A - BT)).$$

As in the proof of Lemma 12, we can prove that H has a positive eigenvector corresponding to the eigenvalue λ_{n+1} and no other positive or negative eigenvectors or Jordan chains corresponding to $\lambda_{n+1}, \ldots, \lambda_{n+m}$.

If M is irreducible nonsingular, or irreducible singular with $\mu \neq 0$, then $\lambda_n > \lambda_{n+1}$, and there are only two linearly independent positive eigenvectors corresponding to real eigenvalues. By Lemma 12, there can be at most two solutions corresponding to $\lambda_n, \lambda_{n-1}, \ldots, \lambda_1$, and to $\lambda_{n+1}, \lambda_{n-1}, \ldots, \lambda_1$, respectively. Since it is know from Theorem 11 that there exist at least two positive solutions, thus (1) has exactly two positive solutions.

If M is irreducible singular with $\mu = 0$, there is only one positive eigenvector corresponding to $\lambda_n = \lambda_{n+1}$, and the unique solution of (1) is obtained by the Jordan chains corresponding to $\lambda_n, \lambda_{n-1}, \ldots, \lambda_1$. □

The next results provide a useful property of the minimal solutions which will be useful in Section 4.

Theorem 14. *Let M be singular and irreducible, and let S and T be the minimal nonnegative solutions of (1) and (4), respectively. Then the following properties hold:*

(a) *if $\mu < 0$, then $Sv_1 = v_2$ and $Tv_2 < v_1$;*
(b) *if $\mu = 0$, then $Sv_1 = v_2$ and $Tv_2 = v_1$;*
(c) *if $\mu > 0$, then $Sv_1 < v_2$ and $Tv_2 = v_1$.*

Proof. From the proof of Theorem 13, it follows that if $\mu \neq 0$, there exist two independent positive eigenvectors a and b of H relative to the central eigenvalues λ_n and λ_{n+1}, respectively. We write $a = \begin{bmatrix} a_1 \\ a_2 \end{bmatrix}$ and $b = \begin{bmatrix} b_1 \\ b_2 \end{bmatrix}$, with $a_1, b_1 \in \mathbb{R}^n$ and $a_2, b_2 \in \mathbb{R}^m$.

Since the solution S is constructed from an invariant subspace containing a, then $Sa_1 = a_2$, since the solution S_+ is constructed from an invariant subspace containing b, then $S_+b_1 = b_2$. Analogously, if T_+ is the second positive solution of the dual equation, then $Tb_2 = b_1$ and $T_+a_2 = a_1$.

The statements (a) and (c) follow from the fact that if $\mu < 0$ then $v = a$ (compare Theorem 9), so $Sv_1 = v_2$ and $Tv_2 < T_+v_2 = v_1$, since $T < T_+$; if $\mu > 0$ then $v = b$, so $Tv_2 = v_1$ and $Sv_1 < S_+v_1 = v_2$, since $S < S_+$.

The statement (b) corresponding to the case $\mu = 0$ can be proved in a similar way. □

Remark 1. When $\mu \geqslant 0$, from Lemma 10 and Theorem 14 we deduce that the minimal nonnegative solution S of (1) is such that $u_2^T S = u_1^T$.

2.8 Perturbation analysis for the minimal solution

We conclude this section with a result of Guo and Higham [24] who perform a qualitative description of the perturbation of the minimal nonnegative solution S of a NARE (1) associated with an M-matrix.

The result is split in two theorems where an M-matrix $\widetilde{M}$ is considered which is obtained by means of a *small* perturbation of M. Here, we denote by $\widetilde{S}$ the minimal nonnegative solution of the perturbed Riccati equation associated with $\widetilde{M}$.

Theorem 15. *If M is a nonsingular M-matrix or an irreducible singular M-matrix with $\mu \neq 0$, then there exist constants $\gamma > 0$ and $\varepsilon > 0$ such that $\|\widetilde{S} - S\| \leqslant \gamma \|\widetilde{M} - M\|$ for all $\widetilde{M}$ with $\|\widetilde{M} - M\| < \varepsilon$.*

Theorem 16. *If M is an irreducible singular M-matrix with $\mu = 0$, then there exist constants $\gamma > 0$ and $\varepsilon > 0$ such that*

(a) $\|\widetilde{S} - S\| \leqslant \gamma \|\widetilde{M} - M\|^{1/2}$ *for all $\widetilde{M}$ with $\|\widetilde{M} - M\| < \varepsilon$;*
(b) $\|\widetilde{S} - S\| \leqslant \gamma \|\widetilde{M} - M\|$ *for all singular $\widetilde{M}$ with $\|\widetilde{M} - M\| < \varepsilon$.*

It is interesting to observe that in the critical case, where $\mu = 0$ or if $\mu \approx 0$, one has to expect poor numerical performances even if the algorithm used for approximating S is backward stable. Moreover, the rounding errors introduced to represent the input values of M in the floating point representation with precision ε may generate an error of the order $\sqrt{\varepsilon}$ in the solution S.

This kind of problems will be overcome in Section 4.1.

3 Numerical methods

We give a brief review of the numerical methods developed so far for computing the minimal nonnegative solution of the NARE (1) associated with an M-matrix.

Here we consider the case where the M-matrix M is nonsingular or is singular, irreducible and $\mu \leqslant 0$. The case $\mu > 0$ can be reduced to the case $\mu < 0$ by means of Lemma 10. The critical case where $\mu = 0$ needs different techniques which will be treated in the next Section 4.

We start with a direct method based on the Schur form of the matrix H then we consider iterative methods based on fixed-point techniques, Newton's iteration and we conclude the section by analyzing a class of doubling algorithms.

The latter class includes methods based on Cyclic Reduction (CR) of [9], and on the Structure-preserving Doubling Algorithm (SDA) of [2].

3.1 Schur method

A classical approach for solving equation (1) is to use the (ordered) Schur decomposition of the matrix M to compute the invariant subspaces of H corresponding to the minimal solution S. This approach for the symmetric algebraic Riccati equation was first presented by Laub in 1979 [40]. Concerning the NARE, a study of that method in the singular and critical case was done by Guo [23] who presented a modified Schur method for the critical or near critical case ($\mu \approx 0$).

As explained in Section 2.4 from

$$H \begin{bmatrix} I_n \\ S \end{bmatrix} = \begin{bmatrix} I_n \\ S \end{bmatrix} (D - CS)$$

it follows that finding the minimal solution S of the NARE (1) is equivalent to finding a basis of the invariant subspace of H relative to the eigenvalues of $D - CS$, i.e., the eigenvalues of H with nonnegative real part.

A method for finding an invariant subspace is obtained by computing a semi-ordered Schur form of H, that is, computing an orthogonal matrix Q and a quasi upper-triangular matrix T such that $Q^*HQ = T$, where T is block upper triangular with diagonal blocks $T_{i,i}$ of size at most 2. The semi-ordering means that if $T_{i,i}$, $T_{j,j}$ and $T_{k,k}$ are diagonal blocks having eigenvalues with positive, null and negative real parts, respectively, then $i < j < k$.

A semi-ordered Schur form can be computed in two steps:

- Compute a real Schur form of H by the customary Hessenberg reduction followed by the application of the QR algorithm as described in [19].
- Swap the diagonal blocks by means of orthogonal transformations as described in [4].

The minimal solution of the NARE can be obtained from the first n columns of the matrix Q partitioned as $\begin{bmatrix} Q_1 \\ Q_2 \end{bmatrix}$ such that Q_1 is an $n \times n$ matrix, that is, $S = Q_2 Q_1^{-1}$.

In the critical case this method does not work, since there is no way to choose an invariant subspace relative to the first n eigenvalues, moreover in the near critical case where $\mu \approx 0$, there is lack of accuracy since the 0 eigenvalue is ill-conditioned. However, the modified Schur method given by C.-H. Guo [24] overcomes these problems.

The cost of this algorithm, following [23], is $200n^3$.

3.2 Functional iterations

In [20] a class of fixed-point methods for (1) is considered.

The fixed-point iterations are based on suitable splittings of A and D, that is $A = A_1 - A_2$ and $D = D_1 - D_2$, with A_1, D_1 chosen to be M-matrices and $A_2, D_2 \geqslant 0$. The form of the iterations is

$$A_1 X_{k+1} + X_{k+1} D_1 = X_k C X_k + X_k D_2 + A_2 X_k + B, \tag{12}$$

where at each step a Sylvester equation of the form $M_1 X + X M_2 = N$ must be solved.

Some possible choices for the splitting are:

1. A_1 and D_1 are the diagonal parts of A and D, respectively;
2. A_1 is the lower triangular part of A and D_1 the upper triangular part of D;
3. $A_1 = A$ and $D_1 = D$.

The solution X_{k+1} of the Sylvester equation can be computed, for instance, by using the Bartels and Stewart method [5], as in MATLAB's `sylvsol` function of the Nick Higham Matrix Function toolbox [28]

The cost of this computation is roughly $60n^3$ ops including the computation of the Schur form of the coefficients A_1 and D_1 [29]. However, observe that for the first splitting, A_1 and D_1 are diagonal matrices and the Sylvester equation can be solved with $O(n^2)$ ops; for the second splitting, the matrices A_1 and D_1 are already in the Schur form. This substantially reduces the cost of the application of the Bartels and Stewart method to $2n^3$. Concerning the third iteration, observe that the matrix coefficients A_1 and D_1 are independent of the iteration. Therefore, the computation of their Schur form must be performed only once.

A monotonic convergence result holds for the three iterations [20].

Theorem 17. *If $\mathcal{R}(X) \leqslant 0$ for some positive matrix X, then for the fixed-point iterations* (12) *with $X_0 = 0$, it holds that $X_k < X_{k+1} < X$ for $k \geqslant 0$. Moreover, $\lim X_k = S$.*

We have also an asymptotic convergence result [20].

Theorem 18. *For the fixed-point iterations* (12) *with $X_0 = 0$, it holds that*

$$\limsup \sqrt[k]{\|X_k - S\|} = \rho((I \otimes A_1 + D_1^T \otimes I)^{-1}(I \otimes (A_2 + SC) + (D_2 + CS)^T \otimes I).$$

These iterations have linear convergence which turns to sublinear in the critical case. The computational cost varies from $8n^3$ arithmetic operations per step for the first splitting, to $64n^3$ for the first step plus $10n^3$ for each subsequent step for the last splitting. The most expensive iteration is the third one which, on the other hand, has the highest (linear) convergence speed.

3.3 Newton's method

Newton's iteration was first applied to the symmetric algebraic Riccati equation by Kleinman in 1968 [37] and later on by various authors. In particular, Benner and Byers [7] complemented the method with an optimization technique (exact line search) in order to reduce the number of steps needed for arriving at convergence. The study of the Newton method for nonsymmetric algebraic Riccati equations was started by Guo and Laub in [26], and a nice convergence result was given by Guo and Higham in [24].

The convergence of the Newton method is generally quadratic except for the critical case where the convergence is observed to be linear with rate $1/2$ [26]. At each step, a Sylvester matrix equation must be solved, so the computational cost is $O(n^3)$ ops per step, but with a large overhead constant.

The Newton method for a NARE [26] consists in the iteration

$$X_{k+1} = \mathcal{N}(X_k) = X_k - (d\mathcal{R}_{X_k})^{-1}\mathcal{R}(X_k), \quad k = 0, 1, \ldots \tag{13}$$

which, in view of (10), can be written explicitly as

$$(A - X_k C)X_{k+1} + X_{k+1}(D - CX_k) = B - X_k C X_k. \tag{14}$$

Therefore, the matrix X_{k+1} is obtained by solving a Sylvester equation. This linear equation is defined by the matrix

$$\Delta_{X_k} = (D - CX_k)^T \otimes I_m + I_n \otimes (A - X_k C)$$

which is nonsingular if $0 \leqslant X_k < S$, as shown in section 2.6. Thus, if $0 \leqslant X_k < S$ for any k, the sequence (13) is well-defined.

In the noncritical case, $d\mathcal{R}_S$ is nonsingular, and the iteration is quadratically convergent in a neighborhood of the minimal nonnegative solution S by the traditional results on Newton's method (see e.g. [36]). Moreover, the following monotonic convergence result holds [24]:

Theorem 19. *Consider Newton's method* (14) *starting from* $X_0 = 0$. *Then for each* $k = 0, 1, \ldots,$ *we have* $0 \leqslant X_k \leqslant X_{k+1} < S$ *and* Δ_{X_k} *is a nonsingular M-matrix. Therefore, the sequence* (X_k) *is well-defined and converges monotonically to* S.

The same result holds when $0 \leqslant X_0 \leqslant S$; the proof in [24] can be easily adapted to this case.

In [26], a hybrid method was suggested, which consists in performing a certain number of iterations of a linearly convergent algorithm, such as the ones of Section 3.2, and then using the computed value as the starting point for Newton's method.

At each step of Newton's iteration, the largest computational work is given by the solution of the Sylvester equation (14). We recall that the solution X_{k+1}, computed by means of the Bartels and Stewart method [5] costs roughly $60n^3$ ops. Therefore the overall cost of Newton's iteration is $66n^3$ ops.

It is worth noting that in the critical and near critical cases, the matrix Δ_k becomes almost singular as X_k approaches the solution S; therefore, some numerical instability is to be expected. Such instability can be removed by means of a suitable technique which we will describe in Section 4.1.

3.4 Doubling algorithms

In this section we report some quadratically convergent algorithms obtained in [13] for solving (1). Quadratically convergent methods for computing the extremal solution of the NARE can be obtained by transforming the NARE into a Unilateral Quadratic Matrix Equation (UQME) of the kind

$$A_2X^2 + A_1X + A_0 = 0 \tag{15}$$

where A_0, A_1, A_2 and X are $p \times p$ matrices. Equations of this kind can be solved efficiently by means of doubling algorithms like Cyclic Reduction (CR) [9,12] or Logarithmic Reduction (LR) [39].

The first attempt to reduce a NARE to a UQME was performed by Ramaswami [46] in the framework of fluid queues. Subsequently, many contributions in this direction have been given by several authors [23, 10, 13, 33, 6] and different reduction techniques have been designed.

Concerning algorithms, Cyclic Reduction and SDA are the most effective computational techniques. The former was applied the first time in [9] by Bini and Meini to solve unilateral quadratic equations. The latter, was first presented by Anderson in 1978 [2] for the numerical solution of discrete-time algebraic Riccati equations. A new interpretation was given by Chu, Fan, Guo, Hwang, Lin, Xu [16, 32, 41], for other kinds of algebraic Riccati equations.

CR applied to (15) generates sequences of matrices defined by the following equations

$$\begin{aligned} V^{(k)} &= (A_1^{(k)})^{-1} \\ A_0^{(k+1)} &= -A_0^{(k)} V^{(k)} A_0^{(k)} \\ A_1^{(k+1)} &= A_1^{(k)} - A_0^{(k)} V^{(k)} A_2^{(k)} - A_2^{(k)} V^{(k)} A_0^{(k)} \quad k = 0, 1, \dots \\ A_2^{(k+1)} &= -A_2^{(k)} V^{(k)} A_2^{(k)} \\ \widehat{A}^{(k+1)} &= \widehat{A}^{(k)} - A_2^{(k)} V^{(k)} A_0^{(k)} \end{aligned} \tag{16}$$

where $A_i^{(0)} = A_i$, $i = 0, 1, 2$, $\widehat{A}^{(0)} = A_1$.

The following result provides convergence properties of CR [12].

Theorem 20. *Let $x_1, \dots, x_{2p}$ be the roots of $a(z) = \det(A_0 + zA_1 + z^2 A_2)$, including roots at the infinity if $\deg a(z) < 2p$, ordered by increasing modulus. Suppose that $|x_p| \leqslant 1 \leqslant |x_{p+1}|$ and $|x_p| < |x_{p+1}|$, and that a solution G exists to (15) such that $\rho(G) = |x_p|$. Then, G is the unique solution to (15) with minimal spectral radius, moreover, if CR (16) can be carried out with no breakdown, the sequence*

$$G^{(k)} = -\left(\widehat{A}^{(k)}\right)^{-1} A_0$$

is such that for any norm

$$\|G^{(k)} - G\| \leqslant \vartheta |x_p / x_{p+1}|^{2^k}$$

where $\vartheta > 0$ is a suitable constant. Moreover, it holds that $\|A_0^{(k)}\| = O(|x_p|^{2^k})$, $\|A_2^{(k)}\| = O(|x_{p+1}|^{-2^k})$.

Observe that, the convergence conditions of the above theorem require that the roots of $a(z)$ have a (p, p) complete splitting with respect to the unit circle. For this reason, before transforming the NARE into a UQME, it is convenient to transform the Hamiltonian H into a new matrix $\widehat{H}$ such that the eigenvalues of $\widehat{H}$ have an (n, m) splitting with respect to the unit circle, i.e., n eigenvalues belong to the closed unit disk and m are outside. This can be obtained by means of one of the two operators: the Cayley transform $\mathcal{C}_\gamma(z) = (z+\gamma)^{-1}(z-\gamma)$, where $\gamma > 0$, or the shrink-and-shift operator $\mathcal{S}_\tau(z) = 1 - \tau z$, where $\tau > 0$. In fact, the Cayley transform maps the right open half-plane into the open unit disk. Similarly, for suitable values of τ, the transformation $\mathcal{S}_\tau$ maps a suitable subset of the right half-plane inside the unit disk. This property is better explained in the following result which has been proved in [13].

Theorem 21. *Let $\gamma, \tau > 0$ and let*

$$H_\gamma = \mathcal{C}_\gamma(H) = (H + \gamma I)^{-1}(H - \gamma I), \quad H_\tau = \mathcal{S}_\tau(H) = I - \tau H.$$

Assume $\mu < 0$, then:

1. H_γ *has eigenvalues* $\xi_i = \mathcal{C}_\gamma(\lambda_i)$, $i = 1, \dots, m+n$, *such that*

$$\max_{i=1,\dots,n} |\xi_i| \leqslant 1 < \min_{i=1,\dots,m} |\xi_{i+n}|;$$

2. if $\tau^{-1} \geqslant \max\{\max_i (A)_{i,i}, \max_i (D)_{i,i}\}$, H_τ *has eigenvalues* $\mu_i = \mathcal{S}_\tau(\lambda_i)$, $i = 1, \dots, m+n$, *such that*

$$\max_{i=1,\dots,n} |\mu_i| \leqslant 1 < \min_{i=1,\dots,m} |\mu_{i+n}|.$$

Moreover, if X *is any solution of* (1) *then*

$$H_\gamma \begin{bmatrix} I \\ X \end{bmatrix} = \begin{bmatrix} I \\ X \end{bmatrix} R_\gamma, \quad H_\tau \begin{bmatrix} I \\ X \end{bmatrix} = \begin{bmatrix} I \\ X \end{bmatrix} R_\tau$$

where $R_\gamma = \mathcal{C}_\gamma(D - CX)$, $R_\tau = \mathcal{S}_\tau(D - CX)$.

In the following we will denote by $\widehat{H} = \begin{bmatrix} \widehat{D} & -\widehat{C} \\ \widehat{B} & -\widehat{A} \end{bmatrix}$ either H_γ or H_τ. Since the transformations $\mathcal{C}_\gamma$ and $\mathcal{S}_\tau$ are invertible, from the above theorem one has that X is a solution of the NARE (1) if and only if X is a solution of the NARE defined by $\widehat{H}$. In particular, the extremal solution S is the solution of the NARE associated with H_γ or H_τ corresponding to the n eigenvalues H_γ or H_τ, respectively, smallest in modulus.

The following result provides a means for reducing a NARE into a UQME:

Theorem 22. *Let* X *be a solution of the NARE* (1). *Then:*

1. $Y = \begin{bmatrix} \widehat{D} - \widehat{C}X & 0 \\ X & 0 \end{bmatrix}$ *is a solution to*

$$\begin{bmatrix} \widehat{D} & 0 \\ \widehat{B} & 0 \end{bmatrix} + \begin{bmatrix} -I & -\widehat{C} \\ 0 & -\widehat{A} \end{bmatrix} Y + \begin{bmatrix} 0 & 0 \\ 0 & -I \end{bmatrix} Y^2 = 0; \tag{17}$$

2. $Y = \begin{bmatrix} \widehat{D} - \widehat{C}X & 0 \\ X(\widehat{D} - \widehat{C}X) & 0 \end{bmatrix}$ *is a solution to*

$$\begin{bmatrix} L_1 & 0 \\ 0 & 0 \end{bmatrix} + \begin{bmatrix} -I & U_1 \\ L_2 & -I \end{bmatrix} Y + \begin{bmatrix} 0 & 0 \\ 0 & U_2 \end{bmatrix} Y^2 = 0, \tag{18}$$

where $U_1 = -\widehat{C}\widehat{A}^{-1}$, $U_2 = -\widehat{A}^{-1}$, $L_1 = \widehat{D} - \widehat{C}\widehat{A}^{-1}\widehat{B}$, $L_2 = -\widehat{A}^{-1}\widehat{B}$.

Conversely,

$$V = \begin{bmatrix} \widehat{D} - \widehat{C}S & 0 \\ S & 0 \end{bmatrix}, \quad W = \begin{bmatrix} \widehat{D} - \widehat{C}S & 0 \\ S(\widehat{D} - \widehat{C}S) & 0 \end{bmatrix}$$

are the unique solutions of UQME (17) *and* (18), *respectively, with* m *eigenvalues equal to 0 and* n *eigenvalues in the closed unit disk.*

A reduction similar to the one provided in equation (17) was proved by Ramaswami in [46] by using probabilistic tools.

The following reduction holds for any NARE (1) provided that $m = n$ and $\det C \neq 0$

Theorem 23. *Let* $m = n$ *and* $\det C \neq 0$. *The matrix* X *is a solution of the NARE* (1) *if and only if* $Y = C^{-1}(D - CX)\widehat{C}$ *is a solution of the UQME*

$$Y^2 + (C^{-1}DC - A)Y + (B - AC^{-1}D)C = 0. \tag{19}$$

Similarly, X *is a solution of the NARE* (1) *if and only if* $Y = D - CX$ *is a solution of the UQME*

$$Y^2 + (D - CAC^{-1})Y + C(B - AC^{-1}D) = 0. \tag{20}$$

If we choose $H = \widehat{H}$, then $Y = \widehat{C}^{-1}(\widehat{D} - \widehat{C}X)\widehat{C}$ is the solution of the (19) with minimal spectral radius. Similarly, $Y = \widehat{D} - \widehat{C}S$ is the solution of (20) with minimal spectral radius.

Observe that if $\det C = 0$, we may replace (1) with a new equation defined by blocks $\widetilde{A}$, $\widetilde{B}$, $\widetilde{C}$, and $\widetilde{D}$ such that $\det \widetilde{C} \neq 0$ according to the following

Lemma 24. *The Riccati equation* (1) *has solution* X *if and only if the Riccati equation*

$$Y\widetilde{C}Y - \widetilde{A}Y - Y\widetilde{D} + \widetilde{B} = 0$$

where $\widetilde{A} = A - BK$, $\widetilde{B} = B$, $\widetilde{C} = \mathcal{R}(K)$, $\widetilde{D} = D - KB$, *has solution* $\widetilde{X} = X(I - KX)^{-1}$ *and* K *is such that* $\det(I - K\widetilde{X}) \neq 0$ *(or equivalently,* $\det(I + XK) \neq 0$*). Moreover,* $\widetilde{D} - \widetilde{C}\widetilde{X} = (I - KX)(D - CX)(I - KX)^{-1}$.

It can be easily verified that if $\widehat{H} = H_\tau$ then

$$\widehat{A} = -I - \tau A, \quad \widehat{B} = -B, \quad \widehat{C} = -C, \quad \widehat{D} = I - \tau D.$$

If $\widehat{H} = H_\gamma$ then a direct calculation shows that

$$\begin{aligned} \widehat{A} &= -I + 2\gamma V^{-1}, & \widehat{B} &= 2\gamma(-A + \gamma I)^{-1}BW^{-1}, \\ \widehat{C} &= 2\gamma(D + \gamma I)^{-1}CV^{-1}, & \widehat{D} &= I - 2\gamma W^{-1}, \end{aligned}$$

with $V = -A + \gamma I + B(D + \gamma I)^{-1}C$ and $W = D + \gamma I + C(-A + \gamma I)^{-1}B$.

Equations (17), (18), (19) and (20) can be solved by means of CR (16), which provides a matrix sequence $G^{(k)}$ that converges, when applicable, to the solution with minimal spectral radius. In view of Theorem 21, and of the subsequent discussion, this solution is the one which is needed for computing the extremal solution S to the NARE (1).

The cost of CR applied to (19) and (20) is about $(38/3)n^3$ ops.

Concerning convergence it follows from Theorem 20 that the approximation error is $O(\sigma^{2^k})$, for $\sigma = \sigma_\gamma$ if $\widehat{H} = H_\gamma$, $\sigma = \sigma_\tau$ if $\widehat{H} = H_\tau$. Here we define

$$\sigma_\tau = \max_{i=1,\dots,n} |\mu_i| / \min_{i=1,\dots,m} |\mu_{n+i}|,$$
$$\sigma_\gamma = \max_{i=1,\dots,n} |\xi_i| / \min_{i=1,\dots,m} |\xi_{n+i}|,$$

where $\sigma_\tau, \sigma_\gamma < 1$ if $\mu < 0$.

Applying CR to (19) and (20) generates blocks of size $m+n$. However, it is possible to verify that the structure of the blocks A_i, $i = 0,1,2$ given in equations (17) and (18) is maintained unchanged by the blocks $A_i^{(k)}$, $i = 0,1,2$. More precisely, it turns out that applying (16) to the equation (17) yields blocks of the kind

$$A_0^{(k)} = \begin{bmatrix} R_1^{(k)} & 0 \\ R_2^{(k)} & 0 \end{bmatrix}, \quad A_1^{(k)} = \begin{bmatrix} -I & R_3^{(k)} \\ R_4^{(k)} & R_5^{(k)} \end{bmatrix},$$
$$A_2^{(k)} = \begin{bmatrix} 0 & 0 \\ 0 & R_6^{(k)} \end{bmatrix}, \quad \widehat{A}^{(k)} = \begin{bmatrix} -I & R_3^{(0)} \\ R_4^{(k)} & R_5^{(0)} \end{bmatrix}.$$

It can be easily verified that the matrices $R_i^{(k)}$, $i = 1,\dots,6$ satisfy the following equations:

$$\begin{aligned}
S^{(k)} &= R_5^{(k)} + R_4^{(k)} R_3^{(k)}, & R_1^{(k+1)} &= -R_1^{(k)} X^{(k)}, \\
Y^{(k)} &= \left(S^{(k)}\right)^{-1} \left(R_2^{(k)} + R_4^{(k)} R_1^{(k)}\right), & R_2^{(k+1)} &= -R_2^{(k)} X^{(k)}, \\
& & R_3^{(k+1)} &= R_3^{(k)} - R_1^{(k)} T^{(k)}, \\
X^{(k)} &= R_3^{(k)} Y^{(k)} - R_1^{(k)}, & R_4^{(k+1)} &= R_4^{(k)} - R_6^{(k)} Y^{(k)}, \\
Z^{(k)} &= \left(S^{(k)}\right)^{-1} R_6^{(k)}, & R_5^{(k+1)} &= R_5^{(k)} - R_2^{(k)} T^{(k)}, \\
T^{(k)} &= R_3^{(k)} Z^{(k)}, & R_6^{(k+1)} &= -R_6^{(k)} Z^{(k)},
\end{aligned} \tag{21}$$

for $k = 0,1,\dots$, starting from the initial values $R_1^{(0)} = \widehat{D}$, $R_2^{(0)} = \widehat{B}$, $R_3^{(0)} = -\widehat{C}$, $R_4^{(0)} = 0$, $R_5^{(0)} = -\widehat{A}$, $R_6^{(0)} = -I$. From Theorem 20 it follows that

$$S = -\left(R_5^{(0)} + R_4^{(k)} R_3^{(0)}\right)^{-1} \left(R_2^{(0)} + R_4^{(k)} R_1^{(0)}\right) + O(\sigma^{2^k}),$$

where $\sigma = \sigma_\tau$ if $\widehat{H} = H_\tau$, while for $\widehat{H} = H_\gamma$ one has $\sigma = \sigma_\gamma$.

The computational cost of this algorithm is $(74/3)n^3$ per step, assuming $m = n$.

Similarly, it turns out that applying (16) to the equation (18) yields blocks of the kind

$$A_0^{(k)} = \begin{bmatrix} E^{(k)} & 0 \\ 0 & 0 \end{bmatrix}, \quad A_1^{(k)} = \begin{bmatrix} -I & G^{(k)} \\ H^{(k)} & -I \end{bmatrix}, \quad A_2^{(k)} = \begin{bmatrix} 0 & 0 \\ 0 & F^{(k)} \end{bmatrix},$$

where the sequences $E^{(k)}$, $F^{(k)}$, $G^{(k)}$, $H^{(k)}$ are given by

$$\begin{aligned}
E^{(k+1)} &= E^{(k)}(I - G^{(k)}H^{(k)})^{-1}E^{(k)}, \\
F^{(k+1)} &= F^{(k)}(I - H^{(k)}G^{(k)})^{-1}F^{(k)}, \\
G^{(k+1)} &= G^{(k)} + E^{(k)}(I - G^{(k)}H^{(k)})^{-1}G^{(k)}F^{(k)}, \\
H^{(k+1)} &= H^{(k)} + F^{(k)}(I - H^{(k)}G^{(k)})^{-1}H^{(k)}E^{(k)},
\end{aligned} \tag{22}$$

for $k \geqslant 0$, starting from the initial values $E^{(0)} = L_1$, $F^{(0)} = U_2$, $G^{(0)} = U_1$, $H^{(0)} = L_2$. The following convergence result holds:

$$S = H^{(k)} + O(\sigma^{2^k})$$

in the noncritical case, where $\sigma = \sigma_\gamma, \sigma_\tau$. Observe that in this case the computation of $\widehat{A}^{(k)}$ is not required.

The cost per step of this iteration is $(64/3)n^3$ for $m = n$.

It is interesting to point out that (22), obtained by applying CR to the solution of the UQME (18), coincides with SDA of [16, 32, 41]. In the critical case where H is singular and $\mu = 0$, the convergence of the doubling algorithms is linear as shown in [15, 25].

4 Exploiting the singularity of H

In this section we assume that M is singular irreducible. Under this assumption, the matrix $H = JM$ has only one independent left and only one independent right eigenvector corresponding to the null eigenvalue. These vectors can be computed easily as already explained in Section 2.5. The knowledge of these eigenvectors can be used for improving the performances of the algorithms by means of two techniques: the shift technique which we deal in Section 4.1, and a suitable choice of the initial approximation in iterative methods which we treat in Section 4.2.

The advantage that one can obtain from these two techniques is twofold: on one hand it is possible to increase the accuracy in the (close to) critical case where the approximation error changes from $O(\sqrt{\varepsilon})$ to $O(\varepsilon)$; on the other hand one can accelerate the convergence speed from the linear convergence to the quadratic convergence in the critical case and improve the quadratic convergence in the close to critical case.

In the rest of the section we consider only the case $\mu \leqslant 0$ in view of Lemma 10.

4.1 The shift technique

The shift technique was introduced by He, Meini and Rhee for a quadratic matrix equation arising in the numerical solution of Markov chains modeling quasi-birth-and-death (QBD) processes [27].

For these problems, the interest is in the computation of the minimal nonnegative solution G of the matrix equation

$$X = A_2X^2 + A_1X + A_0,$$

where $A_i \geqslant 0$, $i = 0, 1, 2$, and $(A_2 + A_1 + A_0)e = e$.

A property generally satisfied in the applications is that the polynomial $\det \varphi(z)$, $\varphi(z) = A_2z^2 + (A_1 - I)z + A_0$, has degree at least $n+1$ and roots $\xi_1, \xi_2, \ldots$, ordered by increasing modulus such that ξ_n and ξ_{n+1} are real and $|\xi_{n-1}| < \xi_n = 1 \leqslant \xi_{n+1}$. Moreover one has $\varphi(1)e = 0$ and $\sigma(G) = \{\xi_1, \ldots, \xi_n\}$ [12].

The conditioning of the minimal nonnegative solution G and the convergence of the available algorithms depend on the ratio $1/\xi_{n+1}$ [39, 9, 27]: the closer is this ratio to 1, the worse conditioned is the solution and the slower is the convergence of the iterative algorithms.

The idea of the shift technique is to consider a new quadratic matrix equation in which the convergence of numerical algorithms and the conditioning of the solution is better, and whose solution easily provides the matrix G. This can be achieved by using the available information of G, that is, $\rho(G) = 1$ and $Ge = e$.

The new UQME is

$$X = B_2X^2 + B_1X + B_0, \tag{23}$$

where

$$\begin{aligned} B_2 &= A_2, \\ B_1 &= A_1 + A_2eu^T, \\ B_0 &= A_0 + (A_1 + A_2 - I)eu^T = A_0 - A_0eu^T, \end{aligned}$$

and u is any positive vector such that $u^Te = 1$. An easy computation shows that the equation (23) has the solution $F = G - eu^T$.

The matrix F has the same eigenvalues as the matrix G except for the eigenvalue 1 of G that becomes the eigenvalue 0 in F, with the same eigenvector e (compare Theorem 25). It can be said that an eigenvalue of G is shifted to 0, this fact gives the name to the technique. Observe that F is the solution with minimal spectral radius of (23).

Concerning the matrix polynomials $\varphi(z)$ and $\psi(z) = B_2z^2 + (B_1 - I)z + B_0$, it holds that

$$\begin{aligned} \varphi(z) &= (A_1 - I + A_2G - zA_2)(zI - G), \\ \psi(z) &= (B_1 - I + B_2F - zB_2)(zI - F) = (A_1 - I + A_2G - zA_2)(zI - F). \end{aligned} \tag{24}$$

The latter equality follows from the fact that $A_2 = B_2$ and $A_1 + A_2G = B_1 + B_2F$ and implies that the determinants of the two matrix polynomials have the same roots except for the root 1 that is replaced by 0. In this way, the ratio between the nth and the $(n+1)$-st root is reduced from $1/\xi_{n+1}$ to $|\xi_{n-1}|/\xi_{n+1}$ (see [27, 22] for further details).

The important case where $\xi_n = \xi_{n+1} = 1$ is critical for the convergence of algorithms since the ratio $1/\xi_{n=1}$ is 1. In fact in this case the convergence of algorithms turns from quadratic to linear or from linear to sublinear. The shift technique transforms this critical equation into another one where the ratio between the nth and the $(n+1)$-st root is $|\xi_{n-1}| < 1$. In this way, the quadratic convergence is preserved.

Even in the case where ξ_n is very close to ξ_{n+1} the shift technique allows one to improve the convergence speed since the ratio between the nth and the $(n+1)$-st root becomes $|\xi_{n-1}|/\xi_{n+1}$ which is smaller than $|\xi_n|/\xi_{n+1}$.

The shift technique has a nice functional interpretation: the matrix polynomial $\psi(z)$ of (24) is obtained by the polynomial $\varphi(z)$ by the simple relation [12]

$$\psi(z)(I - z^{-1}Q) = \varphi(z),$$

where $Q = eu^T$. This characterization has the advantage that the shift technique can be extended to matrix equations of any degree or even to matrix power series [12].

The shift technique can be applied to the UQMEs (17), (18), (19), (20) which derive from NAREs. In particular, in the case of equation (17) this technique has been analyzed in detail in [10]. The cases of (18), (19), (20) can be similarly treated.

Similarly to the case of the quadratic matrix equation, one can directly apply the shift technique to the singular matrix H associated with the NARE [25]. Here the goal is to construct a new matrix $\widetilde{H}$ having the same eigenvalues of H except for the eigenvalue 0 which is moved to a positive eigenvalue η of $\widetilde{H}$. In this way we obtain a new NARE associated with $\widetilde{H}$ having better computational feature and with the same solution S of the original NARE.

The construction of $\widetilde{H}$ is based on the following result of which we give a simpler proof. This result was proved by Brauer in 1952 [14] and it has been rediscovered several times (see [31]).

Theorem 25. *Let A be an $n \times n$ matrix with eigenvalues $\lambda_1, \lambda_2, \dots, \lambda_n$ and let v be a nonnull vector such that $Av = \lambda_1 v$. For any nonnull vector x, set $Q = vx^T$. Then the eigenvalues of $A + Q$ are $\lambda_1 + x^T v, \lambda_2, \dots, \lambda_n$.*

Proof. Since $AQ = \lambda_1 Q$, one has the following identity

$$(\lambda - \lambda_1)(A + Q - \lambda I) = (A - \lambda I)((\lambda - \lambda_1)I - Q).$$

Taking the determinant of both sides and using the formula for the characteristic polynomial of a rank-one matrix, $p_{vx^T}(\lambda) = \det(vx^T - \lambda I) = (\lambda - x^T v)\lambda^{n-1}$, it holds that

$$\begin{aligned} p_{A+Q}(\lambda)(\lambda - \lambda_1)^n &= (-1)^n p_A(\lambda) p_{vx^T}(\lambda - \lambda_1) \\ &= (-1)^n p_A(\lambda)(\lambda - \lambda_1)^{n-1}(\lambda - \lambda_1 - x^T v). \end{aligned}$$

The unique factorization of polynomials completes the proof. □

From the above theorem follows immediately a corollary that will be useful in the following.

Corollary 26. *Let* A *be a singular matrix and* $Aw = 0$ *for a nonzero vector* w*. Assume that* p *is a vector such that* $p^T w = 1$ *and* η *is a scalar. Then the eigenvalues of the matrix*

$$\widetilde{A} = A + \eta w p^A$$

are those of A *except that one zero eigenvalue of* A *is replaced by* η.

We now construct a rank-one modification of the matrix H:

$$\widehat{H} = H + \eta v p^T, \tag{25}$$

where, v is a positive vector such that $Hv = 0$, $\eta > 0$ is a scalar and $p \geqslant 0$ is a vector with $p^T v = 1$. From Corollary 26 the eigenvalues of $\widehat{H}$ are those of H except that one zero eigenvalue of H is replaced by η.

We write $p^T = \begin{bmatrix} p_1 \\ p_2 \end{bmatrix}$ and

$$\widetilde{H} = \begin{bmatrix} \widetilde{D} & -\widetilde{C} \\ \widetilde{B} & -\widetilde{A} \end{bmatrix},$$

where

$$\widetilde{D} = D + \eta v_1 p_1^T, \quad \widetilde{C} = C - \eta v_1 p_2^T,$$

$$\widetilde{B} = B + \eta v_2 p_1^T, \quad \widetilde{A} = A - \eta v_2 p_2^T.$$

Corresponding to $\widetilde{H}$ we define the new NARE

$$X\widetilde{C}X - X\widetilde{D} - \widetilde{A}X + \widetilde{B} = 0, \tag{26}$$

which defines the Riccati operator

$$\widetilde{\mathcal{R}}(X) = X\widetilde{C}X - X\widetilde{D} - \widetilde{A}X + \widetilde{B}. \tag{27}$$

We have the following important property about the NARE (26).

Theorem 27. *If* $\mu \leqslant 0$*, then* S *is a solution of the NARE* (26) *and* $\sigma(\widetilde{D} - \widetilde{C}S) = \{\lambda_1, \ldots, \lambda_{n-1}, \eta\}$*, where* S *is the minimal nonnegative solution of the original NARE* (1).

Computing the minimal nonnegative solution S of the NARE (1) can be achieved by computing the solution S of the new NARE (26) corresponding to eigenvalues with positive real parts. Observe that equation (26) is not associated with an M-matrix, however the algorithms and the techniques of Section 3 can be applied and, if break-down is not encountered, convergence is much faster than for the original equation (1). In particular, in the critical case, the convergence of SDA applied to the new NARE (26) is again quadratic. A detailed convergence analysis of SDA is reported in [25].

When $\mu = 0$, the matrix H has two zero eigenvalues. The above shift technique moves one zero eigenvalue to a positive number. We may use a *double-shift* to move the other zero eigenvalue to a negative number. Recall that $Hv = 0$, where $v = \left[\begin{smallmatrix} v_1 \\ v_2 \end{smallmatrix}\right]$, and $w^T H = 0$, where $w = \left[\begin{smallmatrix} u_1 \\ -u_2 \end{smallmatrix}\right]$. We define the matrix

$$\overline{H} = H + \eta v p^T + \xi q w^T, \tag{28}$$

where $\eta > 0$, $\xi < 0$, p and q are such that $p^T v = q^T w = 1$. Since v and w are orthogonal vectors, the double-shift moves one zero eigenvalue to η and the other to ξ. Indeed, the eigenvalues of $\widetilde{H} = H + \xi q w^T$ are those of $\widetilde{H}^T = H^T + \xi w q^T$, which are the eigenvalues of H except that one zero eigenvalue is replaced by ξ, by Lemma 26. Also, the eigenvalues of $\overline{H} = \widetilde{H} + \eta v p^T$ are the eigenvalues of $\widetilde{H}$ except that the remaining zero eigenvalue is replaced by η, by Lemma 26 again.

From $\overline{H}$ we may define a new Riccati equation

$$X\overline{C}X - X\overline{D} - \overline{A}X + \overline{B} = 0. \tag{29}$$

As before, the minimal nonnegative solution S of (1) is a solution of (29) such that $\sigma(\overline{D} - \overline{C}S) = \{\eta, \lambda_1, \ldots, \lambda_{n-1}\}$. However, it seems very difficult to determine the existence of a solution $\overline{Y}$ of the dual equation of (29) such that $\sigma(\overline{A} - \overline{B}\,\overline{Y}) = \{-\xi, -\lambda_{n+2}, \ldots, -\lambda_{n+m}\}$.

4.2 Choosing a new initial value

If the right eigenvector of H relative to the null eigenvalue is partitioned as $v = \begin{bmatrix} v_1 \\ v_2 \end{bmatrix}$, from Theorem 14 it follows that for the minimal nonnegative solution S, it holds that $Sv_1 = v_2$ (and then $(D - CS)v_1 = 0$).

In the algorithms in which the initial value can be chosen, like Newton's method, the *usual* choice $X_0 = 0$ does not exploit this information, rather it relies only on the positivity of S. Note that in the Riccati equations modeling fluid queues, the condition $Xv_1 = v_2$ is equivalent to the stochasticity of S since $v_1 = v_2 = e$.

A possibly better convergence is expected if one could generate a sequence such that $X_k v_1 = v_2$ for any $k \geqslant 0$. More precisely, one must choose an iteration which preserves the affine subspace $\widehat{W} = \{A \in \mathbb{C}^{n\times n} : Av_1 = v_2\}$ and an initial value $X_0 \in \widehat{W}$ for which the sequence converges to the desired solution.

A similar idea has been used in [45] in order to improve the convergence speed of certain functional iterations for solving nonlinear matrix equations related to special Markov chains.

A nice property of Newton's method is that it is structure-preserving with respect to the affine subspace $\widehat{W}$. To prove this fact consider the following preliminary result which concerns the Newton iteration

Lemma 28. *The Newton method* $X_{k+1} = \mathcal{N}(X_k)$,

$$\mathcal{N}(X_k) = X_k - (d\mathcal{F}_{X_k})^{-1}\mathcal{F}(X_k)$$

applied to the matrix equation $\mathcal{F}(X) = 0$*, when defined, preserves the affine structure* $\widehat{V}$ *if and only if* $\mathcal{F}$ *is a function from* $\widehat{V}$ *to its parallel linear subspace* V.

Proof. Consider the matrix $X \in \widehat{V}$. The matrix $\mathcal{N}(X)$ belongs to $\widehat{V}$ if and only if $\mathcal{N}(X) - X = (d\mathcal{F}_X)^{-1}(-\mathcal{F}(X))$ belongs to V, and that occurs if and only if $\mathcal{F}(X)$ (and then $-\mathcal{F}(X)$) belongs to V. □

Now, we are ready to prove that the Newton method applied to the Riccati operator is structure-preserving with respect to $\widehat{W}$.

Proposition 29. *If* X_0 *is such that* $X_0 v_1 = v_2$*, and the Newton method applied to the Riccati equation* $\mathcal{R}(X) = 0$ *is well defined then* $X_k v_1 = v_2$ *for any* $k \geqslant 0$*. That is, the Newton method preserves the structure* $\widehat{W}$.

Proof. In view of Lemma 28, one needs to prove that $\mathcal{R}$ is a function from $\widehat{W}$ to the parallel linear subspace W.

If $X \in \widehat{W}$, then $\mathcal{R}(X)v_1 = 0$, in fact

$$\mathcal{R}(X)v_1 = XCXv_1 - AXv_1 - XDv_1 + Bv_1 = XCv_2 - Av_2 - XDv_1 + Bv_1,$$

and the last term is 0 since $Cv_2 = Dv_1$ and $Av_2 = Bv_1$. □

A possible choice for the starting value is $(X_0)_{i,j} = (v_2)_i/s$ where $s = \sum_i v_1(i)$. It must be observed that the structured preserving convergence is not anymore monotonic. Since the approximation error has a null component along the subspace W, one should expect a better convergence speed for the sequences obtained with $X_0 \in \widehat{W}$. A proof of this fact and the convergence analysis of this approach is still work in place.

If $\mu = 0$, the differential of $\mathcal{R}$ is singular at the solution S as well as at any point $X \in \widehat{W}$. This makes the sequence X_k undefined. A way to overcome this drawback is considering the shifted Riccati equation described in Section 4.1.

The differential of the shifted Riccati equation (26) at a point X is represented by the matrix

$$\widetilde{\Delta}_X = \Delta_X + I \otimes (\eta(Xv_1 - v_2)p_2^T) + (\eta v_1(p_1^T + p_2^T X))^T \otimes I, \tag{30}$$

where the vector $p \neq 0$ partitioned as $p = \begin{bmatrix} p_1 \\ p_2 \end{bmatrix}$ is an arbitrary nonnegative vector such that $p^T v = 1$. Choosing $p_2 = 0$ provides a nice simplification of the problem, in fact

$$\widetilde{\Delta}_X = \Delta_X - Q^T \otimes I,$$

where $Q = \eta v_1 p_1^T$.

The next result gives more insights on the action of the Newton iteration on the structure $\widehat{V}$.

Proposition 30. *Assume that $p_2 = 0$. If $X \in \widehat{W}$ then $\mathcal{R}(X) = \widetilde{\mathcal{R}}(X)$, where $\widetilde{\mathcal{R}}$ is defined in (27). Moreover the sequences generated by Newton's method, when defined, applied to $\mathcal{R}(X) = 0$ and to $\widetilde{\mathcal{R}}(X) = 0$ with $X_0 \in \widehat{W}$ are the same.*

Proof. The fact $\mathcal{R}(X) = \widehat{\mathcal{R}}(X)$, in the assumption $p_2 = 0$, follows from

$$\widetilde{\mathcal{R}}(X) = \mathcal{R}(X) - \eta(Xv_1 - v_2)p_1^T.$$

Let $\mathcal{N}(X) = X - (d\mathcal{R}_X)^{-1}\mathcal{R}(X)$ and $\widetilde{\mathcal{N}}(X) = X - (d\widetilde{\mathcal{R}}_X)^{-1}\widetilde{\mathcal{R}}(X)$ denote the Newton operator for the original equation and for the shifted one, respectively. To prove that the sequences are the same, it must be shown that

$$(A - XC)\mathcal{N}(X) + \mathcal{N}(X)(\widetilde{D} - CX) = \widetilde{B} - XCX$$

holds for any $X \in \widehat{W}$ and for any η (for which the equation has a unique solution). One has

$$\begin{aligned}&(A - XC)\mathcal{N}(X) + \mathcal{N}(X)(\widetilde{D} - CX)\\ &= B - XCX + \mathcal{N}(X)\eta v_1 p_1^T = B - XCX + \eta v_2 p_1^T = \widetilde{B} - XCX,\end{aligned}$$

where we have used that $\mathcal{N}(X)v_1 = v_2$ since $\mathcal{N}(X) \in \widehat{W}$. This completes the proof. □

Since any starting value $X_0 \in \widehat{V}$ gives the same sequence for the Newton method applied either to the Riccati equation (1) or to the shifted Riccati equation (26), then, choosing such an initial value has the same effect of applying the shift technique.

For the applicability one needs that the matrix Δ_{X_k} is nonsingular at each step. Unfortunately the derivative might be singular for some singular M-matrix and some $X \in \widehat{W}_+ = \{X \in \widehat{W}, X \geqslant 0\}$.

If a breakdown occurs, it is always possible to perform the iteration by using the shifted iteration, with $p_2 = 0$ and for a suitable choice of the parameter η. In fact, the iteration is proved in Proposition 30 to be the same by any choice of p_1 and η.

The convergence is more subtle. Besides the loss of monotonic convergence, one may note that S is not the only solution belonging to $\widehat{W}$, even if it is the only belonging to $\widehat{W}_+$. In fact, in view of Theorem 13, there are at most two positive solutions, and only one of them has the property $Sv_1 = v_2$. The proof of convergence is still work in progress, we conjecture that for each $X_0 \in \widehat{W}_+$, the sequence generated by the Newton method, if defined, converges to S.

A possible improvement of the algorithm could be obtained by implementing the exact line search introduced in [7].

5 Numerical experiments and comparisons

We present some numerical experiments to illustrate the behavior of the algorithms presented in Sections 3 and 4.1 in the critical and noncritical case. To compare the accuracy of the methods we have used the relative error err $= \|X - \widehat{X}\|_1 / \|X\|_1$ on the computed solution $\widehat{X}$, when the exact solution X was provided. Elsewhere, we have used the relative residual error

$$\mathrm{res} = \frac{\|\widehat{X}C\widehat{X} - \widehat{X}D - A\widehat{X} + B\|_1}{\|\widehat{X}C\widehat{X}\|_1 + \|\widehat{X}D\|_1 + \|A\widehat{X}\|_1 + \|B\|_1}.$$

The tests were performed using MATLAB 6 Release 12 on a processor AMD Athlon 64. The code for the different algorithms is available for download at the web page `http://bezout.dm.unipi.it/mriccati/`.

In these tests we consider three methods: the Newton method (N), the SDA, and the Cyclic Reduction (CR) algorithm applied to the UQME (17) (in both SDA and CR we have considered the matrix $\widehat{H}$ obtained by the Cayley transform of H and not the one relying on the shrink-and-shift operator).

We have also considered the improved version of these methods applied to the singular/critical case; we denoted them as IN, ISDA and ICR, respectively, where "I" stands for "Improved". The initial value for IN is chosen as suggested in Section 4.1; the parameter for the shift is chosen as $\eta = \max\{\max(A)_{i,i}, \max(D)_{i,i}\}$ and the vector p is chosen to be $e/\sum_i v_i$.

The iterations are stopped when the relative residual/error ceases to decrease or becomes smaller than 10ε, where ε is the machine precision.

Test 31. A null recurrent case [6, Example 1]. Let

$$M = \left[\begin{array}{cc|cc} 0.003 & -0.001 & -0.001 & -0.001 \\ -0.001 & 0.003 & -0.001 & -0.001 \\ \hline -0.001 & -0.001 & 0.003 & -0.001 \\ -0.001 & -0.001 & -0.001 & 0.003 \end{array}\right]$$

where D is a 2×2 matrix. The minimal positive solution is $X = \frac{1}{2}\begin{bmatrix} 1 & 1 \\ 1 & 1 \end{bmatrix}$.

As suggested by the Theorem 16, the accuracy of the customary algorithms N, SDA and CR is poor in the critical case, and is near to $\sqrt{\varepsilon} \approx 10^{-8}$. We report in Table 1 the number of steps and the relative error for the three algorithms. If one uses the singularity, due to the particular structure of the problem, the solution is achieved in one step by IN, ISDA and ICR with full accuracy.

Algorithm	Steps	Relative error
N	21	$6.0 \cdot 10^{-7}$
SDA	36	$8.6 \cdot 10^{-7}$
CR	31	$4.7 \cdot 10^{-9}$

Table 1. Accuracy of the algorithms in the critical case, Test 31

Test 32. Random choice of a singular M-matrix with $Me = 0$ [20]. To construct M, we generated a 100×100 random matrix R, and set $M = \mathrm{diag}(Re) - R$. The matrices A, B, C and D are 50×50. We generated 5 different matrices M and computed the relative residuals and number of steps needed for the iterations to converge.

All the algorithms (N, IN, SDA, ISDA, CR and ICR) arrive at a relative residual less than 10ε. The number of steps needed by the algorithms are reported in Table 2. As one can see the basic algorithms require the same number of steps, whilst using the singularity the Newton method requires one or two steps less than ISDA and ICR, however, the cost per step of these two methods make their overall cost much lower than the Newton method.

The use of the singularity reduces dramatically the number of steps needed for the algorithms to converge.

Algorithm	Steps needed
N	11–12
IN	3
SDA	11–12
ISDA	4-5
CR	11–13
ICR	4–5

Table 2. Minimum and maximum number of steps needed for algorithms to converge in Test 32

Table 3 summarizes the spectral and computational properties of the solutions of the NARE (1).

Table 4 reports the computational cost of the algorithms for solving (1) with $m = n$, together with the convergence properties in the noncritical case.

	M nonsingular	M singular irreducible		
		$\mu < 0$	$\mu = 0$	$\mu > 0$
splitting	complete	complete	incomplete	complete
	$\lambda_{n+1} < 0 < \lambda_n$	$\lambda_{n+1} < 0 = \lambda_n$	$\lambda_{n+1} = 0 = \lambda_n$	$\lambda_{n+1} = 0 < \lambda_n$
solutions $\geqslant 0$	2	2	1	2
Δ_S	nonsingular	nonsingular	singular	nonsingular
accuracy	ε	ε	$\sqrt{\varepsilon}$	ε

Table 3. Summary of the properties of the NARE

Algorithm	Computational cost	Reference
Schur method	$200n^3$	[23, 40]
Functional iteration	$8n^3$—$14n^3$ (per step)	[20, 26]
Newton's method	$66n^3$ (per step)	[26, 24]
CR applied to (17)	$\frac{74}{3}n^3$ (per step)	[10, 13]
CR applied to (18) (SDA)	$\frac{64}{3}n^3$ (per step)	[16, 25, 13]
CR applied to (19), (20)	$\frac{38}{3}n^3$ (per step)	[33, 13]

Table 4. Comparison of the algorithms

References

1. S. Ahn and V. Ramaswami. Transient analysis of fluid flow models via stochastic coupling to a queue. *Stoch. Models*, 20(1):71–101, 2004.
2. B. D. O. Anderson. Second-order convergent algorithms for the steady-state Riccati equation. *Internat. J. Control*, 28(2):295–306, 1978.
3. S. Asmussen. Stationary distributions for fluid flow models with or without Brownian noise. *Comm. Statist. Stochastic Models*, 11(1):21–49, 1995.
4. Z. Bai and J. W. Demmel. On swapping diagonal blocks in real Schur form. *Linear Algebra Appl.*, 186:73–95, 1993.
5. R. H. Bartels and G. W. Stewart. Solution of the matrix equation $AX + XB = C$. *Commun. ACM*, 15(9):820–826, 1972.

6. N. G. Bean, M. M. O'Reilly, and P. G. Taylor. Algorithms for return probabilities for stochastic fluid flows. *Stochastic Models*, 21(1):149–184, 2005.
7. P. Benner and R. Byers. An exact line search method for solving generalized continuous-time algebraic Riccati equations. *IEEE Trans. Automat. Control*, 43(1):101–107, 1998.
8. A. Berman and R. J. Plemmons. *Nonnegative matrices in the mathematical sciences*, volume 9 of *Classics in Applied Mathematics*. Society for Industrial and Applied Mathematics (SIAM), Philadelphia, PA, 1994. Revised reprint of the 1979 original.
9. D. Bini and B. Meini. On the solution of a nonlinear matrix equation arising in queueing problems. *SIAM J. Matrix Anal. Appl.*, 17(4):906–926, 1996.
10. D. A. Bini, B. Iannazzo, G. Latouche, and B. Meini. On the solution of algebraic Riccati equations arising in fluid queues. *Linear Algebra Appl.*, 413(2-3):474–494, 2006.
11. D. A. Bini, B. Iannazzo, and F. Poloni. A Fast Newton's Method for a Nonsymmetric Algebraic Riccati Equation. *SIAM Journal on Matrix Analysis and Applications*, 30(1):276–290, 2008.
12. D. A. Bini, G. Latouche, and B. Meini. *Numerical methods for structured Markov chains*. Numerical Mathematics and Scientific Computation. Oxford University Press, New York, 2005. Oxford Science Publications.
13. D. A. Bini, B. Meini, and F. Poloni. From algebraic Riccati equations to unilateral quadratic matrix equations: old and new algorithms. Technical Report 1665, Dipartimento di Matematica, Università di Pisa, Italy, July 2007.
14. A. Brauer. Limits for the characteristic roots of a matrix. IV. Applications to stochastic matrices. *Duke Math. J.*, 19:75–91, 1952.
15. C.-Y. Chiang and W.-W. Lin. A structured doubling algorithm for nonsymmetric algebraic Riccati equations (a singular case). Technical report, National Center for Theoretical Sciences, National Tsing Hua University, Taiwan R.O.C., July 2006.
16. E. K.-W. Chu, H.-Y. Fan, and W.-W. Lin. A structure-preserving doubling algorithm for continuous-time algebraic Riccati equations. *Linear Algebra Appl.*, 396:55–80, 2005.
17. A. da Silva Soares and G. Latouche. Further results on the similarity between fluid queues and QBDs. In *Matrix-analytic methods (Adelaide, 2002)*, pages 89–106. World Sci. Publ., River Edge, NJ, 2002.
18. S. Fital and C.-H. Guo. Convergence of the solution of a nonsymmetric matrix Riccati differential equation to its stable equilibrium solution. *J. Math. Anal. Appl.*, 318(2):648–657, 2006.
19. G. H. Golub and C. F. Van Loan. *Matrix computations*. Johns Hopkins Studies in the Mathematical Sciences. Johns Hopkins University Press, Baltimore, MD, third edition, 1996.
20. C.-H. Guo. Nonsymmetric algebraic Riccati equations and Wiener-Hopf factorization for M-matrices. *SIAM J. Matrix Anal. Appl.*, 23(1):225–242, 2001.
21. C.-H. Guo. A note on the minimal nonnegative solution of a nonsymmetric algebraic Riccati equation. *Linear Algebra Appl.*, 357:299–302, 2002.
22. C.-H. Guo. Comments on a shifted cyclic reduction algorithm for quasi-birth-death problems. *SIAM J. Matrix Anal. Appl.*, 24(4):1161–1166, 2003.
23. C.-H. Guo. Efficient methods for solving a nonsymmetric algebraic Riccati equation arising in stochastic fluid models. *J. Comput. Appl. Math.*, 192(2):353–373, 2006.

24. C.-H. Guo and N. J. Higham. Iterative Solution of a Nonsymmetric Algebraic Riccati Equation. *SIAM Journal on Matrix Analysis and Applications*, 29(2):396–412, 2007.
25. C.-H. Guo, B. Iannazzo, and B. Meini. On the Doubling Algorithm for a (Shifted) Nonsymmetric Algebraic Riccati Equation. *SIAM J. Matrix Anal. Appl.*, 29(4):1083–1100, 2007.
26. C.-H. Guo and A. J. Laub. On the iterative solution of a class of nonsymmetric algebraic Riccati equations. *SIAM J. Matrix Anal. Appl.*, 22(2):376–391, 2000.
27. C. He, B. Meini, and N. H. Rhee. A shifted cyclic reduction algorithm for quasi-birth-death problems. *SIAM J. Matrix Anal. Appl.*, 23(3):673–691, 2001/02.
28. N. J. Higham. The Matrix Function Toolbox. `http://www.ma.man.ac.uk/~higham/mftoolbox`.
29. N. J. Higham. *Functions of Matrices: Theory and Computation.* Society for Industrial and Applied Mathematics, Philadelphia, PA, USA, 2008.
30. L. Hogben, editor. *Handbook of linear algebra.* Discrete Mathematics and its Applications (Boca Raton). Chapman & Hall/CRC, Boca Raton, FL, 2007. Associate editors: Richard Brualdi, Anne Greenbaum and Roy Mathias.
31. R. A. Horn and S. Serra Capizzano. Canonical and standard forms for certain rank one perturbations and an application to the (complex) Google pageranking problem. To appear in Internet Mathematics, 2007.
32. T.-M. Hwang, E. K.-W. Chu, and W.-W. Lin. A generalized structure-preserving doubling algorithm for generalized discrete-time algebraic Riccati equations. *Internat. J. Control*, 78(14):1063–1075, 2005.
33. B. Iannazzo and D. Bini. A cyclic reduction method for solving algebraic Riccati equations. Technical report, Dipartimento di Matematica, Università di Pisa, Italy, 2005.
34. J. Juang. Global existence and stability of solutions of matrix Riccati equations. *J. Math. Anal. Appl.*, 258(1):1–12, 2001.
35. J. Juang and W.-W. Lin. Nonsymmetric algebraic Riccati equations and Hamiltonian-like matrices. *SIAM J. Matrix Anal. Appl.*, 20(1):228–243, 1999.
36. L. V. Kantorovich. *Functional analysis and applied mathematics.* NBS Rep. 1509. U. S. Department of Commerce National Bureau of Standards, Los Angeles, Calif., 1952. Translated by C. D. Benster.
37. D. Kleinman. On an iterative technique for riccati equation computations. *IEEE Trans. Automat. Control*, 13(1):114–115, 1968.
38. P. Lancaster and L. Rodman. *Algebraic Riccati equations.* Oxford Science Publications. The Clarendon Press Oxford University Press, New York, 1995.
39. G. Latouche and V. Ramaswami. A logarithmic reduction algorithm for quasi-birth-death processes. *J. Appl. Probab.*, 30(3):650–674, 1993.
40. A. J. Laub. A Schur method for solving algebraic Riccati equations. *IEEE Trans. Automat. Control*, 24(6):913–921, 1979.
41. W.-W. Lin and S.-F. Xu. Convergence analysis of structure-preserving doubling algorithms for Riccati-type matrix equations. *SIAM J. Matrix Anal. Appl.*, 28(1):26–39, 2006.
42. L.-Z. Lu. Newton iterations for a non-symmetric algebraic Riccati equation. *Numer. Linear Algebra Appl.*, 12(2-3):191–200, 2005.
43. L.-Z. Lu. Solution form and simple iteration of a nonsymmetric algebraic Riccati equation arising in transport theory. *SIAM J. Matrix Anal. Appl.*, 26(3):679–685, 2005.

44. V. L. Mehrmann. *The autonomous linear quadratic control problem*, volume 163 of *Lecture Notes in Control and Information Sciences*. Springer-Verlag, Berlin, 1991. Theory and numerical solution.
45. B. Meini. New convergence results on functional iteration techniques for the numerical solution of M/G/1 type Markov chains. *Numer. Math.*, 78(1):39–58, 1997.
46. V. Ramaswami. Matrix analytic methods for stochastic fluid flows. In D. Smith and P. Hey, editors, *Teletraffic Engineering in a Competitive World*, Proceedings of the 16th International Teletraffic Congress, Elsevier Science B.V., Edimburgh, UK, pages 1019–1030, 1999.
47. L. C. G. Rogers. Fluid models in queueing theory and Wiener-Hopf factorization of Markov chains. *Ann. Appl. Probab.*, 4(2):390–413, 1994.
48. D. Williams. A "potential-theoretic" note on the quadratic Wiener-Hopf equation for Q-matrices. In *Seminar on Probability, XVI*, volume 920 of *Lecture Notes in Math.*, pages 91–94. Springer, Berlin, 1982.

A Generalized Conjugate Direction Method for Nonsymmetric Large Ill-Conditioned Linear Systems

Edouard R. Boudinov[1] and Arkadiy I. Manevich[2]

[1] FORTIS Bank, Brussels, Belgium
edouard.boudinov@mail.ru
[2] Department of Computational Mechanics and Strength of Structures,
Dniepropetrovsk National University, Dniepropetrovsk, Ukraine
armanevich@yandex.ru

Abstract. A new version of the generalized conjugate direction (GCD) method for nonsymmetric linear algebraic systems is proposed which is oriented on large and ill-conditioned sets of equations. In distinction from the known Krylov subspace methods for unsymmetrical matrices, the method uses explicitly computed A-conjugate (in generalized sense) vectors, along with an orthogonal set of residuals obtained in the Arnoldi orthogonalization process. Employing entire sequences of orthonormal basis vectors in the Krylov subspaces, similarly to GMRES and FOM, ensures high stability of the method. But instead of solution of a linear set of equations with a Hessenberg matrix in each iteration for determining the step we use A-conjugate vectors and some simple recurrence formulas. The performance of the proposed algorithm is illustrated by the results of extensive numerical experiments with large-scale ill-conditioned linear systems and by comparison with the known efficient algorithms.

Keywords: linear algebraic equations, large-scale problems, iterative methods for linear systems, Krylov subspace methods, conjugate direction methods, orthogonalization.

1 Introduction

The method proposed in this paper is based on the notion of **A**-conjugacy in generalized sense, or "one-sided conjugacy" (in Russian literature term "**A**-pseudo-orthogonality" is also used). We remind the primary definition: vectors d_k are named conjugate direction vectors of a real non-singular matrix **A** (in generalized sense) if the following conditions are satisfied:

$$(d_i, \mathbf{A}d_k) = 0 \quad \text{for} \quad i < k; \qquad (d_i, \mathbf{A}d_k) \neq 0 \quad \text{for} \quad i = k; \tag{1}$$

(in general case $(d_i, \mathbf{A}d_k) \neq 0$ for $i > k$).

The notion of **A**-conjugacy in generalized sense has been introduced and studied already in 1970-s by G. W. Stewart [4], V. V. Voevodin and E. E. Tyrtyshnikov [7], [11], [12], and others.

A few generalized CD-algorithms for non-symmetric systems, based on one-sided conjugacy, have been elaborated already in 1980-s and later (L. A. Hageman, D. M. Young [10] and others, see also [19], [20]). These algorithms relate to different classes of the Krylov subspace methods: minimum residual methods, orthogonal residual methods, orthogonal errors methods. Convergence of these algorithms has been well-studied and, in particular, the finite termination property has been proved. Of course, these results relate to precise arithmetic.

However in practice the generalized CD-algorithms turned out to be less efficient, at whole, than methods based on an orthogonalization procedure, such as elaborated in the same years Full Orhogonalization Method (FOM) [15], Generalized Minimal Residual (GMRES) [16]. It is well known that the convergence of CD-algorithms in finite precision arithmetic essentially differs from its theoretical estimates in exact arithmetic.

In this paper we propose a new generalized conjugate direction algorithm for solving nonsymmetric linear systems (fitting into the class of orthogonal residual methods) which is competitive with the most efficient known methods in the case of large dimension or ill-conditioned systems. Similarly to GMRES and FOM, the algorithm employs entire sequences of orthonormal basis vectors in the Krylov subspaces obtained in the Arnoldi orthogonalization process [1]. This process is also considered as a way of computing residuals, instead of their usual updating.

For simplicity we describe the algorithm in two forms, sequentially introducing new elements. First a "basic algorithm" is presented which determines iterates by employing the one-sided conjugation and some recurrent formulas (but residuals are updated by the usual formula). Then the final algorithm is described which uses the orthogonalization process for deriving residuals. The performance of the proposed algorithm is demonstrated by applying to a set of standard linear problems. The results are compared to that obtained by the classical conjugate gradients method, GMRES and some other efficient methods.

2 Basic algorithm

We solve the problem

$$\mathbf{A}x = b, \qquad x, b \in \Re^N, \tag{2}$$

where $\mathbf{A}$ is an $N \times N$ non-singular real matrix (in general case a nonsymmetric one). Given an initial guess x_1, we compute an initial residual $r_1 = b - \mathbf{A}x_1$ and initial conjugate vector d_1 as a normalized residual r_1: $d_1 = r_1^0 = r_1/\|r_1\|$. The condition $(d_1, \mathbf{A}d_1) \neq 0$ is assumed to be satisfied.

The "basic algorithm" is as follows:

$$x_{k+1} = x_k + \alpha_k d_k, \qquad \alpha_k = \frac{(r_k, d_k)}{(d_k, \mathbf{A}d_k)}, \tag{3}$$

$$r_{k+1} = r_k - \alpha_k \mathbf{A} d_k, \tag{4}$$

$$d_{k+1} = r_{k+1} + \sum_{i=1}^{k} \beta_i^{(k+1)} d_i, \tag{5}$$

((x, y) denotes the scalar product of x and y). Coefficients α_i (3) provide the r_{k+1} to be orthogonal to the d_k.

Coefficients $\beta_i^{(k+1)}$ $(i = 1, \ldots, k)$ are computed from one-sided conjugacy conditions (1), which lead to a triangular set of equations with respect to $\beta_i^{(k+1)}$.

This process can be slightly simplified if to use the following apparent identity which follows from formula (4):

$$\mathbf{A} d_i = \frac{r_i - r_{i+1}}{\alpha_i} \qquad (i = 1, \ldots, k-1). \tag{6}$$

Then the following two-term recurrent formulas for coefficients $\beta_i^{(k)}$ can be easily derived:

$$\beta_i^{(k)} = \alpha_i \left[\frac{\beta_{i-1}^{(k)}}{\alpha_{i-1}} - \frac{(r_i, \mathbf{A} r_k)}{\|r_i\|^2} \right], \qquad \beta_1^{(k)} = -\frac{(d_1, \mathbf{A} r_k)}{(d_1, \mathbf{A} d_1)}. \tag{7}$$

The termination criterion is taken in the form $\|r_k\| \leqslant \varepsilon$ or $\|r_k\| \leqslant \varepsilon \|r_1\|$.

The algorithm constructs the orthogonal set of vectors r_i, $i \leqslant k$, and the **A**-conjugate (in the generalized sense) set of vectors d_i, $i \leqslant k$. Note that this method relates to "long recurrence" algorithms with respect to conjugate vectors (c.v.), because every new c. v. is computed from conditions of **A**-conjugacy with respect to all preceding c. v.'s. But it is a "short recurrence" algorithm with respect to the orthogonal set of residuals.

In the case of symmetric matrix **A** the algorithm is reduced to the classical CG-method: the vector set d_i, $i \leqslant k$ becomes **A**-conjugate in usual sense and all $\beta_i^{(k)}$, $i < k$, vanish.

3 Final algorithm

The basic algorithm is close to a few known algorithms, such as ORTHORES (L.A. Hageman, D.M. Young [10]) and some others. In exact arithmetic it reaches the solution in at most N iterations almost for every initial vector x_1 ([4], [13]). But in practice the efficiency of this algorithm is found to be insufficient for large and/or ill-conditioned systems. The main reason of this shortage, in our opinion, is connected with the updating formula for residuals (4). The updating formula (with α_k (3)) ensures orthogonality of the current residual r_{k+1} to the last conjugate direction d_k with high accuracy, but the orthogonality to all preceding residuals r_i, $i \leqslant k$, is maintained only in exact arithmetic. Round-off errors are not corrected in the next step; they only are accumulated from step to step. This accumulation gradually violates the orthogonality of vectors $\{r_i\}$ and destroys $\mathbf{A}$-conjugacy of vectors $\{d_i\}$. We would like to underline that the basic property of residuals $\{r_i\}$, required for efficiency of the algorithm, is their mutual orthogonality. Accumulation of errors and derangement of the residuals orthogonality is a principal inherent drawback of the basic algorithm (as well as every short recurrence CD-algorithm).

At first sight, the remedy is the direct computation of the residuals by formula $r_k = b - \mathbf{A}x_k$. But this way is wrong. The roundoff errors in computation of step lengths (point x_{k+1}) are again accumulated, so the residuals are computed "exactly", but at "inexact" points! The orthogonality of residuals again is gradually distorted. Besides, the additional matrix-vector multiplication per iteration is required.

We propose another way, which is realized in the final algorithm. Instead of the usual updating residuals we compute r_k simply from the conditions of orthogonality with respect to all preceding r_i^0, $i < k$ (using the modified Gram-Schmidt orthogonalization). Indeed, it is known a priori that the new residual should be orthogonal to all r_i^0, $i < k$, so we need only in proper scaling in order to the normal vector would coincide with the residual (in exact arithmetic). Such a scaling is given by the following formula:

$$r_{k+1} = -\alpha_k \left(\mathbf{A}r_k^0 - \sum_{i=1}^{k} \gamma_{k,i} r_i^0 \right), \quad \gamma_{k,i} = (\mathbf{A}r_k^0, r_i^0), \quad r_k^0 = \frac{r_k}{\|r_k\|}. \tag{8}$$

It can be easily shown that in exact arithmetic formulas (8) and (4) for the residuals are identical (both they determine a vector orthogonal to all $\{r_i\}$, $i \leqslant k$ in the Krylov subspace K_{k+1}, and have equal projections onto the vector r_k^0).

Other formulas of the algorithm remain principally the same, but some changes appears because we introduce the normalized vectors r_i^0 instead of r_i. The vector d_k (5) now is defined as follows:

$$d_{k+1} = r_{k+1}^0 + \sum_{i=1}^{k} \beta_i^{(k+1)} d_i, \tag{9}$$

$$\beta_i^{(k)} = \alpha_i \left[\frac{\beta_{i-1}^{(k)}}{\alpha_{i-1}} - \frac{(r_i^0, \mathbf{A}r_k^0)}{\|r_i\|} \right], \qquad \beta_1^{(k)} = -\frac{(r_1^0, \mathbf{A}r_k^0)}{(r_1^0, \mathbf{A}r_1^0)}. \tag{10}$$

Formula for the iterate x_{k+1} (3) remains the same, but formula for the step length α_k is changed due to new scaling of the vector d_i:

$$x_{k+1} = x_k + \alpha_k d_k, \quad \alpha_k = \frac{\|r_k\|}{(d_k, \mathbf{A}d_k)}. \tag{11}$$

This formula for α_k yields from (4) and the identity

$$(r_k, d_k) = \left(r_k, (r_k^0 + \sum_{i=1}^{k-1} \beta_i^{(k)} d_i) \right) = \|r_k\| \tag{12}$$

(since all vectors d_i, $i \leqslant k-2$, are linear combinations of the vectors r_j, $j \leqslant i$).

It is evident that the orthogonal vector set $\{r_i^0\}$ is less susceptible to degeneracy than the **A**- conjugate vectors set $\{d_i\}$. Hence all computations based on the vectors $\{r_i^0\}$ have higher accuracy than those based on $\{d_i\}$. Therefore it is worthwhile to replace, whenever it possible, operations based on $\{d_i\}$ by the ones based on $\{r_i^0\}$. One has

$$(d_k, \mathbf{A}d_k) = \left((r_k^0 + \sum_{i=1}^{k-1} \beta_i^{(k)} d_i), \mathbf{A}d_k \right) = (r_k^0, \mathbf{A}d_k) = \left(r_k^0, \mathbf{A}r_k^0 + \sum_{i=1}^{k-1} \beta_i^{(k)} \frac{r_i - r_{i+1}}{\alpha_i} \right) = (r_k^0, \mathbf{A}r_k^0) - \frac{\beta_{k-1}^{(k)}}{\alpha_{k-1}} \|r_k\| \tag{13}$$

(here we use formulas (1), (6)).

Thus the coefficients α_k and $\beta_i^{(k)}$ are computed via the vectors $\{r_i^0\}$, and the **A**-conjugate vectors $\{d_i\}$ are used only for computation of the current vector d_k by Eq. (9).

With modification (8) the algorithm becomes "long recurrence" one also with respect to residuals. This property is usually considered as a shortage since it is connected with increased storage requirements and complication of computations. But the long recurrence property makes algorithm more stable and less sensitive to the round-off errors. This was noted already in 80-s ([14]). So in the case of ill-conditioned or large problems long recurrence becomes more likely a merit of an algorithm rather than a drawback.

The final algorithm performs only one matrix-vector multiplication per iteration. We omit here all additional details and options of the algorithm.

It can be easily seen that the final algorithm constructs the same bases in Krylov subspaces as do GMRes and FOM (they use the similar Gram-Schmidt ortogonalization with the same initial vectors). But as for determining steps in these subspaces, computational scheme of our algorithm and that of GMRes (and FOM) are quite different. The GMRES finds the step by solving a linear set of equations with an upper Hessenberg matrix. This process involves the Givens rotations for reducing Hessenberg matrices to the triangular form and/or other computational elements. In our algorithm this subproblem is solved by employing conjugate directions. It is important that no extra matrix-vector product is required per iteration.

4 Numerical experiments

The algorithm has been realized in JAVA programming language and has been tested on a variety of linear algebraic problems (in most cases ill-conditioned).

For comparison we have chosen the following methods: the classical CG [2]; the Bi-CG [3], [6], the Conjugate Gradient Squared (CGS) [17], the Bi-CGSTAB [18] and the GMRES [16].

We used the MATLAB implementations of the Bi-CG, CGS and Bi-CGSTAB methods ([21]). But for CG and GMRES we employed our implementations. In order to reduce the execution time the matrices were first precalculated and then used in the methods implemented in MATLAB. Our implementations of the CG and GMRES methods were benchmarked against the MATLAB implementation of these methods (pcg and gmres functions MATLAB), and it was established that the numbers of iterations were identical in the both implementations, but the running time was less in our implementation.

The termination criterion was taken in the form $\|r_k\| \leqslant \varepsilon\|r_1\|$ (with $\varepsilon = 10^{-13} - 10^{-15}$). [3]

All computations have been performed on PC Pentium 3.2 GHz with RAM of 2000 MB in double precision.

Our main aims were to compare 1) the long reccurence algorithms with the short recurrence ones, 2) the orthogonalization procedure for specifying residuals with usual updating residuals.

First we present the results for symmetric systems with the following matrices (here degree in the denominators is gradually increased, and so the matrices become more degenerate):

$$\text{SYMM1}: \qquad a_{ii} = \tfrac{1}{i} \qquad a_{ij}(i \neq j) = \tfrac{1}{ij} \tag{14}$$

$$\text{SYMM2}: \qquad a_{ii} = \tfrac{1}{i^2} \qquad a_{ij}(i < j) = \tfrac{1}{ij^2} \qquad a_{ij}(i > j) = \tfrac{1}{i^2 j} \tag{15}$$

$$\text{SYMM3}: \qquad a_{ii} = \tfrac{1}{i^3} \qquad a_{ij}(i \neq j) = \tfrac{1}{(ij)^2} \tag{16}$$

$$\text{SYMM4}: \qquad a_{ii} = \tfrac{1}{i^4} \qquad a_{ij}(i < j) = \tfrac{1}{(ij)^2 j} \qquad a_{ij}(i > j) = \tfrac{1}{(ij)^2 i} \tag{17}$$

$$\text{SYMM5}: \qquad a_{ii} = \tfrac{1}{i^5} \qquad a_{ij}(i \neq j) = \tfrac{1}{(ij)^3} \tag{18}$$

In table 1 the results of the calculations for the number of variables $N = 1000$ and $\varepsilon = r/r_1 = 10^{-13}$ (in the termination criterion) are presented. Notations in this and following tables: N is the number of variables, ε_x is the accuracy in arguments, k_{iter} is the number of iterations, t is the running time.

The classical CG and our basic method have successfully solved the relatively simple problems SYMM1 – SYMM3; in the problem SYMM3 accuracy of the CG was very low, and in others problems (SYMM4, SYMM5) these algorithms

[3] The algorithm provides in the k-th iteration the orthogonal residual point x_{k+1}^{or} in which r_{k+1} is orthogonal to all preceding d_i: $(r_{k+1}^{or}, d_i) = 0,\ i = 1, \ldots, k$. Having obtained the conjugate vectors basis in the Krylov subspace K_k, by the cost of a few additional computations we obtained the minimal residual point x_{k+1}^{mr}, which is defined by conditions $(r_{k+1}^{mr}, \mathbf{A}d_i) = 0,\ i = 1, \ldots, k$. For correct comparison with GMRES we used this point in the termination criterion.

Table 1. Number of iterations for symmetric matrices (14)-(18). $N = 1000$, $\varepsilon = 10^{-13}$, "*"—the algorithm failed.

Problem	CG	GMRES	Basic algorithm	Final algorithm GCD
SYMM1	152	88	88	88
SYMM2	1001	177	198	177
SYMM3	1001	239	269	239
SYMM4	*	258	*	258
SYMM5	*	170	*	170

have failed. The GMRES and our final algorithm GCD have successfully solved all the problems with identical accuracy and the number of iterations, and the running time was practically the same. We would like to note that in all the cases the number of iterations (and so the number of stored conjugate vectors) in GMRES and our algorithm was less by several times comparing to the number of variables N (for N=1000 it did not exceed 258).

Table 2 shows the results for larger number of variables $N = 10000$. The CG-algorithm and our basic algorithm have solved with reasonable accuracy only the first problem. The GMRES and our final algorithm have solved all problems, and the numbers of iterations were again identical for both the methods and much less than the dimension of the problem. The running times in both the methods were approximately the same.

Table 2. Number of iterations for symmetric matrices (14)-(18). $N = 10000$, $\varepsilon = 10^{-13}$, "*"—the algorithm failed.

Problem	CG	GMRES	Basic algorithm	Final algorithm GCD
SYMM1	448	186	208	186
SYMM2	*	513	554	513
SYMM3	*	766	*	766

We see that even at solving linear systems with symmetric matrices general algorithms designed for non-symmetric problems turn out to be more stable and efficient comparing to special algorithms for symmetric systems; it is clear that the matrices may remain symmetric in the process of computations only in exact arithmetic.

In the next cycle of numerical experiments we consider the linear problems with nonsymmetric matrices. They were obtained from the matrices of type

(14)–(18) by introducing an asymmetry factor μ:

$$\text{ASYMM1}: \quad a_{ii} = \frac{1}{i}, \quad a_{ij}(i<j) = \frac{1+\mu}{ij}, \quad a_{ij}(i>j) = \frac{1-\mu}{ij} \tag{19}$$

$$\text{ASYMM2}: \quad a_{ii} = \frac{1}{i^2}, \quad a_{ij}(i<j) = \frac{1+\mu}{ij^2}, \quad a_{ij}(i>j) = \frac{1-\mu}{i^2j} \tag{20}$$

$$\text{ASYMM3}: \quad a_{ii} = \frac{1}{i^3}, \quad a_{ij}(i<j) = \frac{1+\mu}{(ij)^2}, \quad a_{ij}(i>j) = \frac{1-\mu}{(ij)^2} \tag{21}$$

$$\text{ASYMM4}: \quad a_{ii} = \frac{1}{i^4}, \quad a_{ij}(i<j) = \frac{1+\mu}{(ij)^2j}, \quad a_{ij}(i>j) = \frac{1-\mu}{(ij)^2i} \tag{22}$$

$$\text{ASYMM5}: \quad a_{ii} = \frac{1}{i^5}, \quad a_{ij}(i<j) = \frac{1+\mu}{(ij)^3}, \quad a_{ij}(i>j) = \frac{1-\mu}{(ij)^3} \tag{23}$$

The following algorithms have been tested, alongside with our algorithm (GCD): Bi-CG, CGS, Bi-CGSTAB and GMRES (the first three algorithms are short recurrence). Results for $N = 1000$ with $\mu = 0.5$ and $\varepsilon = 10^{-13}$ are presented in Table 3.

Table 3. Numbers of iterations in unsymmetrical problems with matrices (19)-(23), solved by various algorithms; $N = 1000$; the asymmetry coefficient $\mu = 0.5$; $\varepsilon = 10^{-13}$; "*"—the algorithm has failed.

Matrix	GMRES	GCD	Bi-CG	CGS	Bi-CGSTAB
ASYMM1	95	95	184	131	92
ASYMM2	183	183	1430	2473	918
ASYMM3	244	244	*	*	*
ASYMM4	264	264	*	*	*
ASYMM5	176	176	*	*	*

The CGS, Bi-CG, Bi-CGSTAB algorithms have solved only problems ASYMM1, ASYMM2. The GMRES and our algorithm have solved all the problems with approximately the same accuracy and numbers of iterations.

The data presented in above tables enable us to draw the following conclusions:

- at solving ill-conditioned problems the short recurrence algorithms compare unfavorably with the long recurrence ones; only long recurrence algorithms are efficient in ill-conditioned problems of moderate and large dimensions;
- algorithms based on usual updating of residuals (CG, our basic algorithm) are at a disadvantage in relation to algorithms based on an orthogonalization procedure (GMRES, our final algorithm);
- the convergence of our final algorithm GCD is identical to that of GMRES.

Therefore in the next computations we dealt only with GMRES and our algorithm GCD. Table 4 shows the results obtained by these algorithms for the same problems with larger number of variables $N = 10000$. Along with the numbers of iterations, here we present also the accuracy in arguments and the execution time. Again the both algorithms have solved all problems with approximately the same numbers of iterations, accuracy and execution times.

Table 4. Results for unsymmetrical problems with matrices (19)-(23); $N = 10000$; the asymmetry coefficient $\mu = 0.5$; $\varepsilon = 10^{-13}$.

	GMRES			Proposed method GCD		
Matrix	ε_x	k_{iter}	t (sec)	ε_x	k_{iter}	t (sec)
ASYMM1	$< 10^{-9}$	200	224	$< 10^{-9}$	200	224
ASYMM2	$< 10^{-5}$	532	621	$< 10^{-6}$	532	619
ASYMM3	< 0.009	788	957	< 0.005	788	956

In order to examine the algorithms in very large scale problems, we considered unsymmetrical problems produced from the matrices (19) with nonzero elements only on five diagonals, i. e., $a_{ij} = 0$ for $j > i+2$ and $j < i-2$. Table 5 presents the results obtained by GMRES and our algorithm. The both methods were very efficient in solving the problems up to N=150000 (on the given PC). We see that the accuracy of the solutions did not decrease as the dimension of the problem increased. The numbers of iterations in the both methods were again identical, and the running time was practically the same.

Table 5. Results for unsymmetrical problems with matrices (19) having only 5 diagonals with non-zero elements: the asymmetry factor $\mu = 0.5$; $\varepsilon = 10^{-10}$.

	GMRES			Proposed method GCD		
N	ε_x	k_{iter}	t (sec)	ε_x	k_{iter}	t (sec)
1000	$< 10^{-7}$	76	0.06	$< 10^{-7}$	76	0.06
10000	$< 10^{-6}$	159	2.02	$< 10^{-6}$	159	1.95
50000	$< 10^{-6}$	264	25.4	$< 10^{-6}$	264	23.0
100000	$< 10^{-6}$	329	83.4	$< 10^{-6}$	329	73.9
150000	$< 10^{-6}$	374	208	$< 10^{-6}$	374	210

Table 6 demonstrates the influence of the asymmetry factor on the efficiency of the GMRES and proposed algorithm, for two problems with $N = 1000$: the matrix (19) and Hilbert matrix modified with the asymmetry factor:

$$a_{ij} = \frac{1+\mu}{i+j-1} \quad (i<j), \qquad a_{ij} = \frac{1-\mu}{i+j-1} \quad (i>j) \tag{24}$$

The matrix asymmetry of the first matrix practically did not affect the performance of the algorithms. In the second problem even a very small matrix asymmetry had an impact on the convergency rate of the both algorithms: they required N iterations for solving the problems. The above conclusion about the comparative efficiency of the both methods holds at any μ-values.

Table 6. Results for unsymmetrical problems with various asymmetry coefficients μ; $N = 1000$; $\varepsilon = 10^{-13}$.

	GMRES			Proposed method GCD		
μ	ε_x	k_{iter}	t (sec)	ε_x	k_{iter}	t (sec)
Matrix Eq. (19)						
0	$< 10^{-9}$	88	1.08	$< 10^{-10}$	89	1.08
0.1	$< 10^{-9}$	90	1.13	$< 10^{-10}$	90	1.13
0.5	$< 10^{-9}$	95	1.14	$< 10^{-10}$	95	1.14
1.0	$< 10^{-9}$	99	1.13	$< 10^{-9}$	99	1.13
2.0	$< 10^{-9}$	106	1.30	$< 10^{-9}$	106	1.30
Hilbert matrix						
0	$< 10^{-3}$	17	0.24	$< 10^{-3}$	17	0.24
0.01	$< 10^{-8}$	1000	20	$< 10^{-8}$	1000	17
0.5	$< 10^{-10}$	1000	20	$< 10^{-10}$	1000	17
1.0	$< 10^{-10}$	1000	20	$< 10^{-10}$	1000	17
2.0	$< 10^{-10}$	1000	20	$< 10^{-10}$	1000	17

We see that the convergence properties of the final GCD algorithm and those of GMRES were the same in all the problems considered. What is the reason for the similarity of these methods? The answer is apparent. Both the methods construct the same orthogonal bases in the Krylov subspace with the same accuracy. Another question may be put: what is the sense to elaborate a new algorithm with the same efficiency as a known method? Computational schemes in these algorithms are different. GMRES does not use conjugate direction. Our algorithm finds iterates employing the conjugate directions. Simultaneously with solution of the linear system one obtains a conjugate vectors set $\{d_i\}$, vectors $\{\mathbf{A}d_i\}$ and quantities $(d_i, \mathbf{A}d_i)$. The knowledge of these quantities, which distinguishes our algorithm from GMRES and FOM, opens new possibilities in solving various problems associated with linear sets of equations, e.g., linear complementarity problems.

5 Conclusions

A generalized conjugate direction method for unsymmetrical linear systems is elaborated, which uses the generalized notion of conjugacy, orthogonalization process for constructing orthogonal basis and computing residuals, and some recurrent formulas for finding iterates. Due to the new algorithmic scheme the method turns out to be very stable and efficient for large scale and/or ill-conditioned problems. The extensive numerical tests show that the efficiency of the proposed method is similar to that of GMRES. In all these methods the efficiency in large scale ill-conditioned problems is maintained, firstly, by employing the long recurrence and, secondly, by replacement of usual updating residuals with the Arnoldi orthogonalization.

The computational schemes in our algorithm GCD and GMRES (or FOM) are different; the knowledge of conjugate vectors set $\{d_i\}$, as well as vectors $\{\mathbf{A}d_i\}$ and quantities $(d_i, \mathbf{A}d_i)$ may be useful at solving various problems, associated with linear sets of equations.

The storage requirements for the proposed method are identical to those of GMRES and other methods employing entire sets of orthogonal or conjugate vectors.

References

1. W. E. ARNOLDI, *The principle of minimized iterations in the solution of the matrix eigenvalue problem*, Quart. Appl. Math., 9, 1951, pp. 17–29.
2. M. R. HESTENES, and E. STIEFEL, *Methods of conjugate gradients for solving linear systems*, J. Research Nat. Bur. Standards, 49, 1952, pp. 409–436.
3. C. LANCZOS, *Solution of linear equations by minimized iterations*, J. Res. Natl. Bur. Stand., 49, 1952, pp. 33–53.
4. G. W. STEWART, *Conjugate Direction Methods for Solving Systems of Linear Equations*, Numer. Math, 21, 1973, pp. 285–297.
5. P. CONCUS, G. H. GOLUB, and D. P. O'LEARY, *A generalized conjugate gradient method for the numerical solution of elliptic partial differential equations*, In: Sparse Matrix Computations (Eds. J. K. Bunch and D. J. Rose), Academic Press, 1976, pp. 309–332.
6. R. FLETCHER, *Conjugate gradients methods for indefinite systems*, Numerical Analysis, Dundee 1975, G. Watson, ed., Springer-Verlag, Berlin, New York, 1976, pp. 73–89.
7. V. V. VOEVODIN, *On methods of conjugate directions*, USSR Comput. Maths and Math. Physics, 19, 1979, N 5, pp. 1313–1317 (in Russian).
8. D. M. YOUNG, and K. C. JEA, *Generalized conjugate gradient acceleration of nonsymmetrizable iterative methods*, Linear Algebra Appl., 34, 1980, pp. 159–194.
9. O. AXELSSON, *Conjugate gradient type methods for unsymmetric and inconsistent systems of linear equations*, Linear Algebra Appl., 29, 1980, pp. 1–16.
10. L. A. HAGEMAN, and D. M. YOUNG, *Applied Iterative Methods*. Academic Press, 1981.
11. V. V. VOEVODIN, and E. E. TYRTYSHNIKOV, *On generalization of conjugate directions methods*, Numerical Methods of Algebra (Chislennye Metody Algebry), Moscow State University Press, Moscow, 1981, pp.3–9 (in Russian).
12. V. V. VOEVODIN, *The problem of a non-selfadjoint generalization of the conjugate gradient method has been closed*, USSR Comput. Math. and Math. Physics, 23, 1983, pp. 143–144 (in Russian).

13. V. V. VOEVODIN, and YU. A. KUZNETSOV, *Matrices and computations.* Moscow, "Nauka", 1984, 320 p. (in Russian).

14. A. I. MANEVICH and P. I. POLYANCHIKOV, *A one-step method of conjugate directions*, Soviet J. Comput. Systems Sci. 23 (1985), No. 2, pp. 14–20; translated from Izv. Akad. Nauk SSSR Tekhn. Kibernet. 1984, No. 6, 41–47 (Russian).

15. Y. SAAD, *Practical use of some Krylov subspace methods for solving indefinite and nonsymmetric linear systems*, SIAM J. Sci. Statist. Comput. 5, 1984, pp. 203–228.

16. Y. SAAD, and M. SCHULTZ, *GMRES, a generalized minimal residual algorithm for solving nonsymmetric linear systems*, SIAM J. Sci. Statist. Comput., 7 (1986), pp. 856–869.

17. P. SONNEVELD, *CGS, a fast Lanczos-type solver for nonsymmetric linear systems*, SIAM J. Sci. Statist. Comput., 10 (1989), pp. 36–52.

18. H. A. VAN DER VORST, *Bi-CGSTAB: a fast and smoothly converging variant to Bi-CG for the solution of nonsymmetric systems*, SIAM J. Sci. Statist. Comput.,13 (1992), pp. 631–644.

19. Y. SAAD, *Iterative methods for sparse linear systems*, PWS Publishing, Boston, 1996, 440 p.

20. Y. SAAD and H. A. VAN DER VORST, *Iterative solution of linear systems in the 20th century*, J. Comput. Appl. Math., 123, 2000, pp. 1–33.

21. *Optimization Toolbox for Use with MATLAB. User's Guide. Version 2*, The MathWorks Inc., 2000.

22. H. A. VAN DER VORST, *Iterative Methods for Large Linear Systems*, Utrecht University, The Netherlands, 2002, 196 p.

There Exist Normal Hankel (ϕ, ψ)-Circulants of Any Order n

V. N. Chugunov[1] and Kh. D. Ikramov[2]

[1] Institute of Numerical Mathematics of Russian Academy of Sciences, Gubkina 8, 119333 Moscow, Russia
vadim@bach.inm.ras.ru

[2] Mechanics and Mathematics Department of Moscow State University, GSP-1, Leninskie gory, 119991 Moscow, Russia
ikramov@cs.msu.su

Abstract. We answer a question motivated by our study of the normal Hankel problem, i.e., the problem of describing normal Hankel matrices. It was shown previously that new solutions to this problem can only be found among the so-called (ϕ, ψ)-circulants. The latter can be described by a system of equations with respect to the real and imaginary parts of their entries. Since the equations are quadratic, it is not at all clear whether this system admits real solutions unless n (the order of the matrix) is three or four (these cases were solved in an earlier publication of the authors). In this note, we construct a class of normal Hankel matrices of an arbitrary order $n \geqslant 5$ that are (ϕ, ψ)-circulants for appropriately chosen values of ϕ and ψ.

1. The issue that we treat in this short paper is motivated by our study of the normal Hankel problem, i.e., the problem of describing normal Hankel matrices. This problem is still open despite a number of available partial results. A detailed account of its present state is given in Section 1 of our paper [1]. We need a shorter version of this account to formulate and then prove our result.

Let

$$H = H_1 + iH_2 \tag{1}$$

be an arbitrary Hankel matrix, H_1 and H_2 being its real and imaginary parts, respectively. Denote by $\mathcal{P}_n$ the backward identity matrix:

$$\mathcal{P}_n = \begin{pmatrix} & & 1 \\ & \cdot^{\cdot^{\cdot}} & \\ 1 & & \end{pmatrix}.$$

Then,

$$T = H\mathcal{P}_n = T_1 + iT_2 \tag{2}$$

is a Toeplitz matrix, T_1 and T_2 being again the real and imaginary parts of T. One can show that, for H to be a normal matrix, it is necessary and sufficient that the associated Toeplitz matrix (2) satisfies the relation

$$\operatorname{Im}(TT^*) = 0. \tag{3}$$

Let $a_1, a_2, \ldots, a_{n-1}$ and $a_{-1}, a_{-2}, \ldots, a_{-n+1}$ be the off-diagonal entries in the first row and the first column of T_1. Denote by $b_1, b_2, \ldots, b_{n-1}$ and $b_{-1}, b_{-2}, \ldots, b_{-n+1}$ the corresponding entries in T_2. Using these entries, we can form the matrices

$$F = \begin{pmatrix} a_{n-1} & b_{n-1} \\ a_{n-2} & b_{n-2} \\ \vdots & \vdots \\ a_1 & b_1 \end{pmatrix}$$

and

$$G = \begin{pmatrix} a_{-1} & b_{-1} \\ a_{-2} & b_{-2} \\ \vdots & \vdots \\ a_{-n+1} & b_{-n+1} \end{pmatrix}.$$

It turns out that all the classes of normal Hankel matrices previously described in the literature correspond to the cases where, for at least one of the matrices F and G, the rank is less than two. Therefore, we hereafter assume that

$$\operatorname{rank} F = \operatorname{rank} G = 2.$$

In this case, the basic equality (3) implies (see details in our paper [2]) that

$$G = FW, \tag{4}$$

where

$$W = \begin{pmatrix} \alpha & \beta \\ \gamma & \delta \end{pmatrix}$$

is a real 2×2 matrix with the determinant

$$\alpha\delta - \beta\gamma = 1. \tag{5}$$

The matrix equality (4) is equivalent to the scalar relations

$$a_{-i} = \alpha a_{n-i} + \gamma b_{n-i}, \qquad b_{-i} = \beta a_{n-i} + \delta b_{n-i}, \qquad 1 \leqslant i \leqslant n-1. \tag{6}$$

Writing the Toeplitz matrix (2) in the form

$$T = \begin{pmatrix} t_0 & t_1 & t_2 & \ldots & t_{n-1} \\ t_{-1} & t_0 & t_1 & \ldots & t_{n-2} \\ t_{-2} & t_{-1} & t_0 & \ldots & t_{n-3} \\ \ldots & \ldots & \ldots & \ldots & \ldots \\ t_{-n+1} & t_{-n+2} & t_{-n+3} & \ldots & t_0 \end{pmatrix}, \tag{7}$$

we can replace real relations (6) by the complex formulas

$$t_{-i} = \phi t_{n-i} + \psi \overline{t_{n-i}}, \qquad 1 \leqslant i \leqslant n-1, \tag{8}$$

where

$$\phi = \frac{\alpha+\delta}{2} + i\frac{\beta-\gamma}{2}, \qquad \psi = \frac{\alpha-\delta}{2} + i\frac{\beta+\gamma}{2}. \tag{9}$$

The complex form of relation (5) is as follows:

$$|\phi|^2 - |\psi|^2 = 1. \tag{10}$$

Let (ϕ, ψ) be a fixed pair of complex numbers obeying condition (10). A Toeplitz matrix T is called a (ϕ, ψ)-circulant if its entries satisfy relations (8). The corresponding Hankel matrix $H = T\mathcal{P}_n$ will be called a Hankel (ϕ, ψ)-circulant.

The case $\psi = 0, |\phi| = 1$ corresponds to the well-known classes of Toeplitz and Hankel ϕ-circulants. However, for $\psi \neq 0$, it is not at all clear whether there exist nontrivial normal Hankel (ϕ, ψ)-circulants. Indeed, if equalities (6) are substituted into our basic relation (3), then the result is a system of $n-1$ real equations with respect to $2n$ real unknowns $a_0, a_1, ..., a_{n-1}$ and $b_0, b_1, ..., b_{n-1}$. Since these equations are quadratic, they need not to have real solutions.

It was shown in [1] that the above system is solvable for $n = 3$ and $n = 4$ for every quadruple $(\alpha, \beta, \gamma, \delta)$ satisfying condition (5). The question of the existence of normal Hankel (ϕ, ψ)-circulants for larger values of n was left open there.

Below, we construct a special class of Toeplitz matrices that generate normal Hankel matrices for any $n \geqslant 5$. These matrices are (ϕ, ψ)-circulants for appropriate values of ϕ and ψ, where $\psi \neq 0$.

2. We seek T as a Toeplitz matrix with the first row of the form

$$0\ 0\ \cdots\ 0\ a\ b\ a.$$

Here, $a = x + iy$ and $b = z + iw$ are complex numbers to be determined. This matrix T must be a (ϕ, ψ)-circulant for appropriate values of ϕ and ψ (that is, for appropriate α, β, γ, and δ).

The Hankel (ϕ, ψ)-circulant corresponding to this T is normal if and only if the basic relation (3) is fulfilled. Now, observe that the property of T to be a (ϕ, ψ)-circulant implies that TT^* is a Toeplitz matrix (see [1] or [2] for explanations of this fact). Moreover, TT^* is obviously a Hermitian matrix. It follows that the matrix relation (3) is equivalent to $n-1$ scalar conditions

$$\operatorname{Im}\{TT^*\}_{1j} = 0, \qquad j = 2, 3, \ldots, n. \tag{11}$$

Due to the "tridiagonal" structure of T, we have

$$\{TT^*\}_{1j} = 0, \qquad j = 4, 5, \ldots, n-2.$$

The remaining conditions in (11) correspond to $j = 2, 3, n-1$ and n. They have the same form for any value of n, beginning from $n = 5$. Thus, to find the desired a and b, it suffices to analyze the case $n = 5$.

Since

$$\{TT^*\}_{12} = b\bar{a} + a\bar{b}, \qquad \{TT^*\}_{13} = |a|^2,$$

the first two conditions in (11) are automatically fulfilled. It remains to satisfy two conditions corresponding to $j = 4$ and $j = 5$. This yields the following system of two equations in four real variables x, y, z and w:

$$\beta x^2 + (\delta - \alpha)xy - \gamma y^2 = 0, \tag{12}$$

$$[2\beta x + (\delta - \alpha)y]z + [(\delta - \alpha)x - 2\gamma y]w = 0. \tag{13}$$

Furthermore, we must keep in mind the relation

$$\operatorname{rank} F = \operatorname{rank} \begin{pmatrix} 0 & 0 \\ x & y \\ z & w \\ x & y \end{pmatrix} = 2,$$

which is equivalent to the inequality

$$yz - xw \neq 0$$

and excludes solutions to system (12), (13) for which $x = y = 0$.

Suppose that (x, y) is a nontrivial solution to equation (12). Substituting x and y into (13), we obtain a linear equation with respect to z and w. However, if at least one of the expressions inside the brackets is nonzero, then this equation is equivalent to the relation

$$yz - xw = 0, \tag{14}$$

signifying that $\operatorname{rank} F = 1$. Indeed, the determinant of the system composed of equations (13) and (14) is given by the formula

$$\begin{vmatrix} 2\beta x + (\delta - \alpha)y & (\delta - \alpha)x - 2\gamma y \\ y & -x \end{vmatrix} = -2[\beta x^2 + (\delta - \alpha)xy - \gamma y^2]$$

and, hence, vanishes in view of (12).

On the other hand, if, for the chosen solution (x, y), we have

$$2\beta x + (\delta - \alpha)y = 0, \tag{15}$$

$$(\delta - \alpha)x - 2\gamma y = 0, \tag{16}$$

then (13) is satisfied by any pair (z, w). Almost all of these pairs satisfy the condition $yz - xw \neq 0$.

By assumption, the homogeneous system (15), (16) has a nontrivial solution (x, y), which means that its determinant

$$\begin{vmatrix} 2\beta & \delta-\alpha \\ \delta-\alpha & -2\gamma \end{vmatrix} = -4\beta\gamma - (\delta-\alpha)^2$$

must be zero. Taking (5) into account, we obtain the condition

$$|\delta+\alpha| = 2. \tag{17}$$

Summing up, we have shown that, for every quadruple $(\alpha, \beta, \gamma, \delta)$ satisfying conditions (5) and (17), there exist complex scalars $a = x + iy$ and $b = z + iw$ specifying the desired Toeplitz matrix T. This matrix is a (ϕ, ψ)-circulant for ϕ and ψ determined by the chosen values of α, β, γ, and δ. The corresponding matrix H (see (2)) is a normal Hankel (ϕ, ψ)-circulant.

V. N. Chugunov acknowledges the support of the Russian Foundation for Basic Research (projects nos. 04-07-90336 and 05-01-00721) and a Priority Research Grant OMN-3 of the Department of Mathematical Sciences of Russian Academy of Sciences.

References

1. V. N. Chugunov, Kh. D. Ikramov, *On normal Hankel matrices of low orders*, Mat. Zametki (accepted for publication).
2. V. N. Chugunov, Kh. D. Ikramov, *On normal Hankel matrices*, Zap. Nauchn. Sem. S.-Peterburg. Otdel. Mat. Inst. Steklov. (POMI) **346** (2007), 63–80.

On the Treatment of Boundary Artifacts in Image Restoration by Reflection and/or Anti-Reflection

Marco Donatelli* and Stefano Serra-Capizzano**

Dipartimento di Fisica e Matematica, Università dell'Insubria - Sede di Como, Via Valleggio 11, 22100 Como, Italy
*marco.donatelli@uninsubria.it,
**stefano.serrac@uninsubria.it, **serra@mail.dm.unipi.it

Abstract. The abrupt boundary truncation of an image introduces artifacts in the restored image. For large image restoration with shift-invariant blurring, it is advisable to use Fast Fourier Transform (FFT)-based procedures for reducing the computational effort. In this direction several techniques manipulate the observed image at the boundary or make some assumptions on the boundary of the true image, in such a way that FFT-based algorithms can be used. We compare the use of reflection with that of anti-reflection, in connection with the choice of the boundary conditions or for extending the observed image, both theoretically and numerically. Furthermore, we combine the two proposals. More precisely we apply anti-reflection, followed by reflection if necessary, to the observed image and we observe that the resulting restoration quality is increased with respect to the case of plain reflection.

Keywords: image deblurring, boundary conditions, fast transforms and matrix algebras.

1 Introduction

The blurred image is expressed as a function of an original scene that is larger than the field of view (FOV) of the blurred image, since pixels from the original scene outside the captured image window contribute to the pixels near the boundaries of the blurred observed image. Indeed the standard observation model can be expressed as

$$g = Af_o + u, \tag{1}$$

where f_o and g, lexicographically ordered, are the true and observed images, and u is the noise. The matrix A represents a convolution of the true image f_o with the point spread function (PSF) that we assume to be known and shift invariant. If the observed image is $n \times n$ and the PSF $m \times m$, then (1) implies that f_o is $(n+m-1)\times(n+m-1)$ and that A is a Toeplitz matrix of matrix size $n^2 \times (n+m-1)^2$. This means that the linear system (1) is underdetermined.

The goal is to recover f_o only in the FOV, i.e., the image f equal to the $n \times n$ middle part of f_o.

A well-established solution to both the problems of nonuniqueness and noise amplification is regularization. A classic approach is the Tikhonov regularization [10], which involves simultaneously minimizing the data error and a measure of the roughness of the solution. This leads to the linear system

$$(A^T A + \mu I)f = A^T g, \tag{2}$$

where $\mu > 0$ is the regularization parameter that should be opportunely chosen and usually satisfies $\mu \ll 1$. In general the solution of the linear system (2) can be computationally expensive, since it is not automatic that an FFT-based algorithm can be applied directly. However an interesting approach is proposed in [8] when $m \ll n$. Indeed, for dealing with the rectangular matrix A while using FFT-based algorithms, it is necessary to resort to iterative methods [2], in which the main task is the application of FFT-based procedures for matrix-vector multiplication. Conversely, for employing FFT-based direct algorithms, the linear system to solve should have coefficient matrix diagonalizable by a suitable fast trigonometric transform, such as sine, cosine, ω-Fourier ($|\omega| = 1$), Hartley transforms (see e.g. [6]). This can be done modifying system (1) in order to obtain a square coefficient matrix. The first approach amounts in imposing boundary conditions (BCs) to f_o an then computing a regularized solution of

$$Bf_b = g, \tag{3}$$

where B is $n^2 \times n^2$, with a structure depending on the shift invariant kernel and on the type of BCs [5]. The second approach is to extend g in some way to obtain g_e of size $2n \times 2n$, and then regularizing

$$Cf_e = g_e, \tag{4}$$

where C is the $(2n)^2 \times (2n)^2$ Circulant matrix obtained by periodically completing A; here the restored image is the $n \times n$ part of f_e corresponding to g [1].

In this paper, we compare the two approaches in the case of reflective pad, i.e. the two proposals in [7] and [1]. We will also consider the use of reflection and anti-reflection in connection as possible choices for boundary conditions. The main results are the following:

- In the case of strongly symmetric (symmetric with respect to each axis independently) PSFs, the considered approaches produce comparable restorations in practical problems.
- Imposing anti-reflective boundary conditions leads to a better restoration quality with respect to the reflective boundary conditions, at least for moderate level of noise [9, 4, 3]. However a direct fast method is available only in the strong symmetric setting.

(a) (b) (c)

Fig. 1. (a) Full reflection of the left top quadrant on right and bottom. (b) Half reflection on each edge of the middle image. (c) Half anti-reflection on each edge of the middle image (scaled image). The edges of the images are emphasized by tiny vertical and horizontal lines.

- To improve the results obtained by image extension as in (4), we use the ideas in [1], but instead of using reflection, we apply anti-reflection or anti-reflection followed by reflection. In this way we obtain a FFT-based algorithm also in the case of a generic PSF (non-necessarily symmetric), so overcoming the limitations in [7, 9] concerning the assumption of a strongly symmetric convolution kernel.

The paper is completed with numerical results to validate the proposals and the related analysis.

2 Reflection for image extension and BCs

In this section, we compare the reflection pad to extend g and the imposition of reflective BCs. The proposal in [1] to extend g, is to form a new image g_e of size $2n \times 2n$ as described in Fig. 1 (a). The observed image g is at the top left quadrant, the top right quadrant is the mirror image of g around the y axis, and the bottom half is the mirror image of the top half around the x axis. After that, the solution of the Tikhonov linear system is computed by circular convolution, because the coefficient matrix C in (4) is Circulant. In [1] it is shown that, for 1-D images and symmetric PSFs, when the truncated image is locally stationary at its boundaries, this approach leads to smaller expected errors in the restored image with respect to apply directly the circular convolution to g. Indeed the circular convolution assumes a circular signal and, independently of g, g_e is always a periodic image, hence it is reasonable to expect that it is obtained from a periodic scene. This clearly reduces the boundary artifact in the restoration, also in the case of non-symmetric PSFs. We note that a reflection of size $n/2$ with respect to each edge can be also used as in Fig. 1 (b), obtaining the same algorithm. Indeed the previous observation means only a translation of the period of the image of $n/2$ in each variable.

The use of reflective or Neumann BCs implies that the true image outside the FOV is a reflection of the image inside the FOV. Therefore f_o is assumed to be an extension by reflection of f like in Fig. 1 (b). The reflection is done with respect to each edge with a bandwidth depending on the support of the PSF, since it is necessary each pixel at the boundary to be well defined. Imposing reflective BCs the square linear system has size n^2 and the matrix B in (3) has a Toeplitz plus Hankel structure. More specifically, if the PSF is strongly symmetric, then B can be diagonalized by the discrete cosine transform of type I (DCT-I) (two dimensional in the case of images).

Now we provide an algebraic formulation of the two approaches in the case of strongly symmetric PSFs and 1-D case. The latter will allow us to give a qualitative comparison of the solution computed from the two strategies applying the Tikhonov regularization to (3) and (4), respectively. Since the PSF is symmetric, we have $h = [h_{-q}, \ldots, h_0, \ldots, h_q]$ with $h_{-i} = h_i$ and $q = (m-1)/2$. Let $\mathcal{T}_k = \{\, \phi_\alpha(x) = \sum_{j=-k}^{k} \alpha_j e^{ijx},\ \alpha_{-j} = \alpha_j\}$ be the set of even trigonometric polynomial of degree at most k, then the symbol

$$\phi_h(x) = \sum_{j=-q}^{q} h_j e^{ijx}, \tag{5}$$

is such that $\phi_h \in \mathcal{T}_q$ and $q \leqslant (n-1)/2$ for $m \leqslant n$.

Imposing reflective BCs, thanks to the symmetry of the PSF, in (3) $B = R_n D R_n^T$ where R_n is the DCT-I matrix (R_n is real and orthogonal), $D = \mathrm{diag}(b)$ with $b = R_n^T(Be_1)/R_n^T e_1$ (the division is component-wise) and e_1 is the first vector of the canonical base. Moreover, since $b_i = \phi_h(i\pi/n)$, $i = 0, \ldots, n-1$, B can be expressed in terms of its symbol ϕ_h and it will be denoted by $B = \mathcal{R}_n(\phi_h)$ (see [7]). Therefore, using the Tikhonov regularization approach (2) for the linear system (3), we obtain

$$f_r = R_n \,\mathrm{diag}\left(\frac{b}{b^2 + \mu}\right) R_n g, \tag{6}$$

where the operations between vectors are intended component-wise. Setting $z = b/(b^2 + \mu)$ and by defining $p_r \in \mathcal{T}_{n-1}$ as the interpolating polynomial in the pairs $\left(\frac{i\pi}{n}, z_i\right)$ for $i = 0, \ldots, n-1$, we find

$$f_r = \mathcal{R}_n(p_r) g. \tag{7}$$

For the other approach in (4), without loss of generality, let $\{1, \ldots, n\}$ be the FOV and n be even. Hence, by reflecting $g = [g_1, \ldots, g_n]$ on both sides, we have

$$g_e = [g_{n/2}, \ldots, g_2, g_1, g_1, g_2, \ldots, g_n, g_n, g_{n-1}, \ldots, g_{n/2+1}], \tag{8}$$

that, as already observed, leads to the same proposal as in [1]. Defining

$$P = \begin{bmatrix} E_l \\ I \\ E_r \end{bmatrix}_{2n \times n}, \tag{9}$$

$E_l = [J\,|\,0]_{n/2\times n}$, $E_r = [0\,|\,J]_{n/2\times n}$ and J being the $n/2 \times n/2$ flip matrix with entries $[J]_{s,t} = 1$ if $s + t = n + 1$ and zero otherwise, we have $g_e = Pg$. While $C = F_{2n}\Lambda F_{2n}^H$, where F_{2n} is the Fourier matrix of order $2n$ and $\Lambda = \mathrm{diag}(c)$, with $c = F_{2n}^H(Ce_1)/F_{2n}^H e_1$. Since $c_i = \phi_h(2\pi i/2n)$, $i = 0, \dots, 2n-1$, we denote $C = \mathcal{C}_{2n}(\phi_h)$. Using the Tikhonov regularization (2) for the linear system (4), the restored signal of size n is

$$f_c = [0\,|\,I\,|\,0]_{n\times 2n}\, C_{2n}(p_c)Pg, \tag{10}$$

where I is the identity of order n and, similarly to the reflective BCs case, $p_c \in \mathcal{T}_{2n-1}$ is the interpolating polynomial in $(\frac{i\pi}{n}, \nu_i)$ for $i = 0, \dots, 2n-1$ with $\nu = c/(c^2 + \mu)$. We show that $p_c \in \mathcal{T}_n$ and it is the interpolating polynomial in $(\frac{i\pi}{n}, \nu_i)$ for $i = 0, \dots, n$, i.e., the points $(\frac{i\pi}{n}, \nu_i)$ for $i = n+1, \dots, 2n-1$ do not add any further information. The interpolation conditions are

$$p_c\left(\frac{i\pi}{n}\right) = \nu_i, \qquad i = 0, \dots, n, \tag{11}$$

$$p_c\left(\frac{(n+i)\pi}{n}\right) = \nu_{n+i}, \qquad i = 1, \dots, n-1. \tag{12}$$

From the trigonometric identity $\cos(\frac{(n+i)\pi}{n}) = \cos(\frac{(n-i)\pi}{n})$, it follows $c_{n+i} = c_{n-i}$ that implies $\nu_{n+i} = \nu_{n-i}$ and $p_c(\frac{(n+i)\pi}{n}) = p_c(\frac{(n-i)\pi}{n})$, for $i = 1, \dots, n-1$. Therefore, conditions (12) can be written as $p_c(\frac{(n-i)\pi}{n}) = \nu_{n-i}$ for $i = 1, \dots, n-1$, that are a subset of (11). Moreover, $c_i = b_i$, and then $\nu_i = z_i$, for $i = 0, \dots, n-1$. Concluding, let $\Omega_n = \{\frac{i\pi}{n}\,|\, i = 0, \dots, n\}$ be the interpolation nodes forming a uniform grid on $[0, \pi]$ and let $\psi = \phi_h/(\phi_h^2 + \mu)$, then

$$p_c \in \mathcal{T}_n \quad \text{interpolating } \psi \text{ in } \quad \Omega_n, \tag{13}$$

$$p_r \in \mathcal{T}_{n-1} \quad \text{interpolating } \psi \text{ in } \quad \Omega_n \setminus \{\pi\}. \tag{14}$$

In order to compare f_r with f_c, it remains to check whether $[0\,|\,I\,|\,0]\, C_{2n}(\phi_\alpha)P$ belongs to the DCT-I algebra. Let $\phi_\alpha \in \mathcal{T}_{\frac{n}{2}}$, then the $n \times 2n$ matrix $T = [0\,|\,I\,|\,0]\, C_{2n}(\phi_\alpha)$ is

$$T = \left[\begin{array}{cc|ccccc|cc} \alpha_{-\frac{n}{2}} & \cdots & \alpha_0 & \cdots & \alpha_{\frac{n}{2}} & & & & \\ & \ddots & \vdots & \ddots & \ddots & \ddots & & & \\ & & \alpha_{-\frac{n}{2}} & & & & \alpha_{\frac{n}{2}} & & \\ & & & \ddots & & & \vdots & \ddots & \\ & & & & \alpha_{-\frac{n}{2}} & \cdots & \alpha_0 & \cdots & \alpha_{\frac{n}{2}} \end{array}\right] \tag{15}$$

and $\mathbb{T}P = \mathcal{R}_n(\phi_\alpha)$. We note that $[0\,|\,I\,|\,0]\,C_{2n}(\phi_\alpha)P = \mathcal{R}_n(\phi_\alpha)$ *holds only if* $\phi_\alpha \in \mathcal{T}_{\frac{n}{2}}$. Therefore it can not be used in (10) since $p_c \in \mathcal{T}_n$, but it generally fails to belong to $\mathcal{T}_{\frac{n}{2}}$. However, from (7) and (10), it holds

$$f_r - f_c = (\mathcal{R}_n(p_r) - [0\,|\,I\,|\,0]\,C_{2n}(p_c)P)g \tag{16}$$

$$= (\mathcal{R}_n(p_r - \phi_\alpha) - [0\,|\,I\,|\,0]\,C_{2n}(p_c - \phi_\alpha)P)g, \tag{17}$$

for $\phi_\alpha \in \mathcal{T}_{\frac{n}{2}}$. We take

$$\phi_\alpha = \arg\min_{p \in \mathcal{T}_{\frac{n}{2}}} \|\psi - p\|_\infty. \tag{18}$$

Therefore

$$\mathcal{C}_{2n}(p_c - \phi_\alpha) = \mathcal{C}_{2n}(p_c - \psi + \psi - \phi_\alpha) \tag{19}$$

$$= \mathcal{C}_{2n}(r_n) + \mathcal{C}_{2n}(a_{\frac{n}{2}}), \tag{20}$$

where r_n is the classical remainder in the trigonometric interpolation with $n+1$ equispaced nodes in $[0, \pi]$ with nodes belonging to Ω_n, while $a_{\frac{n}{2}}$ is the sup-norm optimal remainder of degree $n/2$. Similarly

$$\mathcal{R}_n(p_r - \phi_\alpha) = \mathcal{R}_n(\tilde{r}_{n-1}) + \mathcal{R}_n(a_{\frac{n}{2}}), \tag{21}$$

where $\tilde{r}_{n-1}$ is the remainder of the trigonometric interpolation with n equispaced nodes in $[0, \pi]$ with nodes belonging to $\Omega_n \setminus \{x_n = \pi\}$. As a consequence, since the transforms associated with the circulant and the cosine algebras are unitary, it follows that the spectral norms $\|C_n(s)\|$, $\|\mathcal{R}_n(s)\|$ are bounded by the infinity norm of s. Moreover $\|P\| = \|[0\,|\,I\,|\,0]\| = 1$ and hence, by using (19)–(21) in (17), we find

$$\|f_r - f_c\| \leqslant (\|\mathcal{R}_n(p_r - \phi_\alpha)\| + \|C_{2n}(p_c - \phi_\alpha)\|)\|g\| \tag{22}$$

$$\leqslant (\|r_n\|_\infty + \|\tilde{r}_{n-1}\|_\infty + 2\|a_{n/2}\|_\infty)\|g\| \tag{23}$$

$$\leqslant 2(Kn\|a_n\|_\infty + \|a_{n/2}\|_\infty)\|g\|, \tag{24}$$

with K constant, where the latter inequality follows from the evaluation of the Lebesgue constants in the interpolation operators. In fact, after the change of variable $y = \cos(x)$, the operator behind r_n is the interpolation on $[-1, 1]$ with Chebyshev nodes of second type (the zeros of $\sin(nx)/\sin(x)$) plus the additional endpoints $\{\pm 1\}$: its Lebesgue constant is known to grow as $K\log(n)$. The other Lebesgue constant related to the operator behind $\tilde{r}_{n-1}$ is again related to the Chebyshev nodes of the second type plus only $y = 1$ (i.e. $x = x_0 = 0$); in this case the associated Lebesgue constant is known to grow as Kn. Since $\|a_t\|_\infty$ is exponentially converging to zero as t tends to infinity (due to the C^∞ regularity of ψ), it follows that $\|f_r - f_c\|$ is exponentially converging to zero as n tends to infinity. As a consequence, the vectors f_r and f_c do not coincide in general, but their numerical difference is negligible already for moderate values of n.

Finally, when the PSF is not strongly symmetric, we notice that B can not be diagonalized by DCT-I and it has only a Toeplitz plus Hankel structure. Therefore in general the linear system arising from Tikhonov regularization and reflective BCs can not be solved by a FFT-based algorithm. On the other hand, the other approach based on the extension of g can be again applied without modifications.

3 Image extension by anti-reflection

The reflective pad is effective if the image is locally stationary at its boundaries, but it can still create significant artifacts if the image intensity has a large gradient at the boundary. Reflecting the image will create a cusp that is likely to be highly inconsistent with the original image, since the image beyond the boundary more than likely continues to change according to the gradient at the boundary rather than the negative of that gradient. According to this observation in [9], the author proposed to anti-reflect instead of to reflect the image at the boundary. The considered idea preserves the continuity of the normal derivative at the boundary without creating a cusp. In Fig. 1 (c) is shown how to extend an image by anti-reflection. We note a different scaling with respect to Fig. 1 (a) and Fig. 1 (b) since the anti-reflection produce value outside the original domain and the following visualization requires to scale the image.

We analyze in detail 1-D images. Imposing anti-reflective BCs the images $f = [f_1, \dots, f_n]$ is assumed to be extended as

$$f_{1-j} = 2f_1 - f_{j+1}, \qquad f_{n+j} = 2f_n - f_{n-j}, \tag{25}$$

for $j = 1, 2, \dots$ [9]. Antireflective BCs usually provide restoration better than reflective BCs, also in practical 2-D applications, while, from a computational effort viewpoint, they share the same properties as the reflective BCs [4, 3]. Indeed, when the PSF is strongly symmetric the matrix B in (3) is essentially diagonalized by discrete sine transform of type III (DST-III), in the sense that the first and last equations are decoupled and the inner $(n-2) \times (n-2)$ block can be diagonalized by DST-III. Hence, several computations involving B, like Tikhonov regularization, can be done by FFT-based algorithms.

In the last case, PSF no strongly symmetric, the matrix B is Toeplitz plus Hankel plus a rank two correction and the linear system arising from Tikhonov regularization can not be handled by simply invoking FFT-based algorithms. Therefore, when the PSF is not strongly symmetric, it could be useful to apply the anti-reflection pad to extend g and regularizing (4). The extended image g_e can be easily computed by $g_e = Pg$, with P defined in (9) where now $E_l = [2e| - J|0]$ and $E_r = [0| - J|2e]$, $e = [1, \dots, 1]^T$. We observe that in the case of a strongly symmetric PSF with the anti-reflective pad, differently from

(a) (b) (c)

Fig. 2. (a) Original image where the box indicates observed region. (b) Gaussian blurred and noisy image. (c) Out of focus blurred and noisy image.

the reflective case, the two approaches (BCs on f and extension of g) produce different restorations, usually of comparable quality: indeed the eigenvalues of B are not a subset of the eigenvalues of C, as it happens for the reflective pad, even if they are defined on a uniform grid $\{ i\pi/(n+1) \mid i = 1, \ldots, n\}$ as well.

The main problem extending g by anti-reflection is that g_e is not periodic and then the model (4) could suffer from this. On the other hand the ringing effects are greatly reduced with respect to the application of the circulant deconvolution directly to g, since the boundaries are far away from the portion of the restored image, when compared with the circulant case. However, we can improve the model, and then the restoration, extending g_e by reflection and obtaining a new periodic extended image g_p of size $4n \times 4n$. Clearly this further proposal leads to a moderate increase in the computational effort. In fact, as observed in [1], g_p is real and symmetric and hence only the computation of the real part of a 2D FFT of size $2n \times 2n$ is required.

4 Numerical experiments

For the following experimentation we use Matlab 7.0. The blurred images are contaminated by a mild white Gaussian noise. The restorations are compared visually and the relative restoration error (RRE) is defined as $\|\hat{f} - f\|_2/\|f\|_2$, where $\hat{f}$ and f are the restored and the true image respectively. For the Tikhonov regularization the parameter μ is chosen experimentally such that it minimizes the RRE, in a certain range of μ.

The image in Fig 2 (a) was blurred with a Gaussian PSF (Fig. 2 (b)) and with an out of focus PSF (Fig. 2 (c)). The observed images are $n \times n$ with $n = 195$. Since both the PSFs are strongly symmetric, we can compare the two approaches based on reflective BCs and reflective extension of the observed image respectively. The restored images and the absolute difference of the RREs for the two strategies in Fig. 3 and Fig. 4 validate the theoretical analysis given in Section 2. We note that both strategies reach the minimum RRE for the same

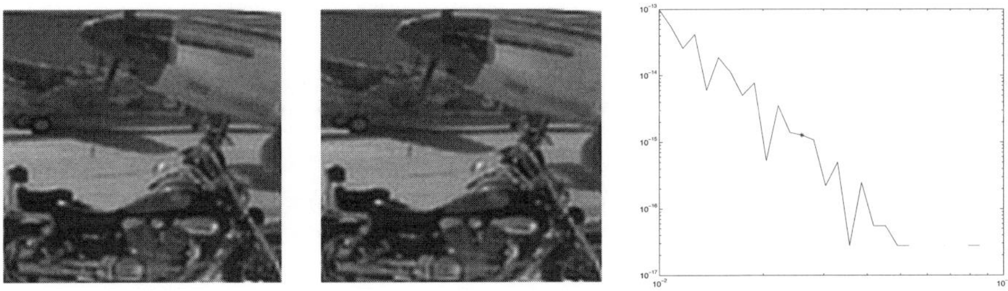

(a) (b) (c)

Fig. 3. Restorations of the image in Fig. 2 (b) (Gaussian blur): (a) restoration by reflective BCs, (b) restoration by reflective extension of the observed image, (c) loglog difference of the RREs for the two approaches ('*' corresponds to the optimal μ equal to 0.026 used in the restored images (a) and (b), absence of line means exactly zero value).

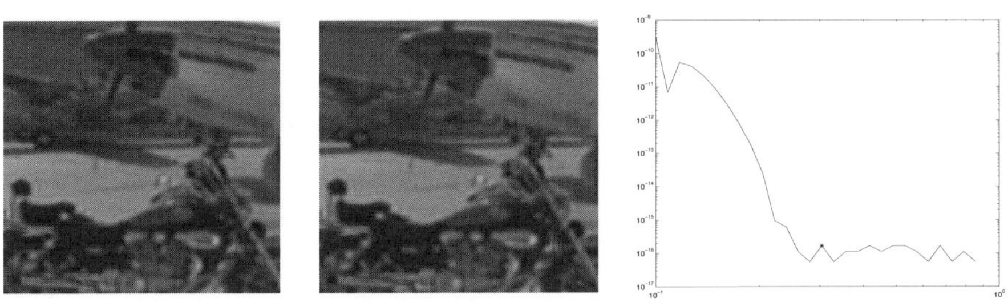

(a) (b) (c)

Fig. 4. Restorations of the image in Fig. 2 (c) (out of focus blur): (a) restoration by reflective BCs, (b) restoration by reflective extension of the observed image, (c) loglog difference of the RREs for the two approaches ('*' corresponds to the optimal μ equal to 0.304 used in the restored images (a) and (b)).

value of μ and we observe that, around this minimum, the absolute difference of the RREs has the same order of the machine precision (10^{-16}).

Now we consider the anti-reflective extension of the observed image described in Section 3 and we compare it only with the reflective extension in the case of a nonsymmetric PSF. Indeed, for strongly symmetric PSFs we have seen that the two approaches based on reflective BCs and reflective extension of the observed image are equivalent. Moreover, in the recent literature, it is widely documented a certain supremacy of the anti-reflective BCs with respect to reflective BCs [4, 3], for moderate levels of noise. On the other hand, when the PSF is not strongly symmetric the BC approach with the Tikhonov regularization leads to a linear system that can not be solved by FFT-based algorithms. Hence, in such case the only fast approach is which based on the extension of the observed image. According to the above comments we choose a PSF representing a motion along

(a) (b) (c) (d)

Fig. 5. (a) Moving blurred and noisy image. (b) Restoration by reflective extension $2n \times 2n$ (RRE $= 0.0932$). (c) Restoration by anti-reflective extension $2n \times 2n$ (RRE $= 0.0807$). (d) Restoration by anti-reflective extension and then reflective extension $4n \times 4n$ (RRE $= 0.0770$).

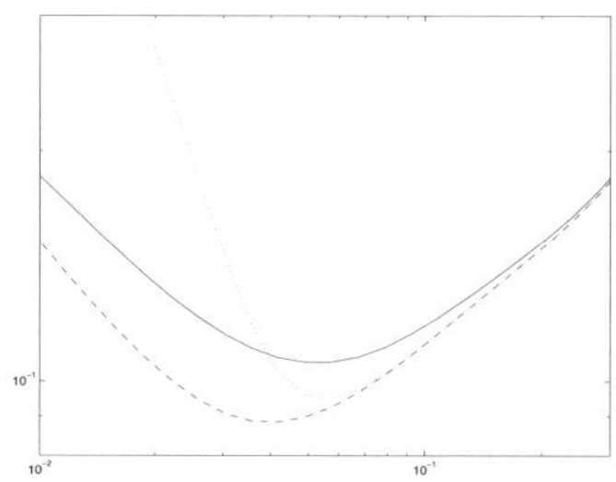

Fig. 6. Loglog RRE vs. μ for the test in Fig. 5 and the three approaches: — reflective extension, $\cdots$ anti-reflective extension $2n \times 2n$, -- anti-reflective extension and then reflective extension $4n \times 4n$.

the x axis. The original image is again that in Fig. 2 (a), while the blurred and noisy image is in Fig. 5 (a). In Fig. 5 (c) the restored image is obtained by anti-reflective extension that, also if the extended image is not periodic, is better than the restored image with reflective extension in Fig. 5 (b). The improvement is especially visible near the right edge, that is in the direction of the motion. If we want further improve the restoration, as described in Section 3, we can extend by reflection the $2n \times 2n$ image obtained by the anti-refletive pad and then apply the circulant de-convolution to the new $4n \times 4n$ problem. Indeed, the restored image in Fig. 5 (d) is better than that in Fig. 5 (c). Moreover the last approach is more stable under perturbations of the parameter μ, as shown in Fig. 6 by the plot of the RREs vs. μ for the considered approaches.

Acknowledgment

The work of the authors was partially supported by MUR, grant №2006017542.

References

1. F. Aghdasi and R. K. Ward, *Reduction of boundary artifacts in image restoration*, IEEE Trans. Image Process., 5 (1996), pp. 611–618.
2. M. Bertero and P. Boccacci, *A simple method for the reduction of the boundary effects in the richardson-lucy approach to image deconvolution*, Astron. Astrophys., 437 (2005), pp. 369–374.
3. M. Christiansen and M. Hanke, *Deblurring methods using antireflective boundary conditions*. manuscript, 2006.
4. M. Donatelli, C. Estatico, A. Martinelli, and S. Serra-Capizzano, *Improved image deblurring with anti-reflective boundary conditions and re-blurring*, Inverse Problems, 22 (2006), pp. 2035–2053.
5. P. C. Hansen, J. G. Nagy, and D. P. O'Leary, *Deblurring Images: Matrices, Spectra, and Filtering*, SIAM, Philadelphia, PA, 2006.
6. T. Kailath and V. Olshevsky, *Displacement structure approach to discrete-trigonometric-transform based preconditioners of g. strang type and t. chan type*, Calcolo, 33 (1996), p. 191208.
7. M. K. Ng, R. H. Chan, and W. C. Tang, *A fast algorithm for deblurring models with Neumann boundary conditions*, SIAM J. Sci. Comput., 21 (1999), pp. 851–866.
8. S. J. Reeves, *Fast image restoration without boundary artifacts*, IEEE Trans. Image Process., 14 (2005), pp. 1448–1453.
9. S. Serra-Capizzano, *A note on anti-reflective boundary conditions and fast deblurring models*, SIAM J. Sci. Comput., 25 (2003), pp. 1307–1325.
10. A. N. Tikhonov, *Solution of incorrectly formulated problems and regularization method*, Soviet Math. Dokl., 4 (1963), pp. 1035–1038.

Zeros of Determinants of λ-Matrices

Walter Gander

Computational Science, ETH, CH-8092 Zurich, Switzerland
gander@inf.ethz.ch

Abstract. Jim Wilkinson discovered that the computation of zeros of polynomials is ill conditioned when the polynomial is given by its coefficients. For many problems we need to compute zeros of polynomials, but we do not necessarily need to represent the polynomial with its coefficients. We develop algorithms that avoid the coefficients. They turn out to be stable, however, the drawback is often heavily increased computational effort. Modern processors on the other hand are mostly idle and wait for crunching numbers so it may pay to accept more computations in order to increase stability and also to exploit parallelism. We apply the method for nonlinear eigenvalue problems.

Keywords: nonlinear eigenvalue problems, Gaussian elimination, determinants, algorithmic differentiation.

1 Introduction

The classical textbook approach to solve an eigenvalue problem $A\mathbf{x} = \lambda\mathbf{x}$ is to first compute the coefficients of the characteristic polynomial $P_n(\lambda) = \det(\lambda I - A)$ by expanding the determinant

$$P_n(\lambda) = c_0 + c_1\lambda + \cdots + c_{n-1}\lambda^{n-1} + \lambda^n.$$

Then second apply some iterative method like e.g. Newton's method to compute the zeros of P_n which are the eigenvalues of the matrix A.

In the beginning of the area of numerical analysis a research focus was to develop reliable solvers for zeros of polynomials. A typical example is e.g. [4]. However, the crucial discovery by Jim Wilkinson [6] was that the zeros of a polynomial can be very sensitive to small changes of the coefficients of the polynomial. Thus the determination of the zeros from the coefficients is ill conditioned. It is easy today to repeat the experiment using a computer algebra system. Executing the following Maple statements

```
p :=1:
for i from 1 by 1 to 20 do  p := p*(x-i) od:
PP := expand(p);
Digits := 7
```

```
PPP := evalf(PP)
Digits := 30
Z := fsolve(PPP, x, complex, maxsols = 20)
```

we can simulate what Jim Wilkinson experienced. We first expand the product

$$\prod_{i=1}^{20}(x-i) = x^{20} - 210x^{19} \pm \cdots + 20!$$

then round the coefficients to floating point numbers with 7 decimal digits.

$$x^{20} - 210.0\,x^{19} + 2.432902 \times 10^{18} \mp \cdots - 8.752948 \times 10^{18}\,x + 20615.0\,x^{18}$$

Continuing now the computation with 30 decimal digits to determine the exact zeros of the polynomial with truncated coefficients we note that we do not obtain the numbers $1, 2, \ldots, 20$. Instead many zeros are complex such as e.g. $17.175 \pm 9.397i$. Thus truncating the coefficients to 7 decimal digits has a very large effect on the zeros. The problem is ill conditioned.

2 Matlab reverses computing

Instead of expanding the determinant to obtain the coefficients of the characteristic polynomial the command `P = poly(A)` in Matlab computes the eigenvalues of A by the QR-Algorithm and expands the linear factors

$$P_n(\lambda) = (\lambda - \lambda_1)(\lambda - \lambda_2)\cdots(\lambda - \lambda_n) = \lambda^n + c_{n-1}\lambda^{n-1} + \cdot + c_0$$

to compute the coefficients.

Given on the other hand the coefficients c_k of a polynomial, the command `lambda = roots(P)` forms the companion matrix

$$A = \begin{pmatrix} -c_{n-1} & -c_{n-2} & \cdots & -c_1 & -c_0 \\ 1 & 0 & \cdots & \cdots & 0 \\ 0 & 1 & 0 & \cdots & 0 \\ \vdots & \ddots & \ddots & \ddots & \vdots \\ 0 & 0 & 0 & 1 & 0 \end{pmatrix}$$

and uses again the QR-Algorithm to find the eigenvalues which are the zeros of the polynomial.

3 Evaluating the characteristic polynomial

How can we evaluate the characteristic polynomial without first computing its coefficients? One way is to use Gaussian elimination and the fact that it is easy to compute the determinant of a triangular matrix. Assume that we have computed the decomposition

$$C = LU$$

with L a lower unit triangular and U an upper triangular matrix. Then

$$\det(C) = \det(L)\det(U) = u_{11}u_{22}\cdots u_{nn}$$

since $\det(L) = 1$. Using partial pivoting for the decomposition we have to change the sign of the determinant each time that we interchange two rows. The program then becomes:

```
function f = determinant(C)
n = length(C);
f = 1;
for i = 1:n
    [cmax,kmax]= max(abs(C(i:n,i)));
    if cmax == 0   % Matrix singular
       f = 0; return
    end
    kmax = kmax+i-1;
    if kmax ~= i
       h = C(i,:); C(i,:) = C(kmax,:); C(kmax,:) = h;
       f = -f;
    end
    f = f*C(i,i);
   % elimination step
    C(i+1:n,i) = C(i+1:n,i)/C(i,i);
    C(i+1:n,i+1:n) = C(i+1:n,i+1:n) - C(i+1:n,i)*C(i,i+1:n);
end
```

Let $C(\lambda) = \lambda I - A$. We would like to use Newton's method to compute zeros of $P(\lambda) = \det(C(\lambda)) = 0$. For this we need the derivative $P'(\lambda)$. It can be computed by *algorithmic differentiation*, that is by differentiating each statement of the program to compute $P(\lambda)$. For instance the statement to update the determinant `f = f*C(i,i);` will be preceded by the statement for the derivative, thus

```
fs =fs*C(i,i)+f*Cs(i,i) ; f = f*C(i,i);
```

We used the variable `Cs` for the matrix $C'(\lambda)$ and `fs` for the derivative of the determinant.

There is, however, for larger matrices the danger that the value of the determinant over- respectively underflows. Notice that for Newton's iteration we do not need both values $f = \det(C(\lambda))$ and $fs = \frac{d}{d\lambda}\det(C(\lambda))$. It is sufficient to compute the *ratio*

$$\frac{\mathsf{P}(\lambda)}{\mathsf{P}'(\lambda)} = \frac{\mathsf{f}}{\mathsf{fs}}.$$

Overflow can be reduced by computing the logarithm. Thus instead of computing `f = f*C(i,i)` we can compute `lf = lf + log(C(i,i))`. Even better is the derivative of the logarithm

$$\mathsf{lfs} := \frac{d}{d\,\lambda}\log(\mathsf{f}) = \frac{\mathsf{fs}}{\mathsf{f}}$$

which yields directly the *inverse Newton correction.*

Thus instead updating the logarithm $\mathsf{lf} = \mathsf{lf} + \log(\mathsf{c}_{\mathsf{ii}})$ we directly compute the derivative

$$\mathsf{lfs} = \mathsf{lfs} + \frac{\mathsf{cs}_{\mathsf{ii}}}{\mathsf{c}_{\mathsf{ii}}}.$$

This considerations lead to

```
function ffs = deta(C,Cs)
% DETA computes Newton correction ffs = f/fs
n = length(C);
lfs = 0;
for i = 1:n
   [cmax,kmax]= max(abs(C(i:n,i)));
   if cmax == 0  % Matrix singular
      ffs = 0; return
   end
   kmax = kmax+i-1;
   if kmax ~= i
      h = C(i,:); C(i,:) = C(kmax,:); C(kmax,:) = h;
      h = Cs(kmax,:); Cs(kmax,:) = Cs(i,:); Cs(i,:) = h;
   end
   lfs = lfs + Cs(i,i)/C(i,i);
  % elimination step
   Cs(i+1:n,i) = (Cs(i+1:n,i)*C(i,i)-Cs(i,i)*C(i+1:n,i))/C(i,i)^2;
   C(i+1:n,i) = C(i+1:n,i)/C(i,i);
   Cs(i+1:n,i+1:n) = Cs(i+1:n,i+1:n) - Cs(i+1:n,i)*C(i,i+1:n)- ...
                        C(i+1:n,i)*Cs(i,i+1:n);
   C(i+1:n,i+1:n) = C(i+1:n,i+1:n) - C(i+1:n,i)*C(i,i+1:n);
end
ffs = 1/lfs;
```

Note that as an alternative to the algorithmic differentiation presented here one could use the *Formula of Jacobi*

$$\frac{d}{d\,\lambda}\det(\mathsf{C}(\lambda)) = \det(\mathsf{C}(\lambda))\ \mathrm{trace}\left(\mathsf{C}^{-1}(\lambda)\mathsf{C}'(\lambda)\right)$$

which gives an explicit expression for the derivative of the determinant.

4 Suppression instead of deflation

If $x_1, \dots, x_k$ are already computed zeros then we would like to continue working with the deflated polynomial

$$P_{n-k}(x) := \frac{P_n(x)}{(x-x_1)\cdots(x-x_k)} \tag{1}$$

of degree $n-k$. However, we cannot explicitly deflate the zeros since we are working with $P(\lambda) = \det(\lambda I - A)$. Differentiating Equation (1) we obtain

$$P'_{n-k}(x) = \frac{P'_n(x)}{(x-x_1)\cdots(x-x_k)} - \frac{P_n(x)}{(x-x_1)\cdots(x-x_k)} \sum_{i=1}^{k} \frac{1}{x-x_i}.$$

Thus the Newton-iteration becomes

$$x_{new} = x - \frac{P_{n-k}(x)}{P'_{n-k}(x)} = x - \frac{P_n(x)}{P'_n(x)} \frac{1}{1 - \dfrac{P_n(x)}{P'_n(x)} \displaystyle\sum_{i=1}^{k} \frac{1}{x-x_i}}.$$

This variant of Newton's Iteration is called *Newton-Maehly Iteration* [2,3].

5 Example

We generate a random symmetric matrix A with eigenvalues $1, 2, \dots, n$:

```
x = [1:n]'; Q = rand(n);
Q = orth(Q); A = Q*diag(x)*Q';
```

respectively a non symmetric matrix with

```
x = [1:n]'; Q = rand(n);
A = Q*diag(x)*inv(Q);
```

Then we compute the solutions of $\det(C(\lambda)) = 0$ with $C(\lambda) = \lambda I - A$ using the Newton-Maehly iteration. We compare the results with the ones obtained by the QR-Algorithm *eig(A)* and with the zeros of the characteristic polynomial *roots(poly(A))*. In Tables 1 and 2 the norm of the difference of the computed eigenvalues to the exact ones is printed. Notice that due to ill-conditioning the roots of the characteristic polynomial differ very much and that for $n = 200$ the coefficients of the characteristic polynomial overflow and the zeros cannot be computed any more. On the other hand we can see that the our method competes in accuracy very well with the standard QR-algorithm.

n	roots(poly(A))	eig(A)	$\det(A-\lambda I)=0$
50	1.3598e+02	3.9436e−13	4.7243e−14
100	9.5089e+02	1.1426e−12	1.4355e−13
150	2.8470e+03	2.1442e−12	3.4472e−13
200	− − −	3.8820e−12	6.5194e−13

Table 1. Norm of difference of the computed to the exact eigenvalues for a symmetric matrix

n	roots(poly(A))	eig(A)	$\det(A-\lambda I)=0$
50	1.3638e+02	3.7404e−12	2.7285e−12
100	9.7802e+02	3.1602e−11	3.5954e−11
150	2.7763e+03	6.8892e−11	3.0060e−11
200	− − −	1.5600e−10	6.1495e−11

Table 2. Norm of difference of the computed to the exact eigenvalues for a non-symmetric matrix

6 Generalization to λ-matrices

Consider a quadratic eigenvalue problem

$$\det(C(\lambda)) = 0, \quad \text{with} \quad C(\lambda) = \lambda^2 M + \lambda C + K.$$

If $\det(M) \neq 0$ then one way to "linearize" the problem is to consider the equivalent general eigenvalue-problem with dimension $2n$:

$$\det\left(\begin{bmatrix} M & 0 \\ 0 & K \end{bmatrix} - \lambda \begin{bmatrix} 0 & M \\ -M & -C \end{bmatrix}\right) = 0$$

Alternatively with our approach we can compute the zeros of $\det(C(\lambda))$ with Newton's iteration. Take the mass-spring system example from [5]. For the nonoverdamped case the matrix is $C(\lambda) = \lambda^2 M + \lambda C + K$ with

$$M = I, \; C = \tau\,\text{tridiag}(-1,3,-1), \; K = \kappa\,\text{tridiag}(-1,3,-1)$$

and with $\kappa = 5$, $\tau = 3$ and $n = 50$. The Matlab program to compute the eigenvalues is

```
% Figure 3.3 in Tisseur-Meerbergen
clear, format compact
n=50
tau = 3, kappa = 5,
e = -ones(n-1,1);
C = (diag(e,-1)+ diag(e,1)+ 3*eye(n));
K = kappa*C;
C = tau*C;
```

```
lam = -0.5+0.1*i;
tic
for k=1:2*n
  ffs = 1; q=0;
  while abs(ffs)>1e-14
    Q = lam*(lam*eye(n)+ C)+K;
    Qs = 2*lam*eye(n)+C;
    ffs = deta(Q,Qs);
    s = 0;
    if k>1
       s = sum(1./(lam-lamb(1:k-1)));
    end
    lam = lam-ffs/(1-ffs*s); q=q+1;
  end
  clc
  k, lam, q, ffs, lamb(k) = lam;
  lam = lam*(1+0.01*i);
end
toc
clf
 plot(real(lamb),imag(lamb),'o')
```

and produces Figure 1. The computation in Matlab needed 13.9 seconds on

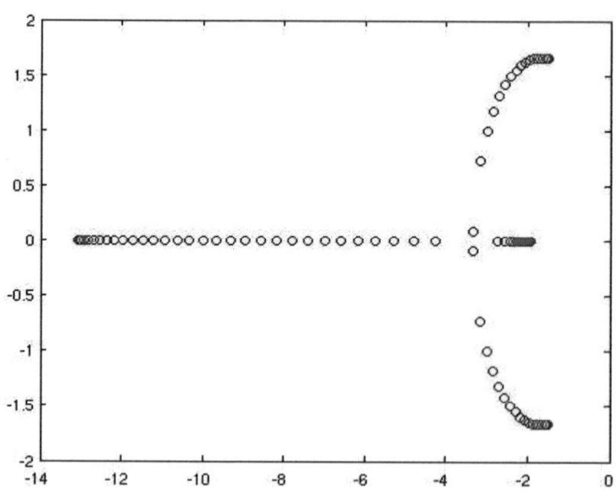

Fig. 1. Eigenvalues in the complex plane for the nonoverdamped case

a IBM X41 laptop. As starting values for the iteration we used the complex number $\lambda(1+0.01i)$ near the last computed eigenvalue λ.

In the second "overdamped" case we have $\kappa = 5$, $\tau = 10$. Since the eigenvalues are all real we can choose real starting values. We chose 1.01λ where again λ was the last eigenvalue found. Figure 2 shows the eigenvalues which are all real and computed with Matlab in 16.3 seconds.

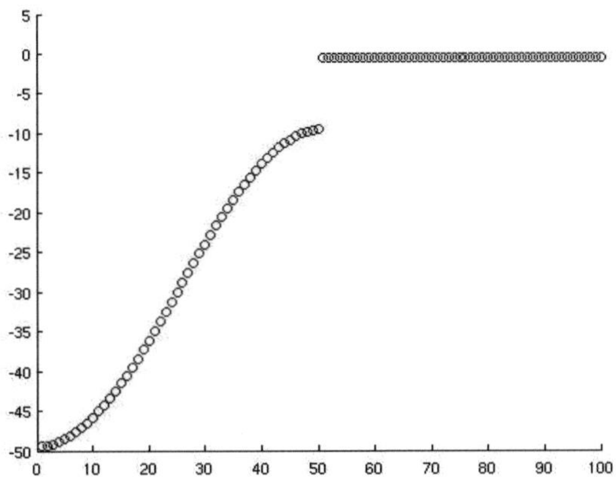

Fig. 2. Real eigenvalues for the overdamped case

Finally we recomputed a cubic eigenvalue problem from [1]. Here we have

$$C(\lambda) = \lambda^3 A_3 + \lambda^2 A_2 + \lambda A_1 + A_0$$

with

$$A_0 = \text{tridiag}(1, 8, 1) \; A_2 = \text{diag}(1, 2, \ldots, n) \text{ and } A_1 = A_3 = I.$$

In [1] the matrix dimension was $n = 20$ thus 60 eigenvalues had to be computed. Using our method we compute these in 1.9 seconds. Figure 3 shows the 150 eigenvalues for $n = 50$ which have been computed in 17.9 seconds.

7 Conclusion

We have demonstrated that computing zeros of polynomials from their coefficients is *ill-conditioned*. However, direct evaluation of the characteristic polynomial is feasible. With this computational intensive method we have shown that medium size nonlinear eigenvalue problems may be solved with a simple program which computes determinants by Gaussian elimination and applies algorithmic differentiation and suppresses already computed zeros. We obtained results in reasonable time in spite that we did not compile the Matlab program

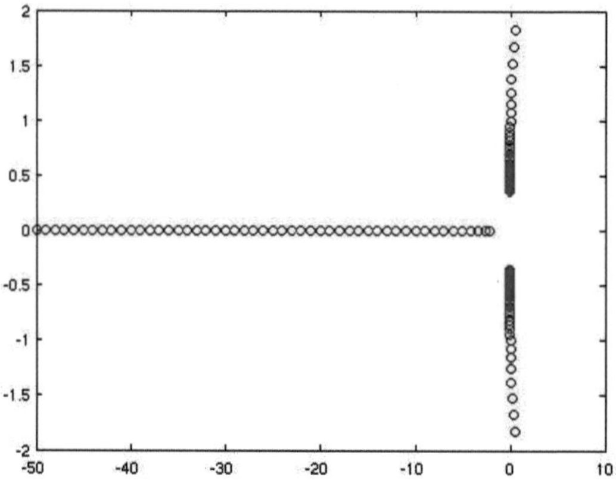

Fig. 3. Cubic Eigenvalue Problem

and that we did not make use of the banded structure of the matrices. This algorithm, though computational expensive, maybe useful for its potential for parallelization on future multicore architectures.

References

1. P. Arbenz and W. Gander, *Solving nonlinear Eigenvalue Problems by Algorithmic Differentiation*, Computing 36, 205-215, 1986.
2. H. J. Maehly, *Zur iterativen Auflösung algebraischer Gleichungen*, ZAMP (Zeitschrift für angewandte Mathematik und Physik), (1954), pp. 260–263.
3. J. Stoer and R. Bulirsch, *Introduction to Numerical Analysis*, Springer, 1991.
4. W. Kellenberger, *Ein konvergentes Iterationsverfahren zur Berechnung der Wurzeln eines Polynoms*, Z. Angew. Math. Phys. 21 (1970) 647–651.
5. F. Tisseur and K. Meerbergen, *The Quadratic Eigenvalue Problem*, SIAM. Rev., 43, pp. 234–286, 2001.
6. J. H. Wilkinson, *Rounding errors in algebraic processes*, Dover Publications, 1994.

How to Find a Good Submatrix*

S. A. Goreinov, I. V. Oseledets, D. V. Savostyanov, E. E. Tyrtyshnikov, and N. L. Zamarashkin

Institute of Numerical Mathematics of Russian Academy of Sciences,
Gubkina 8, 119333 Moscow, Russia
{sergei,ivan,draug,tee,kolya}@bach.inm.ras.ru

Abstract. Pseudoskeleton approximation and some other problems require the knowledge of sufficiently well-conditioned submatrix in a large-scale matrix. The quality of a submatrix can be measured by modulus of its determinant, also known as volume. In this paper we discuss a search algorithm for the maximum-volume submatrix which already proved to be useful in several matrix and tensor approximation algorithms. We investigate the behavior of this algorithm on random matrices and present some of its applications, including maximization of a bivariate functional.

Keywords: maximum volume, low rank, maxvol, pseudoskeleton approximation.

1 Introduction

Several problems in matrix analysis require the knowledge of a good submatrix in a given (supposedly large) matrix. By "good" we mean a sufficiently well-conditioned submatrix. The application that we are particularly interested in is the approximation of a given matrix by a low-rank matrix:

$$A \approx UV^{\top},$$

where A is $m \times n$ and U and V are $m \times r$ and $n \times r$, respectively. Optimal approximation in spectral or Frobenius norm can be computed via singular value decomposition (SVD) which, however, requires too many operations. A much faster way is to use CGR decompositions [1] (later also referred to as CUR by some authors) which in Matlab notation can be written as:

$$A \approx A(:, \mathcal{J})A(\mathcal{I}, \mathcal{J})^{-1}A(\mathcal{I}, :), \tag{1}$$

where $\mathcal{I}, \mathcal{J}$ are appropriately chosen index sets of length r from $1 : n$ and $1 : m$. It can be seen that the right hand side matrix coincides with A in r rows and

* This work was supported by RFBR grant №08-01-00115a and by a Priority Research Grant OMN-3 of the Department of Mathematical Sciences of Russian Academy of Sciences.

r columns. Moreover, if A is strictly of rank r and $\hat{A} = A(\mathcal{I}, \mathcal{J})$ is nonsingular, the exact equality holds. However, in the approximate case the quality of the approximation (1) relies heavily on the "quality" of the submatrix. The question is how to measure this quality and how to find a good submatrix. A theoretical answer (basically, existence theory) [3] is that if $\hat{A}$ has maximal in modulus determinant among all $r \times r$ submatrices of A, then element-wise error estimate is of the form

$$|A - A_r| \leqslant (r+1)\sigma_{r+1},$$

where $|A| = \max_{ij} |a_{ij}|$ denotes Chebyshëv norm, A_r is the right hand side of (1) and σ_{r+1} is the $r+1$–th singular value of the matrix A, i.e. the error of the best rank-r approximation in the spectral norm. That is the theory, but what about a practical algorithm? How to find a good submatrix? That is the topic of the current paper.

As we have seen, the submatrix quality can be measured by its determinant, so we want to find a submatrix with the largest possible determinant. An intermediate step to the solution of that problem is computation of the maximal volume submatrix not in a matrix where both dimensions are large, but in the matrix where only one dimension is large, i.e. in a "tall matrix". Such procedure (called `maxvol`) plays a crucial role in several matrix algorithms we have developed, and it deserves a special description [2, 4]. In this paper we investigate the behavior of the `maxvol` algorithm on random matrices and present some theoretical results and its application for fast search of the maximum entry of large-scale matrix. We also propose a new approach for maximization of a bivariate functional on the base of `maxvol` algorithm.

1.1 Notation

In this article we use Matlab-like notation for defining rows and columns of matrix. Therefore we write i-th row of matrix A as $a_{i,:}$ and j-th column of A as $a_{:,j}$. We will also use columns and rows of identity matrix, denoting them as e_i and e_j^T respectively, using the same notations for different sizes, but the actual size will be always clear by the context.

1.2 Definitions and basic lemmas

Let us give some formal definitions and prove basic lemmas to rely on.

Definition 1. *We refer to the modulus of determinant of square matrix as its* volume.

Definition 2. *We call $r \times r$ submatrix $A_{\blacksquare}$ of rectangular $m \times n$ matrix A* maximum volume submatrix, *if it has maximum determinant in modulus among all possible $r \times r$ submatrices of A.*

Definition 3. *We call* $r \times r$ *submatrix* $A_\square$ *of rectangular* $n \times r$ *matrix* A *of full rank* dominant, *if all the entries of* $AA_\square^{-1}$ *are not greater than* 1 *in modulus.*

The main observation that lays ground for the algorithms for the construction of `maxvol` algorithm is the following lemma.

Lemma 1. *For* $n \times r$ *matrix maximum volume* $r \times r$ *submatrix is dominant.*

Proof. Without loss of generality we can consider that $A_\blacksquare$ *occupies first* r *rows of* A. *Let us refer to them as* upper submatrix. *Then*

$$AA_\blacksquare^{-1} = \begin{bmatrix} I_{r\times r} \\ Z \end{bmatrix} = B. \tag{2}$$

Multiplication by a nonsingular matrix does not change the ratio of determinants of any pair of $r \times r$ *submatrices in* A. *Therefore, the upper submatrix* $I_{r\times r}$ *is a maximum-volume submatrix in* B *and it is dominant in* B *iff* $A_\blacksquare$ *is dominant in* A.

Now, if there is some $|b_{ij}| > 1$ *in* B, *then we can construct a new submatrix with a volume larger than volume of the upper submatrix. To see that, swap rows* i *and* j *in* B, *and it is easy to see that a new upper submatrix*

$$B_{\blacksquare'} = \begin{pmatrix} 1 & & & & \\ & \ddots & & & \\ * & * & b_{ij} & * & * \\ & & & \ddots & \\ & & & & 1 \end{pmatrix} \tag{3}$$

has

$$|\det(B_{\blacksquare'})| = |b_{ij}| > 1 = |\det(I_{r\times r})|.$$

That means that $I_{r\times r}$ *(and hence* $A_\blacksquare$*) is not the maximum volume submatrix.*

The volume of a dominant submatrix can not be very much smaller than the maximum volume, as the following lemma shows.

Lemma 2. *For any nonsingular* $n \times r$ *matrix* A

$$|\det(A_\square)| \geqslant |\det(A_\blacksquare)|/r^{r/2} \tag{4}$$

for all dominant $r \times r$ *submatrices of* A.

Proof. Suppose that $A_\square$ *is the upper submatrix and write*

$$AA_\square^{-1} = \begin{bmatrix} I_{r\times r} \\ Z \end{bmatrix} = B. \tag{5}$$

All entries of B *are not greater than* 1 *in modulus, therefore by Hadamard inequality the volume of any* $r \times r$ *submatrix* $B_{r\times r}$ *of* B *is not greater than*

$$|\det(B_{r\times r})| \leqslant \prod_{i=1}^{r} |b_{\sigma_i,:}| \leqslant r^{r/2},$$

where σ_i *are indices of rows that contain* $B_{r\times r}$.

The inequality is sharp. For example, if Z *contains Fourier, Hadamard or Walsh matrix as a submatrix, it is easy to see that the equality is attained.*

2 Algorithm maxvol

A dominant property of the maximal-volume submatrix allows us to construct a simple and efficient algorithm for the search of maximal volume submatrix.

Algorithm 1. Given: $n \times r$ matrix A. Find: $r \times r$ dominant submatrix $A_{\square}$.

0 Start with arbitrary nonsingular $r \times r$ submatrix $A_{\boxdot}$. Reorder rows in A so that $A_{\boxdot}$ occupies first r rows in A.

1 Compute

$$AA_{\boxdot}^{-1} = B$$

and find its maximal in modulus entry b_{ij}.

2 If $|b_{ij}| > 1$, **then**

swap rows i and j in B. Now, upper submatrix in B has the form (3) and the volume $|b_{ij}| \geqslant 1$. By swapping the rows we have increased the volume of the upper submatrix in B, as well as in A.

Let $A_{\boxdot}$ be the new upper submatrix of A and go to step 1.

If $|b_{ij}| = 1$, return $A_{\square} = A_{\boxdot}$.

On each iterative step of Algorithm 1, volume of $A_{\boxdot}$ increases until the volume of $A_{\square}$ is reached. In practice, we can simplify the stopping criterion in the iterative step to $|b_{ij}| < 1 + \delta$ with sufficiently small parameter δ (we think that $\delta \sim 10^{-2}$ can be a good choice). This dramatically reduces the number of iterative steps but does not change the "good" properties of a submatrix.

If computations proceed in a naive way, then the most expensive part of iterations is step 1, which needs one $r \times r$ matrix inversion and nr^2 operations for the matrix-by-matrix product $AA_{\boxdot}^{-1}$. We can reduce the complexity of this step by a factor of r if we note that on each iteration, $A_{\boxdot}$ is updated by a rank-one matrix, and apply Sherman-Woodbury-Morrison formula for the matrix inverse. Now we describe this in detail.

Swapping of rows i and j of matrix A is equivalent to the following rank-one update.

$$A := A + e_j(a_{i,:} - a_{j,:}) + e_i(a_{j,:} - a_{i,:}) = A + (e_j - e_i)(a_{i,:} - a_{j,:}) = A + pv^T. \quad (6)$$

For the upper submatrix, this update is

$$A_{\square} := A_{\square} + e_j(a_{i,:} - a_{j,:}) = A_{\square} + qv^{\mathsf{T}}. \tag{7}$$

For the inverse of the upper submatrix, we use the SWM formula

$$A_{\square}^{-1} := A_{\square}^{-1} - A_{\square}^{-1} q(1 + v^{\mathsf{T}} A_{\square}^{-1} q)^{-1} v^{\mathsf{T}} A_{\square}^{-1}. \tag{8}$$

Note that

$$v^{\mathsf{T}} A_{\square}^{-1} q = (a_{i,:} - a_{j,:}) A_{\square}^{-1} e_j = \left((AA_{\square}^{-1})_{i,:} - (AA_{\square}^{-1})_{j,:}\right) e_j = b_{ij} - b_{jj} = b_{ij} - 1.$$

We proceed with the formula of fast update of $B = AA_{\square}^{-1}$,

$$\begin{aligned} B = AA_{\square}^{-1} &:= (A + pv^{\mathsf{T}})(A_{\square}^{-1} - A_{\square}^{-1} qv^{\mathsf{T}} A_{\square}^{-1}/b_{ij}) \\ &= AA_{\square}^{-1} - AA_{\square}^{-1} qv^{\mathsf{T}} A_{\square}^{-1}/b_{ij} + pv^{\mathsf{T}} A_{\square}^{-1} - pv^{\mathsf{T}} A_{\square}^{-1} qv^{\mathsf{T}} A_{\square}^{-1}/b_{ij} \\ &= B - \left(Bq - b_{ij}p + pv^{\mathsf{T}} A_{\square}^{-1} q\right) v^{\mathsf{T}} A_{\square}^{-1}/b_{ij}. \end{aligned}$$

Using $v^{\mathsf{T}} A_{\square}^{-1} = b_{i,:} - b_{j,:}$ and $v^{\mathsf{T}} A_{\square}^{-1} q = b_{ij} - 1$, we have

$$B := B - (b_{:,j} - b_{ij}p + (b_{ij} - 1)p)(b_{i,:} - b_{j,:})/b_{ij},$$

and finally

$$B := B - (b_{:,j} - e_j + e_i)(b_{i,:} - e_j^{\mathsf{T}})/b_{ij}. \tag{9}$$

Note also that the upper $r \times r$ submatrix of B remains to be identity after each update, because $b_{1:r,j} = e_j$ for $j \leqslant r$ and $(e_i)_{1:r} = 0$ for $i > r$ that is always the case. So we need to update only the submatrix Z. This can be also done by rank-one update:

$$Z := Z - (b_{:,j} + e_i)(b_{i,:} - e_j^{\mathsf{T}})/b_{ij}. \tag{10}$$

Note that in the last formula we use "old" indexing, i.e. rows of Z are numbered from $r + 1$ to n.

Therefore, each iterative step of the algorithm reduces to a rank-one update of Z which can be done in $(n - r)r$ operations, and a search for a maximum-modulus element in Z, which is of the same complexity. Overall complexity for the algorithm 1 is therefore $\mathcal{O}(nr^2)$ for initialization and $\mathcal{O}(cnr)$ for iterative part, where c is the number of iterations. We can write a rather rough estimate for c as follows. Each iteration step increases volume of $A_{\square}$ by a value $|b_{ij}| \geqslant 1 + \delta$. After k steps

$$|\det(A_{\square}^{\{k\}})| \geqslant |\det(A_{\square}^{\{0\}})|(1 + \delta)^k,$$

therefore

$$c \leqslant \left(\log|\det(A_{\blacksquare})| - \log|\det(A_{\square}^{\{0\}})|\right) / \log(1 + \delta). \tag{11}$$

This shows that good initial guess for $A_{\square}$ can reduce the number of iterations. If no "empirical" guesses are available, it is always safe to apply Gaussian elimination with pivoting to A and use the set of pivoted rows as an initial approximation to the maximal volume submatrix.

3 maxvol-based maximization methods

As an application consider the following simple and interesting problem: find maximum in modulus element of a low-rank matrix $A = UV^T$, given by U and V. This problem arises for example in maximization of two-dimensional separable function on a grid, or as an essential part of the Cross3D algorithm for computation of Tucker approximation of three dimensional tensor in linear time [4]. Direct comparison of all elements requires rnm operations for $m \times n$ matrix of rank r. Is it possible to devise an algorithm with complexity linear in matrix size?

3.1 Theoretical estimates

Our idea is not to search for maximum element in the whole submatrix, but only in the submatrix of maximal volume. Though looking not very natural at the first glance, this algorithm actually works well in many cases. Often the maximal element in the maximal volume submatrix is not necessarily the same as the true maximal element, but it can not be very much smaller (for example, if it zero, then the submatrix of maximal volume is zero and the matrix is also zero, which we hope is not true). But are there any quantitative estimates? In fact, we can replace maximal-volume submatrix by an arbitrary dominant submatrix, which yields the same estimate. But first we need to extend the definition of the dominant submatrix to the case of $m \times n$ matrices. It is done in a very simple manner.

Definition 4. *We call* $r \times r$ *submatrix* $A_\square$ *of rectangular* $m \times n$ *matrix* A dominant, *if it is dominant in columns and rows that it occupies in terms of Definition 3.*

Theorem 1. *If* $A_\square$ *is a dominant* $r \times r$ *submatrix of a* $m \times n$ *matrix* A *of rank* r*, then*

$$|A_\square| \geqslant |A|/r^2. \tag{12}$$

Proof. If maximum in modulus element b *of* A *belongs to* $A_\square$*, the statement is trivial. If not, consider* $(r+1) \times (r+1)$ *submatrix, that contains* $A_\square$ *and* b*,*

$$\hat{A} = \begin{bmatrix} A_\square & c \\ d^T & b \end{bmatrix}. \tag{13}$$

Elements of vectors c *and* d *can be bounded as follows*

$$|c| \leqslant r|A_\square|, \qquad |d| \leqslant r|A_\square|. \tag{14}$$

This immediately follows from $c = A_\square(A_\square^{-1}c) = A_\square\tilde{c}$*, where all elements of* $\tilde{c}$ *are not greater than* 1 *in modulus. Bound for elements of* d *is proved in the same way.*

Now we have to bound $|b|$. *Since* A *has rank* r *and* $A_\square$ *is nonsingular,*

$$b = d^T A_\square^{-1} c, \tag{15}$$

and it immediately follows that

$$|A| = |b| \leqslant |d| r \leqslant |A_\square| r^2,$$

which completes the proof.

The restriction $\operatorname{rank} A = r$ may be removed with almost no change in the bound (12). However, one has to replace $A_\square$ by $A_\blacksquare$.

Theorem 2. *If* $A_\blacksquare$ *is maximum-volume* $r \times r$ *(nonsingular) submatrix of* $m \times n$ *matrix* A*, then*

$$|A_\blacksquare| \geqslant |A|/(2r^2 + r).$$

Proof. Again, consider submatrix $\hat{A}$ *that contains* $A_\blacksquare$ *and* b*, see* (13). *Bound* (14) *follows immediately, because the maximum-volume submatrix is dominant, see Lemma 1. Since* $\operatorname{rank} A$ *is now arbitrary, the equality* (15) *is no longer valid. Instead, we use an inequality from [3],*

$$\left| b - d^T B^{-1} c \right| \leqslant (r+1) \sigma_{r+1}(\hat{A}), \tag{16}$$

where $\sigma_1(\hat{A}) \geqslant \sigma_2(\hat{A}) \geqslant \ldots \geqslant \sigma_{r+1}(\hat{A})$ *are singular values of* $\hat{A}$. *That gives*

$$\begin{aligned} |b| &\leqslant (r+1)\sigma_{r+1}(\hat{A}) + |d^T A_\blacksquare^{-1} c| \leqslant (r+1)\sigma_{r+1}(\hat{A}) + |d^T \tilde{c}| \\ &\leqslant (r+1)\sigma_{r+1}(\hat{A}) + |d| r \leqslant (r+1)\sigma_{r+1}(\hat{A}) + |A_\blacksquare| r^2. \end{aligned} \tag{17}$$

We need an estimate for $\sigma_{r+1}(\hat{A})$ *in terms of values of its elements. Note that*

$$\hat{A}^T \hat{A} = \begin{bmatrix} A_\blacksquare^T & d \\ c^T & b \end{bmatrix} \begin{bmatrix} A_\blacksquare & c \\ d^T & b \end{bmatrix} = \begin{bmatrix} A_\blacksquare^T A_\blacksquare + dd^T & A_\blacksquare^T c + bd \\ c^T A_\blacksquare + bd^T & c^T c + b^2 \end{bmatrix}.$$

From the singular value interlacing theorem,

$$\sigma_r(A_\blacksquare^T A_\blacksquare + dd^T) \geqslant \sigma_{r+1}^2(\hat{A}),$$

and for $r > 1$

$$\sigma_{r-1}(A_\blacksquare^T A_\blacksquare) \geqslant \sigma_r(A_\blacksquare^T A_\blacksquare + dd^T) \geqslant \sigma_{r+1}^2(\hat{A}).$$

Finally we have $\sigma_1(A_\blacksquare) \geqslant \sigma_{r+1}(A)$ *and* $|A_\blacksquare| \geqslant \sigma_1(A_\blacksquare)/r$. *Plugging this into* (17), *we get*

$$|b| \leqslant (r+1) r |A_\blacksquare| + r^2 |A_\blacksquare| = (2r^2 + r)|A_\blacksquare|,$$

which completes the proof.

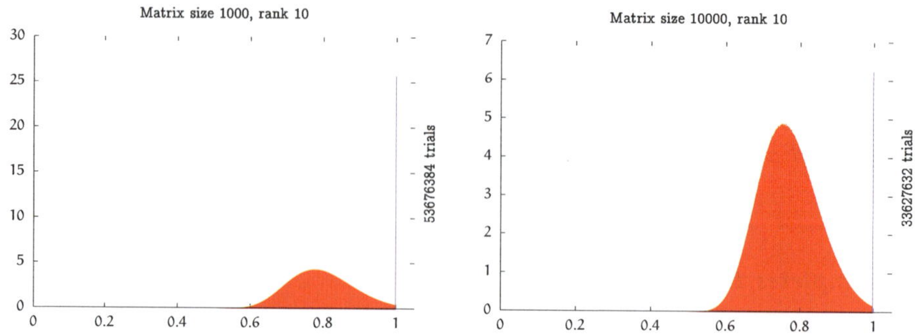

Fig. 1. Distribution of the ratio of `maxvol` over true maximal element

Now it is clear that we can reduce the search to only r^2 elements of the dominant matrix. Then the search time does not depend on matrix size, and the total complexity is just the complexity of finding $A_\square$, which is $\mathcal{O}(nr^2+mr^2)$ operations. Maximum element in $A_\square$ is "sufficiently good" in the sense of proven theorems.

In practical cases, the ratio $|A|/|A_\square|$ is sufficiently smaller than r^2. Consider two examples, which illustrate this fact.

3.2 Search for the maximum element in random low-rank matrices

In order to see how good is the maximal element in our "good" submatrix, we tested it first on random matrices. Given n, m, r, two matrices U and V were generated with elements uniformly distributed in $[-1 : 1]$. Then U, V were replaced with Q-factors of their QR-decompositions and a matrix $A = UDV^\top$ was generated with random positive diagonal D with elements uniformly distributed on $[0, 1]$. We generated a large set of trial matrices, for each of these matrices we computed maximal element using the proposed algorithm.

The actual degradation of the maximal element is presented on the Figure 1, where the histogram of the ratio of maximal element in $A_\square$ over the true maximal element is given. Note that this ratio for certain is not lower than 0.5 for all trials (smooth humps in the middle part of histograms), and in some 5% of cases (sharp peaks in the right part of histograms) we even found a true maximal element, which was much less probable for a random choice.

3.3 Maximization of bivariate functions

There is an interesting application of our algorithm. It can be applied to the problem of global optimization of bivariate functions. Suppose we want to find a maximum of $|f(x, y)|$ in some rectangle $(x, y) \in \Pi = [a_0, a_1] \times [b_0, b_1]$, and f is some given function. "Discretizing" the problem on some sufficiently fine grid (x_i, y_j), $i = 1, 2, \ldots, m$, $j = 1, 2, \ldots, n$ we obtain an $m \times n$ matrix $A = [f(x_i, y_j)]$ to find the maximal in modulus element in. Assume additionally that the function $f(x, y)$ can be sufficiently well approximated by a sum of separable functions:

$$f(x, y) \approx \sum_{\alpha=1}^{r} u_\alpha(x) v_\alpha(y).$$

Then it easy to see that in this case the matrix A admits a rank-r approximation of the form

$$A = f(x_i, y_j) \approx UV^\top,$$

where U, V are $n \times r$ and $m \times r$ matrices, respectively, with elements $U = [u_\alpha(x_i)]$, $V = [v_\alpha(y_j)]$. Thus the "discretized" problem is equivalent to the problem of finding maximal in modulus element in a large low-rank matrix A, so we can apply our method. We have no guarantee that we will find the exact maximum, but we will have an estimate of it. As an example we considered a standard banana function minimization problem:

$$b(x, y) = 100(y - x)^2 + (1 - x)^2.$$

This function has minimum in $(1, 1)$ equal to 0 and is positive in all other points. In order to reformulate the problem as a maximization problem, we introduce an auxiliary function

$$f(x, y) = \frac{1}{b(x, y) + 10^{-6}},$$

the maximum of which is located at the same point $(1, 1)$. A rectangle $[-2, 2] \times [2, 2]$ was chosen and discretized on a 500×500 uniform grid, the corresponding matrix A was approximated by a matrix of rank 10 for which the maximum was found by our `maxvol` algorithm. The extremal point was contained in the grid, and the `maxvol`returned the exact position of the minimum: $(1, 1)$. For other choices of grids the situation was the same, and the approximations to the extremum were very good (the error was $\mathcal{O}(h)$, where h is a grid size).

This result is very encouraging. However, it should not be treated as a universal optimization method, but it can be very useful in global optimization methods, because it gives us an estimate of the value of the global optimum — this can be efficiently used, for example, in branch-and-bound methods, with `maxvol` estimates for the maximal value in a particular domain. Another possibility is to use "local" separable approximations to functions and then minimize this local part by the `maxvol` algorithm. Incorporation of our method into robust optimization methods will be the subject of future research.

4 Conclusion and future work

In this paper we presented a simple iterative method for the search of a submatrix of maximal volume in a given rectangular matrix. This submatrix plays an important role in the theory and algorithms for the approximation by low (tensor) rank matrices. As an application, we constructed an algorithm for the computation of maximal in modulus element in a given low-rank matrix and proved, that the element can not be much smaller than the "true" maximal element. Experiments on random matrices prove that our algorithm performs very good, as well as the experiment with the minimization of the banana function. A future work will be focused on maximizing separable functions by using branch-and-bound method and `maxvol` estimates of the maximal element in each subdomain and by using "local" approximations by separable functions.

References

1. S. A. Goreinov, E. E. Tyrtyshikov, and N. L. Zamarashkin, A theory of pseudo–skeleton approximations, *Linear Algebra Appl.*, **261**: 1–21, 1997.
2. E. E. Tyrtyshnikov, Incomplete cross approximation in the mosaic-skeleton method. *Computing*, **64**(4): 367–380, 2000.
3. S. A. Goreinov and E. E. Tyrtyshnikov, The maximal-volume concept in approximation by low-rank matrices, *Contemporary Mathematics*, **208**: 47–51, 2001.
4. I. V. Oseledets, D. V. Savostyanov, and E. E. Tyrtyshnikov, Tucker dimensionality reduction of three-dimensional arrays in linear time, *SIAM J. Matrix Anal. Appl.*, **30**(3): 939–956, 2008.

Conjugate and Semi-Conjugate Direction Methods with Preconditioning Projectors

V. P. Il'in

Institute of Computational Mathematics and Mathematical Geophysics, Siberian Branch of Russian Academy of Sciences, ak. Lavrentieva 6, 630090 Novosibirsk, Russia
ilin@sscc.ru

Abstract. The acceleration of the original projective iterative methods of multiplicative or additive type for solving systems of linear algebraic equations (SLAEs) by means of conjugate direction approaches is considered. The orthogonal and varitional properties of the preconditioned conjugate gradient, conjugate residual and semi-conjugate residual algorithms, as well as estimations of the number of iterations, are presented. Similar results were obtained for the dynamically preconditioned iterative process in Krylov subspaces. Application of discussed techniques for domain decomposition, Kaczmarz, and Cimmino methods is proposed.

1 Introduction

The aim of this paper is to analyze the iterative algorithms in the Krylov subspaces whose preconditioners are some kinds of projector operators. At first we consider the general approach for acceleration of some convergent iterations with a constant iteration matrix.

Let us have the system of linear algebraic equations:

$$Au = f, \quad u = \{u_i\}, \quad f = \{f_i\} \in R^N, \quad A = \{a_{i,j}\} \in R^{N,N}, \tag{1}$$

and the convergent stationary iterative process

$$u^{k+1} = Bu^k + g, \quad u^k \underset{k\to\infty}{\to} u, \quad g = (I - B)A^{-1}f. \tag{2}$$

Suppose that the iteration matrix B has eigenvalues $\lambda_q(B)$ and spectral radius $\rho = \max_q\{|\lambda_q(B)|\} < 1$. Then the vector u is the solution of system

$$\tilde{A}u \equiv (I - B)u = g, \tag{3}$$

where I is the identity matrix and $\tilde{A}$ is the preconditioned non-singular matrix. If $\tilde{A}$ is a symmetric positive definite (s.p.d) matrix, its spectral condition number is

$$æ = \|\tilde{A}\|_2\|\tilde{A}^{-1}\|_2 = (1 + \rho)/(1 - \rho). \tag{4}$$

and to solve SLAE (3) we can apply some iterative conjugate direction methods (see [1] – [4]):

$$\begin{aligned} &r^0 = g - \tilde{A}u^0, \quad p^0 = r^0; \quad n = 0, 1, \ldots : \\ &u^{n+1} = u^n + \alpha_n p^n, \quad r^{n+1} = r^n - \alpha_n \tilde{A} p^n, \\ &p^{n+1} = r^{n+1} + \beta_n p^n, \end{aligned} \tag{5}$$

which have the optimality property in the Krylov subspaces

$$\mathcal{K}_{n+1}(r^0, \tilde{A}) = \mathrm{Span}\{p^0, p^1, \ldots, p^n\} = \mathrm{Span}\{p^0, \tilde{A}p^0, \ldots, \tilde{A}^n p^0\}.$$

In conjugate direction (CG) and conjugate residual (CR) methods, $s = 0, 1$ respectively, the iterative parameters $\alpha_n^{(s)}$ and $\beta_n^{(s)}$ are defined as follows:

$$\alpha_n^{(s)} = (\tilde{A}^s r^n, r^n)/(\tilde{A}p^n, \tilde{A}^s p^n), \quad \beta_n^{(s)} = (\tilde{A}^s r^{n+1}, r^{n+1})/(\tilde{A}^s r^n, r^n). \tag{6}$$

These algorithms provide the residual and direction (correction) vectors r^n and p^n with the orthogonal peculiarities

$$(\tilde{A}^s r^n, r^k) = (\tilde{A}^s r^n, r^n)\delta_{n,k}, \quad (\tilde{A}p^n, \tilde{A}^s p^k) = (\tilde{A}p^n, \tilde{A}^s p^n)\delta_{n,k}. \tag{7}$$

Also, the functionals $\Phi_n^{(s)}(r^n) = (\tilde{A}^{s-1} r^n, r^n)$, $s = 0, 1$, are minimized in the Krylov subspaces, and the number of iterations necessary for satisfying the condition

$$(\Phi_n^{(s)}(r^n)/\Phi_0^{(s)}(r^0))^{1/2} \leqslant \varepsilon < 1,$$

is estimated by the value

$$n(\varepsilon) \leqslant 1 + \Big(\ln \frac{1 + \sqrt{1 - \varepsilon^2}}{\varepsilon}\Big)/\ln\gamma, \quad \gamma = (\sqrt{æ} - 1)/(\sqrt{æ} - 1). \tag{8}$$

It should be noted that matrix-vector multiplication in (5) presents the implementation of one iteration (2) that does not require explicit forming of matrices $\tilde{A}$ and B, because, for example,

$$\tilde{A}p^n = p^n - Bp^n.$$

If matrix $\tilde{A}$ is nonsymmetric and positive definite, i.e.

$$(\tilde{A}u, u) \geqslant \delta(u, u), \quad \delta > 0, \quad u \neq 0,$$

system (3) can be solved by means of the semi-conjugate residual (SCR) method realizing the stabilized version of the generalized conjugate residual (GCR) algorithm, which is described in [5] and has instability features in terms of truncation errors, see [4].

In SCR, the vectors u^{n+1} and r^{n+1} are computed according to formulas (5), with the coefficients $\alpha_n^{(s)}$ from (6) for $s = 1$, and the direction vectors p^{n+1} are defined as follows:

$$\begin{aligned} &p^{n+1,0} = r^{n+1}, \quad p^{n+1,l} = p^{n+1,l-1} + \beta_{n,l}p^{l-1}, \quad l = 1, ..., n, \\ &\beta_{n,l} = -(\tilde{A}p^l, \tilde{A}p^{n+1,l-1})/(\tilde{A}p^l, \tilde{A}p^l), \quad p^{n+1} = p^{n+1,n}. \end{aligned} \tag{9}$$

Relations (5), (9) realize the construction of A^tA-orthogonal (conjugate) vectors $p^0, p^1, \ldots, p^{n+1}$ by means of modified Gram–Schmidt orthogonalization [6]. In this case, the functional $\Phi_n^{(1)}(r^n) = (r^n, r^n)$ is minimized in the subspace $\mathcal{K}_{n+1}(r^0, \tilde{A})$ and the residual vectors are right semi-conjugate, in the sense that the equalities $(\tilde{A}r^k, r^n) = 0$ are satisfied for $k < n$. Since SCR and GMRES methods (see [4]) have the same variational properties in the Krylov subspaces, similar estimate of the number of iterations $n(\varepsilon)$ is valid for them, and it will be used below.

This paper is organized as follows. In Section 2, we describe projective methods of the multiplicative type using the conjugate direction and semi-conjugate direction approaches. The next Section is devoted to the additive type projective methods in the Krylov subspaces. Also, the application of dynamic preconditioners is discussed. This approach means using variable iteration matrix B_n at different iterations. This is the implementation requirement, for example, in many two-level iterative processes.

2 Multiplicative projector methods

Let $\Omega = \{i = 1, 2, ..., N\}$ denote a set of matrix row numbers and $\Omega_p, p = 1, 2, ..., l$, be its non-intersecting integer subsets, with the numbers m_p of their elements,

$$\Omega = \bigcup_{p=1}^{l} \Omega_p, \quad m_1 + ... + m_l = N.$$

Also, let us introduce subvectors $u_{(p)}, f_{(p)}, p = 1, ..., l$, of dimensions m_p and rectangular matrices $A_{(p)} \in R^{m_p \times N}$:

$$u_{(p)} = \{u_i, i \in \Omega_p\}, \quad f_{(p)} = \{f_i, i \in \Omega_p\}, \quad A_{(p)} = \{A_i, i \in \Omega_p\}, \tag{10}$$

where A_i is the i-th row of matrix A. Then SLAE (1) can be rewritten as

$$A_{(p)}u = f_{(p)}, \quad p = 1, 2, ..., l. \tag{11}$$

To solve (11), we consider an iterative process in which the computing of each n-th approximation step consists of the following stages:

$$u^{n,p} = u^{n,p-1} + \omega A^+_{(p)} r^{n,p-1}_{(p)}, \quad n = 1, 2, ..., \quad p = 1, 2, ..., l, \quad u^n = u^{n,l}. \tag{12}$$

Here $u^{0,0} = \{u_i^0, i = 1, 2, ..., N\}$ is the initial guess, and ω is some iterative parameter,

$$r_{(p)}^{n,p-1} = f_{(p)} - A_{(p)} u^{n,p-1}$$

is the residual subvector of dimension m_p, and A_p^+ is pseudoinverse to matrix $A_{(p)}$ defined by the formula $A_{(p)}^+ = A_{(p)}^t (A_{(p)} A_{(p)}^t)^{-1}$ if $A_{(p)}$ has a full rank.

We have from the above that $I - A_{(p)}^+ A_{(p)}$ is a symmetric positive semi-definite matrix realizing orthogonal projection into the p-th subspace, which is presented geometrically by the union of subspaces described by the i-th equations, $i \in \Omega_p$.

Iterative method (12) can be written in the matrix form,

$$u^n = Bu^{n-1} + g, \quad B = (I - T_l) \cdots (I - T_1), \quad T_p = \omega A_{(p)}^+ A_{(p)}. \tag{13}$$

Projective algorithm (12), (13) for $\omega = 1$ and $m_p = 1$ presents the "point-wise" method published by S.Kaczmarz in [7]. Its different generalizations and investigations were made by many authors, see [8], [9].

In [10] the following assertion was proved for abstract iterative projection method of the multiplicative type, with application to the domain decomposition approach:

Theorem 1. *Let $T_p, p = 1, ..., l,$ be s.p.d. matrices, and the following inequalities be valid for any vector $v \in R^N$:*

$$(T_p v, v)/(v, v) \leqslant \alpha < 2, \quad p = 1, 2, ..., l; \quad \|v\| \leqslant \beta \sum_{p=1}^{l} (T_p v, v).$$

Then the estimate

$$\|B\|_2 \leqslant \rho = 1 - (2 - \alpha)/\{\beta[l + \alpha^2 l(l-1)/2]\}$$

is true for the Euclidian norm $\|B\|_2$. If the matrices $\bar{T}_p = \omega T_p$ for all p satisfy the conditions

$$(\bar{T}_p v, v)/(v, v) \leqslant \bar{\alpha} < 2, \quad \|v\| \leqslant \bar{\beta}[(\bar{T}_1 v, v) + ... + (\bar{T}_l v, v)],$$

then for $\omega = (\bar{\alpha}\sqrt{(l-1)l})^{-1}$ we have $\rho = 1 - (3\ \bar{\alpha}\ \bar{\beta}\ l)^{-1}$.

It should be noted that iteration matrix B in iterative process (13) is non-symmetric, because matrices T_p are not commutative in general.

Now we consider the alternative direction block version of the Kaczmarz method, in which each iteration consists of two stages. The first one realizes conventional formulas (12) or (13), and the second stage implements similar computations but in the backward ordering on the number p:

$$\begin{aligned} &u^{n+1/2,p} = u^{n,p-1} + \omega A_{(p)}^+ r_{(p)}^{n,p-1}, \\ &p = 1, 2, ..., l, \quad u^{n+1/2} = u^{n+1/2,l} = u^{n+1/2,l+1}, \\ &u^{n+1,p} = u^{n+1/2,p+1} + \omega A_{(p)}^+ r_{(p)}^{n+1/2,p+1}, \\ &\quad p = l, ..., 2, 1, \quad u^{n+1} = u^{n+1,1}. \end{aligned} \tag{14}$$

The iteration matrix in iterations (14) is the matrix product $B = B_2 B_1$, where B_1 coincides with B from (13) and B_2 has a similar form. Thus,

$$u^{n+1} = B_2 B_1 u^n + g, \quad B_2 = (I - T_1)(I - T_2) \cdots (I - T_l) = B_1^t. \tag{15}$$

Under conditions of Theorem 1, the estimate $\|B_k\|_2 \leqslant \rho$ is valid for each matrix B_1, B_2, and for the iteration matrix of the alternative direction method we have an inequality $\|B\| \leqslant \|B_1\| \cdot \|B_2\| \leqslant \rho^2 < 1$.

Since method (14), (15) can be presented in the form (2) with s.p.d. matrix B, it is possible to accelerate the convergence of iterations by means of conjugate direction methods, applied formally for solving preconditioned SLAE (3), and the following result is true.

Theorem 2. *The iterations of the alternative direction multiplicative projective conjugate gradient* (ADMPCG) *and conjugate residual* (ADMPCR) *methods defined by relations* (3), (5), *and* (6) *for* $s = 0, 1$ *respectively, are convergent under conditions of Theorem* 1, *and the estimate* (8) *is valid for the number of iterations* $n(\varepsilon)$, *where* $æ = (1+\rho^2)/(1-\rho^2)$ *and the value* ρ *is determined in Theorem* 1.

Now let us consider the successive multiplicative projective semi-conjugate residual (SMPSCR) method in the Krylov subspaces which is an alternative to the above ADMPCR algorithm. The new approach is based on the acceleration of iterative process (13) with non-symmetric iteration matrix B by means of formulas (5) and (9) where preconditioned matrix $\bar{A}$ is described by (3), (13). The SMPSCR procedure requires, for computing u^{n+1}, to save in memory all previous direction vectors $p^0, ..., p^n$, similarly to the GMRES method [4]. These two approach have the same convergent property because they provide minimization of the functional (r^n, r^n) in the subspace $\mathcal{K}_{n+1}(r^0, \tilde{A})$. The following result is true for the successive multiplicative method.

Theorem 3. *Suppose, that the SMPSCR algorithm, defined by formulas* (3), (5),(6) *and* (9),(11)–(13) *for* $s = 1$, *has diagonalizable matrix* $\tilde{A} = X\Lambda X^{-1}$, $\Lambda = \mathrm{diag}(\lambda_1, ..., \lambda_N)$, *where* λ_i *are eigenvalues of* $\tilde{A}$ *and* X *is a square matrix whose columns are corresponding eigenvectors. Then this method is convergent under conditions of Theorem 1, and the following estimate is valid for the number of iterations*:

$$n(\varepsilon) \leqslant 1 + \Big(\ln \frac{1 + \sqrt{1 - \varepsilon_1^2}}{\varepsilon_1}\Big)/\ln\gamma, \quad \varepsilon_1 = \varepsilon/(\|X\|_2 \cdot \|X^{-1}\|_2).$$

Here $\gamma_1 = a + \sqrt{a^2 - d^2}, \gamma_2 = c + \sqrt{c^2 - d^2}$, *where* a, d *are the semi-major axis and the focal distance* $(d^2 < c^2)$ *for the ellipse* $E(a, d, c)$ *which includes all values* λ_i, *excludes origin, and is centered at* c.

It should be noted that for the SMPSCR method, as for GMRES, different reduced versions with a bounded number of saved direction vectors p^n can be constructed. This will decrease the computational resources for the implementation of the algorithm, but the quantities $n(\varepsilon)$ will increase in these cases.

3 Additive projective methods

Let us recall that the Kaczmarz method is based on successive projection of the points from the space R^N onto the hyperplanes which are described by the corresponding equations of the algebraic system. A similar idea is used in the Cimmino algorithm (see [11]–[13] and its references). But here projections of the given point u^n onto all hyperplanes are made simultaneously, and the next step of the iterative approximation is chosen by means of some averaging procedure, or linear combination, with projective points $u^{n,i}$, $i = 1, ..., N$. Such an additive type iterative process to solve SLAE (11) can be presented in a generalized block version as

$$u^{n,p} = u^{n-1} + A^+_{(p)} r^{n-1}_{(p)}, \quad p = 1, 2, ..., l, \quad u^n = (u^{n,1} + u^{n,2} + ... + u^{n,l})/l. \tag{16}$$

These relations can be written in the following matrix form:

$$\begin{aligned} u^n &= Bu^{n-1} + g, \quad B = I - l^{-1} \sum_{p=1}^{l} A^+_{(p)} A_{(p)} \\ &= I - l^{-1} \sum_{p=1}^{l} T_p, \quad g = l^{-1} \sum_{p=1}^{l} A^+_{(p)} f_{(p)}, \end{aligned} \tag{17}$$

where matrices T_p are defined in (13).

Obviously, the limit vector of this sequence $u = \lim_{n \to \infty} u^n$, if it exists, satisfies the preconditioned system of equations

$$\tilde{A}u = \tilde{f}, \quad \tilde{A} = \sum_{p=1}^{l} T_p, \quad \tilde{f} = \sum_{p=1}^{l} A^+_{(p)} f_{(p)}. \tag{18}$$

If matrix $\tilde{A}$ of system (18) is a s.p.d. one, its spectral properties are obtained from the following result [10].

Theorem 4. *Let the quantities $0 < \alpha < 2$ and $0 < \rho < 1$ be defined from Theorem 1. Then the spectral radius $\lambda(\tilde{A})$ of s.p.d. matrix $\tilde{A}$ from system (18) satisfies the inequalities*

$$(2 - \alpha)(1 - \rho)/4 \leqslant \lambda(\tilde{A}) \leqslant \alpha l.$$

Now we can estimate the convergence rate of the additive projective approach.

Theorem 5. *Estimate* (8) *for the number of iterations* $n(\varepsilon)$ *is valid for the conjugate gradient and conjugate residual methods to solve the SLAE* (18), *i.e. to accelerate the additive projective algorithm* (17). *In this case the condition number satisfies the estimate* $\text{æ}(\tilde{A}) \leqslant 4\alpha l(2-\alpha)^{-1}(1-\rho)^{-1}$.

Remark 1. It follows from Theorems 1 and 5 that the multiplicative method is faster, in comparison to a similar additive procedure. However the latter has a considerable advantage for parallel implementation on a multi-processor computer, because the calculation of each projection at the subspace can be done independently.

Remark 2. Theorems 1 and 4 were proved in [10] to analyse convergence properties of the multiplicative and additive domain decomposition methods. It is evident that Theorems 2, 3 and 5 on the accelerations of projective iterative methods by means of conjugate direction or semi-conjugate direction algorithms in the Krylov subspaces can be used successively in these applications. Thus, the block variant of SLAE (11) can be interpreted as a matrix representation of the algebraic domain decomposion (ADD) formulation.

4 Iterations in Krylov subspaces with dynamic preconditioning

If we have a large problem, i.e. the original algebraic system (1) has a dimensionality of several millions or hundreds of millions, then it is natural to use some iterative procedure for solving auxiliary SLAEs at each step of block projection method (12) or (17).

In this case we obtain a two level iterative approach: at the external level we have iterative method of the form

$$u^{n+1} = B_n u^n + g^n = u^n + C_n^{-1}(f - Au^n), \quad B_n = I - C_n^{-1}A, \tag{19}$$

with variable (dynamic) iteration matrices B_n and preconditioning matrices C_n, and at the internal level the subsystems of dimensionality m_p are solved iteratively.

The acceleration of iterative process (19) in the Krylov subspaces

$$\mathcal{K}_{n+1}(r^0, C_n^{-1}A) = \text{span}\{C_0^{-1}r^0, AC_1^{-1}r^0, \ldots, A^n C_n^{-1}r^0\}$$

can be done by the following dynamically preconditioned semi-conjugate residual (DPSCR) method:

$$
\begin{aligned}
&r^0 = f - Au^0, \quad p^0 = C_0^{-1} r^0, \quad n = 0, 1, ..: \\
&u^{n+1} = u^n + \alpha_n p^n, \quad r^{n+1} = r^n - \alpha_n A p^n, \\
&p^{n+1} = C_{n+1}^{-1} r^{n+1} + \sum_{k=0}^{n} \beta_{n,k} p^k = p^{n+1,l} + \sum_{k=l}^{n} \beta_{n,k} p^k, \\
&p^{n+1,l} = p^{n+1,l-1} + \beta_{n,l-1} p^{l-1}, \quad p^{n+1,0} = C_{n+1}^{-1} r^{n+1}, \quad p^{n+1} = p^{n+1,n}, \\
&\alpha_n = (AC_n^{-1} r^n, r^n)/(Ap^n, Ap^n), \quad \beta_{n,k} = -(Ap^k, Ap^{n,k})/(Ap^k, Ap^k).
\end{aligned} \tag{20}
$$

The algorithm DPSCR provides minimization of the residual norm $\|r^{n+1}\|$ in the subspace $\mathcal{K}_{n+1}(r^0, C_n^{-1}A)$, and the following equality is true:

$$
\|r^{n+1}\|^2 = (r^0, r^0) - \frac{(AC_0^{-1} r^0, r^0)^2}{(Ap^0, Ap^0)} - \ldots - \frac{(AC_n^{-1} r^n, r^n)^2}{(Ap^n, Ap^n)}. \tag{21}
$$

Thus, this method converges if matrices $C_n^{-1}A$ are positive definite. In order to decrease the computational complexity of the algorithm, for large n two reduced versions of method (20) can be applied. The first one is based on the procedure of periodical restarting after each m iterations. This means that for $n = ml, l = 1, 2, ...$, the residual vector r^n is computed not from the recurrent relation but from the original equation ($r^{ml} = f - Au^{ml}$), and subsequent calculations are performed in the conventional form. The second way consists in truncated orthogonalization, i.e. for $n > m$ only the last m direction vectors $p^n, ..., p^{n-m+1}$ and $Ap^n, ..., Ap^{n-m+1}$ are saved in the memory and used in the recursion.

The following combination of these two approaches can be proposed. Let m_1 be the restart period, m_2 be the number of saved orthogonal direction vectors, and $n' = n - \left[\frac{n}{m_2}\right] m_2$, where $[b]$ is the integer part of b. Then the unified reduced recursion for p^n is written as

$$
p^{n+1} = C_{n+1}^{-1} r^{n+1} + \sum_{k=n-m+1}^{n} \beta_{n,k} p^k, \quad m = \min\{n', m_1\}. \tag{22}
$$

It is easy to show from (21) that the reduced versions of DPSCR converge also, if matrices $C_n^{-1}A$ are positive definite for all n.

References

1. Golub G., Van Loan C. Matrix computations. The Johns Hopkins Univ. Press, Baltimore, 1989.
2. O.Axelsson. Iterative solution methods. Cambridge Univ. Press, New York, 1994.
3. V.P.Il'in. Iterative Incomplete Factorization Methods, World Scientific Publ., Singapore, 1992.

4. Y.Saad. Iterative methods for sparse linear systems, PWS Publ., New York, 1996.
5. S.C.Eisenstat, H.C.Elman, M.H.Schultz. Variational iterative methods for non-symmetric systems of linear equations, SIAM J. Num. Anal., 20, (1983), pp. 345-357.
6. C.L.Lawson, R.J.Hanson. Solving Least Square Problems, Prentice-Hall, Inc., New Jersey, 1974.
7. S.Kaczmarz. Angenäherte Auflösung von Systemen linearer Gleichungen, Bull. Internat. Acad. Polon. Sci. Lettres A, 335–357 (1937). Translated into English: Int. J. Control 57(6): 1269–1271 (1993).
8. K.Tanabe. Projection method for solving a singular system of linear equation and its applications, Number Math., 17, (1971), pp. 203-214.
9. V.P.Il'in. On the iterative Kaczmarz method and its generalizations (in Russian), Sib.J.Industr. Math., 9, (2006), pp. 39-49.
10. J.H.Bramble, J.E.Pasciak, J.Wang, J.Xu. Convergence estimates for product iterative methods with applications to domain decomposition, Math. of Comput., 57, (1991), 195, pp. 1-21.
11. G.Cimmino. Calcolo approssimato per le soluzioni dei sistemi di equazioni lineari, La Ricerca Scientifica, II, 9 (1938), pp. 326-333.
12. R.Bramley, A.Sameh. Row projection methods for large nonsymmetric linear systems, SIAM J. Sci. Stat. Comput., 13, (1992), pp. 168-193.
13. G.Appleby, D.C.Smolarski. A linear acceleration row action method for projecting onto subspaces, Electronic Transactions on Num. Anal., 20, (2005), pp. 243-275.

Some Relationships between Optimal Preconditioner and Superoptimal Preconditioner

Jian-Biao Chen[1,*], Xiao-Qing Jin[2,**], Yi-Min Wei[3,***], and Zhao-Liang Xu[1,†]

[1] Department of Mathematics, Shanghai Maritime University, Shanghai 200135, P. R. China

[2] Department of Mathematics, University of Macau, Macao, P. R. China
xqjin@umac.mo

[3] Institute of Mathematics, School of Mathematical Sciences, Fudan University, Shanghai 200433, P. R. China
ymwei@fudan.edu.cn

Abstract. For any given n-by-n matrix A_n, a specific circulant preconditioner $t_F(A_n)$ proposed by E. Tyrtyshnikov [SIAM J. Matrix Anal. Appl., Vol. 13 (1992), pp. 459–473] is defined to be the solution of

$$\min_{C_n} \|I_n - C_n^{-1}A_n\|_F$$

over all n-by-n nonsingular circulant matrices C_n. The $t_F(A_n)$, called the superoptimal circulant preconditioner, has been proved to be a good preconditioner for a large class of structured systems including some ill-conditioned problems from image processing. In this paper, we study this preconditioner from an operator viewpoint. We will give some relationships between the optimal preconditioner (operator) proposed by T. Chan [SIAM J. Sci. Statist. Comput., Vol. 9 (1988), pp. 766–771] and superoptimal preconditioner (operator).

Keywords: optimal preconditioner, superoptimal preconditioner.

1 Introduction

In 1986, circulant preconditioners were proposed for solving Toeplitz systems [18, 22] by the preconditioned conjugate gradient method. Since then, the use

* The research of this author is partially sponsored by the Hi-Tech Research and Development Program of China (grant number: 2007AA11Z249).

** The research of this author is supported by the research grant RG-UL/07-08S/Y1/JXQ/FST from University of Macau.

*** The research of this author is supported by the National Natural Science Foundation of China under the grant 10871051 and Shanghai Science and Technology Committee under grant 08511501703.

† The research of this author is supported by the National Natural Science Foundation of China under the grant 10871051.

of circulant preconditioners for solving structured systems has been studied extensively [4, 10–12, 16, 17, 19, 20]. In 1988, T. Chan [6] proposed a specific circulant preconditioner as follows. For any arbitrary matrix A_n, T. Chan's circulant preconditioner $c_F(A_n)$ is defined to be the minimizer of the Frobenius norm

$$\min_{C_n} \|A_n - C_n\|_F$$

where C_n runs over all circulant matrices. The $c_F(A_n)$ is called the optimal circulant preconditioner in [6]. A generalization of the optimal circulant preconditioner is defined in [9]. More precisely, given a unitary matrix $U \in \mathbb{C}^{n \times n}$, let

$$\mathcal{M}_U \equiv \{U^* \Lambda_n U \mid \Lambda_n \text{ is any } n\text{-by-}n \text{ diagonal matrix}\}. \tag{1}$$

The optimal preconditioner $c_U(A_n)$ is defined to be the minimizer of

$$\min_{W_n} \|A_n - W_n\|_F$$

where W_n runs over $\mathcal{M}_U$. We remark that in (1), when $U = F$, the Fourier matrix, $\mathcal{M}_F$ is the set of all circulant matrices [8], and then $c_U(A_n)$ turns back to $c_F(A_n)$. The matrix U can also take other fast discrete transform matrices such as the discrete Hartley matrix, the discrete sine matrix or the discrete cosine matrix, etc., and then $\mathcal{M}_U$ is the set of matrices that can be diagonalized by a corresponding fast transform [2, 4, 10, 17]. We refer to [14] for a survey of the optimal preconditioner.

Now we introduce the superoptimal circulant preconditioner proposed by Tyrtyshnikov in 1992. For any arbitrary matrix A_n, the superoptimal circulant preconditioner $t_F(A_n)$ is defined to be the minimizer of

$$\min_{C_n} \|I_n - C_n^{-1} A_n\|_F$$

where C_n runs over all nonsingular circulant matrices. The generalized superoptimal preconditioner $t_U(A_n)$ is defined to be the minimizer of

$$\min_{W_n} \|I_n - W_n^{-1} A_n\|_F$$

where W_n runs over all nonsingular matrices in $\mathcal{M}_U$ given by (1). Again, $t_U(A_n)$ turns back to $t_F(A_n)$ when $U = F$.

In this paper, we study the superoptimal preconditioner from an operator viewpoint. We will give some relationships between the optimal preconditioner (operator) and superoptimal preconditioner (operator). Now, we introduce some lemmas which will be used later. Let $\delta(E_n)$ denote the diagonal matrix whose diagonal is equal to the diagonal of the matrix E_n.

Lemma 1. **([3])** *Let* $A_n \in \mathbb{C}^{n\times n}$. *Then*

$$c_U(A_n) \equiv U^*\delta(UA_nU^*)U.$$

For a relationship between $c_U(A_n)$ and $t_U(A_n)$, we have

Lemma 2. **([3])** *Let* $A_n \in \mathbb{C}^{n\times n}$ *such that* A_n *and* $c_U(A_n)$ *are invertible. Then*

$$t_U(A_n) \equiv c_U(A_nA_n^*)c_U(A_n^*)^{-1}.$$

Lemma 3. **([3, 7])** *For any matrix* $A_n \in \mathbb{C}^{n\times n}$,

$$\delta(UA_nA_n^*U^*) - \delta(UA_nU^*)\cdot\delta(UA_n^*U^*)$$

is a positive semi-definite diagonal matrix.

2 Relationships between c_U and t_U

The optimal preconditioner was studied from an operator viewpoint in [3]. Let the Banach algebra of all n-by-n matrices over the complex field, equipped with a matrix norm $\|\cdot\|$, be denoted by $(\mathbb{C}^{n\times n}, \|\cdot\|)$. Let $(\mathcal{M}_U, \|\cdot\|)$ be the subalgebra of $(\mathbb{C}^{n\times n}, \|\cdot\|)$. We note that $\mathcal{M}_U$ is an inverse-closed, commutative algebra. Let t_U be an operator from $(\mathbb{C}^{n\times n}, \|\cdot\|)$ to $(\mathcal{M}_U, \|\cdot\|)$ such that for any A_n in $\mathbb{C}^{n\times n}$, $t_U(A_n)$ is the minimizer of $\|I_n - W_n^{-1}A_n\|_F$ over all nonsingular $W_n \in \mathcal{M}_U$. Before we discuss the operator t_U in details, we introduce the following theorem which is concerned with the operator norms of c_U.

Theorem 1. **([2, 3])** *For all* $n \geqslant 1$, *we have*

(i) $\|c_U\|_F \equiv \sup\limits_{\|A_n\|_F=1} \|c_U(A_n)\|_F = 1.$
(ii) $\|c_U\|_2 \equiv \sup\limits_{\|A_n\|_2=1} \|c_U(A_n)\|_2 = 1.$

The following theorem includes some properties of $t_U(A_n)$.

Theorem 2. *Let* $A_n \in \mathbb{C}^{n\times n}$ *with* $n \geqslant 1$ *such that* A_n *and* $c_U(A_n)$ *are invertible. We have*

(i) $t_U(\alpha A_n) = \alpha t_U(A_n)$, *for all* $\alpha \in \mathbb{C}$.
(ii) $t_U(A_n^*) = t_U(A_n)^*$ *for the normal matrix* A_n.
(iii) $t_U(B_nA_n) = B_nt_U(A_n)$ *for* $B_n \in \mathcal{M}_U$ *if* B_nA_n *and* $c_U(B_nA_n)$ *are invertible.*

(iv) $t_U(A_n)$ *is stable for any normal and stable matrix* A_n. *We recall that a matrix is stable if all the real parts of its eigenvalues are negative.*

Proof. For (i), if $\alpha = 0$, (i) holds obviously. If $\alpha \neq 0$, it follows from Lemma 2 that

$$t_U(\alpha A_n) = c_U(\alpha A_n \bar{\alpha} A_n^*) c_U(\bar{\alpha} A_n^*)^{-1} = \alpha\bar{\alpha} c_U(A_n A_n^*)[\bar{\alpha} c_U(A_n^*)]^{-1}$$

$$= \alpha c_U(A_n A_n^*) c_U(A_n^*)^{-1} = \alpha t_U(A_n).$$

For (ii), we have by Lemma 2 again,

$$\begin{aligned} t_U(A_n)^* &= [c_U(A_n A_n^*) c_U(A_n^*)^{-1}]^* = [c_U(A_n^*)^{-1}]^* c_U(A_n A_n^*) \\ &= c_U(A_n)^{-1} c_U(A_n A_n^*) \end{aligned}$$

and then by Lemma 1,

$$t_U(A_n^*) = c_U(A_n^* A_n) c_U(A_n)^{-1} = U^*\delta(UA_n^* A_n U^*)UU^*\delta(UA_n U^*)^{-1}U$$

$$= U^*\delta(UA_n U^*)^{-1}UU^*\delta(UA_n^* A_n U^*)U = c_U(A_n)^{-1} c_U(A_n^* A_n).$$

Since A_n is normal, we obtain $t_U(A_n^*) = t_U(A_n)^*$. For (iii), we have

$$t_U(B_n A_n) = c_U(B_n A_n A_n^* B_n^*) c_U(A_n^* B_n^*)^{-1} = B_n c_U(A_n A_n^*) B_n^* c_U(A_n^* B_n^*)^{-1}$$

$$= B_n c_U(A_n A_n^*) c_U(A_n^*)^{-1} = B_n t_U(A_n).$$

For (iv), it follows from [15] that $\delta(UA_n U^*)$ and $\delta(UA_n^* U^*)$ are stable. Since $\delta(UA_n A_n^* U^*)$ is a positive diagonal matrix, we know that $\delta(UA_n A_n^* U^*) \cdot \delta(UA_n^* U^*)^{-1}$ is also stable. □

In general, we remark that (ii) is not true. For example, Let $U = I_2$ and

$$A_2 = \begin{bmatrix} 1 & 1 \\ 0 & 2 \end{bmatrix}.$$

It is easy to verify that $t_U(A_n^*) \neq t_U(A_n)^*$.

Theorem 3. *Let* $A_n \in \mathbb{C}^{n\times n}$ *with* $n \geqslant 1$ *such that* A_n *and* $c_U(A_n)$ *are invertible. We have*

(i) $\sup_{\|A_n\|_F = 1} \|t_U(A_n)\|_F \geqslant 1.$

(ii) $\sup_{\|A_n\|_2 = 1} \|t_U(A_n)\|_2 \geqslant 1.$

Proof. For (i), we have by Lemmas 1 and 2,

$$t_U(A_n) = U^*\delta(UA_nA_n^*U^*)\delta(UA_n^*U^*)^{-1}U.$$

Notice that from Lemma 3 and the invertibility of A_n and $c_U(A_n)$,

$$\delta(UA_nA_n^*U^*) \geqslant \delta(UA_nU^*)\delta(UA_n^*U^*) > 0,$$

where $M \geqslant N$ for any matrices M and N means that the all entries of $M - N$ are non-negative. We obtain

$$|\delta(UA_nA_n^*U^*)\delta(UA_n^*U^*)^{-1}| \geqslant |\delta(UA_nU^*)| > 0 \tag{2}$$

where $|Q| = [|q_{ij}|]$ for any matrix $Q = [q_{ij}]$. Thus we have by (2) and Theorem 1,

$$\sup_{\|A_n\|_F=1} \|t_U(A_n)\|_F = \sup_{\|A_n\|_F=1} \|\delta(UA_nA_n^*U^*)\delta(UA_n^*U^*)^{-1}\|_F$$

$$\geqslant \sup_{\|A_n\|_F=1} \|\delta(UA_nU^*)\|_F = \sup_{\|A_n\|_F=1} \|c_U(A_n)\|_F = \|c_U\|_F = 1.$$

For (ii), it follows by (2) that

$$\|t_U(A_n)\|_2 = \|\delta(UA_nA_n^*U^*)\delta(UA_n^*U^*)^{-1}\|_2 \geqslant \|\delta(UA_nU^*)\|_2 = \|c_U(A_n)\|_2.$$

Hence by Theorem 1 again,

$$\sup_{\|A_n\|_2=1} \|t_U(A_n)\|_2 \geqslant \|c_U\|_2 = 1. \qquad \square$$

Finally, we give a relationship of the unitarily invariant norm between $c_U(A_n)^{-1}A_n$ and $t_U(A_n)^{-1}A_n$.

Theorem 4. *Let $A_n \in \mathbb{C}^{n\times n}$ with $n \geqslant 1$ such that A_n and $c_U(A_n)$ are invertible. For every unitarily invariant norm $\|\cdot\|$, we have*

$$\|t_U(A_n)^{-1}A_n\| \leqslant \|c_U(A_n)^{-1}A_n\|.$$

Proof. It follows from [13, Theorem 2.2]: if the singular values are ordered in the following decreasing way: $\sigma_1 \geqslant \sigma_2 \geqslant \cdots \geqslant \sigma_n$, then we have

$$\sigma_k[t_U(A_n)^{-1}A_n] \leqslant \sigma_k[c_U(A_n)^{-1}A_n], \quad k = 1, 2, \ldots, n.$$

Thus, for every unitarily invariant norm $\|\cdot\|$, the result holds from [21, p.79, Theorem 3.7]. $\square$

References

1. D. Bertaccini, *A Circulant Preconditioner for the Systems of LMF-Based ODE Codes*, SIAM J. Sci. Comput., Vol. 22 (2000), pp. 767–786.
2. R. Chan and X. Jin, *An Introduction to Iterative Toeplitz Solvers*, SIAM, Philadelphia, 2007.
3. R. Chan, X. Jin and M. Yeung, *The Circulant Operator in the Banach Algebra of Matrices*, Linear Algebra Appl., Vol. 149 (1991), pp. 41–53.
4. R. Chan and M. Ng, *Conjugate Gradient Methods for Toeplitz Systems*, SIAM Review, Vol. 38 (1996), pp. 427–482.
5. R. Chan, M. Ng and X. Jin, *Strang-Type Preconditioners for Systems of LMF-Based ODE Codes*, IMA J. Numer. Anal., Vol. 21 (2001), pp. 451–462.
6. T. Chan, *An Optimal Circulant Preconditioner for Toeplitz Systems*, SIAM J. Sci. Statist. Comput., Vol. 9 (1988), pp. 766–771.
7. C. Cheng, X. Jin, S. Vong and W. Wang, *A Note on Spectra of Optimal and Superoptimal Preconditioned Matrices*, Linear Algebra Appl., Vol. 422 (2007), pp. 482–485.
8. P. Davis, *Circulant Matrices*, 2nd ed., Chelsea Publishing, New York, 1994.
9. T. Huckle, *Circulant and Skew Circulant Matrices for Solving Toeplitz Matrix Problems*, SIAM J. Matrix Anal. Appl., Vol. 13 (1992), pp. 767–777.
10. X. Jin, *Developments and Applications of Block Toeplitz Iterative Solvers*, Science Press, Beijing; and Kluwer Academic Publishers, Dordrecht, 2002.
11. X. Jin, *Three Useful Preconditioners in Structured Matrix Computations*, Proceedings of the 4th ICCM (2007), Vol. III, pp. 570–591. Eds: L.-Z. Ji, K.-F Liu, L. Yang and S.-T. Yau, Higher Education Press, Beijing, 2007.
12. X. Jin and Y. Wei, *Numerical Linear Algebra and Its Applications*, Science Press, Beijing, 2004.
13. X. Jin and Y. Wei, *A Short Note on Singular Values of Optimal and Superoptimal Preconditioned Matrices*, Int. J. Comput. Math., Vol. 84 (2007), pp. 1261–1263.
14. X. Jin and Y. Wei, *A Survey and Some Extensions of T. Chan's Preconditioner*, Linear Algebra Appl., Vol. 428 (2008), pp. 403–412.
15. X. Jin, Y. Wei and W. Xu, *A Stability Property of T. Chan's Preconditioner*, SIAM J. Matrix Aanal. Appl., Vol. 25 (2003), pp. 627–629.
16. T. Ku and C. Kuo, *Design and Analysis of Toeplitz Preconditioners*, IEEE Trans. Signal Process., Vol. 40 (1992), pp. 129–141.
17. M. Ng, *Iterative Methods for Toeplitz Systems*, Oxford University Press, Oxford, 2004.
18. J. Olkin, *Linear and Nonlinear Deconvolution Problems*, Ph.D thesis, Rice University, Houston, 1986.
19. D. Potts and G. Steidl, *Preconditioners for Ill-Conditioned Toeplitz Matrices*, BIT, Vol. 39 (1999), pp. 579–594.
20. S. Serra, *Preconditioning Strategies for Asymptotically Ill-Conditioned Block Toeplitz Systems*, BIT, Vol. 34 (1994), pp. 579–594.
21. G. Stewart and J. Sun, *Matrix Perturbation Theory*, Academic Press, Boston, 1990.
22. G. Strang, *A Proposal for Toeplitz Matrix Calculations*, Stud. Appl. Math., Vol. 74 (1986), pp. 171–176.

23. E. Tyrtyshnikov, *Optimal and Super-Optimal Circulant Preconditioners*, SIAM J. Matrix Anal. Appl., Vol. 13 (1992), pp. 459–473.

Scaling, Preconditioning, and Superlinear Convergence in GMRES-Type Iterations*

Igor Kaporin

Computing Center of Russian Academy of Sciences,
Vavilova 40, Moscow 119991, Russia
kaporin@ccas.ru

Abstract. A theoretical justification is found for several standard techniques related to ILU preconditioning, such as pre-scaling and pivot modification, with implications for practical implementation. An improved estimate for the reduction of the GMRES residual is obtained within the general framework of two-stage preconditioning. In particular, an estimate in terms of a conditioning measure of the scaled coefficient matrix and the Frobenius norm of the scaled ILU residual is presented.

Keywords: unsymmetric sparse matrix, two-side scaling, incomplete LU preconditioning, two-stage preconditioning, superlinear convergence.

1 Introduction

In the present paper we address certain theoretical issues related to the construction of computational methods for the numerical solution of large linear systems with general nonsingular unsymmetris sparse coefficient matrices.

As is known, *direct solvers* (which are based on the "exact" sparse triangular factorization of the matrix) represent a quite robust, advanced and well-established piece of numerical software. As an example, one can refer to the UMFPACK solver [5], which implements an unsymmetric multifrontal sparse Gauss elimination. However, the sparsity structure inherent to many important classes of problems (such as fully three-dimensional discrete models) is rather unsuitable to such methods. This is due to huge volumes of intermediate data generated by a direct solver (namely, arrays presenting nonzero elements of the triangular factors) which are many orders of magnitude larger than the order of the system. Moreover, the corresponding computation time grows even faster than the storage space as the linear system size increases.

An alternative to direct solvers is represented by iterative methods. Unfortunately, any "classical" fixed-storage simplistic schemes (for instance, the

* This work was partially supported through the Presidium of Russian Academy of Sciences program P-14 and the program "Leading Scientific Schools" (project NSh-2240.2006.1).

ILU(0) preconditioned GMRES(m) method) are completely unreliable for general unsymmetric linear systems. More promising are the Incomplete LU-type Preconditioned Krylov subspace iterative solvers based on the approximate factorization "by value" without any restrictions on the sparsity of the triangular factors. An appropriate use of the "approximate" triangular factorization makes it possible to generate much more compact triangular factors as compared to those arising in direct solvers. Is should be stressed that almost all results and techniques developed for the "exact" LU-factorization need to be essentially revisited and reformed in order to be useful for the purpose of efficient preconditioning of the Krylov subspace iterations. In this case, for instance, a careful pivoting (strictly targeted at "as good as possible" diagonal dominance in the approximate triangular factors) appears to be much more important than any near-optimum pre-ordering or even the dynamic account for the local fill-in [12].

It can be definitely stated that the currently available software products implementing preconditioned iterative sparse linear solvers still suffer from the following deficiences:
(a) their reliability is still worse than that of direct solvers;
(b) in order to provide satisfactory reliability and efficiency, they require quite complicated tuning of the solver control parameters (which are related to the numerical algorithm itself rather than to the problem solved).

Below we present a superlinear convergence estimate for Preconditioned GMRES-type iterative linear equation solvers. The formulation of the result is specifically adjusted to the case when the preconditioning is based on approximate triangular factorization applied to a pre-scaled coefficient matrix. Hence, in addition to many empirical observations (see, e.g., [2]), a certain theoretical evidence is found for the considered robust iterative solvers.

2 Problem setting

Consider the linear algebraic system

$$Ax = b \tag{1}$$

with a general unsymmetric nonsingular sparse $n \times n$ matrix A. The Incomplete LU (ILU) preconditioned GMRES-type iterative methods use the *preconditioner* matrix $C \approx A$ of the form $C = PLUQ$ which is obtained from the ILU equation

$$A = PLUQ + E,$$

where L and U are nonsingular lower and upper triangular matrices, respectively, while P and Q are permutation matrices. Hence, the preconditioner is given by

$$C = PLUQ, \tag{2}$$

which is obviously an "easily invertible" matrix. The additive term E is the ILU *error matrix*, a standard assumption for which is

$$|(E)_{ij}| = O(\tau), \tag{3}$$

where $\tau \ll 1$ is a prescribed threshold parameter.

Note that a more general structure of the error matrix is admissible in preconditioned GMRES-type methods, namely,

$$|(E - X)_{ij}| = O(\tau), \qquad \text{rank}(X) = O(1), \tag{4}$$

which was proposed in [19] in the context of preconditioned Toeplitz-like system solvers. The low-rank term in ILU error matrix may arise due to the use of *pivot correction*, which technique can be helpful in the case of *diagonal pivoting*, see [12] for more detail.

We consider the preconditioned Krylov subspace iterative solver for unsymmetric linear system (1) as an application of GMRES iterations [16] to the *right preconditioned* system

$$AC^{-1}y = b, \tag{5}$$

so that the solution of (1) is obtained as $x = C^{-1}y$.

Note: Under a proper choice of permutation matrices P and Q (mainly aimed at the improvement of diagonal dominance of L and U), one can observe that

$$\|E\|_F^2 \equiv \text{trace}(E^T E) = O(n\tau^2), \tag{6}$$

i.e., only relatively few nonzero entries of E attain their maximum allowed magnitude. At the same time, the stability of the triangular factors is often improved (more precisely, the ratio $\text{cond}(C)/\text{cond}(A)$ is not large), which is desirable from the numerical stability viewpoint.

3 Scaling techniques

It was noted by many authors (see, e.g. [2, 13] and references therein) that the ILU factorization 'by value' applied to a properly *two-side scaled* coefficient matrix

$$A_S = D_L A D_R \tag{7}$$

may yield much better preconditioning compared to similar algorithms applied to the original coefficient matrix A (especially in several hard-to-solve cases, see also [12]). The incomplete triangular factorization (now applied to scaled matrix (7)) yields the equation

$$A_S = P_S L_S U_S Q_S + E_S,$$

where P_S and Q_S are permutation matrices arising due to pre-ordering and pivoting applied to the scaled matrix. Hence, according to (7) the resulting preconditioner is

$$D_L^{-1} P_S L_S U_S Q_S D_R^{-1} \equiv C \approx A.$$

Note that in actual implementation the latter preconditioning can readily be transformed to the same form (2) (though with different triangular factors, even if the permutations would be the same).

Next we will construct a preconditioning quality measure via
(i) a special condition number of A_S (presenting the scaling quality measure) and
(ii) Frobenius norm of the scaled ILU error matrix E_S.
The corresponding ILU-GMRES convergence estimate can be referred to as the *constructive* one, because the residual norm bound is expressed literally via the very functionals which are expected to be directly optimized in the procedures of scaling and approximate factorization. Moreover, the improvement of scaling quality and the attained value of the factorization quality criteria can be readily evaluated *a posteriori* (numerically).

Note: Below in Section 5 we present a convergence estimate for the GMRES method which *does not depend* on the quantities of the type $\|\text{cond}(D_L)\|$, $\|E\|$, or $\|(LU)^{-1}\|$. Taking into account that our result holds in 'exact arithmetics', one can conclude that "bad" (i.e., too large) values of quality indicators (which are often associated with ILU preconditioning) such as
(a) condition numbers of the scaling matrices D_L and D_R,
(b) size of the elements of the unscaled 'original' error matrix E, and
(c) norm of the inverse of the scaled preconditioner,
may have their destructive effect on the GMRES convergence only in the presence of round-off errors.

4 How to estimate GMRES convergence

From now on, let $\|\cdot\|$ denote the matrix spectral norm

$$\|B\| = \max_{z\neq 0} \|Bz\|/\|z\|, \qquad \|z\| = \sqrt{z^T z}. \tag{8}$$

For the kth residual

$$r_k = b - Ax_k \tag{9}$$

in the preconditioned minimum residual method (also known as GMRES(∞), cf. [16]) one has, by the construction,

$$\|r^k\| = \min_{P_k(0)=1} \|P_k(M) r^0\| = \|P_k^*(M) r^0\|. \tag{10}$$

Here

$$M = AC^{-1} \tag{11}$$

is the right preconditioned matrix and $P_k^*(\cdot)$ is the polynomial determined at the kth step of the minimum residual method; this polynomial has the degree not greater than k and is normalized by the condition $P_k^*(0) = 1$.

For the sake of simplicity, let M be diagonalizable, that is,

$$M = V\Lambda V^{-1}. \tag{12}$$

Here, the columns of V are the (normalized) eigenvectors $\nu_1, \nu_2, \ldots, \nu_n$ of M and the entries of the diagonal matrix Λ are the corresponding eigenvalues $\lambda_1, \lambda_2, \ldots, \lambda_n$ of M. Using (10) and (12), one finds

$$\begin{aligned} \|r^k\| &= \|P_k^*(M)r^0\| \leqslant \|\widetilde{P}_k(M)r^0\| = \|\widetilde{P}_k(V\Lambda V^{-1})r^0\| = \|V\widetilde{P}_k(\Lambda)V^{-1}r^0\| \\ &= \|VD\widetilde{P}_k(\Lambda)D^{-1}V^{-1}r^0\| \leqslant \|VD\|\|\widetilde{P}_k(\Lambda)\|\|(VD)^{-1}\|\|r^0\| \\ &= \kappa\, \|\widetilde{P}_k(\Lambda)\|\|r^0\| = \kappa \max_{1\leqslant i\leqslant n} |\widetilde{P}_k(\lambda_i)|\; \|r^0\| \end{aligned} \tag{13}$$

which holds for an arbitrary polynomial $\widetilde{P}_k$ of a degree not greater than k and normalized by the condition $\widetilde{P}_k(0) = 1$. Note that hereafter, the notation

$$\kappa = \min_{D=\text{diag.}} \|VD\|\|(VD)^{-1}\| = \min_{D=\text{diag.}} \text{cond}(VD) \tag{14}$$

is used to denote the condition number of VD, where D is an arbitrary nonsingular diagonal matrix.

Nontrivial choices of $\widetilde{P}_k(\cdot)$ and upper bounds for $\max_{1\leqslant i\leqslant n} |\widetilde{P}_k(\lambda_i)|$ are typically obtained via the separation of the spectrum of M into the cluster part and the outlying part, see, for instance [7, 4, 3] for the case of SPD matrix M, and [6, 15, 16, 21] for the general case. In [18], an alternative technique is used, which allows to relax diagonalizability condition (12).

A standard approach to the analysis of preconditioned iterations is to use the general theory of Krylov subspace methods for a preconditioned system (5) using substitution (11). Unfortunately, one can hardly estimate and/or control any related properties of the preconditioned matrix M even *a posteriori*. It is not known how one can effectively relate any characteristics of the "localization" or "distribution" of the eigenvalue spectrum of M to the result of preconditioning. For instance, in general even the tr(M) is very hard to estimate (e.g., its exact evaluation seems like n times the solution cost of the original linear system).

Therefore, we actually reject the use of the preconditioned spectrum as an "interface" between the preconditioning and iterations. Instead of that, we separately use some properties of the two factors E_S and A_S^{-1} in the multiplicative splitting such as

$$D_L(I - M^{-1})D_L^{-1} = (A_S - C_S)A_S^{-1} = E_S A_S^{-1}.$$

This well conforms to the two-stage preconditioning scheme, where at the first stage one improves some conditioning measure for matrix A_S by the choice of D_L and D_R, and at the second stage seeks for an easily invertible C_S which *directly* approximates A_S.

5 Superlinear GMRES convergence via scaled error matrix

Let us denote the singular values of a real-valued $n \times n$-matrix Z as

$$\sigma_1(Z) \geqslant \sigma_2(Z) \geqslant \cdots \geqslant \sigma_n(Z) \geqslant 0.$$

Recalling the definition of the Frobenius matrix norm given in (6) and taking into account that $\sigma_i(Z)^2$ is exactly the ith eigenvalue of $Z^T Z$, one has

$$\|Z\|_F^2 = \mathrm{tr}(Z^T Z) = \sum_{i=1}^{n} \sigma_i(Z)^2. \tag{15}$$

Moreover, by $(\det Z)^2 = \det(Z^T Z)$ it follows

$$|\det\ Z| = \prod_{i=1}^{n} \sigma_i(Z). \tag{16}$$

5.1 Main result

Next we present a scale-invariant generalization of the preconditioned GMRES convergence result earlier presented in [12].

Theorem 1. *Let C_S be a preconditioner for the scaled matrix A_S as defined in (7) and the iterates x_k be generated by the* GMRES(∞) *method with the preconditioner $C = D_L^{-1} C_S D_R^{-1}$. Then the kth residual $r_k = b - Ax_k$ satisfies*

$$\frac{\|r_k\|}{\|r_0\|} \leqslant \kappa\ \mathcal{K}(A_S) \left(4e\frac{n}{k} \sin^2[C_S, A_S]\right)^{k/2}, \qquad k = 1, 2, \dots, n, \tag{17}$$

where $e = \exp(1)$, the quantity κ was defined in (14),

$$\mathcal{K}(Z) = \left(n^{-1/2}\|Z\|_F\right)^n \Big/ |\det Z| \tag{18}$$

denotes the unsymmetric *K-condition number of a nonsingular matrix Z, and*

$$\sin^2[Y, Z] = 1 - \frac{(\mathrm{tr} Z^T Y)^2}{\|Y\|_F^2 \|Z\|_F^2} \tag{19}$$

denotes the squared sine of the Euclidean acute angle between the matrices Y and Z.

Proof. Let us define the scalar

$$\xi = \frac{\mathrm{trace}(A_S^T C_S)}{\|C_S\|^2}. \tag{20}$$

(Note that if $C_S \approx A_S$, then $\xi \approx 1$.) Let the eigenvalues of the preconditioned matrix $M = AC^{-1}$ (recall that $C = D_L^{-1} C_S D_R^{-1}$) be numbered by the decrease of the distance to ξ:

$$|\xi - \lambda_1| \geqslant |\xi - \lambda_2| \geqslant \ldots |\xi - \lambda_n| \geqslant 0. \tag{21}$$

Following the techniques introduced in [24] (see also [11]), let us consider the polynomial $\widetilde{P}_k$ of the form

$$\widetilde{P}_k(\lambda) = \prod_{i=1}^{k} \left(1 - \frac{\lambda}{\lambda_i}\right).$$

Taking into account that $\widetilde{P}_k(\lambda_j) = 0$ for $1 \leqslant j \leqslant k$ and using the above ordering of the eigenvalues one can deduce from (13) the following residual norm estimate:

$$\begin{aligned}
\frac{1}{\kappa}\frac{\|r^k\|}{\|r^0\|} &\leqslant \max_{1\leqslant j\leqslant n} |\widetilde{P}_k(\lambda)| = \max_{k<j\leqslant n} |\widetilde{P}_k(\lambda)| = \max_{k<j\leqslant n} \left|\prod_{i=1}^{k}\left(1 - \frac{\lambda_j}{\lambda_i}\right)\right| \\
&= \max_{k<j\leqslant n} \prod_{i=1}^{m} \frac{|(\xi - \lambda_i) - (\xi - \lambda_j)|}{|\lambda_i|} \\
&\leqslant \max_{k<j\leqslant n} \prod_{i=1}^{k} \frac{|\xi - \lambda_i| + |\xi - \lambda_j|}{|\lambda_i|} \\
&\leqslant 2^k \prod_{i=1}^{k} \frac{|\xi - \lambda_i|}{|\lambda_i|} = 2^k \prod_{i=1}^{k} \left|1 - \frac{\xi}{\lambda_i}\right| \\
&= 2^k \prod_{i=1}^{k} |\lambda_{\pi(i)}(I - \xi M^{-1})| \leqslant 2^k \prod_{i=1}^{k} |\lambda_i(I - \xi M^{-1})|,
\end{aligned}$$

where the index permutation $\pi(i)$ corresponds to the reordering of the original numbering (21) according to the decrease of eigenvalue modules of the matrix $I - \xi M^{-1}$:

$$|\lambda_1(I - \xi M^{-1})| \geqslant |\lambda_2(I - \xi M^{-1})| \geqslant \cdots \geqslant |\lambda_n(I - \xi M^{-1})|.$$

Next we use the identity

$$I - \xi M^{-1} = I - \xi C A^{-1} = (A - \xi C)A^{-1} = D_L^{-1}(A_S - \xi C_S)A_S^{-1} D_L$$

and apply the classical inequalities

$$\prod_{i=1}^{k} |\lambda_i(XY)| \leqslant \prod_{i=1}^{k} \sigma_i(XY) \leqslant \prod_{i=1}^{k} \sigma_i(X)\sigma_i(Y),$$

the left of which is known as the Weyl inequality (written for $Z = XY$), while the right one was found by Horn, see [14], with $X = A_S - \xi C_S$ and $Y = A_S^{-1}$. This yields the following estimate:

$$\begin{aligned}
\frac{1}{\kappa}\frac{\|r^k\|}{\|r^0\|} &\leqslant 2^k \prod_{i=1}^{k} |\lambda_i(I - \xi M^{-1})| \\
&= 2^k \prod_{i=1}^{k} |\lambda_i(D_L^{-1}(A_S - \xi C_S)A_S^{-1}D_L)| \\
&= 2^k \prod_{i=1}^{k} |\lambda_i((A_S - \xi C_S)A_S^{-1})| \leqslant 2^k \prod_{i=1}^{k} \sigma_i((A_S - \xi C_S)A_S^{-1}) \\
&\leqslant 2^k \prod_{i=1}^{k} \sigma_i(A_S - \xi C_S)\sigma_i(A_S^{-1}) \\
&= 2^k \left(\prod_{i=1}^{k} (\sigma_i(A_S - \xi C_S))^2\right)^{1/2} \left(\prod_{i=1}^{k} \sigma_i(A_S^{-1})\right). \qquad (22)
\end{aligned}$$

The first product can be bounded using the inequality between the arithmetic and geometric mean values

$$\left(\prod_{i=1}^{m} \eta_i\right)^{1/m} \leqslant \frac{1}{m}\sum_{i=1}^{m} \eta_i, \qquad \eta_i > 0, \qquad (23)$$

taken with $m = k$ and $\eta_i = (\sigma_i(A_S - \xi C_S))^2$:

$$\begin{aligned}
\left(\prod_{i=1}^{k} (\sigma_i(A_S - \xi C_S))^2\right)^{1/2} &\leqslant \left(\frac{1}{k}\sum_{i=1}^{k} (\sigma_i(A_S - \xi C_S))^2\right)^{k/2} \\
&\leqslant \left(\frac{1}{k}\sum_{i=1}^{n} (\sigma_i(A_S - \xi C_S))^2\right)^{k/2} \\
&= \left(\frac{1}{k}\|A_S - \xi C_S\|_F^2\right)^{k/2} \\
&= \left(\frac{1}{k}\|A_S\|_F^2 \sin^2[A_S, C_S]\right)^{k/2}. \qquad (24)
\end{aligned}$$

Here the last equality follows from (19) and (20).

The second product in (22) can also be estimated using inequality (23), this time taken with $m = n - k$ and $\eta_i = \sigma_i(A_S)$:

$$\begin{aligned}
\prod_{i=1}^{k} \sigma_i(A_S^{-1}) &= \left(\prod_{i=1}^{k} \sigma_{n+1-i}(A_S) \right)^{-1} \\
&= \frac{1}{|\det A_S|} \left(\prod_{i=1}^{n} \sigma_i(A_S) \right) \left(\prod_{i=1}^{k} \sigma_{n+1-i}(A_S) \right)^{-1} \\
&= \frac{1}{|\det A_S|} \prod_{i=1}^{n-k} \sigma_i(A_S) \\
&= \frac{1}{|\det A_S|} \left(\prod_{i=1}^{n-k} (\sigma_i(A_S))^2 \right)^{1/2} \\
&\leqslant \frac{1}{|\det A_S|} \left(\frac{1}{n-k} \sum_{i=1}^{n-k} (\sigma_i(A_S))^2 \right)^{\frac{n-k}{2}} \\
&\leqslant \frac{1}{|\det A_S|} \left(\frac{1}{n-k} \sum_{i=1}^{n} (\sigma_i(A_S))^2 \right)^{\frac{n-k}{2}} \\
&= \frac{1}{|\det A_S|} \left(\frac{1}{n-k} \|A_S\|_F^2 \right)^{\frac{n-k}{2}} \\
&\leqslant \frac{\exp(k/2)}{|\det A_S|} \left(\frac{1}{n} \|A_S\|_F^2 \right)^{\frac{n-k}{2}} .
\end{aligned} \tag{25}$$

The latter inequality follows from

$$\begin{aligned}
\left(\frac{n}{n-k} \right)^{\frac{n-k}{2}} &= \exp\left(\frac{n-k}{2} \log \frac{n}{n-k} \right) \\
&\leqslant \exp\left(\frac{n-k}{2} \left(\frac{n}{n-k} - 1 \right) \right) = \exp\left(\frac{k}{2} \right),
\end{aligned}$$

where we have used $\log \eta \leqslant \eta - 1$.

Substituting now the above two ineqialities (24) and (25) into (22), one gets

$$\begin{aligned}
\frac{1}{\kappa} \frac{\|r^k\|}{\|r^0\|} &\leqslant 2^k \left(\frac{1}{k} \|A_S\|_F^2 \sin^2[A_S, C_S] \right)^{\frac{k}{2}} \frac{\exp(k/2)}{|\det A_S|} \left(\frac{1}{n} \|A_S\|_F^2 \right)^{\frac{n-k}{2}} \\
&= \left(4e \frac{n}{k} \sin^2[A_S, C_S] \right)^{k/2} \frac{\left(n^{-1} \|A_S\|_F^2 \right)^{n/2}}{|\det A_S|}.
\end{aligned}$$

Finally, it only remains to recall definition (18), and the required inequality (17) follows.

Hence, Theorem 1 actually gives a theoretical basis for two-stage preconditionings. For instance, at the first stage one chooses the scaling matrices D_L and D_R (subject to the condition of near minimization of $\mathcal{K}(D_L A D_R)$, see Section 5.3 below and [12] for more detail), and at the second stage one constructs an easily invertible approximation for the scaled matrix $A_S = D_L A D_R$, e.g., with the use of an approximate triangular factorization with permutations as in [12].

Note that the earlier supelinear GMRES convergence estimate [11] was formulated in terms of the quantities $\|I_n - AC^{-1}\|_F$ and $|\lambda(AC^{-1})|_{\min}$, which, in general, can hardly be estimated even *a posteriori*. It turns out that simplistic upper bounds like

$$\|I_n - AC^{-1}\|_F = \|(A - C)C^{-1}\|_F \leqslant \|C^{-1}\|\|E\|_F$$

are often senseless due to occasionally huge values of $\|C^{-1}\|$, see for instance the data in Tables 2–7 below. At the same time, one can see there that "reasonably huge" values of the norm of the inverse preconditioner *may not* destroy the GMRES convergence.

Also, the above preconditioning quality measure (19) satisfies a natural condition of being a scale-invariant functional of its matrix arguments, that is,

$$\sin^2[\gamma C_S, \alpha A_S] = \sin^2[C_S, A_S], \qquad \alpha \neq 0, \quad \gamma \neq 0.$$

This well conforms with the obvious fact that the GMRES residual norm is invariant with respect to any re-scaling of the preconditioner (i.e. $C := \beta C$, $\beta \neq 0$).

Certainly, the particular value of the constant $4e$ in (17) is somewhat overestimated due to rather rough techniques used in the proof of Theorem 1. Based on special analytical examples, it can be conjectured that the unimprovable value for this constant equals to one.

Note: Starting with a sufficiently large iteration number k, the right-hand sides of the above estimate (17) decrease faster than any geometric progression. In this sense, these estimates confirm the superlinear GMRES convergence, which is often observed when the preconditioning is good enough.

5.2 The corresponding GMRES iteration number bound

Using the techniques developed in [11] one can readily find an upper bound for the iteration number needed to attain the specified residual norm reduction $\varepsilon \ll 1$.

We will use the following auxiliary result (for the proof see [11]).

Lemma 1. *Let* $t > 0$ *and*

$$s \geqslant \frac{1 + (1 + e^{-1})t}{\log(e + t)}, \tag{26}$$

where $e = \exp(1)$. *Then the inequality*

$$s \log s \geqslant t \tag{27}$$

holds.

As was mentioned in [11], for any $t > 0$ it holds $t < s \log s < 1.064t$, i.e. the relative overestimation in (27) is never larger than 6.5%.

Now we can prove a GMRES iteration number bound similar to the ones presented in [11],[12].

Theorem 2. *The iteration number* k *sufficient for the* ε *times reduction of the residual norm in the minimum residual method satisfies*

$$k \leqslant \left\lceil \frac{4en \sin^2[C_S, A_S] + (2 + 2e^{-1}) \log\left(\frac{\kappa}{\varepsilon}\mathcal{K}(A_S)\right)}{\log\left(e + \left(2en \sin^2[C_S, A_S]\right)^{-1} \log\left(\frac{\kappa}{\varepsilon}\mathcal{K}(A_S)\right)\right)} \right\rceil \tag{28}$$

with κ *determined in (14) and* $e = \exp(1)$.

Proof. By the result of Theorem 1, a sufficient condition to satisfy the required inequality $\|r^k\|/\|r^0\| \leqslant \varepsilon$ is

$$\kappa\, \mathcal{K}(A_S) \left(4e\frac{n}{k} \sin^2[C_S, A_S]\right)^{k/2} \leqslant \varepsilon,$$

which can be rewritten as

$$\frac{k}{2} \log\left(\frac{k}{4en \sin^2[C_S, A_S]}\right) \geqslant \log\left(\frac{\kappa}{\varepsilon}\mathcal{K}(A_S)\right).$$

Multiplying the latter inequality by $\left(2en \sin^2[C_S, A_S]\right)^{-1}$ and denoting

$$s = \frac{k}{4en \sin^2[C_S, A_S]}, \qquad t = \frac{1}{2en \sin^2[C_S, A_S]} \log\left(\frac{\kappa}{\varepsilon}\mathcal{K}(A_S)\right),$$

one can see that the resulting inequality is equivalent to condition (27). By Lemma 1, a sufficient condition for (27) to hold is (26), which yields exactly the required estimate (28). The use of the closest integer from above is valid, since the function $s \log s$ increases for $s \geqslant 1/e$, and by (26) it holds $s \geqslant 1$.

5.3 Relating the new estimate to scaling

In view of (17), it is natural to require that the scaling should minimize functional (18) with $Z = A_S = D_L A D_R$. As is shown in [12], the minimizer satisfies exactly the requirement that A_S have the Euclidean norms of each row and column equal to the same number, e.g.,

$$\sum_{j=1}^{n}(D_L)_i^2(A)_{ij}^2(D_R)_j^2 = 1, \qquad \sum_{i=1}^{n}(D_L)_i^2(A)_{ij}^2(D_R)_j^2 = 1$$

exactly as was recommended in [2, 13]. The diagonal matrices D_L and D_R can be evaluated as an approximate solution of the above nonlinear system of equations using the RAS (Row-Alternating Scaling) iterations (see, e.g. [17] and references therein). Each RAS half-iteration consists in one-side re-scaling of the current matrix to normalize all its rows or all the columns at odd and even steps, respectively. The RAS algorithm and its "symmetrized" version are investigated in [12] from the viewpoint of $\mathcal{K}(A_S)$ reduction.

Note that both the convergence theory above and the numerical examples given later (cf. also [12]) clearly indicate that it makes sense to invest a considerable fraction of computational efforts into the evaluation of scaling matrices D_L and D_R for which the factor $\mathcal{K}(A_S)$ in the right hand side of the GMRES convergence estimate (17) is reduced considerably. In this respect, one can even use sparse triangular matrices instead of diagonal D_L and D_R, as it was done in the two-side explicit preconditioning proposed and investigated in [8]. There, a general unsymmetric matrix A was preconditioned using the two-side transformation

$$\widehat{A} = G_L A G_U,$$

with G_L and G_U chosen as the sparse lower and upper triangular matrices, respectively. The positions and values of their entries were determined from *the same condition* of $\mathcal{K}(G_L A G_U)$ minimization. To this end, a RAS-type procedure was used, where at each half-step one evaluates the K-condition number minimizer G_L or G_U, where $\mathbf{K}(M) = (n^{-1}\mathrm{tr}M)^n/\det M$ and $M = \widehat{A}\widehat{A}^T$ or $M = \widehat{A}^T\widehat{A}$, respectively (cf. also [9]). The strategy considered in [8] was as follows: allowing the matrices G_L and G_U to have a sufficiently large number of nonzeroes, one can assume that the matrix $M = \widehat{A}\widehat{A}^T$ comes close enough to the identity matrix I_n to make the *explicit* Conjugate Gradient iterations efficient in solving the two-side preconditioned system $My = f$. Since such a construction is completely free of necessity of solving systems with large sparse triangular matrices, this method is considered suitable for the parallel implementation.

In the context of the present paper, even the use of G_L and G_U containing not more than 2 nonzeroes in each row and column instead of diagonal matrices D_L and D_R, respectively, may result in a further considerable reduction of $\mathcal{K}(G_L A G_U)$. Moreover, one can expect that approximate triangular factorization of the type

$$G_L A G_U = \widehat{P}\widehat{L}\widehat{U}\widehat{Q} + \widehat{E}$$

will possess even better preconditioning quality than that obtained with simple diagonal scaling. In this case, the convergence estimate (17) of Theorem 1 will take the form

$$\frac{\|r_k\|}{\|r_0\|} \leqslant \kappa\, \mathcal{K}(G_L A G_U) \left(4e\frac{n}{k}\sin^2[\widehat{P}\widehat{L}\widehat{U}\widehat{Q}, G_L A G_U]\right)^{k/2}.$$

Hence, the resulting *two-level* preconditioner takes the form $C = G_L^{-1}\widehat{P}\widehat{L}\widehat{U}\widehat{Q}G_U^{-1}$ and its application additionaly requires two matrix-vector multiplications with the sparse matrices G_L and G_U. Of course, such a scheme would involve certain additional algorithmic complications; however, the expected gain in preconditioning quality should prevail.

5.4 Relating the new estimate to ILU preconditioning

If the matrix $A_S = D_L A D_R$ is scaled to satisfy

$$\|A_S\|_F^2 = n, \tag{29}$$

(note that (29) holds for scalings obtained using any number of RAS iterations) then the following upper bound holds:

$$\sin^2[C_S, A_S] \equiv 1 - \frac{(\text{trace}A_S^T C_S)^2}{\|C_S\|_F^2\|A_S\|_F^2} = \frac{\min_\sigma \|A_S - \sigma C_S\|_F^2}{\|A_S\|_F^2}$$
$$\leqslant \|A_S - C_S\|_F^2 \|A_S\|_F^{-2} = n^{-1}\|A_S - C_S\|_F^2 = n^{-1}\|E_S\|_F^2.$$

Hence, under condition (29) the result of Theorem 1 coincides exactly with the one presented in [12]:

$$\frac{\|r^k\|}{\|r^0\|} \leqslant \frac{\kappa}{|\det A_S|}\left(\frac{3.3}{\sqrt{k}}\|E_S\|_F\right)^k, \tag{30}$$

where we have also used the numerical inequality $2\sqrt{e} < 3.3$.

It should be noted that if the ILU threshold parameter is chosen sufficiently small, e.g. $\tau = 0.001$, and the ILU factors are stable enough, then the typical values of $\|E_S\|_F$ are *not big* (one can often observe $\|E_S\|_F < 1$ even for realistic large-scale problems, cf. numerical data in [12]). As was noted above, the quantity $\|E_S\|_F$ can be easily evaluated in the process of the approximate factorization of A_S, which allows us to use it as an *a posteriori* indicator of ILU preconditioning quality.

Turning back to the low-rank modified form of error term (4), one can generalize the main result to take into account the case when the *pivot modification* rule is used in the ILU factorization (see [12] for more detail). Setting $\xi = 1$ in (22), one finds, for any integer $1 \leqslant m \ll k$, the following estimate:

$$\frac{1}{\kappa}\frac{\|r^k\|}{\|r^0\|} \leqslant 2^k \left(\prod_{i=1}^{m}\sigma_i(E_S)\right)\left(\prod_{i=m+1}^{k}\sigma_i(E_S)^2\right)^{1/2}\left(\prod_{i=1}^{k}\sigma_i(A_S^{-1})\right). \tag{31}$$

Estimating these three products separately, one has

$$\left(\prod_{i=1}^{m}\sigma_i(E_S)\right) \leqslant \|E_S\|^m,$$

$$\left(\prod_{i=m+1}^{k}\sigma_i(E_S)^2\right)^{1/2} \leqslant \left(\frac{1}{k-m}\sum_{i=m+1}^{n}\sigma_i(E_S)^2\right)^{(k-m)/2}$$

$$= \left(\frac{1}{k-m}\min_{\mathrm{rank}(X)=m}\|E_S - X\|_F^2\right)^{(k-m)/2}$$

where we have used the well known result of Eckart and Young, see, e.g. Theorem B5 in [14], Section 10. Finally, by (25) and (29), it follows

$$\left(\prod_{i=1}^{k}\sigma_i(A_S^{-1})\right) \leqslant \frac{\exp(k/2)}{|\det A_S|}.$$

Substituting the latter three inequalities in (31) gives the needed generalization of (30):

$$\frac{\|r^k\|}{\|r^0\|} \leqslant \frac{\kappa}{|\det A_S|}\Big(3.3\|E_S\|\Big)^m \left(\frac{3.3}{\sqrt{k-m}}\min_{\mathrm{rank}(X)=m}\|E_S - X\|_F\right)^{k-m}. \quad (32)$$

One can readily apply the techniques of Section 5.2 and find that the corresponding iteration number bound will differ only by an additive term of the type $m + o(m)$. However, in certain cases, for some moderate value of m, it may hold

$$\min_{\mathrm{rank}(X)=m}\|E_S - X\|_F \ll \|E_S\|_F.$$

For instance, the use of pivot modifications in ILU algorithms is equivalent to the approximate triangular decomposition of a diagonally perturbed input matrix,

$$A_S + \widetilde{D} = P_S L_S U_S Q_S + \widetilde{E}_S,$$

where $\widetilde{D}$ is a diagonal matrix having only m nonzero elements (which may have considerably larger magnitudes compared to the ILU threshold parameter τ), and the entries of $\widetilde{E}_S$ satisfy the bound (3). Clearly, $\mathrm{rank}(\widetilde{D}) = m$, and therefore one finds

$$\min_{\mathrm{rank}(X)=m}\|E_S - X\|_F = \min_{\mathrm{rank}(X)=m}\|\widetilde{E}_S - \widetilde{D} - X\|_F \leqslant \|\widetilde{E}_S\|_F,$$

which quantity may really be considerably smaller than the Frobenius norm of the total residual $E_S = \widetilde{E}_S - \widetilde{D}$.

Hence, one can expect that m pivot modifications in ILU preconditioning may cost m additional GMRES iterations. It should be noted that the complete diagonal pivoting in ILU described in [12] usually requires a rather small, if any, number of pivot modifications.

Table 1. RAS(δ) scaling statistics for 18 test problems with $\delta = 0.8$ and $\delta = 0.1$

Problem	Size n	#Nonzeroes $nz(A)$	$\log \mathcal{K}(A)$	$\delta = 0.8$: #RAS iters. and $\log \mathcal{K}(A_S)$		$\delta = 0.1$: #RAS iters. and $\log \mathcal{K}(A_S)$	
gre_1107	1107	5664	4.487+02	7	4.067+02	35	3.828+02
qh1484	1484	6110	3.562+04	19	1.296+03	39	1.286+03
west2021	2021	7310	1.974+04	21	3.177+02	131	2.006+02
nnc1374	1374	8588	1.409+04	41	9.204+02	79	6.190+02
sherman3	5005	20033	8.789+04	3	1.177+03	4	1.177+03
sherman5	3312	20793	1.140+04	6	2.164+02	20	1.769+02
saylr4	3564	22316	5.760+03	3	4.985+03	4	4.984+03
lnsp3937	3937	25407	7.792+04	29	1.272+03	95	1.216+03
gemat12	4929	33044	1.107+04	13	3.214+03	92	3.154+03
dw8192	8192	41746	2.532+04	3	5.493+03	10	5.486+03
circuit3	12127	48137	3.583+04	22	7.603+03	62	7.545+03
cryg10K	10000	49699	4.562+04	4	6.393+03	27	6.389+03
fd18	16428	63406	2.173+05	17	4.490+03	93	4.015+03
bayer10	13436	71594	1.312+05	24	2.412+03	173	1.917+03
lhr04c	4101	82682	3.635+03	26	9.399+02	193	8.253+02
utm5940	5940	83842	5.625+03	12	3.406+03	54	3.332+03
bayer04	20545	85537	4.193+05	45	2.214+03	209	1.648+03
orani678	2529	90158	1.219+03	11	1.639+02	88	1.026+02

Table 2. RAS(0.1)+ILU(0.001) preconditioning statistics for the 18 test problems

Problem	Precond. density	Lower est. for $\|C_S^{-1}\|$	$\|E_S\|_F$	#GMRES iterations	#Estimated iterations
gre_1107	15.87	1.216+06	1.962−02	10	90
qh1484	2.72	2.108+05	4.889−03	3	221
west2021	2.73	2.824+11	1.235−02	4	47
nnc1374	8.18	1.395+07	1.268−02	21	129
sherman3	4.82	4.675+01	4.360−02	15	280
sherman5	2.15	3.809+00	2.801−02	6	49
saylr4	0.80	9.194+02	1.388−02	60	889
lnsp3937	5.56	1.346+04	4.833−02	8	294
gemat12	2.52	8.304+05	2.504−02	12	631
dw8192	4.01	6.982+01	3.978−02	15	1126
circuit3	1.56	6.375+03	1.688−02	12	1343
cryg10K	3.72	3.344+03	4.379−02	34	1315
fd18	11.73	2.278+30	6.855−02	30	921
bayer10	3.55	7.903+38	5.448−02	7	452
lhr04c	2.04	6.725+03	6.300−02	18	218
utm5940	5.65	4.073+04	9.949−02	30	830
bayer04	2.97	1.105+38	5.123−02	5	390
orani678	0.95	3.656+00	5.503−02	6	37

Table 3. RAS(0.1)+ILU(0.01) preconditioning statistics for the 18 test problems

Problem	Precond. density	Lower est. for $\|C_S^{-1}\|$	$\|E_S\|_F$	#GMRES iterations	#Estimated iterations
gre_1107	13.73	3.281+03	2.708−01	21	158
qh1484	2.24	2.243+06	8.372−02	21	341
west2021	2.37	9.970+02	1.103−01	8	73
nnc1374	7.27	2.489+04	1.775−01	48	212
sherman3	2.62	3.451+01	3.554−01	35	438
sherman5	1.42	4.039+00	2.624−01	11	85
saylr4	0.78	9.163+02	4.895−02	59	1064
lnsp3937	3.65	3.564+05	4.329−01	16	475
gemat12	1.74	9.887+17	2.667−01	48	963
dw8192	2.73	2.662+01	3.506−01	45	1670
circuit3	1.23	1.483+03	1.952−01	56	1969
cryg10K	2.34	1.046+03	3.828−01	78	1949
fd18	9.33	1.826+45	7.006−01	93	1507
bayer10	2.52	5.266+50	5.658−01	12 †	755
lhr04c	1.01	1.590+05	6.587−01	55	392
utm5940	2.52	2.457+02	8.077−01	68	1338
bayer04	2.28	4.151+32	5.299−01	9	651
orani678	0.38	2.854+00	4.440−01	8	70

Table 4. RAS(0.1)+ILU(0.07) preconditioning statistics for the 18 test problems

Problem	Precond. density	Lower est. for $\|C_S^{-1}\|$	$\|E_S\|_F$	#GMRES iterations	#Estimated iterations
gre_1107	9.30	9.996+02	2.291+00	50	409
qh1484	1.73	2.163+05	7.746−01	62	596
west2021	1.87	7.695+12	1.172+00	18	177
nnc1374	5.70	8.907+04	1.574+00	610 †	452
sherman3	1.56	3.132+01	1.919+00	73	800
sherman5	1.00	1.770+00	1.234+00	22	168
saylr4	0.76	8.869+02	1.444−01	60	1279
lnsp3937	2.23	5.106+04	2.628+00	36	966
gemat12	1.12	1.932+10	2.104+00	174	1780
dw8192	1.42	5.544+01	1.808+00	205	2627
circuit3	0.98	2.001+02	1.653+00	163	3321
cryg10K	1.55	1.627+02	2.352+00	175	3272
fd18	5.61	9.233+59	4.458+00	222	3048
bayer10	1.66	7.786+28	3.764+00	47	1637
lhr04c	0.39	5.661+03	3.397+00	92	874
utm5940	0.96	6.650+01	4.138+00	137	2557
bayer04	1.53	7.189+51	3.625+00	104 †	1442
orani678	0.15	2.408+00	2.180+00	14	191

Table 5. RAS(0.8)+ILU(0.001) preconditioning statistics for the 18 test problems

Problem	Precond. density	Lower est. for $\|C_S^{-1}\|$	$\|E_S\|_F$	#GMRES iterations	#Estimated iterations
gre_1107	14.98	1.442+07	2.157−02	14	96
qh1484	3.10	1.665+13	6.356−03	4	230
west2021	2.94	1.236+06	1.079−02	6	69
nnc1374	9.18	4.938+04	6.882−03	36	169
sherman3	4.81	6.704+01	4.341−02	15	280
sherman5	2.19	3.916+00	2.643−02	6	58
saylr4	.79	9.445+02	1.384−02	60	889
lnsp3937	5.29	6.959+02	4.531−02	8	302
gemat12	2.55	1.932+04	2.515−02	13	642
dw8192	3.96	4.462+01	3.933−02	16	1125
circuit3	1.57	4.101+03	1.666−02	12	1350
cryg10K	3.73	4.367+04	4.396−02	41	1316
fd18	13.61	1.745+41	6.887−02	24	1021
bayer10	4.32	1.735+27	4.986−02	8	549
lhr04c	2.21	9.265+08	6.237−02	19	244
utm5940	6.14	3.131+03	1.012−01	32	849
bayer04	3.74	3.142+31	4.947−02	10	507
orani678	1.06	1.589+01	5.903−02	5	54

Table 6. RAS(0.8)+ILU(0.01) preconditioning statistics for the 18 test problems

Problem	Precond. density	Lower est. for $\|C_S^{-1}\|$	$\|E_S\|_F$	#GMRES iterations	#Estimated iterations
gre_1107	12.28	4.509+05	2.640−01	28	165
qh1484	2.49	5.968+10	8.875−02	19	348
west2021	2.40	2.354+04	1.440−01	14	114
nnc1374	8.56	5.503+06	1.594−01	168 †	291
sherman3	2.62	3.515+01	3.559−01	35	438
sherman5	1.46	4.191+00	2.577−01	11	99
saylr4	.78	9.378+02	4.929−02	59	1065
lnsp3937	3.45	1.648+02	4.110−01	15	487
gemat12	1.78	2.472+14	2.602−01	35	973
dw8192	2.72	4.546+01	3.481−01	48	1669
circuit3	1.23	1.281+03	2.654−01	44	2106
cryg10K	2.35	7.093+02	3.858−01	78	1953
fd18	10.41	7.823+41	7.219−01	139	1673
bayer10	3.41	1.119+39	4.974−01	28 †	886
lhr04c	1.26	3.094+06	6.223−01	41	428
utm5940	2.80	5.501+03	8.158−01	70	1367
bayer04	3.05	1.450+54	5.359−01	24	840
orani678	.41	5.877+01	4.694−01	8	99

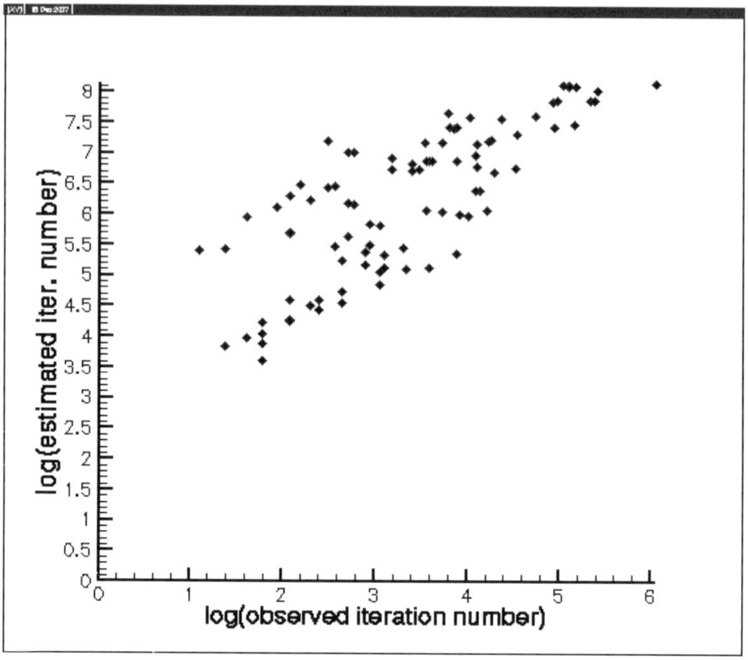

Fig. 1. Set of points $(\log k, \log k_{est})$ depicting the correlation between the observed and estimated iteration numbers

6 Numerical experiments

The correctness of the above convergence estimate (30) has also been tested numerically using several small-sized "hard" test matrices taken from the University of Florida Sparse Matrix Collection [1]. The limitation on the sizes of the matrices was set in order to make easier the "exact" LU-factorization of the coefficient matrix A which was used for the evaluation of $\log|\det A|$.

The linear systems were solved with an artificial right-hand side $b = Ax_*$, where the components of the exact solution were chosen as $x_*(i) = i/n$, $i = 1,2,\ldots,n$. The initial guess was always chosen as $x_0 = 0$ and the stopping criterion in GMRES iteration was set as $\|\tilde{r}_k\| \leqslant \varepsilon\|r_0\|$ with $\varepsilon = 10^{-8}$, where $\|\tilde{r}_k\|$ is the *estimated* GMRES residual norm. If the matrix A is very ill-conditioned and the preconditioning is not sufficiently strong (e.g. if the ILU threshold parameter τ is set too large), the true residual norm can be much larger than the estimated one (due to the calculations in finite precision). In the cases of a complete failure when $\|r_k\| > \|r_0\|$, we put the "†" mark after the GMRES iteration number in Tables 2-7. In the GMRES(m) scheme, we took $m = 900$ and used approximate LU preconditioning with the "best" default tuning of the pre-ordering and pivoting (see [12] for more detail).

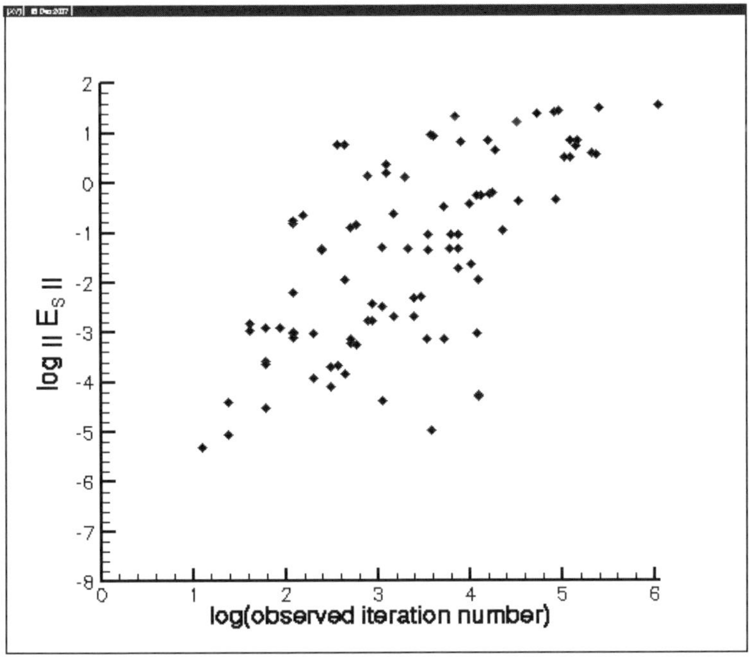

Fig. 2. Set of points $(\log k, \log \|E_S\|_F)$ depicting the correlation between the observed iteration number and the Frobenius norm of the scaled ILU error

Note: It has been observed (especially in calculations with "nnc1374" matrix), that much better results, in the sense of closeness between the "iterative" and the "true" residual (the latter is $r_k = b - Ax_k$), are obtained using the BiCGStab iterations [22]. Probably, an improved GMRES implementation [23] (where the plain rotations are replaced by elementary reflections) would be more competitive.

In the scaling procedure, the RAS stopping criterion was

$$\max\left(\frac{\max_i \sum_j (A_S)_{ij}^2}{\min_i \sum_j (A_S)_{ij}^2}, \frac{\max_j \sum_i (A_S)_{ij}^2}{\min_j \sum_i (A_S)_{ij}^2}\right) \leqslant 1 + \delta$$

with $\delta = 0.1,\ 0.8$, and the ILU threshold parameter τ was set to $\tau = 0.001$, 0.01, 0.07.

We present numerical results for 18 sample problems from the above mentioned collection. The problems are taken form the subset of 60 matrices which has been used in [12] for testing of ILU preconditionings. Hence, the statistics on the total of $2 \times 3 \times 18 = 108$ test runs are reported in Tables 2- 7.

In Table 1 we list the names of the test matrices with their sizes and number of nonzeroes, and present values for the quality measure $\mathcal{K}(A_S)$ which charac-

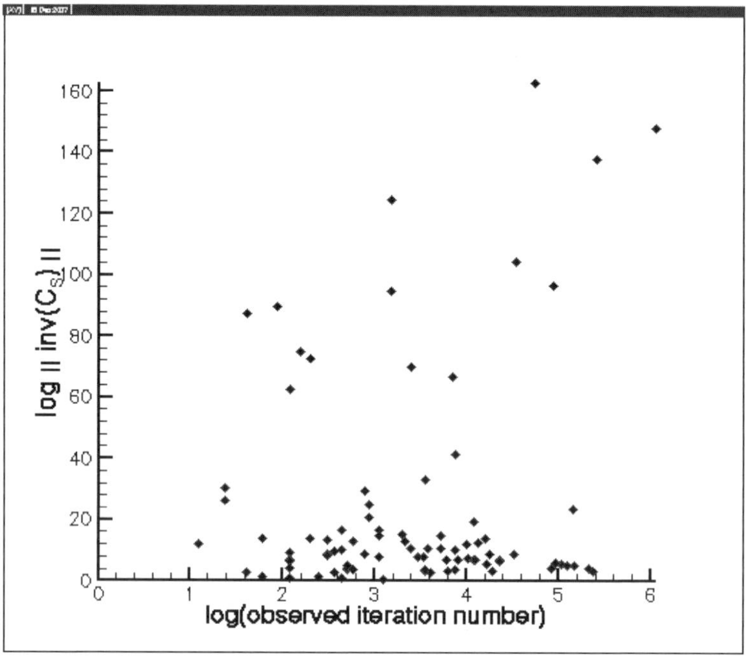

Fig. 3. Set of points $(\log k, \log \|C_S^{-1}\|)$ depicting the (absence of) correlation between the observed iteration number and the (lower bound for) spectral norm of the inverse scaled preconditioner

terize the result of scaling. Clearly, the smaller δ, the smaller is $\mathcal{K}(A_S)$, which corresponds to better scaling. However, the number of RAS iterations increases considerably when refining the precision from $\delta = 0.8$ to $\delta = 0.1$.

In GMRES(m) we took $m = 900$ and used approximate LU preconditioning as in [12]. All computing was done in double precision. The iteration number counts and other related data are given in Tables 2-7. For each test run we give:

1. The resulting preconditioner density $nz(L+U)/nz(A)$;
2. The lower bound on the spectral norm of C_S^{-1} (taken as $(v^T U_S^{-1})(L_S^{-1}u)/n$, where the components of the vectors u and v are 1 or -1 with signs determined in the course of back substitutions to obtain a local maximum at each step);
3. The Frobenius norm of the scaled ILU residual E_S;
4. The actual number k of GMRES iterations;
5. The upper bound k_{est} for the iteraton number k obtained from estimate (30) with $\kappa = 1$ in the same way as in the proof of Theorem 2.

First of all, the results presented give another confirmation that good prescaling can be useful for the improvement of the ILU-GMRES performance.

Table 7. RAS(0.8)+ILU(0.07) preconditioning statistics for the 18 test problems

Problem	Precond. density	Lower est. for $\|C_S^{-1}\|$	$\|E_S\|_F$	#GMRES iterations	#Estimated iterations
gre_1107	8.25	9.027+05	2.353+00	67	433
qh1484	1.92	2.938+08	7.703−01	59	598
west2021	1.85	4.015+06	1.124+00	27	236
nnc1374	7.41	1.838+06	1.650+00	720†	619
sherman3	1.56	3.129+01	1.922+00	73	801
sherman5	1.01	1.706+00	1.436+00	22	210
saylr4	.76	9.742+02	1.440−01	60	1279
lnsp3937	2.17	1.929+01	2.523+00	37	975
gemat12	1.16	5.290+15	2.142+00	>900†	1820
dw8192	1.40	2.341+01	1.777+00	215	2614
circuit3	1.00	2.395+02	1.649+00	153	3340
cryg10K	1.56	1.694+02	2.350+00	163	3273
fd18	7.07	2.029+64	4.708+00	420	3405
bayer10	2.26	3.880+70	4.046+00	114	2006
lhr04c	.46	3.720+21	3.887+00	>900†	1044
utm5940	1.07	4.046+02	4.218+00	145	2626
bayer04	2.03	7.703+49	3.686+00	121†	1789
orani678	.15	1.685+01	2.152+00	13	238

Next we adress consistency analysis for the above presented GMRES convergence theory. One can see that for the cases considered, upper bound (28) is, on average, a twenty times overestimation of the actual iteration count. However, the relative variations of the upper bound (from one problem to another) correlate with the actual iteration numbers rather well, as is illustrated in Figure 1. (In Figs. 1–3 we have used only the data on 99 out of 108 test runs, thus ignoring the breakdown occasions marked by "†".) A much weaker correlation is observed between the Frobenius norm of the scaled ILU residual $\|E_S\|_F$ and the actual GMRES iteration number, see Figure 2. Furthermore, the conventional indicator $\|C_S^{-1}\|$ does not demonstrate any correlation with the GMRES iteration number. Note that, if there is a hidden dependence, for instance, of the form $k = \alpha k_{est}^{\beta}$, then the points $(\log k, \log k_{est})$ lie at the corresponding straight line. The reader may clearly observe that only the discrete set shown in Figure 1 can safely be interpreted as a "linear function plus noise". More precisely, one can find two intersecting straight lines in Figure 1 which, in fact, correspond to two different classes of test problems.

7 Conclusion

First, a theoretical justification is found for the standard pre-scaling technique related to the ILU factorization, with implications for practical implementation. (Namely, a more accurate evaluation of D_L and D_R may be useful, or even sparse matrices with more than n nonzeroes can be used instead of the diagonal ones.) Second, an estimate for the reduction of the *original* (unscaled) residual is obtained in terms of the *scaled* ILU error. These results can readily be used as a working tool for the construction of efficient two-stage preconditionings for Krylov subspace methods.

Acknowledgments

The author thanks Eugene Tyrtyshnikov for his kind interest in this research and for his valuable assistance in related presentations.

References

1. *Univ. of Florida Sparse Matrix Collection.* http://www.cise.ufl.edu/research/sparse/matrices/
2. V.F. de Almeida, A.M. Chapman, and J.J. Derby, *On Equilibration and Sparse Factorization of Matrices Arising in Finite Element Solutions of Partial Differential Equations*, Numer. Methods Partial Different. Equ., 16 (2000), pp. 11–29.
3. O. Axelsson and I. Kaporin, *On the sublinear and superlinear rate of convergence of conjugate gradient methods*, Numerical Algorithms, 25 (2000), pp. 1–22.
4. O. Axelsson and G. Lindskog, *On the rate of convergence of the preconditioned conjugate gradient method*, Numerische Mathematik, 48 (1986), pp. 499–523.
5. T. Davis, http://www.cise.ufl.edu/research/sparse/umfpack/
6. S.L. Campbell, I.C. Ipsen, C.T. Kelley, and C.D. Meyer, *GMRES and the Minimal Polynomial*, BIT, 36 (1996), pp. 664–675.
7. A. Jennings, *Influence of the eigenvalue spectrum on the convergence rate of the conjugate gradient method*, Journal of the Institute of Mathematics and Its Applications, 20 (1977), pp. 61–72.
8. I. Kaporin, *Explicitly preconditioned conjugate gradient method for the solution of nonsymmetric linear systems*, Int. J. Computer Math., 40 (1992), pp. 169–187.
9. I. Kaporin, *New convergence results and preconditioning strategies for the conjugate gradient method*, Numer. Linear Algebra with Appls., 1 (1994), pp. 179–210.
10. I. Kaporin, *High quality preconditioning of a general symmetric positive matrix based on its* $U^TU + U^TR + R^TU$*-decomposition*, Numerical Linear Algebra Appl., 5 (1998), pp. 484–509.
11. I. Kaporin, *Superlinear convergence in minimum residual iterations*, Numerical Linear Algebra Appl., 12 (2005), pp. 453–470.

12. I. KAPORIN, *Scaling, Reordering, and Diagonal Pivoting in ILU Preconditionings*, Russian Journal of Numerical Analysis and Mathematical Modelling, 22 (2007), pp. 341–375.
13. O. E. LIVNE AND G. H. GOLUB, *Scaling by Binormalization*, Numer. Alg., 35 (2004), pp. 97–120.
14. A. W. MARSHALL AND I. OLKIN, *Inequalities: Theory of Majorization and its Applications*, Academic Press, New York, 1979.
15. I. MORET *A note on the superlinear convergence of GMRES*, SIAM Journal on Numerical Analysis, 34 (1997), pp. 513–516.
16. Y. SAAD AND M. H. SCHULTZ, GMRES*: A generalized minimal residual method for solving nonsymmetric linear systems*, SIAM J. Sci. Statist. Comput., 7 (1986), pp. 856–869.
17. M. H. SCHNEIDER AND S. A. ZENIOS, *A comparative study of algorithms for matrix balancing*, Operations Research, 38 (1990), pp. 439–455.
18. V. SIMONCINI AND D.B. SZYLD, *On the Occurrence of Superlinear Convergence of Exact and Inexact Krylov Subspace Methods*, Dept. Math., Temple University Report 03-3-13; Philadelphia, Pennsylvania, March 2003, 25pp.
19. E. E. TYRTYSHNIKOV, *A unifying approach to some old and new theorems on distribution and clustering*, Linear Algebra and Applications, 232 (1996), pp. 1–43.
20. E. E. TYRTYSHNIKOV, *Krylov subspace methods and minimal residuals*, J. Numer. Math. (2007, submitted).
21. H. A. VAN DER VORST AND C. VUIK, *The superlinear convergence behaviour of GMRES*, Journal of Computational and Applied Mathematics, 48 (1993), pp. 327–341.
22. H. A. VAN DER VORST, *Bi-CGStab: a fast and smoothly converging variant of Bi-CG for the solution of nonsymmetric linear systems*, SIAM J. Sci. Statist. Comput., 13 (1992), pp. 631–644.
23. H. F. WALKER, *Implementation of the* GMRES *method using Householder transformations*, SIAM J. Sci. Statist. Comput., 9 (1988), pp. 152–163.
24. R. WINTHER, *Some superlinear convergence results for the conjugate gradient method*, SIAM J. Numer. Analysis, 17 (1980), pp. 14–17.

Toeplitz and Toeplitz-Block-Toeplitz Matrices and Their Correlation with Syzygies of Polynomials

Houssam Khalil[1,*], Bernard Mourrain[2], and Michelle Schatzman[1,†]

[1] Institut Camille Jordan, 43 boulevard du 11 novembre 1918, 69622 Villeurbanne Cedex, France
[*] khalil@math.univ-lyon1.fr
[†] schatz@math.univ-lyon1.fr
[2] INRIA, GALAAD team, 2004 route des Lucioles, BP 93, 06902 Sophia Antipolis Cedex, France
mourrain@sophia.inria.fr

Abstract. In this paper, we re-investigate the resolution of Toeplitz systems $T u = g$, from a new point of view, by correlating the solution of such problems with syzygies of polynomials or moving lines. We show an explicit connection between the generators of a Toeplitz matrix and the generators of the corresponding module of syzygies. We show that this module is generated by two elements of degree n and the solution of $T u = g$ can be reinterpreted as the remainder of an explicit vector depending on g, by these two generators.

This approach extends naturally to multivariate problems and we describe for Toeplitz-block-Toeplitz matrices, the structure of the corresponding generators.

Keywords: toeplitz matrix, rational interpolation, syzygie.

1 Introduction

Structured matrices appear in various domains, such as scientific computing, signal processing, ... They usually express, in a linearize way, a problem which depends on less parameters than the number of entries of the corresponding matrix. An important area of research is devoted to the development of methods for the treatment of such matrices, which depend on the actual parameters involved in these matrices.

Among well-known structured matrices, Toeplitz and Hankel structures have been intensively studied [5, 6]. Nearly optimal algorithms are known for the multiplication or the resolution of linear systems, for such structure. Namely, if A is a Toeplitz matrix of size n, multiplying it by a vector or solving a linear system with A requires $\tilde{\mathcal{O}}(n)$ arithmetic operations (where $\tilde{\mathcal{O}}(n) = \mathcal{O}(n \log^c(n))$ for some $c > 0$) [2, 12]. Such algorithms are called super-fast, in opposition with fast algorithms requiring $\mathcal{O}(n^2)$ arithmetic operations.

The fundamental ingredients in these algorithms are the so-called generators [6], encoding the minimal information stored in these matrices, and on which the matrix transformations are translated. The correlation with other types of structured matrices has also been well developed in the literature [10, 9], allowing to treat so efficiently other structures such as Vandermonde or Cauchy-like structures.

Such problems are strongly connected to polynomial problems [4, 1]. For instance, the product of a Toeplitz matrix by a vector can be deduced from the product of two univariate polynomials, and thus can be computed efficiently by evaluation-interpolation techniques, based on FFT. The inverse of a Hankel or Toeplitz matrix is connected to the Bezoutian of the polynomials associated to their generators.

However, most of these methods involve univariate polynomials. So far, few investigations have been pursued for the treatment of multilevel structured matrices [11], related to multivariate problems. Such linear systems appear for instance in resultant or in residue constructions, in normal form computations, or more generally in multivariate polynomial algebra. We refer to [8] for a general description of such correlations between multi-structured matrices and multivariate polynomials. Surprisingly, they also appear in numerical scheme and preconditionners. A main challenge here is to devise super-fast algorithms of complexity $\tilde{\mathcal{O}}(n)$ for the resolution of multi-structured systems of size n.

In this paper, we consider block-Toeplitz matrices, where each block is a Toeplitz matrix. Such a structure, which is the first step to multi-level structures, is involved in many bivariate problems, or in numerical linear problems.We re-investigate first the resolution of Toeplitz systems $T\,u = g$, from a new point of view, by correlating the solution of such problems with syzygies of polynomials or moving lines. We show an explicit connection between the generators of a Toeplitz matrix and the generators of the corresponding module of syzygies. We show that this module is generated by two elements of degree n and the solution of $T\,u = g$ can be reinterpreted as the remainder of an explicit vector depending on g, by these two generators.

This approach extends naturally to multivariate problems and we describe for Toeplitz-block-Toeplitz matrices, the structure of the corresponding generators. In particular, we show the known result that the module of syzygies of k non-zero bivariate polynomials is free of rank $k-1$, by a new elementary proof.

Exploiting the properties of moving lines associated to Toeplitz matrices, we give a new point of view to resolve a Toeplitz-block-Toeplitz system.

In the next section we study the scalar Toeplitz case. In Section 3 we consider the Toeplitz-block-Toeplitz case.

Let $R = \mathbb{K}[x]$. For $n \in \mathbb{N}$, we denote by $\mathbb{K}[x]_n$ the vector space of polynomials of degree $\leqslant n$. Let $L = \mathbb{K}[x, x^{-1}]$ be the set of Laurent polynomials in the variable x. For any polynomial $p = \sum_{i=-m}^{n} p_i\, x^i \in L$, we denote by p^+ the sum of terms with positive exponents: $p^+ = \sum_{i=0}^{n} p_i\, x^i$ and by p^-, the sum of terms with strictly negative exponents: $p^- = \sum_{i=-m}^{-1} p_i\, x^i$. We have $p = p^+ + p^-$.

For $n \in \mathbb{N}$, we denote by $\mathfrak{U}_n = \{\omega; \omega^n = 1\}$ the set of roots of unity of order n.

2 Univariate case

We begin by the univariate case and the following problem:

Problem 1. Given a Toeplitz matrix $T = (t_{i-j})_{i,j=0}^{n-1} \in \mathbb{K}^{n\times n}$ ($T = (T_{ij})_{i,j=0}^{n-1}$ with $T_{ij} = t_{i-j}$) of size n and $g = (g_0, \ldots, g_{n-1}) \in \mathbb{K}^n$, find $u = (u_0, \ldots, u_{n-1}) \in \mathbb{K}^n$ such that

$$Tu = g. \tag{1}$$

Let $E = \{1, \ldots, x^{n-1}\}$, and Π_E be the projection of R on the vector space generated by E, along $\langle x^n, x^{n+1}, \ldots\rangle$.

Definition 1. *We define the following polynomials:*

- $T(x) = \sum_{i=-n+1}^{n-1} t_i x^i$,
- $\tilde{T}(x) = \sum_{i=0}^{2n-1} \tilde{t}_i x^i$ *with* $\tilde{t}_i = \begin{cases} t_i & \text{if } i < n \\ t_{i-2n} & \text{if } i \geqslant n \end{cases}$,
- $u(x) = \sum_{i=0}^{n-1} u_i x^i$, $g(x) = \sum_{i=0}^{n-1} g_i x^i$.

Notice that $\tilde{T} = T^+ + x^{2n} T^-$ and $T(w) = \tilde{T}(w)$ if $w \in \mathfrak{U}_{2n}$. We also have (see [8])

$$Tu = g \Leftrightarrow \Pi_E(T(x)u(x)) = g(x).$$

For any polynomial $u \in \mathbb{K}[x]$ of degree d, we denote it as $u(x) = \underline{u}(x) + x^n \overline{u}(x)$ with $\deg(\underline{u}) \leqslant n-1$ and $\deg(\overline{u}) \leqslant d-n$ if $d \geqslant n$ and $\overline{u} = 0$ otherwise. Then, we have

$$\begin{aligned} T(x)\,u(x) &= T(x)\underline{u}(x) + T(x)x^n\overline{u}(x) \\ &= \Pi_E(T(x)\underline{u}(x)) + \Pi_E(T(x)x^n\overline{u}(x)) \\ &\quad +(\alpha_{-n+1}x^{-n+1} + \cdots + \alpha_{-1}x^{-1}) \\ &\quad +(\alpha_n x^n + \cdots + \alpha_{n+m}x^{n+m}) \\ &= \Pi_E(T(x)\underline{u}(x)) + \Pi_E(T(x)x^n\overline{u}(x)) \\ &\quad +x^{-n+1}A(x) + x^n B(x), \end{aligned} \tag{2}$$

with $m = \max(n-2, d-1)$,

$$\begin{aligned} A(x) &= \alpha_{-n+1} + \cdots + \alpha_{-1}x^{n-2}, \\ B(x) &= \alpha_n + \cdots + \alpha_{n+m}x^m. \end{aligned} \tag{3}$$

See [8] for more details, on the correlation between structured matrices and (multivariate) polynomials.

2.1 Moving lines and Toeplitz matrices

We consider here another problem, related to interesting questions in Effective Algebraic Geometry.

Problem 2. Given three polynomials $a, b, c \in R$ respectively of degree $< l, < m, < n$, find three polynomials $p, q, r \in R$ of degree $< \nu - l, < \nu - m, < \nu - n$, such that

$$a(x)\,p(x) + b(x)\,q(x) + c(x)\,r(x) = 0. \tag{4}$$

We denote by $\mathcal{L}(a, b, c)$ the set of $(p, q, r) \in \mathbb{K}[x]^3$ which are solutions of (4). It is a $\mathbb{K}[x]$-module of $\mathbb{K}[x]^3$. The solutions of the problem (2) are $\mathcal{L}(a, b, c) \cap \mathbb{K}[x]_{\nu-l-1} \times \mathbb{K}[x]_{\nu-m-1} \times \mathbb{K}[x]_{\nu-n-1}$.

Given a new polynomial $d(x) \in \mathbb{K}[x]$, we denote by $\mathcal{L}(a, b, c; d)$ the set of $(p, q, r) \in \mathbb{K}[x]^3$ such that

$$a(x)\,p(x) + b(x)\,q(x) + c(x)\,r(x) = d(x).$$

Theorem 1. *For any non-zero vector of polynomials $(a, b, c) \in \mathbb{K}[x]^3$, the $\mathbb{K}[x]$-module $\mathcal{L}(a, b, c)$ is free of rank 2.*

Proof. By the Hilbert's theorem, the ideal I generated by (a, b, c) has a free resolution of length at most 1, that is of the form:

$$0 \to \mathbb{K}[x]^p \to \mathbb{K}[x]^3 \to \mathbb{K}[x] \to \mathbb{K}[x]/I \to 0.$$

As $I \neq 0$, for dimensional reasons, we must have $p = 2$.

Definition 2. *A μ-base of $\mathcal{L}(a, b, c)$ is a basis (p, q, r), (p', q', r') of $\mathcal{L}(a, b, c)$, with (p, q, r) of minimal degree μ.*

Notice if μ_1 is the smallest degree of a generator and μ_2 the degree of the second generator (p', q', r'), we have $d = \max(\deg(a), \deg(b), \deg(c)) = \mu_1 + \mu_2$. Indeed, we have

$$\begin{aligned} 0 \to \mathbb{K}[x]_{\nu-d-\mu_1} \oplus \mathbb{K}[x]_{\nu-d-\mu_2} \to \\ \mathbb{K}[x]^3_{\nu-d} \to \mathbb{K}[x]_\nu \to \mathbb{K}[x]_\nu/(a, b, c)_\nu \to 0, \end{aligned}$$

for $\nu >> 0$. As the alternate sum of the dimension of the $\mathbb{K}$-vector spaces is zero and $\mathbb{K}[x]_\nu/(a, b, c)_\nu$ is 0 for $\nu >> 0$, we have

$$\begin{aligned} 0 &= 3\,(d - \nu - 1) + \nu - \mu_1 - d + 1 + \nu - \mu_2 - d + 1 + \nu + 1 \\ &= d - \mu_1 - \mu_2. \end{aligned}$$

For $\mathcal{L}(\tilde{T}(x), x^n, x^{2n} - 1)$, we have $\mu_1 + \mu_2 = 2n$. We are going to show now that in fact $\mu_1 = \mu_2 = n$:

Proposition 1. *The* $\mathbb{K}[x]$*-module* $\mathcal{L}(\tilde{T}(x), x^n, x^{2n}-1)$ *has a* n*-basis.*

Proof. Consider the map

$$\begin{aligned} \mathbb{K}[x]_{n-1}^3 &\to \mathbb{K}[x]_{3n-1} \\ (p(x), q(x), r(x)) &\mapsto \tilde{T}(x)p(x) + x^n q(x) + (x^{2n}-1)r(x) \end{aligned} \tag{5}$$

which $3n \times 3n$ matrix is of the form

$$S := \begin{pmatrix} T_0 & \mathbf{0} & -\mathbb{I}_n \\ T_1 & \mathbb{I}_n & \mathbf{0} \\ T_2 & \mathbf{0} & \mathbb{I}_n \end{pmatrix} \tag{6}$$

where T_0, T_1, T_2 are the coefficient matrices of $(\tilde{T}(x), x\,\tilde{T}(x), \ldots, x^n \tilde{T}(x))$, respectively for the list of monomials $(1, \ldots, x^{n-1})$, $(x^n, \ldots, x^{2n-1})$, $(x^{2n}, \ldots, x^{3n-1})$. Notice in particular that $T = T_0 + T_2$

Reducing the first rows of $(T_0|\mathbf{0}| - \mathbb{I}_n)$ by the last rows $(T_2|\mathbf{0}|\mathbb{I}_n)$, we replace it by the block $(T_0 + T_2|\mathbf{0}|\mathbf{0})$, without changing the rank of S. As $T = T_0 + T_2$ is invertible, this shows that the matrix S is of rank $3n$. Therefore, there is no syzygies in degree $n-1$. As the sum $2n = \mu_1 + \mu_2$ and $\mu_1 \leqslant n, \mu_2 \leqslant n$ where μ_1, μ_2 are the smallest degree of a pair of generators of $\mathcal{L}(\tilde{T}(x), x^n, x^{2n}-1)$ of degree $\leqslant n$, we have $\mu_1 = \mu_2 = n$. Thus there exist two linearly independent syzygies (u_1, v_1, w_1), (u_2, v_2, w_2) of degree n, which generate $\mathcal{L}(\tilde{T}(x), x^n, x^{2n}-1)$.

A similar result can also be found in [12], but the proof much longer than this one, is based on interpolation techniques and explicit computations. Let us now describe how to construct explicitly two generators of $\mathcal{L}(\tilde{T}(x), x^n, x^{2n}-1)$ of degree n (see also [12]).

As $\tilde{T}(x)$ is of degree $\leqslant 2n-1$ and the map (5) is a surjective function, there exists $(u, v, w) \in \mathbb{K}[x]_{n-1}^3$ such that

$$\tilde{T}(x)u(x) + x^n v(x) + (x^{2n}-1)\,w = \tilde{T}(x)x^n, \tag{7}$$

we deduce that $(u_1, v_1, w_1) = (x^n - u, -v, -w) \in \mathcal{L}(\tilde{T}(x), x^n, x^{2n}-1)$.

As there exists $(u', v', w') \in \mathbb{K}[x]_{n-1}^3$ such that

$$\tilde{T}(x)u'(x) + x^n v'(x) + (x^{2n}-1)\,w' = 1 = x^n\,x^n - (x^{2n}-1) \tag{8}$$

we deduce that $(u_2, v_2, w_2) = (-u', x^n - v', -w' - 1) \in \mathcal{L}(\tilde{T}(x), x^n, x^{2n}-1)$.

Now, the vectors (u_1, v_1, w_1), (u_2, v_2, w_2) of $\mathcal{L}(\tilde{T}(x), x^n, x^{2n}-1)$ are linearly independent since by construction, the coefficient vectors of x^n in (u_1, v_1, w_1) and (u_2, v_2, w_2) are respectively $(1, 0, 0)$ and $(0, 1, 0)$.

Proposition 2. *The vector* u *is solution of* (1) *if and only if there exist* $v(x) \in \mathbb{K}[x]_{n-1}, w(x) \in \mathbb{K}[x]_{n-1}$ *such that*

$$(u(x), v(x), w(x)) \in \mathcal{L}(\tilde{T}(x), x^n, x^{2n}-1; g(x)).$$

Proof. The vector u is solution of (1) if and only if we have

$$\Pi_E(T(x)u(x)) = g(x).$$

As $u(x)$ is of degree $\leqslant n-1$, we deduce from (2) and (3) that there exist polynomial $A(x) \in \mathbb{K}[x]_{n-2}$ and $B(x) \in \mathbb{K}[x]_{n-1}$ such that

$$T(x)u(x) - x^{-n+1}A(x) - x^n B(x) = g(x).$$

By evaluation at the roots $\omega \in \mathfrak{U}_{2n}$, and since $\omega^{-n} = \omega^n$ and $\tilde{T}(\omega) = T(\omega)$ for $\omega \in \mathfrak{U}_n$, we have

$$\tilde{T}(\omega)u(\omega) + \omega^n v(\omega) = g(\omega), \forall \omega \in \mathfrak{U}_{2n}(\omega),$$

with $v(x) = -x\,A(x) - B(x)$ of degree $\leqslant n-1$. We deduce that there exists $w(x) \in \mathbb{K}[x]$ such that

$$\tilde{T}(x)u(x) + x^n v(x) + (x^{2n}-1)w(x) = g(x).$$

Notice that $w(x)$ is of degree $\leqslant n-1$, because $(x^{2n}-1)\,w(x)$ is of degree $\leqslant 3n-1$.

Conversely, a solution $(u(x), v(x), w(x)) \in \mathcal{L}(\tilde{T}(x), x^n, x^{2n}-1; g(x)) \cap \mathbb{K}[x]^3_{n-1}$ implies a solution $(u, v, w) \in \mathbb{K}^{3\,n}$ of the linear system:

$$S \begin{pmatrix} u \\ v \\ w \end{pmatrix} = \begin{pmatrix} g \\ 0 \\ 0 \end{pmatrix}$$

where S is has the block structure (6), so that $T_2\,u + w = 0$ and $T_0\,u - w = (T_0 + T_2)u = g$. As we have $T_0 + T_2 = T$, the vector u is a solution of (1), which ends the proof of the proposition.

2.2 Euclidean division

As a consequence of proposition 1, we have the following property:

Proposition 3. *Let* $\{(u_1, v_1, w_1), (u_2, v_2, w_2)\}$ *a* n*-basis of* $\mathcal{L}(\tilde{T}(x), x^n, x^{2n}-1)$, *the remainder of the division of* $\begin{pmatrix} 0 \\ x^n\,g \\ g \end{pmatrix}$ *by* $\begin{pmatrix} u_1 & u_2 \\ v_1 & v_2 \\ w_1 & w_2 \end{pmatrix}$ *is the vector solution given in the proposition* (2).

Proof. The vector $\begin{pmatrix} 0 \\ x^n\,g \\ -g \end{pmatrix} \in \mathcal{L}(\tilde{T}(x), x^n, x^{2\,n}-1; g)$ (a particular solution). We divide it by $\begin{pmatrix} u_1 & u_2 \\ v_1 & v_2 \\ w_1 & w_2 \end{pmatrix}$ we obtain

$$\begin{pmatrix} u \\ v \\ w \end{pmatrix} = \begin{pmatrix} 0 \\ x^n\,g \\ g \end{pmatrix} - \begin{pmatrix} u_1 & u_2 \\ v_1 & v_2 \\ w_1 & w_2 \end{pmatrix} \begin{pmatrix} p \\ q \end{pmatrix}$$

(u,v,w) is the remainder of division, thus $(u,v,w) \in \mathbb{K}[x]^3_{n-1} \cap \mathcal{L}(\tilde{T}(x), x^n, x^{2n} - 1; g)$. However (u,v,w) is the unique vector $\in \mathbb{K}[x]^3_{n-1} \cap \mathcal{L}(\tilde{T}(x), x^n, x^{2n} - 1; g)$ because if there is another vector then their difference is in $\mathcal{L}(\tilde{T}(x), x^n, x^{2n} - 1) \cap \mathbb{K}[x]^3_{n-1}$ which is equal to $\{(0,0,0)\}$.

Problem 3. Given a matrix and a vector of polynomials $\begin{pmatrix} e(x) & e'(x) \\ f(x) & f'(x) \end{pmatrix}$ of degree n, and $\begin{pmatrix} p(x) \\ q(x) \end{pmatrix}$ of degree $m \geqslant n$, such that $\begin{pmatrix} e_n & e'_n \\ f_n & f'_n \end{pmatrix}$ is invertible; find the remainder of the division of $\begin{pmatrix} p(x) \\ q(x) \end{pmatrix}$ by $\begin{pmatrix} e(x) & e'(x) \\ f(x) & f'(x) \end{pmatrix}$.

Proposition 4. *The first coordinate of remainder vector of the division of* $\begin{pmatrix} 0 \\ x^n g \end{pmatrix}$ *by* $\begin{pmatrix} u & u' \\ r & r' \end{pmatrix}$ *is the polynomial* $v(x)$ *solution of* (1).

We describe here a generalized Euclidean division algorithm to solve problem (3).

Let $E(x) = \begin{pmatrix} p(x) \\ q(x) \end{pmatrix}$ of degree m, $B(x) = \begin{pmatrix} e(x) & e'(x) \\ f(x) & f'(x) \end{pmatrix}$ of degree $n \leqslant m$. $E(x) = B(x)Q(x) + R(x)$ with $\deg(R(x)) < n$, and $\deg(Q(x)) \leqslant m - n$. Let $z = \frac{1}{x}$

$$\begin{aligned} E(x) &= B(x)Q(x) + R(x) \\ \Leftrightarrow\ E(\frac{1}{z}) &= B(\frac{1}{z})Q(\frac{1}{z}) + R(\frac{1}{z}) \\ \Leftrightarrow z^m E(\frac{1}{z}) &= z^n B(\frac{1}{z}) z^{m-n} Q(\frac{1}{z}) + z^{m-n+1} z^{n-1} R(\frac{1}{z}) \\ \Leftrightarrow\ \hat{E}(z) &= \hat{B}(z)\hat{Q}(z) + z^{m-n+1}\hat{R}(z) \end{aligned} \tag{9}$$

with $\hat{E}(z), \hat{B}(z), \hat{Q}(z), \hat{R}(z)$ are the polynomials obtained by reversing the order of coefficients of polynomials $E(z), B(z), Q(z), R(z)$.

$$\begin{aligned} (9) \Rightarrow \frac{\hat{E}(z)}{\hat{B}(z)} &= \hat{Q}(z) + z^{m+n-1}\frac{\hat{R}(z)}{\hat{B}(z)} \\ \Rightarrow \hat{Q}(z) &= \frac{\hat{E}(z)}{\hat{B}(z)} \bmod z^{m-n+1} \end{aligned}$$

$\frac{1}{\hat{B}(z)}$ exists because its coefficient of highest degree is invertible. Thus $\hat{Q}(z)$ is obtained by computing the first $m - n + 1$ coefficients of $\frac{\hat{E}(z)}{\hat{B}(z)}$.

To find $W(x) = \frac{1}{\hat{B}(x)}$ we will use Newton's iteration: Let $f(W) = \hat{B} - W^{-1}$. $f'(W_l).(W_{l+1} - W_l) = -W_l^{-1}(W_l + 1 - W_l)W_l^{-1} = f(W_l) = \hat{B} - W_l^{-1}$, thus

$$W_{l+1} = 2W_l - W_l\hat{B}W_l.$$

and $W_0 = \hat{B}_0^{-1}$ which exists.

$$\begin{aligned} W - W_{l+1} &= W - 2W_l + W_l \hat{B} W_l \\ &= W(\mathbb{I}_2 - \hat{B} W_l)^2 \\ &= (W - W_l)\hat{B}(W - W_l). \end{aligned}$$

Thus $W_l(x) = W(x) \bmod x^{2l}$ for $l = 0, \ldots, \lceil \log(m-n+1) \rceil$.

Proposition 5. *We need* $\mathcal{O}(n \log(n) \log(m-n) + m \log m)$ *arithmetic operations to solve problem* (3)

Proof. We must do $\lceil \log(m-n+1) \rceil$ Newton's iteration to obtain the first $m-n+1$ coefficients of $\frac{1}{\hat{B}} = W(x)$. And for each iteration we must do $\mathcal{O}(n \log n)$ arithmetic operations (multiplication of polynimials of degree n). And then we need $\mathcal{O}(m \log m)$ aritmetic operations to do the multiplication $\hat{E}.\frac{1}{\hat{B}}$.

2.3 Construction of the generators

The canonical basis of $\mathbb{K}[x]^3$ is denoted by $\sigma_1, \sigma_2, \sigma_3$. Let ρ_1, ρ_2 the generators of $\mathcal{L}(\tilde{T}(x), x^n, x^{2n} - 1)$ of degree n given by

$$\begin{aligned} \rho_1 &= x^n \sigma_1 - (u, v, w) = (u_1, v_1, w_1) \\ \rho_2 &= x^n \sigma_2 - (u', v', w') = (u_2, v2, w_2) \end{aligned} \tag{10}$$

with (u, v, w), (u', v', w') are the vector given in (7) and (8).

We will describe here how we compute (u_1, v_1, w_1) and (u_2, v_2, w_2). We will give two methods to compute them, the second one is the method given in [12]. The first one uses the Euclidean gcd algorithm:

We will recall firstly the algebraic and computational properties of the well known extended Euclidean algorithm (see [13]): Given $p(x), p'(x)$ two polynomials in degree m and m' respectively, let

$$\begin{aligned} r_0 &= p, & r_1 &= p', \\ s_0 &= 1, & s_1 &= 0, \\ t_0 &= 0, & t_1 &= 1 \end{aligned}$$

and define

$$\begin{aligned} r_{i+1} &= r_{i-1} - q_i r_i, \\ s_{i+1} &= s_{i-1} - q_i s_i, \\ t_{i+1} &= t_{i-1} - q_i t_i, \end{aligned}$$

where q_i results when the division algorithm is applied to r_{i-1} and r_i, i.e. $r_{i-1} = q_i r_i + r_{i+1}$. $\deg r_{i+1} < \deg r_i$ for $i = 1, \ldots, l$ with l is such that $r_l = 0$, therefore $r_{l-1} = \gcd(p(x), p'(x))$.

Proposition 6. *The following relations hold:*

$$s_i p + t_i p' = r_i \quad \textit{and} \quad (s_i, t_i) = 1 \quad \textit{for } i = 1, \dots, l$$

and

$$\begin{cases} \deg r_{i+1} < \deg r_i, & i = 1, \dots, l-1 \\ \deg s_{i+1} > \deg s_i & \textit{and} \quad \deg t_{i+1} > \deg t_i, \\ \deg s_{i+1} = \deg(q_i . s_i) = \deg v - \deg r_i, \\ \deg t_{i+1} = \deg(q_i . t_i) = \deg u - \deg r_i. \end{cases}$$

Proposition 7. *By applying the Euclidean gcd algorithm in* $p(x) = x^{n-1}T$ *and* $p'(x) = x^{2n-1}$ *in degree* $n-1$ *and* $n-2$ *we obtain* ρ_1 *and* ρ_2 *respectively.*

Proof. We saw that $Tu = g$ if and only if there exist $A(x)$ and $B(x)$ such that

$$\bar{T}(x)u(x) + x^{2n-1}B(x) = x^{n-1}b(x) + A(x)$$

with $\bar{T}(x) = x^{n-1}T(x)$ a polynomial of degree $\leqslant 2n-2$. In (7) and (8) we saw that for $g(x) = 1$ $(g = e_1)$ and $g(x) = x^n T(x)$ $(g = (0, t_{-n+1}, \dots, t_{-1})^T)$ we obtain a base of $\mathcal{L}(\tilde{T}(x), x^n, x^{2n} - 1)$. $Tu_1 = e_1$ if and only if there exist $A_1(x)$, $B_1(x)$ such that

$$\bar{T}(x)u_1(x) + x^{2n-1}B_1(x) = x^{n-1} + A_1(x) \tag{11}$$

and $Tu_2 = (0, t_{-n+1}, \dots, t_{-1})^T$ if and only if there exist $A_2(x)$, $B_2(x)$ such that

$$\bar{T}(x)(u_2(x) + x^n) + x^{2n-1}B_2(x) = A_2(x) \tag{12}$$

with $\deg A_1(x) \leqslant n-2$ and $\deg A_2(x) \leqslant n-2$. Thus By applying the extended Euclidean algorithm in $p(x) = x^{n-1}T$ and $p'(x) = x^{2n-1}$ until we have $\deg r_l(x) = n-1$ and $\deg r_{l+1}(x) = n-2$ we obtain

$$u_1(x) = \frac{1}{c_1}s_l(x), \quad B_1(x) = \frac{1}{c_1}t_l(x), \quad x^{n-1} + A_1(x) = \frac{1}{c_1}r_l(x)$$

and

$$x^n + u_2(x) = \frac{1}{c_2}s_{l+1}(x), \quad B_2(x) = \frac{1}{c_2}t_{l+1}(x), \quad A_2(x) = \frac{1}{c_2}r_{l+1}(x)$$

with c_1 and c_2 are the highest coefficients of $r_l(x)$ and $s_{l+1}(x)$ respectively, in fact: The equation (11) is equivalent to

$$
\begin{array}{r}
n-1\left\{\begin{array}{c}\\ \\ \end{array}\right.\\
n\left\{\begin{array}{c}\\ \\ \\ \end{array}\right.\\
n-1\left\{\begin{array}{c}\\ \\ \\ \end{array}\right.
\end{array}
\left(\begin{array}{ccc|ccc}
\multicolumn{3}{c|}{\overbrace{\hphantom{t_{n-1}\ \cdots\ t_{-n+1}}}^{n}} & \multicolumn{3}{c}{\overbrace{\hphantom{1\quad \ddots \quad 1}}^{n-1}}\\
t_{-n+1} & & & & & \\
\vdots & \ddots & & & & \\
\hline
t_0 & \cdots & t_{-n+1} & & & \\
\vdots & \ddots & \vdots & & & \\
t_{n-1} & \cdots & t_0 & & & \\
\hline
 & \ddots & \vdots & 1 & & \\
 & & & & \ddots & \\
 & & t_{n-1} & & & 1
\end{array}\right)
\begin{pmatrix} u_1 \\ \\ B_1 \end{pmatrix}
=
\left(\begin{array}{c} A_1 \\ \hline 1 \\ 0 \\ \vdots \\ 0 \end{array}\right)
$$

since T is invertible then the $(2n-1)\times(2n-1)$ block at the bottom is invertible and then u_1 and B_1 are unique, therefore u_1, B_1 and A_1 are unique. And, by proposition (6), $\deg r_l = n-1$ $(r_l = c_1(x^n + A_1(x))$ then $\deg s_{l+1} = (2n-1)-(n-1) = n$ and $\deg t_{l+1} = (2n-2)-(n-1) = n-1$ thus, by the same proposition, $\deg s_l \leqslant n-1$ and $\deg t_l \leqslant n-2$. Therfore $\frac{1}{c_1}s_l = u_1$ and $\frac{1}{c_1}t_l = B_1$.

Finaly, $Tu = e_1$ if and only if there exist $v(x)$, $w(x)$ such that

$$\tilde{T}(x)u(x) + x^n v(x) + (x^{2n}-1)w(x) = 1 \tag{13}$$

$\tilde{T}(x) = T^+ + x^{2n}T^- = T + (x^{2n}-1)T^-$ thus

$$T(x)u(x) + x^n v(x) + (x^{2n}-1)(w(x) + T^-(x)u(x)) = 1 \tag{14}$$

of another hand $T(x)u(x) - x^{-n+1}A_1(x) + x^n B_1(x) = 1$ and $x^{-n+1}A_1(x) = x^n(xA_1) - x^{-n}(x^{2n}-1)xA_1$ thus

$$T(x)u(x) + x^n(B(x) - xA(x)) + (x^{2n}-1)x^{-n+1}A(x) = 1. \tag{15}$$

By comparing (14) and (15), and as $1 = x^n x^n - (x^{2n}-1)$ we have the proposition and we have $w(x) = x^{-n+1}A(x) - T_-(x)u(x) + 1$ which is the part of positive degree of $-T_-(x)u(x)+1$.

Remark 1. A superfast Euclidean gcd algorithm, which uses no more than $\mathcal{O}(n\log^2 n)$, is given in [13] chapter 11.

The second method to compute (u_1, v_1, w_1) and (u_2, v_2, w_2) is given in [12]. We are interested in computing the coefficients of σ_1, σ_2, the coefficients of σ_3 correspond to elements in the ideal $(x^{2n}-1)$ and thus can be obtain by reduction of $(\tilde{T}(x)\,x^n).B(x)$ by $x^{2n}-1$, with $B(x) = \begin{pmatrix} x^n - u_0 & -v_0 \\ -u_1 & x^n - v_1 \end{pmatrix} = \begin{pmatrix} u(x) & v(x) \\ u'(x) & v'(x) \end{pmatrix}$.

A superfast algorithm to compute $B(x)$ is given in [12]. Let us describe how to compute it.

By evaluation of (10) at the roots $\omega_j \in \mathfrak{U}_{2n}$ we deduce that $(u(x)\, v(x))^T$ and $(u'(x)\, v'(x))^T$ are the solution of the following rational interpolation problem:

$$\begin{cases} \tilde{T}(\omega_j)u(\omega_j) + \omega_j^n v(\omega_j) = 0 \\ \tilde{T}(\omega_j)u'(\omega_j) + \omega_j^n v'(\omega_j) = 0 \end{cases} \text{with}$$

$$\begin{cases} u_n = 1,\ v_n = 0 \\ u'_n = 0,\ v'_n = 1 \end{cases}.$$

Definition 3. *The τ-degree of a vector polynomial $w(x) = (w_1(x)\, w_2(x))^T$ is defined as*

$$\tau - \deg w(x) := \max\{\deg w_1(x), \deg w_2(x) - \tau\}.$$

$B(x)$ is a n–reduced basis of the module of all vector polynomials $r(x) \in \mathbb{K}[x]^2$ that satisfy the interpolation conditions

$$f_j^T r(\omega_j) = 0, \quad j = 0, \ldots, 2n-1$$

with $f_j = \begin{pmatrix} \tilde{T}(\omega_j) \\ \omega_j^n \end{pmatrix}$.

$B(x)$ is called a τ–reduced basis (with $\tau = n$) that corresponds to the interpolation data (ω_j, f_j), $j = 0, \ldots, 2n-1$.

Definition 4. *A set of vector polynomial in $\mathbb{K}[x]^2$ is called τ-reduced if the τ-highest degree coefficients are lineary independent.*

Theorem 2. *Let $\tau = n$. Suppose J is a positive integer. Let $\sigma_1, \ldots, \sigma_J \in \mathbb{K}$ and $\phi_1, \ldots, \phi_J \in \mathbb{K}^2$ wich are $\neq (0\,0)^T$. Let $1 \leqslant j \leqslant J$ and $\tau_J \in \mathbb{Z}$. Suppose that $B_j(x) \in \mathbb{K}[x]^{2\times 2}$ is a τ_J-reduced basis matrix with basis vectors having τ_J–degree δ_1 and δ_2, respectively, corresponding to the interpolation data $\{(\sigma_i, \phi_i); i = 1, \ldots, j\}$.*

Let $\tau_{j\to J} := \delta_1 - \delta_2$. Let $B_{j\to J}(x)$ be a $\tau_{j\to J}$-reduced basis matrix corresponding to the interpolation data $\{(\sigma_i, B_j^T(\sigma_j)\phi_i); i = j+1, \ldots, J\}$.

Then $B_J(x) := B_j(x)B_{j\to J}(x)$ is a τ_J-reduced basis matrix corresponding to the interpolation data $\{(\sigma_i, \phi_i); i = 1, \ldots, J\}$.

Proof. For the proof, see [12].

When we apply this theorem for the $\omega_j \in \mathfrak{U}_{2n}$ as interpolation points, we obtain a superfast algorithm ($\mathcal{O}(n \log^2 n)$) wich compute $B(x)$.[12]

We consider the two following problems:

3 Bivariate case

Let $m \in \mathbb{N}, m \in \mathbb{N}$. In this section we denote by $E = \{(i,j);\ 0 \leqslant i \leqslant m-1,\ 0 \leqslant j \leqslant n-1\}$, and $R = \mathbb{K}[x,y]$. We denote by $\mathbb{K}[x,y]_{\substack{m\\n}}$ the vector space of bivariate polynomials of degree $\leqslant m$ in x and $\leqslant n$ in y.

Notation. For a block matrix M, of block size n and each block is of size m, we will use the following indication :

$$M = \left(M_{(i_1,i_2),(j_1,j_2)}\right)_{\substack{0\leqslant i_1,j_1\leqslant m-1\\ 0\leqslant i_2,j_2\leqslant n-1}} = (M_{\alpha\beta})_{\alpha,\beta\in E}. \tag{16}$$

(i_2, j_2) gives the block's positions, (i_1, j_1) the position in the blocks.

Problem 4. Given a Toeplitz block Toeplitz matrix $T = (t_{\alpha-\beta})_{\alpha\in E,\beta\in E} \in \mathbb{K}^{mn\times mn}$ ($T = (T_{\alpha\beta})_{\alpha,\beta\in E}$ with $T_{\alpha\beta} = t_{\alpha-\beta}$) of size mn and $g = (g_\alpha)_{\alpha\in E} \in \mathbb{K}^{mn}$, find $u = (u_\alpha)_{\alpha\in E}$ such that

$$Tu = g. \tag{17}$$

Definition 5. *We define the following polynomials:*

- $T(x,y) := \sum_{(i,j)\in E-E} t_{i,j} x^i y^j,$
- $\tilde{T}(x,y) := \sum_{i,j=0}^{2n-1,2m-1} \tilde{t}_{i,j} x^i y^j$ *with*

$$\tilde{t}_{i,j} := \begin{cases} t_{i,j} & si\ i < m, j < n \\ t_{i-2m,j} & si\ i \geqslant m, j < n \\ t_{i,j-2n} & si\ i < m, j \geqslant n \\ t_{i-2m,j-2n} & si\ i \geqslant m, i \geqslant n \end{cases},$$

- $u(x,y) := \sum_{(i,j)\in E} u_{i,j}\, x^i y^j,\ g(x,y) := \sum_{(i,j)\in E} g_{i,j} x^i y^j.$

3.1 Moving hyperplanes

For any non-zero vector of polynomials $\mathbf{a} = (a_1, \ldots, a_n) \in \mathbb{K}[x,y]^n$, we denote by $\mathcal{L}(\mathbf{a})$ the set of vectors $(h_1, \ldots, h_n) \in \mathbb{K}[x,y]^n$ such that

$$\sum_{i=1}^{n} a_i\, h_i = 0. \tag{18}$$

It is a $\mathbb{K}[x,y]$-module of $\mathbb{K}[x,y]^n$.

Proposition 8. *The vector* u *is solution of* (17) *if and only if there exist* $h_2, \ldots, h_9 \in \mathbb{K}[x,y]_{\substack{m-1\\n-1}}$ *such that* $(u(x,y), h_2(x,y), \ldots, h_9(x,y))$ *belongs to* $\mathcal{L}(\tilde{T}(x,y), x^m, x^{2m}-1, y^n, x^m y^n, (x^{2m}-1)\, y^n, y^{2n}-1, x^m(y^{2n}-1), (x^{2m}-1)\,(y^{2n}-1))$.

Proof. Let $L = \{x^{\alpha_1}y^{\alpha_2}, 0 \leqslant \alpha_1 \leqslant m-1, 0 \leqslant \alpha_2 \leqslant n-1\}$, and Π_E the projection of R on the vector space generated by L. By [8], we have

$$Tu = g \Leftrightarrow \Pi_E(T(x,y)\,u(x,y)) = g(x,y) \tag{19}$$

which implies that

$$\begin{aligned} T(x,y)u(x,y) = {} & g(x,y) + x^m y^n A_1(x,y) + x^m y^{-n} A_2(x,y) + \\ & x^{-m} y^n A_3(x,y) + x^{-m} y^{-n} A_4(x,y) + \\ & x^m A_5(x,y) + x^{-m} A_6(x,y) + y^n A_7(x,y) + y^{-n} A_8(x,y), \end{aligned} \tag{20}$$

where the $A_i(x,y)$ are polynomials of degree at most $m-1$ in x and $n-1$ in y. Since $\omega^m = \omega^{-m}$, $\upsilon^n = \upsilon^{-n}$, $\tilde{T}(\omega,\upsilon) = T(\omega,\upsilon)$ for $\omega \in \mathfrak{U}_{2m}$, $\upsilon \in \mathfrak{U}_{2n}$, we deduce by evaluation at the roots $\omega \in \mathfrak{U}_{2m}$, $\upsilon \in \mathfrak{U}_{2n}$ that

$$\begin{aligned} R(x,y) := {} & \tilde{T}(x,y)u(x,y) + x^m h_2(x,y) + y^n h_4(x,y) + \\ & x^m y^n h_5(x,y) - g(x,y) \in (x^{2m}-1, y^{2n}-1) \end{aligned}$$

with $h_2 = -(A_5+A_6)$, $h_4 = -(A_7+A_8)$, $h_5 = -(A_1(x,y)+A_2(x,y)+A_3(x,y)+A_4(x,y))$.

By reduction by the polynomials $x^{2m}-1$, $y^{2n}-1$, and as $R(x,y)$ is of degree $\leqslant 3m-1$ in x and $\leqslant 3n-1$ in y, there exist $h_3(x,y), h_6(x,y), \ldots, h_8(x,y) \in \mathbb{K}[x,y]_{\substack{m-1\\n-1}}$ such that

$$\begin{aligned} & \tilde{T}(x,y)u(x,y) + x^m h_2(x,y) + (x^{2m}-1)h_3(x,y) + y^n h_4(x,y) + \\ & x^m y^n h_5(x,y) + (x^{2m}-1)y^n h_6(x,y) + (y^{2n}-1)h_7(x,y) + \\ & x^m (y^{2m}-1)h_7(x,y) + (x^{2n}-1)(y^{2n}-1)h_8(x,y) = g(x,y). \end{aligned} \tag{21}$$

Conversely a solution of (21) can be transformed into a solution of (20), which ends the proof of the proposition.

In the following, we are going to denote by $\mathbf{T}$ the vector $\mathbf{T} = (\tilde{T}(x,y), x^m, x^{2m}-1, y^n, x^m y^n, (x^{2m}-1)\,y^n, y^{2n}-1, x^m(y^{2n}-1), (x^{2m}-1)\,(y^{2n}-1))$.

Proposition 9. *There is no elements of* $\mathbb{K}[x,y]_{\substack{m-1\\n-1}}$ *in* $\mathcal{L}(\mathbf{T})$.

Proof. We consider the map

$$\mathbb{K}[x,y]^9_{\substack{m-1\\n-1}} \to \mathbb{K}[x,y]_{\substack{3m-1\\3n-1}} \tag{22}$$

$$p(x,y) = (p_1(x,y), \ldots, p_9(x,y)) \mapsto \mathbf{T}.p \tag{23}$$

which $9mn \times 9mn$ matrix is of the form

$$S := \left(\begin{array}{c|c|c|c}
 & \begin{matrix} E_{21} & -E_{11}+E_{31} \\ \vdots & \vdots \\ E_{2n} & -E_{1n}+E_{3n} \end{matrix} & & \begin{matrix} -E_{11} & -E_{21} & E_{11}-E_{31} \\ \vdots & \vdots & \vdots \\ -E_{1n} & -E_{2n} & E_{1n}-E_{3n} \end{matrix} \\ T_0 & & & \\
\hline
T_1 & & \begin{matrix} E_{11} & E_{21} & -E_{11}+E_{31} \\ \vdots & \vdots & \vdots \\ E_{1n} & E_{2n} & -E_{1n}+E_{3n} \end{matrix} & \\
\hline
T_2 & & & \begin{matrix} E_{11} & E_{21} & -E_{11}+E_{31} \\ \vdots & \vdots & \vdots \\ E_{1n} & E_{2n} & -E_{1n}+E_{3n} \end{matrix}
\end{array}\right) \tag{24}$$

with E_{ij} is the $3m \times mn$ matrix $e_{ij} \otimes I_m$ and e_{ij} is the $3 \times n$ matrix with entries equal zero except the (i,j)th entrie equal 1. And the matrix $\begin{pmatrix} T_0 \\ T_1 \\ T_2 \end{pmatrix}$ is the following $9mn \times m$ matrix

$$\left(\begin{array}{cccc}
t_0 & 0 & \dots & 0 \\
t_1 & t_0 & \dots & 0 \\
\vdots & \ddots & \ddots & \vdots \\
t_{n-1} & \dots & t_1 & t_0 \\
\hline
0 & t_{n-1} & \dots & t_1 \\
t_{-n+1} & 0 & \ddots & \vdots \\
\vdots & \ddots & \ddots & t_{-n+1} \\
t_{-1} & \dots & t_{-n+1} & 0 \\
\hline
0 & t_{-1} & \dots & t_{-n+1} \\
\vdots & \ddots & \ddots & \vdots \\
\vdots & & \ddots & t_{-1} \\
0 & \dots & \dots & 0
\end{array}\right) \text{ and } t_i = \left(\begin{array}{cccc}
t_{i,0} & 0 & \dots & 0 \\
t_{i,1} & t_{i,0} & \dots & 0 \\
\vdots & \ddots & \ddots & \vdots \\
t_{i,n-1} & \dots & t_{i,1} & t_{i,0} \\
\hline
0 & t_{i,m-1} & \dots & t_{i,1} \\
t_{i,-m+1} & 0 & \ddots & \vdots \\
\vdots & \ddots & \ddots & t_{i,-m+1} \\
t_{i,-1} & \dots & t_{i,-m+1} & 0 \\
\hline
0 & t_{i,-1} & \dots & t_{i,-m+1} \\
\vdots & \ddots & \ddots & \vdots \\
\vdots & & \ddots & t_{i,-1} \\
0 & \dots & \dots & 0
\end{array}\right).$$

For the same reasons in the proof of proposition (1) the matrix S is invertible.

Theorem 3. *For any non-zero vector of polynomials* $\mathbf{a} = (a_i)_{i=1,\dots,n} \in \mathbb{K}[x,y]^n$, *the* $\mathbb{K}[x,y]$*-module* $\mathcal{L}(a_1, \dots, a_n)$ *is free of rank* $n-1$.

Proof. Consider first the case where a_i are monomials.

$a_i = x^{\alpha_i} y^{\beta_i}$ that are sorted in lexicographic order such that $x < y$, a_1 being the biggest and a_n the smallest. Then the module of syzygies of $\mathbf{a}$ is generated by the S-polynomials:

$$S(a_i, a_j) = \operatorname{lcm}(a_i, a_j)\left(\frac{\sigma_i}{a_i} - \frac{\sigma_j}{a_j}\right),$$

where $(\sigma_i)_{i=1,\ldots,n}$ is the canonical basis of $\mathbb{K}[x,y]^n$ [3]. We easily check that $S(a_i,a_k) = \frac{\mathrm{lcm}(a_i,a_k)}{\mathrm{lcm}(a_i,a_j)} S(a_i,a_j) - \frac{\mathrm{lcm}(a_i,a_k)}{\mathrm{lcm}(a_j,a_k)} S(a_j,a_k)$ if $i \neq j \neq k$ and $\mathrm{lcm}(a_i,a_j)$ divides $\mathrm{lcm}(a_i,a_k)$. Therefore $\mathcal{L}(\mathbf{a})$ is generated by the $S(a_i,a_j)$ which are minimal for the division, that is, by $S(a_i,a_{i+1})$ (for $i = 1,\ldots,n-1$), since the monomials a_i are sorted lexicographically. As the syzygies $S(a_i,a_{i+1})$ involve the basis elements σ_i, σ_{i+1}, they are linearly independent over $\mathbb{K}[x,y]$, which shows that $\mathcal{L}(\mathbf{a})$ is a free module of rank $n-1$ and that we have the following resolution:

$$0 \to \mathbb{K}[x,y]^{n-1} \to \mathbb{K}[x,y]^n \to (\mathbf{a}) \to 0.$$

Suppose now that a_i are general polynomials $\in \mathbb{K}[x,y]$ and let us compute a Gröbner basis of a_i, for a monomial ordering refining the degree [3]. We denote by $m_1,\ldots,m_s$ the leading terms of the polynomials in this Gröbner basis, sorted by lexicographic order.

The previous construction yields a resolution of $(m_1, \ldots, m_s)$:

$$0 \to \mathbb{K}[x,y]^{s-1} \to \mathbb{K}[x,y]^s \to (m_i)_{i=1,\ldots,s} \to 0.$$

Using [7] (or [3]), this resolution can be deformed into a resolution of $(\mathbf{a})$, of the form

$$0 \to \mathbb{K}[x,y]^p \to \mathbb{K}[x,y]^n \to (\mathbf{a}) \to 0,$$

which shows that $\mathcal{L}(\mathbf{a})$ is also a free module. Its rank p is necessarily equal to $n-1$, since the alternate sum of the dimensions of the vector spaces of elements of degree $\leqslant \nu$ in each module of this resolution should be 0, for $\nu \in \mathbb{N}$.

3.2 Generators and reduction

In this section, we describe an explicit set of generators of $\mathcal{L}(\mathbf{T})$. The canonical basis of $\mathbb{K}[x,y]^9$ is denoted by $\sigma_1,\ldots,\sigma_9$.

First as $\tilde{T}(x,y)$ is of degree $\leqslant 2m-1$ in x and $\leqslant 2n-1$ in y and as the function (22) in surjective, there exists $u_1, u_2 \in \mathbb{K}[x,y]^9_{\substack{m-1\\n-1}}$ such that $\mathbf{T}\cdot u_1 = \tilde{T}(x,y)x^m$, $\mathbf{T}\cdot u_2 = \tilde{T}(x,y)y^n$. Thus,

$$\begin{aligned}\rho_1 &= x^m\sigma_1 - u_1 \in \mathcal{L}(\mathbf{T}),\\ \rho_2 &= y^n\sigma_1 - u_2 \in \mathcal{L}(\mathbf{T}).\end{aligned}$$

We also have $u_3 \in \mathbb{K}[x,y]_{\substack{m-1\\n-1}}$, such that $\mathbf{T}\cdot u_3 = 1 = x^m x^m - (x^{2m}-1) = y^n y^n - (y^{2n}-1)$. We deduce that

$$\begin{aligned}\rho_3 &= x^m\sigma_2 - \sigma_3 - u_3 \in \mathcal{L}(\mathbf{T}),\\ \rho_4 &= y^n\sigma_4 - \sigma_7 - u_3 \in \mathcal{L}(\mathbf{T}).\end{aligned}$$

Finally, we have the obvious relations:

$$\begin{aligned}\rho_5 &= y^n\sigma_2 - \sigma_5 \in \mathcal{L}(\mathbf{T}),\\ \rho_6 &= x^m\sigma_4 - \sigma_5 \in \mathcal{L}(\mathbf{T}),\\ \rho_7 &= x^m\sigma_5 - \sigma_6 + \sigma_4 \in \mathcal{L}(\mathbf{T}),\\ \rho_8 &= y^n\sigma_5 - \sigma_8 + \sigma_2 \in \mathcal{L}(\mathbf{T}).\end{aligned}$$

Proposition 10. *The relations* $\rho_1, \ldots, \rho_8$ *form a basis of* $\mathcal{L}(\mathbf{T})$.

Proof. Let $\mathbf{h} = (h_1, \ldots, h_9) \in \mathcal{L}(\mathbf{T})$. By reduction by the previous elements of $\mathcal{L}(\mathbf{T})$, we can assume that the coefficients h_1, h_2, h_4, h_5 are in $\mathbb{K}[x,y]_{\substack{m-1\\n-1}}$. Thus, $\tilde{T}(x,y)h_1 + x^m h_2 + y^n h_4 + x^m y^n h_5 \in (x^{2n} - 1, y^{2m} - 1)$. As this polynomial is of degree $\leqslant 3m-1$ in x and $\leqslant 3n-1$ in y, by reduction by the polynomials, we deduce that the coefficients $h_3, h_6, \ldots, h_9$ are in $\mathbb{K}[x,y]_{\substack{m-1\\n-1}}$. By proposition 9, there is no non-zero syzygy in $\mathbb{K}[x,y]^9_{\substack{m-1\\n-1}}$. Thus we have $\mathbf{h} = 0$ and every element of $\mathcal{L}(\mathbf{T})$ can be reduced to 0 by the previous relations. In other words, $\rho_1, \ldots, \rho_8$ is a generating set of the $\mathbb{K}[x,y]$-module $\mathcal{L}(\mathbf{T})$. By theorem 3, the relations ρ_i cannot be dependent over $\mathbb{K}[x,y]$ and thus form a basis of $\mathcal{L}(\mathbf{T})$.

3.3 Interpolation

Our aim is now to compute efficiently a system of generators of $\mathcal{L}(\mathbf{T})$.

More precisely, we are interested in computing the coefficients of $\sigma_1, \sigma_2, \sigma_4, \sigma_5$ of ρ_1, ρ_2, ρ_3. Let us call $B(x,y)$ the corresponding coefficient matrix, which is of the form:

$$\begin{pmatrix} x^m & y^n & 0\\ 0 & 0 & x^m\\ 0 & 0 & 0\\ 0 & 0 & 0 \end{pmatrix} + \mathbb{K}[x,y]^{4,3}_{\substack{m-1\\n-1}} \tag{25}$$

Notice that the other coefficients of the relations ρ_1, ρ_2, ρ_3 correspond to elements in the ideal $(x^{2m} - 1, y^{2n} - 1)$ and thus can be obtained easily by reduction of the entries of $(\tilde{T}(x,y), x^m, y^n, x^m y^n) \cdot B(x,y)$ by the polynomials $x^{2m} - 1, y^{2n} - 1$.

Notice also that the relation ρ_4 can be easily deduced from ρ_3, since we have $\rho_3 - x^m\sigma_2 + \sigma_3 + y^n\sigma_4 - \sigma_7 = \rho_4$. Since the other relations ρ_i (for $i > 4$) are explicit and independent of $\tilde{T}(x,y)$, we can easily deduce a basis of $\mathcal{L}(\mathbf{T})$ from the matrix $B(x,y)$.

As in $\mathcal{L}(\mathbf{T}) \cap \mathbb{K}[x,y]_{\substack{m-1\\n-1}}$ there is only one element, thus by computing the basis given in proposition (10) and reducing it we can obtain this element in $\mathcal{L}(\mathbf{T}) \cap \mathbb{K}[x,y]_{\substack{m-1\\n-1}}$ which gives us the solution of $Tu = g$. We can give a fast algorithm to do these two step, but a superfast algorithm is not available.

4 Conclusions

We show in this paper a correlation between the solution of a Toeplitz system and the syzygies of polynomials. We generalized this way, and we gave a correlation between the solution of a Toeplitz-block-Toeplitz system and the syzygies of bivariate polynomials. In the univariate case we could exploit this correlation to give a superfast resolution algorithm. The generalization of this technique to the bivariate case is not very clear and it remains an important challenge.

References

1. D. Bini and V. Y. Pan. *Polynomial and matrix computations. Vol. 1.* Progress in Theoretical Computer Science. Birkhäuser Boston Inc., Boston, MA, 1994. Fundamental algorithms.
2. R. Bitmead and B. Anderson. Asymptotically fast solution of Toeplitz and related systems of equations. *Linear Algebra and Its Applications*, 34:103–116, 1980.
3. D. Eisenbud. *Commutative algebra*, volume 150 of *Graduate Texts in Mathematics.* Springer-Verlag, New York, 1995. With a view toward algebraic geometry.
4. P. Fuhrmann. *A polynomial approach to linear algebra.* Springer-Verlag, 1996.
5. G. Heinig and K. Rost. *Algebraic methods for Toeplitz-like matrices and operators*, volume 13 of *Operator Theory: Advances and Applications.* Birkhäuser Verlag, Basel, 1984.
6. T. Kailath and A. H. Sayed. Displacement structure: theory and applications. *SIAM Rev.*, 37(3):297–386, 1995.
7. H. M. Möller and F. Mora. New constructive methods in classical ideal theory. *J. Algebra*, 100(1):138–178, 1986.
8. B. Mourrain and V. Y. Pan. Multivariate polynomials, duality, and structured matrices. *J. Complexity*, 16(1):110–180, 2000.
9. V. Y. Pan. Nearly optimal computations with structured matrices. In *Proceedings of the Eleventh Annual ACM-SIAM Symposium on Discrete Algorithms (San Francisco, CA, 2000)*, pages 953–962, New York, 2000. ACM.
10. V. Y. Pan. *Structured matrices and polynomials.* Birkhäuser Boston Inc., Boston, MA, 2001. Unified superfast algorithms.
11. E. Tyrtyshnikov. Fast algorithms for block Toeplitz matrices. *Sov. J. Numer. Math. Modelling*, 1(2):121–139, 1985.
12. M. Van Barel, G. Heinig, and P. Kravanja. A stabilized superfast solver for nonsymmetric Toeplitz systems. *SIAM J. Matrix Anal. Appl.*, 23(2):494–510 (electronic), 2001.
13. J. von zur Gathen and J. Gerhard. *Modern computer algebra.* Cambridge University Press, Cambridge, second edition, 2003.

Concepts of Data-Sparse Tensor-Product Approximation in Many-Particle Modelling

Heinz-Jürgen Flad[1], Wolfgang Hackbusch[2], Boris N. Khoromskij[2], and Reinhold Schneider[1]

[1] Institut für Mathematik, Technische Universität Berlin, Straße des 17. Juni 137, D-10623 Berlin, Germany
{flad,schneidr}@math.tu-berlin.de

[2] Max-Planck-Institute for Mathematics in the Sciences, Inselstr. 22-26, D-04103 Leipzig, Germany
{wh,bokh}@mis.mpg.de

Abstract. We present concepts of data-sparse tensor approximations to the functions and operators arising in many-particle models of quantum chemistry. Our approach is based on the systematic use of structured tensor-product representations where the low-dimensional components are represented in hierarchical or wavelet based matrix formats. The modern methods of tensor-product approximation in higher dimensions are discussed with the focus on analytically based approaches. We give numerical illustrations which confirm the efficiency of tensor decomposition techniques in electronic structure calculations.

Keywords: Schrödinger equation, Hartree-Fock method, density functional theory, tensor-product approximation.

1 Introduction

Among the most challenging problems of scientific computing nowadays are those of high dimensions, for instance, multi-particle interactions, integral or differential equations on $[0,1]^d$ and the related numerical operator calculus for $d \geqslant 3$. Many standard approaches have a computational complexity that grows exponentially in the dimension d and thus fail because of the well known "curse of dimensionality". To get rid of this exponential growth in the complexity one can use the idea of tensor-product constructions (cf. [86]) on all stages of the solution process. Hereby we approximate the quantity of interest in tensor-product formats and use other approximation methods for the remaining low-dimensional components. Depending on the specific properties of the problem, these low-dimensional components are already in a data-sparse format, like band structured matrices, or can be approximated via hierarchical (low-rank) matrix and wavelet formats, respectively. In order to obtain low-rank tensor-product approximations it is convenient to start already with a separable approximation of possibly large separation rank. This is the case e.g. for hyperbolic cross approximations in tensor-product wavelet bases or for Gaussian-type and plane wave basis sets which are frequently used in quantum chemistry and solid state physics. With such a representation at hand it is possible to apply algebraic recompression methods to generate the desired low-rank approximations. We want to stress, however, that these recompression methods in multi-linear algebra lead to severe computational problems since they are, in fact, equivalent to some kind of nonlinear approximation in $d \geqslant 3$. Despite these computational difficulties, such kind of procedure is especially favourable for smooth functions with few singularities which are actually typical for our envisaged applications to be discussed below.

A large class of translation invariant kernels of integral operators can be represented via integral transformations of a separable function, e.g. Gaussian function. Using exponentially convergent quadrature rules for the parametric integrals it is possible to derive low-rank tensor-product approximations for these integral operators. In a similar manner it is possible to derive such representations for matrix-valued functions in the tensor-product format.

It is the purpose of the present paper to discuss possible applications of the afore outlined approach to electronic structure calculations with applications in quantum chemistry and solid state physics. It will be shown in the following how to combine the different techniques, which complement each other nicely, to provide a feasible numerical operator calculus for some standard many-particle models in quantum chemistry. Within the present work, we focus on the *Hartree-Fock* method and the *Kohn-Sham* equations of *density functional theory* (DFT). We present a brief survey on existing approximation methods, and give some numerical results confirming their efficiency. Our approach aims towards a numerical solution of the Hartree-Fock and Kohn-Sham equations with computational complexity that scales almost linearly in the number of particles (atoms). In particular, large molecular systems such as biomolecules, and nanostructures, reveal severe limitations of the standard numerical algorithms and tensor-product approximations might help to overcome at least some of them.

The rest of the paper is organised as follows. Section 2 gives a brief outline of electronic structure calculations and of the Hartree-Fock method in particular. This is followed by a discussion of best N-term approximation and its generalization to tensor product wavelet bases. We present an application of this approach to the Hartree-Fock method. In Section 4, we first introduce various tensor product formats for the approximation of functions and matrices in higher dimensions. Thereafter we consider a variety of methods to obtain separable approximations of multivariate functions. These methods center around the *Sinc* interpolation and convenient integral representations for these functions. Section 5 provides an overview on different data sparse formats for the univariate components of tensor products. Finally, we discuss in Section 6 possible applications of these tensor-product techniques in order to obtain linear scaling methods for Hartree-Fock and Kohn-Sham equations.

2 Basic principles of electronic structure calculations

The physics of *stationary states*, i.e. time harmonic, *quantum mechanical systems of* N *particles*, is completely described by a single *wave function*

$$(\mathbf{r}_1, s_1, ..., \mathbf{r}_N, s_N) \mapsto \Psi(\mathbf{r}_1, s_1, ..., \mathbf{r}_N, s_N) \in \mathbb{C} \ , \ \mathbf{r}_i \in \mathbb{R}^3 \ , \ s_i \in S \ ,$$

which is a function depending on the spatial coordinates $\mathbf{r}_i \in \mathbb{R}^3$ of the particles $i = 1, \ldots, N$ together with their spin degrees of freedom s_i. Since identical quantum mechanical particles, e.g. electrons, cannot be distinguished, the wave function must admit a certain symmetry with respect to the interchange of particles. The *Pauli exclusion principle* states that for electrons, the spin variables can take only two values $s_i \in S = \{\pm\frac{1}{2}\}$, and the wave function has to be antisymmetric with respect to the permutation of particles

$$\Psi(\mathbf{r}_1, s_1, \ldots, \mathbf{r}_i, s_i, \ldots \mathbf{r}_j, s_j, \ldots, \mathbf{r}_N, s_N) = -\Psi(\mathbf{r}_1, s_1, \ldots, \mathbf{r}_j, s_j, \ldots \mathbf{r}_i, s_i, \ldots, \mathbf{r}_N, s_N) \ .$$

The *Born Oppenheimer approximation* considers a quantum mechanical ensemble of N electrons moving in an exterior electrical field generated by the nuclei of K atoms. Therein the wave function is supposed to be a solution of the stationary electronic Schrödinger equation

$$\mathbf{H}\Psi = E\Psi \ ,$$

with the many-particle Schrödinger operator (non-relativistic Hamiltonian) $\mathbf{H}$ given by

$$\mathbf{H} := -\frac{1}{2}\sum_{i=1}^{N} \Delta_i - \sum_{a=1}^{K}\sum_{i=1}^{N} \frac{Z_a}{|\mathbf{r}_i - \mathbf{R}_a|} + \sum_{i<j\leqslant N} \frac{1}{|\mathbf{r}_i - \mathbf{r}_j|} + \sum_{a<b\leqslant K} \frac{Z_a Z_b}{|\mathbf{R}_a - \mathbf{R}_b|}, \quad (1)$$

where $Z_a, \mathbf{R}_a$ are charges and positions of the nuclei, respectively.

The basic problem in *wave function methods* is to calculate (approximately) the wave functions $\Psi(\mathbf{r}_1, s_1, ..., \mathbf{r}_N, s_N)$, $\mathbf{r}_i \in \mathbb{R}^3$, $s_i = \pm\frac{1}{2}$ as an eigenfunction of the non-relativistic Hamiltonian $\mathbf{H}$. The wave function can be assumed to be real valued, which we will pursue in the sequel, for sake of simplicity. Perhaps in case of translation symmetry, e.g. for bulk crystals, a complex setting might be helpful. One is mostly interested in the ground energy. Due to the Ritz or Courant-Fischer min-max principle [76] this problem can be casted in a variational formulation

$$E = \min\{\langle \mathbf{H}\Phi, \Phi\rangle \ : \langle \Phi, \Phi\rangle = 1\} \ , \ \ \Psi = \operatorname{argmin}\{\langle \mathbf{H}\Phi, \Phi\rangle \ : \langle \Phi, \Phi\rangle = 1\} \ . \quad (2)$$

The wave functions can be approximated by antisymmetric tensor products $\Psi = \sum c_k \Psi_k$ where each Ψ_k denotes a *Slater determinant*

$$\Psi_k(\mathbf{r}_1, s_1, \ldots, \mathbf{r}_N, s_N) = \frac{1}{\sqrt{N!}} \det(\varphi_{k_i}(\mathbf{r}_{k_j}, s_{k_j}))_{i,j=1}^N . \tag{3}$$

Here the functions $\varphi_\nu : \mathbb{R}^3 \times \{\pm\frac{1}{2}\} \to \mathbb{R}$ are supposed to be pairwise orthogonal

$$\langle \varphi_\nu, \varphi_\mu \rangle = \sum_s \int_{\mathbb{R}^3} \varphi_\nu(\mathbf{r}, s) \varphi_\mu(\mathbf{r}, s) d^3 r = \delta_{\mu,\nu} .$$

An approximation by a single Slater determinant (3)

$$\Psi \approx \Psi_{SL} := \frac{1}{\sqrt{N!}} \det(\varphi_i(\mathbf{r}_j, s_j))_{i,j=1}^N ,$$

which is a kind of rank one approximation by an antisymmetric tensor-product gives the Hartree-Fock energy functional

$$\mathcal{E}_{HF}(\varphi_1, \ldots, \varphi_N) := \langle \mathbf{H}\Psi_{SL}, \Psi_{SL} \rangle .$$

Which inserted into (2) yields the following constraint minimization problem

$$E_{HF} = \min\{\mathcal{E}_{HF}(\varphi_1, \ldots, \varphi_N) = \langle \mathbf{H}\Phi_{SL}, \Phi_{SL} \rangle : \ \langle \varphi_i, \varphi_j \rangle = \delta_{i,j} \ , \ i, j = 1, \ldots, N\}$$

with N^2 constraint conditions $\langle \varphi_i, \varphi_j \rangle = \delta_{i,j}$, $i, j = 1, \ldots, N$.

An additional simplification can be made for even number of electrons, restricting pairs of orbitals with opposite spin to the same spatial behaviour $\varphi_i(\mathbf{r}, \frac{1}{2}) = \varphi_{N/2+i}(\mathbf{r}, -\frac{1}{2}) =: \phi_i(\mathbf{r})$, $i = 1, \ldots N/2$. This gives the so called *restricted Hartree-Fock* model for close shell systems. The corresponding Hartree-Fock energy functional can be calculated explicitly

$$\mathcal{E}_{HF}(\phi_1, \ldots, \phi_{N/2}) = 2 \sum_{i=1}^{N/2} \int_{\mathbb{R}^3} \left(\frac{1}{2} |\nabla \phi_i(\mathbf{r})|^2 + V_c(\mathbf{r}) |\phi_i(\mathbf{r})|^2 \right) d^3 r +$$
$$2 \sum_{i=1}^{N/2} \sum_{j=1}^{N/2} \int_{\mathbb{R}^3} \int_{\mathbb{R}^3} \left(\frac{\phi_i(\mathbf{r})\phi_i(\mathbf{r})\phi_j(\mathbf{r}')\phi_j(\mathbf{r}')}{|\mathbf{r} - \mathbf{r}'|} - \frac{\phi_i(\mathbf{r})\phi_i(\mathbf{r}')\phi_j(\mathbf{r})\phi_j(\mathbf{r}')}{2|\mathbf{r} - \mathbf{r}'|} \right) d^3 r' d^3 r, \tag{4}$$

where V_c denotes the Coulomb potential due to the nuclei. Variational calculus applied to the restricted Hartree-Fock energy functional

$$E_{HF} = \min_{\langle \phi_i, \phi_j \rangle = \delta_{i,j}} \mathcal{E}_{HF}(\phi_1, \ldots, \phi_{N/2})$$

yields the Hartree-Fock equations

$$\mathcal{F}\phi_i(\mathbf{r}) = \varepsilon_i \, \phi_i(\mathbf{r}), \tag{5}$$

with the Fock operator $\mathcal{F}$, as a necessary condition, that for a minimizer there exist pairwise orthogonal orbitals ϕ_i, $i = 1, \ldots, N/2$. Defining the *reduced one-electron spin density matrix*, or simply "density matrix" in the following, as the kernel function

$$\rho(\mathbf{r}, \mathbf{r}') := \sum_{i=1}^{N/2} \phi_i^*(\mathbf{r})\phi_i(\mathbf{r}') \tag{6}$$

of the corresponding spectral projection operator P_ρ, and introducing the total electron density

$$n(\mathbf{r}) := 2\rho(\mathbf{r}, \mathbf{r}), \tag{7}$$

the Fock operator is given by

$$\mathcal{F}\phi(\mathbf{r}) = -\frac{1}{2}\Delta\phi(\mathbf{r}) + V_c(\mathbf{r})\,\phi(\mathbf{r}) + V_H\phi(\mathbf{r}) + (\mathcal{K}\phi)\,(\mathbf{r}), \tag{8}$$

with *Hartree potential*

$$V_H(\mathbf{r}) := \int_{\mathbb{R}^3} \frac{n(\mathbf{r}')}{|\mathbf{r} - \mathbf{r}'|} d^3r', \tag{9}$$

and *exchange operator*

$$(\mathcal{K}\phi)\,(\mathbf{r}) := -\int_{\mathbb{R}^3} \frac{\rho(\mathbf{r}, \mathbf{r}')}{|\mathbf{r} - \mathbf{r}'|}\,\phi(\mathbf{r}')d^3r'. \tag{10}$$

This is a coupled nonlinear eigenvalue problem, since the Fock operator depends on the density matrix (6), where ϕ_i, $i = 1, \ldots, N/2$ are the eigenfunctions corresponding to the $N/2$ lowest eigenvalues $\varepsilon_1 \leqslant \ldots \varepsilon_{N/2} < \varepsilon_{N/2+1} \leqslant \ldots$.

A rank one approximation as in the Hartree-Fock model seems to be a rather poor approximation to the wave function. One way to pursue is to use more than one Slater determinant to approximate the wave function. This results in methods like *configuration interaction* (CI), *multi configuration self-consistent-field* (MCSCF), *coupled cluster* (CC) methods and other. All these methods have in common that they require a much larger computational effort. As a general reference for these methods, we refer to the monograph [63].

In contrast to these wave function methods, DFT [75] tries to replace the linear, but high-dimensional Schrödinger operator by nonlinear but low dimensional partial differential equations with the same ground state energy and which look formally like the Hartree-Fock equations. The major difference is that in the Kohn-Sham equations, the nonlocal exchange term $\mathcal{K}$ is replaced by a local *exchange-correlation potential* V_x depending only on the electron density n. However the functional dependence turns out to be very complicated and is not known explicitly. The most simple form is called the *local density approximation* (LDA) [75] where the exchange correlation potential is a nonlinear function of the density

$$V_x(\mathbf{r}) = -\frac{1}{\pi}(3\pi^2 n(\mathbf{r}))^{\frac{1}{3}} + \text{correction terms}.$$

Electron correlation can be incorporated into the correction term, which is usually taken from quantum Monte Carlo calculations for a homogeneous electron gas. The corresponding Euler-Lagrange equations take the form (5) with Fock operator

$$\mathcal{F}\phi(\mathbf{r}) := -\frac{1}{2}\Delta\phi(\mathbf{r}) + \Big(V_c(\mathbf{r}) + V_H(\mathbf{r}) + V_x[n](\mathbf{r})\Big)\phi(\mathbf{r}).$$

Since there are no nonlocal terms, apparently the computation of these density functional models becomes more simpler and efficient as Hartree-Fock computations. Nevertheless, the exact functional is still unknown and all known approximate functionals have certain deficiencies. Therefore a model error is still intrinsical in all of these methods. It should be mentioned, that actually so-called hybride models, i.e. convex combinations of Kohn-Sham and Hartree-Fock models, have shown the best performance in benchmark calculations for molecules.

Usually the orbitals ϕ_i are approximated either by atomic centered basis functions e.g. *Gaussian-type orbitals* (GTO) or even harmonic polynomials (plane waves). A new alternative approach is based on multi-scale wavelet bases, cf. [2, 34, 43, 62] and references therein. Due to the Coulomb singularity of the nuclear potential, the orbitals are only smooth away from the nuclei. To get rid of these electron nuclear cusps the core electrons are added to smooth the nuclear potential. The Coulomb potential is thereby replaced by a pseudopotential and only the valence electrons are considered explicitly in an external field generated by the pseudopotentials [30].

2.1 Density matrix formulation of Hartree-Fock and DFT models

Solving the spectral problem corresponding to the Hartree-Fock equations (5) leads to $\mathcal{O}(N^3)$ complexity at least. In order to achieve algorithms which scale linear in the number of electrons N it therefore turns out to be necessary to circumvent the computation of the eigenvalue problem. Instead it is possible to reformulate the Hartree-Fock method in terms of the density matrix (6). It has been already mentioned before that the nonlinear part of the Fock operator (8) can be expressed in terms of the density matrix. Furthermore it is not hard to see that the Hartree-Fock energy functional (4) can be rewritten as a functional of the density matrix $\mathcal{E}_{HF}(\rho)$ where the orthogonality constraints of the orbitals have to be replaced by the idempotency condition of the density matrix considered as a spectral projector, i.e.,

$$\int_{\mathbb{R}^3} \rho(\mathbf{r}, \mathbf{r}'')\rho(\mathbf{r}'', \mathbf{r}')\, d^3r'' - \rho(\mathbf{r}, \mathbf{r}') = 0.$$

The density matrix plays the key role in order to achieve linear scaling in Hartree-Fock and DFT methods. For instance it is well known that the density matrix exhibits exponential decay $\rho(\mathbf{r},\mathbf{r}') \sim \exp(-\lambda|\mathbf{r}-\mathbf{r}'|)$ for nonmetallic systems. This so-called shortsightedness of the density matrix [68] enables e.g. an efficient treatment of the nonlocal exchange term (10) in the Fock operator. Various computational schemes, entirely based on the density matrix, exist to perform Hartree-Fock or DFT calculations, cf. [4, 7, 70, 71, 73] and references therein. In Section 6, we discuss an approach using the *sign* function of an operator which seems to be especially suitable for our purposes.

3 Hyperbolic cross approximation in wavelet bases

The idea of *sparse grids* or *hyperbolic cross approximation* is based on the following observation. Let us consider a complete orthonormal basis in $L_2(\mathbb{R})$, $\{\psi_{l,k} : l \in \mathbb{N} \cup \{-1,0\},\ k \in \mathcal{J}_l\}$ with $\sharp\mathcal{J}_l \sim 2^l$, which obeys the following approximation property

$$\|u - \sum_{l=-1}^{L-1} \sum_{k} u_{l,k}\psi_{l,k}\| \leqslant C_R 2^{-Ls}\|u\|_s \quad \text{with } u_{l,k} = \langle \psi_{l,k}, u\rangle,$$

with respect to the Sobolev spaces $H^s(\mathbb{R})$, $s > 0$. Typical bases are e.g. wavelet bases or trigonometric polynomials $\psi_{l,k}(x) = e^{2\pi i(2^l+k)x}$, $l \geqslant 0$. The orthogonality can be relaxed to a biorthogonal setting, which also includes hierarchical nodal basis functions as originally proposed by Zenger. We refer to [49, 81] and references therein for the detailed exposition of this approach.

Setting $W_l := \operatorname{span}\{\psi_{l,k} : k \in \mathcal{J}_l\}$, the tensor-products $\Psi_{\mathbf{l},\mathbf{k}}(\mathbf{x}) = \psi_{l_1,k_1}(x_1) \cdots \psi_{l_d,k_d}(x_d)$, $l_i \geqslant -1$, $k_i \in \mathcal{J}_{l_i}$, $i = 1,\ldots,d$, form an orthogonal basis in $L_2(\mathbb{R}^d)$. Taking the spaces $Z_L = \sum_{l_1+\cdots+l_d \leqslant L} W_{l_1} \otimes \cdots \otimes W_{l_d}$, we then get $\dim Z_L = L^d 2^L$ and

$$\|u - u_{Z_l}\| \lesssim 2^{-Ls}\|u\|_{s,\ldots,s}.$$

This means that one obtains the convergence rate 2^{-Ls} with $L^d 2^L$ degrees of freedom, instead of the usual complexity 2^{Ld}. The price to pay is that one has to require a slightly higher regularity in terms of the mixed or tensor-product Sobolev norms $\|.\|_{s,\ldots,s}$ in $H^{s,\ldots,s} = \bigotimes_{i=1}^{d} H^s(\mathbb{R})$. Therefore, for functions satisfying this regularity requirement we can get rid of the curse of dimensions. Up to a logarithmic term, we can achieve asymptotically the same complexity as in one dimension. It has been proven by Yserentant [88] that eigenfunctions of the many-particle Schrödinger operator (1) belong to Sobolev spaces of mixed partial derivatives. This result enables the construction of sparse grid approximations for the entire wavefunction [47, 48, 89].

Practically the sparse grid approach is limited because the Riesz constant C_R of the basis enters by C_R^d. Usually this becomes rather large with increasing d. Recent experiences demonstrate that hyperbolic cross approximations or sparse grid approximations can be applied successfully for a moderate number of dimensions $d \leqslant 10, ..., 30$. In the application we have in mind, namely the numerical solution of the Hartree-Fock or Kohn-Sham equation, the orbitals are functions in $\mathbb{R}^3$ and the operators have kernels in $\mathbb{R}^6$. This makes the sparse grid approach highly attractive for the present problem.

Hyperbolic cross approximations can be used also in an adaptive setting. Results for graded meshes have been obtained by Griebel et al. (c.f. [47]). Schwab and Nitsche (cf. [72]) have considered point singularities, and demonstrated that an adaptive sparse grid approximation works well. In fact, wavelet bases are highly advantageous for local adaptive approximation. This can be explained best in the framework of a *best N-term approximation* [16, 17, 20, 28]. For $\tau \leqslant 2$, the space $l_\tau^w(\mathcal{J})$ is the collection of sequences, respectively infinite vectors, $\mathbf{u} \in l_2(\mathcal{J})$, satisfying

$$\sharp \{\lambda \in \mathcal{J} : |u_\lambda| > \eta\} \lesssim \eta^{-\tau}$$

for all $\eta > 0$. The quantity

$$\|\mathbf{u}\|_{l_\tau^w(\mathcal{J})} := \|\mathbf{u}\|_{l_2(\mathcal{J})} + |\mathbf{u}|_{l_\tau^w(\mathcal{J})} \quad \text{with } |\mathbf{u}|_{l_\tau^w(\mathcal{J})} := \sup_{\eta>0}(\, \eta\, [\sharp\{\lambda \in \mathcal{J} : |u_\lambda| > \eta\}]^{1/\tau})$$

defines a quasi-norm in l_τ^w. Rearranging $\mathbf{u} \in l_2(\mathcal{J})$ by a non-increasing sequence $\mathbf{u}^* = (u_k^*)_{k\in\mathbb{N}}$, i.e. $|u_k^*| \leqslant |u_{k-1}^*|$, we have an alternative representation of this quasi-norm via

$$|\mathbf{u}|_{l_\tau^w(\mathcal{J})} = \sup_{k>0}\, (k^{1/\tau}\, |u_k^*|)$$

and, if $\tau < 2$,

$$\|\mathbf{u}\| = \|\mathbf{u}\|_{l_2(\mathcal{J})} \leqslant \|\mathbf{u}\|_{l_\tau^w(\mathcal{J})}.$$

The quantity

$$\sigma_N(\mathbf{u}) := \inf_{\sharp\, \mathrm{supp}\, \mathbf{v} \leqslant N} \|\mathbf{u} - \mathbf{v}\| = (\sum_{k>N} |u_k^*|^2)^{1/2}$$

denotes the error of the *best* N*-term approximation of* $\mathbf{u}$. An approximation $\mathbf{v}$ satisfying $\|\mathbf{u} - \mathbf{v}\| \lesssim \sigma_N(\mathbf{u})$ is given by $\mathbf{v}^*$ where $v_k^* = u_k^*$ if $k = 1, \ldots, N$ and $v_k^* = 0$, for $k > N$.

Proposition 1. [16, 21]

1. *For* $\mathbf{u} \in l_2(\mathcal{J})$ *and* $s > 0$ *the estimate* $\sigma_N(\mathbf{u}) \lesssim N^{-s}$ *holds if and only if* $\mathbf{u} \in l_\tau^w(\mathcal{J})$ *with* $\frac{1}{\tau} = s + \frac{1}{2}$ *and* $\sigma_N(\mathbf{u}) \lesssim N^{-s}\|\mathbf{u}\|_{l_\tau^w(\mathcal{J})}$.
2. *For* $0 < \tau < \tau' \leqslant 2$ *there holds*

$$l_\tau(\mathcal{J}) \subset l_\tau^w(\mathcal{J}) \subset l_{\tau'}(\mathcal{J}).$$

3. *The wavelet expansion of a function* $u = \sum_{l,k} u_{l,k}\psi_{l,k}$ *belongs to the Besov space* $B^s_{\tau,\tau}(\mathbb{R})$ *if and only if* $(u_{l,k}) \in l_\tau(\mathcal{J})$ *where* $\frac{1}{\tau} = s + \frac{1}{2}$.

In a recent paper, Nitsche [72] has shown that the quasi Banach spaces $l_\tau(\mathbb{Z}^d)$, $0 < \tau < 1$, are also tensor-product spaces.

Theorem 1. [72] *For* $0 < \tau \leqslant 2$ *there holds* $l_\tau(\mathbb{Z}^d) = \bigotimes_{i=1}^d l_\tau(\mathbb{Z})$.

In the case of a tensor-product wavelet expansion of a function $v = \sum v_\lambda \psi_\lambda$ in the Besov space $\bigotimes_{i=1}^d B^{s^*}_{\tau,\tau}(\mathbb{R})$ we therefore obtain the following rate for the best N-term approximation

$$\inf_{\sharp \operatorname{supp} \mathbf{v}_N \leqslant N} \|v - v_N\| \lesssim N^{-s} \|v\|_{\bigotimes_{i=1}^d B^{s^*}_{\tau,\tau}(\mathbb{R})}$$

for all $0 \leqslant s < s^*$. Note that due to $s < s^*$ no logarithmic term is present in the above estimate. A similar but more complicated result is true for the energy space $H^1(\mathbb{R}^3)$. Let us note that this means that asymptotically the number of degrees of freedoms required to obtain an accuracy ε behaves (almost) like the corresponding number for a one-dimensional problem.

In order to apply Nitsche's theorem to solutions of Hartree-Fock and Kohn-Sham equations it becomes necessary to establish an asymptotic smoothness property for the behaviour of the solutions at the atomic nuclei. This has been achieved in the Hartree-Fock case by applying the calculus of pseudodifferential operators on manifolds with conical singularities. Our results can be summarized in the following theorem:

Theorem 2. [35, 37] *The self-consistent-field solutions* ϕ_i *of the Hartree-Fock equations, obtained via the level-shifting algorithm, satisfy the asymptotic smoothness property*

$$|\partial^\beta_{\mathbf{x}} \phi_i(\mathbf{x})| \leqslant C_\beta |\mathbf{x} - \mathbf{A}|^{1-|\beta|} \qquad \text{for } \mathbf{x} \neq \mathbf{A} \text{ and } |\beta| \geqslant 1$$

in a bounded neighbourhood $\Omega \subset \mathbb{R}^3$ *of a nucleus at* $\mathbf{A}$. *This is sufficient to ensure* $\phi_i \in \bigotimes_{i=1}^3 B^{s^*}_{\tau,\tau}(\Omega)$ *for each* $s^* > 0$ *and* $\frac{1}{\tau} = s^* + \frac{1}{2}$.

Here we have used the standard short-hand notation for mixed partial derivatives

$$\partial^\beta := \frac{\partial^{\beta_1}}{\partial x_1^{\beta_1}} \frac{\partial^{\beta_2}}{\partial x_2^{\beta_2}} \frac{\partial^{\beta_3}}{\partial x_3^{\beta_3}},$$

with the absolute value of the multi-index $|\beta| := \beta_1 + \beta_2 + \beta_3$. The iterative solutions of the Hartree-Fock equations to which the theorem applies refer to the so-called level-shifting algorithm with an appropriate initial guess. This algorithm corresponds to a fixed point iteration scheme for which convergence has been proven by Cancès and Le Bris [13]. It should be mentioned however that so far no proof exists that every solution of the Hartree-Fock equations can be obtained via the level-shifting algorithm. In view of (6), this theorem can be immediately extended to the density matrix.

Corollary 1. *The one-electron reduced density matrix* $\rho(\mathbf{x}, \mathbf{y})$ *belongs to* $\bigotimes_{i=1}^{6} B^{s^*}_{\tau,\tau}(\Omega)$ *for each* $s^* > 0$ *and* $\frac{1}{\tau} = s^* + \frac{1}{2}$.

For further applications of the best N-term approximation to post Hartree-Fock methods, we refer to [36].

Since the univariate basis functions and therefore the tensor-products are fixed from the beginning, a sparse grid approximation will be by far not an optimal tensor-product approximation with respect to the separation rank. In general we expect that the separation rank of an "optimal" tensor-product approximation [3] is much smaller than for a sparse grid approximation of comparable accuracy. For example, let $f, g \in H^s(\mathbb{R})$, and $\|f - f_L\| \lesssim 2^{-Ls}$, $\|g - g_L\| \lesssim 2^{-Ls}$. Then $F(x, y) = f(x)g(y)$ has separation rank $r_{opt} = 1$. The sparse grid approximation F_{Z_L} has the same approximation rate 2^{-Ls} and $\mathcal{O}(L^2 2^L)$ degrees of freedom, but a relatively large separation rank $L2^L >> 1$. We will see in a moment that the sparse grid approximation is not too bad. Because, to store both functions f_L and g_L with respect to the given basis requires $2 \cdot 2^L$ coefficients, whereas the sparse grid approximation requires $\mathcal{O}(L^2 2^L)$ nonzero coefficients in contrast to $\mathcal{O}(2^{dL})$ for the full product. Keeping in mind that a really optimal tensor-product approximation for $d > 2$ is still an unsolved problem, and in general it might be quite expensive, the sparse grids approximation is simple and cheap from the algorithmic point of view. It achieves also an almost optimal complexity for storage requirements. It is a trivial task to convert an "optimal" tensor-product representation into a sparse grid approximation. The opposite direction is a highly nontrivial task and requires fairly sophisticated compression algorithms.

It is worthwhile to mention that previous wavelet matrix compression approaches are based on some Calderón-Zygmund type estimates for the kernels. The sparse grid approximation is intimately related to *wavelet matrix compression* of integral operators with globally smooth kernels. The kernel functions of *Calderón-Zygmund operators* are not globally smooth. Nevertheless, it can be shown that they can be approximated within *linear* or *almost linear complexity* by means of wavelet Galerkin methods see e.g. [8, 17–19, 77], since they are smooth in the far field region. This result is proved, provided that the Schwartz kernel $K(\mathbf{x}, \mathbf{y})$ in $\mathbb{R}^d \times \mathbb{R}^d$ is approximated by tensor-product bases $\Psi \otimes \Psi$, where Ψ is an isotropic wavelet basis in $\mathbb{R}^d$. Recently developed fast methods like wavelet matrix compression and hierarchical matrices are working well for isotropic basis functions or isotropic clusters. Corresponding results for sparse grid approximations with $\bigotimes_{i=1}^{2d} \Psi^i$ have not been derived so far. Tensor-product bases in the framework of sparse grids do not have this geometric

[3] It should be mentioned that in our applications at best almost optimal tensor-product approximations can be achieved. This is not of particular significance since we are aiming at a certain accuracy and small variations of the separation rank, required in order to achieve this accuracy, do not cause much harm.

isotropy, which might spoil the efficiency of these methods. This is not the case for more general tensor-product approximations of these operators discussed in Sections 4.2.2 and 4.2.3 below. Therefore tensor-product approximations will provide an appropriate and efficient tool handling nonlocal operators acting on functions which are represented by means of tensor-product (sparse grid) bases. The development of such a tool will play a fundamental role for dealing with operators in high dimensions.

4 Toolkit for tensor-product approximations

The numerical treatment of operators in higher dimensions arising in traditional *finite element methods* (FEM) and *boundary element methods* (BEM) as well as in quantum chemistry, material sciences and financial mathematics all have in common the fundamental difficulty that the computational cost of traditional methods usually has an exponential growth in d even for algorithms with linear complexity $\mathcal{O}(N)$ in the problem size N (indeed, N scales exponentially in d as $N = n^d$, where n is the "one dimensional" problem size).

There are several approaches to remove the dimension parameter d from the exponent (cf. [5, 41, 49, 53, 58]). For the approximation of functions, such methods are usually based on different forms of the separation of variables. Specifically, a multivariate function $F : \mathbb{R}^d \to \mathbb{R}$ can be approximated in the form

$$F_r(x_1, ..., x_d) = \sum_{k=1}^{r} s_k \Phi_k^{(1)}(x_1) \cdots \Phi_{(k)}^{d}(x_d) \approx F,$$

where the set of functions $\{\Phi_k^{(\ell)}(x_\ell)\}$ can be fixed, like the best N-term approximation discussed in Section 3, or chosen adaptively. The latter approach tries to optimize the functions $\{\Phi_k^{(\ell)}(x_\ell)\}$ in order to achieve for a certain *separation rank* r at least the almost optimal approximation property. By increasing r, the approximation can be made as accurate as desired. In the case of globally analytic functions there holds $r = \mathcal{O}(|\log \varepsilon|^{d-1})$, while for analytic functions with point singularities one can prove $r = \mathcal{O}(|\log \varepsilon|^{2(d-1)})$ (cf. [53]).

In the following we want to give a short overview of various approaches to generate separable approximations with low separation rank. We first introduce in Section 4.1 two different tensor-product formats which have been used in the following. Section 4.2 provides a succint discussion of low rank tensor-product approximations of special functions, including the Coulomb and Yukawa potential, for which a certain type of "seperable" integral representation exists. This integral representation can be used to obtain separable approximations either by applying the *Sinc* approximation (Section 4.2.1) or directly through a best N-term approximation of exponential sums (Section 4.2.2).

4.1 Tensor-product representations in higher dimension

Let a d-th order tensor $\mathcal{A} = [a_{i_1 \dots i_d}] \in \mathbb{C}^{\mathcal{I}}$ be given, defined on the product index set $\mathcal{I} = I_1 \times \dots \times I_d$. It can be approximated via the *canonical decomposition* (CANDECOMP) or *parallel factors* (PARAFAC) model (shortly, canonical model) in the following manner

$$\mathcal{A} \approx \mathcal{A}_{(r)} = \sum_{k=1}^{r} b_k V_k^{(1)} \otimes \dots\dots \otimes V_k^{(d)}, \quad b_k \in \mathbb{C}, \tag{11}$$

where the Kronecker factors $V_k^{(\ell)} \in \mathbb{C}^{I_\ell}$ are unit-norm vectors which are chosen such that for a certain approximation only a minimal number r of components in the representation (11) are required. The minimal number r is called the *Kronecker rank* of a given tensor $\mathcal{A}_{(r)}$. Here and in the following we use the notation $\otimes$ to represent the canonical tensor

$$\mathcal{U} \equiv [u_{\mathbf{i}}]_{\mathbf{i} \in \mathcal{I}} = b\, U^{(1)} \otimes \dots \otimes U^{(d)} \in \mathbb{C}^{\mathcal{I}},$$

defined by $u_{i_1 \dots i_d} = b \cdot u_{i_1}^{(1)} \cdots u_{i_d}^{(d)}$ with $U^{(\ell)} \equiv [u_{i_\ell}^{(\ell)}]_{i_\ell \in I_\ell} \in \mathbb{C}^{I_\ell}$. We make use of the multi-index notation $\mathbf{i} := (i_1, \dots, i_d) \in \mathcal{I}$.

The Tucker model deals with the approximation

$$\mathcal{A} \approx \mathcal{A}_{(\mathbf{r})} = \sum_{k_1=1}^{r_1} \dots \sum_{k_d=1}^{r_d} b_{k_1 \dots k_d} V_{k_1}^{(1)} \otimes \dots \otimes V_{k_d}^{(d)}, \tag{12}$$

where the Kronecker factors $V_{k_\ell}^{(\ell)} \in \mathbb{C}^{I_\ell}$ ($k_\ell = 1, \dots, r_\ell$, $\ell = 1, \dots, d$) are complex vectors of the respective size $n_\ell = |I_\ell|$, $\mathbf{r} = (r_1, \dots, r_d)$ (the Tucker rank) and $b_{k_1 \dots k_d} \in \mathbb{C}$. Without loss of generality, we assume that the vectors $\{V_{k_\ell}^{(\ell)}\}$ are orthonormal, i.e.,

$$\left\langle V_{k_\ell}^{(\ell)}, V_{m_\ell}^{(\ell)} \right\rangle = \delta_{k_\ell, m_\ell}, \quad k_\ell, m_\ell = 1, \dots, r_\ell;\ \ell = 1, \dots, d,$$

where δ_{k_ℓ, m_ℓ} is Kronecker's delta.

On the level of operators (matrices) we distinguish the following tensor-product structures. Given a matrix $A \in \mathbb{C}^{N \times N}$ with $N = n^d$, we approximate it with the canonical model by a matrix $A_{(r)}$ of the form

$$A \approx A_{(r)} = \sum_{k=1}^{r} V_k^{(1)} \otimes \cdots \otimes V_k^{(d)}, \tag{13}$$

where the $V_k^{(\ell)}$ are hierarchically structured matrices of order $n \times n$. Again the important parameter r is denoted as the Kronecker rank.

We also introduce the following rank-$(r_1, ..., r_d)$ Tucker-type tensor-product matrix format

$$\mathcal{A} = \sum_{k_1=1}^{r_1} \cdots \sum_{k_d=1}^{r_d} b_{k_1 \dots k_d} V_{k_1}^{(1)} \otimes \dots \otimes V_{k_d}^{(d)} \in \mathbb{R}^{I_1^2 \times \dots \times I_d^2}, \tag{14}$$

where the Kronecker factors $V_{k_\ell}^{(\ell)} \in \mathbb{R}^{I_\ell \times I_\ell}$, $k_\ell = 1, ..., r_\ell$, $\ell = 1, ..., d$, are matrices of a certain structure (say, $\mathcal{H}$-matrix, wavelet based format, Toeplitz/circulant, low-rank, banded, etc.). The matrix representation in the form (14) is a model reduction which is a generalisation of the low-rank approximation of matrices, corresponding to the case $d = 2$. For a class of matrix-valued functions (cf. [53, 58] and Section 6.1 below) it is possible to show that $r = \mathcal{O}(|\log \varepsilon|^{2(d-1)})$. Further results on the tensor-product approximation to certain matrix-valued functions can be found in [41, 54].

Note that *algebraic recompression methods* based on the *singular value decomposition* (SVD) cannot be directly generalised to $d \geqslant 3$. We refer to [5, 6, 25–27, 33, 58, 59, 64, ?,67, 74, 90] and references therein for detailed description of the methods of numerical multi-linear algebra. In the following, we stress the significance of analytical methods for the separable approximation of multivariate functions and related function-generated matrices/tensors.

4.2 Separable approximation of functions

Separable approximation of functions plays an important role in the design of effective tensor-product decomposition methods. For a large class of functions (cf. [84, 85]) it is possible to show that tensor-product approximations with low separation rank exist. In this section, we overview the most commonly used methods to construct separable approximations of multivariate functions.

4.2.1 ***Sinc*** **interpolation methods** *Sinc-approximation* methods provide the efficient tools for interpolating C^∞ functions on $\mathbb{R}$ having exponential decay as $|x| \to \infty$ (cf. [80]). Let

$$S_{k,\mathfrak{h}}(x) = \frac{\sin[\pi(x - k\mathfrak{h})/\mathfrak{h}]}{\pi(x - k\mathfrak{h})/\mathfrak{h}} \qquad (k \in \mathbb{Z},\ \mathfrak{h} > 0,\ x \in \mathbb{R})$$

be the k-th *Sinc* function with step size $\mathfrak{h}$, evaluated at x. Given f in the Hardy space $H^1(D_\delta)$ with respect to the strip $D_\delta := \{z \in \mathbb{C} : |\Im z| \leqslant \delta\}$ for a $\delta < \frac{\pi}{2}$. Let $\mathfrak{h} > 0$ and $M \in \mathbb{N}_0$, the corresponding *Sinc*-interpolant (cardinal series representation) and quadrature read as

$$C_M(f, \mathfrak{h}) = \sum_{k=-M}^{M} f(k\mathfrak{h}) S_{k,\mathfrak{h}}, \quad T_M(f, \mathfrak{h}) = \mathfrak{h} \sum_{k=-M}^{M} f(k\mathfrak{h}),$$

where the latter approximates the integral

$$I(f) = \int_{\mathbb{R}} f(x)dx.$$

For the interpolation error, the choice $\mathfrak{h} = \sqrt{\pi\delta/bM}$ implies the exponential convergence rate

$$\|f - C_M(f, \mathfrak{h})\|_\infty \leqslant CM^{1/2}e^{-\sqrt{\pi\delta bM}}.$$

Similarly, for the quadrature error, the choice $\mathfrak{h} = \sqrt{2\pi\delta/bM}$ yields

$$|I(f) - T_M(f, \mathfrak{h})| \leqslant Ce^{-\sqrt{2\pi\delta bM}}.$$

If f *has a double-exponential decay* as $|x| \to \infty$, i.e.,

$$|f(\xi)| \leqslant C\exp(-be^{a|\xi|}) \qquad \text{for all } \xi \in \mathbb{R} \text{ with } a, b, C > 0,$$

the convergence rate of both *Sinc*-interpolation and *Sinc*-quadrature can be improved up to $\mathcal{O}(e^{-cM/\log M})$.

For example, let $d = 2$. Given a function $F(\zeta, \eta)$ defined in the product domain $\Omega := [0, 1] \times [a, b]$, $a, b \in \mathbb{R}$, we assume that for each fixed $\eta \in [a, b]$, the univariate function $F(\cdot, \eta)$ belongs to $C^\infty(0, 1]$ and allows a certain holomorphic extension (with respect to ζ) to the complex plane $\mathbb{C}$ (cf. [53] for more details). Moreover, the function $F(\cdot, \eta)$ restricted onto $[0, 1]$ is allowed to have a singularity with respect to ζ at the end-point $\zeta = 0$ of $[0, 1]$. Specifically, it is assumed that there is a function $\phi : \mathbb{R} \to (0, 1]$ such that for any $\eta \in [a, b]$ the composition $f(x) = F(\phi(x), \eta)$ belongs to the class $H^1(D_\delta)$. For this class of functions a separable approximation is based on the transformed *Sinc*-interpolation [41, 80] leading to

$$F_M(\zeta, \eta) = \sum_{k=-M}^{M} F(\phi(k\mathfrak{h}), \eta) S_{k,\mathfrak{h}}(\phi^{-1}(\zeta)) \approx F(\zeta, \eta).$$

The following error bound

$$\sup_{\zeta \in [a,b]} |F(\zeta, \eta) - F_M(\zeta, \eta)| \leqslant Ce^{-sM/\log M} \tag{15}$$

holds with $\phi^{-1}(\zeta) = \text{arsinh}(\text{arcosh}(\zeta^{-1}))$. In the case of a multivariate function in $[0, 1]^{d-1} \times [a, b]$, one can adapt the corresponding tensor-product approximation by successive application of the one-dimensional interpolation (cf. [53]). In the numerical example shown in Fig. 1), we approximate the Euclidean distance $|x - y|$ in $\mathbb{R}^3$ on the domain $|x_i - y_i| \leqslant 1$ $(i = 1, 2, 3)$, by the *Sinc*-interpolation. To that end, the approximation (15) applies to the function $F(\zeta, \eta, \vartheta) = \sqrt{\zeta^2 + \eta^2 + \vartheta^2}$ in $\Omega := [0, 1]^3$.

4.2.2 Integral representation methods *Integral representation methods* are based on the quadrature approximation of integral Laplace-type transforms representing spherically symmetric functions. In particular, some functions of the Euclidean distance in $\mathbb{R}^d$, say,

$$1/|x-y|, \quad |x-y|^\beta, \quad e^{-|x-y|}, \quad e^{-\lambda|x-y|}/|x-y|, \quad x,y \in \mathbb{R}^d,$$

can be approximated by *Sinc*-quadratures of the corresponding Gaussian integral on the semi-axis [41, 53, 54, 65].

For example, in the range $0 < a \leqslant |x-y| \leqslant A$, one can use the integral representation

$$\frac{1}{|x-y|} = \frac{1}{\sqrt{\pi}} \int_{\mathbb{R}} \exp(-|x-y|^2 t^2)dt = \int_{\mathbb{R}} F(\rho;t)dt, \quad x,y \in \mathbb{R}^d \tag{16}$$

of the Coulomb potential with

$$F(\rho;t) = \frac{1}{\sqrt{\pi}} e^{-\rho^2 t^2}, \quad \rho = |x-y|, \quad d=3.$$

After the substitution $t = \log(1+e^u)$ and $u = \sinh(w)$ in the integral (16), we apply the quadrature to obtain

$$T_M(F,\mathfrak{h}) := \mathfrak{h} \sum_{k=-M}^{M} \cosh(k\mathfrak{h}) G(\rho, \sinh(k\mathfrak{h})) \approx \int_{\mathbb{R}} F(\rho,t)dt = \frac{1}{\rho} \tag{17}$$

with $G(\rho,u) = \frac{2}{\sqrt{\pi}} \frac{e^{-\rho^2 \log^2(1+e^u)}}{1+e^u}$ and with $\mathfrak{h} = C_0 \log M/M$. The quadrature (17) is proven to converge exponentially in M,

$$E_M := \left| \frac{1}{\rho} - T_M(F,\mathfrak{h}) \right| \leqslant Ce^{-sM/\log M},$$

where C, s do not depend on M (but depend on ρ), see [53]. With the proper scaling of the Coulomb potential, one can apply this quadrature in the reference interval $\rho \in [1,R]$. A numerical example for this quadrature with values $\rho \in [1,R]$, $R \leqslant 5000$, is presented in Fig. 2. We observe almost linear error growth in ρ.

In electronic structure calculations, the Galerkin discretisation of the Coulomb potential in tensor-product wavelet bases is of specific interest. For simplicity, we consider an isotropic 3d-wavelet basis

$$\gamma_{j,\mathbf{a}}^{(\mathbf{s})}(\mathbf{x}) := \psi_{j,a_1}^{(s_1)}(x_1)\, \psi_{j,a_2}^{(s_2)}(x_2)\, \psi_{j,a_3}^{(s_3)}(x_3),$$

where the functions $\psi_{j,a}^{(0)}(x) := 2^{j/2}\psi^{(0)}(2^j x - a)$, $\psi_{j,a}^{(1)}(x) := 2^{j/2}\psi^{(1)}(2^j x - a)$, with $j, a \in \mathbb{Z}$, correspond to univariate scaling functions and wavelets, respectively. The *nonstandard* representation of the Coulomb potential (cf. [8, 34])

requires integrals of the form

$$\int_{\mathbb{R}^3}\int_{\mathbb{R}^3} \gamma_{j,\mathbf{a}}^{(\mathbf{p})}(\mathbf{x}) \frac{1}{|\mathbf{x}-\mathbf{y}|} \gamma_{j,\mathbf{b}}^{(\mathbf{q})}(\mathbf{y})\, d^3x d^3y = \frac{2^{-2j+1}}{\sqrt{\pi}} \int_0^\infty \mathcal{I}^{(\mathbf{p},\mathbf{q})}(t, \mathbf{a}-\mathbf{b}) dt,$$

with

$$\mathcal{I}^{(\mathbf{p},\mathbf{q})}(t,\mathbf{a}) = G^{(p_1,q_1)}(a_1,t)\, G^{(p_2,q_2)}(a_2,t)\, G^{(p_3,q_3)}(a_3,t),$$

and

$$G^{(p,q)}(a,t) = \int_{\mathbb{R}}\int_{\mathbb{R}} \psi^{(p)}(x-a)\, e^{-(x-y)^2 t^2}\, \psi^{(q)}(y)\, dx dy.$$

In order to benefit from the tensor-product structure, it is important to have a uniform error bound with respect to the spatial separation $|\mathbf{a}-\mathbf{b}|$ of the wavelets. Recently, the following theorem was proven by Schwinger [79]

Theorem 3. *Given a univariate wavelet basis $\psi_{j,a}^{(p)}$ which satisfies*

$$\left| \int \psi^{(p)}(x-y)\, \psi^{(q)}(y)\, dy \right| \lesssim e^{-c|x|} \text{ for } c > 0.$$

Then for any $\delta < \frac{\pi}{4}$, the integration error of the exponential quadrature rule (cf. [80]) with $h = \sqrt{\frac{\pi\delta}{M}}$ ($h = \sqrt{\frac{2\pi\delta}{M}}$ pure scaling functions, i.e., $\mathbf{p} = \mathbf{q} = (0,0,0)$) satisfies

$$\left| \int_0^\infty \mathcal{I}^{(\mathbf{p},\mathbf{q})}(t,\mathbf{a})dt - h \sum_{m=-M}^{M} e^{mh} \mathcal{I}^{(\mathbf{p},\mathbf{q})}(e^{mh},\mathbf{a}) \right| \leqslant C e^{-\alpha\sqrt{M}} \tag{18}$$

for $\alpha = 2\sqrt{\pi\delta}$ ($\alpha = \sqrt{2\pi\delta}$ pure scaling functions) with constant C independent of the translation parameter $\mathbf{a}$.

We illustrate the theorem for the case of pure scaling functions in Fig. 8. Similar results for wavelets are presented in [14].

4.2.3 On the best approximation by exponential sums Using integral representation methods, the *Sinc*-quadrature can be applied, for example, to the integrals

$$\frac{1}{\rho} = \int_0^\infty e^{-\rho\xi} d\xi, \quad \text{and} \quad \frac{1}{\rho} = \frac{1}{\sqrt{\pi}} \int_{-\infty}^{\infty} e^{-\rho^2 t^2} dt$$

to obtain an exponentially convergent sum of exponentials approximating the inverse function $\frac{1}{\rho}$. Instead, one can directly determine the best approximation of a function with respect to a certain norm by exponential sums $\sum_{\nu=1}^{n} \omega_\nu e^{-t_\nu x}$

or $\sum_{\nu=1}^{n} \omega_\nu e^{-t_\nu x^2}$, where $\omega_\nu, t_\nu \in \mathbb{R}$ are to be chosen optimally. For some applications in quantum chemistry of approximation by exponential sums we refer e.g. to [1, 60, 62].

We recall some facts from the approximation theory by exponential sums (cf. [10] and the discussion in [53]). The existence result is based on the fundamental *Big Bernstein Theorem*: If f is completely monotone for $x \geqslant 0$, i.e.,

$$(-1)^n f^{(n)}(x) \geqslant 0 \qquad \text{for all } n \geqslant 0,\ x \geqslant 0,$$

then it is the restriction of the Laplace transform of a measure to the half-axis:

$$f(z) = \int_{\mathbb{R}_+} e^{-tz} d\mu(t).$$

For $n \geqslant 1$, consider the set E_n^0 of exponential sums and the extended set E_n:

$$E_n^0 := \left\{ u = \sum_{\nu=1}^{n} \omega_\nu e^{-t_\nu x} : \ \omega_\nu,\ t_\nu \in \mathbb{R} \right\},$$

$$E_n := \left\{ u = \sum_{\nu=1}^{\ell} p_\nu(x) e^{-t_\nu x} : \ t_\nu \in \mathbb{R}, \right.$$

$$\left. p_\nu \text{ polynomials with } \sum_{\nu=1}^{\ell} (1 + \text{degree}(p_\nu)) \leqslant n \right\}.$$

Now one can address the problem of finding the best approximation to f over the set E_n characterised by the best N-term approximation error

$$d_\infty(f, E_n) := \inf_{v \in E_n} \|f - v\|_\infty.$$

We recall the complete elliptic integral of the first kind with modulus κ,

$$\mathbf{K}(\kappa) = \int_0^1 \frac{dt}{\sqrt{(1-t^2)(1-\kappa^2 t^2)}} \qquad (0 < \kappa < 1)$$

(cf. [12]), and define $\mathbf{K}'(\kappa) := \mathbf{K}(\kappa')$ by $\kappa^2 + (\kappa')^2 = 1$.

Theorem 4. [4] ([10]) *Assume that f is completely monotone and analytic for $\Re e\, z > 0$, and let $0 < a < b$. Then for the uniform approximation on the interval $[a, b]$,*

$$\lim_{n\to\infty} d_\infty(f, E_n)^{1/n} \leqslant \frac{1}{\omega^2}, \quad \textit{where } \omega = \exp \frac{\pi \mathbf{K}(\kappa)}{\mathbf{K}'(\kappa)} \quad \textit{with } \kappa = \frac{a}{b}.$$

[4] The same result holds for E_n^0, but the best approximation may belong to the closure E_n of E_n^0.

In the case discussed below, we have $\kappa = 1/R$ for possibly large R. Applying the asymptotics

$$\mathbf{K}(\kappa') = \ln\tfrac{4}{\kappa} + C_1\kappa + \dots \qquad \text{for } \kappa' \to 1,$$
$$\mathbf{K}(\kappa) = \tfrac{\pi}{2}\{1 + \tfrac{1}{4}\kappa^2 + C_1\kappa^4 + \dots\} \qquad \text{for } \kappa \to 0,$$

of the complete elliptic integrals (cf. [44]), we obtain

$$\frac{1}{\omega^2} = \exp\left(-\frac{2\pi\mathbf{K}(\kappa)}{\mathbf{K}(\kappa')}\right) \approx \exp\left(-\frac{\pi^2}{\ln(4R)}\right) \approx 1 - \frac{\pi^2}{\ln(4R)}.$$

The latter expression indicates that the number n of different terms to achieve a tolerance ε is asymptotically

$$n \approx \frac{|\log \varepsilon|}{|\log \omega^{-2}|} \approx \frac{|\log \varepsilon| \ln(4R)}{\pi^2}.$$

This result shows the same asymptotical convergence in n as the corresponding bound in the *Sinc*-approximation theory.

Optimisation with respect to the maximum norm leads to the nonlinear minimisation problem $\inf_{v \in E_n^0} \|f - v\|_{L^\infty[1,R]}$ involving $2n$ parameters $\{\omega_\nu, t_\nu\}_{\nu=1}^n$. The numerical implementation is based on the Remez algorithm (cf. [12]). For the particular application with $f(x) = x^{-1}$, we have the same asymptotical dependence $n = n(\varepsilon, R)$ as in the *Sinc*-approximation above, however, the numerical results [5] indicate a noticeable improvement compared with the quadrature method, at least for $n \leqslant 15$.

The best approximation to $1/\rho^\mu$ in the interval $[1, R]$ with respect to a W-weighted L^2-norm can be reduced to the minimisation of an explicitly given differentiable functional

$$d_2(f, E_n) := \inf_{v \in E_n} \|f - v\|_{L^2_W}.$$

Given $R > 1$, $\mu > 0$, $n \geqslant 1$, find the $2n$ real parameters $t_1, \omega_1, \dots, t_n, \omega_n \in \mathbb{R}$, such that

$$F_\mu(R; t_1, \omega_1, \dots, t_n, \omega_n) := \int_1^R W(x)\Big(\frac{1}{x^\mu} - \sum_{i=1}^n \omega_i e^{-t_i x}\Big)^2 dx = \min. \qquad (19)$$

[5] Numerical results for the best approximation of x^{-1} by sums of exponentials can be found in [10] and [11]; a full list of numerical data is presented in www.mis.mpg.de/scicomp/EXP_SUM/1_x/tabelle.

In the particular case of $\mu = 1$ and $W(x) = 1$, the integral (19) can be calculated in a closed form[6]:

$$F_1(R; t_1, \omega_1, \ldots, t_n, \omega_n) = 1 - \frac{1}{R} - 2\sum_{i=1}^{n} \omega_i \left[\mathrm{Ei}(-t_i) - \mathrm{Ei}(-t_i R)\right]$$

$$+\frac{1}{2}\sum_{i=1}^{n} \frac{\omega_i^2}{t_i}\left[e^{-2t_i} - e^{-2t_i R}\right] + 2\sum_{1\leqslant i<j\leqslant n} \frac{\omega_i \omega_j}{t_i + t_j}\left[e^{-(t_i+t_j)} - e^{-(t_i+t_j)R}\right]$$

with the integral exponential function $\mathrm{Ei}(x) = \int_{-\infty}^{x} \frac{e^t}{t} dt$ (cf. [12]). In the special case $R = \infty$, the expression for $F_1(\infty; \ldots)$ even simplifies. Gradient or Newton type methods with a proper choice of the initial guess can be used to obtain the minimiser of F_1 (cf. [56]).

5 Data sparse formats for univariate components

5.1 Hierarchical matrix techniques

The hierarchical matrix ($\mathcal{H}$-matrix) technique [46, 50, 51, 55] (see also the mosaic-skeleton method [83]) allows an efficient treatment of dense matrices arising, e.g., from BEM, evaluation of volume integrals and multi-particle interactions, certain matrix-valued functions, etc. In particular, it provides matrix formats which enable the computation and storage of inverse FEM stiffness matrices corresponding to elliptic problems as well as of BEM matrices.

The hierarchical matrices are represented by means of a certain block partitioning. Fig. 4 shows typical admissible block structures.

Each block is filled by a submatrix of a rank not exceeding k. Then, for the mentioned class of matrices, it can be shown that the exact dense matrix A and the approximating hierarchical matrix $A_{\mathcal{H}}$ differ by $\|A - A_{\mathcal{H}}\| \leqslant \mathcal{O}(\eta^k)$ for a certain number $\eta < 1$. This exponential decrease allows to obtain an error ε by the choice $k = \mathcal{O}(\log(1/\varepsilon))$. It is shown (cf. [50–52]) that the $\mathcal{H}$-matrix arithmetic exhibits almost linear complexity in N:

- *Data compression.* The storage of $N \times N$ $\mathcal{H}$-matrices as well as the matrix-by-vector multiplication and matrix-matrix addition have a cost $\mathcal{O}(kN \log N)$, where the local rank k is the parameter determining the approximation error.
- *Matrix-by-matrix and matrix-inverse complexity.* The approximate matrix-matrix multiplication and the approximate inversion both take $\mathcal{O}(k^2 N \log^2 N)$ operations.
- *The Hadamard (entry-wise) matrix product.* The exact Hadamard product of two rank-k $\mathcal{H}$-matrices leads to an $\mathcal{H}$-matrix of the block-rank k^2 (see Section 5.2 below).

[6] In the general case, the integral (19) may be approximated by certain quadratures.

5.2 Hierarchical Kronecker tensor-product approximations

Since n is much smaller than N, one can apply the hierarchical (or low-rank) matrix structure to represent the Kronecker factors V_k^ℓ in (13) with the complexity $\mathcal{O}(n \log^q n)$ or even $\mathcal{O}(n)$ that finally leads to $\mathcal{O}(rn) = \mathcal{O}(rN^{1/d})$ data to represent the compressed matrix A_r. We call by $\mathrm{HKT}(r, s)$ the class of Kronecker rank-r matrices, where the Kronecker factors V_k^ℓ are represented by the block-rank s $\mathcal{H}$-matrices (shortly, HKT-matrices). It was shown in [58] that the advantages of replacing A with A_r (cf. (13)), where all the Kronecker factors possess the structure of general $\mathcal{H}$-matrices, are the following:

- *Data compression.* The storage for the V_k^ℓ matrices of (13) is only $\mathcal{O}(rn) = \mathcal{O}(rN^{1/d})$ while that for the original (dense) matrix A is $\mathcal{O}(N^2)$, where $r = \mathcal{O}(\log^\alpha N)$ for some $\alpha > 0$. Consequently, we enjoy a *linear-logarithmic complexity* of $\mathcal{O}(n \log^\alpha n)$ in the univariate problem size n.
- *Matrix-by-vector complexity.* Instead of $\mathcal{O}(N^2)$ operations to compute Ax, $x \in \mathbb{C}^N$, we now need only $\mathcal{O}(rkn^d \log n) = \mathcal{O}(rkN \log n)$ operations. If the vector can be represented in a tensor-product form (say, $x = x^1 \otimes \ldots \otimes x^d$, $x^i \in \mathbb{C}^n$) the corresponding cost is reduced to $\mathcal{O}(rkn \log n) = \mathcal{O}(rkN^{1/d} \log n)$ operations.
- *Matrix-by-matrix complexity.* Instead of $\mathcal{O}(N^3)$ operations to compute AB, we now need only $\mathcal{O}(r^2 n^3) = \mathcal{O}(r^2 N^{3/d})$ operations for rather general structure of the Kronecker factors. Remarkably, this result is much better than the corresponding matrix-by-vector complexity for a general vector x.
- *Hadamard product.* The Hadamard (entry-wise) product of two HKT-matrices $A * B$ is presented in the same format: $(U_1 \times V_1) * (U_2 \times V_2) = (U_1 * U_2) \times (V_1 * V_2)$. In turn, the exact Hadamard product $U_1 * U_2$ (same for $V_1 * V_2$) of two rank-k $\mathcal{H}$-matrices results in an $\mathcal{H}$-matrix of the block-rank k^2 and with the corresponding "skeleton" vectors defined by the Hadamard products of those in the initial factors (since there holds $(a \otimes b) * (a_1 \otimes b_1) = (a * a_1) \otimes (b * b_1)$).

Therefore, basic linear algebra operations can be performed in the tensor-product representation using one-dimensional operations, thus avoiding an exponential scaling in the dimension d.

The exact product of two HKT-matrices can be represented in the same format, but with squared Kronecker rank and properly modified block-rank [58]. If $A, B \in \mathrm{HKT}(r, s)$, where s corresponds to the block-rank of the $\mathcal{H}$-matrices involved, then in general $AB \notin \mathrm{HKT}(r, s)$. However,

$$A = \sum_{k=1}^{r} U_k^A \otimes V_k^A, \quad B = \sum_{l=1}^{r} U_l^B \otimes V_l^B, \quad U_k^A, V_k^A, U_l^B, V_l^B \in \mathbb{C}^{n \times n}, \tag{20}$$

leads to

$$AB = \sum_{k=1}^{r}\sum_{l=1}^{r}(U_k^A U_l^B)\otimes(V_k^A V_l^B).$$

It can be proven that the $U_k^A U_l^A$ and $V_k^A V_l^B$ matrices possess the same hierarchical partitioning as the initial factors in (20) with blocks of possibly larger (than s) rank bounded, nevertheless, by $s_{AB} = \mathcal{O}(s \log N)$. Thus, $AB \in \mathrm{HKT}(r^2, s_{AB})$ with $s_{AB} = \mathcal{O}(s \log N)$.

5.3 Wavelet Kronecker tensor-product approximations

Wavelet matrix compression was introduced in [8]. This techniques has been considered by one of the authors during the past decade in a series of publications (cf. [77]). The compression of the Kronecker factors $V_i \in \mathbb{R}^{n\times n}$ is not so obvious, since it is not clear to what extend they satisfy a Calderón-Zygmund condition. It is more likely that they obey more or less a hyperbolic cross structure. An underlying truncation criterion based on the size of the coefficients will provide an automatic way to find the optimal structure independent of an a priori assumption. A basic thresholding or a posteriori criterion has been formulated by Harbrecht [61] and in [22]. With this criterion at hand, we expect linear scaling with respect to the size of the matrices.

- *Data compression.* The matrices in (13) V_k^ℓ can be compressed requiring total storage size about $\mathcal{O}(rn) = \mathcal{O}(rN^{1/d})$, where $r = \mathcal{O}(\log^\alpha N)$ is as above. The data vector requires at most $\mathcal{O}(n \log^d n)$ nonzero coefficients.
- *Matrix-by-vector complexity.* Instead of $\mathcal{O}(N^2)$ operations to compute Ax, $x \in \mathbb{C}^N$, we now need only $\mathcal{O}(rn^d) = \mathcal{O}(rN)$ operations. If the vector is represented in a tensor-product form (say, $x = x^1 \otimes \ldots \otimes x^d$, $x^i \in \mathbb{C}^n$) or in sparse grid representation, then the corresponding cost is reduced to $\mathcal{O}(rn)$, resp. $\mathcal{O}(rn \log^d n)$ operations .
- *Matrix-by-matrix complexity.* Using the compression of the Lemarié algebra [82], instead of $\mathcal{O}(N^3)$ operations to compute AB, we need only $\mathcal{O}(r^2 n \log^q n) = \mathcal{O}(r^2 N^{1/d} \log^q N)$, or even $\mathcal{O}(r^2 n)$ operations.

Adaptive wavelet schemes for nonlinear operators have been developed in [3, 24] and for nonlocal operators in [23]. Corresponding schemes for hyperbolic cross approximations have not been worked out up to now. Perhaps basic ideas can be transfered immediately to the tensor-product case.

6 Linear scaling methods for Hartree-Fock and Kohn-Sham equations

Operator-valued functions $\mathcal{G}(\mathcal{L})$ of elliptic operators $\mathcal{L}$ play a prominent role in quantum many-particle theory. A possible representation of the operator $\mathcal{G}(\mathcal{L})$

is given by the Dunford-Cauchy integral (cf. [38–41])

$$\mathcal{G}(\mathcal{L}) = \frac{1}{2\pi i}\int_\Gamma \mathcal{G}(z)(zI - \mathcal{L})^{-1}\,dz,$$

where Γ envelopes the spectrum spec$(\mathcal{L})$ of the operator $\mathcal{L}$ in the complex plane. This kind of representation is especially suitable for tensor-product approximation using *Sinc* or Gauss-Lobatto quadratures for the contour integral to get an approximate operator of the form

$$\mathcal{G}(\mathcal{L}) \approx \sum c_k \mathcal{G}(z_k)(z_k I - \mathcal{L})^{-1}. \tag{21}$$

An important example for an operator valued function is the *sign* function of the shifted Fock operator which can be directly related to the spectral projector P_ρ associated with the density matrix ρ. This relation

$$P_\rho = \frac{1}{2}\left[I - \text{sign}(\mathcal{F} - \mu I)\right] = -\frac{1}{2\pi i}\int_\Gamma (\mathcal{F} - zI)^{-1}\,dz,$$

where $\Gamma \cap \text{spec}(\mathcal{F}) = \varnothing$ encloses the $N/2$ lowest eigenvalues of the Fock operator, has been first noticed by Beylkin, Coult and Mohlenkamp [7]. In order to be applicable, the method requires a finite gap between the highest occupied $\varepsilon_{N/2}$ and lowest unoccupied $\varepsilon_{N/2+1}$ eigenvalue to adjust the parameter $\varepsilon_{N/2} < \mu < \varepsilon_{N/2+1}$. This constraint, in particular, excludes metallic systems.

In general, the approximability of inverse matrices, required in (21), within the HKT format is still an open problem. First results on fast approximate algorithms to compute inverse matrices in the HKT format for the case $d \geqslant 2$ can be found in [41]. In Fig. 8, we consider the HKT representation to the discrete Laplacian inverse $(-\Delta_h)^{-1}$ (homogeneous Dirichlet boundary conditions) in $\mathbb{R}^d$, which can be obtained with $\mathcal{O}(dn \log^q n)$ cost. Numerical examples for still higher dimensions $d \leqslant 1024$ are presented in [45]. For comparison, the following numerical example manifests the optimal Kronecker rank of the discrete elliptic inverse in $d = 2$. Let $-\Delta_h$ now correspond to a five-point stencil discretization of the Laplacian on a uniform mesh in the unit rectangle in $\mathbb{R}^2$ (Dirichlet boundary conditions). It is easy to see that the Kronecker rank of $-\Delta_h$ is 2. The Kronecker ranks of $(-\Delta_h)^{-1}$ for different relative approximation accuracies (in the Frobenius norm) are given in Table 6. Our results indicate a logarithmic bound $\mathcal{O}(\log \varepsilon^{-1})$ for the approximate Kronecker rank r.

6.1 Matrix-valued functions approach for density matrices

Let $\mathbb{F} \in \mathbb{R}^{M\times M}$ be the Fock matrix that represents the Fock operator $\mathcal{F}$ (cf. (8)) in an orthogonal basis $\{\varphi_i\}_{i=1}^{M}$, $M \geqslant N/2$. There exist two different approaches to compute the Galerkin discretization $\mathbb{D} \in \mathbb{R}^{M\times M}$ of the density matrix (6) via the matrix *sign* of the shifted Fock matrix

$$\mathbb{D} = \frac{1}{2}[\mathbb{I} - \text{sign}(\mathbb{F} - \mu\mathbb{I})], \quad \text{with } \mu \in (\varepsilon_{N/2}, \varepsilon_{N/2+1}).$$

The first approach uses an exponentially convergent quadrature for the integral to obtain an expansion into resolvents (21) whereas the second approach is based on a Newton-Schultz iteration scheme. Concerning the tensor-product approximation of resolvents in the HKT format we refer to our discussion in Section 5.2. For the Newton-Schultz iteration scheme proposed in [7]

$$S^{(n+1)} = S^{(n)} + \frac{1}{2}\left[I - (S^{(n)})^2\right] S^{(n)} \quad , \quad S^{(0)} = (\mathbb{F} - \mu\mathbb{I}) / \|\mathbb{F} - \mu\mathbb{I}\|_2, \tag{22}$$

the sequence $S^{(n)}$ converges to $\mathrm{sign}(\mathbb{F} - \mu\mathbb{I})$. First applications in quantum chemistry by Németh and Scuseria [71] demonstrate the practicability of this approach. Iterations schemes of the form (22) seem to be especially favourable for tensor-product formats. Starting from an initial approximation of the Fock matrix $\mathbb{F}$, with low separation rank one has to perform matrix-matrix multiplications which can be handled in an efficient manner in the tensor-product format, cf. our discussion in Section 5.2. After each iteration step a recompression of the tensor-product decomposition of $S^{(n+1)}$ becomes necessary. For the recompression one can apply the simple *alternating least squares* (ALS) method [5, 87, 90] or Newton-type and related algebraic iterative methods [33]. The ALS algorithm starts with an initial decomposition of $S^{(n+1)}$ with separation rank r and obtains the best approximation with separation rank $\tilde{r} \leqslant r$ by iteratively solving an optimisation problem for each coordinate separately. Assume that r is actually much larger than necessary, i.e., $\tilde{r} << r$, then the ALS algorithm costs $\mathcal{O}(d\,\tilde{r}(\tilde{r}^2 + rn^2))$. We refer to [78] for the discussion of wavelet methods for density matrix computation in electronic structure calculation.

6.2 Computation of Hartree potentials in tensor-product formats

A common bottleneck for the numerical solution of Hartree-Fock and Kohn-Sham equations is the computation of the Hartree-potential (9). Traditionally,

Table 1. Canonical tensor product approximation with optimal separation rank of the inverse Laplacian $(-\Delta_h)^{-1}$ on a uniform rectangular grid in $[0,1]^2$. Here $n = 64$ and, hence, $-\Delta_h \in \mathbb{R}^{N\times N}$ with $N = 4096$.

Kronecker rank for $(-\Delta_h)^{-1}$	6	8	9	11	12	14
Relative Frobenius error	10^{-3}	10^{-4}	10^{-5}	10^{-6}	10^{-7}	10^{-8}

highly adapted GTO basis sets are used for the approximation of electron densities which enable an analytic evaluation of the subsequent convolution with the Coulomb potential. This kind of approach became widely known as *density-fitting* or *resolution of the identity* method (cf. [31, 32, 69]) and turned out to be an essential ingredient for computational efficiency. We want to discuss two alternative approaches based on canonical (11) and Tucker (12) tensor-product decompositions which have been presented in [14] and [15], respectively. Both approaches have to be considered as a generalization of the density-fitting method with respect to the fact that they provide tensor-product approximations for both, the electron density and the Hartree potential. The latter becomes possible via separable approximations of the Coulomb potential obtained via *Sinc* interpolation or best approximation by exponential sums discussed in Sections 4.2.2 and 4.2.3, respectively.

The canonical tensor-product approach is based on approximations for the electron density (7) in the format

$$n(\mathbf{x}) = \sum_{k=1}^{K} n_k^{(1)}(x_1) n_k^{(2)}(x_2) n_k^{(3)}(x_3) \approx \sum_{k=1}^{\kappa} \eta_k^{(1)}(x_1)\, \eta_k^{(2)}(x_2)\, \eta_k^{(3)}(x_3),$$

which can be obtained from separable approximations with large separation rank K, e.g. expansions in terms of GTO basis sets, via algebraic compression methods [33]. Using a separable approximation of the Coulomb potential with separation rank $2M+1$, cf. (17), we obtain from an intermediate tensor-product representation with separation rank $\kappa(2M+1)$, after a further compression step, the Hartree potential in the canonical format with low separation rank

$$V_H(\mathbf{x}) = \int_{\mathbb{R}^3} \frac{1}{|\mathbf{x}-\mathbf{y}|}\, n(\mathbf{y}) d^3y \approx \sum_{k=1}^{\kappa'} \nu_k^{(1)}(x_1)\, \nu_k^{(2)}(x_2)\, \nu_k^{(3)}(x_3).$$

With such an approximation at hand, it is straightforward to assemble the Galerkin matrix with respect to an arbitrary tensor-product basis $\{G_\Lambda(\mathbf{x}) := g_{\lambda_1}(x_1) g_{\lambda_2}(x_2) g_{\lambda_3}(x_3)\}$, $\Lambda := (\lambda_1, \lambda_2, \lambda_3) \in \mathcal{I}$, in an efficient manner by utilising the tensor-product decomposition of the constituents

$$\int_{\mathbb{R}^3}\int_{\mathbb{R}^3} G_\Lambda V_H(\mathbf{x}) G_\Omega\, d^3x d^3y = \sum_{k=1}^{\kappa} \langle g_{\lambda_1} \nu_k^{(1)}, g_{\omega_1}\rangle \langle g_{\lambda_2} \nu_k^{(2)}, g_{\omega_2}\rangle \langle g_{\lambda_3} \nu_k^{(3)}, g_{\omega_3}\rangle.$$

This approach is therefore especially suitable in combination with conventional GTO or tensor-product wavelet bases. It turned out that significant improvements beyond standard GTO bases can be achieved, however the unconstraint optimization of univariate components required for the compression steps represents a generically ill-posed problem and rather sophisticated optimization techniques are necessary in order to achieve the required high accuracies.

The second approach uses the Tucker format (12) to obtain data sparse representations of electron densities. In contrast to the canonical approach, the Tucker approximation is typically a well posed problem. We have studied Tucker approximations of electron densities and Hartree potentials represented on regular cartesian grids. This kind of representation underlies the recently developed BIGDFT code [9] for DFT based large scale electronic structure calculations. The presently most efficient approach to compute the Hartree potential on a cartesian grid uses the *Fast Fourier Transform* (FFT) to perform the convolution with the Coulomb potential [42]. Concerning the computational complexity FFT scales $\mathcal{O}(n^3 \log n)$ on a cubic grid, where n is the number of grid points in each direction. Within tensor-product formats it is possible to perform this step with sublinear, i.e., $\mathcal{O}(n^\alpha)$, $1 \leqslant \alpha < 3$ complexity. The Tucker format is not as convenient for the convolution as the canonical format. Therefore it is favourable to simply rearange the Tucker tensor with Tucker rank r into a canonical tensor with separation rank r^2 and perform the convolution in the canonical format as described above. In a subsequent step it is possible to compress the resulting Hartree potential of rank $r^2(2M+1)$ again within the Tucker format.

We have studied canonical and Tucker type tensor-product approximations of electron densities and Hartree potentials for a series of small molecules. We refer to [14, 15] for a detailed discussion of the canonical and Tucker approach including benchmark calculations for some small molecules. As an illustrative example we present in Fig. 6 results for the C_2H_6 molecule. The relative errors of the tensor-product approximations refer to the $L^2(\mathbb{R}^3)$ and discrete Frobenius norm for the canonical and Tucker format, respectively. It can be seen from Fig. 6 that the error in the canonical format decreases approximately like $e^{-c\sqrt{\kappa}}$ whereas the Tucker format shows an exponential convergence with respect to the Tucker rank. As already noticed in the previous paragraph it is always possible to rearange the Tucker into the canonical format where the Tucker rank r corresponds to a canonical separation rank $\kappa = r^2$. Here we observe a similar behaviour for electron densities and Hartree potentials.

7 Conclusions

The importance of tensor-product approximations for electronic structure calculations in quantum chemistry and solid state physics can be hardly overestimated. Their scope of applications ranges from basic issues related to the problem to find a convenient ansatz for the many-particle wavefunction in terms of Slater determinants up to more technical issues concerning the efficient computation of integrals involving the Coulomb potential. Within the present work we reviewed some recent developments in numerical analysis comprising best N-term approximation in tensor-product wavelet bases as well as more general canonical and Tucker type tensor-product formats, which can be combined with data sparse representations for the low dimensional components using $\mathcal{H}$ matrices or wavelets. Furthermore, separable approximations of certain kernel functions, like the Coulomb or Yukawa potential, enable fast convolutions in tensor-product formats. For large scale Hartree-Fock and DFT electronic structure calculations, based on GTO bases or cartesian grids, the computation of the nonlinear Hartree potential becomes a dominant step. Tensor-product formats provide possible improvements with respect to conventional approaches based on density-fitting schemes and FFT for GTO and grid based methods, respectively. An essential prerequisite, however, is the availability of accurate and fast compression algorithms, which have to be successively applied in order to avoid a disproportionate increase of the separation rank e.g. within tensor-product convolutions. Such kind of algorithms are presently under development [33, 74].

8 Acknowledgments

The authors gratefully acknowledge S. R. Chinnamsetty, M. Espig, Dr. L. Grasedyck, V. Khoromskaia and S. Schwinger (Leipzig) for useful discussions. This work was supported by the Deutsche Forschungsgemeinschaft (SPP 1145).

References

1. J. Almlöf, *Elimination of energy denominators in Møller-Plesset perturbation theory by a Laplace transform approach*, Chem. Phys. Lett., 181 (1991), pp. 319–320.
2. T. A. Arias, *Multiresolution analysis of electronic structure: Semicardinal and wavelet bases*, Rev. Mod. Phys., 71 (1999), pp. 267–312.
3. A. Barinka, W. Dahmen and R. Schneider, *Fast computation of adaptive wavelet expansions*, Numer. Math., 105 (2007), pp. 549–589.
4. K. R. Bates, A. D. Daniels and G. E. Scuseria, *Comparison of conjugate gradient density matrix search and Chebyshev expansion methods for avoiding diagonalization in large-scale electronic structure calculations*, J. Chem. Phys., 109 (1998), pp. 3308–3312.
5. G. Beylkin and M. J. Mohlenkamp, *Numerical operator calculus in higher dimensions*, Proc. Natl. Acad. Sci. USA, 99 (2002), pp. 10246–10251.
6. G. Beylkin and M. J. Mohlenkamp, *Algorithms for numerical analysis in high dimension*, SIAM J. Sci. Comput., 26 (2005), pp. 2133–2159.
7. G. Beylkin, N. Coult and M. J. Mohlenkamp, *Fast spectral projection algorithms for density-matrix computations*, J. Comp. Phys., 152 (1999), pp. 32–54.
8. G. Beylkin, R. Coifman and V. Rokhlin, *Fast wavelet transforms and numerical algorithms I*, Comm. Pure Appl. Math., 44 (1991), pp. 141–183.
9. BigDFT is a wavelet based DFT program by T. Deutsch, S. Goedecker, X. Gonze and R. Schneider, with contributions from P.-M. Anglade, A. Bergman, D. Caliste, L. Genovese, A. Ghasemi, F. Krüger, F. Lancȩn, A. Neelov, J. Pivonski, and M. Rayson.

10. D. BRAESS, *Nonlinear Approximation Theory*, Springer-Verlag, Berlin, 1986.
11. D. BRAESS AND W. HACKBUSCH, *Approximation of* $1/x$ *by exponential sums in* $[1, \infty)$, IMA J. Numer. Anal., 25 (2005), pp. 685–697.
12. I. N. BRONSHTEIN, K. A. SEMENDYAYEV, G. MUSIOL AND H. MUEHLIG, *Handbook of Mathematics, 4th ed.* Springer-Verlag, Berlin, 2004.
13. E. CANCÈS AND C. LE BRIS, *On the convergence of SCF algorithms for the Hartree-Fock equations*, ESAIM: M2AN, 34 (2000), pp. 749–774.
14. S. R. CHINNAMSETTY, M. ESPIG, W. HACKBUSCH, B. N. KHOROMSKIJ AND H.-J. FLAD, *Tensor product approximation with optimal rank in quantum chemistry*, J. Chem. Phys., 127 (2007), 084110 (14 pages).
15. S. R. CHINNAMSETTY, H.-J. FLAD, V. KHOROMSKAIA AND B. N. KHOROMSKIJ, *Tensor decomposition in electronic structure calculations on Cartesian grids*, MPI MIS 65, Leipzig 2007 (submitted).
16. A. COHEN, *Numerical analysis of wavelet methods*, Elsevier, Amsterdam, 2003.
17. A. COHEN, W. DAHMEN AND R. A. DEVORE, *Adaptive wavelet methods for elliptic operator equations – convergence rates*, Math. Comp., 70 (2001), pp. 27–75.
18. W. DAHMEN, S. PRÖSSDORF AND R. SCHNEIDER, *Wavelet approximation methods for pseudodifferantial equations II: Matrix compression and fast solution*, Adv. Comput. Math., 1 (1993), pp. 259–335.
19. W. DAHMEN, S. PRÖSSDORF AND R. SCHNEIDER, *Multiscale methods for pseudo-differential equations on smooth closed manifolds*, Wavelet Anal. Appl., 5, Academic Press, San Diego, 1994.
20. W. DAHMEN, *Wavelet and multiscale methods for operator equations*, Acta Numerica, 6 (1997), pp. 55–228.
21. W. DAHMEN, *Multiscale and wavelet methods for operator equations*, in Multiscale problems and methods in numerical simulations, Lecture Notes in Math., 1825, Springer–Verlag, Berlin, 2003, pp. 31–96.
22. W. DAHMEN, H. HARBRECHT AND R. SCHNEIDER, *Compression techniques for boundary integral equations—asymptotically optimal complexity estimates*, SIAM J. Numer. Anal., 43 (2006), pp. 2251–2271.
23. W. DAHMEN, H. HARBRECHT AND R. SCHNEIDER, *Adaptive methods for boundary integral equations: complexity and convergence estimates*, Math. Comp., 76 (2007), pp. 1243–1274.
24. W. DAHMEN, R. SCHNEIDER AND Y. XU, *Nonlinear functionals of wavelet expansions – Adaptive reconstruction and fast evaluation*, Numer. Math., 86 (2000), pp. 40–101.
25. L. DE LATHAUWER, B. DE MOOR AND J. VANDEWALLE, *On the best rank-1 and rank-*$(R_1, R_2, \ldots, R_N)$ *approximation of higher-order tensors*, SIAM J. Matrix Anal. Appl., 21 (2000), pp. 1324–1342.
26. L. DE LATHAUWER, B. DE MOOR AND J. VANDEWALLE, *A multilinear singular value decomposition*, SIAM J. Matrix Anal. Appl., 21 (2000), pp. 1253–1278.
27. L. DE LATHAUWER, B. DE MOOR AND J. VANDEWALLE, *Computation of the canonical decomposition by means of a simultaneous generalized Schur decomposition*, SIAM J. Matrix Anal. Appl., 26 (2004), pp. 295–327.
28. R. A. DEVORE, *Nonlinear Approximation*, Acta Numerica, 7 (1998), pp. 51–150.
29. G. DESLAURIERS AND S. DUBUC, *Symmetric iterative interpolation processes*, Constr. Approx., 5 (1989), pp. 49–68.

30. M. Dolg, *Effective core potentials* in: Modern Methods and Algorithms of Quantum Chemistry, J. Grotendorst, ed., John Neumann Institute for Computing, NIC Series Vol. 1, Jülich, 2000 , pp. 479–508.
31. B. I. Dunlap, *Robust and variational fitting*, Phys. Chem. Chem. Phys., 2 (2000), pp. 2113–2116.
32. K. Eichkorn, O. Treutler, H. Öhm, M. Häser and R. Ahlrichs, *Auxiliary basis sets to approximate Coulomb potentials*, Chem. Phys. Lett., 240 (1995), pp. 283–290.
33. M. Espig, *Efficient Best-Approximation with Sums of Elementary-Tensors in High Dimensions*, PhD thesis (in German), MPI-MIS Leipzig 2007.
34. H.-J. Flad, W. Hackbusch, D. Kolb and R. Schneider, *Wavelet approximation of correlated wavefunctions. I. Basics*, J. Chem. Phys., 116 (2002), pp. 9641–9657.
35. H.-J. Flad, W. Hackbusch and R. Schneider, *Best N-term approximation in electronic structure calculation. I. One-electron reduced density matrix*, ESAIM: M2AN, 40 (2006), pp. 49–61.
36. H.-J. Flad, W. Hackbusch and R. Schneider, *Best N-term approximation in electronic structure calculation. II. Jastrow factors*, ESAIM: M2AN, 41 (2007), pp. 261–279.
37. H.-J. Flad, R. Schneider and B.-W. Schulze, *Asymptotic regularity of solutions of Hartree-Fock equations with Coulomb potential*, preprint 5, Universität Potsdam 2007.
38. I. P. Gavrilyuk, W. Hackbusch and B. N. Khoromskij, *$\mathcal{H}$-Matrix approximation for the operator exponential with applications*, Numer. Math., 92 (2002), pp. 83–111.
39. I. P. Gavrilyuk, W. Hackbusch and B. N. Khoromskij, *Data-sparse approximation to operator-valued functions of elliptic operators*, Math. Comp. 73 (2004), pp. 1297–1324.
40. I. P. Gavrilyuk, W. Hackbusch and B.N. Khoromskij, *Data-sparse approximation to a class of operator-valued functions*, Math. Comp. 74 (2005), pp. 681–708.
41. I. P. Gavrilyuk, W. Hackbusch, and B.N. Khoromskij, *Tensor-product approximation to elliptic and parabolic solution operators in higher dimensions*, Computing 74 (2005), pp. 131–157.
42. L. Genovese, T. Deutsch, A. Neelov, S. Goedecker and G. Beylkin, *Efficient solution of Poisson's equation with free boundary conditions*, arXiv:cond-mat/0605371v1 (2006).
43. S. Goedecker, *Wavelets and their Application for the Solution of Differential Equations*. Presses Polytechniques Universitaires et Romandes, Lausanne, 1998.
44. I. S. Gradshteyn and I.M. Ryzhik, *Tables of integrals, series and products, 6th edition*, Academic Press, San Diego, 2000.
45. L. Grasedyck, *Existence and computation of low Kronecker-rank approximations for large linear systems of tensor product structure*, Computing 72 (2004), pp. 247–265.
46. L. Grasedyck and W. Hackbusch, *Construction and arithmetics of $\mathcal{H}$-matrices*, Computing 70 (2003), pp. 295–334.
47. J. Garcke and M. Griebel, *On the computation of the eigenproblems of hydrogen and helium in strong magnetic and electric fields with the sparse grid combination technique*, J. Comp. Phys., 165 (2000), pp. 694–716.

48. M. Griebel and J. Hamaekers, *Sparse grids for the Schrdinger equation*, ESAIM: M2AN, 41 (2007), pp. 215–247.
49. M. Griebel and S. Knapek, *Optimized tensor-product approximation spaces*, Constr. Approx., 16 (2000), pp. 303–332.
50. *W. Hackbusch, A sparse matrix arithmetic based on $\mathcal{H}$-matrices. I. Introduction to $\mathcal{H}$-matrices*, Computing, 62 (1999), pp. 89–108.
51. W. Hackbusch and B.N. Khoromskij, *A sparse $\mathcal{H}$-matrix arithmetic. II. Application to multi-dimensional problems*, Computing, 64 (2000), pp. 21–47.
52. W. Hackbusch and B. N. Khoromskij, *A sparse $\mathcal{H}$-matrix arithmetic: General complexity estimates*, J. Comp. Appl. Math., 125 (2000), pp. 479–501.
53. *W. Hackbusch and B. N. Khoromskij, Low-rank Kronecker product approximation to multi-dimensional nonlocal operators. Part I. Separable approximation of multi-variate functions*, Computing, 76 (2006), pp. 177–202.
54. W. Hackbusch and B. N. Khoromskij, *Low-rank Kronecker product approximation to multi-dimensional nonlocal operators. Part II. HKT representations of certain operators*, Computing **76** (2006), pp. 203–225.
55. W. Hackbusch, B. N. Khoromskij and R. Kriemann, *Hierarchical matrices based on a weak admissibility criterion*, Computing, 73 (2004), pp. 207–243.
56. W. Hackbusch, B. N. Khoromskij and A. Klaus, *Approximation of functions by exponential sums based on Newton-type optimisation*, Techn. Rep. 3, MIP MIS, Leipzig 2005.
57. W. Hackbusch and B.N. Khoromskij: *Tensor-Product Approximation to Operators and Functions in High Dimension.* Journal of Complexity 23 (2007), 697-714.
58. W. Hackbusch, B. N. Khoromskij and E. E. Tyrtyshnikov, *Hierarchical Kronecker tensor-product approximations.* J. Numer. Math., 13 (2005), pp. 119–156.
59. W. Hackbusch, B. N. Khoromskij and E. E. Tyrtyshnikov, *Approximate iterations for structured matrices*, Numer. Math., to appear.
60. M. Häser and J. Almlöf, *Laplace transform techniques in Møller-Plesset perturbation theory*, J. Chem. Phys., 96 (1992), pp. 489–494.
61. sc H. Harbrecht, *Wavelet Galerkin schemes for the boundary element method in three dimensions*, PhD thesis, Technische Universität Chemnitz, 2001.
62. R. J. Harrison, G. I. Fann, T. Yanai, Z. Gan and G. Beylkin, *Multiresolution quantum chemistry: Basic theory and initial applications*, J. Chem. Phys., 121 (2004), pp. 11587–11598.
63. T. Helgaker, P. Jørgensen and J. Olsen, *Molecular Electronic-Structure Theory*, Wiley, New York, 1999.
64. B. N. Khoromskij, *Structured rank-*$(\mathrm{r}_1, ..., \mathrm{r}_\mathrm{d})$ *decomposition of function-related tensors in* $\mathbb{R}^\mathrm{d}$, Comp. Meth. in Applied Math., 6 (2006), pp. 194–220.
65. B. N. Khoromskij, *Structured data-sparse approximation to high-order tensors arising from the deterministic Boltzmann equation*, Math. Comp., 76 (2007), pp. 1291–1315.
66. B. N. Khoromskij, *Multi-linear approximation of higher-dimensional convolution in linear cost*, preprint MPI MIS Leipzig, 2007.
67. B. N. Khoromskij and V. Khoromskaia, *Low Rank Tucker Tensor Approximation to the Classical Potentials*, Cent. Eur. J. Math., 5 (2007), pp. 1–28.

68. W. Kohn, *Density functional theory for systems of very many atoms*, Int. J. Quantum Chem., 56 (1995), pp. 229–232.
69. F. R. Manby, P. J. Knowles and A. W. Lloyd, *The Poisson equation in density fitting for the Kohn-Sham Coulomb problem*, J. Chem. Phys., 115 (2001), pp. 9144–9148.
70. J. M. Millam and G. E. Scuseria, *Linear scaling conjugate gradient density matrix search as an alternative to diagonalization for first principles electronic structure calculations*, J. Chem. Phys., 106 (1997), pp. 5569–5577.
71. K. Németh and G. E. Scuseria, *Linear scaling density matrix search based on sign matrices*, J. Chem. Phys., 113 (2000), pp. 6035–6041.
72. P.-A. Nitsche, *Best* N *term approximation spaces for tensor product wavelet bases*, Constr. Approx., 24 (2006), pp. 49–70.
73. C. Ochsenfeld and M. Head-Gordon, *A reformulation of the coupled perturbed self-consistent field equations entirely within a local atomic orbital density matrix-based scheme*, Chem. Phys. Lett., 270 (1997), pp. 399-405.
74. I. V. Oseledets, D. V. Savostianov and E. E. Tyrtyshnikov, *Tucker dimensionality reduction of three-dimensional arrays in linear time*, SIAM J. Matrix Anal. Appl., 2007, to appear.
75. R. G. Parr and W. Yang, *Density-functional Theory of Atoms and Molecules*, Oxford University Press, 1989.
76. M. Reed and B. Simon, *Methods of Modern Mathematical Physiscs. IV: Analysis of Operators*, Academic Press, San Diego, 1978.
77. R. Schneider, *Multiskalen- und Wavelet-Matrixkompression*, Teubner, Stuttgart, 1998.
78. R. Schneider and T. Weber, *Wavelets for density matrix computation in electronic structure calculation.* Appl. Numer. Math., 56 (2006), pp. 1383-1396.
79. S. Schwinger, Diploma thesis, MPI Leipzig and TU Chemnitz, 2005.
80. F. Stenger, *Numerical Methods Based on Sinc and Analytic Functions*, Springer-Verlag, Heidelberg, 1993.
81. V. N. Temlyakov, *Approximation of functions with bounded mixed derivative*, Proc. Steklov Inst. Math., 1989 (1989), pp. 275-293.
82. P. Tchamitchian, *Wavelet functions and operators*, in Wavelets, Theory and Applications, G. Erlebacher, Y. Hussaini, L. Jameson, eds., Oxford University Press, 1996, pp. 87–178.
83. E. E. Tyrtyshnikov, *Incomplete cross approximation in the mosaic-skeleton method*, Computing, 64 (2000), pp. 367–380.
84. E. E. Tyrtyshnikov, *Tensor approximations of matrices generated by asymptotically smooth functions*, Sb. Math., 194 (2003), pp. 941–954.
85. E. Tyrtyshnikov, *Kronecker-product approximations for some function-related matrices*, Linear Algebra Appl., 379 (2004), pp. 423-437.
86. C. F. Van Loan, *The ubiquitous Kronecker product*, J. Comput. Appl. Math., 123 (2000), 85–100.
87. C. F. Van Loan and N. P. Pitsianis, *Approximation with Kronecker products*, NATO Adv. Sci. Ser E Appl. Sci., 232, Kluwer, Dordrecht, 1993, pp. 293–314.
88. H. Yserentant, *On the regularity of the electronic Schrödinger equation in Hilbert spaces of mixed derivatives*, Numer. Math., 98 (2004), pp. 731-759.
89. H. Yserentant, *Sparse grid spaces for the numerical solution of the electronic Schrödinger equation*, Numer. Math., 101 (2005), pp. 381-389.

90. T. Zhang and G.H. Golub, *Rank-one approximation to high order tensors*, SIAM J. Matrix Anal. Appl., 23 (2001), pp. 534-550.

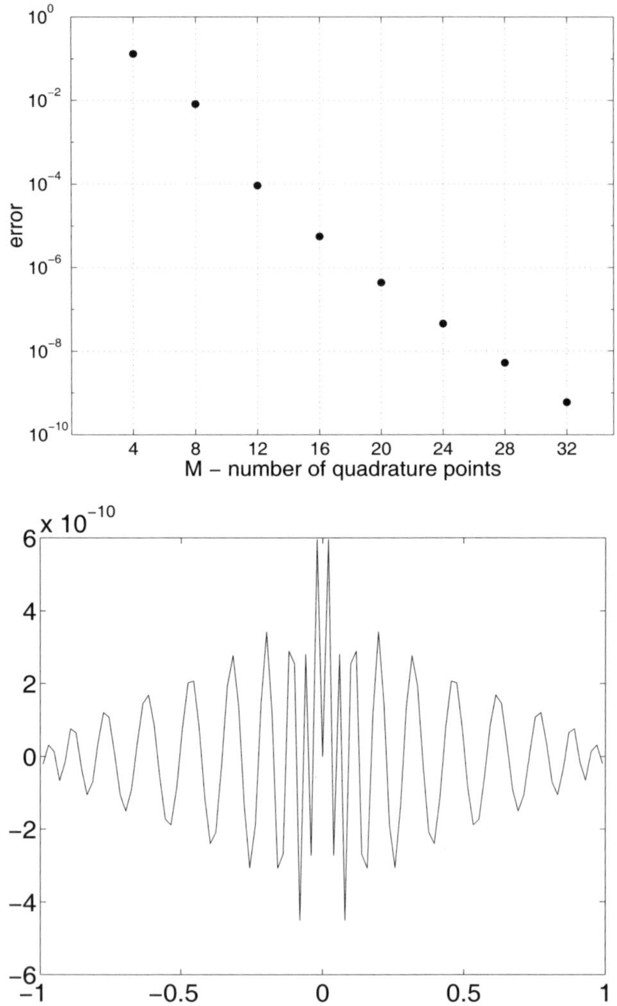

Fig. 1. Convergence of the *Sinc*-approximation of $F(\zeta,\eta,\vartheta) = \sqrt{\zeta^2+\eta^2+\vartheta^2}$ on the slice $\eta = \vartheta = 0$ (left), and the pointwise error for $M = 32$ on $(\zeta,0,0)$, $\zeta \in [-1,1]$ (right).

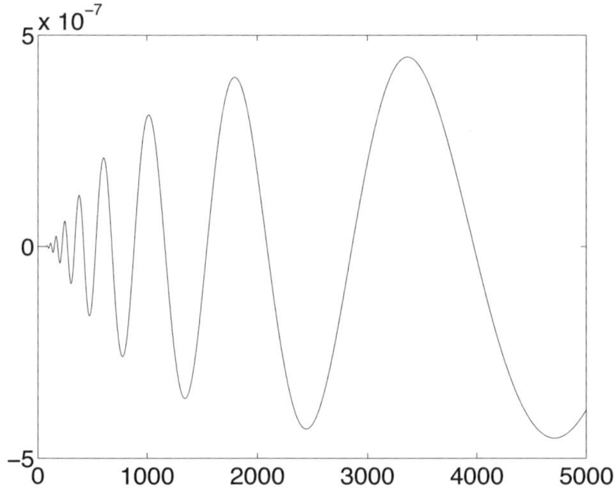

Fig. 2. Error of the quadrature formula (17) for the Coulomb potential on the interval $[1, 5000]$ for $M = 64$.

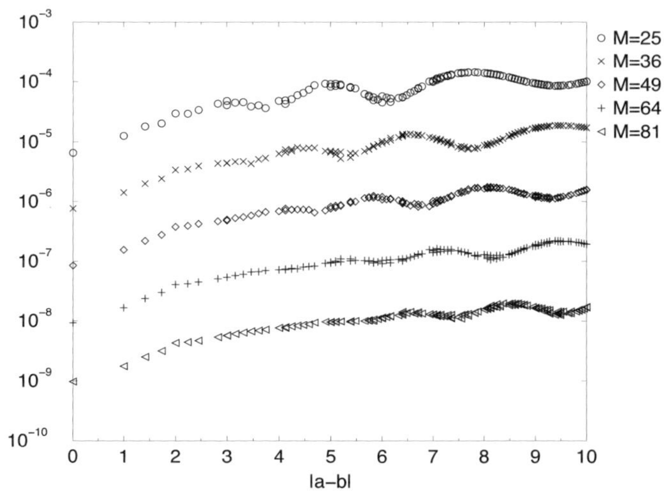

Fig. 3. Relative error of Coulomb integrals (18) for the case of pure scaling functions, i.e., $\mathbf{p} = \mathbf{q} = (0, 0, 0)$, for different number of quadrature points M. The isotropic 3d-scaling functions were generated from the univariate Deslauriers Dubuc interpolating scaling function [29] which is exact up to 5'th order.

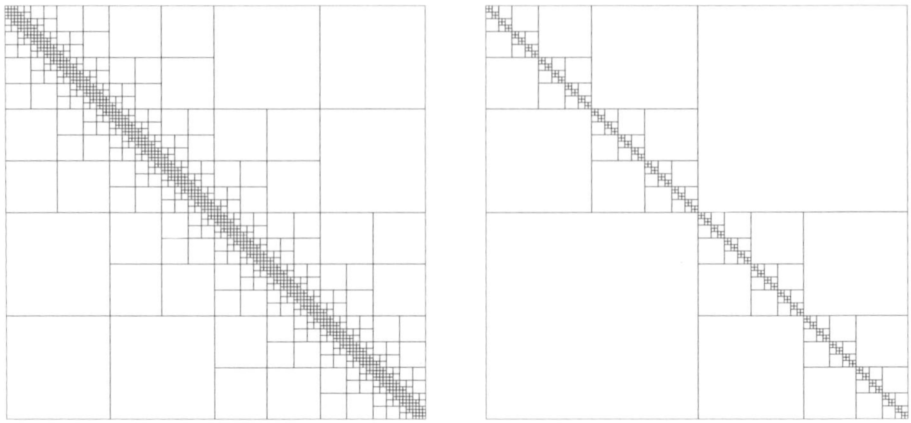

Fig. 4. Hierarchical partitioning by the standard and weak admissibility conditions

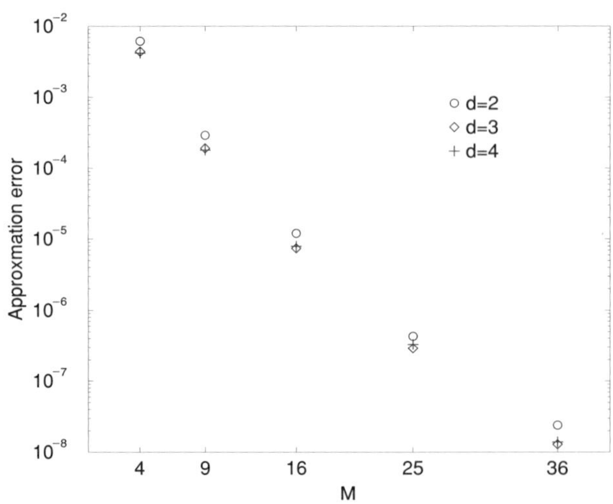

Fig. 5. Error for canonical tensor product approximations of $(-\Delta_h)^{-1}$ in $[0,1]^d$ with $d = 2, 3$ and 4. Here M denotes the number of Sinc-quadrature points so that the separation rank is given by $r = 2M + 1$, cf. [41] for further details.

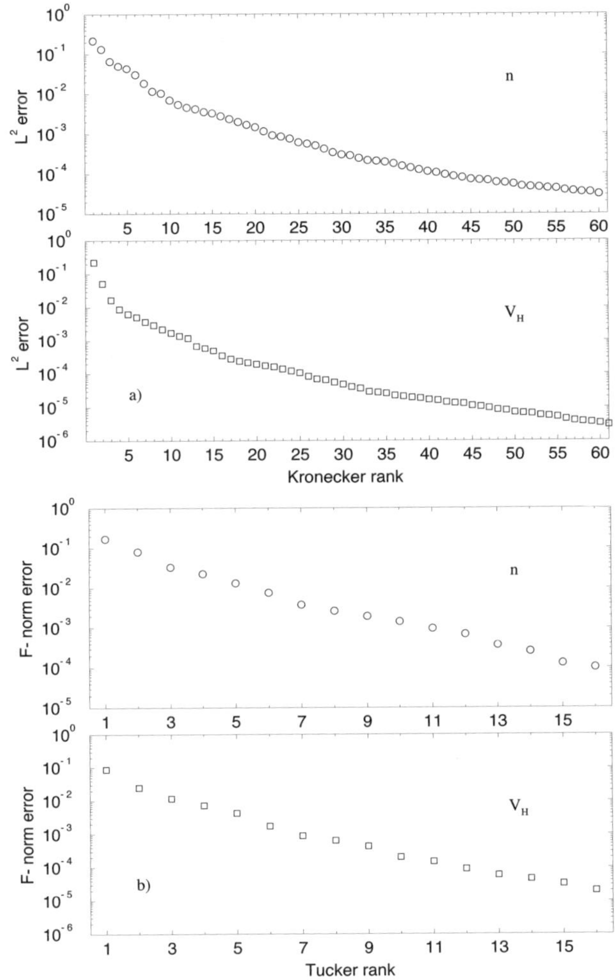

Fig. 6. Errors for tensor product approximations of the electron density n and Hartree potential V_H of C_2H_6. a) Kronecker rank of the canonical format versus error in the L^2 norm. b) Tucker rank of the Tucker format versus error in the Frobenius norm.

Separation of Variables in Nonlinear Fermi Equation

Yu. I. Kuznetsov

Institute of Computational Mathematics and Mathematical Geophysics,
Siberian Branch of Russian Academy of Sciences, ak. Lavrentieva 6, 630090
Novosibirsk, Russia
sur@ommfao1.sscc.ru

Abstract. The transformation of the discrete nonlinear oscillatory system separating the Fourier coefficients and the harmonics of a linear oscillator system is obtained. The numerical experiments show that an initial impulse of the nonlinear system is localized in those harmonics whose numbers are divisible by the numbers of nonzero initial harmonics. Obtained separation of the variables is connected with symmetric Toeplitz and persymmetric Hankel matrices.

Such characteristics of seismic activity as clustering in space, in time, in seismic energy and so on, which are observed in the last decades, provide evidence of the wave nature of seismic processes.

Problem of modelling the elastic chain of rotating spherical blocks of equal size is reduced to solving 1D nonlinear sine-Gordon equation:

$$\frac{\partial^2 \vartheta}{\partial \xi^2} - \frac{\partial^2 \vartheta}{\partial \eta^2} = \sin(\vartheta).$$

Variables $\xi = k_0 x$ and $\eta = c_0 k_0 t$ are dimensionless coordinates, x is a coordinate in the chain, t is time, c_0 and k_0 are characteristic velocity and wave number. The solution has the character of a soliton. Computations based on the sine-Gordon equation have a rather good agreement with real data corresponding to a seismic foci belt.

In this work we consider another nonlinear equation with a soliton solution, considered in [1]:

$$\ddot{z}_k = (z_{k+1} - 2z_k + z_{k-1})\Big(1 + \alpha(z_{k+1} - z_{k-1})\Big), \tag{1}$$

$z_0 = z_{N+1} = 0$, $k = 1(1)N$, describing describing N uniformly spaced equal masses fixed on a thread. Here $z_k = z_k(t)$ is a deviation of k-th mass from the equilibrium state. Discrete systems have more rich dynamics as compared to the continuous ones [2]. They allow to separate spacial and temporal variables. It is known [1] that the oscillatory system described by (1) is not similar to the linear one: the initial momentum energy does not dissipate in the phase space with time, but is focused in a group of harmonics.

By analogy to [1] the basis of the linear problem (1) at $\alpha = 0$ is used. However a special transformation is used for the coefficients of decomposition, which allows to separate the Fourier coefficients from the considered harmonics. As a result the new conclusions are obtained.

The boundary conditions of the system (1) are determined by the rigid fixing of the endpoints. Let $Z(t) = (z_1(t), \ldots, z_N(t))^T$ and

$$Z(t) = \sum_{i=1}^{N} c_i(t) Y^{(i)}, \tag{2}$$

where $Y^{(i)} = (y_1^{(i)}, \ldots, y_N^{(i)})^T$, $y_0^{(i)} = 0$, $y_{N+1}^{(i)} = 0$ is the solution of the homogeneous equation

$$y_{k+1}^{(i)} - 2y_k^{(i)} + y_{k-1}^{(i)} = -\lambda_i y_k^{(i)}, \qquad i = 1(1)N. \tag{3}$$

Here λ_i is the squared frequency, $Y^{(i)}$ is the amplitude vector of the linear oscillator.

1. The representation (2) has the component form:

$$z_k(t) = \sum_{i=1}^{N} c_i(t) y_k^{(i)}. \tag{4}$$

Theorem 1. *The linear system (1) has solution (2). The Fourier coefficients $c_i(t)$ are defined by the equations*

$$\begin{aligned} -\frac{1}{\alpha}\Big(\ddot{c}_l(t) + \lambda_l c_l(t)\Big) = \sum_{i=1}^{l-1} \lambda_i c_i(t) t_{l-i} - \sum_{i=N-l+2}^{N} \lambda_i c_i(t) t_{2N-l-i+2} + \\ \sum_{i=l+1}^{N} \lambda_i c_i(t) t_{i-l} - \sum_{i=1}^{N-l} \lambda_i c_i(t) t_{i+l}, \end{aligned} \tag{5}$$

$l = 1(1)N$, *where*

$$t_k = c_k \sin\frac{k\pi}{N+1}, \qquad k = 1(1)N, \tag{6}$$

and

$$c_{k+1} t_{N-k} + c_{N-k} t_{k+1} = 0, \quad k = 0(1)N-1. \tag{7}$$

Proof. The system (1) can be transformed as follows:

$$J = -\frac{1}{\alpha}\sum_{l=1}^{N}\Big(\ddot{c}_l(t) + \lambda_l c_l(t)\Big) y_k^{(l)} = \sum_{i=1}^{N} \lambda_i c_i(t) \sum_{j=1}^{N} c_j(t) y_k^{(i)} \Big(y_{k+1}^{(j)} - y_{k-1}^{(j)}\Big). \tag{8}$$

Evidently,

$$\lambda_i = 4\sin^2\frac{i\pi}{2(N+1)}, \quad y_k^{(i)} = \sin\frac{ik\pi}{N+1}, \quad i, k = 1(1)N. \tag{9}$$

Let us transform the right-hand of the equality (5). As far as

$$y_{k+1}^{(j)} - y_{k-1}^{(j)} = \sin\frac{j(k+1)\pi}{N+1} - \sin\frac{j(k-1)\pi}{N+1} = 2\sin\frac{j\pi}{N+1}\cos\frac{jk\pi}{N+1},$$

then

$$y_k^{(i)}\left(y_{k+1}^{(j)} - y_{k-1}^{(j)}\right) = 2\sin\frac{ik\pi}{N+1}\cos\frac{jk\pi}{N+1}\sin\frac{j\pi}{N+1} = \left(\sin\frac{(i+j)k\pi}{N+1} + \sin\frac{(i-j)k\pi}{N+1}\right)\sin\frac{j\pi}{N+1},$$

and the system (8) has the form

$$J = J_+ + J_-,$$

where

$$J_+ = \sum_{i=1}^{N}\lambda_i c_i(t)\sum_{j=1}^{N} c_j \sin\frac{j\pi}{N+1}\sin\frac{(i+j)k\pi}{N+1},$$

$$J_- = \sum_{i=1}^{N}\lambda_i c_i(t)\sum_{j=1}^{N} c_j \sin\frac{j\pi}{N+1}\sin\frac{(i-j)k\pi}{N+1}.$$

Let us introduce in J_+ and J_- the new variables $l = i+j$ and $l = i-j$ respectively. Then with the notation $c_j = 0$, if $j \leqslant 0$ or $j > N$, we obtain:

$$J_+ = \sum_{l=2}^{N}\sin\frac{lk\pi}{N+1}\sum_{i=1}^{N}\lambda_i c_i(t)c_{l-i}(t)\sin\frac{(l-i)\pi}{N+1}$$

$$+ \sum_{l=N+2}^{2N}\sin\frac{lk\pi}{N+1}\sum_{i=1}^{N}\lambda_i c_i(t)c_{l-i}(t)\sin\frac{(l-i)\pi}{N+1},$$

$$J_- = \sum_{l=1-N}^{N-1}\sin\frac{lk\pi}{N+1}\sum_{i=1}^{N}\lambda_i c_i(t)c_{i-l}(t)\sin\frac{(i-l)\pi}{N+1}$$

$$= \sum_{l=1}^{N-1}\sin\frac{lk\pi}{N+1}\sum_{i=1}^{N}\lambda_i c_i(t)\left(c_{i-l}(t)\sin\frac{(i-l)\pi}{N+1} - c_{i+l}(t)\sin\frac{(i+l)\pi}{N+1}\right).$$

In the second equality for J_-, the terms with indices l and $-l$ were combined. In the second term in J_+, was taken by $2N - l + 2$ instead of l. Respectively the correction of the limits of the summation on i is made. From the inequality

$1 \leqslant j \leqslant N$ and the equalities $j = l - i$ and $j = i - l$, we have $i \leqslant l - 1$, $i \geqslant l - N$, and $i \geqslant l + 1$ respectively. Then

$$J_+ = \sum_{l=2}^{N} \sin \frac{lk\pi}{N+1} (\beta_l(t) - \gamma_l(t)),$$

$$\beta_l(t) = \sum_{i=1}^{l-1} \lambda_i c_i(t) c_{l-i}(t) \sin \frac{(l-i)\pi}{N+1},$$

$$\gamma_l(t) = \sum_{i=N-l+2}^{N} \lambda_i c_i(t) c_{2N-l-i+2}(t) \sin \frac{(2N-l-i+2)\pi}{N+1},$$

$$J_- = \sum_{l=1}^{N-1} \sin \frac{lk\pi}{N+1} (\delta_l(t) - \varepsilon_l(t)),$$

$$\delta_l(t) = \sum_{i=l+1}^{N} \lambda_i c_i(t) c_{i-l}(t) \sin \frac{(i-l)\pi}{N+1},$$

$$\varepsilon_l(t) = \sum_{i=1}^{N-l} \lambda_i c_i(t) c_{i+l}(t) \sin \frac{(i+l)\pi}{N+1}.$$

Now let us use the orthogonality relations:

$$\left(Y^{(j)}, Y^{(l)}\right) = \sum_{k=1}^{N} \sin \frac{jk\pi}{N+1} \sin \frac{lk\pi}{N+1} = \frac{N+1}{2} \delta_{jl}, \quad j, l = 1(1)N. \tag{10}$$

The transformed relation (8) has the following vector form:

$$-\frac{1}{\alpha} \sum_{l=1}^{N} (\ddot{c}_l(t) + \lambda_l c_l(t)) Y^{(l)} = \sum_{l=2}^{N} (\beta_l(t) - \gamma_l(t)) Y_k^{(l)} + \sum_{l=1}^{N-1} (\delta_l(t) - \varepsilon_l(t)) Y_k^{(l)}.$$

Finally, by (10) we find (5). □

2.

Corollary 1. *The vector* $C = \left(c_1(t), \ldots, c_N(t)\right)^T$ *is bound by the following relation:*

$$\Lambda^{1/2} \ddot{C} = -B \Lambda^{1/2} C. \tag{11}$$

$$B_T = \begin{bmatrix} 0 & t_1 & t_2 & \cdots & & t_{N-2} & t_{N-1} \\ t_1 & 0 & t_1 & & & & t_{N-2} \\ t_2 & t_1 & \ddots & \ddots & & & \vdots \\ \vdots & & \ddots & & & & \\ & & & \ddots & t_1 & & t_2 \\ t_{N-2} & & & & t_1 & 0 & t_1 \\ t_{N-1} & t_{N-2} & \cdots & & t_2 & t_1 & 0 \end{bmatrix}, \tag{12}$$

$$B_H = \begin{bmatrix} t_2 & t_3 & & \cdots & t_{N-1} & t_N & 0 \\ t_3 & & & & & 0 & t_N \\ & & & & \iddots & & t_{N-1} \\ \vdots & & \iddots & & & & \vdots \\ t_{N-1} & & \iddots & \iddots & & & \\ t_N & 0 & & & & & t_3 \\ 0 & t_N & t_{N-1} & \cdots & & t_3 & t_2 \end{bmatrix}, \tag{13}$$

the values t_k *are defined in (6).*

Proof. Let us represent the equations (5) for the vector C in the vector form. The first and the third terms in the right-hand side of equation (5) determine the symmetric Toeplitz matrix B_T. The second and the fourth terms of the same equation form the persymmetric Hankel matrix B_H. □

Let us now define the vector $C = (c_1(t), \ldots, c_N(t))^T$ and the symmetric matrix

$$B = \Lambda^{1/2}(I + \alpha(B_T - B_H))\Lambda^{1/2},$$

where I is the identity matrix, $\Lambda = \mathrm{diag}(\lambda_1, \ldots, \lambda_N)$.

If

$$\mathcal{C} = \begin{bmatrix} \Lambda^{1/2} C \\ \Lambda^{1/2} \dot{C} \end{bmatrix}, \tag{14}$$

then

$$\dot{\mathcal{C}} = A\mathcal{C}, \tag{15}$$

where

$$A = \begin{bmatrix} 0 & I \\ -B & 0 \end{bmatrix} \in \mathcal{R}^{2N \times 2N}.$$

The vector $\mathcal{C}$ determines the coordinates in the Lagrange space. The total energy of the linear oscillator (1) at $\alpha = 0$ is the Hamilton function:

$$H = \frac{1}{2}(\dot{Z}(t), \dot{Z}(t)) + \frac{1}{2}(\Lambda Z(t), Z(t)). \tag{16}$$

At $\alpha = 0$ the energy (16) is conserved. The eigenvalue problem for the matrix A is of the following form

$$\begin{bmatrix} 0 & I \\ -B & 0 \end{bmatrix} \begin{bmatrix} U_j \\ V_j \end{bmatrix} = \mu_j \begin{bmatrix} U_j \\ V_j \end{bmatrix}, \quad BU_j = x_j U_j, \tag{17}$$

where $x_j = -\mu_j^2$ are the real numbers. The vectors U_j form the orthonormal basis in $\mathcal{R}^N$, $(U_j, U_l) = \delta_{jl}$, $j, l = 1(1)N$. By choosing α one can ensure $x_j > 0$, hence

$$\mu_j = \pm \mathbf{i} p_j, \quad p_j = \sqrt{x_j}, \tag{18}$$

where $\mathbf{i} = \sqrt{-1}$. A pair of eigenvectors of A correspond to this pair of eigenvalues

$$\mathcal{U}_{\pm j} = \begin{bmatrix} U_j \\ \pm \mathbf{i} p_j U_j \end{bmatrix}, \quad \mathcal{V}_{\pm j} = \frac{1}{2} \begin{bmatrix} U_j \\ \pm \mathbf{i} \frac{1}{p_j} U_j \end{bmatrix}. \tag{19}$$

The vectors $\mathcal{U}_{\pm j}, \mathcal{V}_{\pm j}$ form the biorthogonal system. As $(\mathcal{U}_{\mp j}, \mathcal{V}_{\mp k}) = \delta_{jk}$, $j, k = 1(1)2N$, then

$$\mathcal{Z}(t) = \sum_{j=1}^{n} \Big(\frac{\phi_j + \mathbf{i}\varphi_j}{2} e^{-\mathbf{i} p_j t} \begin{bmatrix} U_j \\ -\mathbf{i} p_j U_j \end{bmatrix} + \frac{\phi_j - \mathbf{i}\varphi_j}{2} e^{\mathbf{i} p_j t} \begin{bmatrix} U_j \\ \mathbf{i} p_j U_j \end{bmatrix} \Big),$$

where

$$\phi_j = \Big(DU_j, Z(0) \Big), \quad \varphi_j = \frac{1}{p_j} \Big(DU_j, \dot{Z}(0) \Big).$$

The motion determined by the vector $Z(t)$ is a periodic one and is a superposition of the harmonics of the linear oscillator.

3. The systems of the equations (15) is solved by RK-method of Radaux. In the numerical experiments the calculations begin with $t = 0$, when the system is at rest.

At $\alpha = 0$, the energy (16) is preserved in the initial harmonics. For $N = 31$ on 100000 steps ($\tau = 0.001$) the relative error of the total energy H is about 4.10^{-4}.

The purpose of the numerical experiments was to show that some localization took place for $\alpha \neq 0$. If $C^{(j)}(0) = e_j$, $j = 1,..,31$, where $C = \big(c_1(t), \dots, c_N(t)\big)^T$, $\alpha = 1$, $N = 31$ on 20000 iterations ($\tau = 0.001$) and

$$\sqrt{\sum_{i=1}^{N} (c_i^{(j)}(t))^2} = \sqrt{\sum_{i=1}^{[N/j]} (c_{ji}^{(j)}(t))^2} + \varepsilon_j(t),$$

where $\sum_{i=1}^{[N/j]} (c_{ji}^{(j)}(t))^2$ contain only the coefficients whose number is divisible by j. Especially expressive is the case for $j = 2^k$, $k = 1,..,4$: $\varepsilon_j(t) \equiv 0$.

We get also: $j = 3$, $\varepsilon_3 = 10^{-6}$; $j = 5$, $\varepsilon_5 = 3 \cdot 10^{-3}$; $j = 6$, $\varepsilon_6 = 5 \cdot 10^{-3}$; $j = 7$, $\varepsilon_7 = 4 \cdot 10^{-2}$.

References

1. E. Fermi, *Collected papers (Note e memorie)*, University of Chicago Press, 1965. V. 2.
2. V. K. Mezentsev, S. L. Musher, I. V. Ryzhenkova, S. K. Turitsyn, *Two-dimensional solitons in discrete systems*, JETP Letters, **60** (11) (1994), 815–821.

Faster Multipoint Polynomial Evaluation via Structured Matrices

B. Murphy* and R. E. Rosholt†

Department of Mathematics and Computer Science, Lehman College, City University of New York, Bronx, NY 10468, USA

*brian.murphy@lehman.cuny.edu

†rhys.rosholt@lehman.cuny.edu

Abstract. We accelerate multipoint polynomial evaluation by reducing the problem to structured matrix computation and transforming the resulting matrix structure.

Keywords: algorithm design and analysis, multipoint polynomial evaluation, Vandermonde matrices, Hankel matrices.

Exploiting the links between computations with polynomials and structured matrices and transformation of matrix structure are two effective means for enhancing the efficiency of algorithms in both areas [P89/90], [P92], [BP94], [GKO95], [P01]. We demonstrate the power of these techniques by accelerating multipoint evaluation of univariate polynomials.

Multipoint polynomial evaluation is a classical problem of algebraic computations. Given the coefficient vector $\mathbf{p} = (p_j)_{j=0}^{N-1}$ of a polynomial

$$p(x) = p_0 + p_1 x + \cdots + p_{N-1} x^{N-1}$$

and n distinct points $x_1, \ldots, x_n$, one seeks the vector $\mathbf{v} = (v_i)_{i=1}^{n}$ of the values $v_i = p(x_i)$, $i = 1, \ldots, n$.

Hereafter "ops" stands for "arithmetic operations", m_M (resp. i_M) denotes the number of ops required for multiplication of a matrix M (resp. the inverse of matrix M^{-1}) by a vector, and we assume that $N \geqslant n$. (N is large, e.g., for univariate polynomials obtained from multivariate polynomials via Kronecker's map.) One can compute the vector $\mathbf{v}$ in $2(N-1)n$ ops, by applying Horner's algorithm n times, whereas the Moenck–Borodin algorithm [MB72] uses $O((N/n)m(n)\log n)$ ops provided a pair of polynomials in x can be multiplied modulo x^k in $m(k)$ ops, $m(k) = O(k \log k)$ where the field of constants supports FFT and $m(k) = O((k\log k)\log\log k)$ over any field of constants [CK91].

We take advantage of shifting to the equivalent problem of multiplication of the $n \times N$ Vandermonde matrix

$$V_{n,N}(\mathbf{x}) = (x_i^j)_{i=1,j=0}^{n,N-1}$$

by the vector $\mathbf{p}$. This enables us to exploit matrix structure to decrease the upper bound to $O(((N/n) + \log n)m(n))$, thus yielding some acceleration of these classical computations.

Our techniques may be of interest as a sample of the structure transformation for the acceleration of computations with structured matrices. In our case we rely on the transformation of the matrix $V_{n,N}(\mathbf{x})$ into the Hankel matrix $H(\mathbf{x}) = V^T_{n,n}(\mathbf{x})V_{n,N}(\mathbf{x})$.

We use the following auxiliary results (see, e.g., [P01, Chapters 2 and 3]).

Fact 1.

$$H(\mathbf{x}) = V^T_{n,n}(\mathbf{x})V_{n,N}(\mathbf{x})$$

is an $n \times N$ Hankel matrix

$$\left(\sum_{i=1}^{n} x_i^{k+j}\right)_{k=1,j=0}^{n,N-1}.$$

Fact 2.

$$m_H = O((N/n)m(n)) \text{for} H = H_{n,N}(\mathbf{x}).$$

Fact 3.

$$m_V = O(m(n)\log n)$$

for an $n \times n$ Vandermonde matrix V and

$$i_V = O(m(n)\log n)$$

if this matrix is nonsingular.

We compute the vector $\mathbf{v}$ as follows.

Algorithm 1.

1. Compute the $N + n$ entries of the Hankel matrix $H_{n,N}(\mathbf{x})$ by using
$$O((N/n)m(n) + m(n)\log n) \quad \text{ops}.$$
2. Compute the vector $\mathbf{z} = H_{n,N}(\mathbf{x})\mathbf{p}$ by using
$$O((N/n)m(n)) \quad \text{ops}.$$
3. Apply $O(m(n)\log n)$ ops to compute and output the vector $\mathbf{v} = V^{-T}_{n,n}(\mathbf{x})\mathbf{z}$.

The matrices $V_{n,n}(\mathbf{x})$ and their transposes $V^T_{n,n}(\mathbf{x})$ are nonsingular because the n points $x_1, \ldots, x_n$ are distinct.

The cost bound on Stages 2 and 3 follow from Facts 2 and 3 respectively. To perform Stage 1 we first apply $O(m(n) \log n)$ ops to compute the coefficients of the polynomial

$$q(x) = \prod_{i=1}^{n} (x - x_i)$$

(cf., e.g. [P01, Section 3.1]) and then apply $O((N/n)m(n))$ ops to compute the power sums

$$\sum_{i=1}^{n} x_i^k, \quad k = 1, 2, \ldots, N + n$$

of its zero (cf. [BP94, page 34]).

References

[BP94] D. Bini, V. Y. Pan, *Polynomial and Matrix Computations, Volume 1: Fundamental Algorithms*, Birkhäuser, Boston, 1994.

[CK91] D. G. Cantor, E. Kaltofen, On Fast Multiplication of Polynomials over Arbitrary Rings, *Acta Informatica*, **28(7)**, 697–701, 1991.

[GKO95] I. Gohberg, T. Kailath, V. Olshevsky, Fast Gaussian Elimination with Partial Pivoting for Matrices with Displacement Structure, *Math. of Computation*, **64**, 1557–1576, 1995.

[MB72] R. Moenck, A. Borodin, Fast Modular Transform via Division, *Proc. of 13th Annual Symposium on Switching and Automata Theory*, 90–96, IEEE Computer Society Press, Washington, DC, 1972.

[P89/90] V. Y. Pan, On Computations with Dense Structured Matrices, *Math. of Computation*, **55(191)**, 179–190, 1990. Proceedings version in *Proc. ISSAC89*, 34-42, ACM Press, New York, 1989.

[P92] V. Y. Pan, Complexity of Computations with Matrices and Polynomials, *SIAM Review*, **34**, **2**, 225–262, 1992.

[P01] V. Y. Pan, *Structured Matrices and Polynomials: Unified Superfast Algorithms*, Birkhäuser/Springer, Boston/New York, 2001.

Testing Pivoting Policies in Gaussian Elimination*

Brian Murphy[1,**], Guoliang Qian[2,***], Rhys Eric Rosholt[1,†], Ai-Long Zheng[3], Severin Ngnosse[2,‡], and Islam Taj-Eddin[2,§]

[1] Department of Mathematics and Computer Science, Lehman College, City University of New York, Bronx, NY 10468, USA
**brian.murphy@lehman.cuny.edu
†rosholt@lehman.cuny.edu

[2] Ph.D. Program in Computer Science, The City University of New York, New York, NY 10036 USA
***guoliangqian@yahoo.com
‡sngnosse@msn.com
§itaj-eddin@gc.cuny.edu

[3] Ph.D. Program in Mathematics, The City University of New York, New York, NY 10036 USA
azheng 1999@yahoo.com

Abstract. We begin with specifying a class of matrices for which Gaussian elimination with partial pivoting fails and then observe that both rook and complete pivoting easily handle these matrices. We display the results of testing partial, rook and complete pivoting for this and other classes of matrices. Our tests confirm that rook pivoting is an inexpensive but solid backup wherever partial pivoting fails.

Keywords: Gaussian elimination, pivoting.

1 Introduction

Hereafter we write GEPP, GECP, and GERP to denote Gaussian elimination with partial, complete, and rook pivoting. GEPP and GPPP are Wilkinson's classical algorithms [1], [2], [3], whereas GERP is a more recent and much less known invention [4], [5], [6]. Each of the three algorithms uses $(2/3)n^3 + O(n^2)$ flops to yield triangular factorization of an $n \times n$ matrix, but they differ in the number of comparisons involved, and GEPP has slightly weaker numerically. Namely, both GERP and GECP guarantee numerical stability [7], [5], whereas GEPP is statistically stable for most of the input instances in computational practice but fails for some rare but important classes of inputs [8], [9], [10]. Nevertheless GEPP is omnipresent in modern numerical matrix computations, whereas GECP is rarely used. The reason is simple: GEPP involves $(1/2)n^2 + O(n)$ comparisons versus $(1/3)n^3 + O(n^2)$ in GECP, that is the computational

* Supported by PSC CUNY Award 69350–0038.

cost of pivoting is negligible versus arithmetic cost for GEPP but is substantial for GECP.

GERP combines the advantages of both GECP and GEPP. According to the theory and extensive tests, GERP is stable numerically almost as as GECP and is likely to use about $2n^2$ comparisons for random input matrices (see [4], [5], [6], and our Remark 1), although it uses the order of n^3 comparisons in the worst case [3, page 160]. Each of GEPP, GECP, and GERP can be combined with initial scaling for additional heuristic protection against instability, which requires from about n^2 to about $2n^2$ comparisons and as many flops [1, Section 3.5.2], [2, Section 3.4.4], [3, Section 9.7], so that the overall computational cost is still strongly dominated by the elimination flops.

The customary examples of well conditioned matrices for which GEPP fails numerically are rather complicated, but in the next section we give a simple example, which should provide clearer insight into this problem. Namely, we specify a class of input matrices for which already the rounding errors at the first elimination step of GEPP completely corrupt the output. The results of our numerical tests in Section 3 show that both GECP and GERP have no problems with this class. We also include the test results for six other input classes. For each class we present the number of comparisons, growth factor, and the norms of the error and residual vectors, which gives a more complete picture versus [4], [5], and [6] (cf. our concluding Remark 2). Our tests confirm that GERP is an inexpensive but solid backup wherever GEPP fails.

2 A hard input class for GEPP

Already the first step of Gaussian elimination tends to magnify the input errors wherever the pivot entry is absolutely smaller than some other entries in the same row and column. For example, represent an input matrix M as follows,

$$M = \begin{pmatrix} 1 & \mathbf{v}^T \\ \mathbf{u} & B \end{pmatrix} = (m_{ij})_{i,j=0}^{n-1}, \quad B = (m_{ij})_{i,j=1}^{n-1}, \tag{1}$$

let ε denote the machine epsilon (also called unit roundoff), and suppose that

$$\mathbf{u} = s\mathbf{e}, \quad \mathbf{v} = t\mathbf{e}, \quad \mathbf{e} = (1, 1, \ldots, 1)^T, \quad |m_{ij}| \leqslant 1 \ \text{ for } \ i, j > 0, \tag{2}$$

$$s < 2/\varepsilon, \quad t = 1.$$

Then the first elimination step, performed error-free, produces an $(n-1) \times (n-1)$ matrix $B_s = B + s\mathbf{e}\mathbf{e}^T$, which turns into a rank-one matrix $\text{fl}(s)\mathbf{e}\mathbf{e}^T$ in the result of rounding. Here and hereafter $\text{fl}(a)$ denotes the floating-point representations of a real number a.

Partial pivoting fixes the latter problem for this matrix but does not help against exactly the same problem where the input matrix M satisfies equations (1) and (2) and where

$$s = 1, \quad t > 2/\varepsilon. \tag{3}$$

In this case the first elimination step, performed error-free, would produce the $(n-1) \times (n-1)$ matrix $B_t = B + t\mathbf{e}\mathbf{e}^T$. Rounding would turn it into the rank-one matrix $fl(t)\mathbf{e}\mathbf{e}^T$.

We refer the reader to [8] and [9] (cf. also [10]) on some narrow but important classes of linear systems of equations coming from computational practice on which GEPP fails to produce correct output.

3 Experimental results

Tables 1–4 show the results of tests by Dr. Xinmao Wang at the Department of Mathematics, University of Science and Technology of China, Hefei, Anhui 230026, China. He implemented GEPP, GECP, and GERP in C++ under the 64-bit Fedore Core 7 Linux with AMD Athlon64 3200+ uniprocessor and 1 GB memory. In his implementation he used n comparisons for computing the maximum of n numbers. He tested the algorithms for $n \times n$ matrices M of the following seven classes.

1. Matrices with random integer entries uniformly and independently of each other distributed in the range $(-10^l, 10^l)$.

2. Matrices $M = PLU$ for $n \times n$ permutation matrices P that define n interchanges of random pairs of rows and for lower unit triangular matrices L and U^T with random integer entries in the range $(-10^b, 10^b)$.

3. Matrices $M = S\Sigma T$ for random orthogonal matrices S and T (computed as the Q-factors in the QR factorization of matrices with random integer entries uniformly and independently of each other distributed in the range $(-10^c, 10^c)$) and for the diagonal matrix $\Sigma = \mathrm{diag}(\sigma_i)_{i=1}^n$ where $\sigma_1 = \sigma_2 = \cdots = \sigma_{n-\rho} = 1$ and $\sigma_{n-\rho+1} = \sigma_n = 10^{-q}$ (cf. [3, Section 28.3]).

4. Matrices M satisfying equations (1)–(3) where B denotes an $(n-1) \times (n-1)$ matrix from matrix class 1 above.

5. Matrices $M = \begin{pmatrix} I & O & & \dots & I \\ -M_1 & I & O & \dots & O \\ & -M_1 & I & & \vdots \\ & & \ddots & \ddots & O \\ & & & -M_1 & I \end{pmatrix}$ from [8, page 232], where

$$M_1 = \exp\begin{pmatrix} -0.05 & 0.3 \\ 0.3 & -0.05 \end{pmatrix} \approx \begin{pmatrix} 0.994357 & 0.289669 \\ 0.289669 & 0.994357 \end{pmatrix}.$$

6. Matrices $M = \begin{pmatrix} 1 & 0 & 0 & \cdots & 0 & -1/C \\ -\frac{kh}{2} & 1-\frac{kh}{2} & 0 & \cdots & 0 & -1/C \\ -\frac{kh}{2} & -kh & 1-\frac{kh}{2} & \ddots & \vdots & \vdots \\ \vdots & \vdots & \ddots & \ddots & 0 & -1/C \\ -\frac{kh}{2} & -kh & \cdots & -kh & 1-\frac{kh}{2} & -1/C \\ -\frac{kh}{2} & -kh & \cdots & -kh & -kh & 1-1/C-\frac{kh}{2} \end{pmatrix}$ from [9, page 1360], where $kh = \frac{2}{3}$, $C = 6$.

7. Matrices $M = \begin{pmatrix} 1 & 0 & \cdots & 0 & 1 \\ -1 & 1 & \ddots & \vdots & \vdots \\ -1 & -1 & \ddots & 0 & 1 \\ \vdots & \vdots & \ddots & 1 & 1 \\ -1 & -1 & \cdots & -1 & 1 \end{pmatrix}$ from [10, page 156].

$n = 128$	minimal	maximal	average
Class 1	31371	37287	34147
Class 2	35150	40904	38168
Class 3, $\rho = 1$	30189	36097	32995
Class 3, $\rho = 2$	30597	36561	32960
Class 3, $\rho = 3$	29938	35761	32967
Class 4	31342	36333	33648
Class 5	24318		
Class 6	32258		
Class 7	32764		

$n = 256$	minimal	maximal	average
Class 1	131692	146780	139419
Class 2	147123	161971	153559
Class 3, $\rho = 1$	127911	143706	136361
Class 3, $\rho = 2$	129228	144226	136427
Class 3, $\rho = 3$	129945	145882	136508
Class 4	131533	146014	138392
Class 5	97790		
Class 6	130050		
Class 7	131068		

Table 1. Numbers of comparisons in GERP.

For each matrix of classes 1–4 the tests were performed for $m = 1000$ input instances M for each of the two values $n = 128$ and $n = 256$, for $b = c = l = 4$, and for $q = 10$. For class 3 the tests were performed for each of the three values

$n = 128$	GEPP	GECP	GERP	$n = 256$	GEPP	GECP	GERP
Class 1	13.8 ± 2.5	6.4 ± 0.4	8.4 ± 0.8	Class 1	21.8 ± 3.8	9.5 ± 0.6	12.8 ± 1.3
Class 2	2.5 ± 0.5	1.5 ± 0.2	1.8 ± 0.2	Class 2	3.4 ± 0.6	1.9 ± 0.2	2.4 ± 0.3
Cl. 3, $\rho = 1$	17.4 ± 4.0	8.7 ± 1.0	11.6 ± 1.8	Cl. 3, $\rho = 1$	32.2 ± 7.4	15.5 ± 1.7	20.6 ± 2.9
Cl. 3, $\rho = 2$	15.6 ± 3.6	7.7 ± 0.8	10.2 ± 1.4	Cl. 3, $\rho = 2$	29.2 ± 6.7	13.8 ± 1.4	18.6 ± 2.9
Cl. 3, $\rho = 3$	14.3 ± 3.5	7.0 ± 0.7	9.3 ± 1.3	Cl. 3, $\rho = 3$	27.0 ± 6.1	12.5 ± 1.3	16.7 ± 2.3
Class 4	FAIL	1	1	Class 4	FAIL	1	1
Class 5	3.4e6	2	2	Class 5	3.1e13	2	2
Class 6	6.6e36	1.33	1.33	Class 6	8.6e74	1.33	1.33
Class 7	1.7e38	2	2	Class 7	5.8e76	2	2

Table 2. Growth factor in GEPP/GECP/GERP.

$\rho = 1, 2, 3$. Besides the results of these tests, Tables 1–4 also cover the test results for matrices M of classes 5–7 (from the papers [8], [9], and [10], respectively), for which GEPP produced corrupted outputs.

To every matrix GEPP, GECP, and GERP were applied. As was expected, for matrix classes 1–3 numerical performance of GEPP, GECP, and GERP was similar but for classes 4–7 GEPP either failed or lost many more correct input bits versus GECP and GERP.

Table 1 shows the maximum, minimum and average numbers of comparisons used in GERP for every input class of matrices.

Table 2 shows the average growth factor

$$\phi = \max{}_{i,j,k=0}^{n-1} \; |m_{ij}^{(k)}| / \max{}_{i,j=0}^{n-1} \; |m_{ij}|$$

(as well as its standard deviation from the average) where $M^{(k)} = (m_{i,j}(k))_{i,j=k}^{n-1}$ denotes the matrix computed in k steps of Gaussian elimination with the selected pivoting policy and $M = M^{(0)} = (m_{ij})_{i,j=0}^{n-1}$ denotes the input matrix.

Tables 3 and 4 show the average norms of the error and residual vectors, respectively, as well as the standard deviations from the average, where the linear systems $M\mathbf{y} = \mathbf{f}$ were solved by applying GECP, GEPP, and GERP. The vectors $\mathbf{f}$ were defined according to the following rule: first generate vectors $\mathbf{y}$ with random components from the sets $\{-1, 0, 1\}$ or $\{-1, 1\}$, then save these vectors for computing the errors vectors, and finally compute the vectors $\mathbf{f} = M\mathbf{y}$.

Remark 1. Table 1 shows the results of testing GERP where n comparisons were used for computing the maximum of n numbers. Extensive additional tests with random matrices (of class 1) for $n = 2^h$ and for h ranging from 5 to 10 were performed in the Graduate Center of the City University of New York. In these tests the modification GERP was run where no tested row or column is examined again until the next elimination step. Furthermore, the tests used $k-1$ comparisons for computing the maximum of k numbers. The observed numbers of comparisons slightly decreased versus Table 1 and always stayed below $2n^2$.

$n = 128$	GEPP	GECP	GERP
Class 1	6.8e-13 ± 3.4e-12	5.2e-13 ± 2.8e-12	4.8e-13 ± 2.2e-12
Class 2	1.7e7 ± 2.6e8	8.7e5 ± 4.6e6	6.6e5 ± 3.7e6
Class 3, $\rho = 1$	1.1e-5 ± 8.4e-6	7.4e-6 ± 5.7e-6	8.7e-6 ± 6.7e-6
Class 3, $\rho = 2$	1.7e-5 ± 8.8e-6	1.2e-5 ± 6.1e-6	1.3e-5 ± 7.0e-6
Class 3, $\rho = 3$	2.1e-5 ± 9.2e-6	1.5e-5 ± 6.2e-6	1.7e-5 ± 7.5e-6
Class 4	FAIL	5.7e-13 ± 6.3e-12	5.7e-13 ± 3.5e-12
Class 5	1.0e-9	2.7e-15	2.7e-15
Class 6	3.1e3	2.7e-15	2.7e-15
Class 7	6.5	0.0	0.0
$n = 256$	GEPP	GECP	GERP
Class 1	3.8e-12 ± 3.7e-11	2.8e-12 ± 4.0e-11	2.6e-12 ± 2.0e-11
Class 2	3.9e7 ± 5.0e8	1.1e6 ± 4.1e6	2.2e6 ± 1.3e7
Class 3, $\rho = 1$	2.0e-5 ± 1.5e-5	1.3e-5 ± 9.3e-6	1.5e-5 ± 1.1e-5
Class 3, $\rho = 2$	3.1e-5 ± 1.6e-5	2.0e-5 ± 1.1e-5	2.4e-5 ± 1.2e-5
Class 3, $\rho = 3$	3.9e-5 ± 1.7e-5	2.5e-5 ± 1.1e-5	2.9e-5 ± 1.2e-5
Class 4	FAIL	3.6e-12 ± 4.0e-11	3.6e-12 ± 2.5e-11
Class 5	1.4e-2	3.7e-15	3.7e-15
Class 6	7.2e57	3.6e-14	3.6e-14
Class 7	11.3	0.0	0.0

Table 3. Norms of the error vectors in GEPP/GECP/GERP.

Remark 2. Similar test results for class 1 were presented earlier in [5] and [6] and for classes 3 and 5–7 in [5], but [5] shows no norms of the error and residual vectors. It seems that GEPP, GECP, and GERP have not been tested earlier for classes 2 and 4.

Acknowledgement

We are happy to acknowledge valuable experimental support of our work by Dr. Xinmao Wang.

References

1. G. H. Golub and C. F. Van Loan, *Matrix Computations*, 3rd edition, The Johns Hopkins University Press, Baltimore, Maryland, 1996.
2. G. W. Stewart, *Matrix Algorithms, Vol I: Basic Decompositions*, SIAM, Philadelphia, 1998.
3. N. J. Higham, *Accuracy and Stability in Numerical Analysis*, SIAM, Philadelphia, 2002 (second edition).
4. L. Neal and G. Pool, A Geometric Analysis of Gaussian Elimination, II, *Linear Algebra and Its Applications*, **173**, 239–264, 1992.

$n = 128$	GEPP	GECP	GERP
Class 1	1.6e-9 ± 3.0e-10	1.1e-9 ± 1.7e-10	1.2e-9 ± 2.1e-10
Class 2	2.2e-4 ± 1.6e-3	1.2e-4 ± 4.7e-4	1.1e-4 ± 6.3e-4
Class 3, $\rho = 1$	3.1e-14 ± 5.1e-15	2.0e-14 ± 2.9e-15	2.3e-14 ± 3.6e-15
Class 3, $\rho = 2$	3.0e-14 ± 5.0e-15	1.9e-14 ± 2.8e-15	2.3e-14 ± 3.6e-15
Class 3, $\rho = 3$	3.0e-14 ± 5.3e-15	1.9e-14 ± 2.8e-15	2.3e-14 ± 3.5e-15
Class 4	FAIL	3.3e2 ± 3.3e2	3.5e2 ± 3.3e2
Class 5	1.1e-9	1.9e-15	1.9e-15
Class 6	2.9e3	1.7e-14	1.7e-14
Class 7	14.5	0.0	0.0
$n = 256$	GEPP	GECP	GERP
Class 1	7.1e-9 ± 1.1e-9	4.4e-9 ± 5.8e-10	5.2e-9 ± 7.2e-10
Class 2	2.1e-3 ± 3.7e-2	6.2e-4 ± 2.1e-3	1.5e-3 ± 1.6e-2
Class 3, $\rho = 1$	9.8e-14 ± 1.5e-14	5.7e-14 ± 6.8e-15	7.4e-14 ± 9.3e-15
Class 3, $\rho = 2$	9.7e-14 ± 1.4e-14	5.7e-14 ± 7.0e-15	7.1e-14 ± 9.2e-15
Class 3, $\rho = 3$	3.9e-5 ± 1.7e-5	5.7e-14 ± 6.9e-15	7.0e-14 ± 9.1e-15
Class 4	FAIL	6.7e2 ± 6.5e2	6.6e2 ± 6.3e2
Class 5	9.0e-3	2.6e-15	2.6e-15
Class 6	2.1e58	1.0e-13	1.0e-13
Class 7	41.1	0.0	0.0

Table 4. Norms of the residual vectors in GEPP/GECP/GERP.

5. L. V. Foster, The Growth Factor and Efficiency of Gaussian Elimination with Rook Pivoting, *J. of Comp. and Applied Math.*, **86**, 177–194, 1997. Corringendum in *J. of Comp. and Applied Math.*, **98**, 177, 1998.
6. G. Pool and L. Neal, The Rook's Pivoting Strategy, *J. of Comp. and Applied Math.*, **123**, 353–369, 2000.
7. J. H. Wilkinson, Error Analysis of Direct Methods of Matrix Inversion, *J. of ACM*, **8**, 281–330, 1961.
8. S. J. Wright, A Collection of Problem for Which Gaussian Elimination with Partial Pivoting Is Unstable, *SIAM J. on Sci. Stat. Computing*, **14**, **1**, 231–238, 1993.
9. L. V. Foster, Gaussian Elimination with Partial Pivoting Can Fail in Practice, *SIAM J. on Matrix Analisys and Applications*, **15**, **4**, 1354–1362, 1994.
10. N. J. Higham and D. J. Higham, Large Growth Factors in Gaussian Elimination with Pivoting, *SIAM J. on Matrix Analisys and Applications*, **10**, **2**, 155–164, 1989.

Newton's Iteration for Matrix Inversion, Advances and Extensions*

Victor Y. Pan

Department of Mathematics and Computer Science,
Lehman College of CUNY, Bronx, NY 10468, USA
victor.pan@lehman.cuny.edu
http://comet.lehman.cuny.edu/vpan/

Abstract. We first cover Newton's iteration for generalized matrix inversion, its ameliorations, recursive compression of its iterates in the case of structured inputs, some techniques of continuation via factorization, and extension to splitting the Singular Value Decomposition. We combine the latter extension with our recent fast algorithms for the null space bases (prompted by our progress in randomized preconditioning). We applied these combinations to compute the respective spaces of singular vectors and to arrive at divide-and-conquer algorithms for matrix inversion and computing determinants. Our techniques promise to be effective for computing other matrix functions in the case of ill conditioned inputs.

Keywords: matrix inversion, Newton's iteration, matrix structure, continuation (homotopy), divide-and-conquer algorithms, null spaces.

1 Introduction

Newton's iteration for generalized matrix inversion amounts mostly to performing a sequence of matrix multiplications. This level-three BLAS performance is particularly effective on systolic arrays and parallel computers.

Newton's iteration for generalized inverse is important in its own right but also as a sample technique for computing various other matrix functions such as the square root, matrix sign function, and the solution of Riccati equation. We survey and advance this approach, show its acceleration in the case of structured input matrices, its combination with our new techniques of homotopic continuation, factorization, and preconditioning, as well as its extension to divide-and-conquer algorithms for splitting the Singular Value Decomposition, that is for computing the respective subspaces generated by singular vectors (hereafter we refer to such subspaces as *singular subspaces* and invoke the usual abbreviation SVD). The latter extensions employ our recent techniques for the computation of bases for null spaces, which should enhance the power of the approach.

* Supported by PSC CUNY Award 69330–0038.

We recall some basic definitions in the next section and then, in Section 3, the convergence analysis from [1] and [2] and some recipes for the initialization. In Section 4 we describe three techniques that exploit input structure to save running time and computer memory by compressing the computed approximations. All three techniques usually require reasonably good initialization (in spite of some interesting phenomenon of autocorrection in compression), and in Section 5 we cover a general recipe for initialization by means of homotopy (continuation), effective for both general and structured inputs. We improve conditioning of continuation by representing it as recursive factorization. These preconditioning techniques can be of interest on its own right, independently of the considered iterative processes. In Section 6 we describe a modified iteration directed to splitting the SVD and its generalizations. This technique produces bases for the respective singular subspaces and can be extended to divide-and-conquer algorithms for the inverses, determinants, square roots, and other matrix functions. The technique is proposed for general Hermitian input matrices. (It does not preserve matrix structure except for symmetry.) We cover this direction in Section 7, where we also employ our recent effective algorithms for computing null space bases of general non-Hermitian matrices. We briefly recall these algorithms in Section 8 and point out their natural extension to randomized preconditioning of ill conditioned inputs. In Section 9 we discuss some directions for further study.

2 Basic definitions

We rely on the customary definitions for matrix computations in [3]–[8]. M^H denotes the Hermitian transpose of a matrix M. I_k is the $k \times k$ identity matrix. I is the identity matrix of an unspecified size. (A, B) is the 1×2 block matrix with blocks A and B. $\mathrm{diag}(a_i)_i$ (resp. $\mathrm{diag}(B_i)_i$) is the diagonal (resp. block diagonal) matrix with the diagonal entries a_i (resp. diagonal blocks B_i). U is a unitary matrix if $U^H U = I$. $N(M)$ denotes the (right) null space of a matrix M. $\mathrm{range}(M)$ is the range of a matrix M, that is its column span. A matrix M is a *matrix basis* for a space $\mathbb{S}$ if its columns form a basis for this space, that is if the matrix has full column rank and if $\mathrm{range}(M) = \mathbb{S}$. A matrix basis for the null space $N(M)$ is a null matrix basis for a matrix M. $\rho = \mathrm{rank}(M)$ is its rank. $\sigma_i(M)$ is its ith largest singular value, $i = 1, 2, \ldots, \rho$. $\mathrm{cond}_2 M = \sigma_1(M)/\sigma_\rho(M) \geqslant 1$ is the condition number of a matrix M of a rank ρ. A matrix is well conditioned if its condition number is not large (relatively to the computational task and computer environment) and is ill conditioned otherwise. C^+ and C^- denote the Moore–Penrose generalized inverse of a matrix C, so that $C^+ = C^- = C^{-1}$ for a nonsingular matrix C.

3 Newton's iteration for matrix inversion. Its initialization and acceleration

Newton's iteration

$$x_{k+1} = x_k - f(x_k)/f'(x_k), \quad k = 0, 1, \ldots, \tag{1}$$

rapidly improves a crude initial approximation $x = x_0$ to the solution $x = r$ of an equation $f(x) = 0$ provided $f(x)$ is a smooth nearly linear function on an open line interval that covers two points r and x_0. Equation (1) can be obtained by truncating all terms of the orders of at least two in Taylor's expansion of the function $f(x)$ at $x = r$.

Hotelling [9] and Schultz [10] extended Newton's iteration (1) to the case where $x = X$, $x_k = X_k$, and $f(x_k) = f(X_k)$ are matrices and $f(X) = M - X^{-1}$ for two matrices M and X. In this case Newton's iteration rapidly improves a crude initial approximation X_0 to the inverse of a nonsingular $n \times n$ matrix M,

$$X_{k+1} = X_k(2I - MX_k), \quad k = 0, 1, \ldots. \tag{2}$$

Indeed, define the error and residual matrices

$$E_k = M^{-1} - X_k, \quad e_k = \|E_k\|, \quad R_k = ME_k = I - MX_k, \quad \rho_k = \|R_k\|$$

for all k, assume a matrix norm $\|\cdot\|$ satisfying the submultiplicative property $\|AB\| \leqslant \|A\| \ \|B\|$, and deduce from equation (2) that

$$R_k = R_{k-1}^2 = R_0^{2^k}, \quad \rho_k \leqslant \rho_0^{2^k}, \tag{3}$$

$$ME_k = (ME_{k-1})^2 = (ME_0)^{2^k}, \quad e_k \leqslant e_0^{2^k} \|M\|^{2^k-1}. \tag{4}$$

The latter equations show quadratic convergence of the approximations X_k to the inverse matrix M^{-1} provided $\rho_0 < 1$. Each step (2) amounts essentially to performing matrix multiplication twice. Iteration (2) is numerically stable for nonsingular matrices M, but numerical stability has been proved in [2] for its extensions (16) and (17) in Section 6 even where the matrix M is singular.

Ben-Israel in [11] and Ben-Israel and Cohen in [12] proved that the iteration converges where $X_0 = aM^H$ for a sufficiently small positive scalar a. Søderstrøm and Stewart [1] analyzed Newton's iteration based on the SVDs of the involved matrices. This study was continued by Schreiber in [13] and then in [2]. We outline this work by using *Generalized SVDs* (hereafter to be referred to as *GSVDs*), that is nonunique representations of matrices as $U\Sigma V^H$ where U and V are unitary matrices and Σ is a diagonal matrix. They turn into SVDs wherever Σ denotes diagonal matrices filled with nonnegative entries in nonincreasing order. Assume that the matrices M and X_0 have GSVDs

$$M = U\Sigma V^H, \ X_0 = V\Phi_0 U^H \tag{5}$$

for some unitary matrices U and V and diagonal matrices $\Sigma = \text{diag}(\sigma_i)_i$ and $\Phi_0 = \text{diag}(\phi_{i,0})_i$. In particular this is the case where

$$X_0 = f(M^H) \tag{6}$$

is a matrix function in M^H, e.g.,

$$X_0 = aM^H + bI \tag{7}$$

for two scalars a and b. Then we have

$$X_k M = VS_k V^H, \quad S_k = \text{diag}(s_i^{(k)})_i, \quad 1 - s_i^{(k+1)} = (1 - s_i^{(k)})^2 \tag{8}$$

for all i and k. Furthermore, we have

$$s_i^{(0)} = \sigma_i \phi_{i,0} = \sigma_i f(\sigma_i) \tag{9}$$

for all i under (6), so that iteration (2) converges to the generalized inverse M^- if $0 < s_i^{(0)} = \sigma_i \phi_{i,0} < 2$ for all i. Convergence is locally quadratic but can be slow initially if the values $s_i^{(0)}$ are close to zero or two for some subscripts i. More precisely, assume the choice (7) for $b = 0$ and $a = 1/(\|M\|_1 \|M\|_\infty)$ [11]. Then it can be proved that $\rho_0 \leqslant 1 - 1/((\text{cond}_2 M)^2 n)$ (cf. [14]). By choosing $a = y/(\|M\|_1 \|M\|_\infty)$ for any value of y such that $1 \leqslant y \leqslant 2(\text{cond}_2 M)^2 n/(1 + (\text{cond}_2 M)^2 n)$ we obtain the slightly improved bound $\rho_0 \leqslant 1 - y/((\text{cond}_2 M)^2 n)$. In particular for $y = 2n/(1+n)$ we obtain that $\rho_0 \leqslant 1 - 2n/((\text{cond}_2 M)^2 (1+n))$. Under these choices we need about $\nu = 2\log_2 \text{cond}_2 M$ steps (2) to decrease the residual norm ρ_k below $1/e = 0.3678781\ldots$. Then in the order of $l = \log_2 \ln h$ additional steps (2) we would yield the bound $\rho_{\nu+l} \leqslant e^{-2^h}$, $e = 2.7182981\ldots$.

The bound on the number ν of initial steps is critical for ill conditioned matrices. It was decreased roughly by twice in [2] by means of replacing iteration (2) by its scaled version

$$X_{k+1} = a_k X_k (2I - MX_k), \quad k = 0, 1, \ldots, l \tag{10}$$

for appropriate scalars a_k.

Clearly, the inversion of a nonsingular matrix M can be reduced to the inversion of either of the Hermitian positive definite matrices $M^H M$ or MM^H because $M^{-1} = (M^H M)^{-1} M^H = M^H (MM^H)^{-1}$ or of the Hermitian matrix $\begin{pmatrix} 0 & M \\ M^H & 0 \end{pmatrix}$, having the inverse $\begin{pmatrix} 0 & M^{-H} \\ M^{-1} & 0 \end{pmatrix}$.

Now suppose M is a Hermitian matrix. Then one can further accelerate the computations by twice by choosing the initial approximation $X_0 = yI/\|M\|_1$ for any value y such that $1 \leqslant y \leqslant 2\sqrt{n}(\text{cond}_2 M)/(1 + \sqrt{n}(\text{cond}_2 M))$. This yields the bound $\rho_0 \leqslant 1 - 2\sqrt{n}/((\text{cond}_2 M)(1 + \sqrt{n}))$.

The paper [2] obtains some acceleration for a large class of inputs by means of replacing iteration (2) with cubic iteration $X_{k+1} = (cX_k^2 + dX_k + eI)X_k$,

$k = 0, 1, \ldots$ for appropriate scalars c, d, and e. The latter resource was employed again in [15] in the case of sructured input matrices. For more narrow input classes one can try to yield further acceleration of convergence by applying more general iteration schemes. For example, recall the following two-stage iteration [16]–[18], having effective applications to integral equations via the associated tensor computations,

$$X_{k+1} = X_k(2I - X_k), \quad Y_{k+1} = Y_k(2I - X_k).$$

Here $Y_0 = I$ and $M = X_0$ is a nonsingular matrix such that $\sigma_1(I - X_0) = \|I - X_0\|_2 < 1$. It is readily verified that $X_k = X_0 Y_k$ for all k and that the matrices X_k converge to the identity matrix I. Consequently the matrices Y_k converge to the inverse $M^{-1} = X_0^{-1}$.

4 Structured iteration, recursive compressions, and autocorrection

Next, assuming that the input matrix M is structured and is given with its short displacement generator, we modify Newton's iteration to perform its steps faster. We begin with recalling some background on the displacement representation of matrices (cf. [19]–[21]). We rely on the Sylvester displacement operators $\nabla_{A,B}(M) \leftarrow AM - MB$, defined by the pairs of the associated $n \times n$ operator matrices A and B. The next simple fact relates them to the Stein operators $\Delta_{A,B}(M) = M - AMB$.

Theorem 1. $\nabla_{A,B} = A\Delta_{A^{-1},B}$ *if* A *is nonsingular.* $\nabla_{A,B} = -\Delta^{-1}_{A,B^{-1}}B$ *if* B *is nonsingular.*

$\nabla_{A,B}(M)$ is the *displacement* of M, its rank is the *displacement rank* of M. The matrix pair $\{S, T\}$ is a *displacement generator* of length l for M if $\nabla_{A,B}(M) = ST^H$ and if S and T are $n \times l$ matrices. If a matrix M has displacement rank $r = \operatorname{rank} \nabla_{A,B}(M)$ and is given with its displacement generator of a length l, then one can readily compute its displacement generator of length r in $O(l^2 n)$ ops [21, Sections 4.6].

Most popular structures of Toeplitz, Hankel, Vandermonde and Cauchy types are associated with the operators $\nabla_{A,B}$ where each of the operator matrices A and B is diagonal or unit f-circulant. For such operators simple l-term bilinear or trilinear expressions of an $n \times n$ matrix M via the entries of its displacement generator $\{S, T\}$ of length l can be found in [20], [21, Sections 4.4.4 and 4.4.5], and [22]. If $l << n$, then M can be called $\nabla_{A,B}$-structured.

One can perform arithmetic operations with matrices in term of the associated displacement generators provided the respective operator matrices match properly [20], [21]. In this way one can multiply such $n \times n$ matrices pairwise in

$O(nl^2(\log n)^d)$ flops, can multiply them by a vector in $O(nl(\log n)^d)$ flops for $d = 1$ or $d = 2$, and needs no flops to sum them. In summation and multiplication, the displacement rank of the output can reach the sum of the displacement ranks of the input matrices. It increases respectively in a single step of Newton's iteration (2) but stays invariant in inversion. Here are some specifics (see other relevant results in [21, Section 1.5]).

Theorem 2. $\nabla_{B,A}(M^{-1}) = -M^{-1}\nabla_{A,B}(M)M^{-1}$.

Theorem 3. *Let* $Y = 2X - XMX$. *Then we have*

$$\nabla_{B,A}(Y) = 2\nabla_{B,A}(X) - \nabla_{B,A}(X)MX - XM\nabla_{B,A}(X) - X\nabla_{A,B}(M)X.$$

It follows that $\text{rank}(\nabla_{B,A}(Y)) \leqslant 2\,\text{rank}(\nabla_{B,A}(X)) + \text{rank}(\nabla_{A,B}(M))$. This bound is sharp.

As the iteration progresses, the displacement rank of the iterates X_i can steadily grow, and soon this can make the iteration step as expensive as in the case of general matrices. In Newton's structured iteration such a problem is countered by means of periodic compression of the displacement generators of the iterates X_k towards the displacement rank of the output (which is the same as for the input, due to Theorem 2). The approximation to the inverse can be spoiled in the compression, but if this deviations is minor it is usually more than compensated in the next iteration step. In [21, Sections 6.3–6.8], [22]–[30] such Compression Subroutines (we denote them **CS**s) have been devised and analyzed, including upper estimates for the initial residual norm that ensure the overall rapid convergence of the iteration. Below we show some details. We begin our iterative process with an initial approximation Y_0 to the inverse M^{-1} and proceed as follows,

$$X_{i+1} = f(Y_i, M), \quad Y_{i+1} = g(X_{i+1}) \text{ for } i = 0, 1, \ldots. \tag{11}$$

Here $f(Y_i, M)$ defines the iterative inversion process, e.g.,

$$f(Y_i, M) = Y_i(2I - MY_i), \tag{12}$$

and $g(X_{i+1})$ defines a Compression Subroutine **CS**.

The first subroutine (due to [23]) compresses displacement by truncating its smallest singular values. Note that the SVD $W = U\Sigma V^H$ defines the generators $\{U\Sigma, V\}$ and $\{U, \Sigma V\}$ for a matrix W, both having length $l = \text{rank}(W)$.

Subroutine CS1. Compute the SVD-based generator of the displacement $\nabla_{B,A}(X_i)$ and decrease its length to a fixed smaller value r (not exceeded by $r_- = \text{rank}(\nabla_{A,B}(M))$) by setting to zero all singular values σ_i of the displacement for $i \geqslant r+1$. Output the resulting displacement generator that defines the matrix Y_i.

Here are the respective estimates for the convergence rate [21, Corollary 6.4.3].

Theorem 4. *Let Algorithm (11), (12) be applied with the function* $g(X_{i+1})$ *defined by the Subroutine* **CS1** *for a nonsingular operator* $\nabla_{B,A}$ *and for* $r = \text{rank}(\nabla_{A,B}(M))$. *Then we have*

$$\mu\rho_i \leqslant (\mu\rho_0)^{2^i}, \quad i = 1, 2, \ldots,$$

where $\rho_i = \|I - MY_i\|_2$, $\mu = ((\|A\|_2 + \|B\|_2 + 1)\nu^-)\text{cond}_2 M$, *and*

$$\nu^- = \|\nabla_{A,B}^{-1}\|_2 = \sup_M(\|M\|_2/\|\nabla_{A,B}(M)\|_2) \tag{13}$$

is the norm of the inverse of the operator $\nabla_{A,B}(M)$.

The norm ν^- is reasonably bounded for the operators $\nabla_{A,B}$ associated with matrices that have structures of Toeplitz, Hankel, Vandermonde and Cauchy types (see [21, Section 6.6], [22]), but not so for the operators associated with rank structured matrices, also studied as semiseparable, quasiseparable, close to Hankel, etc. (cf. [31]). Thus we cannot readily ensure effectiveness of Newton's structured iteration for these matrices (see, however, our comments preceding Theorem 7 in the next section).

The second Compression Subroutine is based on substitution. It first appeared in [32], [33] in algebraic form (over the fields) and then in [26], [34] in numerical form. Given an approximate inverse X_0 and a displacement generator $\{S, T\}$ of a length r for the input matrix M, the subroutine recursively computes displacement generators of the same length r for approximate inverses.

Subroutine CS2. Assume iteration (11) and let $\nabla_{A,B}(M) = ST^H$ for two given matrices S and T. Compute and output the two matrices

$$S_{i+1} = Y_i g(Y_i, M)S, \tag{14}$$
$$T_{i+1}^H = T^H Y_i g(Y_i, M). \tag{15}$$

The respective estimates for the convergence rate are from [21, Theorem 6.5.2].

Theorem 5. *Under the definitions and assumptions of Theorem 4, apply Subroutine* **CS2** *(rather than* **CS1***), let* $\rho_0 \leqslant 1$, $e_i = \|M^{-1} - Y_i\|_2 \leqslant \|M^{-1}\|_2$ *for all* i. *Write* $C = 3\nu^-\|\nabla_{A,B}(M)\|_2 \ \|Y_0\|_2/(1-\rho_0)$, $\mu = C^2\|M\|_2$, $e = \rho_0\|Y_0\|_2/(1-\rho_0)$. *Let* $\mu e_1 \leqslant (Ce\|M\|_2)^2$. *Then* $\mu e_{i+1} \leqslant (\mu e_1)^{2^i}$, $i = 0, 1, \ldots$.

The third Compression Subroutine **CS3** is due to [30]. It compresses the displacement and simultaneously improves the current approximation to the inverse.

Subroutine CS3. Given a shortest displacement generator $\{S, T\}$ of a length r_- for a nonsingular $n \times n$ matrix M, a displacement generator $\{\tilde{S}, \tilde{T}\}$ of a length $r_+ > r$ for its approximate inverse X, and an integer r such that $r_- \leqslant r < r_+$, the subroutine computes a (shortest) generator $\{S_-, T_-\} = \{\tilde{S}Y_S, -\tilde{T}Y_T\}$ of length r where Y_S and Y_T are $n \times r$ matrices that minimize the 2-norms of all columns of the matrices $S - M\tilde{S}Y_S$ and $T + M\tilde{T}Y_T$.

Let us motivate the construction of this subroutine. Observe that

$$\nabla_{B,A}(M^{-1}) = S_-T_-^H = -M^{-1}\nabla_{A,B}(M)M^{-1} = -M^{-1}ST^HM^{-1}.$$

Thus we can choose a displacement generator $\{S_-, T_-\}$ for the inverse M^{-1} such that $S_- = M^{-1}S$ and $-T_-^H = T^HM^{-1}$ or equivalently $S = MS_-$ and $T = -M^HT_-$. Therefore, by minimizing the norms of all columns in the matrices $S - MS_X$ and $T + M^HT_X$ for $S_X = \tilde{S}Y_S$ and $T_X = \tilde{T}Y_T$, Subroutine **CS3** minimizes the 2-norms of all columns of the matrices $MS_- - MS_X$ and $M^HT_- - M^HT_X$.

The Subroutine **CS3** can be applied recursively just as the function $g(X_{i+1})$ in equation (11), to compress and refine the approximations X_k computed in the iteration, but can be alternatively combined with Subroutines **CS1** or **CS2**.

In the case of structured inputs, compression by means of all the above techniques has the computational cost dominated at the iteration stage.

Extensive numerical tests for the inversion of Toeplitz matrices in the City University of New York, University of Pisa, Italy, and Katholieke Universiteit Leuven, Belgium, confirmed the results of the above analysis and showed slightly faster convergence with Subroutine **CS1** versus **CS2** and with Subroutine **CS3** versus **CS1** (cf. [21, Section 6.11], [28], [30]). The study of Subroutine **CS1** by Bini and Meini in [28] has naturally lead them to introducing the intersting general concept of approximate displacement rank.

The test results in [21, Tables 6.4–6.21] and [30] show that the iteration with compression via Subroutines **CS1**, **CS2**, and **CS3** frequently diverges if $\text{cond}_2 M$ is close to one. Table 6.21 in [21] shows an interesting phenomenon in such tests based on Subroutine **CS1** where the iteration initially stumbled, producing residuals with the norms near one or a little above one. Then it diverged in about 75 percents of cases but in about 25 percents of cases still converged. Explaining this phenomenon of *autocorrection* could have helped to employ it for advancing the approach. (Perhaps, compression of the displacement of a matrix X_k removes the disturbance by the extraneous singular vectors.) Our techniques in the next section, however, enable convergence of Newton structured iteration independently of autocorrection.

5 Homotopic (continuation) approach

Homotopic (continuation) techniques are a natural means of the initialization of iterative processes. In the case of Newton's iteration for matrix inversion they were first studied in [23] and then in [21, Section 6.9], [35], and [36]. (In particular it was proved in [23] that homotopic processes with Newton's structured iteration enable inversion of an $n \times n$ well conditioned Toeplitz or Toeplitz-like matrix in $O(n \log^2 n)$ arithmetic operations.) The idea of this approach is to invert at first a readily invertible matrix M_0, e.g., $M_0 = M + t_0 I$ for a large scalar t_0. Then one can recursively invert the matrices $M_i = t_i I + M$, $i = 1, 2, \dots, l$, where $t_0 > t_1 > \dots > t_l = 0$ and the matrix M_{i-1}^{-1} serves as the initial approximation to the matrix M_i^{-1}. By choosing proper scalars t_i, we can yield smaller initial residual norms and therefore faster convergence at every step. The condition numbers of the matrices M_i grow approaching $\text{cond}_2 M$ as i grows, which is disturbing where M is an ill conditioned matrix. We, however, alleviate the problem by representing our homotopic process as recursive factorization of the matrix M. We rely on the following result [37].

Theorem 6. *For an $n \times n$ matrix M define some scalars $t = t_0 \neq 0$, $u_k > 1$, and $t_{k+1} = t_k(1 - 1/u_k) = t\prod_{j=0}^{k}(1 - 1/u_j)$ for $k = 0, 1, \dots$. Write $M_k = M + t_k I$, $P_k = I - t_k(u_k M_k)^{-1}$, $k = 0, 1, \dots$. Suppose the matrices M_k are nonsingular for all k. Then we have*

a) $M_{k+1} = P_k M_k = P_k P_{k-1} \cdots P_0 M_0$,
b) $M = P_{k-1} \cdots P_0 M_0 V_k$ *where* $V_k = M_k^{-1} M$, *and*
c) $I - V_k = a_k M_k^{-1}$

for $k = 0, 1, \dots$.

Theorem 6 enables us to invert the matrix M by recursively computing the matrices M_0, M_0^{-1}, P_k, P_k^{-1}, M_{k+1}, M_{k+1}^{-1}, for $k = 0, 1, \dots, l-1$ and a fixed integer l, and then computing $V_l = M_l^{-1} M$. Then if M is nonsingular, we compute the inverses V_l^{-1} and

$$M^{-1} = V_l^{-1} M_l^{-1} = V_l^{-1} M_0^{-1} P_0^{-1} \cdots P_l^{-1}.$$

We can invert the matrices M_0, P_0, ..., P_{l-1}, and V_l by applying Newton's iteration, its extensions, or any other iteration that converges faster where the initial residual norm is smaller. Alternatively we can apply any algorithm whose performance is enhanced where the input matrix is better conditioned. In this case the homotopic process plays the role of preconditioning. E. g., in the case of structured input matrices the inversion is reduced to solving a smaller number of linear systems of equations with the same input matrix. This can be efficiently handled by algorithms of the Conjugate Gradient type if the matrix is well conditioned. *The latter approach promises to be effective also for well conditioned rank structured input matrices.* Next we estimate the residual norms and condition numbers for the matrices M_0, P_0, ..., P_{l-1}, and V_l.

Theorem 7. *(Cf. [37].) Let M have the spectrum $\Lambda(M) = \{\lambda_1, \dots, \lambda_n\}$. Then under the assumptions of Theorem 6, we have*

$$\begin{aligned}
\Lambda(I - M_0/t) &= \{-\lambda_i/t\}_{i=1}^n, && t = t_0,\\
\Lambda(I - P_k) &= \{(t_k/u_k)/(t_k + \lambda_i)\}_{i=1}^n, && t_{k+1} = t_k(1 - 1/u_k),\\
& && k = 0, 1, \ldots, l-1,\\
\Lambda(I - V_l) &= \{t_l/(\lambda_i + t_l)\}_{i=1}^n.
\end{aligned}$$

Corollary 1. *Under the assumptions of Theorem 6, let* M *be a Hermitian and positive definite matrix and let*

$$\lambda^+ \geqslant \lambda_1 \geqslant \lambda_2 \geqslant \cdots \geqslant \lambda_n \geqslant \lambda_- > 0,$$

$$t = t_0 > 0, \quad u_k > 1, \quad t_{k+1} = t_k(1 - 1/u_k), \quad k = 0, 1, \ldots.$$

Then

a) *the matrices* M_0, $M_0/t - I$, P_k, $I - P_k$ *for* $k = 0, 1, \ldots, l-1$, V_l *and* $I - V_l$ *are Hermitian and positive definite,*

b) $\|I - M_0/t\|_2 \leqslant \lambda_1/t \leqslant \lambda^+/t$, $\|I - P_k\|_2 < 1/u_k$, $k = 0, 1, \ldots, l-1$, $\|I - V_l\|_2 \leqslant 1/(1 + \lambda_n/t_l) \leqslant 1/(1 + \lambda_-/t_l)$,

c) $\text{cond}_2(M_0) = 1 + \frac{\lambda_1 - \lambda_n}{t + \lambda_n} < 1 + \lambda_1/t \leqslant 1 + \lambda^+/t$,
$\text{cond}_2(P_k) = \frac{1 - (t_k/u_k)/(t_k + \lambda_1)}{1 - (t_k/u_k)/(t_k + \lambda_n)} < 1/(1 - 1/u_k) = t_k/t_{k+1}$, $k = 0, 1, \ldots, l-1$,
$\text{cond}_2(V_l) = \frac{1 - t_l/(\lambda_1 + t_l)}{1 - t_l/(\lambda_n + t_l)} < 1 + t_l/\lambda_n \leqslant 1 + t_l/\lambda_-$.

Corollary 2. *For a scalar* $u > 1$, *write* $\kappa^+ = \lambda^+/\lambda_-$, $t = \lambda^+ u$, $l = l(u) = 1 + \lceil \log((u-1)u\kappa^+)/\log(1 + 1/(u-1)) \rceil$, $u_k = u$, $k = 0, 1, \ldots, l-1$. *Then* $\text{cond}_2(M_0) \leqslant 1 + 1/u < 2$, $\text{cond}_2(P_k) < 1/(1 - 1/u)$ *for all* k, $\text{cond}_2(V_l) < 1 + t(1 - 1/u)^l/\lambda_-$, *and none of the residual norms in Corollary 1b exceeds* $1/u$.

The above results can be readily applied to bounding the overall number of Newton's or, say, conjugate gradient steps in the homotopic process and to guide us in choosing the parameters $t, l, u_0, \ldots, u_{l-1}$. For an example of other useful applications, Corollary 1b implies that the matrices P_k and V_k for all k are diagonally dominant. Indeed their 2-norm distance from the identity matrix I is less than one, and the same property holds for the matrix M_0/t if $t > \lambda_1$.

To extend our study to the case of indefinite Hermitian matrices M we just need to modify the matrices M_k and P_k by replacing $t_k \leftarrow t_k\sqrt{-1}$ for all k. We refer the reader to [36, Section 7] on some extensions of homotopic techniques to the case of non-Hermitian input matrices.

If the input matrix M has structure of Toeplitz type or has rank structure, then so do the matrices M_k, P_k, and V_k for all k, and we can accelerate the computations respectively. We can extend the structures of other types from the matrix M to the matrices M_k for all k (and consequently also to the matrices P_k and V_k for all k) simply by redefining the matrices: $M_k \leftarrow M + t_k N$ where the matrix N shares the structure with the matrix M. E.g., for a Hankel-like matrix

M, we can choose N being the reflection matrix, which has entries ones on its antidiagonal and has zero entries elsewhere. For matrices M having structure of Vandermonde or Cauchy type, we can choose N being a Vandermonde or Cauchy matrix, respectively, associated with the same operator $\nabla_{A,B}$.

Alternatively, to invert a matrix M having the structures of Vandermonde or Cauchy types we can first compute the matrix $N = VMW$ where each of V and W is an appropriate Vandermonde matrix or the inverse or transpose of such a matrix. This would reduce the original inversion problem to the case of a Toeplitz-like matrix N because $M^{-1} = W^{-1}N^{-1}V^{-1}$. (This technique of *displacement transformation* is due to [38], was extensively used by G. Heinig, and is most widely known because of its effective application to practical solution of Toeplitz and Toepitz-like linear sytems of equations in [39].)

We have the following lower bound on the number l of homotopic steps,

$$l + 1 > \log_\kappa \operatorname{cond}_2(M)$$

for every scalar κ exceeding the condition numbers of the matrices M_0, P_0, ..., P_{l-1}, and V_l. This bound is implied by the inequality

$$\operatorname{cond}_2(M) \leqslant \operatorname{cond}_2(M_0)\operatorname{cond}_2(V_l)\prod_{k=0}^{l-1}\operatorname{cond}_2(P_k).$$

With an appropriate choice of step sizes one only needs $O(\log \operatorname{cond}_2 M + \log_2 \ln h)$ Newton's steps overall to approximate M^{-1} with the residual norm below $1/e^{2^h}$, $e = 2.718291\ldots$ (cf. [36]).

6 Splitting GSVDs

We keep using the definitions in equations (5) and (8) and at first recall the following iteration from [13],

$$Y_k = X_k(2I - MX_k),\ X_{k+1} = Y_kMY_k,\ \ k = 0, 1, \ldots, \tag{16}$$

such that

$$X_{k+1}M = ((2I - X_kM)X_kM)^2,\ \ k = 0, 1, \ldots,$$

and for $X_0 = aM^H$ the singular values $s_i^{(k)}$ of the matrices X_kM satisfy the quartic equations

$$s_i^{(k+1)} = ((2 - s_i^{(k)})s_i^{(k)})^2,\ i = 1, 2, \ldots, n;\ k = 0, 1, \ldots$$

The basic quartic polynomial mapping $s \leftarrow (2-s)^2s^2$ for this iteration has four fixed points $\tilde{s}_0 = 1$, $\tilde{s}_1 = (3-\sqrt{5})/2 = 0.3819\ldots$, $\tilde{s}_2 = 1$, and $\tilde{s}_3 = (3+\sqrt{5})/2 = 2.618\ldots$. The iteration sends the singular values $s_i^{(0)}$ of the matrix X_0M to zero if they lie in the interval $\{s : 0 < s < \tilde{s}_1\}$ and sends them to one if they lie in the interval $\{s : \tilde{s}_1 < s < 2 - \tilde{s}_1 = (1+\sqrt{5})/2 = 1.618\ldots\}$.

If all singular values of the matrix X_0M lie in these two intervals, then under (6) the matrices X_k converge to the generalized inverse $(M_{<s})^-$ of the matrix $M_{<s}$ where $sf(s) = \tilde{s}_1$ under (6). Here and hereafter we write $M_{<s} = U\Sigma_{<s}V^H$ where $\Sigma_{<s} = \mathrm{diag}(\sigma_i^{(<s)})$, $\sigma_i^{(<s)}$ equals σ_i if $\sigma_i \geqslant s$ and equals zero otherwise, so that $M_{<s}$ and $M_{\geqslant s} = M - M_{<s}V^H$ denote the two matrices obtained by setting to zeros all singular values σ_i of the matrix M exceeded by s and greater than s, respectively.

The convergence is locally quadratic but initially is slow for matrices M having singular values σ_i such that the values $s_i^{(0)} = \sigma_i\phi_i^{(0)}$ (equal to $\sigma_i f(\sigma_i)$ under (6)) lie near the points $\tilde{s}_1$ and/or $2 - \tilde{s}_1$.

The iteration can be directed towards the matrix $M_{<s}$ for any fixed smaller positive s (cf. [2, Section 7]). At first one should choose appropriate scalars $c, d, a_0, a_1, \ldots$, define the initial approximation $X_0 = cI + dM^H$, and apply iteration (10). For appropriate scalars a_k and sufficiently large l one yields that $0 \leqslant s_i^{(l)} < \tilde{s}_1$ if $\sigma_i < s$ and $\tilde{s}_1 \leqslant s_i^{(l)} < 2 - \tilde{s}_1$ otherwise. Then one writes $X_0 \leftarrow X_l$ and shifts to iteration (16).

Similar results are obtained in [2] for the iteration

$$X_{k+1} = (3I - 2X_kM)X_kMX_k, \quad k = 0, 1, \ldots, \tag{17}$$

such that $X_{k+1}M = (3I - 2X_kM)(X_kM)^2$, $k = 0, 1, \ldots$. This iteration is associated with the cubic polynomial mapping $s \leftarrow s^2(3 - 2s)$, which has non-negative fixed points 0, 1/2, and 1. The iteration sends the singular values s_i of the matrix X_0M towards zero where $0 \leqslant s_i^{(0)} < 1/2$ and towards one where $1/2 < s_i^{(0)} \leqslant \tilde{s}_4 = (1+\sqrt{3})/2 = 1.366\ldots$. The convergence to zero and one is locally quadratic but initially is slow near the points 1/2 and $\tilde{s}_4$. Then again this can be readily extended to the iteration for which the iterates X_k converge to the matrix $(M_{<s})^-$ for a selected smaller positive s.

Both iterations (16) and (17) are proved to be numerically stable in [2].

Having the matrices M and $(M_{<s})^-$, one can readily compute the matrices $M_{<s} = M(M_{<s})^-M$ and $M_{\geqslant s} = M - M_{<s}$.

The paper [2] also extends iteration (17) to yield convergence to the projection matrices $P_s = M(M_{<s})^- = U\mathrm{diag}(I_{r(s)}, 0)U^H$ and $P^{(s)} = (M_{<s})^-M = V\mathrm{diag}(I_{r(s)}, 0)V^H$ where the integer $r(s)$ is defined by the threshold value s. In this case it is sufficient to choose two scalars a and b satisfying

$$a > 0, \ b \geqslant 0, \ as_1^2 + b = 1/2, \ \ a\sigma_1^2 + b < \tilde{s}_4 = 1.37\ldots,$$

to set $X_0 = aP + bI$, and to apply the iteration

$$X_{k+1} = (I - 2(X_k - I))X_k^2, \ \ k = 0, 1, \ldots.$$

The iteration converges to the matrices P_s and $P^{(s)}$ for $P = MM^H$ and $P = M^HM$, respectively.

7 Divide-and-conquer algorithms and computing singular subspaces

Let us briefly comment on some applications of the splitting techniques from the previous section. Clearly, the SVD and GSVD computation for a matrix M can be reduced to the computation of the SVDs or GSVDs of the pair of matrices $M_{<s}$ and $M_{\geqslant s}$, and this can be recursively extended. Similar divide-and-conquer process can be applied to computing the generalized inverse $M^{-} = (M_{<s})^{-} + (M_{\geqslant s})^{-}$ and the solutions $M^{-}\mathbf{f} = (M_{<s})^{-}\mathbf{f} + (M_{\geqslant s})^{-}\mathbf{f}$ of linear systems $M\mathbf{x} = \mathbf{f}$. This process can rely on computing the pairs of matrices $(M_{<s})^{-}$ and $(M_{\geqslant s})^{-}$ or $M_{<s}$ and $M_{\geqslant s}$. In the latter case the vectors $(M_{<s})^{-}\mathbf{f}$ and $(M_{\geqslant s})^{-}\mathbf{f}$ can be computed as the least-squares solutions of the linear systems $M_{<s}\mathbf{x}_{<s} = \mathbf{f}$ and $M_{\geqslant s}\mathbf{x}_{\geqslant s} = \mathbf{f}$, respectively, that have the minimum Euclidean norms (see [3, Section 5.5] on the respective theory and algorithms). The computations with the matrices $M_{<s}$ and $M_{\geqslant s}$ are simpler than with the matrix M because $\text{cond}M = (\text{cond}M_{<s})\text{cond}M_{\geqslant s}$ and $\text{rank}(M) = \text{rank}(M_{<s}) + \text{rank}(M_{\geqslant s})$. Splitting is more effective where the treshold value s balances it, that is where the ratios $(\text{cond}M_{<s})/\text{cond}M_{\geqslant s}$ and/or $\text{rank}(M_{<s})/\text{rank}(M_{\geqslant s})$ are close to one.

In a sample application of splitting SVD or GSVD to computing the determinant $\det M$, we can compute some unitary matrix bases $U^H_{<s}$, $V_{<s}$, $U^H_{\geqslant s}$, and $V_{\geqslant s}$, respectively, for the left and right null spaces of the matrices $M_{<s}$ and $M_{\geqslant s}$, respectively, such that

$$\begin{pmatrix} U^H_{<s} \\ U^H_{\geqslant s} \end{pmatrix} M(V_{<s}, V_{\geqslant s}) = \text{diag}(\tilde{U}_1, \tilde{U}_2)\Sigma\text{diag}(\tilde{V}_1, \tilde{V}_2) = \text{diag}(\tilde{M}_1, \tilde{M}_2).$$

Here $(\tilde{U}_1, \tilde{U}_2)$ and $(\tilde{V}_1, \tilde{V}_2)$ are unitary matrices, whereas the diagonal blocks $\tilde{U}_i$, $\tilde{V}_i$, and $\tilde{M}_i$ are $n_i \times n_i$ matrices for $i = 1, 2$; $n_1 = r(s)$ and $n_2 = n - n_1$. Therefore

$$\det M = (\det \tilde{M}_1)(\det \tilde{M}_2)/((\det(U_{<s}, U_{\geqslant s})(\det(V_{<s}, V_{\geqslant s})),$$

where the matrices $(U_{<s}, U_{\geqslant s})$ and $(V_{<s}, V_{\geqslant s})$ are unitary and the matrices M_1 and M_2 have both sizes and condition numbers decreased versus the matrix M. Indeed $\text{cond}M = (\text{cond}\tilde{M}_1)(\text{cond}\tilde{M}_2)$.

We can apply the same techniques wherever we can zero some singular values of the input matrix M and preserve the respective singular subspaces of the matrix M. We only need to compute the respective matrix bases for the left and right null spaces of the resulting matrix and for their complements (cf. the next section), which are the respective singular subspaces of the input matrix. The desired suppression of some small singular values can be achieved in smaller numbers of steps (16) or (17), as soon as the respective singular values nearly vanish, whereas the other singular values remain bounded and separated from zero and do not necessarily move close to one.

Clearly, the approach can be applied to computing any polynomial or rational function in a Hermitian matrix M having a GSVD $M = U^H \Sigma U$. Consequently it can be extended to polynomial and rational approximation of irrational functions in Hermitian matrices.

8 Computation of null matrix bases

Computation of null matrix bases can rely on factorizations of input matrices (say, on their QR or PLUP* factorizations) but also on the recent alternatives in [40], [41]. Here is a relevant basic result from [40], [41], which for simplicity we state only for square input matrices M and for the right null spaces. (Recall that the left null space of a matrix M is the right null space of its Hermitian transpose M^H.)

Theorem 8. *Assume an $n \times n$ matrix M of a rank ρ, a pair of two matrices U and V of sizes $n \times r$, and the nonsingular matrix $C = M + UV^H$. Then*

$$r \geqslant \text{rank}(U) \geqslant n - \rho, \tag{18}$$

$$N(M) = \text{range}(C^+ UY) \tag{19}$$

provided Y is a matrix basis for the null space $N(MC^+U)$. Furthermore if

$$r = \text{rank}(U) = n - \rho, \tag{20}$$

then we have

$$N(M) = \text{range}(C^+ U), \tag{21}$$

$$V^H C^+ U = I_r. \tag{22}$$

One can choose the matrices U and V at random based on the following simple results from [42].

Theorem 9. *For a finite set Δ of cardinality $|\Delta|$ in a ring $\mathbb{R}$ and four matrices $M \in \mathbb{R}^{n\times n}$ of a rank ρ, U and V in $\Delta^{r\times n}$, and $C = M + UV^T$, we have*

a) $\text{rank}(C) \leqslant r + \rho$,
b) $\text{rank}(C) = n$ *with a probability of at least* $1 - \frac{2r}{|\Delta|}$ *if* $r + \rho \geqslant n$ *and either the entries of both matrices U and V have been randomly sampled from the set Δ or $U = V$ and the entries of the matrix U have been randomly sampled from this set,*
c) $\text{rank}(C) = n$ *with a probability of at least* $1 - \frac{r}{|\Delta|}$ *if* $r + \rho \geqslant n$, *the matrix U (respectively V) has full rank r, and the entries of the matrix V (respectively U) have been randomly sampled from the set Δ.*

With weakly random generation of the matrices U and V (allowing to endow them with the desired patterns of structure and sparseness) our null space computations are expected to be numerically stable according to the theoretical and experimental study in [42, Sections 4, 6, and 8]. In particular this study shows that under the assumptions of the previous theorem and under weakly random choice of sparse and structured matrices U and V of a rank $r \geqslant n - \rho$, the ratio $(\mathrm{cond}_2 C)/\mathrm{cond}_2 M$ is likely to be neither large nor small provided the matrices M, U, and V are scaled so that the ratio $\|M\|_2/\|UV^H\|_2$ is neither large nor small. We refer the reader to [40]–[44] on such a randomized additive preprocessing $M \to M + UV^H$ and its applications to some fundamental matrix computations (such as eigen-solving and linear system solving).

9 Discussion

1. One can enhance the power of our techniques in various ways by combining them with the available software and hardware. E.g., the iterations of Sections 6 and 7 can be applied concurrently to a number of initial approximations X_0, thus producing matrices $M_{<s(j)}$ for a number of treshold values $s(j)$, $j = 1, 2, \ldots$. The matrix $M_{<s} - M_{<t}$ for $t > s$ is obtained by zeroing all singular values of the matrix M that are less than s or not less than t. Thus we can employ the power of parallel processing to accelerate splitting the given computational problem into subproblems having smaller sizes and condition numbers.

2. Our divide-and-conquer processes employ various algorithms for computing null space bases numerically, that is some bases for the spaces of singular vectors associated with the smallest singular values. It is a challenge to refine these algorithms, particularly where these values form clusters not clearly separated from each other.

3. Another challenge is to define new polynomial mappings that would modify iterations (16) and (17) to accelerate the convergence of some singular values to zero while keeping sufficiently many of them away from zero.

4. Many of the techniques in this paper can be applied to the computation of various other matrix functions besides the inverse, e.g., the square roots, matrix sign function, and the solution of Riccati's equation (cf. [45] and the bibliography therein). Newton's iteration is fundamental for such tasks. Wherever the output matrix is structured, our techniques of recursive compression in Section 4 can support acceleration, at least locally. Our techniques in Section 5 for continuation via recursive factorization can treat the paramount problem of initialization, in both cases of general and structured input matrices. Finally the techniques in Sections 6–8 for splitting GSVDs and computing bases for the respective singular subspaces can be readily extended from the case of inversion to computing the square roots and other functions in Hermitian matrices. A

natural challenge is the transition from GSVDs to eigen-decompositions, which would involve non-Hermitian input matrices.

References

1. T. Søderstrøm, G. W. Stewart, On the Numerical Properties of an Iterative Method for Computing the Moore–Penrose Generalized Inverse, *SIAM Journal on Numerical Analysis*, **11**, 61–74, 1974.
2. V. Y. Pan, R. Schreiber, An Improved Newton Iteration for the Generalized Inverse of a Matrix, with Applications, *SIAM Journal on Scientific and Statistical Computing*, **12**, **5**, 1109–1131, 1991.
3. G. H. Golub, C. F. Van Loan, *Matrix Computations*, 3rd edition, The Johns Hopkins University Press, Baltimore, Maryland, 1996.
4. G. W. Stewart, *Matrix Algorithms, Vol I: Basic Decompositions*, SIAM, Philadelphia, 1998.
5. G. W. Stewart, *Matrix Algorithms, Vol II: Eigensystems*, SIAM, Philadelphia, 1998 (first edition), 2001 (second edition).
6. J. W. Demmel, *Applied Numerical Linear Algebra*, SIAM, Philadelphia, 1997.
7. L. N. Trefethen, D. Bau, III, *Numerical Linear Algebra*, SIAM, Philadelphia, 1997.
8. N. J. Higham, *Accuracy and Stability in Numerical Analysis*, SIAM, Philadelphia, 2002 (second edition).
9. H. Hotelling, Analysis of a Complerx Statistocal Variable into Principal Components, *J. of Educational Psysh.*, **24**, 417–441, 498–520, 1933.
10. G. Schultz, Iterative Berechnung der Reciproken Matrix, *Z. Angew. Meth. Mech.*, **13**, 57–59, 1933.
11. A. Ben-Israel, A Note on Iterative Method for Generalized Inversion of Matrices, *Mathematics of Computation*, **20**, 439–440, 1966.
12. A. Ben-Israel, D. Cohen, On Iterative Computation of Generalized Inverses and Associated Projections, *SIAM Journal on Numerical Analysis*, **3**, 410–419, 1966.
13. R. Schreiber, Computing Generalized Inverses and Eigenvalues of Symmetric Matrices Using Systolic Arrays, in *Computing Methods in Applied Scence and Engeneering* (edited by R. Glowinski and J.-L. Lions), North–Holland, Amsterdam, 1984.
14. V. Y. Pan, J. Reif, Fast and Efficient Parallel Solution of Dense Linear Systems, *Computers and Math. (with Applications)*, **17**, **11**, 1481–1491, 1989.
15. G. Codevico, V. Y. Pan, M. Van Barel, Newton-like Iteration Based on Cubic Polynomials for Structured Matrices, *Numerical Algorithms*, **36**, 365–380, 2004.
16. I. V. Oceledetz, E. E. Tyrtyshnikov, Approximate Inversion of Matrices in the Process of Solving a Hypersingular Integral Equation, *Computational Math. and Math. Physics*, **45**, **2**, 302–313, 2005 (Translated from *JVM i MF*, **45**, **2**, 315–326, 2005).
17. V. Olshevsky, I.V. Oseledets, and E.E. Tyrtyshnikov, Tensor Properties of Multilevel Toeplitz and Related Matrices, *Linear Algebra and Its Applications*, **412**, 1–21, 2006.
18. V. Olshevsky, I.V. Oseledets, and E.E. Tyrtyshnikov, Superfast Inversion of Two-Level Toeplitz Matrices Using Newton Iteration and Tensor-Displacement Structure, *Operator Theory: Advances and Applications*, **179**, 229–240, 2008.

19. T. Kailath, S.-Y. Kung, M. Morf, Displacement Ranks of Matrices and Linear Equations, *J. Math. Anal. Appl.,* **68(2)**, 395–407, 1979.
20. I. Gohberg, V. Olshevsky, Complexity of Multiplication with Vectors for Structured Matrices, *Linear Algebra and Its Applications*, **202**, 163–192, 1994.
21. V. Y. Pan, *Structured Matrices and Polynomials: Unified Superfast Algorithms*, Birkhäuser/Springer, Boston/New York, 2001.
22. V. Y. Pan, X. Wang, Inversion of Displacement Operators, *SIAM Journal on Matrix Analysis and Applications*, **24**, **3**, 660–677, 2003.
23. V. Y. Pan, Parallel Solution of Toeplitz-like Linear Systems, *J. of Complexity*, **8**, 1–21, 1992.
24. V. Y. Pan, Concurrent Iterative Algorithm for Toepliz-like Linear Systems, *IEEE Transactions on Parallel and Distributed Systems*, **4**, **5**, 592–600, 1993.
25. V. Y. Pan, Decreasing the Displacement Rank of a Matrix, *SIAM Journal on Matrix Analysis and Applications*, **14**, **1**, 118–121, 1993.
26. V. Y. Pan, S. Branham, R. Rosholt, A. Zheng, Newton's Iteration for Structured Matrices and Linear Systems of Equations, *SIAM volume on Fast Reliable Algorithms for Matrices with Structure* (edited by T. Kailath and A. H. Sayed), 189–210, SIAM Publications, Philadelphia, 1999.
27. V. Y. Pan, Y. Rami, Newton's Iteration for the Inversion of Structured Matrices, in *Structured Matrices: Recent Developments in Theory and Computation*, (edited by D. Bini, E. Tyrtyshnikov and P. Yalamov), 79–90, Nova Science Publishers, USA, 2001.
28. D. A. Bini, B. Meini, Approximate Displacement Rank and Applications, in *AMS Conference "Structured Matrices in Operator Theory, Control, Signal and Image Processing"*, Boulder, 1999 (edited by V. Olshevsky), *American Math. Society*, 215–232, Providence, RI, 2001.
29. V. Y. Pan, Y. Rami, X. Wang, Structured Matrices and Newton's Iteration: Unified Approach, *Linear Algebra and Its Applications*, **343–344**, 233–265, 2002.
30. V. Y. Pan, M. Van Barel, X. Wang, G. Codevico, Iterative Inversion of Structured Matrices, *Theoretical Computer Science*, **315**, **2–3** (Special Issue on Algebraic and Numerical Computing), 581–592, 2004.
31. R. Vandebril, M. Van Barel, G. Golub, N. Mastronardi, A Bibliography on Semiseparable Matrices, *Calcolo*, **42**, **3–4**, 249–270, 2005.
32. D. Bini, V. Y. Pan, Improved Parallel Computations with Toeplitz-like and Hankel-like Matrices, *Linear Algebra and Its Applications*, **188/189**, 3–29, 1993.
33. D. Bini, V. Y. Pan, *Polynomial and Matrix Computations, Vol.1: Fundamental Algorithms*, Birkhäuser, Boston, 1994.
34. V. Y. Pan, A. Zheng, X. Huang, O. Dias, Newton's Iteration for Inversion of Cauchy-like and Other Structured Matrices, *Journal of Complexity,* **13**, 108–124, 1997.
35. V. Y. Pan, A Homotopic Residual Correction Process, *Proc. of the Second Conference on Numerical Analysis and Applications* (edited by L. Vulkov, J. Wasniewsky and P. Yalamov), *Lecture Notes in Computer Science*, **1988**, 644–649, Springer, Berlin, 2001.
36. V. Y. Pan, M. Kunin, R. Rosholt, H. Kodal, Homotopic Residual Correction Processes, *Math. of Computation*, **75**, 345–368, 2006.
37. V. Y. Pan, New Homotopic/Factorization and Symmetrization Techniques for Newton's and Newton/Structured Iteration, *Computer and Math. with Applications*, **54**, 721–729, 2007.

38. V. Y. Pan, Computations with Dense Structured Matrices, *Math. of Comp.*, **55(191)**, 179–190, 1990. Proc. version in *Proc. Annual ACM-SIGSAM International Symposium on Symbolic and Algebraic Computation (ISSAC '89)*, 34-42, ACM Press, New York, 1989.
39. I. Gohberg, T. Kailath, V. Olshevsky, Fast Gaussian Elimination with Partial Pivoting for Matrices with Displacement Structure, *Math. of Comp.*, **64(212)**, 1557–1576, 1995.
40. V. Y. Pan, Computations in the Null Spaces with Additive Preconditioning, Technical Report TR 2007009, *CUNY Ph.D. Program in Computer Science, Graduate Center, City University of New York*, April 2007. Available at http://www.cs.gc.cuny.edu/tr/techreport.php?id=352
41. V. Y. Pan, G. Qian, Solving Homogeneous Linear Systems with Weakly Random Additive Preprocessing, Technical Report TR 2008009, *CUNY Ph.D. Program in Computer Science, Graduate Center, the City University of New York*, 2008. Available at http://www.cs.gc.cuny.edu/tr/techreport.php?id=352
42. V. Y. Pan, D. Ivolgin, B. Murphy, R. E. Rosholt, Y. Tang, X. Yan, Additive Preconditioning for Matrix Computations, Tech. Report TR 2008004, *Ph.D. Program in Computer Science, Graduate Center, the City University of New York*, 2008. Available at http://www.cs.gc.cuny.edu/tr/techreport.php?id=352
Proceedings version in *Proc. of the Third International Computer Science Symposium in Russia (CSR 2008), Lecture Notes in Computer Science (LNCS)*, **5010**, 372–383, 2008.
43. V. Y. Pan, X. Yan, Additive Preconditioning, Eigenspaces, and the Inverse Iteration, *Linear Algebra and Its Applications*, in press.
44. V. Y. Pan, D. Grady, B. Murphy, G. Qian, R. E. Rosholt, A. Ruslanov, Schur Aggregation for Linear Systems and Determinants, *Theoretical Computer Science, Special Issue on Symbolic–Numerical Algorithms* (D.A. Bini, V.Y. Pan, and J. Verschelde, editors), in press.
45. N. J. Higham, *Functions of Matrices: Theory and Appllications*, SIAM, Philadelphia, 2008.

Truncated Decompositions and Filtering Methods with Reflective/Anti-Reflective Boundary Conditions: A Comparison

C. Tablino Possio*

Dipartimento di Matematica e Applicazioni, Università di Milano Bicocca, via Cozzi 53, 20125 Milano, Italy
cristina.tablinopossio@unimib.it

Abstract. The paper analyzes and compares some spectral filtering methods as truncated singular/eigen-value decompositions and Tikhonov/Reblurring regularizations in the case of the recently proposed Reflective [18] and Anti-Reflective [21] boundary conditions. We give numerical evidence to the fact that spectral decompositions (SDs) provide a good image restoration quality and this is true in particular for the Anti-Reflective SD, despite the loss of orthogonality in the associated transform. The related computational cost is comparable with previously known spectral decompositions, and results substantially lower than the singular value decomposition. The model extension to the cross-channel blurring phenomenon of color images is also considered and the related spectral filtering methods are suitably adapted.

Keywords: filtering methods, spectral decompositions, boundary conditions.

1 Introduction

In this paper we deal with the classical image restoration problem of blurred and noisy images in the case of a space invariant blurring. Under such assumption the image formation process is modelled according to the following integral equation with space invariant kernel

$$g(x) = \int h(x - \tilde{x}) f(\tilde{x}) d\tilde{x} + \eta(x), \;\; x \in \mathbb{R}^2, \tag{1}$$

where f denotes the true physical object to be restored, g is the recorded blurred and noisy image, η takes into account unknown errors in the collected data, e.g. measurement errors and noise.
As customary, we consider the discretization of (1) by means of a standard

* The work of the author was partially supported by MIUR 2006017542.

2D generalization of the rectangle quadrature formula on an equispaced grid, ordered row-wise from the top-left corner to the bottom-right one. Hence, we obtain the relations

$$g_i = \sum_{j \in \mathbb{Z}^2} h_{i-j} f_j + \eta_i, \quad i \in \mathbb{Z}^2, \tag{2}$$

in which an infinite and a shift-invariant matrix $\widetilde{A}_\infty = [h_{i-j}]_{(i,j)=((i_1,i_2),(j_1,j_2))}$, i.e., a two-level Toeplitz matrix, is involved.
In principle, (2) presents an infinite summation since the true image scene does not have a finite boundary. Nevertheless, the data g_i are clearly collected only at a finite number of values, so representing only a finite region of such an infinite scene. In addition, the blurring operator typically shows a finite support, so that it is completely described by a Point Spread Function (PSF) mask such as

$$h_{PSF} = [h_{i_1,i_2}]_{i_1=-q_1,\ldots,q_1,i_2=-q_2,\ldots,q_2} \tag{3}$$

where $h_{i_1,i_2} \geqslant 0$ for any i_1, i_2 and $\sum_{i=-q}^{q} h_i = 1$, $i = (i_1, i_2)$, $q = (q_1, q_2)$ (normalization according to a suitable conservation law).
Therefore, relations (2) imply

$$g_i = \sum_{s=-q}^{q} h_s f_{i-s} + \eta_i, \quad i_1 = 1, \ldots, n_1, i_2 = 1, \ldots, n_2, \tag{4}$$

where the range of collected data defines the so called Field of View (FOV).
Once again, we are assuming that all the involved data in (5), similarly to (2), are reshaped in a row-wise ordering. In such a way we obtain the linear system

$$\widetilde{A}\tilde{f} = g - \eta \tag{5}$$

where $\widetilde{A} \in \mathbb{R}^{N(n)\times N(n+2q)}$ is a finite principal sub-matrix of $\widetilde{A}_\infty$, with main diagonal containing $h_{0,0}$, $\tilde{f} \in \mathbb{R}^{N(n+2q)}$, $g, \eta \in \mathbb{R}^{N(n)}$ and with $N(m) = m_1 m_2$, for any two-index $m = (m_1, m_2)$.
Such a reshape is considered just to perform the theoretical analysis, since all the deblurring/denoising methods are able to deal directly with data in matrix form. For instance, it is evident that the blurring process in (4) consists in a discrete convolution between the PSF mask, after a rotation of $180°$, and the proper true image data in

$$\widetilde{F} = [f_{i_1,i_2}]_{i_1=-q_1+1,\ldots,n_1+q_1,i_2=-q_2+1,\ldots,n_2+q_2}.$$

Hereafter, with a two-index notation, we denote by $F = [f_{i_1,i_2}]_{i_1=1,\ldots,n_1,i_2=1,\ldots,n_2}$ the true image inside the FOV and by $G = [g_{i_1,i_2}]_{i_1=1,\ldots,n_1,i_2=1,\ldots,n_2}$ the recorded image. Thus, assuming the knowledge of PSF mask in (3) and of some statistical properties of η, the deblurring problem is defined as to restore, as best as possible, the true image F on the basis of the recorded image G. As evident from (4), the problem is undetermined since the number of unknowns involved in the convolution exceeds the number of recorded data. Boundary conditions (BCs) are introduced to artificially describe the scene outside the FOV: the values of unknowns outside the FOV are fixed or are defined as linear combinations of the unknowns inside the FOV, the target being to reduce (5) into a square linear system

$$A_n f = g - \eta \tag{6}$$

with $A_n \in \mathbb{R}^{N(n)\times N(n)}$, $n = (n_1, n_2)$, $N(n) = n_1 n_2$ and $f, g, \eta \in \mathbb{R}^{N(n)}$.
The choice of the BCs does not affect the global spectral behavior of the matrix. However, it may have a valuable impact both with respect to the accuracy of the restored image and to the computational costs for recovering f from the blurred datum, with or without noise.
Notice also that, typically, the matrix A is very ill-conditioned and there is a significant intersection between the subspace related to small eigen/singular values and the high frequency subspace. Such a feature requires the use of suitable regularization methods that allow to properly restore the image F with controlled noise levels [12–14, 24], among which we can cite truncated SVD, Tikhonov, and total variation [12, 14, 24].
Hereafter, we focus our attention on special case of PSFs satisfying a strong symmetry property, i.e., such that

$$h_{|i|} = h_i \quad \text{for any } i = -q, \ldots, q. \tag{7}$$

This assumption is fulfilled in the majority of models in real optical applications. For instance, in most 2D astronomical imaging with optical lens [5] the model of the PSF is circularly symmetric, and hence, strongly symmetric; in the multi-image deconvolution of some recent interferometric telescopes, the PSF is strongly symmetric too [6]. Moreover, in real applications when the PSF is obtained by measurements (like a guide star in astronomy), the influence of noise leads to a numerically nonsymmetric PSF, also when the kernel of the PSF is strongly (or centro) symmetric. In such a case, by employing a symmetrized version of the measured PSF, comparable restorations are observed [15, 1].

The paper is organized as follows. In Section 2 we focus on two recently proposed BCs, i.e., the Reflective [18] and Anti-Reflective BCs [21] and their relevant properties. Section 3 summarizes some classical filtering techniques as the truncated singular/eigen-values decomposition and the Tikhonov method. The Re-blurring method [11, 9] is considered in the case of Anti-Reflective BCs and its re-interpretation in the framework of the classical Tikhonov regularization is given. In Section 4 the model is generalized for taking into account the cross-channel blurring phenomenon and the previous filtering methods are suitable adapted. Lastly, Section 5 deals with some computational issues and reports several numerical tests, the aim being to compare the quoted filtering methods and the two type of BCs, both in the case of gray-scale and color images. In Section 6 some conclusions and remarks end the paper.

2 Boundary conditions

In this section we summarize the relevant properties of two recently proposed type of BCs, i.e., the Reflective [18] and Anti-Reflective BCs [21]. Special attention is given to the structural and spectral properties of the arising matrices. In fact, though the choice of the BCs does not affect the global spectral behavior of the matrix A, it can have a valuable impact with respect both to the accuracy of the restoration (especially close to the boundaries where ringing effects can appear), and the computational costs for recovering the image from the blurred one, with or without noise.
Moreover, tanking into account the scale of the problem, the regularization methods analysis can be greatly simplified whenever a spectral (or singular value) decomposition of A is easily available. This means that the target is to obtain the best possible approximation properties, keeping unaltered the fact that the arising matrix shows an exploitable structure. For instance, the use of periodic BCs enforces a circulant structure, so that the spectral decomposition can be computed efficiently with the fast Fourier transform (FFT) [8]. Despite these computational facilities, they give rise to significant ringing effects when a significant discontinuity is introduced into the image.
Hereafter, we focus on two recently proposed boundary conditions, that more carefully describe the scene outside the FOV.
Clearly, several other methods deal with this topic in the image processing literature, e.g. local mean value [22] or extrapolation techniques (see [17] and references therein). Nevertheless, the penalty of their good approximation properties could lie in a linear algebra problem more difficult to cope with.

2.1 Reflective boundary conditions

In [18] Ng *et al.* analyze the use of Reflective BCs, both from model and linear algebra point of view. The improvement with respect to Periodic BCs is due to the preservation of the continuity of the image. In fact, the scene outside the FOV is assumed to be a reflection of the scene inside the FOV. For example, with a boundary at $x_1 = 0$ and $x_2 = 0$ the reflective condition is given by $f(\pm x_1, \pm x_2) = f(x_1, x_2)$.
More precisely, along the borders, the BCs impose

$$f_{i_1,1-i_2} = f_{i_1,i_2}, f_{i_1,n_2+i_2} = f_{i_1,n_2+1-i_2}, \text{ for any } i_1 = 1, \dots, n_1,\ i_2 = 1, \dots, q_2$$
$$f_{1-i_1,i_2} = f_{i_1,i_2}, f_{n_1+i_1,i_2} = f_{n_1+1-i_1,i_2}, \text{ for any } i_1 = 1, \dots, q_1,\ i_2 = 1, \dots, n_2,$$

and, at the corners, the BCs impose for any $i_1 = 1, \dots, q_1$, $i_2 = 1, \dots, q_2$

$$f_{1-i_1,1-i_2} = f_{i_1,i_2}, \qquad f_{n_1+i_1,n_2+i_2} = f_{n_1+1-i_1,n_2+1-i_2},$$
$$f_{1-i_1,n_2+i_2} = f_{i_1,n_2+1-i_2}, \qquad f_{n_1+i_1,1-i_2} = f_{n_1+1-i_1,i_2},$$

i.e., a double reflection, first with respect to one axis and after with respect to the other, no matter about the order.

As a consequence the rectangular matrix $\widetilde{A}$ is reduced to a square Toeplitz-plus-Hankel block matrix with Toeplitz-plus-Hankel blocks, i.e., A_n shows the two-level Toeplitz-plus-Hankel structure. Moreover, if the blurring operator satisfies the strong symmetry condition (7) then the matrix A_n belongs to DCT-III matrix algebra. Therefore, its spectral decomposition can be computed very efficiently using the fast discrete cosine transform (DCT-III) [23].

More in detail, let $\mathcal{C}_n = \{A_n \in \mathbb{R}^{N(n)\times N(n)}, n = (n_1, n_2), N(n) = n_1 n_2 \mid A_n = R_n \Lambda_n R_n^T\}$ be the two-level DCT-III matrix algebra, i.e., the algebra of matrices that are simultaneously diagonalized by the orthogonal transform

$$R_n = R_{n_1} \otimes R_{n_2}, \quad R_m = \left[\sqrt{\frac{2-\delta_{t,1}}{m}} \cos\left\{ \frac{(s-1)(t-1/2)\pi}{m} \right\} \right]_{s,t=1}^{m}, \tag{8}$$

with $\delta_{s,t}$ denoting the Kronecker symbol.

Thus, the explicit structure of the matrix is A_n = Toeplitz(V) + Hankel$(\sigma(V)$, $J\sigma(V))$, with $V = [V_0\ V_1\ \dots\ V_{q_1}\ 0 \dots 0]$ and where each V_{i_1}, $i_1 = 1, \dots, q_1$ is the unilevel DCT-III matrix associated to the i_1^{th} row of the PSF mask, i.e., V_{i_1} = Toeplitz(v_{i_1}) + Hankel$(\sigma(v_{i_1}), J\sigma(v_{i_1}))$, with $v_{i_1} = [h_{i_1,0}, \dots, h_{i_1,q_2}, 0, \dots, 0]$. Here, we denote by σ the shift operator such that $\sigma(v_{i_1}) = [h_{i_1,1}, \dots, h_{i_1,q_2}$, $0, \dots, 0]$ and by J the usual flip matrix; at the block level the same operations are intended in block-wise sense.

Beside this structural characterization, the spectral description is completely known. In fact, let f be the bivariate generating function associated to the PSF mask (3), that is

$$\begin{aligned} f(x_1, x_2) = h_{0,0} &+ 2\sum_{s_1=1}^{q_1} h_{s_1,0} \cos(s_1 x_1) + 2\sum_{s_2=1}^{q_2} h_{0,s_2} \cos(s_2 x_2) \\ &+ 4\sum_{s_1=1}^{q_1}\sum_{s_2=1}^{q_2} h_{s_1,s_2} \cos(s_1 x_1)\cos(s_2 x_2), \end{aligned} \tag{9}$$

then the eigenvalues of the corresponding matrix $A_n \in \mathcal{C}_n$ are given by

$$\lambda_s(A_n) = f\left(x_{s_1}^{[n_1]}, x_{s_2}^{[n_2]}\right), \quad s = (s_1, s_2), \quad x_r^{[m]} = \frac{(r-1)\pi}{m},$$

where $s_1 = 1, \dots, n_1$, $s_2 = 1, \dots, n_2$, and where the two-index notation highlights the tensorial structure of the corresponding eigenvectors.

Lastly, notice that standard operations like matrix-vector products, resolution of linear systems and eigenvalues evaluations can be performed by means of FCT-III [18] within $O(n_1 n_2 \log(n_1 n_2))$ arithmetic operations (ops). For example, by multiplying by $e_1 = [1, 0, \ldots, 0]^T$ both the sides of $R_n^T A_n = \Lambda_n R_n^T$, it holds that

$$[\Lambda_n]_{(i_1,i_2)} = [R_n^T(A_n e_1)]_{(i_1,i_2)}/[R_n^T e_1]_{(i_1,i_2)}, \quad i_1 = 1, \ldots, n_1, i_2 = 1, \ldots, n_2,$$

i.e., it is enough to consider an inverse FCT-III applied to the first column of A_n, with a computational cost of $O(n_1 n_2 \log(n_1 n_2))$ ops.

2.2 Anti-reflective boundary conditions

More recently, Anti-reflective boundary conditions (AR-BCs) have been proposed in [21] and studied [2–4, 9, 10, 19]. The improvement is due to the fact that not only the continuity of the image, but also of the normal derivative, are guaranteed at the boundary. This regularity, which is not shared with Dirichlet or periodic BCs, and only partially shared with reflective BCs, significantly reduces typical ringing artifacts.

The key idea is simply to assume that the scene outside the FOV is the anti-reflection of the scene inside the FOV. For example, with a boundary at $x_1 = 0$ the anti-reflexive condition impose $f(-x_1, x_2) - f(x_1^*, x_2) = -(f(x_1, x_2) - f(x_1^*, x_2))$, for any x_2, where x_1^* is the center of the one-dimensional anti-reflection, i.e.,

$$f(-x_1, x_2) = 2f(x_1^*, x_2) - f(x_1, x_2), \text{for any } x_2.$$

In order to preserve a tensorial structure, at the corners, a double anti-reflection, first with respect to one axis and after with respect to the other, is considered, so that the BCs impose

$$f(-x_1, -x_2) = 4f(x_1^*, x_2^*) - 2f(x_1^*, x_2) - 2f(x_1, x_2^*) + f(x_1, x_2),$$

where (x_1^*, x_2^*) is the center of the two-dimensional anti-reflection.

More precisely, by choosing as center of the anti-reflection the first available data, along the borders, the BCs impose

$$\begin{aligned} f_{1-i_1,i_2} = {} & 2f_{1,i_2} - f_{i_1+1,i_2}, \ f_{n_1+i_1,i_2} = 2f_{n_1,i_2} - f_{n_1-i_1,i_2}, \\ & \qquad i_1 = 1, \ldots, q_1, \quad i_2 = 1, \ldots, n_2, \\ f_{i_1,1-i_2} = {} & 2f_{i_1,1} - f_{i_1,i_2+1}, \ f_{i_1,n_2+i_2} = 2f_{i_1,n_2} - f_{i_1,n_2-i_2}, \\ & \qquad i_1 = 1, \ldots, n_1, \quad i_2 = 1, \ldots, q_2. \end{aligned}$$

At the corners, the BCs impose for any $i_1 = 1, \ldots, q_1$ and $i_2 = 1, \ldots, q_2$,

$$\begin{aligned} f_{1-i_1,1-i_2} &= 4f_{1,1} - 2f_{1,i_2+1} - 2f_{i_1+1,1} + f_{i_1+1,i_2+1}, \\ f_{1-i_1,n_2+i_2} &= 4f_{1,n_2} - 2f_{1,n_2-i_2} - 2f_{i_1+1,n_2} + f_{i_1+1,n_2-i_2}, \\ f_{n_1+i_1,1-i_2} &= 4f_{n_1,1} - 2f_{n_1,i_2+1} - 2f_{n_1-i_1,1} + f_{n_1-i_1,i_2+1}, \\ f_{n_1+i_1,n_2+i_2} &= 4f_{n_1,n_2} - 2f_{n_1,n_1-i_2} - 2f_{n_1-i_1,n_2} + f_{n_1-i_1,n_2-i_2}. \end{aligned}$$

As a consequence the rectangular matrix $\widetilde{A}$ is reduced to a square Toeplitz-plus-Hankel block matrix with Toeplitz-plus-Hankel blocks, plus an additional structured low rank matrix.

Moreover, under the assumption of strong symmetry of the PSF and of a mild finite support condition (more precisely $h_i = 0$ if $|i_j| \geqslant n-2$, for some $j \in \{1,2\}$), the resulting linear system $A_n f = g$ is such that A_n belongs to the $\mathcal{AR}_n^{2D}$ commutative matrix algebra [3]. This new algebra shares some properties with the τ (or DST-I) algebra [7].

Going inside the definition, a matrix $A_n \in \mathcal{AR}_n^{2D}$ has the following block structure

$$A_n = \left[\begin{array}{c|c|c} D_0 + Z^{[1]} & 0^T & 0 \\ \hline D_1 + Z^{[2]} & & 0 \\ \vdots & & \vdots \\ D_{q_1-1} + Z^{[q_1]} & & 0 \\ D_{q_1} & \tau(D_0,\ldots,D_{q_1}) & D_{q_1} \\ 0 & & D_{q_1-1} + Z^{[q_1]} \\ \vdots & & \vdots \\ 0 & & D_1 + Z^{[2]} \\ \hline 0 & 0^T & D_0 + Z^{[1]} \end{array}\right],$$

where $\tau(D_0,\ldots,D_{q_1})$ is a block τ matrix with respect to the $\mathcal{AR}^{1D}$ blocks D_{i_1}, $i_1 = 1,\ldots,q_1$ and $Z^{[k]} = 2\sum_{t=k}^{q_1} D_t$ for $k = 1,\ldots,q_1$. In particular, the $\mathcal{AR}^{1D}$ block D_{i_1} is associated to i_1^{th} row of the PSF, i.e., $h_{i_1}^{[1D]} = [h_{i_1,i_2}]_{i_2=-q_2,\ldots,q_2}$ and it is defined as

$$D_{i_1} = \left[\begin{array}{c|c|c} h_{i_1,0} + z_{i_1}^{[1]} & 0^T & 0 \\ \hline h_{i_1,1} + z_{i_1}^{[2]} & & 0 \\ \vdots & & \vdots \\ h_{i_1,q_2-1} + z_{i_1}^{[q_2]} & & 0 \\ h_{i_1,q_2} & \tau(h_{i_1,0},\ldots,h_{i_1,q_2}) & h_{i_1,q_2} \\ 0 & & h_{i_1,q_2-1} + z_{i_1}^{[q_2]} \\ \vdots & & \vdots \\ 0 & & h_{i_1,1} + z_{i_1}^{[2]} \\ \hline 0 & 0^T & h_{i_1,0} + z_{i_1}^{[1]} \end{array}\right],$$

where $z_{i_1}^{[k]} = 2\sum_{t=k}^{q_2} h_{i_1,t}$ for $k = 1,\ldots,q_2$ and $\tau(h_{i_1,0},\ldots,h_{i_1,q_2})$ is the unilevel τ matrix associated to the one-dimensional PSF $h_{i_1}^{[1D]}$ previously defined.

Notice that the rank-1 correction given by the elements $z_{i_1}^{[k]}$ pertains to the contribution of the anti-reflection centers with respect to the vertical borders, while

the low rank correction given by the matrices $Z_{[k]}$ pertains to the contribution of the anti-reflection centers with respect to the horizontal borders.
It is evident from the above matrix structure that favorable computational properties are guaranteed also by virtue of the τ structure. Therefore, firstly we recall the relevant properties of the two-level τ algebra [7].
Let $\mathcal{T}_n = \{A_n \in \mathbb{R}^{N(n)\times N(n)}, n = (n_1, n_2), N(n) = n_1 n_2 \mid A_n = Q_n \Lambda_n Q_n\}$ be the two-level τ matrix algebra, i.e., the algebra of matrices that are simultaneously diagonalized by the symmetric orthogonal transform

$$Q_n = Q_{n_1} \otimes Q_{n_2}, \quad Q_m = \left[\sqrt{\frac{2}{m+1}} \sin\left\{\frac{st\pi}{m+1}\right\}\right]_{s,t=1}^{m}. \tag{10}$$

With the same notation as the DCT-III algebra case, the explicit structure of the matrix is two level Toeplitz-plus-Hankel. More precisely,

$$A_n = \text{Toeplitz}(V) - \text{Hankel}(\sigma^2(V), J\sigma^2(V))$$

with $V = [V_0\ V_1\ \dots\ V_{q_1}\ 0 \dots 0]$, where each V_{i_1}, $i_1 = 1, \dots, q_1$ is a the unilevel τ matrix associated to the i_1^{th} row of the PSF mask, i.e., $V_{i_1} =$Toeplitz$(v_{i_1}) -$ Hankel$(\sigma^2(v_{i_1}), J\sigma^2(v_{i_1}))$ with $v_{i_1} = [h_{i_1,0}, \dots, h_{i_1,q_2}, 0, \dots, 0]$. Here, we denote by σ^2 the double shift operator such that $\sigma^2(v_{i_1}) = [h_{i_1,2}, \dots, h_{i_1,q_2}, 0, \dots, 0]$; at the block level the same operations are intended in block-wise sense. Once more, the spectral characterization is completely known since for any $A_n \in \mathcal{T}_n$ the related eigenvalues are given by

$$\lambda_s(A_n) = f\left(x_{s_1}^{[n_1]}, x_{s_2}^{[n_2]}\right), s = (s_1, s_2), \quad x_r^{[m]} = \frac{r\pi}{m+1},$$

where $s_1 = 1, \dots, n_1$, $s_2 = 1, \dots, n_2$, and f is the bivariate generating function associated to the PSF defined in (9).
As in the DCT-III case, standard operations like matrix-vector products, resolution of linear systems and eigenvalues evaluations can be performed by means of FST-I within $O(n_1 n_2 \log(n_1 n_2))$ (ops). For instance, it is enough to consider a FST-I applied to the first column of A_n to obtain the eigenvalues

$$[\Lambda_n]_{(i_1,i_2)} = [Q_n(A_n e_1)]_{(i_1,i_2)}/[Q_n e_1]_{(i_1,i_2)}, \quad i_1 = 1, \dots, n_1, i_2 = 1, \dots, n_2.$$

Now, with respect to the $\mathcal{AR}_n^{2D}$ matrix algebra, a complete spectral characterization is given in [3, 4]. A really useful fact is the existence of a transform T_n that simultaneously diagonalizes all the matrices belonging to $\mathcal{AR}_n^{2D}$, although the orthogonality property is partially lost.

Theorem 1. [4] *Any matrix* $A_n \in \mathcal{AR}_n^{2D}$, $n = (n_1, n_2)$, *can be diagonalized by* T_n, *i.e.,*

$$A_n = T_n \Lambda_n \widetilde{T}_n, \quad \widetilde{T}_n = T_n^{-1}$$

where $T_n = T_{n_1} \otimes T_{n_2}$, $\widetilde{T}_n = \widetilde{T}_{n_1} \otimes \widetilde{T}_{n_2}$, *with*

$$T_m = \begin{bmatrix} \alpha_m^{-1} & 0^T & 0 \\ \alpha_m^{-1} p & Q_{m-2} & \alpha_m^{-1} Jp \\ 0 & 0^T & \alpha_m^{-1} \end{bmatrix} \quad \text{and} \quad \widetilde{T}_m = \begin{bmatrix} \alpha_m & 0^T & 0 \\ -Q_{m-2}p & Q_{m-2} & -Q_{m-2}Jp \\ 0 & 0^T & \alpha_m \end{bmatrix}.$$

The entries of the vector $p \in \mathbb{R}^{m-2}$ *are defined as* $p_j = 1 - j/(m-1)$, $j = 1, \ldots, m-2$, $J \in \mathbb{R}^{m-2 \times m-2}$ *is the flip matrix, and* α_m *is a normalizing factor chosen such that the Euclidean norm of the first and last column of* T_m *will be equal to* 1.

Theorem 2. [3] *Let* $A_n \in \mathcal{AR}_n^{2D}$, $n = (n_1, n_2)$, *the matrix related to the PSF* $h_{PSF} = [h_{i_1,i_2}]_{i_1=-q_1,\ldots,q_1, i_2=-q_2,\ldots,q_2}$. *Then, the eigenvalues of* A_n *are given by*

- 1 *with algebraic multiplicity* 4,
- *the* $n_2 - 2$ *eigenvalues of the unilevel* τ *matrix related to the one-dimensional PSF* $h^{\{r\}} = [\sum_{i_1=-q_1}^{q_1} h_{i_1,-q_2}, \ldots, \sum_{i_1=-q_1}^{q_1} h_{i_1,q_2}]$, *each one with algebraic multiplicity* 2,
- *the* $n_1 - 2$ *eigenvalues of the unilevel* τ *matrix related to the one-dimensional PSF* $h^{\{c\}} = [\sum_{i_2=-q_2}^{q_2} h_{-q_1,i_2}, \ldots, \sum_{i_2=-q_2}^{q_2} h_{q_1,i_2}]$, *each one with algebraic multiplicity* 2,
- *the* $(n_1-2)(n_2-2)$ *eigenvalues of the two-level* τ *matrix related to the two-dimensional PSF* h_{PSF}.

Notice that the three sets of multiple eigenvalues are exactly related to the type of low rank correction imposed by the BCs through the centers of the anti-reflections. More in detail, the eigenvalues of $\tau_{n_2-2}(h^{\{r\}})$ and of $\tau_{n_1-2}(h^{\{c\}})$ take into account the condensed PSF information considered along the horizontal and vertical borders respectively, while the eigenvalue equal to 1 takes into account the condensed information of the whole PSF at the four corners.
In addition, it is worth noticing that the spectral characterization can be completely described in terms of the generating function associated to the PSF defined in (9), simply by extending to 0 the standard τ evaluation grid, i.e., it holds

$$\lambda_s(A_n) = f\left(x_{s_1}^{[n_1]}, x_{s_2}^{[n_2]}\right), s = (s_1, s_2), s_j = 0, \ldots, n_j, \quad x_r^{[m]} = \frac{r\pi}{m+1},$$

where the 0–index refers to the first/last columns of the matrix T_m [3].
See [2, 4] for some algorithms related to standard operations like matrix-vector products, resolution of linear systems and eigenvalues evaluations with a computational cost of $O(n_1 n_2 \log(n_1 n_2))$ ops.

It is worthwhile stressing that the computational cost of the inverse transform is comparable with that of the direct transform and, at least at first sight, the very true penalty is the loss of orthogonality due to the first/last column of the matrix T_m.

3 Filtering methods

Owing to the ill-conditioning, the standard solution $f = A_n^{-1}g$ is not physically meaningful since it is completely corrupted by the noise propagation from data to solution, i.e., by the so called inverted noise. For this reason, restoration methods look for an approximate solution with controlled noise levels: widely considered regularization methods are obtained through spectral filtering [14, 16]. Hereafter, we consider the truncated Singular Values Decompositions (SVDs) (or Spectral Decompositions (SDs)) and the Tikhonov (or Re-blurring) regularization method.

3.1 Truncated SVDs and truncated SDs

The Singular Values Decomposition (SVD) highlights a standard perspective for dealing with the inverted noise. More precisely, if

$$A_n = U_n \Sigma_n V_n^T \in \mathbb{R}^{N(n) \times N(n)}$$

is the SVD of A_n, i.e., U_n and V_n are orthogonal matrices and Σ_n is a diagonal matrix with entries $\sigma_1 \geqslant \sigma_2 \geqslant \ldots \sigma_{N(n)} \geqslant 0$, then the solution of the linear system $A_n f = g$ can be written as

$$f = \sum_{k=1}^{N(n)} \left(\frac{u_k^T g}{\sigma_k} \right) v_k,$$

where u_k and v_k denote the kth column of the matrix U_n and V_n, respectively. With regard to the image restoration problem, the idea is to consider a sharp filter, i.e., to take in the summation only the terms corresponding to singular values greater than a certain threshold value δ, so damping the effects caused by division by the small singular values. Therefore, by setting the filter factors as

$$\phi_k = \begin{cases} 1, & \text{if } \sigma_k \geqslant \delta, \\ 0, & \text{otherwise,} \end{cases}$$

the filtered solution is defined as

$$f_{\text{filt}} = \sum_{k=1}^{N(n)} \left(\phi_k \frac{u_k^T g}{\sigma_k} \right) v_k = \sum_{k \in I_\delta} \left(\phi_k \frac{u_k^T g}{\sigma_k} \right) v_k, \quad I_\delta = \{k \mid \sigma_k \geqslant \delta\}.$$

Due to scale of the problem, the SVD of the matrix A_n is in general an expensive computational task (and not negligible also in the case of a separable PSF). Thus, an "a priori" known spectral decomposition, whenever available, can give rise to a valuable simplification. More precisely, let

$$A_n = V_n \Lambda_n \widetilde{V}_n \in \mathbb{R}^{N(n)\times N(n)}, \quad \widetilde{V}_n = V_n^{-1}$$

be a spectral decomposition of A_n, then the filtered solution is defined as

$$f_{\text{filt}} = \sum_{k=1}^{N(n)} \left(\phi_k \frac{\tilde{v}_k g}{\lambda_k} \right) v_k = \sum_{k\in I_\delta} \left(\phi_k \frac{\tilde{v}_k g}{\lambda_k} \right) v_k, \quad I_\delta = \{k \mid |\lambda_k(A)| \geqslant \delta\},$$

where v_k and $\tilde{v}_k$ denote the kth column of V_n and the kth row of $\widetilde{V}_n$, respectively, and where $\phi_k = 1$ if $k \in I_\delta$, 0 otherwise.

3.2 Tikhonov and re-blurring regularizations

In the classical Tikhonov regularization method, the image filtering is obtained by looking for the solution of the following minimization problem

$$\min_f \|A_n f - g\|_2^2 + \mu\|D_n f\|_2^2, \tag{11}$$

where $\mu > 0$ is the regularization parameter and D_n is a carefully chosen matrix (typically $D_n = I_n$ or represents the discretization of a differential operator, properly adapted with respect to the chosen BCs).
The target is to minimize the Euclidean norm of the residual $\|A_n f - g\|_2$ without explosions with respect to the quantity $\|D_n x\|_2$. As well known, (11) is equivalent to the solution to the damped least square problem

$$(A_n^T A_n + \mu D_n^T D_n) f = A_n^T g. \tag{12}$$

In addition, the regularization Tikhonov method can be reinterpreted in the framework of classical spectral filtering method. For instance, in the case of $D_n = I_n$, by making use of the SVD of $A_n = U_n \Sigma_n V_n^T$, the solution of (12) can be rewritten as

$$f_{\text{filt}} = V_n \Phi_n \Sigma_n^{-1} U_n^T g,$$

where $\Phi_n = \text{diag}(\phi_k)$ with $\phi_k = \sigma_k^2/(\sigma_k^2 + \mu)$, $k = 1, \ldots, N(n)$.
A severe drawback in adopting the Tikhonov regularization approach in the case of $A_n \in \mathcal{AR}_n^{2D}$ is due to the fact that $A_n^T \notin \mathcal{AR}_n^{2D}$, so that all the favorable computational properties are substantially spoiled. An alternative approach, named Re-blurring, has been proposed in [11, 9]: the proposal is to replace A_n^T by $A_n^{'}$ in (12), where $A_n^{'}$ is the blurring matrix related to the current BCs with a

PSF rotated by 180°. This approach is completely equivalent to (12) in the case of Dirichlet and Periodic BCs, while the novelty concerns both Reflective BCs and Anti-Reflective BCs, where in general $A_n' \neq A_n^T$. The authors show that the Re-blurring with anti-reflective BCs is computationally convenient and leads to a larger reduction of the ringing effects arising in classical deblurring schemes. From the modelling point of view, the authors motivation relies upon the fact that Re-blurring smoothes the noise in the right hand side of the system, in the same manner as this happens in the case of Dirichlet, Periodic and Reflective BCs.

Hereafter, we consider an explanation of the observed approximation results. As previously claimed, we focus our attention on the case of a strongly symmetric PSF, so that the matrix A_n' equals the matrix A_n. Moreover, also in this case it is evident that the linear system

$$(A_n^2 + \mu D_n^2)f = A_n g \tag{13}$$

is not equivalent to a minimization problem, again because the matrix $A \in \mathcal{AR}_n^{2D}$ is not symmetric. Nevertheless, the symmetrization of (13) can be performed by diagonalization, so obtaining

$$(\Lambda_{A,n}^2 + \mu\Lambda_{D,n}^2)\hat{f} = \Lambda_{A,n}\hat{g}, \tag{14}$$

where $\hat{f} = \widetilde{T}_n f$ and $\hat{g} = \widetilde{T}_n g$. In such a way (14) is again equivalent to the minimization problem

$$\min_f \|\Lambda_{A,n}\widetilde{T}_n f - \widetilde{T}_n g\|_2^2 + \mu\|\Lambda_{D,n}\widetilde{T}_n f\|_2^2, \tag{15}$$

or equivalently, again by making use of the diagonalization result, to

$$\min_f \|\widetilde{T}_n(A_n f - g)\|_2^2 + \mu\|\widetilde{T}_n D_n f\|_2^2. \tag{16}$$

Clearly, the last formulation in (16) is the most natural and it allows to claim that the Re-blurring method can be interpreted as a standard Tikhonov regularization method in the space transformed by means of $\widetilde{T}_n$.

Recalling that $\widetilde{T}_n$ is not an orthogonal transformation, the goal becomes to compare $\|\widetilde{T}_n f\|_2$ and $\|f\|_2$, that is to bound $\|\widetilde{T}_n\|_2 = \|\widetilde{T}_{n_1}\|_2\|\widetilde{T}_{n_1}\|_2$, being $\|\widetilde{T}_n f\|_2 \leqslant \|\widetilde{T}_n\|_2\|f\|_2$.

A quite sharp estimate of such a norm can be found by exploiting the structure of the unilevel matrix $\widetilde{T}_m \in \mathbb{R}^{m\times m}$. Let $\check{f} = [f_2, \ldots, f_{m-1}]$, it holds that

$$\begin{aligned}
\|\widetilde{T}_m f\|_2^2 &= \alpha_m^2 f_1^2 + \|Q_{m-2}(-f_1 p + \check{f} - f_n Jp)\|_2^2 + \alpha_m^2 f_m^2 \\
&= \alpha_m^2(f_1^2 + f_m^2) + \| - f_1 p + \check{f} - f_n Jp\|_2^2 \\
&\leqslant \alpha_m^2(f_1^2 + f_m^2) + (\|\check{f}\|_2 + (|f_1| + |f_n|)\|p\|_2)^2 \\
&\leqslant \alpha_m^2(f_1^2 + f_m^2) + \|\check{f}\|_2^2 + 3\|p\|_2^2\|f\|_2^2 + 4\|p\|_2\|f\|_2^2 \\
&\leqslant (1 + 2\|p\|_2)^2\|f\|_2^2,
\end{aligned}$$

being $\alpha_m^2 = 1 + \|p\|_2^2$. Since, by definition, $\|p\|_2^2 \simeq m$, we have

$$\|\widetilde{T}_m\|_2 \leqslant 1 + 2\|p\|_2 \simeq 2\sqrt{m}. \tag{17}$$

Notice that the bound given in (17) is quite sharp, since for instance $\|\widetilde{T}_m e_1\|_2^2$ equals $1 + 2\|p\|_2^2$.

4 Cross-channel blurring

Hereafter, we extend the analysis of the deblurring problem to the case of color images digitalized, for instance, according to the standard RGB system. Several techniques can be used for recording color images, but the main problem concerns the fact that light from one color channel can end up on a pixel assigned to another color. The consequence of this phenomenon is called cross-channel blurring among the three channels of the image and it sums up to the previously analyzed blurring of each one of the three colors, named within-channel blurring.

By assuming that the cross-channel blurring takes place after the within-channel blurring of the image, that it is spatially invariant and by assuming that the same within-channel blurring occurs in all the three color channels, the problem can be modelled [16] as

$$(A_{\text{color}} \otimes A_n)f = g - \eta \tag{18}$$

with $A_n \in \mathbb{R}^{N(n)\times N(n)}$, $n = (n_1, n_2)$, $N(n) = n_1 n_2$, and

$$A_{\text{color}} = \begin{bmatrix} a_{rr} & a_{rg} & a_{rb} \\ a_{gr} & a_{gg} & a_{gb} \\ a_{br} & a_{bg} & a_{bb} \end{bmatrix}.$$

The row-entries denote the amount of within-channel blurring pertaining to each color channel; a normalized conservation law prescribes that $A_{\text{color}}e = e$, $e = [1\ 1\ 1]^T$. Lastly, the vectors $f, g, \eta \in \mathbb{R}^{3N(n)}$ are assumed to collect the three color channels in the RGB order.

Clearly, if $A_{\text{color}} = I_3$, i.e., the blurring is only of within-channel type, the problem is simply decoupled into three independent gray-scale deblurring problems. In the general case, taking into account the tensorial structure of the whole blurring matrix $A_{\text{color}} \otimes A_n$ is evident that the truncated SVDs and SDs can be formulated as the natural extension of those considered in the within-blurring case. Notice that in the case of SDs, we will consider a SVD for the matrix A_{color}, since it naturally assures an orthogonal decomposition, no matter about the specific matrix, while its computational cost is negligible with respect to the scale of the problem. In addition, we tune the filtering strategy with respect the spectral information given only by the matrix A_n, i.e., for any fixed σ_k (or

λ_k) we simultaneously sum, or discard, the three contribution on f related to the three singular values of A_{color}.
With respect to the Tikhonov regularization method, the approach is a bit more involved. Under the assumption $A_n = A_n^T = V_n \Lambda_n \widetilde{V}_n$, the damped least square problem

$$[(A_{\text{color}} \otimes A_n)^T (A_{\text{color}} \otimes A_n) + \mu I_{3n}] f = (A_{\text{color}} \otimes A_n)^T g$$

can be rewritten as

$$[(A_{\text{color}}^T A_{\text{color}}) \otimes V_n \Lambda_n^2 \widetilde{V}_n + \mu (I_3 \otimes I_n)] f = (A_{\text{color}} \otimes V_n \Lambda_n \widetilde{V}_n)^T g. \tag{19}$$

Thus, by setting $S_{3n} = I_3 \otimes \widetilde{V}_n$, $\hat{f} = S_{3n} f$, $\hat{g} = S_{3n} g$, (19) can be transformed in

$$S_{3n}[(A_{\text{color}}^T A_{\text{color}}) \otimes V_n \Lambda_n^2 \widetilde{V}_n + \mu (I_3 \otimes I_n)] S_{3n}^{-1} \hat{f} = S_{3n} (A_{\text{color}} \otimes V_n \Lambda_n \widetilde{V}_n)^T S_{3n}^{-1} \hat{g},$$

so obtaining the linear system

$$[(A_{\text{color}}^T A_{\text{color}}) \otimes \Lambda_n^2 + \mu (I_3 \otimes I_n)] \hat{f} = (A_{\text{color}}^T \otimes \Lambda_n) \hat{g},$$

that can easily be decoupled into $n_1 n_2$ linear systems of dimension 3.
Clearly, in the case of any matrix $A_n \in \mathcal{C}_n$, all these manipulations can be performed by means of an orthogonal transformation S_{3n}. Notice also that the computational cost is always $O(n_1 n_2 \log n_1 n_2)$ ops.
With respect to $A_n = T_n \Lambda_n \widetilde{T}_n \in \mathcal{AR}_n^{2D}$, we can consider the same strategy by referring to the Re-blurring regularization method. More precisely, the linear system

$$[(A_{\text{color}}^T A_{\text{color}}) \otimes A_n^2 + \mu (I_3 \otimes I_n)] f = (A_{\text{color}}^T \otimes A_n) g$$

can be transformed in

$$[(A_{\text{color}}^T A_{\text{color}}) \otimes \Lambda_n^2 + \mu (I_3 \otimes I_n)] \hat{f} = (A_{\text{color}}^T \otimes \Lambda_n) \hat{g}.$$

Though the transformation $S_{3n} = I_3 \otimes \widetilde{T}_n$ is not orthogonal as in the Reflective case, the obtained restored image are fully comparable with the previous ones and the computational cost is still $O(n_1 n_2 \log n_1 n_2))$ ops.

5 Numerical tests

5.1 Some computational issues

Before analyzing the image restoration results, we discuss how the methods can work without reshaping the involved data. In fact, the tensorial structure of the matrices, obtained by considering Reflective and Anti-Reflective BCs, can be exploited in depth, so that the algorithms can deal directly, and more naturally, with the data collected in matrix form. Hereafter, we consider a two-index notation in the sense of the previously adopted row-wise ordering.

In the SD case considered in Section 3.1, since $\tilde{v}_k = \tilde{v}_{k_1}^{[n_1]} \otimes \tilde{v}_{k_2}^{[n_2]}$ is represented in matrix form as $(\tilde{v}_{k_1}^{[n_1]})^T \tilde{v}_{k_2}^{[n_2]}$, the required scalar product can be computed as

$$\tilde{v}_k g = \left[\left(\tilde{v}_{k_1}^{[n_1]} \right)^T \tilde{v}_{k_2}^{[n_2]} \right] \odot G,$$

where $\odot$ denotes the summation of all the involved terms after a element-wise product. Clearly, $v_k = v_{k_1}^{[n_1]} \otimes v_{k_2}^{[n_2]}$ is represented in matrix form as $v_{k_1}^{[n_1]} (\tilde{v}_{k_2}^{[n_2]})^T$. In a similar manner, in the case of the SVD of A_n with separable PSF $h = h_1 \otimes h_2$, we can represent $v_k = v_{k_1}^{[n_1]} \otimes v_{k_2}^{[n_2]}$ in matrix form as $v_{k_1}^{[n_1]} (v_{k_2}^{[n_2]})^T$ and $u_k^T = (u_{k_1}^{[n_1]} \otimes u_{k_2}^{[n_2]})^T$ as $u_{k_1}^{[n_1]} (u_{k_2}^{[n_2]})^T$.
The eigenvalues required for the SD can be stored into a matrix $\Lambda^* \in \mathbb{R}^{n_1 \times n_2}$. In the case of $A_n \in \mathcal{C}_n$ this matrix can be evaluated as

$$\Lambda^* = \left(\widetilde{V}_{n_2} A^* \widetilde{V}_{n_1}^T \ ./\ \widetilde{V}_{n_2} E_1^* \widetilde{V}_{n_1}^T \right)^T$$

where $A^* \in \mathbb{R}^{n_2 \times n_1}$ denotes the first column of A_n and E_1^* the first canonical basis vector, reshaped as matrices in column-wise order. In addition, the two-level direct and inverse transform $y = V_n x$ and $y = \widetilde{V}_n x$ can be directly evaluated on a matrix data as

$$Y = V_{n_1} X V_{n_2}^T = (V_{n_2} (V_{n_1} X)^T)^T \quad \text{and} \quad Y = \widetilde{V}_{n_1} X \widetilde{V}_{n_2} = (\widetilde{V}_{n_2} (\widetilde{V}_{n_1} X)^T)^T$$

by referring to the corresponding unilevel transforms.
In the same way, the eigenvalues required in the case of $A_n \in \mathcal{AR}_n^{2D}$ can be suitably stored as

$$\Lambda^* = \left[\begin{array}{c|c|c} 1 & \Lambda^*(\tau_{n_2-2}(h^r)) & 1 \\ \hline \Lambda^*(\tau_{n_1-2}(h^c)) & \Lambda^*(\tau_{n-2}(h)) & \Lambda^*(\tau_{n_1-2}(h^c)) \\ \hline 1 & \Lambda^*(\tau_{n_2-2}(h^r)) & 1 \end{array} \right] \in \mathbb{R}^{n_1 \times n_2},$$

with reference to the notations of Theorem 2, where the eigenvalues of the unilevel and two-level τ matrices are evaluated as outlined in Section 2.2.
Lastly, the linear systems obtained, for any fixed μ, in the case of Tikhonov and Re-blurring regularization methods can be solved with reference to the matrix Φ_n of the corresponding filter factors by applying the Reflective and Anti-Reflective transforms with a computational cost $O(n_1 n_2 \log n_1 n_2)$ ops.

5.2 Truncated decompositions

In this section we compare the effectiveness of truncated spectral decompositions (SDs) with respect to the standard truncated SVDs both in the case of

Reflective and Anti-Reflective BCs. Due to scale of the problem, the SVD of the matrix A_n is in general an expensive computational task (and not negligible also in the case of a separable PSF). Thus, a spectral decomposition, whenever available as in these cases, leads to a valuable simplification.

Firstly, we consider the case of the separable PSF caused by atmospheric turbulence

$$h_{i_1,i_2} = \frac{1}{2\pi\sigma_{i_1}\sigma_{i_2}} \exp\left(-\frac{1}{2}\left(\frac{i_1}{\sigma_{i_1}}\right)^2 - \frac{1}{2}\left(\frac{i_2}{\sigma_{i_2}}\right)^2\right),$$

where σ_{i_1} and σ_{i_2} determine the width of the PSF itself. Since the Gaussian function decays exponentially away from its center, it is customary to truncate the values in the PSF mask after an assigned decay $|i_1|, |i_2| \leqslant l$. It is evident from the quoted definition that the Gaussian PSF satisfies the strong symmetry condition (7). Another example of strongly symmetric PSF is given by the PSF representing the out-of-focus blur

$$h_{i_1,i_2} = \begin{cases} \frac{1}{\pi r^2}, & \text{if } i_1^2 + i_2^2 \leqslant r^2, \\ 0, & \text{otherwise,} \end{cases}$$

where r is the radius of the PSF.

In the reported numerical tests, the blurred image g has been perturbed by adding a Gaussian noise contribution $\eta = \eta_n \nu$ with ν fixed noise vector, $\eta_n = \rho\|g\|_2/\|\nu\|_2$, and ρ assigned value. In such a way the Signal Noise Ratio (SNR) [5] is given by

$$\text{SNR} = 20\log_{10}\frac{\|g\|_2}{\|\eta\|_2} = 20\log_{10}\rho^{-1} \text{ (dB)}.$$

5.2.1 Gray-scale images In Figure 1 we report the template true image (the FOV is delimited by a white frame), together with the blurred image with the Gaussian PSF with support 15×15 and $\sigma_{i_1} = \sigma_{i_2} = 2$ and the reference perturbation ν, reshaped in matrix form.

We consider the optimal image restoration with respect to the relative restoration error (RRE), i.e., $\|f_{\text{filt}} - f_{\text{true}}\|_2/\|f_{\text{true}}\|_2$, where f_{filt} is the computed approximation of the true image f_{true} by considering spectral filtering. More in detail, the RRE is analyzed by progressively adding a new basis element at a time, according to the non-decreasing order of the singular/eigen-values (the eigenvalues are ordered with respect to their absolute value).

In the case of SDs (or SVDs related to a separable PSF) this can be done as described in Section 5.1 and, beside the preliminary cost related to the computation of the decomposition, the addition of a new term has a computational cost equal to $4n_1n_2$ ops. The algorithm proposed in [4], that makes use of the Anti-Reflective direct and inverse transforms, is less expensive in the case of tests with few threshold values.

Hereafter, the aim is to compare the truncated SVD with the truncated SD

Fig. 1. True image (FOV is delimited by a white frame), reference noise perturbation, blurred image with the Gaussian PSF with support 15×15 and $\sigma_{i_1} = \sigma_{i_2} = 2$, and blurred image with the Out-of-Focus PSF with support 15×15.

Table 1. Optimal RREs of truncated SVD and SD with reference to the true image in Figure 1 (Gaussian blur $\sigma_{i_1} = \sigma_{i_2} = 2$).

Reflective BCs				
PSF	5x5	11x11	15x15	21x21
$\rho = 0$				
SVD	0.059164	0.087402	0.090742	0.093856
SD	0.043754	0.087400	0.090746	0.093867
$\rho = 0.001$				
SVD	0.060278	0.091964	0.094468	0.097034
SD	0.060278	0.091964	0.094476	0.097034
$\rho = 0.01$				
SVD	0.091151	0.11214	0.11307	0.11495
SD	0.091152	0.11214	0.11307	0.11495
$\rho = 0.05$				
SVD	0.11635	0.13356	0.13508	0.13739
SD	0.11635	0.13356	0.13510	0.13739
$\rho = 0.1$				
SVD	0.13024	0.14607	0.14746	0.15047
SD	0.13024	0.14607	0.14746	0.15047

Anti-Reflective BCs				
PSF	5x5	11x11	15x15	21x21
$\rho = 0$				
SVD	0.039165	0.064081	0.086621	0.087237
SD	0.038316	0.063114	0.083043	0.083521
$\rho = 0.001$				
SVD	0.062182	0.094237	0.098897	0.10042
SD	0.059617	0.089105	0.092814	0.094343
$\rho = 0.01$				
SVD	0.096049	0.12231	0.12403	0.12536
SD	0.091383	0.11230	0.11343	0.11495
$\rho = 0.05$				
SVD	0.12791	0.15070	0.15188	0.15492
SD	0.11666	0.13414	0.13570	0.13816
$\rho = 0.1$				
SVD	0.14399	0.16756	0.16964	0.17225
SD	0.13083	0.14709	0.14852	0.15162

restorations both in the case of Reflective and Anti-Reflective BCs. Periodic BCs are not analyzed here, since Reflective and Anti-Reflective BCs give better performances with respect to the approximation of the image at the boundary. In Table 1 and 2 we report the results obtained by varying the dimension of the PSF support, the parameter ρ related to the amount of the noise perturbation and the variance of the considered Gaussian blur. As expected the optimal RRE worses as the parameter ρ increases and the Anti-Reflective BCs show better performances in the case of low noise levels. In fact, for low ρ values, the reduction of ringing artifacts is significant, while the quality of the restoration for higher ρ values is essentially driven by the goal of noise filtering. Therefore, in such a case, the choice of the BCs becomes more an more meaningless since it is not able to influence the image restoration quality. Some examples of restored images are reported in Figure 2.

Table 2. Optimal RREs of truncated SVD and SD with reference to the true image in Figure 1 (Gaussian blur $\sigma_{i_1} = \sigma_{i_2} = 5$).

Reflective BCs				
PSF	5x5	11x11	15x15	21x21
$\rho = 0$				
SVD	0.063387	0.081274	0.097351	0.14634
SD	0.045365	0.081274	0.096387	0.14634
$\rho = 0.001$				
SVD	0.063915	0.096243	0.11449	0.15217
SD	0.063915	0.096274	0.11449	0.15217
$\rho = 0.01$				
SVD	0.089032	0.13343	0.14947	0.17397
SD	0.089032	0.13343	0.14946	0.17397
$\rho = 0.05$				
SVD	0.12203	0.16002	0.17339	0.18335
SD	0.12203	0.16002	0.17339	0.18335
$\rho = 0.1$				
SVD	0.13412	0.16793	0.17963	0.19057
SD	0.13412	0.16793	0.17963	0.19057

Anti-Reflective BCs				
PSF	5x5	11x11	15x15	21x21
$\rho = 0$				
SVD	0.040214	0.079543	0.088224	0.13686
SD	0.039437	0.078970	0.088832	0.13129
$\rho = 0.001$				
SVD	0.068197	0.095808	0.11522	0.15767
SD	0.063575	0.093247	0.1127	0.14893
$\rho = 0.01$				
SVD	0.09412	0.14482	0.16825	0.21148
SD	0.089038	0.13611	0.15270	0.17446
$\rho = 0.05$				
SVD	0.13553	0.18563	0.21006	0.22962
SD	0.12253	0.16269	0.17439	0.18414
$\rho = 0.1$				
SVD	0.15010	0.20164	0.22256	0.23960
SD	0.13487	0.16916	0.18088	0.19218

More impressive is the fact that SDs give better, or equal, results with respect to those obtained by considering SVDs. This numerical evidence is really interesting in the case of Anti-Reflective BCs: despite the loss of the orthogonality property in the spectral decomposition, the restoration results are better than those obtained by considering SVD. Moreover, the observed trend with respect to the Reflective BCs is also conserved.

A further analysis refers to the so-called Picard plots (see Figure 3), where the coefficients $|u_k^T g|$, or $|\tilde{v}_k g|$, (black dots) are compared with the singular values σ_k, or the absolute values of the eigenvalues $|\lambda_k|$, (red line). As expected, initially these coefficients decrease faster than σ_k, or $|\lambda_k|$, while afterwards they level off at a plateau determined by the level of the noise in the image.

The threshold of this change of behavior is in good agreement with the optimal k value obtained in the numerical test by monitoring the RRE.

Moreover, notice that the Picard plots related to the SDs are quite in agreement with those corresponding to SVDs. In the case of the Anti-Reflective SD we observe an increasing data dispersion with respect to the plateau, but the correspondence between the threshold and the chosen optimal k is still preserved.

The computational relevance of this result is due to the significant lower computational cost required by the Anti-Reflective SDs with respect to the corresponding SVDs.

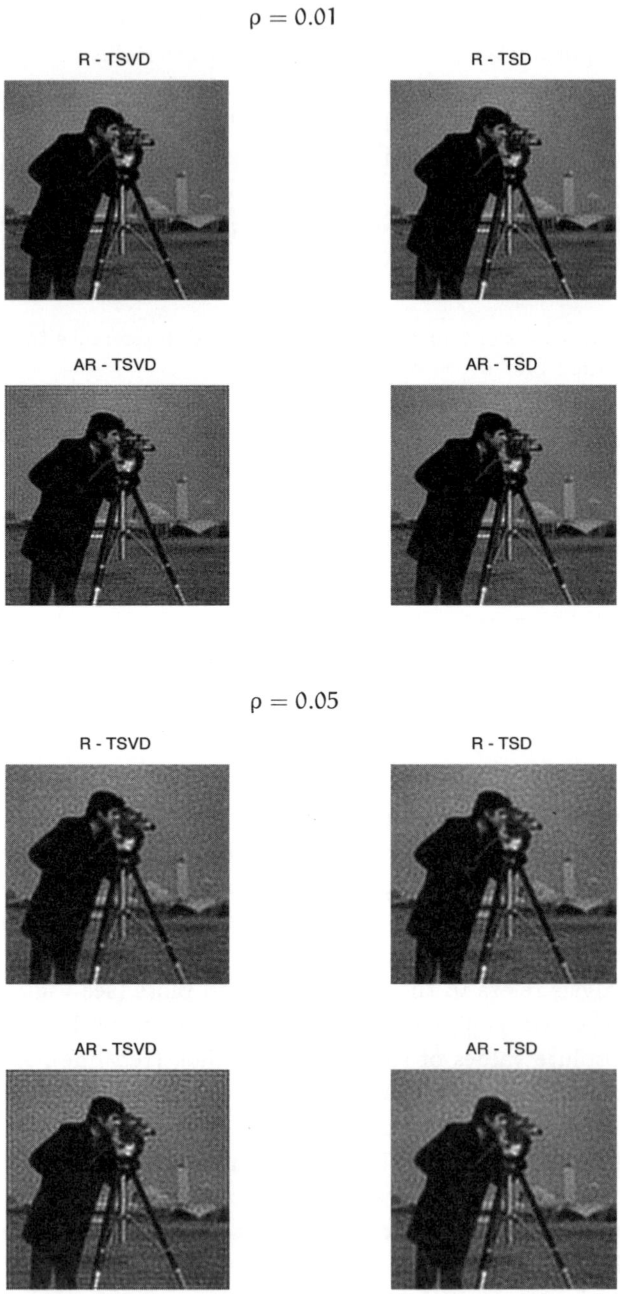

Fig. 2. Optimal restorations of truncated SVD and SD in the case of Reflective and Anti-Reflective BCs with reference to Figure 1 (Gaussian blur $\sigma_{i_1} = \sigma_{i_2} = 2$).

Lastly, Table 3 reports the spectral filtering results obtained in the case of Out-

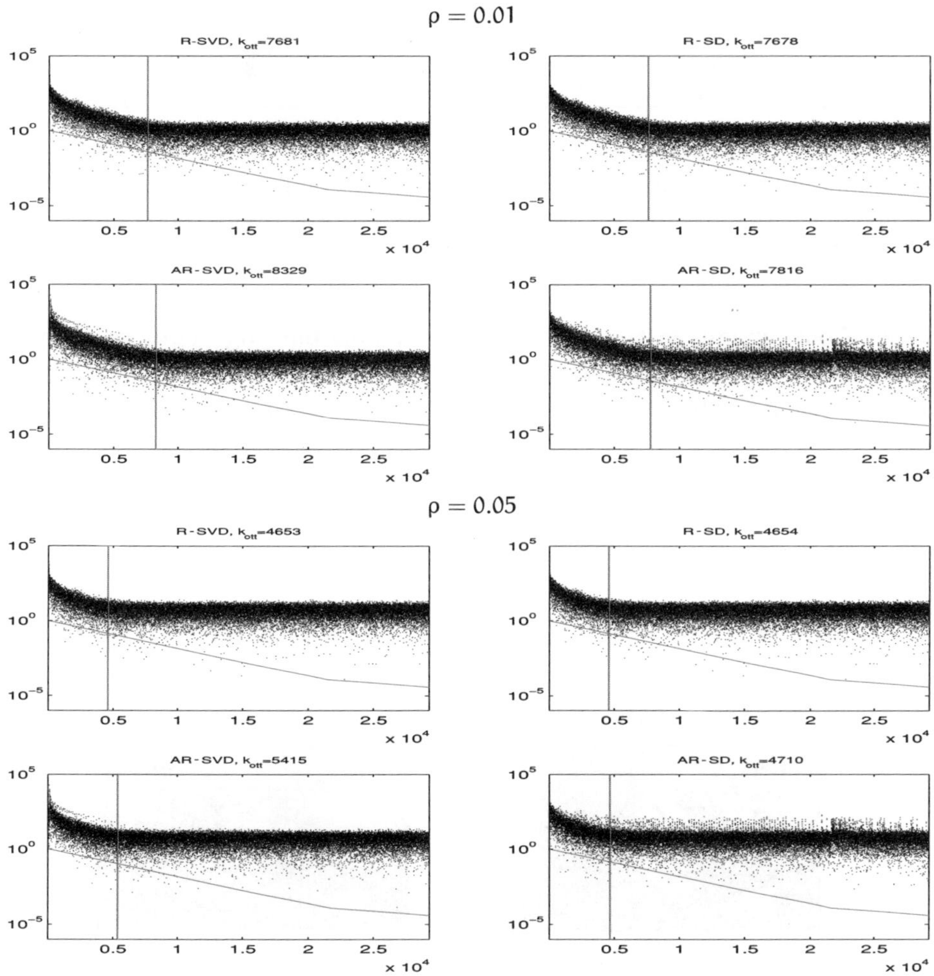

Fig. 3. Picard plot of truncated SVD and SD in the case of Reflective and Anti-Reflective BCs with reference to Figure 1 (Gaussian blur $\sigma_{i_1} = \sigma_{i_2} = 2$).

of-Focus blur by varying the dimension of the PSF support and the parameter ρ related to the noise perturbation. The RRE follows the same trend observed in the case of Gaussian blur. Other image restoration tests with different gray-scale images have been considered in [20].

A more interesting remark again pertains the computational cost. Since the Out-of-Focus PSF is not separable, but the transforms are, the use of SDs related to Reflective or Anti-Reflective BCs allows to exploit the tensorial nature of the

Table 3. Optimal RREs of truncated SDs with reference to the true image in Figure 1 (Out-of-Focus blur).

	Reflective BCs				Anti-Reflective BCs			
PSF	5x5	11x11	15x15	21x21	5x5	11x11	15x15	21x21
ρ =0	0.072593	0.084604	0.088323	0.096479	0.072821	0.085366	0.091252	0.099293
ρ =0.001	0.072671	0.085809	0.091035	0.10436	0.072904	0.086643	0.093929	0.10752
ρ =0.01	0.080016	0.12255	0.13569	0.15276	0.080427	0.12316	0.13803	0.15683
ρ =0.05	0.10645	0.15365	0.16810	0.18777	0.10685	0.15571	0.17147	0.19172
ρ =0.1	0.12089	0.16314	0.17836	0.20471	0.12147	0.16482	0.17987	0.20829

corresponding transforms, both with respect to the computation of the eigenvalues and of the eigenvectors (or of the Reflective and Anti-Reflective transforms).

5.2.2 Color images in the case of cross-channel blurring Here, we analyze some restoration tests in the case of the template color image reported in Figure 4, by assuming the presence of a cross-channel blurring phenomenon modelled as in (18). The entity of this mixing effect is chosen according to the matrix

$$A_{\text{color}} = \begin{bmatrix} 0.7 & 0.2 & 0.1 \\ 0.25 & 0.5 & 0.25 \\ 0.15 & 0.1 & 0.75 \end{bmatrix}. \tag{20}$$

In Figure 4 the cross-channel blurred image with Gaussian PSF with support

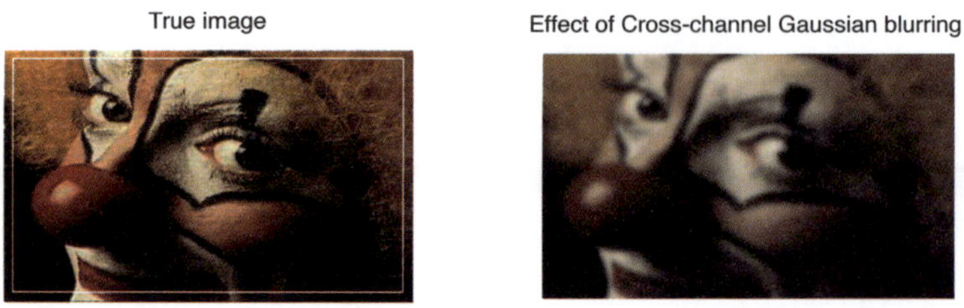

Fig. 4. True image (FOV is delimited by a white frame) and cross-channel blurred image with the Gaussian PSF with support 15×15 and $\sigma_{i_1} = \sigma_{i_2} = 2$ and matrix A_{color} in (20).

15×15 and $\sigma_{i_1} = \sigma_{i_2} = 2$ is also reported. Notice that the entity of the cross-channel blurring is not negligible, since the whole image results to be darkened and the color intensities of the additive RGB system are substantially altered. Table 4 reports the optimal RREs of truncated SVDs and SDs obtained by varying the dimension of the Gaussian PSF support and the parameter ρ related to the amount of the noise perturbation. It is worth stressing that we tune the filtering strategy with respect the spectral information given just by the matrix A_n, i.e., for any fixed σ_k (or λ_k) we simultaneously sum, or discard, the three contribution on f related to the three singular values of A_{color}. In fact, the magnitude of singular values of the considered matrix A_{color} does not differ enough to dramatically change the filtering information given just by A_n. Nevertheless, also the comparison with the restoration results obtained by considering a global ordering justifies this approach.

The color case behaves as the gray-scale one: as expected the optimal RRE becomes worse as the parameter ρ increases and the Anti-Reflective SD shows better performances in the case of low noise levels.

Table 4. Optimal RREs of truncated SVD and SD with reference to the true image in Figure 4 (Cross-channel and Gaussian Blur $\sigma_{i_1} = \sigma_{i_2} = 2$).

Reflective BCs				
PSF	5x5	11x11	15x15	21x21
ρ =0				
SVD	0.078276	0.12114	0.11654	0.1178
SD	0.078276	0.12114	0.11654	0.1178
ρ =0.001				
SVD	0.078992	0.1212	0.11663	0.11792
SD	0.078992	0.12119	0.11663	0.11792
ρ =0.01				
SVD	0.10152	0.12396	0.12088	0.12198
SD	0.10152	0.12396	0.12088	0.12198
ρ =0.05				
SVD	0.12102	0.13853	0.13743	0.13844
SD	0.12102	0.13853	0.13743	0.13844
ρ =0.1				
SVD	0.13437	0.14898	0.14854	0.14947
SD	0.13437	0.14898	0.14854	0.14947

Anti-Reflective BCs				
PSF	5x5	11x11	15x15	21x21
ρ =0				
SVD	0.076646	0.1006	0.1098	0.10646
SD	0.074953	0.098474	0.10508	0.10216
ρ =0.001				
SVD	0.077394	0.10639	0.1111	0.11002
SD	0.075727	0.10233	0.10612	0.10443
ρ =0.01				
SVD	0.10431	0.12695	0.12624	0.12779
SD	0.10087	0.11737	0.11805	0.118
ρ =0.05				
SVD	0.13017	0.15075	0.15063	0.15166
SD	0.12127	0.13699	0.13756	0.13795
ρ =0.1				
SVD	0.1456	0.16516	0.16626	0.16647
SD	0.13507	0.14796	0.14955	0.15018

In addition, by referring to Figure 5, we note that the truncated SVD in the case of Anti-Reflective BCs shows a little more 'freckles' than the corresponding truncated SVD in the case of Reflective BCs. Nevertheless, for low noise levels, is just the Anti-Reflective SD that exhibits less 'freckles' than the Reflective SD.

5.3 Tikhonov and re-blurring regularizations

By considering a Gaussian blurring of the true image reported in Figure 1, Table 5 compares the optimal RRE obtained in the case of the Tikhonov method for Reflective BCs and of the Re-blurring method for Anti-Reflective BCs. In addition, in Table 6, the same comparison refers to the case of the Out-of-Focus PSF.

As expected, the RRE deteriorates as the dimension of the noise level or the dimension of the PSF support increases. Notice also that the gap between the Reflective and Anti-Reflective BCs is reduced also for low noise levels. Further numerical tests can be found in [9, 2].

Lastly, we focus our attention on the case of the color image in Figure 4. The

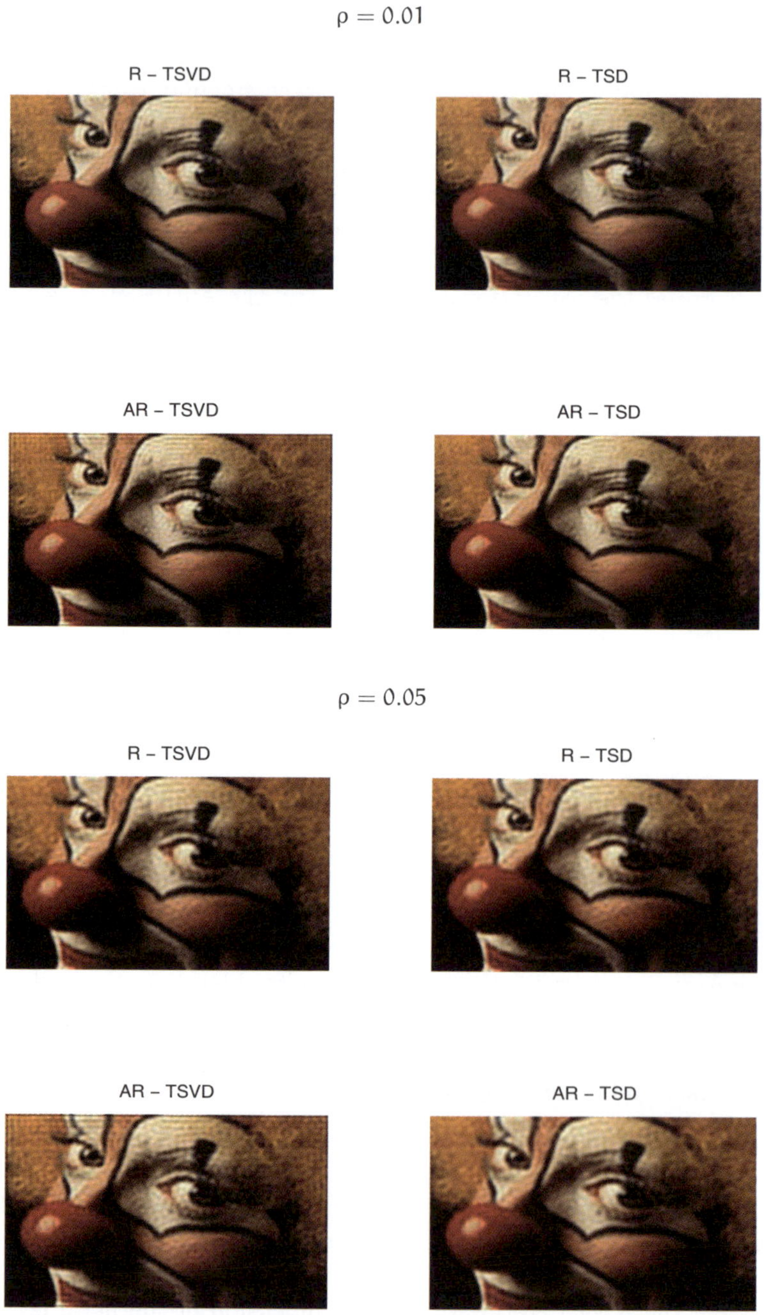

Fig. 5. Optimal restorations of truncated SVD and SD in the case of Reflective and Anti-Reflective BCs with reference to Figure 4 (Cross-channel and Gaussian Blur $\sigma_{i_1} = \sigma_{i_2} = 2$).

Table 5. Optimal RREs of Tikhonov and Re-blurring methods and corresponding μ_{ott} with reference to the true image in Figure 1 (Gaussian Blur $\sigma_{i_1} = \sigma_{i_2} = 2$).

PSF	5x5		11x11		15x15		21x21	
$\rho = 0$								
R	0.041015	4.1e-005	0.079044	9e-006	0.086386	1.1e-005	0.089556	1.6e-005
AR	0.034237	1.1e-005	0.059465	1e-006	0.078963	1e-006	0.079805	1e-006
$\rho = 0.001$								
R	0.050155	0.000188	0.087482	5.7e-005	0.090825	4.3e-005	0.093071	4.9e-005
AR	0.048556	0.000163	0.085279	4.6e-005	0.089388	3.3e-005	0.090821	3.3e-005
$\rho = 0.01$								
R	0.083456	0.005555	0.10748	0.001786	0.10863	0.001678	0.11023	0.001573
AR	0.083436	0.005536	0.10744	0.001792	0.10868	0.001691	0.11019	0.001575
$\rho = 0.05$								
R	0.12024	0.038152	0.12982	0.01929	0.13071	0.018417	0.13307	0.017892
AR	0.12049	0.038379	0.13006	0.01957	0.13096	0.018669	0.1333	0.018105
$\rho = 0.1$								
R	0.14767	0.06587	0.14721	0.039231	0.14822	0.038181	0.15097	0.037893
AR	0.14813	0.066251	0.14766	0.039707	0.14866	0.038644	0.15144	0.038296

Table 6. Optimal RREs of Tikhonov and Re-blurring methods and corresponding μ_{ott} with reference to the true image in Figure 1 (Out-of-Focus blur).

PSF	5x5		11x11		15x15		21x21	
$\rho = 0$								
R	0.031422	0.000172	0.05346	6.9e-005	0.060954	3.5e-005	0.074785	2.7e-005
AR	0.036213	0.000302	0.051236	6.8e-005	0.06683	5.7e-005	0.084482	5.8e-005
$\rho = 0.001$								
R	0.034441	0.000271	0.061465	0.000145	0.073751	0.000101	0.09074	7.9e-005
AR	0.038313	0.000402	0.059957	0.000138	0.076695	0.000126	0.095274	0.000106
$\rho = 0.01$								
R	0.069647	0.008493	0.11361	0.004117	0.12881	0.003037	0.14914	0.001873
AR	0.070384	0.008923	0.11404	0.00422	0.12982	0.003139	0.15061	0.001969
$\rho = 0.05$								
R	0.12204	0.053687	0.1532	0.030719	0.16614	0.022121	0.18769	0.01346
AR	0.12256	0.05423	0.15402	0.031574	0.16739	0.023213	0.18933	0.014472
$\rho = 0.1$								
R	0.16366	0.092379	0.17357	0.055919	0.1829	0.042944	0.20323	0.028803
AR	0.16433	0.093069	0.17485	0.057326	0.18457	0.044901	0.20511	0.031011

Table 7. Optimal RREs of Tikhonov and Re-blurring methods and corresponding μ_{ott} with reference to the true image in Figure 4 (Cross-channel and Gaussian Blur $\sigma_{i_1} = \sigma_{i_2} = 2$).

PSF	5x5		11x11		15x15		21x21	
$\rho = 0$								
R	0.069148	0.000203	0.11508	0.001204	0.1123	0.000717	0.11335	0.000726
AR	0.062854	0.000102	0.091232	7e-006	0.1014	4.4e-005	0.098266	1.5e-005
$\rho = 0.001$								
R	0.071259	0.000312	0.11515	0.001228	0.11239	0.000744	0.11347	0.000755
AR	0.066734	0.000209	0.098658	5.8e-005	0.10276	7.7e-005	0.10111	4.5e-005
$\rho = 0.01$								
R	0.094871	0.004975	0.11896	0.002919	0.11712	0.002421	0.1182	0.002459
AR	0.094458	0.004841	0.1144	0.001884	0.11481	0.00184	0.11507	0.001755
$\rho = 0.05$								
R	0.13209	0.029798	0.13662	0.015305	0.13599	0.014896	0.13669	0.014824
AR	0.13239	0.029944	0.13561	0.014992	0.13593	0.014772	0.13611	0.014595
$\rho = 0.1$								
R	0.16281	0.051315	0.15543	0.029068	0.15547	0.028822	0.15586	0.02868
AR	0.16341	0.051659	0.15526	0.029213	0.15602	0.029029	0.15588	0.02872

image restorations have been obtained by considering the transformation procedure outlined at the end of Section 4. Despite the RREs in Table 7 are bigger than in the gray-scale case, the perception of the image restoration quality is very satisfying and a little less 'freckles' than in the corresponding SDs and SVDs are observed (see Figure 6). Notice, also that the lack of orthogonality in the S_{3n} transform related to the Anti-reflective BCs does not deteriorate the performances of the restoration.

6 Conclusions

In this paper we have analyzed and compared SD and SVD filtering methods in the case both of Reflective and Anti-Reflective BCs. Numerical evidence is given of the good performances achievable through SDs and with a substantially lower computational cost with respect to SVDs. In addition, the tensorial structure of the Reflective and Anti-Reflective SDs can be exploited in depth also in the case of not separable PSFs.
A special mention has to be done to the fact that the loss of orthogonality of the Anti-Reflective transform does not seems to have any consequence on the trend of the image restoration results. The analysis in the case of cross-channel blurring in color images allows to confirm the quoted considerations. Finally, the Re-blurring regularizing method has been re-interpreted as a standard Tikhonov regularization method in the space transformed by means of $\widetilde{T}_n$. Some numerical tests highlight the image restoration performances, also in the case of cross-channel blurring.
Future works will concern the analysis of effective strategies allowing to properly choose the optimal regularizing parameters in the Anti-Reflective BCs case.

References

1. B. Anconelli, M. Bertero, P. Boccacci, M. Carbillet, and H. Lanteri, *Reduction of boundary effects in multiple image deconvolution with an application to LBT LINC-NIRVANA*, Astron. Astrophys., 448 (2006), pp. 1217–1224.
2. A. Aricò, M. Donatelli, and S. Serra-Capizzano, *The Antireflective Algebra: Structural and Computational Analyses with Application to Image Deblurring and Denoising*, Calcolo, 45–3 (2008), pp. 149–175.
3. A. Aricò, M. Donatelli, and S. Serra Capizzano, *Spectral analysis of the anti-reflective algebra*, Linear Algebra Appl., 428 (2008), pp. 657–675.
4. A. Aricò, M. Donatelli, J. Nagy, and S. Serra-Capizzano, *The anti-reflective transform and regularization by filtering*, Numerical Linear Algebra in Signals, Systems, and Control., in Lecture Notes in Electrical Engineering edited by S. Bhattacharyya, R. Chan, V. Olshevsky, A. Routray, and P. Van Dooren, Springer Verlag, in press.

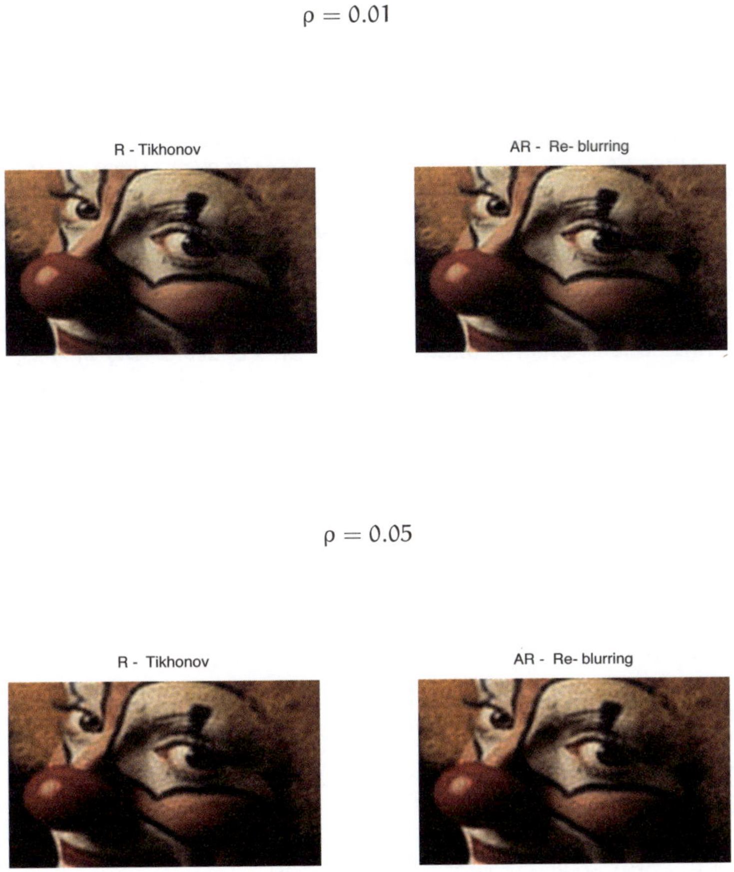

Fig. 6. Optimal RREs of Tikhonov and Re-blurring methods with reference to the true image in Figure 4 (Cross-channel and Gaussian blur $\sigma_{i_1} = \sigma_{i_2} = 2$ - $\rho = 0.05$).

5. M. Bertero and P. Boccacci, *Introduction to inverse problems in imaging*, Inst. of Physics Publ. London, UK, 1998.
6. M. Bertero and P. Boccacci, *Image restoration for Large Binocular Telescope (LBT)*, Astron. Astrophys. Suppl. Ser., 147 (2000), pp. 323–332.
7. D. Bini and M. Capovani, *Spectral and computational properties of band symmetric Toeplitz matrices*, Linear Algebra Appl., 52/53 (1983), pp. 99–125.
8. P. J. Davis, *Circulant Matrices*, Wiley, New York, 1979.
9. M. Donatelli, C. Estatico, A. Martinelli, and S. Serra Capizzano, *Improved image deblurring with anti-reflective boundary conditions and reblurring*, Inverse Problems, 22 (2006), pp. 2035–2053.
10. M. Donatelli, C. Estatico, J. Nagy, L. Perrone, and S. Serra Capizzano, *Anti-reflective boundary conditions and fast 2D deblurring models*, Proceeding to SPIE's 48th Annual Meeting, San Diego, CA USA, F. Luk Ed, 5205 (2003), pp. 380–389.
11. M. Donatelli and S. Serra Capizzano, *Anti-reflective boundary conditions and re-blurring*, Inverse Problems, 21 (2005), pp. 169–182.
12. H. Engl, M. Hanke, and A. Neubauer, *Regularization of Inverse Problems*, Kluwer Academic Publishers, Dordrecht, The Netherlands, 2000.
13. C. W. Groetsch, *The Theory of Tikhonov Regularization for Fredholm Integral Equations of the First Kind*, Pitman, Boston, 1984.
14. P. C. Hansen, *Rank-deficient and discrete ill-posed problems*, SIAM, Philadelphia, PA, 1997.
15. M. Hanke and J. Nagy, *Restoration of atmospherically blurred images by symmetric indefinite conjugate gradient technique*, Inverse Problems, 12 (1996), pp. 157–173.
16. P. C. Hansen, J. Nagy, and D. P. O'Leary, *Deblurring Images Matrices, Spectra and Filtering*, SIAM Publications, Philadelphia, 2006.
17. R. L. Lagendijk and J. Biemond, *Iterative Identification and Restoration of Images* , Springer-Verlag New York, Inc., 1991.
18. M. K. Ng, R. H. Chan, and W. C. Tang, *A fast algorithm for deblurring models with Neumann boundary conditions*, SIAM J. Sci. Comput., 21 (1999), no. 3, pp. 851–866.
19. L. Perrone, *Kronecker Product Approximations for Image Restoration with Anti-Reflective Boundary Conditions*, Numer. Linear Algebra Appl., 13–1 (2006), pp. 1–22.
20. F. Rossi, *Tecniche di filtraggio nella ricostruzione di immagini con condizioni al contorno antiriflettenti*, (in Italian), Basic Degree Thesis, University of Milano-Bicocca, Milano, 2006.
21. S. Serra Capizzano, *A note on anti-reflective boundary conditions and fast deblurring models*, SIAM J. Sci. Comput., 25–3 (2003), pp. 1307–1325.
22. Y. Shi and Q. Chang, *Acceleration methods for image restoration problem with different boundary conditions*, Appl. Numer. Math., 58–5 (2008), pp. 602–614.
23. G. Strang, *The Discrete Cosine Transform*, SIAM Review, 41–1 (1999), pp. 135–147.
24. C. R. Vogel, *Computational Methods for Inverse Problems*, SIAM, Philadelphia, PA, 2002.

Discrete-Time Stability of a Class of Hermitian Polynomial Matrices with Positive Semidefinite Coefficients

Harald K. Wimmer

Mathematisches Institut, Universität Würzburg, D-97074 Würzburg, Germany
wimmer@mathematik.uni-wuerzburg.de

Abstract. Polynomial matrices $G(z) = Iz^m - \sum_{i=0}^{m-1} C_i z^i$ with positive semidefinite coefficients C_i are studied. If C_0 is positive definite and $\sum C_i = I$ then all characteristic values of $G(z)$ are in the closed unit disc and those lying on the unit circle are m-th roots of unity having linear elementary divisors. The result yields a stability and convergence criterion for a system of difference equations.

Keywords: polynomial matrix, zeros of polynomials, root location, block companion matrix, difference equation, stability.

1 Introduction

In this note we deal with a theorem on polynomials and its extension to polynomial matrices. The following result can be found in [1], and to some extent also in [3], [4, p. 92]) and [5, p. 3].

Theorem 1. *Let* $g(z) = z^m - \left(c_{m-1}z^{m-1} + \cdots + c_1 z + c_0\right)$ *be a real polynomial such that*

$$c_i \geqslant 0, \ i = 0, \ldots, m-1, \ c_0 > 0, \quad \text{and} \quad \sum_{i=0}^{m-1} c_i = 1. \tag{1}$$

(i) *Then all zeros of* $g(z)$ *are in the closed unit disc.*
(ii) *The zeros of* $g(z)$ *lying on the unit circle are simple and they are* m*-th roots of unity.*
(iii) *The number of zeros of* $g(z)$ *on the unit circle is equal to*

$$k = \gcd\bigl\{\{m\} \cup \{i; c_i \neq 0\}\bigr\}.$$

Moreover $g(z) = (z^k - 1)p(z^k)$ *with* $p(\lambda^k) \neq 0$ *if* $|\lambda| = 1$.

It is the purpose of this note to extend the preceding theorem to polynomial matrices and to derive a stability and convergence result for a system of difference equations. We consider matrices $G(z) = Iz^m - \sum_{i=0}^{m-1} C_i z^i$ where the coefficients $C_i \in \mathbb{C}^{n\times n}$ are hermitian and positive semidefinite ($C_i \geqslant 0$) and C_0 is positive definite ($C_0 > 0$), and we assume $\sum_{i=0}^{m-1} C_i = I$. The following notation will be used. We define $\sigma(G) = \{\lambda;\ \det G(\lambda) = 0\}$. In accordance with [2, p. 341] the elements of $\sigma(G)$ will be called the characteristic values of $G(z)$. If $G(\lambda)v = 0$ and $v \in C^n$, $v \neq 0$, then v is said to be an eigenvector corresponding to λ. An r-tuple of vectors $(v_0, v_1, \dots, v_{r-1})$, $v_i \in \mathbb{C}^n$, $v_0 \neq 0$, is called a Jordan chain (or Keldysh chain) of length r if

$$G(\lambda)v_0 = 0,\ G'(\lambda)v_0 + G(\lambda)v_1 = 0, \cdots,$$
$$\tfrac{1}{(r-1)!}G^{(r-1)}(\lambda)v_0 + \tfrac{1}{(r-2)!}G^{(r-2)}(\lambda)v_1 + \cdots + G(\lambda)v_{r-1} = 0.$$

Let $\mathbb{D} = \{z \in \mathbb{C}; |z| < 1\}$ be the open unit disc and $\partial\mathbb{D} = \{z \in \mathbb{C}; |z| = 1\}$ the unit circle of the complex plane, and let $\mathbb{R}_{\geqslant}$ be the set of nonnegative real numbers. Let $E_m = \{\zeta \in \mathbb{C};\ \zeta^m = 1\}$ be the group of m-th roots of unity. If $\zeta \in E_m$ then $\operatorname{ord}\zeta$ will denote the order of ζ, i.e. if $\operatorname{ord}\zeta = s$ then s is the smallest positive divisor of m such that $\zeta^s = 1$.

2 Polynomial matrices

Let us have a closer look at Theorem 1. It is not difficult to show that the zeros of the polynomial $g(z)$ lie in the closed unit disc. But it is remarkable that the unimodular zeros of $g(z)$ should be roots of unity. In the theorem below we shall encounter this property in a more general setting. Accordingly, the focus of this section will be on characteristic values on the unit circle and corresponding eigenvectors. To make the exposition self-contained we dont take advantage of Theorem 1 in the subsequent proof.

Theorem 2. *Let*

$$G(z) = Iz^m - \sum_{i=0}^{m-1} C_i z^i$$

be an $n \times n$ polynomial matrix with hermitian coefficients C_i such that

$$C_i \geqslant 0,\ i = 0, \dots, m-1,\ C_0 > 0,\ \text{and}\ \sum_{i=0}^{m-1} C_i = I. \tag{2}$$

(i) *Then $|\lambda| \leqslant 1$ for all $\lambda \in \sigma(G)$.*
(ii) *If $\lambda \in \sigma(G)$ and $|\lambda| = 1$ then $\lambda^m = 1$. The elementary divisors of $G(z)$ corresponding to λ are linear. If $\operatorname{ord}\lambda = s$ then $E_s \subseteq \sigma(G)$.*
(iii) *Let $v \in \mathbb{C}^n$, $v \neq 0$. Define*

$$k(v) = \gcd\big\{\{m\} \cup \{i;\ C_i v \neq 0,\ i = 0, 1, \dots, m-1\}\big\}. \tag{3}$$

Suppose the set

$$M(v) = \{\lambda \in \sigma(G);\ |\lambda| = 1,\ G(\lambda)v = 0\} \tag{4}$$

is nonempty. Then $M(v) = E_{k(v)}$. *If* $m = k(v)\ell$ *then*

$$G(z)v = \Big[Iz^{k(v)\ell} - \sum_{j=0}^{\ell-1} C_{k(v)j}\, z^{k(v)j}\Big]v = (z^{k(v)} - 1)\, p\big(z^{k(v)}\big) \tag{5}$$

where $p(z) \in \mathbb{C}^n[z]$ *and*

$$p\big(\lambda^{k(v)}\big) \neq 0 \quad \text{if} \quad |\lambda| = 1. \tag{6}$$

Proof. Note that $\det C_0 \neq 0$ implies $0 \notin \sigma(G)$. Let λ be a characteristic value of $G(z)$ and v a corresponding eigenvector with $v^*v = 1$. Set $g_v(z) = v^*G(z)v$ and $c_i = v^*C_iv$. Then $g_v(z) = z^m - \sum_{i=0}^{m-1} c_i z^i$, and the assumptions (2) imply (1). We have $g_v(\lambda) = 0$ or equivalently

$$1 = \sum_{i=0}^{m-1} \frac{c_i}{\lambda^{m-i}}. \tag{7}$$

Hence

$$1 = \left| \sum_{i=0}^{m-1} \frac{c_i}{\lambda^{m-i}} \right| \leqslant \sum_{i=0}^{m-1} \frac{c_i}{|\lambda^{m-i}|}. \tag{8}$$

(i) Suppose $|\lambda| > 1$, i.e. $|\frac{1}{\lambda^{m-i}}| < 1$, $0 \leqslant i \leqslant m-1$. Then (8) implies the strict inequality $1 < \sum c_i$, in contradiction to (1).
(ii) Put

$$\beta_i = \frac{c_i}{\lambda^{m-i}}, \quad i = 0, \ldots, m-1. \tag{9}$$

If $|\lambda| = 1$ then (1) and (7) yield

$$\left| \sum\nolimits_{i=0}^{m-1} \beta_i \right| = \sum\nolimits_{i=0}^{m-1} |\beta_i| = 1.$$

Hence $\beta_i = u\alpha_i$, $i = 0, \ldots, m-1$, with $\alpha_i \in \mathbb{R}_{\geqslant}$, $u \in \mathbb{C}$, $|u| = 1$. From (7) we obtain $1 = u \sum \alpha_i$. Therefore $u \in \mathbb{R}$, $u > 0$. Thus $u = 1$, and we have (9) with $\beta_i \in \mathbb{R}_{\geqslant}$. Take $i = 0$. Then $c_0 > 0$ yields $\lambda^m \in \mathbb{R}_{>}$. Because of $|\lambda| = 1$ we obtain $\lambda^m = 1$, i.e. $\lambda \in E_m$. We rewrite (9) as

$$\beta_i = \lambda^i c_i, \quad \beta_i \in \mathbb{R}_{\geqslant}, \; i = 0, \ldots, m-1. \tag{10}$$

Let $\operatorname{ord} \lambda = s$ and $m = \ell s$. Suppose $c_i \neq 0$, i.e. $c_i > 0$. Then (10) implies $\lambda^i = 1$, that is $i \in \{0, s, 2s, \ldots, (\ell-1)s\}$. Therefore $c_i = 0$ if $i \notin s\mathbb{Z}$. Because of $C_i \geqslant 0$ we have $c_i = v^*C_iv = 0$ if and only if $C_iv = 0$. Hence

$$G(z)v = \Big\{I\,(z^s)^\ell - \Big[C_{(\ell-1)s}\,(z^s)^{\ell-1} + \cdots + C_s z^s + C_0\Big]\Big\}v. \tag{11}$$

Let $\mu^s = 1$. Then (11) and $G(1) = 0$ imply $G(\mu)v = 0$. Thus $E_s \subseteq \sigma(G)$. Moreover $v^* G(\lambda) = 0$, and

$$g_v(z) = z^{\ell s} - \sum_{j=0}^{\ell-1} c_{js} z^{js} \quad \text{and} \quad \sum_{j=0}^{\ell-1} c_{js} = 1.$$

Let us show that the elementary divisors corresponding to λ are linear. It suffices to prove (see e.g. [2, p. 342]) that the vector v cannot be extended to a Jordan chain of length greater than 1. Suppose there exists a vector $w \in \mathbb{C}^n$ such that $G'(\lambda) v + G(\lambda) w = 0$. Then $v^* G(\lambda) = 0$ and $\lambda^s = 1$ imply

$$0 = v^* [G(\lambda) w + G'(\lambda) v] = v^* G'(\lambda) v = g_v'(\lambda) =$$
$$\ell s \lambda^{\ell s - 1} - \sum_{j=0}^{\ell-1} js\, c_{js} \lambda^{js-1} = \lambda^{-1} \Big[\ell s - \sum\nolimits_{j=0}^{\ell-1} js\, c_{js} \Big].$$

Thus we would obtain $\ell s = \sum_{j=0}^{\ell-1} js\, c_{jk}$, which is incompatible with $\sum_{j=0}^{\ell-1} c_{js} = 1$.

(iii) Let $M(v)$ be defined by (4) and let $D(v)$ denote the set of common positive divisors of $\{m \cup \{i;\ C_i v \neq 0,\ i = 0, \dots, m-1\}\}$. Then $s \in D(v)$ with $m = \ell s$ is equivalent to (11). We know that $M(v) \subseteq E_m$, and we have seen that $s \in D(v)$ if $E_s \subseteq M(v)$. Since (11) implies $E_s \subseteq M(v)$ it is obvious that

$$s \in D(v) \quad \text{if and only if} \quad E_s \subseteq M(v). \tag{12}$$

Now let $\lambda, \mu \in M(v)$ and ord $\lambda = s$, ord $\mu = t$. Set $q = \mathrm{lcm}\{s, t\}$, and $r = m/q$. Then

$$G(z)v = \Big[I z^{rq} - \sum\nolimits_{j=0}^{r-1} C_{jq} z^{jq} \Big] v.$$

Hence $E_q \subseteq M(v)$. In particular, we have $\lambda\mu \in M(v)$. Therefore $M(v)$ is a subgroup of E_m. Hence $M(v) = E_{\hat{k}}$ for some divisor $\hat{k}$ of m. Note that $E_s \subseteq E_{\hat{k}}$ is equivalent to $s | \hat{k}$. Therefore it follows from (12) that $\hat{k}$ is the greatest element of $D(v)$. Thus, if $k(v)$ is given by (3) then $\hat{k} = k(v)$.

It remains to show that the polynomial vector $p(z^{k(v)})$ in (5) satisfies the condition (6). Suppose $p(\lambda^{k(v)}) = 0$ for some $\lambda \in \partial\mathbb{D}$. Then $\lambda \in M(v)$ and therefore $\lambda \in E_{k(v)}$, i.e. $\lambda^{k(v)} - 1 = 0$. Hence $G(\lambda)v = G'(\lambda)v = 0$. But then there would exist an an elementary divisor $(z - \lambda)^t$ with $t \geqslant 2$. Therefore we have (6). □

From $\sum C_i = I$ follows $G(1) = 0$. Thus $1 \in \sigma(G)$. More precisely, $\det G(z) = (z-1)^n f(z)$, $f(1) \neq 0$. To check whether $G(z)$ has characteristic values on $\partial\mathbb{D}$ different from 1 we introduce the following matrices. Let s, $s \neq 1$, $s \neq m$, be a positive divisor of m such that $m = s\ell$. Define

$$T_s = I - \sum\nolimits_{j=0}^{\ell-1} C_{js}.$$

Corollary 1. *For each nontrivial divisor* s *of* m *the matrix* T_s *is nonsingular if and only if* $\lambda = 1$ *is only characteristic value of* $G(z)$ *on the unit circle.*

Proof. Suppose

$$G(\lambda)v = 0 \quad \text{and} \quad v \neq 0, \quad |\lambda| = 1, \ \text{ord}\, \lambda = s,\ s > 1.$$

Then $C_i v = 0$ for $i \notin s\mathbb{Z}$. Hence $\sum_{j=0}^{\ell-1} C_{js} v = \sum_{i=0}^{m-1} C_i v = v$, and therefore $T_s v = 0$, and rank $T_s < n$. Conversely, suppose rank $T_s < n$ for some s. Let $T_s v = 0, v \neq 0$. Then $\sum C_i = I$ and $C_i \geqslant 0$ imply $G(z)v = \left[Iz^m - \sum_{j=0}^{\ell-1} C_{js} z^{js}\right] v$ and we conclude that $\{1\} \subsetneqq E_s \subseteq \sigma(G)$. □

3 A difference equation

Theorem 2 deals with the location of characteristic values with respect to the unit circle. Therefore it can be applied to stability problems of systems of difference equations.

Theorem 3. *Let* $C_0, \ldots, C_{m-1} \in \mathbb{C}^{n \times n}$ *be hermitian and such that* (2), *i.e.*

$$C_i \geqslant 0,\ i = 0, \ldots, m-1,\ C_0 > 0, \ \text{ and } \ \sum_{i=0}^{m-1} C_i = I,$$

holds. Then all solutions $\left(x(t)\right)_{t \in \mathbb{N}_0}$ *of the difference equation*

$$x(t+m) = C_{m-1}x(t+m-1) + \cdots + C_1 x(t+1) + C_0 x(t), \tag{13a}$$

$$x(0) = x_0, \ldots, x(m-1) = x_{m-1}, \tag{13b}$$

are bounded for $t \to \infty$. *The sequence* $\left(x(jm)\right)_{j \in \mathbb{N}_0}$ *is convergent.*

Proof. It is well known that the solutions of (13) are bounded if and only if all characteristic values of the associated polynomial matrix $G(z) = Iz^m - \sum C_i z^i$ are in the closed unit disc and if those which lie on the unit circle have linear elementary divisors. To prove convergence of $\left(x(jm)\right)$ we consider the block companion matrix

$$F = \begin{pmatrix} 0 & I & 0 & \ldots & 0 \\ 0 & 0 & I & \ldots & 0 \\ \cdot & \cdot & \cdot & \ldots & \cdot \\ \cdot & \cdot & \cdot & \ldots & \cdot \\ C_0 & C_1 & C_2 & \ldots & C_{m-1} \end{pmatrix}$$

associated with $G(z)$. Note that $\det G(z) = \det(zI - F)$. Moreover $G(z)$ and F have the same elementary divisors. Set $y(t) = \left(x^T(t), x^T(t+1), \cdots, x^T(t+m-1)\right)^T$ and define y_0 conforming to (13b). Then (13) is equivalent to

$$y(t+1) = Fy(t),\ y(0) = y_0.$$

The corresponding equation for $w(j) = x(jm)$ is

$$w(j+1) = F^m w(j).$$

We know that $\sigma(G) \subseteq \overline{\mathbb{D}}$ and that $\lambda \in \sigma(G) \cap \partial\mathbb{D}$ implies $\lambda^m = 1$. Therefore $\sigma(F^m) \subseteq \{1\} \cup \mathbb{D}$, and F^m is similar to $\operatorname{diag}(I, \hat{F})$ with $\sigma(\hat{F}) \subseteq \mathbb{D}$. Hence $(w(j))$ is convergent. □

References

1. N. Anderson, E. B. Saff, and R. S. Varga, *An extension of the Eneström-Kakeya theorem and its sharpness*, SIAM J. Math. Anal. 12 (1981), pp. 10–22.
2. H. Baumgärtel, *Analytic Perturbation Theory for Matrices and Operators*, Operator Theory, Advances and Applications, Vol. 15, Birkhäuser, Basel, 1985.
3. A. Hurwitz, *Über einen Satz des Herrn Kakeya*, Tôhoku Math. J. 4 (1913), pp. 89–93; in: Mathematische Werke von A. Hurwitz, 2. Band, pp. 627–631, Birkhäuser, Basel, 1933.
4. A. M. Ostrowski, *Solutions of Equations in Euclidean and Banach Spaces*, 3rd ed., Academic Press, New York, 1973.
5. V. V. Prasolov, *Polynomials*, Algorithms and Computation in Mathematics, Vol. 11, Springer, New York, 2004.

MATRICES AND APPLICATIONS

Splitting Algorithm for Solving Mixed Variational Inequalities with Inversely Strongly Monotone Operators*

Ildar Badriev[†] and Oleg Zadvornov

Department of Computational Mathematics and Mathematical Cybernetics, 420008
Kazan State University, Kazan, Kremlevskya, 18, Russia
[†]Ildar.Badriev@ksu.ru

Abstract. We consider a boundary value problem whose generalized statement is formulated as a mixed variational inequality in a Hilbert space. The operator of this variational inequality is a sum of several inversely strongly monotone operators (which are not necessarily potential operators). The functional occurring in this variational inequality is also a sum of several lower semi-continuous convex proper functionals. For solving of the considered variational inequality a decomposition iterative method is offered. The suggested method does not require the inversion of original operators. The convergence of this method is investigated.

Keywords: variational inequality, inversely strongly monotone operator, variational inequality, iterative method.

1 Statement of the problem

Let $\Omega \subset \mathbb{R}^n$, $n \geqslant 1$ be a bounded domain with a Lipschitz continuous boundary Γ. We consider the following boundary value problem with respect to the function $u = (u_1, u_2, \dots, u_n)$:

$$\sum_{j=1}^{n} \left(\frac{\partial}{\partial x_j} v_j^{(i)}(x) + d_{ij}(x) u_j(x) \right) = f_i(x),\ x \in \Omega,\ i = 1, \dots, n, \tag{1}$$

$$u(x) = 0, \quad x \in \Gamma, \tag{2}$$

$$-v_j^{(i)}(x) \in \frac{g_j(|\partial u(x)/\partial x_j|)}{|\partial u(x)/\partial x_j|} \frac{\partial u_i(x)}{\partial x_j},\ x \in \Omega,\ i, j = 1, \dots, n, \tag{3}$$

where $f = (f_1, f_2, \dots, f_n)$ is a given function, $D = \{d_{ij}\}$ is an unsymmetric matrix such that

$$(D\xi,\ \xi) \geqslant \alpha_0\, (D\xi,\ D\xi) \qquad \forall\, \xi \in \mathbb{R}^n, \quad \alpha_0 > 0. \tag{4}$$

* This work was supported by the Russian Foundation for Basic Research, project №06-01-00633 and 07-01-00674.

We assume that the multi-valued functions g_j can be represented in the form

$$g_j(\xi) = g_{0j}(\xi) + \vartheta_j\, h(\xi - \beta_j),$$

where ϑ_j, β_j are the given non negative constants, h is the multi-valued and g_{0j} are the single-valued functions given by the formulas

$$h(\xi) = \begin{cases} 0, & \xi < 0, \\ [0,1], & \xi = 0, \\ 1, & \xi > 0, \end{cases} \qquad g_{0j}(\xi) = \begin{cases} 0, & \xi \leqslant \beta_j, \\ g_j^*(\xi - \beta_j), & \xi \geqslant \beta_j, \end{cases}$$

$g_j^* : [0, +\infty) \to [0, +\infty)$ are the continuous functions which satisfy the following conditions:

$$g_j^*(0) = 0, \; g_j^*(\xi) > g_j^*(\zeta) \quad \forall\, \xi > \zeta \geqslant 0, \tag{5}$$

$$\exists\, \sigma_j > 0 : \; |\, g_j^*(\xi) - g_j^*(\zeta)\,| \leqslant \frac{1}{\sigma_j} |\, \xi - \zeta\,| \quad \forall\, \xi, \zeta \geqslant 0, \tag{6}$$

$$\exists\, k_j > 0, \; \xi_j^* \geqslant 0 : \; g_j^*(\xi_j^*) \geqslant k_j\, \xi_j^*, \; g_j^*(\xi) - g_j^*(\zeta) \geqslant k_j\,(\xi - \zeta) \quad \forall\, \xi \geqslant \zeta \geqslant \xi_j^*. \tag{7}$$

Let us introduce the notations: $V = \left[\overset{\circ}{W}{}_2^{(1)}(\Omega)\right]^n$, $H = [L_2(\Omega)]^n$, $B_j = \partial/\partial x_j : V \to H$, $j = 1, 2, \dots, n$.

A generalized solution of the problem (1)–(3) is defined as the function $u \in V$ satisfying for all $\eta \in V$ the variational inequality

$$\begin{aligned} &(A_0 u, \eta - u)_V + \sum_{j=1}^{n} (A_j \circ B_j(u), B_j(\eta - u))_H + F_0(\eta) - F_0(u) + \\ &\sum_{j=1}^{n} [F_j(B_j \eta) - F_j(B_j u)] \geqslant 0. \end{aligned} \tag{8}$$

Here $B_j^* : H \to V$, $j = 1, 2, \dots, n$ are operators conjugate to B_j. The operators $A_0 : V \to V$ and $A_j : H \to H$, $j = 1, 2, \dots, n$ are generated by the forms

$$(A_0 u, \eta)_V = \int_\Omega (Qu, \eta)\, dx, \; u, \eta \in V, \quad (A_j\, y,\, z)_H = \int_\Omega (G_j\,(y),\, z)\, dx, \; y, z \in H.$$

The operators $G_j : \mathbb{R}^n \to \mathbb{R}^n$ and the functionals $F_0 : V \to \mathbb{R}^1$, $F_j : H \to \mathbb{R}^1$, $j = 1, 2, \dots, n$ are defined by the formulas:

$$G_j\,(y) = g_{0j}\,(|y|)\, |y|^{-1}\, y\,, y \neq 0, \; G_j\,(0) = 0, \quad F_0(\eta) = -\int_\Omega (f, \eta)\, dx, \; \eta \in V,$$

$$F_j\,(z) = \vartheta_j \int_\Omega \mu(|z| - \beta_j)\, dx, \; z \in H, \quad \mu(\zeta) = \begin{cases} 0, & \zeta < 0, \\ \zeta, & \zeta \geqslant 0. \end{cases}$$

The following result is valid.

Lemma 1. *Let the condition (4) be satisfied. Then* A_0 *is an inversely strongly monotone operator, i.e.,*

$$(A_0\eta - A_0 u, \eta - u)_V \geqslant \sigma_0 \, \| A_0\eta - A_0 u \|_V^2, \quad \sigma_0 > 0 \quad \forall u, \eta \in V. \qquad (9)$$

Proof. It follows from (4) that

$$|Q\xi| \leqslant \alpha_0^{-1/2} (Q\xi, \xi)^{1/2} \quad \forall \xi \in \mathbb{R}^n,$$

and hence

$$|(Q\xi, \zeta)| \leqslant \alpha_0^{-1/2} (Q\xi, \xi)^{1/2} |\zeta| \quad \forall \xi, \zeta \in \mathbb{R}^n.$$

Because of this

$$|(A_0 u, \eta)_V| \leqslant \int_\Omega |(Q u, \eta)| \, dx \leqslant \alpha_0^{-1/2} \int_\Omega (Qu, Qu)^{1/2} |\eta| \, dx \leqslant$$

$$\alpha_0^{-1/2} \left(\int_\Omega (Qu, u) \, dx \right)^{1/2} \left(\int_\Omega |\eta|^2 \, dx \right)^{1/2} = \alpha_0^{-1/2} (A_0 u, u)_V^{1/2} \, \| \eta \|_H \leqslant$$

$$\alpha_0^{-1/2} c_H (A_0 u, u)_V^{1/2} \, \| \eta \|_V = \sigma_0^{-1/2} (A_0 u, u)_V^{1/2} \, \| \eta \|_V,$$

where c_H is the Friedrichs constant (the constant of embedding V into H), $\sigma_0 = \alpha_0 / c_H^2$.

Therefore,

$$\| A_0 u \|_V = \sup_{\eta \neq 0} \frac{|(A_0 u, \eta)_V|}{\| \eta \|_V} \leqslant \sigma_0^{-1/2} (A_0 u, u)_V^{1/2},$$

whence by virtue of linearity of A_0 it follows required inequality. □

By analogy with [4] we obtain that the following results are valid.

Lemma 2. *Let the conditions (5)–(7) be satisfied. Then* A_j *are coercive and inversely strongly monotone operators, i.e.,*

$$(A_j y - A_j z, y - z)_H \geqslant \sigma_j \, \| A_j y - A_j z \|_H^2, \ \sigma_j > 0, \quad \forall y, z \in H. \qquad (10)$$

Lemma 3. *The functionals* $F_0 : V \to \mathbb{R}^1$, $F_j : H \to \mathbb{R}^1$, $j = 1, 2, \ldots, n$ *are convex and Lipschitz continuous ones.*

It follows from these results that the variational inequality (8) has at least one solution (see e.g. [5]).

2 The iterative process

In the following, we will consider the abstract variational inequality (8) postulating the properties (9), (10) and assuming that $B_j : V \to H$, $j = 1, 2, \dots, n$ are linear continuous operators and F_j, $j = 0, 1, 2, \dots, n$ are proper convex and Lipschitz continuous functionals. In addition, we assume that the operator $\sum_{j=1}^{n} B_j^* B_j : V \to V$ is a canonical isomorphism, i.e.,

$$\sum_{j=1}^{n} \left(B_j^* B_j u, \eta \right)_V = (u, \eta)_V \qquad \forall u, \eta \in V. \tag{11}$$

To solve the variational inequality (8) we consider the following splitting algorithm.

Let $u^{(0)} \in V$, $y_j^{(0)} \in H$, $\lambda_j^{(0)} \in H$, $j = 1, 2, \dots, n$ be arbitrary elements. For $k = 0, 1, 2, \dots$ and for known $y_j^{(k)}$, $\lambda_j^{(k)}$, $j = 1, 2, \dots, n$ we define $u^{(k+1)}$ as a solution of the variational inequality:

$$\begin{aligned} &\frac{1}{\tau_0} \left(u^{(k+1)} - u^{(k)}, \eta - u^{(k+1)} \right)_V + \\ &F_0(\eta) - F_0\left(u^{(k+1)}\right) + \left(A_0 u^{(k)}, \eta - u^{(k+1)} \right)_V + \\ &\left(\sum_{j=1}^{n} B_j^* \lambda_j^{(k)} + r \sum_{j=1}^{n} B_j^* \left(B_j u^{(k)} - y_j^{(k)} \right), \eta - u^{(k+1)} \right)_V \geqslant 0 \quad \forall \eta \in V. \end{aligned} \tag{12}$$

Then we find $y_j^{(k+1)}$, $j = 1, 2, \dots, n$, by solving the variational inequalities

$$\begin{aligned} &\frac{1}{\tau_j} \left(y_j^{(k+1)} - y_j^{(k)}, z - y_j^{(k+1)} \right)_H + \\ &F_j(z) - F_j\left(y_j^{(k+1)}\right) + \left(A_j y_j^{(k)} - \lambda_j^{(k)}, z - y_j^{(k+1)} \right)_H + \\ &r \left(y_j^{(k)} - B_j u^{(k+1)}, z - y_j^{(k+1)} \right)_H \geqslant 0 \quad \forall z \in H, \quad j = 1, 2, \dots, n. \end{aligned} \tag{13}$$

Finally, we set

$$\lambda_j^{(k+1)} = \lambda_j^{(k)} + r \left(B_j u^{(k+1)} - y_j^{(k+1)} \right), \quad j = 1, 2, \dots, n. \tag{14}$$

Here $\tau_j > 0$, $j = 0, 1, 2, \dots, n$ and $r > 0$ are the iterative parameters.

To analyze the convergence of the method (12)–(14) we formulate it via the transition operator $T : V \times H^n \times H^n \to V \times H^n \times H^n$ that takes each vector $q = (q_0, q_1, \dots, q_{2n}) = (u, Y, \Lambda)$, $Y \in H^n$, $\Lambda \in H^n$ to the element $Tq = (T_0 q, T_1 q, \dots, T_{2n} q)$ as follows

$$T_0 q = \mathrm{Prox}_{\tau_0 F_0} \left(q_0 - \tau_0 \left[A_0 q_0 + \sum_{j=1}^{n} B_j^* q_{n+j} + r \sum_{j=1}^{n} B_j^* \left(B_j q_0 - q_j \right) \right] \right), \tag{15}$$

$$T_j q = \mathrm{Prox}_{\tau_j F_j} \Big(q_j - \tau_j \Big[A_j q_j - q_{n+j} + r \Big(q_j - B_j T_0 q \Big) \Big] \Big), \; j = 1, 2, \ldots, n, \tag{16}$$

$$T_{n+j} q = q_{n+j} + r \Big(B_j T_0 q - T_j q \Big), \qquad j = 1, 2, \ldots, n. \tag{17}$$

Here Prox_G is a proximal mapping (see e.g. [5]). Recall that a mapping $\mathrm{Prox}_G : Z \to Z$ is said to be proximal if it takes each element p of Hilbert space Z to the element $v = \mathrm{Prox}_G(p)$ that is the solution of the minimization problem

$$\frac{1}{2} \| v - p \|_Z^2 + G(v) = \min_{z \in P} \Big\{ \frac{1}{2} \| z - p \|_Z^2 + G(z) \Big\}.$$

This problem is equivalent (if G is a convex proper lower semi-continuous functional) to a variational inequality

$$(v - p, \, z - v)_Z + G(z) - G(v) \geqslant 0 \quad \forall z \in Z. \tag{18}$$

It is easy to show that a proximal mapping is a firmly nonexpansive; i.e.,

$$\| \mathrm{Prox}_G(p) - \mathrm{Prox}_G(z) \|_Z^2 \leqslant (\mathrm{Prox}_G(p) - \mathrm{Prox}_G(z) \, , \, p - z)_Z \qquad \forall p, \, z \in Z.$$

We introduce the notations

$$Y^{(k)} = \Big(y_1^{(k)}, y_2^{(k)}, \ldots, y_n^{(k)} \Big), \quad \Lambda^{(k)} = \Big(\lambda_1^{(k)}, \lambda_2^{(k)}, \ldots, \lambda_n^{(k)} \Big).$$

Then using the definition of a proximal mapping by the variational inequality (18) it is easy to verify that the iterative process (12)–(14) can be represented in the form

$$\begin{cases} q^{(0)} \quad \text{is an arbitrary element,} \\ q^{(k+1)} = T q^{(k)}, \; q^{(k)} = \big(u^{(k)}, Y^{(k)}, \Lambda^{(k)} \big), \; k = 0, 1, 2, \ldots, \end{cases} \tag{19}$$

i.e., T is the transition operator of this iterative process.

Let us now obtain a relationship between the solution of the original variational inequality (8) and the components of the fixed point of the transition operator T.

The following result is true.

Theorem 1. *Let the operator* $T : V \times H^n \times H^n \to V \times H^n \times H^n$ *is defined by the relationships (15) – (17). Then the point* $q = (u, Y, \Lambda)$ *where* $u \in V$, $Y = (y_1, y_2, \ldots, y_n) \in H^n$, $\Lambda = (\lambda_1, \lambda_2, \ldots, \lambda_n) \in H^n$, *is a fixed point of the operator* T *if and only if*

$$y_j = B_j u, \quad j = 1, 2, \ldots, n, \tag{20}$$

$$\lambda_j \in \partial F_j(y_j) + A_j y_j, \quad j = 1, 2, \ldots, n, \tag{21}$$

$$-\sum_{j=1}^{n} B_j^* \lambda_j \in \partial F_0(u) + A_0 u. \tag{22}$$

Moreover, the first component u *of each fixed point* q *of the operator* T *is a solution of the problem (8).*

Proof. Let $q = (u, Y, \Lambda)$ be fixed point of the operator T, i.e., according to (15) – (17)

$$u = \mathrm{Prox}_{\tau_0 F_0}\left(u - \tau_0 \left[A_0 u + \sum_{j=1}^{n} B_j^* \lambda_j + r \sum_{j=1}^{n} B_j^* \left(B_j u - y_j \right) \right] \right), \tag{23}$$

$$y_j = \mathrm{Prox}_{\tau_j F_j}\left(y_j - \tau_j \left[A_j y_j - \lambda_j + r \left(y_j - B_j u \right) \right] \right), \quad j = 1, 2, \ldots, n, \tag{24}$$

$$\lambda_j = \lambda_j + r\,(B_j u - y_j), \quad j = 1, 2, \ldots, n. \tag{25}$$

Obviously, the relations (25) are equivalent to (20).

By the (20) and definition (18) of a proximal mapping the relations (24) are equivalent to variational inequalities

$$\tau_j\, (A_j y_j - \lambda_j, z - y_j)_H + \tau_j F_j(z) - \tau_j F_j(y_j) \geqslant 0 \quad \forall z \in H, \quad j = 1, 2, \ldots, n,$$

or

$$(A_j y_j - \lambda_j, z - y_j)_H + F_j(z) - F_j(y_j) \geqslant 0 \quad \forall z \in H, \quad j = 1, 2, \ldots, n, \tag{26}$$

each of these is equivalent to $-A_j y_j - \lambda_j \in \partial F_j(y_j)$, $j = 1, 2, \ldots, n$, i.e., inclusions (21) hold.

In an analogous way we have that relation (23) is equivalent to the variational inequality

$$\left(A_0 u + \sum_{j=1}^{n} B_j^* \lambda_j, \eta - u \right)_V + F_0(\eta) - F_0(u) \geqslant 0 \qquad \forall \eta \in V, \tag{27}$$

i.e., to the inclusion (22).

We have thereby shown that the equality $Tq = q$ is equivalent to relations (20) – (22).

Let us now verify that the first component u of each fixed point q of the operator T is a solution of the problem (8). To this end, in inequalities (26), we use relations (20) to replace y_j by $B_j u$, $j = 1, 2, \ldots, n$ and set $z = B_j \eta$, where η is an arbitrary element of V. By adding resulting inequalities and using the definition of conjugate operator we have

$$\sum_{j=1}^{n} (B_j^* \circ A_j \circ B_j\,(u), \eta - u)_V + \sum_{j=1}^{n} [F_j\,(B_j \eta) - F_j\,(B_j u)] \geqslant 0 \quad \forall \eta \in V. \tag{28}$$

By adding inequalities (27)–(28) we have that u is a solution of the problem (8). The proof of the theorem is complete. □

Theorem 2. *Suppose that there exists a solution of problem (8) and*

$$\begin{aligned}&\exists\, u^* \in \operatorname{dom} F_0 : \; B_j\, u^* \in \operatorname{dom} F_j;\\ &F_j \textit{ is continuous at the point } B_j\, u^*, \; j = 1, 2, \dots, n.\end{aligned} \tag{29}$$

Then the set of fixed points of the operator T *is nonempty.*

Proof. Let u be a solution of the problem (8), $y_j = B_j\, u$, $j = 1, 2, \dots, n$. The variational inequality (8) is equivalent to the following inclusion

$$-A_0\, u - \sum_{j=1}^{n} B_j^*\, A_j\, y_j \in \partial \left[F_0 + \sum_{j=1}^{n} F_j \circ B_j \right] (u). \tag{30}$$

If conditions (29) are satisfied, then it follows from Propositions 5.6 and 5.7 [5] that

$$\partial \left[F_0 + \sum_{j=1}^{n} F_j \circ B_j \right] (u) = \partial\, F_0(u) + \sum_{j=1}^{n} \partial\, (F_j \circ B_j\,)\, (u) = \partial\, F_0(u) + \sum_{j=1}^{n} B_j^*\, \partial\, F_j(y_j). \tag{31}$$

Relations (31) and (30) imply that there exist elements $\nu \in \partial\, F_0(u)$, $z_j \in \partial F_j(y_j\,)$, $j = 1, 2, \dots, n$, such that

$$-A_0\, u - \sum_{j=1}^{n} B_j^*\, A_j\, y_j = \nu + \sum_{j=1}^{n} B_j^*\, z_j,$$

or

$$-A_0\, u - \sum_{j=1}^{n} B_j^*\, (\, A_j\, y_j + z_j\,) = \nu.$$

Let $\lambda_j = A_j\, y_j + z_j$, $j = 1, 2, \dots, n$; then we have the inclusions

$$-A_0\, u - \sum_{j=1}^{n} B_j^*\, \lambda_j = \nu \in \partial\, F_0(u); \quad -A_j\, y_j + \lambda_j = z \in \partial F_j(y_j\,), \quad j = 1, 2, \dots, n,$$

i.e., the relations (21), (22) hold.

Next relations (20) are valid by virtue of the definition of y_j. Therefore by Theorem 1, the operator T has a fixed point, namely the point $q = (\, u, Y, \Lambda\,)$ where $Y = (\, y_1, y_2, \dots, y_n\,) \in H^n$, $\Lambda = (\, \lambda_1, \lambda_2, \dots, \lambda_n\,) \in H^n$. The proof of the theorem is complete. □

Thus the convergence analysis of the iterative process (12)–(14) can be reduced to that of the successive approximation method for finding a fixed point of T.

3 The investigation of the convergence of the iterative process

Let introduce the Hilbert space $Q = V \times H^n \times H^n$ with the inner product

$$(\cdot,\cdot)_Q = \frac{1-\tau_0 r}{\tau_0}(\cdot,\cdot)_V + \sum_{j=1}^{n} \frac{1}{\tau_j}(\cdot,\cdot)_H + \frac{1}{r}\sum_{j=1}^{n}(\cdot,\cdot)_H,$$

where r, τ_j, $j = 0,1,2,\dots,n$, are positive constants; moreover, $\tau_j r < 1$, $j = 0,1,2,\dots,n$.

The investigation of the convergence of the iterative process (19) is based on the following

Theorem 3. *Let conditions (9)–(11) be satisfied, and let*

$$\tau_j < \frac{2\sigma_j}{2\sigma_j r + 1}, \qquad j = 0,1,2,\dots,n. \tag{32}$$

Then operator T *is nonexpansive. Moreover, the inequality*

$$\begin{aligned}
&\| Tq - Tp \|_Q^2 + \delta_0\,(A_0 q_0 - A_0 p_0\,,\, q_0 - p_0)_V + \sum_{j=1}^{n} \delta_j\,(A_j q_j - A_j p_j\,,\, q_j - p_j)_H + \\
&\frac{1}{\tau_0(1-\tau_0 r)}\,\|(1-\tau_j r)[(q_0 - T_0 q) - (p_0 - T_0 p)] - \tau_0(A_0 q_0 - A_0 p_0)\|_V^2 + \\
&\sum_{j=1}^{n}\frac{1}{\tau_j(1-\tau_j r)}\,\|(1-\tau_j r)[(q_j - T_j q) - (p_j - T_j p)] - \tau_j(A_j q_j - A_j p_j)\|_H^2 + \\
&r\sum_{j=1}^{n}\|(q_j - B_j T_0 q) - (p_j - B_j T_0 p)\|_H^2 \leqslant \|q - p\|_Q^2,
\end{aligned} \tag{33}$$

is valid for arbitrary $q, p \in Q$, *where* $\delta_j = 2 - \tau_j/[\sigma_j(1-\tau_j r)]$, $j = 0,1,2,\dots,n$.

Proof. First, note that by virtue of conditions (32) we have $\tau_j r < 1$, $\delta_j > 0$, $j = 0,1,2,\dots,n$; therefore it follows from (9), (10) and (33) that T is nonexpansive operator.

We rewrite relation (15) in view of (11) in the form

$$\begin{aligned}
T_0 q &= \mathrm{Prox}_{\tau_0 F_0}\left(q_0 - \tau_0 A_0 q_0 - \tau_0 r q_0 - \tau_0 \sum_{j=1}^{n} B_j^*\,(q_{n+j} - r q_j)\right) \\
&= \mathrm{Prox}_{\tau_0 F_0}\left(S_0 q_0 - \tau_0 \sum_{j=1}^{n} B_j^*\,(q_{n+j} - r q_j)\right),
\end{aligned}$$

where $S_0 : V \to V$ is the operator given by the formula $S_0 = (1-\tau_0 r)\,I - \tau_0 A_0$.

According to [3] by using (9) we obtain

$$\| S_0\, p_0 - S_0\, q_0 \|_V^2 = (1 - \tau_0\, r)^2 \, \| q_0 - p_0 \|_V^2 -$$
$$2\tau_0\, (1 - \tau_0\, r)\, (A_0\, q_0 - A_0\, p_0\,,\, q_0 - p_0\,)_V + \tau_0^2 \, \|A_0\, q_0 - A_0\, p_0 \|_V^2 \leqslant$$
$$(1 - \tau_0\, r)^2 \, \| q_0 - p_0\|_V^2 - 2\tau_0 \left(1 - \tau_0 \, \frac{2\,\sigma_0\, r + 1}{2\,\sigma_0}\right) (A_0\, q_0 - A_0\, p_0,\, q_0 - p_0\,)_V\,,$$

i.e.,

$$\begin{aligned} &\| S_0 p_0 - S_0 q_0 \|_V^2 \leqslant \\ &\quad (1 - \tau_0\, r)^2 \, \| q_0 - p_0\|_V^2 - \tau_0\, (1 - \tau_0\, r)\, \delta_0 \, (A_0\, q_0 - A_0\, p_0,\, q_0 - p_0\,)_V\,, \end{aligned} \tag{34}$$

for any $q_0, p_0 \in V$. Further, by using the firmly nonexpansiving property of proximal mapping $\mathrm{Prox}_{\tau_0\, F_0}$ we obtain

$$\| T_0\, q - T_0\, p \|_V^2 \leqslant (T_0\, q - T_0\, p,\, S_0\, q_0 - S_0\, p_0\,)_V -$$
$$\tau_0 \sum_{j=1}^{n} \left(B_j^*\, (\, q_{n+j} - p_{n+j}\,) - r\, B_j^*\, (q_j - p_j\,)\,,\, T_0\, q - T_0\, p\,\right)_V .$$

Let us transform the first term in the right side by the relation

$$(v, w)_Z = \frac{1}{2\,\varepsilon}\, \|v\|_Z^2 - \frac{1}{2\,\varepsilon}\, \|v - \varepsilon\, w\,\|_Z^2 + \frac{\varepsilon}{2}\, \|w\|_Z^2 \quad \forall\, v, w \in Z,\ \forall\, \varepsilon > 0 \tag{35}$$

with $Z = V$, $v = S_0\, q_0 - S_0\, p_0$, $w = T_0\, q - T_0\, p$. We have

$$\| T_0\, q - T_0\, p \|_V^2 \leqslant \frac{1}{2\,\varepsilon}\, \|S_0\, q_0 - S_0\, p_0 \|_V^2 + \frac{\varepsilon}{2}\, \|T_0\, q - T_0\, p \|_V^2 -$$
$$\frac{1}{2\,\varepsilon}\, \|(S_0\, q_0 - S_0\, p_0\,) - \varepsilon\, (T_0\, q - T_0\, p\,)\, \|_V^2 -$$
$$\tau_0 \sum_{j=1}^{n} \left(B_j^*\, (\, q_{n+j} - p_{n+j}\,) - r\, B_j^*\, (q_j - p_j\,)\,,\, T_0\, q - T_0\, p\,\right)_V .$$

Therefore, by virtue of (34) we obtain

$$\frac{2 - \varepsilon}{2}\, \| T_0\, q - T_0\, p \|_V^2 \leqslant \frac{(1 - \tau_0\, r)^2}{2\,\varepsilon}\, \|q_0 - p_0 \|_V^2 -$$
$$\frac{\tau_0\, (1 - \tau_0\, r)\, \delta_0}{2\,\varepsilon}\, (A_0\, q_0 - A_0\, p_0,\, q_0 - p_0\,)_V -$$
$$\frac{1}{2\,\varepsilon}\, \|(1 - \tau_0\, r)\, (q_0 - p_0\,) - \tau_0\, (A_0\, q_0 - A_0\, p_0\,) - \varepsilon\, (T_0\, q - T_0\, p\,)\, \|_V^2 -$$
$$\tau_0 \sum_{j=1}^{n} \left(B_j^*\, (\, q_{n+j} - p_{n+j}\,) - r\, B_j^*\, (q_j - p_j\,)\,,\, T_0\, q - T_0\, p\,\right)_V .$$

After division by τ_0 by choosing $\varepsilon = 1 - \tau_0 r$ we have

$$\frac{1+\tau_0 r}{2\tau_0} \| T_0 q - T_0 p \|_V^2 + \frac{\delta_0}{2} (A_0 q_0 - A_0 p_0, q_0 - p_0)_V +$$
$$\frac{1}{2(1-\tau_0 r)\tau_0} \|(1-\tau_0 r) [(q_0 - T_0 q) - (p_0 - T_0 p)] - \tau_0 (A_0 q_0 - A_0 p_0) \|_V^2 \leqslant$$
$$\frac{1-\tau_0 r}{2\tau_0} \|q_0 - p_0\|_V^2 - \sum_{j=1}^{n} (q_{n+j} - p_{n+j}, B_j (T_0 q - T_0 p))_H +$$
$$r \sum_{j=1}^{n} (q_j - p_j, B_j (T_0 q - T_0 p))_H .$$

From this inequality after the transformation of the terms $(q_j - p_j, B_j (T_0 q - T_0 p))_H$ by the (34) with $Z = H$, $\varepsilon = 1$, $v = q_j - p_j$, $w = B_j (T_0 q - T_0 p)$ it follows that

$$\frac{1+\tau_0 r}{2\tau_0} \| T_0 q - T_0 p \|_V^2 + \frac{\delta_0}{2} (A_0 q_0 - A_0 p_0, q_0 - p_0)_V +$$
$$\frac{1}{2(1-\tau_0 r)\tau_0} \|(1-\tau_0 r) [(q_0 - T_0 q) - (p_0 - T_0 p)] - \tau_0 (A_0 q_0 - A_0 p_0) \|_V^2 +$$
$$\frac{r}{2} \sum_{j=1}^{n} \| (q_j - B_j T_0 q) - (p_j - B_j T_0 p) \|_H^2 \leqslant$$
$$\frac{1-\tau_0 r}{2\tau_0} \|q_0 - p_0\|_V^2 + \frac{r}{2} \sum_{j=1}^{n} \| q_j - p_j \|_H^2 +$$
$$\frac{r}{2} \sum_{j=1}^{n} \| B_j (T_0 q - T_0 p) \|_H^2 - \sum_{j=1}^{n} (q_{n+j} - p_{n+j}, B_j (T_0 q - T_0 p))_H . \tag{36}$$

For each $j = 1, 2, \ldots, n$ we rewrite the (16) in the form

$$\begin{aligned} T_j q &= \mathrm{Prox}_{\tau_j F_j} (q_j - \tau_j r q_j - \tau_j A_j q_j + \tau_j q_{n+j} + \tau_j B_j T_0 q) \\ &= \mathrm{Prox}_{\tau_j F_j} (S_j q_j + \tau_j q_{n+j} + \tau_j B_j T_0 q) , \end{aligned}$$

where the operators $S_j : H \to H$ are defined by the relationships $S_j = (1 - \tau_j r) I - \tau_j A_j$.

By virtue of (10) by the analogous with (34) we obtain the estimates

$$\| S_j p_j - S_j q_j \|_H^2 \leqslant (1-\tau_j r)^2 \|q_j - p_j\|_H^2 - \tau_j (1-\tau_j r) \delta_j (A_j q_j - A_j p_j, q_j - p_j)_H , \tag{37}$$

and take in account the firmly non expanding of the proximal mapping $\mathrm{Prox}_{\tau_j F_j}$ and the equality (34) with an arbitrary $\varepsilon > 0$, $Z = H$, $v = S_j q_j - S_j p_j$,

$w = T_j q - T_j p$ we have

$$\| T_j q - T_j p \|_H^2 \leqslant (T_j q - T_j p , S_j q_j - S_j p_j)_H +$$
$$\tau_j r (T_j q - T_j p , B_j (T_0 q - T_0 p))_H + \tau_j (T_j q - T_j p , q_{n+j} - p_{n+j})_H =$$
$$\frac{1}{2\varepsilon} \| S_j q_j - S_j p_j \|_H^2 + \frac{\varepsilon}{2} \| T_j q - T_j p \|_H^2 -$$
$$\frac{1}{2\varepsilon} \| (S_j q_j - S_j p_j) - \varepsilon (T_j q - T_j p)\|_H^2 +$$
$$\tau_j r (T_j q - T_j p , B_j (T_0 q - T_0 p))_H + \tau_j (T_j q - T_j p , q_{n+j} - p_{n+j})_H .$$

By setting in the last inequality $\varepsilon = 1 - \tau_j r$ and using the estimation (37) for $\|S_j q_j - S_j p_j \|_H^2$ we obtain the inequality

$$\frac{1 + \tau_j r}{2\tau_j} \| T_j q - T_j p \|_H^2 + \frac{\delta_j}{2} (A_j q_j - A_j p_j , q_j - p_j)_H +$$
$$\frac{1}{2(1 - \tau_j r)\tau_j} \| (1 - \tau_j r) [(q_j - T_j q) - (p_j - T_j p)] - \tau_j (A_j q_j - A_j p_j) \|_H^2 \leqslant$$
$$\frac{1 - \tau_j r}{2\tau_j} \| q_j - p_j \|_H^2 - (T_j q - T_j p , q_{n+j} - p_{n+j})_H +$$
$$r (T_j q - T_j p , B_j (T_0 q - T_0 p))_H ,$$

which after using the relation (34) with $\varepsilon = 1$, $Z = H$, $v = T_j q - T_j p$, $w = B_j(T_0 q - T_0 p)$ for the transformation the last term implies

$$\frac{1 + \tau_j r}{2\tau_j} \| T_j q - T_j p \|_H^2 + \frac{\delta_j}{2} (A_j q_j - A_j p_j , q_j - p_j)_H +$$
$$\frac{r}{2} \| (T_j q - B_j T_0 q) - (T_j p - B_j T_0 p) \|_H^2 +$$
$$\frac{1}{2(1 - \tau_j r)\tau_j} \| (1 - \tau_j r) [(q_j - T_j q) - (p_j - T_j p)] - \tau_j (A_j q_j - A_j p_j) \|_H^2 \leqslant$$
$$\frac{1 - \tau_j r}{2\tau_j} \| q_j - p_j \|_H^2 + (q_{n+j} - p_{n+j} , T_j q - T_j p)_H +$$
$$\frac{r}{2} \| T_j q - T_j p \|_H^2 + \frac{r}{2} \| B_j (T_0 q - T_0 p) \|_H^2 . \tag{38}$$

Further for each $j = 1, 2, \ldots, n$ by virtue of (17) we have

$$\frac{1}{2r} \| T_{n+j} q - T_{n+j} p \|_H^2 = \frac{1}{2r} \| q_{n+j} - p_{n+j} \|_H^2 +$$
$$(q_{n+j} - p_{n+j} , B_j (T_0 q - T_0 p))_H -$$
$$(q_{n+j} - p_{n+j} , T_j q - T_j p)_H + \frac{r}{2} \| B_j (T_0 q - T_0 p) - (T_j q - T_j p) \|_H^2 . \tag{39}$$

By adding relations (38), (39) with $j = 1, 2, \ldots, n$, and the relations (36) after multiplying by 2 we have

$$\begin{aligned}
&\frac{1+\tau_0 r}{\tau_0} \| T_0 q - T_0 p \|_V^2 + \delta_0 (A_0 q_0 - A_0 p_0, q_0 - p_0)_V + \\
&\frac{1}{(1-\tau_0 r)\tau_0} \|(1-\tau_0 r)[(q_0 - T_0 q) - (p_0 - T_0 p)] - \tau_0 (A_0 q_0 - A_0 p_0)\|_V^2 + \\
&r \sum_{j=1}^{n} \| (q_j - B_j T_0 q) - (p_j - B_j T_0 p) \|_H^2 + \\
&\sum_{j=1}^{n} \frac{1+\tau_j r}{\tau_j} \| T_j q - T_j p \|_H^2 + \sum_{j=1}^{n} \delta_j (A_j q_j - A_j p_j, q_j - p_j)_H + \\
&r \sum_{j=1}^{n} \| (T_j q - B_j T_0 q) - (T_j p - B_j T_0 p) \|_H^2 + \\
&\sum_{j=1}^{n} \frac{1}{(1-\tau_j r)\tau_j} \| (1-\tau_j r)[(q_j - T_j q) - (p_j - T_j p)] - \tau_j (A_j q_j - A_j p_j) \|_H^2 + \\
&\frac{1}{r} \sum_{j=1}^{n} \| T_{n+j} q - T_{n+j} p \|_H^2 \leqslant \frac{1-\tau_0 r}{\tau_0} \|q_0 - p_0\|_V^2 + r \sum_{j=1}^{n} \| q_j - p_j \|_H^2 + \\
&r \sum_{j=1}^{n} \| B_j (T_0 q - T_0 p) \|_H^2 - 2 \sum_{j=1}^{n} (q_{n+j} - p_{n+j}, B_j (T_0 q - T_0 p))_H + \\
&\sum_{j=1}^{n} \frac{1-\tau_j r}{\tau_j} \| q_j - p_j \|_H^2 + 2 \sum_{j=1}^{n} (q_{n+j} - p_{n+j}, T_j q - T_j p)_H + \\
&r \sum_{j=1}^{n} \| T_j q - T_j p \|_H^2 + r \sum_{j=1}^{n} \| B_j (T_0 q - T_0 p) \|_H^2 + \\
&\frac{1}{r} \sum_{j=1}^{n} \| q_{n+j} - p_{n+j} \|_H^2 + 2 \sum_{j=1}^{n} (q_{n+j} - p_{n+j}, B_j (T_0 q - T_0 p))_H - \\
&2 \sum_{j=1}^{n} (q_{n+j} - p_{n+j}, T_j q - T_j p)_H + r \sum_{j=1}^{n} \| B_j (T_0 q - T_0 p) - (T_j q - T_j p) \|_H^2 .
\end{aligned}$$

Then by virtue of (11)

$$\sum_{j=1}^{n} \| B_j \eta \|_H^2 = \| \eta \|_V^2 \qquad \forall \eta \in V. \tag{40}$$

Taking in account this equation we have

$$\frac{1+\tau_0 r}{\tau_0}\, \| T_0 q - T_0 p \|_V^2 + \sum_{j=1}^{n} \frac{1+\tau_j r}{\tau_j}\, \| T_j q - T_j p \|_H^2 + \frac{1}{r} \sum_{j=1}^{n} \| T_{n+j} q - T_{n+j} p \|_H^2 +$$

$$\delta_0\, (A_0 q_0 - A_0 p_0, q_0 - p_0)_V + \sum_{j=1}^{n} \delta_j\, (A_j q_j - A_j p_j, q_j - p_j)_H +$$

$$\sum_{j=1}^{n} \frac{1}{(1-\tau_j r)\tau_j}\, \| (1-\tau_j r)\, [(q_j - T_j q) - (p_j - T_j p)] - \tau_j (A_j q_j - A_j p_j) \|_H^2 +$$

$$\frac{1}{(1-\tau_0 r)\,\tau_0}\, \|(1-\tau_0 r)\, [(q_0 - T_0 q) - (p_0 - T_0 p)] - \tau_0 (A_0 q_0 - A_0 p_0) \|_V^2 +$$

$$r \sum_{j=1}^{n} \| (q_j - B_j T_0 q) - (p_j - B_j T_0 p) \|_H^2 \leqslant$$

$$\frac{1-\tau_0 r}{\tau_0}\, \|q_0 - p_0 \|_V^2 + \sum_{j=1}^{n} \frac{1}{\tau_j}\, \| q_j - p_j \|_H^2 + \frac{1}{r} \sum_{j=1}^{n} \| q_{n+j} - p_{n+j} \|_H^2 ,$$

i.e., the inequality (33) is true. The proof of the theorem is complete. □

Recall (see [6]), that the operator $T : Q \to Q$ is named the asymptotically regular if $T^{k+1} q - T^k q \to 0$ as $k \to +\infty$ for any $q \in Q$.

It is valid the following

Theorem 4. *Let the operator* T *has at least one fixed point and let the conditions (9)–(11), (32) are hold. Then the iterative sequence* $\{q^{(k)}\}_{k=0}^{+\infty}$, *constructed according to (19), converges weakly to* q^* *in* Q *as* $k \to +\infty$, q^* *is a fixed point of the operator* T, *the relation*

$$\lim_{k\to+\infty} \left\| y_j^{(k)} - B_j\, u^{(k)} \right\|_H = 0, \qquad j = 1, 2, \ldots, n, \tag{41}$$

is valid and the operator $T : Q \to Q$ *is an asymptotically regular; i.e.,*

$$\lim_{k\to+\infty} \left\| q^{(k+1)} - q^{(k)} \right\|_Q = 0. \tag{42}$$

Proof. We use the inequality (33) with $q = q^{(k)}$ assuming that p is a fixed point of the operator T (the existence of at least one fixed point is provided by the assumptions of the theorem). Since $Tq^{(k)} = q^{(k+1)}$ by the definition of the iterative sequence, $p_j = T_j p$, $j = 0, 1, 2, \ldots, n$, for a fixed point, and, by Theorem 1 $p_j = B_j T_0 p = B_j p_0$, $j = 1, 2, \ldots, n$, we have

$$\left\| q^{(k+1)} - p \right\|_Q^2 + \delta_0 \left(A_0 u^{(k)} - A_0 p_0 \, , \; u^{(k)} - p_0 \right)_V +$$
$$\sum_{j=1}^{n} \delta_j \left(A_j y_j^{(k)} - A_j p_j \, , \, y_j^{(k)} - p_j \right)_H +$$
$$\frac{1}{\tau_0 (1 - \tau_0 r)} \left\| (1 - \tau_0 r) (u^{(k)} - u^{(k+1)}) - \tau_0 (A_0 u^{(k)} - A_0 p_0) \right\|_V^2 +$$
$$\sum_{j=1}^{n} \frac{1}{\tau_j (1 - \tau_j r)} \left\| (1 - \tau_j r) (y_j^{(k)} - y_j^{(k+1)}) - \tau_j (A_j y_j^{(k)} - A_j p_j) \right\|_H^2 +$$
$$r \sum_{j=1}^{n} \left\| y_j^{(k)} - B_j u^{(k+1)} \right\|_H^2 \leqslant \left\| q^{(k)} - p \right\|_Q^2 ,$$

This inequality implies that the numerical sequence $\left\{ \| q^{(k)} - p \|_Q \right\}_{k=0}^{+\infty}$ is non increasing and hence have a finite limit:

$$\lim_{k \to +\infty} \left\| q^{(k)} - p \right\|_Q < +\infty,$$

therefore

$$\lim_{k \to +\infty} \left(A_0 u^{(k)} - A_0 p_0 \, , \; u^{(k)} - p_0 \right)_V = 0, \tag{43}$$

$$\lim_{k \to +\infty} \left(A_j y_j^{(k)} - A_j p_j \, , \; y_j^{(k)} - p_j \right)_H = 0, \; j = 1, 2, \ldots, n, \tag{44}$$

$$\lim_{k \to +\infty} \left\| y_j^{(k)} - B_j u^{(k+1)} \right\|_H = 0, \quad j = 1, 2, \ldots, n, \tag{45}$$

$$\lim_{k \to +\infty} \left\| (1 - \tau_0 r) \left(u^{(k)} - u^{(k+1)} \right) - \tau_0 \left(A_0 u^{(k)} - A_0 p_0 \right) \right\|_V = 0, \tag{46}$$

$$\lim_{k \to \infty} \left\| (1 - \tau_j r) \left(y_j^{(k)} - y_j^{(k+1)} \right) - \tau_j \left(A_j y_j^{(k)} - A_j p_j \right) \right\|_H = 0, \; j = 1, 2, \ldots, n. \tag{47}$$

By using (9), (10), (43) and (44), we obtain

$$\lim_{k \to \infty} \left\| A_0 u^{(k)} - A_0 p_0 \right\|_V = 0, \; \lim_{k \to \infty} \left\| A_j y_j^{(k)} - A_j p_j \right\|_H = 0, \; j = 1, 2, \ldots, n. \tag{48}$$

It follows from (46) – (48) that

$$\lim_{k \to +\infty} \left\| u^{(k)} - u^{(k+1)} \right\|_V = 0, \quad \lim_{k \to +\infty} \left\| y_j^{(k)} - y_j^{(k+1)} \right\|_H = 0, \quad j = 1, 2, \ldots, n. \tag{49}$$

Further by using (40), (45), (49), from the inequality

$$\left\| y_j^{(k)} - B_j\, u^{(k)} \right\|_H \leqslant \left\| y_j^{(k)} - B_j\, u^{(k+1)} \right\|_H + \left\| B_j\, (u^{(k)} - u^{(k+1)}) \right\|_H \leqslant$$
$$\left\| y_j^{(k)} - B_j\, u^{(k+1)} \right\|_H + \left(\sum_{i=1}^{n} \left\| B_i\, (u^{(k)} - u^{(k+1)}) \right\|_H^2 \right)^{1/2} =$$
$$\left\| y_j^{(k)} - B_j\, u^{(k+1)} \right\|_H + \left\| u^{(k)} - u^{(k+1)} \right\|_V, \quad j = 1, 2, \dots, n,$$

we obtain (41). It follows from (17) and (41) that

$$\lim_{k\to\infty} \left\| \lambda_j^{(k)} - \lambda_j^{(k+1)} \right\|_H = r \lim_{k\to\infty} \left\| y_j^{(k+1)} - B_j u^{(k+1)} \right\|_H = 0, \ j = 1, 2, \dots, n. \tag{50}$$

Relations (49), (50) imply that the condition (42) is satisfied, i.e., T is an asymptotical regular operator. Since, by addition, by the assumptions of the Theorem, the operator T have a non empty set of fixed points and, by Theorem 3, is non expanding operator, it follows from [7] that the iterative sequence $\{q^{(k)}\}_{k=0}^{+\infty}$ constructed by (19) is weakly converges in Q as $k \to +\infty$. Its limit q^* is the fixed point of the operator T. The proof of the theorem is complete. □

Note that if the assumptions of the Theorem 1 are valid, then it follows from Theorems 2, 4 that the sequences $\{u^{(k)}\}_{k=0}^{+\infty}$ and $\{y_j^{(k)}\}_{k=0}^{+\infty}$, constructed by (12)–(14) are is weakly converge to u and $B_j\, u$, $j = 1, 2, \dots, n$ in V and H, respectively, as $k \to +\infty$, where u is a solution of variational inequality (8).

4 Application of the iterative method to the problems (1)–(3)

Let us apply the suggested iterative methods (12)–(14) to the problems (1)–(3). Since in (14) calculations are performed by explicit formulas, it is sufficient to consider only the problems (12), (13).

Since F_0 is a linear functional, the variational inequality (12) by standard way can be rewritten in the form

$$\frac{1}{\tau_0} \left(u^{(k+1)} - u^{(k)}, \eta \right)_V +$$
$$\left(A_0\, u^{(k)} - \widehat{f} + r u^{(k)} + \sum_{j=1}^{n} B_j^* \left(\lambda_j^{(k)} - r y_j^{(k)} \right), \eta \right)_V = 0 \quad \forall \eta \in V,$$

where the element $\widehat{f} \in V$ is defined by the formula

$$(\widehat{f}, \eta)_V = \int_\Omega (f, \eta)\, dx, \quad \eta \in V.$$

Thus the first step of the iterative process can be reduced to solving of n Dirichlet problems for Poisson equation.

Further, for each $j = 1, 2, \dots, n$, let us rewrite variational inequality (13) in the form

$$\left(y_j^{(k+1)}, z - y_j^{(k+1)}\right)_H + G_j(z) - G_j\left(y_j^{(k+1)}\right) \geqslant 0 \quad \forall z \in H, \tag{51}$$

where

$$G_j(z) = \tau_j F_j(z) - \left(y_j^{(k)} - \tau_j \left[A_j y_j^{(k)} - \lambda_j^{(k)} + r\left(y_j^{(k)} - B_j u^{(k+1)}\right)\right], z\right)_H .$$

By using the definition of a proximal mapping we obtain that the variational inequality (51) is equivalent to a following minimization problem

$$\frac{1}{2}\|z\|_H^2 + G_j(z) \geqslant \frac{1}{2}\left\|y_j^{(k+1)}\right\|_H^2 + G_j\left(y_j^{(k+1)}\right) \quad \forall z \in H,$$

or

$$\frac{1}{2\tau_j}\|z\|_H^2 + F_j(z) - \frac{1}{2\tau_j}\left\|y_j^{(k+1)}\right\|_H^2 - F_j\left(y_j^{(k+1)}\right) \geqslant$$
$$\left(\frac{1}{\tau_j} y_j^{(k)} - \left[A_j y_j^{(k)} - \lambda_j^{(k)} + r\left(y_j^{(k)} - B_j u^{(k+1)}\right)\right], z - y_j^{(k+1)}\right)_H \quad \forall z \in H,$$

i.e.,

$$\frac{1}{\tau_j} y_j^{(k)} - \left[A_j y_j^{(k)} - \lambda_j^{(k)} + r\left(y_j^{(k)} - B_j u^{(k+1)}\right)\right] \in \partial \widehat{F}_j\left(y_j^{(k+1)}\right), \tag{52}$$

where

$$\widehat{F}_j(z) = \frac{1}{2\tau_j}\|z\|_H^2 + F_j(z).$$

It is known (see [5]), that $p \in \partial \widehat{F}_j(z)$ if and only if $z \in \partial \widehat{F}_j^*(q)$, where $\widehat{F}_j^*$ is a functional conjugate to $\widehat{F}_j$ (see, e.g., [5]). So the inclusion (52) is equivalent to the following one:

$$y_j^{(k+1)} \in \partial \widehat{F}_j^*\left(\frac{1}{\tau_j} y_j^{(k)} - \left[A_j y_j^{(k)} - \lambda_j^{(k)} + r\left(y_j^{(k)} - B_j u^{(k+1)}\right)\right]\right). \tag{53}$$

Since

$$\widehat{F}_j(z) = \int_\Omega \int_0^{|z|} g_{\tau_j}(\xi)\, d\xi\, dx, \quad g_{\tau_j}(\xi) = \begin{cases} \xi/\tau_j, & \xi \leqslant \beta_j, \\ \xi/\tau_j + \vartheta_j, & \xi \geqslant \beta_j, \end{cases}$$

then it is not difficult to check that

$$\widehat{F}_j^*(z) = \int_\Omega \int_0^{|z|} \varphi_{\tau_j}(\xi)\, d\xi\, dx, \quad \varphi_{\tau_j}(\xi) = \begin{cases} \tau_j \xi, & \xi \leqslant \beta_j/\tau_j, \\ \beta_j, & \beta_j/\tau_j < \xi \leqslant \beta_j/\tau_j + \vartheta_j, \\ \tau_j(\xi + \vartheta_j), & \xi \geqslant \beta_j/\tau_j + \vartheta_j. \end{cases}$$

Then we obtain that functional $\widehat{F}_j^*$ is Gato differentiable, moreover,

$$(\widehat{F}_j^*)'(z) = \frac{\varphi_{\tau_j}(|z|)}{|z|}\,z,$$

hence by virtue of the Proposition 5.3 [5] the subdifferential $\partial\widehat{F}_j^*(z)$ contains unique element coinciding with $(\widehat{F}_j^*)'(z)$. So calculations by (53) are performed by explicit formulas.

References

1. I.B. Badriev, O.A. Zadvornov *A Decomposition Method for Variational Inequalities of the Second Kind with Strongly Inverse-Monotone Operators*, Differential inequalities, Pleaiades Publishing, Inc, 2003, 39, pp. 936–944.
2. O.A. Zadvornov *On the Convergence of the Semi-implicite Method for Solving the Variational Inqualities of the Second Kind*, Izvestiya Vyzov. Matematica, 2005, 6, pp. 61–70 (in Russian).
3. I.B. Badriev, O.A. Zadvornov *On the convergence of Dual-Type Iterative Method for Mixed Variational Inqualities*, Differential inequalities, Pleaiades Publishing, Inc, 2006, 42(8), pp. 1180–1188.
4. I.B. Badriev, O.A. Zadvornov, A.M. Saddek *Convergence Analysis of Iterative Methods for Some Variational Inqualities with Pseudomonotone Operators*, Differential inequalities, Pleaiades Publishing, Inc, 2001, 37(7), pp. 934–942.
5. Ekeland I., Temam R. *Convex Analysis and Variational Problems*, North-Holland Publishing Company, Amsterdam, 1976.
6. Browder F.E., Petryshin W.V. *The solution by iteration of nonlinear functional equations in Banach spaces*, Bull. Amer. Math. Soc., 1966, V. 72, pp. 571-575.
7. Opial Z. *Weak convergence of the sequence of successive approximations for nonexpansive mappings*, Bull. Amer. Math. Soc., 1967, V. 73, pp. 591–597.
8. Gaewskii H., Gröger K., Zacharias K. *Nichtlineare Operatorgleichungen und Operatordifferentialgleichungen*, Berlin: Academie-Verlag, 1974.

Multilevel Algorithm for Graph Partitioning

N. S. Bochkarev, O. V. Diyankov*, and V. Y. Pravilnikov

Neurok TechSoft LLC, Russia
*diyankov@aconts.com

Abstract. A class of multilevel algorithms for partitioning of a sparse matrix prior to parallel solution of a system of linear equations is described. This matrix partitioning problem can be described in terms of a graph partitioning problem which is known to be NP-hard, so several heuristics for its solution have been proposed in the past decades. For this purpose we use the multilevel algorithm proposed by B. Hendrickson and R. Leland [2] and further developed by G. Karypis and V. Kumar [3]. This algorithm is very efficient and tends to produce high quality partitioning for a wide range of matrices arising in many practical applications.

Keywords: graph partitioning, parallel computations, load balancing.

1 Introduction

Efficient algorithms for graph partitioning are critical for scientific simulations on high-performance parallel computers. For example, parallel iterative solution of a linear system of equations

$$Ax = b$$

where A is a large sparse matrix, b is a right-hand side and x is a vector of unknowns is based on the partitioning of the matrix A. The main purpose of the partitioning procedure is to divide the matrix A into required number of parts (stripes) in such a way that each part has approximately the same number of rows and the number of interprocess communications performed during the parallel solution is kept as small as possible. This class of problems can be strictly described in terms of the *graph partitioning problem* (see sec. 2), which is known to be NP-hard, so several heuristics for its solution have been developed in the past decades. They can be subdivided into three main categories. The first one contains so-called *spectral algorithms.* While achieving partitioning of a very good quality, they require a large amount of hardware resources (CPU cycles and memory) because of the necessity to find eigenvector corresponding to the second largest eigenvalue of the Laplacian matrix of the adjacency graph of A. The second group contains *greedy algorithms* which find graphs partitioning by sequentially adding nodes to growing subsets following some greedy strategy such as minimizing the number of cut edges (see sec. 2) at each step.

The third group contains *multilevel (ML) algorithms* which are among the best ones in terms of partitioning quality and computational resources requirements, which is very important as problems become larger. The *ML* approach itself was originally proposed by B. Hendrickson and R. Leland [2] and further developed by G. Karypis and V. Kumar [3] in their *METIS* package. In this paper we describe its analogue and present numerical tests results. The remainder of the paper is organized as follows. In section 2 we define the graph partitioning problem. In section 3 the main idea behind multilevel techniques is demonstrated. In sections 4, 5, and 6 we describe in details different phases of the multilevel approach - coarsening, initial partitioning and uncoarsening respectively. In section 7 we present a variant of the original *ML* approach which is called *Cell-Based Multilevel (CBML)* approach and describe its advantages. Section 8 presents numerical tests results. Section 9 provides a summary of the tests.

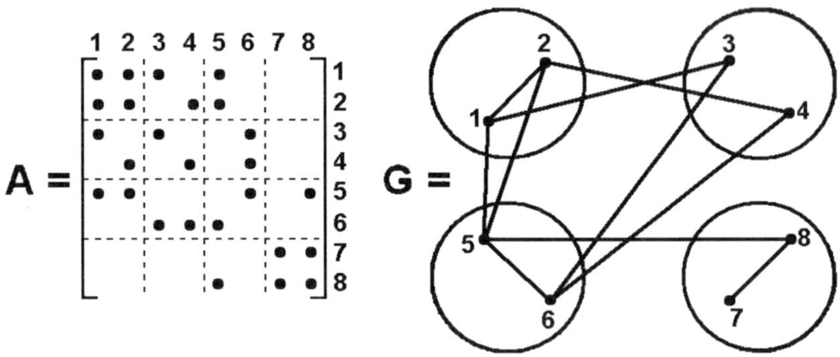

Fig. 1. Sparse matrix and its adjacency graph.

2 The problem statement

It is well known that nonzero pattern of a sparse matrix can be represented by its adjacency graph (see Fig. 1). Namely, given a square $n \times n$ sparse matrix A containing nz nonzero entries, its adjacency graph is $G = (V, E)$, where V is the set of nodes corresponding to the rows of A ($|V| = n$), and E is the set of edges corresponding to the nonzero entries of A ($|E| = nz$). The graph is undirected when A is symmetric and directed otherwise. In the following paragraphs we assume that A is symmetric. This restriction is easy to fulfil by considering the matrix $A^* = A + A^T$ and its adjacency graph instead of A since the exact values of the matrix's nonzero entries are unimportant. The k-*way graph partitioning problem* is formulated as follows: partition V into k disjoint subsets $\{V_1, V_2, ..., V_n\}$ such that $|V_i| \approx \left|\frac{V}{k}\right|$ for $i = \overline{1..k}$ *(load-balancing condition)*, while minimizing the number of edges whose incident nodes belong to different partitions *(cut-size minimization condition)* (see Fig. 1). These edges are called *cut edges* and their number is called *cut size*. This problem can be trivially extended to graphs with weights assigned to the nodes and edges (see [3]). When the number of parts is a power of 2, i.e. $k = 2^p$, the problem is frequently solved in a *recursive bisection* fashion. Namely, we first obtain 2-way partitioning of our graph: $V = V_1 \bigcup V_2$. Then, we recursively apply the same procedure to subgraphs of G induced by V_1 and V_2. After p steps the original graph is partitioned into k parts. It's worth to say that this approach often works worse than the original k-way partitioning approach when $k > 2$, but still frequently used due to its simplicity. In this paper we describe the original k-way partitioning approach and propose some improvements. We estimate *partitioning quality* depending on the degree the load-balancing and cut size minimization conditions are fulfilled.

3 Multilevel k-way graphs partitioning

The whole procedure may be depicted by Fig. 2. It consists of the following main phases:

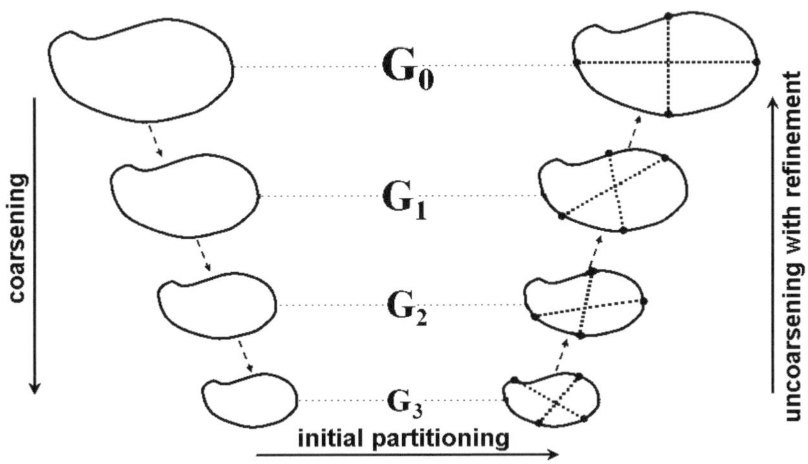

Fig. 2. Multilevel graph partitioning algorithm.

1. **Coarsening phase.** During the coarsening phase, a sequence of smaller graphs $\{G_1, G_2, ..., G_m\}$ is constructed until the number of nodes in

the coarsest graph G_m becomes less than some predefined value (around a few hundreds). The number of graphs in this sequence is called *coarsening depth.* We use special parameter ν to control the coarsening depth. Namely, we try to build a coarser graph G_{i+1} from a finer one G_i until $|V_i| \geqslant \nu * |V_1|$. Each graph G_i forms a layer of coarsening. That is why this approach is called "multilevel". At each layer, possibly except the first one, weights are assigned to the nodes and edges of the graphs (see sec. 4) in order to partitioning of the coarsest graph be good with respect to the original one. There are many possibilities to construct a coarser graph from a finer one. But we use edges collapsing technique that is based on *matchings* (see sec. 4).

2. **Initial partitioning phase.** During the initial partitioning phase, high quality partitioning of the coarsest graph G_m is computed. Since the number of nodes of G_m is small comparing with that in G_1 this phase can be accomplished very quickly. Actually, it takes about 10% of the total partitioning time. There exists many algorithms to do this (see [3], [5]). For this we use *Restarted Greedy Graph Growing (RGGG)* algorithm which is described in details in sec. 5.
3. **Uncoarsening phase with refinement.** During this phase, just found partitioning of the coarsest graph G_m is projected back to the original graph G_1 by going through the set of intermediate graphs. On each layer just projected partitioning is refined. There are many *local refinement algorithms* intended to do this. It is worth to mention *Kernigan-Lin refinement algorithm* [5] and it's linear-time variant - *Fiduccia-Mattheyses refinement algorithm* [1]. In our multilevel approach we use a variant of the original Fiduccia-Mattheyses local refinement algorithm which is called *boundary Fiduccia-Mattheyses local refinement algorithm.*

4 Coarsening phase

Given a weighted graph G_i with weights assigned to the nodes and the edges, the next level coarser graph G_{i+1} is constructed from it by merging together some subsets of its nodes $\{v_{j_1}^{(i)}, v_{j_2}^{(i)}, \dots, v_{j_k}^{(i)}\}$ (ancestors) into multinodes $v_j^{(i+1)}$ (descendants). The weight of $v_j^{(i+1)}$ equals to the sum of weights of $\{v_{j_1}^{(i)}, v_{j_2}^{(i)}, \dots, v_{j_k}^{(i)}\}$. In the case when more than one node of $\{v_{j_1}^{(i)}, v_{j_2}^{(i)}, \dots, v_{j_k}^{(i)}\}$ contain edges incident to the same node $u \not\subseteq \{v_{j_1}^{(i)}, v_{j_2}^{(i)}, \dots, v_{j_k}^{(i)}\}$, the weight of the edge $(v_j^{(i+1)}, u)$ equals to the sum of the weights of these edges. It is obvious that a coarser graph can be constructed from a finer one in many different ways. For matrices with unstructured nonzero patterns it seems reasonable to use coarsening procedure based on collapsing together the edges of G that form a *matching*, because of the necessity to preserve connectivity structure of the original graph in the coarsest one. Remind that a *matching* in a graph (weighted or unweighted) is a subset of its edges with the following property: no two of which are incident to the same node. A matching is called *maximal* if it is impossible to add one more edge to it such that the resulting subset of edges forms a matching too. The maximal matching that has the maximum number of edges is called *maximum matching.* Since the goal of the coarsening procedure is to decrease the size of the graph, matching should contain a large number of edges. But the complexity of computing maximum matching is higher than that of computing maximal matching. That is why we construct maximal matchings during the coarsening phase. They can be generated very quickly using depth-first search [6] or randomized algorithm. We implemented the following three types of matchings:

- Random matching (RM). This type of matchings is very popular because of its simplicity and often gives good results. It is demonstrated by the following pseudocode:

  Algorithm 3. Random Matching Algorithm
  INPUT: graph G(V,E)
  OUTPUT: matching M

  ```
  1.    matching M = ∅;
  2.    foreach( u ∈ V ) mask[u] = 0;
  3.    foreach( u ∈ V ) {
  4.       if( 0 == mask[u] ) {
  5.          mask[u] = 1;
  6.          if( exist v ∈ adj[u] such that 0 == mask[v] ) {
  7.             mask[v] = 1;
  8.             M ⟵ (u, v);
  9.          }
  10.      }
  11.  }
  ```

  Initially, the matching is empty (line 1) and all nodes are unmasked (line 2). Then, the nodes are visited in random order (line 3). If node u is already masked it is skipped. Otherwise, it is masked (lines 4, 5) and then we arbitrary select its adjacent unmasked node v if such a node exists (line 6), mask it (line 7) and add the edge (u, v) to the matching. Obviously, that this algorithm has a linear time complexity with respect to the number of nodes, i.e $O(|V|)$.
- Heavy-Edge Matching (HEM). As in the previous algorithm, the nodes are visited in random order. But now we select unmasked node v adjacent to u in such a way that the weight of the edge (u, v) is maximal over all

unmatched adjacent edges. The algorithm can be illustrated by the following pseudocode:

Algorithm 4. Heavy-Edge Matching Algorithm

```
INPUT: graph G(V,E)
OUTPUT: matching M
matching M = ∅;
foreach( u ∈ V ) mask[u] = 0;
foreach( u ∈ V ) {
   if( 0 == mask[u] ) {
      mask[u] = 1;
      if( exist v ∈ adj[u] such that 0 == mask[v] and w(u,v) ⟶ max ) {
         mask[v] = 1;
         M ⟵ (u, v);
      }
   }
}
```

This algorithm has linear time complexity with respect to the number of edges, i.e $O(|E|)$.

- Heavy-Clique Matching (HCM). In this section we describe our version of heavy-clique matching algorithm. The algorithm can be effective for graphs with a few highly-connected components [6]. In [3] one can find a variant of HCM algorithm based on the concept of *edges density*. In contrast to this algorithm, we developed our own one. Remind that for undirected graph G one can define the concept of *degree of a node* [6], which gives the number of edges incident to the node. The algorithm can be demonstrated by the following pseudocode:

Algorithm 5. Heavy-Clique Matching Algorithm

```
INPUT: graph G(V,E)
OUTPUT: matching M
matching M = ∅;
for( i = 1; G ≠ ∅; i++ ) {
   if( min_{u∈V_i}(deg[u]) == 1 ) break;
   build G_i from G such that min_{u⊂V_i}(deg[u]) is as large as possible;
   G = G\G_i;
}
foreach( G_i ) M ⟵ HEM(G_i);
M ⟵ HEM(G);
```

Initially, the matching is empty (line 1). In line 4 we try to build subgraph G_i from a given one G which has the following property: $\min\limits_{u \subset V_i}(deg[u])$ *is as large as possible*, where deg[u] is the degree of node u. In other words, we try to extract subgraph from G in which the minimal degree of a node is as large as possible. The operation in line 4 can be implemented in $O(|E|)$ by the algorithm which is described below (see *Maximal Minimum Degree Subgraph Extraction* algorithm). Then we perform the same procedure for the subgraph of G which is induced by the set of nodes $\{V \backslash V_i\}$ until the condition in line 3 is satisfied or the subgraph becomes empty (loop in lines 2-6). As a result, we obtain the sequence of graphs $\{G_1, G_2, ..., G_q$ $\}$ and the remaining part of input graph which consists of isolated nodes or isolated pairs of nodes. After that we build required matching M as a conjunction of heavy-edge matchings for all graphs G_i and G. Let's consider the algorithm which demonstrates how we can build a subgraph of G with maximal minimum degree (the operation in line 4) in $O(|E|)$ time:

Algorithm 6. Maximal Minimum Degree Subgraph Extraction
INPUT: graph G(V,E)
OUTPUT: subgraph $G^* \in G$

```
1.  sort nodes of G in degrees-ascending order;
2.  for( u = 0; u < |V|; u++ ) {
3.     save D[u] <--- deg[u];
5.     G = G\u;
6.     maintain nodes degrees-ascending order;
7.  }
8.  find u* : D[u*] = max_u(D[u]);
9.  while( u < u* ) G = G\{u};
10. G* = G;
```

In line 1 we sort the nodes of G in degrees ascending order. In the loop in lines 2-7 we visit sorted nodes one at a time, take the node with minimum degree, save its degree and then exclude it with its incident edges from G. It is necessary to recalculate the degrees of the remaining nodes and maintain their degrees ascending order. In line 8 we find maximal value of all degrees that were saved in line 3 and corresponding node u^*. Then the output of the algorithm is obtained by removing the nodes of G which were visited before u^* in the loop in lines 2-7.

We considered three algorithms for matchings generation. In [3] one can find other ones. When the edges are unweighted (or have the same weight) it seems reasonable to use RM. In order to construct a coarser graph from a finer one we need to collapse together matched edges. This procedure can also be implemented in $O(|E|)$. It is worth to note that coarsening phase usually takes about 80% of the total partitioning time.

5 Initial partitioning phase

During this phase high quality partitioning of the coarsest graph G_m is constructed. Since G_m has quite a small number of nodes, this phase takes quite a small amount of time. For this we use *Restarted Greedy Graph Growing* algorithm. The algorithm can be outlined by the following pseudocode:

Algorithm 7. Restarted Greedy Graph Growing Algorithm
INPUT: graph G, the number of parts k, the number of restarts rests
OUTPUT: partitioning of G into k parts

```
1.  while( rests-- ) {
2.     put all nodes in partition P0;
3.     for( j = 1; j < k; j++ ) {
4.        randomly select u ∈ P0 and put it in Pi;
5.        while( size[Pi] < V/k ) {
6.           select u ∈ P0 such that cutsize ⟶ min;
7.           move u from P0 to Pk;
8.        }
9.     }
10.    save partitioning;
11. }
```

At the beginning of each restart we put all nodes in partition P_0 (line 2). In order to construct partition P_i, we first randomly select a node from partition P_0 and put it in partition P_i (*growing subset*) which was empty before it (line 4). Then we sequentially move nodes from P_0 to P_i in such a way that each movement results in the smallest possible increase in the cut size (lines 6,7). We continue this until the size of P_i becomes more than or equal to $|\frac{V}{k}|$. Then we try to construct partition P_{i+1} in the same way (loop in lines 3-9). It is obvious that in order to construct k-way partitioning of the graph we must construct $k-1$ partitions. After that, remaining nodes in partition P_0 form missing k-th partition. After the required number of restarts is finished, we use the best partitioning as the result. This procedure can be used as a standalone partitioner, but greedy algorithms often give partitioning of a poor quality and the required amount of time often exceeds the amount of time required by multilevel partitioner.

6 Uncoarsening with refinement phase

This phase consists of the following two steps. First, partitioning of the coarsest graph is projected back to the original graph by going through intermediate graphs. Since each node of G_{i+1} is formed by a distinct subset of nodes of G_i, the projection is trivial to realize. Namely, we can derive partitioning of G_i from partitioning of G_{i+1} by assigning to the set of nodes $\{v_{j_1}^{(i)}, v_{j_2}^{(i)}, \ldots, v_{j_k}^{(i)}\}$ that collapsed into $v_j^{(i+1)}$ the partition number that holds $v_j^{(i+1)}$. The next step is a refinement procedure for just found partitioning of G_i. We use a modification of the original *local Fiduccia-Mattheyses refinement algorithm* [1] which we call *boundary Fiduccia-Mattheyses refinement algorithm*. The central concept behind refinement algorithms is the concept of *gain of a node*. Given a node u which belongs to the partition P_i, the gain of movement of node u from P_i to P_j $(i \neq j)$ is given by the following formula:

$$\text{gain}(P_i \underset{u}{\longrightarrow} P_j) = \sum_{v \subset P_j} w_{(u,v)} - \sum_{v \subset P_i} w_{(u,v)},$$

where $w_{(u,v)}$ is the weight of the edge (u, v). In other words, $\text{gain}(P_i \underset{u}{\longrightarrow} P_j)$ gives the decrease in the cut size we obtain after the movement is performed. The boundary Fiduccia-Mattheyses refinement algorithm may be outlined by the followig pseudocode:

Algorithm 8. Boundary Fiduccia-Mattheyses Refinement Algorithm
INPUT: graph G(V,E), the number of parts k, the number of restarts rests
OUTPUT: partitioning of G into k parts

```
while( rests-- ) {
    unlock all nodes V;
    for( p = 0; p < k; p++ ) {
        put all boundary nodes from P_i to PQ_i;
    }
    while( all PQs are not empty ) {
        from all PQs find i, j and u ⊂ P_i such that gain(P_i -u-> P_j) --> max;
        if( gain(P_i -u-> P_j) < 0 ) break;
        if( after the movement P_i and P_j are still balanced ) {
            move P_i -u-> P_j;
            adjust gains of all unlocked nodes v ⊂ adj[u];
        }
        else {
            remove u from PQ_i;
        }
    }
}
```

The boundary Fiduccia-Mattheyses refinement algorithm is iterative in nature. The number of iterations (or restarts) is controlled by the $rests$ parameter. We maintain k priority queues to hold boundary nodes from each partition that are allowed to move, i.e. unlocked (see [1] for explanation). Initially all the nodes are unlocked (line 2). At lines 3-5 we initialize all queues with boundary nodes from corresponding partitions. As a key of a node we use the maximum gain from all gains considered to allowable movements of that node, i.e. movements from P_i to P_j, $i \neq j$. Then, in the loop in lines 6-16 we look at the tops of all queues and select the node that has the maximum value of key (line 7). After that we know all information that is necessary to perform just found movement. If the gain of the movement is negative we exit from the loop in lines 6-16 because only movements with positive gains can refine partitioning. It is worth to note that due to the load-balancing condition such a movement may be not allowed (it is controlled by line 9). In this case we simply remove the node from its queue and perform this procedure again. We exit from the loop in lines 6-16 only if there is no movements that preserve load-balancing and decrease the cut size. An advantage of this algorithm over the one described in [1] is its time complexity. It can be approximated by the formula $O(|N^* * S|)$ where N^* is the number of *boundary nodes* (nodes which have at least one adjacent node that belongs to another partition) and S is the average sparsity.

7 Cell-based multilevel approach

In this section we introduce a new multilevel technique for sparse matrix partitioning. We call it *cell-based multilevel (CBML)* partitioning algorithm. It is effective for large sparse matrices arising from such discretization of PDEs in which several unknowns are related to each grid cell. Let us consider such a matrix A. In the case when A is a multiblock matrix we consider one of its blocks. The adjacency graph G of A has special connectivity structure. Namely, the set of its nodes V can be represented as $V = \bigcup V_\mu$ in which each subset V_μ has the following property: all the nodes $u \subset V_\mu$ are *indistinguishable* (remind that two nodes v and u are called indistinguishable if $adj[u] \bigcup u = adj[v] \bigcup v$). This information about connectivity pattern may be employed to find better partitioning compared with that generated by the original algorithm. Namely we can consider so-called *reduced* graph G^* of G in which each node $v \subset V^*$ corresponds to subset V_μ, has the same connectivity structure as any node in V_μ and the weight $w[v] = |V_\mu|$. Then we apply multilevel technique to reduced graph G^*. After partitioning of G^* is generated, partitioning of G can be derived from it since each node of G^* is formed from distinct subset of nodes of G. In the case when $\frac{|V^*|}{|V|} \ll 1$ this approach seems to produce better partitioning compared with that generated by the original multilevel algorithm applied to G (see Fig. 8 in the numerical test results section).

8 Numerical test results

In this section we present the results of comparison of our partitioner with *METIS* package which can be downloaded from

http://www.glaros.dtc.umn.edu/gkhome/metis/metis/download.

All tests are performed on Opteron 2.0 GHz, with 2 Gb RAM running under SLES 9. We use a public XOM matrix collection which can be downloaded from http://www.aconts.com/XOMMatrices. Table 1 demonstrates test matrices properties.

Table 1. Summary of public XOM matrix collection

Problem	N	Z	Z/N	ZD	ND	SD	POD
CI-1	113465	1654732	14,58	0	2	23619	477535
CI-2	62449	460319	7,37	0	0	128	202
CIT-1	17436	344245	19,74	0	4207	7388	98092
CIT-2	249428	5613978	22,51	30	1323	16106	1024619
SBO-1	21700	145122	6,69	1	0	1	7
SBO-2	111756	888190	7,95	0	0	8	12
SBO-3	216051	1849317	8,56	0	0	0	0
SBO-4	93264	667882	7,16	0	0	0	0
SEO-1	22421	204784	9,13	0	180	94	849

Here N is the number of rows, Z is the number of nonzero entries, $S = Z/N$ is the average sparsity, ZD is the number of zero diagonal entries, ND is the number of negative diagonal entries, SD is the number of "small" diagonal entries (i.e. $|a_{ii}| < 0.01 * \sum_{j \neq i} |a_{ij}|$), POD is the number of positive off-diagonal entries.

- **Matching algorithms impact on partitioning quality.** The main obsevation behind these tests is that all algorithms for matching generation give good results with respect to partitioning quality and time requrements. But it seems reasonable to use "heavy" matchings (i.e. *HEM* or *HCM*) on "deep" layers of coarsing where weights are assigned to the nodes and the edges. As the experiments show, HEM is the best algorithm which results in very good partitioning of the coarsest graph. In the tests described below HEM is used as default matching strategy. In Fig. 3 comparison of deferent matching strategies for different numbers of parts (2,4,6,8,10,12,16,20,24,28,32) is presented for the problem CI-1.
- **Coarsening depth parameter impact on partitioning quality.** As was described in sec. 3, the depth of coarsening is controlled by special parameter

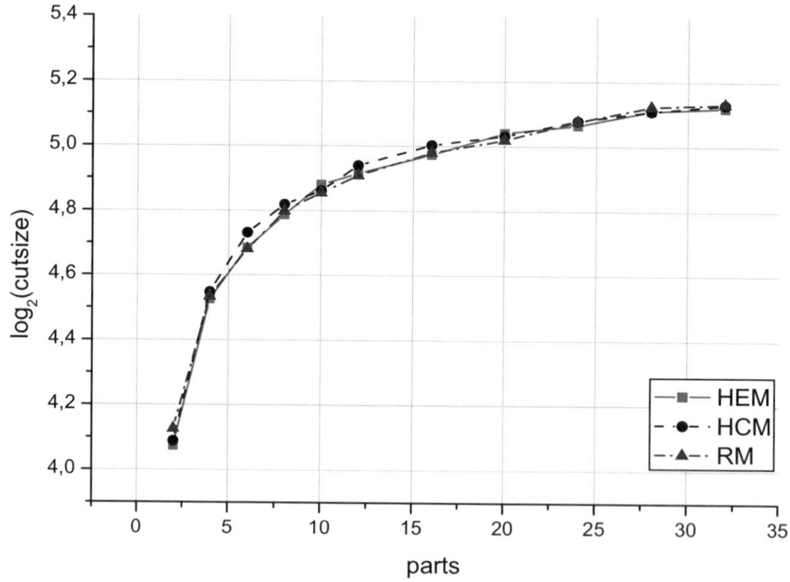

Fig. 3. Comparison of different matching algorithms and their impact on partitioning quality for different numbers of parts (2,4,6,8,10,12,16,20,24,28,32) for the problem CI-1.

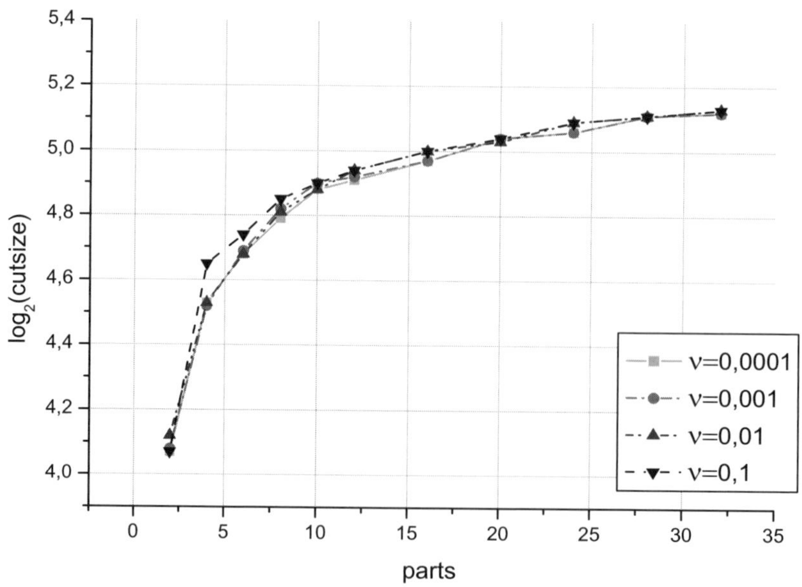

Fig. 4. Influence of coarsening depth parameter ν on partitioning quality for different numbers of parts (2,4,6,8,10,12,16,20,24,28,32) for the problem CI-1.

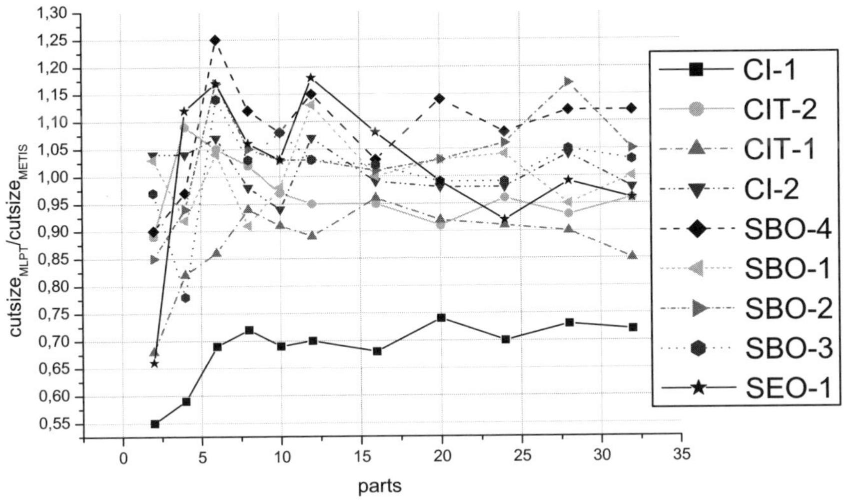

Fig. 5. Comparison of partitioning quality generated by *MLPT* with that generated by *METIS* package for different numbers of parts (2,4,6,8,10,12,16,20,24,28,32). Here $\mathrm{cutsize}_{\mathrm{MLPT}}$ is the cut size for MLPT and $\mathrm{cutsize}_{\mathrm{METIS}}$ is the cut size for METIS.

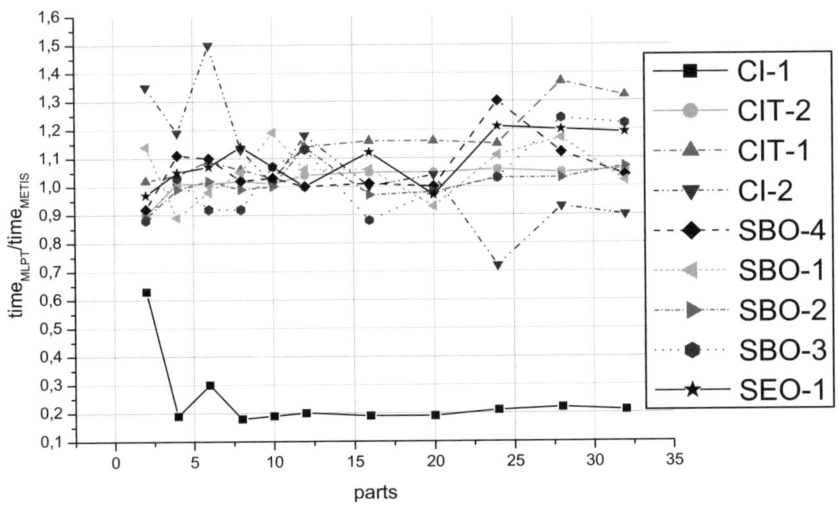

Fig. 6. Comparison of partitioning time required by *MLPT* with that required by *METIS* package for different numbers of parts (2,4,6,8,10,12,16,20,24,28,32). Here $\mathrm{time}_{\mathrm{MLPT}}$ is the time requred by MLPT and $\mathrm{time}_{\mathrm{METIS}}$ is the time required by METIS.

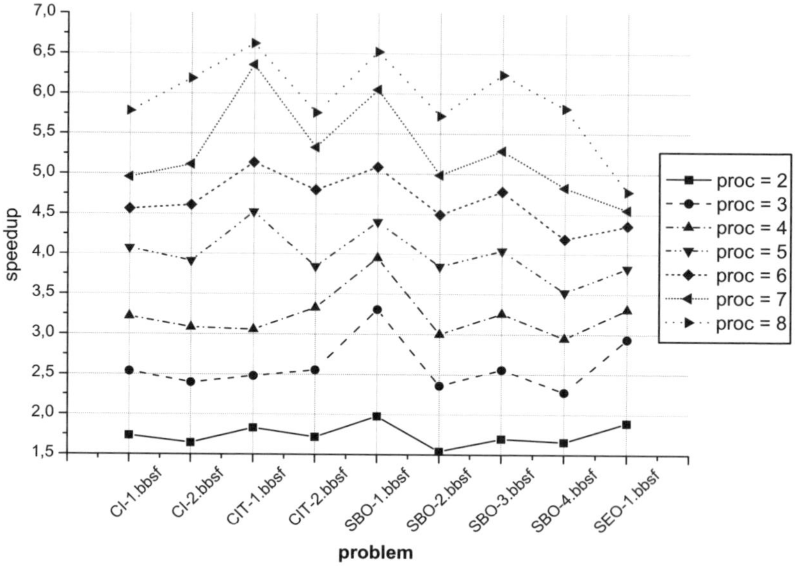

Fig. 7. Influence of partitioning quality on the performance of matrix-vector product (*MVP*) operation on MPI architecture for different numbers of parts (2,3,4,5,6,7,8). Here speedup is the ratio of serial time to parallel time required by the operation.

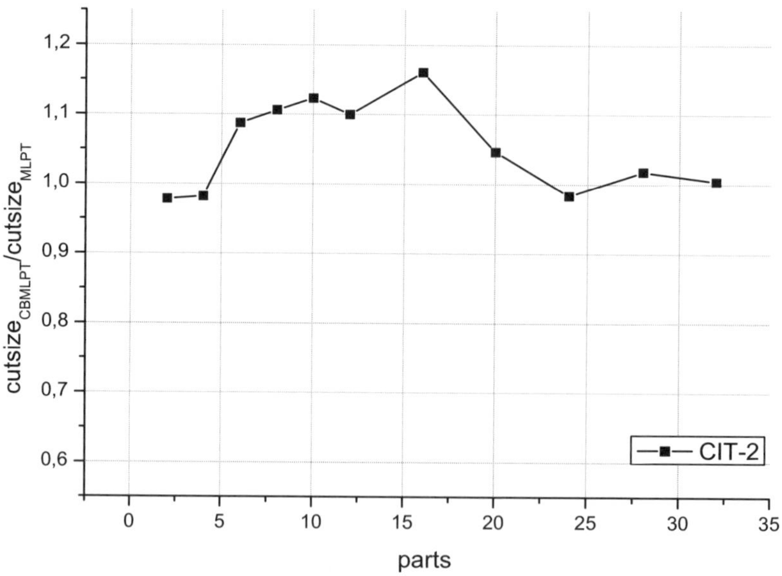

Fig. 8. Comparison of *Cell-Based Multilevel Algorithm* with the original one for the problem CIT-2 for different numbers of parts (2,4,6,8,10,12,16,20,24,28,32). Here $\mathrm{cutsize}_{\mathrm{CBMLPT}}$ is the cut size for CBMLPT and $\mathrm{cutsize}_{\mathrm{MLPT}}$ is the cut size for MLPT.

ν, i.e. we try to build a coarser graph G_{i+1} from a finer one G_i until $|V_i| \geqslant \nu*|V_1|$, where $|V_1|$ is the number of nodes of G_1. The smaller the parameter's value the smaller the number of nodes in the coarsest graph and then the better partitioning we can obtain after the initial partitioning phase. As our experiments show, small values of ν result in better initial partitionig. In this section we present the results of tests where the coarsening depth parameter is varying for the problem CI-1 for different numbers of parts (2,4,6,8,10,12,16,20,24,28,32).

- **Comparison with METIS package.** In this section the results of comparison of our partitioner (*MLPT*) with *METIS* package are presented for all matrices for different numbers of parts. In the tests we use HEM algorithm to find matching and the value of the coarsening depth parameter is $\nu = 0.0001$. We can conclude that our variant of multilevel algorithm generates partitioning competetive with that generated by *METIS* package. In Fig. 5 we compare the quality of partitioning generated by our partitioner with that generated by *METIS* package. In Fig. 6 we compare the required time.
- **MPI matrix-vector product** . While solving a large sparse linear system of equations via some Krylov-like iterative methods on a machine with distributed memory architecture, it is very important to perform *matrix-vector product (mvp)* operation as fast as possible. In Fig. 7 we present the impact of partitioning quality on the performance of MVP operation on MPI architecture for 2, 3, 4, 5, 6, 7, and 8 numbers of parts. We can conclude that there is significant speedup for most matrices.
- **Comparison of *CBML* approach with the original one**. In this section we present the advantages of cell-based multilevel approach over the original one for the problem CIT−1. As it was mentioned earlier, this approach tends to generate good partitionings for the paroblems arising from descretization of PDEs in which several unknowns are related to each grid cell. The Fig. 8 presents the result of comparison. One can conclude that *CBML* algorithm is preferable over the original one for such systems.

9 Conclusion

We evaluated the performance of our multilevel partitioner for a range of matrices arising from discretization of PDEs. One can conclude that the multilevel technique work quite well. As it was mentioned earlier, the coarsening phase requires more than half of the total partitioning time. This fact demonstrates that in order to effectively parallelize the whole algorithm some tricks must be employed to parallelize the coarsening phase. In [4] a parallel multilevel algorithm is proposed which is based on the graph coloring. Comparing partitioning quality one can conclude that the best partitionings are generated when HEM algorithm is used to find the edges to contract. In addition, obtained partitionings are competitive with those generated by METIS package.

References

1. C. M. Fiduccia and R. M. Mattheyses, A linear time heuristic for improving network partitions. In: *Proc. 19th IEEE Design Automation Conference*, 1982, pp. 175-181.
2. B. Hendrickson and R. Leland, A Multilevel Algorithm for Partitioning Graphs. *Tech. report SAND93-1301*, Sandia National Laboratories, Albuquerque, NM, 1993.
3. G. Karypis and V. Kumar, Multilevel Graph Partition and Sparse Matrix Ordering. In: *Intl. Conf. on Parallel Processing*, 1995.
4. G. Karypis and V. Kumar, A parallel algorithm for multilevel graph partitioning and sparse matrix ordering. In: *J. Parallel and Distributed Computing*, 1998, No.48, pp. 71-95.
5. B. W. Kernigan and S. Lin, An efficient heuristic procedure for Partitioning graphs. In: *Bell Sys. Tech. J.*, 1970, No. 49, pp. 291-307.
6. O. Ore, Theory of Graphs. *AMS Colloquium Publications 38. AMS*, 1962.

2D-Extension of Singular Spectrum Analysis: Algorithm and Elements of Theory

N. E. Golyandina* and K. D. Usevich**

Mathematical Department, St. Petersburg State University, Universitetskij pr. 28, St. Petersburg Petrodvorets 198504, Russia
*nina@gistatgroup.com, **usevich.k.d@gmail.com

Abstract. Singular Spectrum Analysis is a nonparametric method, which allows one to solve problems like decomposition of a time series into a sum of interpretable components, extraction of periodic components, noise removal and others. In this paper, the algorithm and theory of the SSA method are extended to analyse two-dimensional arrays (e.g. images). The 2D-SSA algorithm based on the SVD of a Hankel-block-Hankel matrix is introduced. Another formulation of the algorithm by means of Kronecker-product SVD is presented. Basic SSA notions such as separability are considered. Results on ranks of Hankel-block-Hankel matrices generated by exponential, sine-wave and polynomial 2D-arrays are obtained. An example of 2D-SSA application is presented.

Keywords: Singular Spectrum Analysis, image analysis, Hankel-block-Hankel matrix, separability, finite rank, Singular Value Decomposition, Kronecker-product SVD.

1 Introduction

The purpose of this paper is to extend the SSA (Singular Spectrum Analysis) algorithm and theory developed in [7] to the case of *two-dimensional arrays* of data (i.e. real-valued functions of two variables defined on Cartesian grid). The monochrome digital images are a standard example here. Singular Spectrum Analysis is a well-known model-free technique for analysis of real-valued time series. Basically, SSA is an exploratory method intended to perform decomposition of a time series into a sum of interpretable components, such as trend, periodicities and noise (see [3, 4, 7] for more details). SSA has proved to be successful for such tasks. Moreover, there are several SSA extensions for time series forecasting, change-point detection, missing values imputation and so on. These are the reasons to believe that the two-dimensional extension of SSA (2D-SSA, first presented in [6]) has similar capabilities. However, its application was hampered by lack of theory, which this paper is intended to reduce.

Suppose we observe a 2D-array of data (a real matrix) being a sum of unknown components $\mathrm{F} = \mathrm{F}^{(1)} + \ldots + \mathrm{F}^{(m)}$. The general task of the 2D-SSA algorithm is to produce a decomposition

$$\mathsf{F} = \widetilde{\mathsf{F}}^{(1)} + \ldots + \widetilde{\mathsf{F}}^{(m)}, \tag{1}$$

where the terms approximate the initial components.

In §2 we present the algorithm of 2D-SSA. First of all, the algorithm is formulated basing on the SVD of the Hankel-block-Hankel (HbH for short) matrix generated by the input 2D-array. However, another equivalent representation of the algorithm fits better for examination and analysis. It is based on the decomposition of a matrix into a sum of Kronecker products.

The key step of the algorithm is grouping of terms of the SVD. This step governs the resulting decomposition (1). Main problems of grouping are: possibility of proper grouping and identification of terms in the SVD. These problems are discussed in §2.4 and investigated in §3 and §4.

In §3 we study the notion of *separability* inherited from the 1D case. Separability means possibility to extract constituents from their sum by 2D-SSA. We also provide a brief review of results on one-dimensional separability as the basis for results in the 2D case.

Section 4 deals with the so-called *2D-SSA rank* of a 2D-array defined as the number of SVD terms corresponding to the 2D-array and equal to the rank of a Hankel-block-Hankel matrix generated by the 2D-array. This number is important, as it should be taken into account when performing identification. We provide rank calculations for different 2D-arrays: exponents, polynomials and sine-waves.

In §5 we demonstrate 2D-SSA notions by an example of periodic noise removal.

General definitions

First of all, let us review definitions that will be used throughout this paper.

The following operator is widely used in the SSA theory and is quite helpful for the 2D-SSA algorithm formulation.

Definition 1. *Let* $\mathsf{A} = (a_{ij})_{i,j=1}^{m,n} \in \mathsf{M}_{m,n}(\mathsf{Q})$ *be a matrix over Euclidean space* Q. *The* hankelization operator $\mathcal{H}^{\mathsf{Q}} : \mathsf{M}_{m,n}(\mathsf{Q}) \mapsto \mathsf{M}_{m,n}(\mathsf{Q})$ *is defined by*

$$\mathcal{H}^{\mathsf{Q}}\mathsf{A} = \begin{pmatrix} \widetilde{a}_1 & \widetilde{a}_2 & \ldots & \widetilde{a}_n \\ \widetilde{a}_2 & \widetilde{a}_3 & \ldots & \widetilde{a}_{n+1} \\ \vdots & \vdots & \ddots & \vdots \\ \widetilde{a}_m & \widetilde{a}_{m+1} & \ldots & \widetilde{a}_{m+n-1} \end{pmatrix}, \quad \widetilde{a}_k = \left(\sum_{(i,j)\in\mathcal{D}_k} a_{ij} \right) \Big/ \#\mathcal{D}_k,$$

where $\mathcal{D}_k = \{(i,j) : 1 \leqslant i \leqslant m,\ 1 \leqslant j \leqslant n,\ i+j = k+1\}$.

Further, we will denote by $\mathcal{M}_{m,n} \stackrel{\text{def}}{=} M_{m,n}(\mathbb{R})$ the space of real matrices with Frobenius inner product:

$$\langle X, Y \rangle_{\mathcal{M}} = \sum_{i=1}^{m} \sum_{j=1}^{n} x_{ij} y_{ij}, \tag{2}$$

where $X = (x_{ij})_{i,j=1}^{m,n}$, $Y = (y_{ij})_{i,j=1}^{m,n} \in \mathcal{M}_{m,n}$.

Introduce an isomorphism between $\mathcal{M}_{m,n}$ and $\mathbb{R}^{mn}$.

Definition 2. The vectorization *(see, for instance, [8]) of* $A = (a_{ij})_{i,j=1}^{m,n} \in \mathcal{M}_{m,n}$ *is given by*

$$\operatorname{vec} A \stackrel{\text{def}}{=} (a_{11}, \ldots, a_{m1}; a_{12}, \ldots, a_{m2}; \ldots; a_{1n}, \ldots, a_{mn})^T. \tag{3}$$

Definition 3. *The* (m,n)-matricizing *of* $X \in \mathbb{R}^{mn}$ *denoted by* $\operatorname{matr}_{m,n}(X)$ *is defined to be* $A \in \mathcal{M}_{m,n}$ *satisfying* $\operatorname{vec} A = X$.

Then, recall the operation of Kronecker product [8, 9].

Definition 4. *For* $A = (a_{ij})_{i,j=1}^{m,n} \in \mathcal{M}_{m,n}$ *and* $B = (b_{kl})_{k,l=1}^{p,q} \in \mathcal{M}_{p,q}$ *their* Kronecker product *is, by definition,*

$$A \otimes B \stackrel{\text{def}}{=} \begin{pmatrix} a_{11}B & \ldots & a_{1n}B \\ \vdots & & \vdots \\ a_{m1}B & \ldots & a_{mn}B \end{pmatrix}. \tag{4}$$

Finally, we need an isomorphism between classes of block matrices.

Definition 5. *The* rearrangement $\mathcal{R}: \mathcal{M}_{mp,nq} \mapsto \mathcal{M}_{pq,mn}$ *is defined as*

$$\begin{aligned} &\mathcal{R}(C) \stackrel{\text{def}}{=} D \in \mathcal{M}_{pq,mn}, \quad \textit{where} \\ &(D)_{i+(j-1)p,k+(l-1)m} = (C)_{i+(k-1)p,j+(l-1)q} \end{aligned} \tag{5}$$

for $1 \leqslant i \leqslant p, 1 \leqslant j \leqslant q, 1 \leqslant k \leqslant m, 1 \leqslant l \leqslant n$.

Note that the introduced rearrangement of a matrix is the transpose of the rearrangement defined in [2]. The following properties of the rearrangement are quite useful, despite being easily checked.

- Let $A = (a_{ij})_{i,j=1}^{m,n} \in \mathcal{M}_{m,n}$ and $B = (b_{kl})_{k,l=1}^{p,q} \in \mathcal{M}_{p,q}$. Then

$$\mathcal{R}(A \otimes B) = \operatorname{vec} B (\operatorname{vec} A)^T. \tag{6}$$

- For any $C \in \mathcal{M}_{mp,nq}$

$$\|\mathcal{R}(C)\|_{\mathcal{M}} = \|C\|_{\mathcal{M}}. \tag{7}$$

2 2D-SSA

2.1 Basic algorithm

Consider a *2D-array* of data

$$F = \begin{pmatrix} f(0,0) & f(0,1) & \dots & f(0, N_y - 1) \\ f(1,0) & f(1,1) & \dots & f(1, N_y - 1) \\ \vdots & \vdots & \ddots & \vdots \\ f(N_x - 1, 0) & f(N_x - 1, 1) & \dots & f(N_x - 1, N_y - 1) \end{pmatrix}.$$

The algorithm is based on the SVD of a Hankel-block-Hankel (HbH) matrix constructed from the 2D-array. The dimensions of the HbH matrix are defined by the *window sizes* (L_x, L_y), which are restricted by $1 \leqslant L_x \leqslant N_x$, $1 \leqslant L_y \leqslant N_y$ and $1 < L_x L_y < N_x N_y$. Let $K_x = N_x - L_x + 1$ and $K_y = N_y - L_y + 1$ for convenience of notation.

Embedding

At this step, the input 2D-array is arranged into a Hankel-block-Hankel matrix of size $L_x L_y \times K_x K_y$:

$$\mathbf{W} = \begin{pmatrix} \mathbf{H}_0 & \mathbf{H}_1 & \mathbf{H}_2 & \dots & \mathbf{H}_{K_y - 1} \\ \mathbf{H}_1 & \mathbf{H}_2 & \mathbf{H}_3 & \dots & \mathbf{H}_{K_y} \\ \mathbf{H}_2 & \mathbf{H}_3 & .\cdot^{\cdot} & .\cdot^{\cdot} & \vdots \\ \vdots & \vdots & .\cdot^{\cdot} & .\cdot^{\cdot} & \vdots \\ \mathbf{H}_{L_y - 1} & \mathbf{H}_{L_y} & \dots & \dots & \mathbf{H}_{N_y - 1} \end{pmatrix}, \tag{8}$$

where

$$\mathbf{H}_j = \begin{pmatrix} f(0,j) & f(1,j) & \dots & f(K_x - 1, j) \\ f(1,j) & f(2,j) & \dots & f(K_x, j) \\ \vdots & \vdots & \ddots & \vdots \\ f(L_x - 1, j) & f(L_x, j) & \dots & f(N_x - 1, j) \end{pmatrix}.$$

Obviously, there is the one-to-one correspondence between 2D-arrays of size $N_x \times N_y$ and HbH matrices (8). Let us call the matrix $\mathbf{W}$ a *Hankel-block-Hankel matrix generated by* the 2D-array F.

SVD

Then, the SVD is applied to the Hankel-block-Hankel matrix (8):

$$\mathbf{W} = \sum_{i=1}^{d} \sqrt{\lambda_i} U_i V_i^{T}. \tag{9}$$

Here λ_i $(1 \leqslant i \leqslant d)$ are the non-zero *eigenvalues* of the matrix $\mathbf{W}\mathbf{W}^T$ arranged in decreasing order $\lambda_1 \geqslant \lambda_2 \geqslant \dots \geqslant \lambda_d > 0$; $\{U_1, \dots, U_d\}$ is a system of orthonormal in $\mathbb{R}^{L_x L_y}$ *eigenvectors* of the matrix $\mathbf{W}\mathbf{W}^T$; $\{V_1, \dots, V_d\}$

is an orthonormal system of vectors in $\mathbb{R}^{K_x K_y}$, hereafter called *factor vectors*. The factor vectors can be expressed as follows: $V_i = \mathbf{W}^T U_i/\sqrt{\lambda_i}$. The triple $(\sqrt{\lambda_i}, U_i, V_i)$ is said to be the *ith eigentriple*. Note that $\sqrt{\lambda_i}$ is called a *singular value* of the matrix $\mathbf{W}$.

Grouping

After specifying m disjoint subsets of indices I_k (groups of eigentriples),

$$I_1 \cup I_2 \cup \cdots \cup I_m = \{1, \ldots, d\}, \tag{10}$$

one obtains the decomposition of the HbH matrix

$$\mathbf{W} = \sum_{k=1}^{m} \mathbf{W}_{I_k}, \quad \text{where } \mathbf{W}_I = \sum_{i \in I} \sqrt{\lambda_i} U_i {V_i}^T. \tag{11}$$

This is the most important step of the algorithm as it controls the resulting decomposition of the input 2D-array. The problem of proper grouping of the eigentriples will be discussed further (in §2.4).

Projection

Projection step is necessary in order to obtain a decomposition (1) of the input 2D-array from the decomposition (11) of the HbH matrix. Firstly, matrices $\mathbf{W}_{I_k}$ are reduced to Hankel-block-Hankel matrices $\widetilde{\mathbf{W}}_{I_k}$. Secondly, 2D-arrays $\widetilde{F}_{I_k}$ are obtained from $\widetilde{\mathbf{W}}_{I_k}$ by the one-to-one correspondence.

The matrices $\widetilde{\mathbf{W}}_{I_k}$, in their turn, are obtained by orthogonal projection of matrices $\mathbf{W}_{I_k}$ in Frobenius norm (2) onto the linear space of block-Hankel $L_x L_y \times K_x K_y$ matrices with Hankel $L_x \times K_x$ blocks. The orthogonal projection of

$$\mathbf{Z} = \begin{pmatrix} Z_{1,1} & Z_{1,2} & \ldots & Z_{1,K_y} \\ Z_{2,1} & Z_{2,2} & \ldots & Z_{2,K_y} \\ \vdots & \vdots & \ddots & \vdots \\ Z_{L_y,1} & Z_{L_y,2} & \ldots & Z_{L_y,K_y} \end{pmatrix}, \quad Z_{i,j} \in \mathcal{M}_{L_x,K_x},$$

can be expressed as a two-step hankelization

$$\widetilde{\mathbf{Z}} = \mathcal{H}^{\mathcal{M}_{L_x,K_x}} \begin{pmatrix} \mathcal{H}^{\mathbb{R}} Z_{1,1} & \mathcal{H}^{\mathbb{R}} Z_{1,2} & \ldots & \mathcal{H}^{\mathbb{R}} Z_{1,K_y} \\ \mathcal{H}^{\mathbb{R}} Z_{2,1} & \mathcal{H}^{\mathbb{R}} Z_{2,2} & \ldots & \mathcal{H}^{\mathbb{R}} Z_{2,K_y} \\ \vdots & \vdots & \ddots & \vdots \\ \mathcal{H}^{\mathbb{R}} Z_{L_y,1} & \mathcal{H}^{\mathbb{R}} Z_{L_y,2} & \ldots & \mathcal{H}^{\mathbb{R}} Z_{L_y,K_y} \end{pmatrix}.$$

In other words, the hankelization is applied at first to the blocks (*within-block* hankelization) and then to the whole matrix, i.e. the blocks on secondary diagonals are averaged between themselves (*between-block* hankelization). Certainly, the hankelization operators can be applied in the reversed order.

Thus, the result of the algorithm is

$$F = \sum_{k=1}^{m} \widetilde{F}_{I_k}. \tag{12}$$

A component $\widetilde{F}_{I_k}$ is said to be the *reconstructed by eigentriples with indices* I_k 2D-array.

2.2 Algorithm: Kronecker products

Let us examine the algorithm in terms of tensors and matrix Kronecker products.

Embedding

Columns of the Hankel-block-Hankel matrix $\mathbf{W}$ generated by the 2D-array F can be treated as vectorized $L_x \times L_y$ submatrices (moving 2D windows) of the input 2D-array F (see Fig. 1).

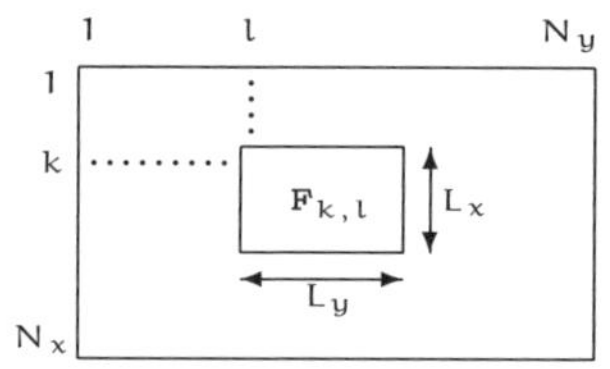

Fig. 1. Moving 2D windows

More precisely, if W_m stands for the mth column of the Hankel-block-Hankel matrix $\mathbf{W} = [W_1 : \ldots : W_{K_x K_y}]$, then

$$W_{k+(l-1)K_x} = \mathrm{vec}(F_{k,l}) \quad \text{for} \quad 1 \leqslant k \leqslant K_x,\ 1 \leqslant l \leqslant K_y, \tag{13}$$

where $F_{k,l}$ denotes the $L_x \times L_y$ submatrix beginning from the entry (k, l)

$$F_{k,l} = \begin{pmatrix} f(k-1, l-1) & \ldots & f(k-1, l+L_y-2) \\ \vdots & \ddots & \vdots \\ f(k+L_x-2, l-1) & \ldots & f(k+L_x-2, l+L_y-2) \end{pmatrix}. \tag{14}$$

An analogous equality holds for the rows of the Hankel-block-Hankel matrix $\mathbf{W}$. Let W^n be the nth row of the matrix $\mathbf{W} = [W^1 : \ldots : W^{L_x L_y}]^T$. Then

$$W^{i+(j-1)L_x} = \mathrm{vec}(F^{i,j}) \quad \text{for} \quad 1 \leqslant i \leqslant L_x,\ 1 \leqslant j \leqslant L_y, \tag{15}$$

where $F^{i,j}$ denotes the $K_x \times K_y$ submatrix beginning from the entry (i, j).

Basically, the HbH matrix is a 2D representation of the 4-order tensor $\mathcal{X}^{ij}_{kl}$

$$\mathcal{X}^{ij}_{kl} = (F_{k,l})_{i,j} = (F^{i,j})_{k,l} = f(i+k-2, j+l-2) \tag{16}$$

and the SVD of the matrix $\mathbf{W}$ is an orthogonal decomposition of this tensor. Another 2D representation of the tensor $\mathcal{X}^{ij}_{kl}$ can be obtained by the rearrangement

(5) of $\mathbf{W}$:

$$\mathbf{X} = \mathcal{R}(\mathbf{W}) = \begin{pmatrix} F_{1,1} & F_{1,2} & \dots & F_{1,K_y} \\ \vdots & \vdots & \ddots & \vdots \\ F_{K_x,1} & F_{K_x,2} & \dots & F_{K_x,K_y} \end{pmatrix}. \tag{17}$$

Let us call this block $L_x K_x \times L_y K_y$ matrix the *2D-trajectory matrix* and formulate the subsequent steps of the algorithm in terms of 2D-trajectory matrices.

SVD

First of all, recall that the eigenvectors $\{U_i\}_{i=1}^d$ form an orthonormal basis of $\text{span}(W_1, \dots, W_{K_x K_y})$ and the factor vectors $\{V_i\}_{i=1}^d$ form an orthonormal basis of $\text{span}(W^1, \dots, W^{L_x L_y})$. Consider matrices

$$\Psi_i = \text{matr}_{L_x,L_y}(U_i) \in \mathcal{M}_{L_x,L_y},$$
$$\Phi_i = \text{matr}_{K_x,K_y}(V_i) \in \mathcal{M}_{K_x,K_y},$$

and call Ψ_i and Φ_i *eigenarrays* and *factor arrays* respectively. It is easily seen that systems $\{\Psi_i\}_{i=1}^d$ and $\{\Phi_i\}_{i=1}^d$ form orthogonal bases of $\text{span}(\{F_{k,l}\}_{k,l=1}^{K_x,K_y})$ and $\text{span}(\{F^{i,j}\}_{i,j=1}^{L_x,L_y})$ (see (13) and (15)). Moreover, by (6) one can rewrite the SVD step of the algorithm as a decomposition of the 2D-trajectory matrix

$$\mathbf{X} = \sum_{i=1}^{d} \mathbf{X}_i = \sum_{i=1}^{d} \sqrt{\lambda_i} \Phi_i \otimes \Psi_i. \tag{18}$$

The decomposition is biorthogonal and has the same optimality properties as the SVD (see [2]). We will call it *Kronecker-product SVD* (*KP-SVD* for short).

Grouping

Grouping step in terms of Kronecker products has exactly the same form as (11). Choosing m disjoint subsets I_k (10) one obtains the grouped expansion

$$\mathbf{X} = \sum_{k=1}^{m} \mathbf{X}_{I_k}, \quad \text{where } \mathbf{X}_I = \sum_{i \in I} \sqrt{\lambda_i} \Phi_i \otimes \Psi_i. \tag{19}$$

Note that it is more convenient in practice to perform the grouping step on the base of Ψ_i and Φ_i (instead of U_i and V_i), since they are two-dimensional as well as the input 2D-array.

Projection

It follows from (18) and (6) that matrices $\mathbf{X}_{I_k}$ are rearrangements of corresponding matrices $\mathbf{W}_{I_k}$. Since the rearrangement $\mathcal{R}$ preserves Frobenius inner product, the resulting 2D-arrays $\widetilde{F}_{I_k}$ in (12) can be expressed through orthogonal projections in Frobenius norm of the matrices $\mathbf{X}_{I_k}$ onto the linear subspace of 2D-trajectory matrices (17) and the one-to-one correspondence between 2D-arrays and matrices like (17).

2.3 Special cases

Here we will consider some special cases of 2D-SSA. It happens that these special cases describe most of well-known SSA-like algorithms.

2.3.1 1D sequences: SSA for time series. The first special case occurs when the input array has only one dimension, namely it is a one-dimensional finite real-valued sequence (*1D-sequence* for short):

$$F = (f(0,0), \dots, f(N_x - 1, 0))^T. \tag{20}$$

In this case, the 2D-SSA algorithm coincides with the original SSA algorithm [7] applied to the same data. Let us briefly describe the SSA algorithm in its standard notation denoting $f(i,0)$ by f_i and N_x by N.

The only parameter $L = L_x$ is called the *window length*. Let $K = N - L + 1 = K_x$. Algorithm consists of four steps (the same as those of 2D-SSA). The result of Embedding step is the Hankel matrix

$$\mathbf{W} = \begin{pmatrix} f_0 & f_1 & f_2 & \dots & f_{K-1} \\ f_1 & f_2 & f_3 & \dots & f_K \\ f_2 & f_3 & f_4 & \dots & f_{K+1} \\ \vdots & \vdots & \vdots & \ddots & \vdots \\ f_{L-1} & f_L & f_{L+1} & \dots & f_{N-1} \end{pmatrix}. \tag{21}$$

This matrix is called the *trajectory matrix*[1]. SVD and Decomposition steps are exactly the same as in the 2D case. Projection in the 1D case is formulated as one-step hankelization $\mathcal{H}^{\mathbb{R}}$.

2.3.2 Extreme window sizes. Let us return to a general 2D-array case when $N_x, N_y > 1$. Consider extreme window sizes: (a) $L_x = 1$ or $L_x = N_x$; (b) $L_y = 1$ or $L_x = N_y$.

1. If conditions (a) and (b) are met both, then due to condition $1 < L_x L_y < N_x N_y$ we get $(L_x, L_y) = (N_x, 1)$ or $(L_x, L_y) = (1, N_y)$. In this case, the HbH matrix $\mathbf{W}$ coincides with the 2D-array F itself or with its transpose. Thus, the algorithm of 2D-SSA is reduced to a grouping of the SVD components of the 2D-array F. This technique is used in image processing and it works well for 2D-arrays that are products of 1D-sequences ($f(i,j) = p_i q_j$).
2. Consider the case when either (a) or (b) is met. Let it be (b). Without loss of generality, we can assume that $L_y = 1$ and $1 < L_x < N_x$. Then the HbH matrix $\mathbf{W}$ generated by F consists of stacked Hankel matrices

$$\mathbf{W} = [\mathbf{H}_0 : \mathbf{H}_1 : \dots : \mathbf{H}_{N_y - 1}]$$

[1] In the SSA literature, the trajectory matrix is usually denoted by $\mathbf{X}$

and we come to the algorithm of MSSA [4,6,10] for simultaneous decomposition of multiple time series. More precisely, we treat the 2D-array as a set of time series arranged into columns and apply the MSSA algorithm with parameter L_x to this set of series.

Practically, MSSA is more preferred than the general 2D-SSA if we expect only one dimension of the input 2D-array to be 'structured'.

2.3.3 Product of 1D sequences. In §2.3.1, we have shown that SSA for time series can be considered as a special case of the 2D-SSA. However, we can establish another relation between SSA and 2D-SSA. Consider the outer product of 1D-sequences as an important particular case of 2D-arrays: $f(i,j) = p_i q_j$. Products of 1D-sequences are of great importance for the general case of 2D-SSA as we can study properties (e.g. separability) of sums of products of 1D-sequences based on properties of the factors. The main fact here is that a 2D-SSA decomposition of the 2D-array $F = \big(f(i,j)\big)_{i,j=0}^{N_x-1,N_y-1}$ can be expressed through SSA decompositions of the 1D-sequences $(p_i)_{i=0}^{N_x-1}$ and $(q_j)_{j=0}^{N_y-1}$.

In matrix notation, the product of two 1D-sequences $P = (p_0, \ldots, p_{N_x-1})^T$ and $Q = (q_0, \ldots, q_{N_y-1})^T$ is $F = PQ^T$. Let us fix window sizes (L_x, L_y) and denote by $\mathbf{W}^{(p)}$ and $\mathbf{W}^{(q)}$ the Hankel matrices generated by P and Q respectively:

$$\mathbf{W}^{(p)} = \begin{pmatrix} p_0 & p_1 & \cdots & p_{K_x-1} \\ p_1 & p_2 & \cdots & p_{K_x} \\ \vdots & \vdots & \ddots & \vdots \\ p_{L_x-1} & p_{L_x} & \cdots & p_{N_x-1} \end{pmatrix}, \quad \mathbf{W}^{(q)} = \begin{pmatrix} q_0 & q_1 & \cdots & q_{K_y-1} \\ q_1 & q_2 & \cdots & q_{K_y} \\ \vdots & \vdots & \ddots & \vdots \\ q_{L_y-1} & q_{L_y} & \cdots & q_{N_y-1} \end{pmatrix}.$$

Then the Hankel-block-Hankel matrix $\mathbf{W}$ generated by the 2D-array F is

$$\mathbf{W} = \mathbf{W}^{(q)} \otimes \mathbf{W}^{(p)}.$$

Thus, the following theorem holds.

Theorem 1 ([9, Th. 13.10]). *Let $\mathbf{W}^{(p)}$ and $\mathbf{W}^{(q)}$ have singular value decompositions*

$$\mathbf{W}^{(p)} = \sum_{m=1}^{d_p} \sqrt{\lambda_m^{(p)}} U_m^{(p)} {V_m^{(p)}}^T, \quad \mathbf{W}^{(q)} = \sum_{n=1}^{d_q} \sqrt{\lambda_n^{(q)}} U_n^{(q)} {V_n^{(q)}}^T. \tag{22}$$

Then

$$\mathbf{W} = \sum_{m=1}^{d_p} \sum_{n=1}^{d_q} \sqrt{\lambda_m^{(p)} \lambda_n^{(q)}} \Big(U_n^{(q)} \otimes U_m^{(p)}\Big) \Big(V_n^{(q)} \otimes V_m^{(p)}\Big)^T \tag{23}$$

yields a singular value decomposition of the matrix $\mathbf{W}$, after rearranging of its terms (in decreasing order of $\lambda_m^{(p)} \lambda_n^{(q)}$).

2.4 Comments on grouping step

Let us now discuss perhaps the most sophisticated point of the algorithm: grouping of the eigentriples. Rules for grouping are not defined within the 2D-SSA algorithm and this step is supposed to be performed by hand, on the base of theoretical results. The way of grouping depends on the task one has to solve. The general task of 2D-SSA is to extract additive components from the observed 2D-array. Let us try to formalize this task.

Suppose we observe a sum of 2D-arrays: $\mathrm{F} = \mathrm{F}^{(1)} + \ldots + \mathrm{F}^{(m)}$. For example, F is a sum of a smooth surface, regular fluctuations and noise. When applying the 2D-SSA algorithm to F, we have to group somehow the eigentriples (i.e. to group the terms of (9) or (18)) at Grouping step. The problems arising here are:

- Is it possible to group the eigentriples providing the initial decomposition of F into $\mathrm{F}^{(k)}$?
- How to identify the eigentriples corresponding to a component $\mathrm{F}^{(k)}$?

In order to answer the first question, we introduce the notion of *separability of the 2D-arrays* $\mathrm{F}^{(1)}, \ldots, \mathrm{F}^{(m)}$ *by 2D-SSA* (following the 1D case [7]) as the possibility to extract them from their sum. In other words, we call the set of 2D-arrays separable if the answer to the first question is positive. In §3.1 we present the strict definition of separability and study its properties. In §3.2 we review some facts on separability of time series (the 1D-SSA case), establish a link between the 1D-SSA and 2D-SSA cases and deduce several important examples of 2D-SSA separability (§3.3). For practical reasons, we discuss approximate and asymptotic separability.

If components are separable, then we come to the second question: how to perform an appropriate grouping? The main idea is based on the following fact: the eigenarrays $\{\Psi_i\}_{i \in \mathrm{I}_k}$ and factor arrays $\{\Phi_i\}_{i \in \mathrm{I}_k}$ corresponding to a component $\mathrm{F}^{(k)}$ can be expressed as linear combinations of submatrices of the component. We can conclude that they repeat the form of the component $\mathrm{F}^{(k)}$. For example, smooth surfaces produce smooth eigenarrays (factor arrays), periodic components generate periodic eigenarrays, and so on. In §3.4 we also describe a tool of *weighted correlations* for checking separability a-posteriori. This tool can be an additional guess for grouping.

Another matter of concern is the number of eigentriples we have to gather to obtain a component $\mathrm{F}^{(k)}$. This number is called the *2D-SSA rank* of the 2D-array $\mathrm{F}^{(k)}$ and is equal to the rank of the HbH matrix generated by $\mathrm{F}^{(k)}$. Actually, we are interested in separable 2D-arrays. Clearly, they have rank-deficient HbH matrices in non-trivial case. This class of 2D-arrays has an important subclass: the 2D-arrays keeping their 2D-SSA rank constant within a range of window sizes. In the 1D case (see §2.3.1) the HbH matrices are Hankel and the subclass coincides with the whole class. For the general 2D case it is not so. However, 2D-arrays from the defined above subclass are of considerable interest since the number of eigentriples they produce does not depend on the choice of window sizes. §4 contains several examples of such 2D-arrays and rank calculations for them.

3 2D separability

This section deals with the problem of separability stated in §2.4 as a possibility to extract terms from the observed sum. We consider the problem of separability for two 2D-arrays, $F^{(1)}$ and $F^{(2)}$. Let us fix window sizes (L_x, L_y) and consider the SVD of the HbH matrix $\mathbf{W}$ generated by $F = F^{(1)} + F^{(2)}$:

$$\mathbf{W} = \sum_{i=1}^{d} \sqrt{\lambda_i} U_i {V_i}^T.$$

If we denote $\mathbf{W}^{(1)}$ and $\mathbf{W}^{(2)}$ the Hankel-block-Hankel matrices generated by $F^{(1)}$ and $F^{(2)}$, then the problem of separability can be formulated as follows: does there exist such a grouping $\{I_1, I_2\}$ that

$$\mathbf{W}^{(1)} = \sum_{i \in I_1} \sqrt{\lambda_i} U_i {V_i}^T \quad \text{and} \quad \mathbf{W}^{(2)} = \sum_{i \in I_2} \sqrt{\lambda_i} U_i {V_i}^T. \tag{24}$$

The important point to note here is that if $\mathbf{W}$ has equal singular values, then the SVD of $\mathbf{W}$ is not unique. For this reason, we introduce two notions (in the same fashion as in [7]): *strong* and *weak separability*. Strong separability means that **any** SVD of the matrix $\mathbf{W}$ allows the desired grouping, while weak separability means that there **exists** such an SVD.

3.1 Basic definitions

Let $\mathfrak{L}^{(m,n)} = \mathfrak{L}^{(m,n)}(G)$ denote the linear space spanned by the $m \times n$ submatrices of a 2D-array G. Particulary, for fixed window sizes (L_x, L_y), we have

$$\mathfrak{L}^{(L_x,L_y)}(F) = \text{span}(\{F_{k,l}\}) \quad \text{and} \quad \mathfrak{L}^{(K_x,K_y)}(F) = \text{span}(\{F^{i,j}\}).$$

Definition 6. *Two 2D-arrays* $F^{(1)}$ *and* $F^{(2)}$ *with equal sizes are* weakly (L_x, L_y)-separable *if*

$$\mathfrak{L}^{(L_x,L_y)}(F^{(1)}) \perp \mathfrak{L}^{(L_x,L_y)}(F^{(2)}) \quad \textit{and} \quad \mathfrak{L}^{(K_x,K_y)}(F^{(1)}) \perp \mathfrak{L}^{(K_x,K_y)}(F^{(2)}).$$

Due to properties of SVDs, Definition 6 means that if $F^{(1)}$ and $F^{(2)}$ are weakly separable, then the sum of SVDs of $\mathbf{W}^{(1)}$ and $\mathbf{W}^{(2)}$ (24) is an SVD of the $\mathbf{W}$. We also introduce the definition of strong separability.

Definition 7. *We call two 2D-arrays* $F^{(1)}$ *and* $F^{(2)}$ strongly separable *if they are weakly separable and the sets of singular values of their Hankel-block-Hankel matrices do not intersect.*

Hereafter we will speak mostly about the weak separability and will say 'separability' for short.

Remark 1. The set of 2D-arrays separable from a fixed 2D-array F is a linear space.

Since the exact separability is not feasible, let us introduce the approximate separability as almost orthogonality of the corresponding subspaces. Consider 2D-arrays F and G and fix window sizes (L_x, L_y). As in (14), F_{k_1,l_1}, G_{k_2,l_2} stand for $L_x \times L_y$ submatrices of F and G and F^{i_1,j_1}, G^{i_2,j_2} do for $K_x \times K_y$ submatrices. Let us introduce a distance between two 2D-arrays in order to measure the approximate separability:

$$\rho^{(L_x,L_y)}(F,G) \stackrel{\text{def}}{=} \max(\rho_L, \rho_K), \tag{25}$$

where

$$\rho_K = \max_{(k_1,l_1),(k_2,l_2)\in J_K} \left| \frac{\langle F_{k_1,l_1}, G_{k_2,l_2}\rangle_{\mathcal{M}}}{\|F_{k_1,l_1}\|_{\mathcal{M}}\|G_{k_2,l_2}\|_{\mathcal{M}}} \right|, \quad J_K = \{1,\dots,K_x\}\times\{1,\dots,K_y\};$$

$$\rho_L = \max_{(i_1,j_1),(i_2,j_2)\in J_L} \left| \frac{\left\langle F^{i_1,j_1}, G^{i_2,j_2}\right\rangle_{\mathcal{M}}}{\|F^{i_1,j_1}\|_{\mathcal{M}}\|G^{i_2,j_2}\|_{\mathcal{M}}} \right|, \quad J_L = \{1,\dots,L_x\}\times\{1,\dots,L_y\}.$$

Remark 2. The 2D-arrays F and G are separable iff $\rho^{(L_x,L_y)}(F,G) = 0$.

A quite natural way to deal with approximate separability is studying asymptotic by array sizes separability of 2D-arrays, namely 'good' approximate separability for relatively big 2D-arrays. Consider two infinite 2D-arrays $\mathcal{F} = (f_{ij})_{i,j=0}^{\infty,\infty}$ and $\mathcal{G} = (g_{ij})_{i,j=0}^{\infty,\infty}$. Let $\mathcal{F}|_{m,n}$ and $\mathcal{G}|_{m,n}$ denote finite submatrices of infinite 2D-arrays $\mathcal{F}$ and $\mathcal{G}$: $\mathcal{F}|_{m,n} = (f_{ij})_{i,j=0}^{m-1,n-1}$, $\mathcal{G}|_{m,n} = (g_{ij})_{i,j=0}^{m-1,n-1}$.

Definition 8. *$\mathcal{F}$ and $\mathcal{G}$ are said to be* asymptotically separable *if*

$$\lim_{N_x,N_y\to\infty} \rho^{(L_x,L_y)}(\mathcal{F}|_{N_x,N_y}, \mathcal{G}|_{N_x,N_y}) = 0 \tag{26}$$

for any $L_x = L_x(N_x,N_y)$ and $L_y = L_y(N_x,N_y)$ such that $L_x, K_x, L_y, K_y \to \infty$ as $N_x, N_y \to \infty$.

3.2 Separability of 1D sequences

As well as the original 1D-SSA algorithm can be treated as a special case of 2D-SSA, the notion of L-*separability of time series* (originally introduced in [7]) is a special case of (L_x, L_y)-separability.

Remark 3. Time series $F^{(1)} = (f_0^{(1)},\dots,f_{N-1}^{(1)})^T$ and $F^{(2)} = (f_0^{(2)},\dots,f_{N-1}^{(2)})^T$ are L-separable if they are $(L,1)$-separable as 2D-arrays.

Let us now give several examples of the (weak) L-separability, which is thoroughly studied in [7].

Example 1. The sequence $F = (f_0, \dots, f_{N-1})^T$ with $f_n = \cos(2\pi\omega n + \varphi)$ is L-separable from a non-zero constant sequence $(c, \dots, c)^T$ if $L\omega$ and $K\omega$, where $K = N - L + 1$, are integers.

Example 2. Two cosine sequences of length N given by

$$f_n^{(1)} = \cos(2\pi\omega_1 n + \varphi_1) \quad \text{and} \quad f_n^{(2)} = \cos(2\pi\omega_2 n + \varphi_2)$$

are L-separable if $\omega_1 \neq \omega_2$, $0 < \omega_1, \omega_2 \leqslant 1/2$ and $L\omega_1, L\omega_2, K\omega_1, K\omega_2$ are integers.

In general, there are only a small number of exact separability examples. Hence, we come to consideration of approximate separability. It is studied with the help of *asymptotic separability of time series* first introduced in [7]. Asymptotic separability is defined in the same fashion as that in the 2D case (see Definition 8). The only difference is that we let just one dimension (and parameter) tend to infinity (because another dimension is fixed).

Example 3. Two cosine sequences given by

$$f_n^{(l)} = \sum_{k=0}^{m} c_k^{(l)} \cos(2\pi\omega_k^{(l)} n + \varphi_k^{(l)}), \quad 0 < \omega_k^{(l)} \leqslant 1/2, \ l = 1, 2, \tag{27}$$

with different frequencies are asymptotically separable.

In Table 1, one can see a short summary on asymptotic separability of time series.

Table 1. Asymptotic separability

	const	cos	exp	exp cos	poly
const	−	+	+	+	−
cos	+	+	+	+	+
exp	+	+	+	+	+
exp cos	+	+	+	+	+
poly	−	+	+	+	−

In this table, **const** stands for non-zero constant sequences, **cos** does for cosine sequences (27), **exp** denotes sequences $\exp(\alpha n)$, **exp cos** stands for $e^{\alpha n} \cos(2\pi\omega n + \phi)$ and **poly** does for polynomial sequences. Note that conditions of separability are omitted in the table. For more details, such as conditions, convergence rates, and other types of separability (e.g. stochastic separability of a deterministic signal from the white noise), see [7].

3.3 Products of 1D sequences

Let us study separability properties for products of 1D-sequences (introduced in §2.3.3). Consider four 1D-sequences

$$P^{(1)} = (p_0^{(1)}, \dots, p_{N_x-1}^{(1)})^T, \quad Q^{(1)} = (q_0^{(1)}, \dots, q_{N_y-1}^{(1)})^T,$$
$$P^{(2)} = (p_0^{(2)}, \dots, p_{N_x-1}^{(2)})^T, \quad Q^{(2)} = (q_0^{(2)}, \dots, q_{N_y-1}^{(2)})^T.$$

Proposition 1. *If $P^{(1)}$ and $P^{(2)}$ are L_x-separable or sequences $Q^{(1)}$ and $Q^{(2)}$ are L_y-separable, then their products $F^{(1)} = P^{(1)}(Q^{(1)})^T$ and $F^{(2)} = P^{(2)}(Q^{(2)})^T$ are (L_x, L_y)-separable.*

Proof. First of all, let us notice that submatrices of the 2D-arrays are products of subvectors of 1D-sequences

$$\begin{aligned} F_{k_1,l_1}^{(1)} &= (p_{k_1-1}^{(1)}, \dots, p_{k_1+L_x-2}^{(1)})^T (q_{l_1-1}^{(1)}, \dots, q_{l_1+L_y-2}^{(1)}), \\ F_{k_2,l_2}^{(2)} &= (p_{k_2-1}^{(2)}, \dots, p_{k_2+L_x-2}^{(2)})^T (q_{l_2-1}^{(2)}, \dots, q_{l_2+L_y-2}^{(2)}). \end{aligned} \tag{28}$$

Let us recall an important feature of Frobenius inner product:

$$\langle AB^T, CD^T \rangle_{\mathcal{M}} = \langle A, C \rangle_2 \langle B, D \rangle_2, \tag{29}$$

where A, B, C, and D are vectors.

Applying (29) to (28), we obtain the orthogonality of all $L_x \times L_y$ submatrices of 2D-arrays:

$$\left\langle F_{k_1,l_1}^{(1)}, F_{k_2,l_2}^{(2)} \right\rangle_{\mathcal{M}} = 0.$$

Likewise, all their $K_x \times K_y$ submatrices are orthogonal too. According to Remark 2, we conclude that the 2D-arrays $F^{(1)}$ and $F^{(2)}$ are separable, and the proof is complete. □

Furthermore, we can generalize Proposition 1 to approximate and asymptotic separability.

Lemma 1. *Under the assumptions of Proposition 1,*

$$\rho^{(L_x,L_y)}(F^{(1)}, F^{(2)}) \leqslant \rho^{L_x}(P^{(1)}, P^{(2)}) \rho^{L_y}(Q^{(1)}, Q^{(2)}).$$

Proof. Equalities (28) and (29) make the proof obvious. □

Proposition 2. *Let $\mathcal{F}^{(1)}$ and $\mathcal{F}^{(2)}$ be products of infinite 1D-sequences:*

$$\mathcal{F}^{(1)} = \mathcal{P}^{(1)}(\mathcal{Q}^{(1)})^T, \quad \mathcal{F}^{(2)} = \mathcal{P}^{(2)}(\mathcal{Q}^{(2)})^T,$$
$$\mathcal{P}^{(j)} = (p_0^{(j)}, \dots, p_n^{(j)}, \dots)^T \quad \text{and} \quad \mathcal{Q}^{(j)} = (q_0^{(j)}, \dots, q_n^{(j)}, \dots)^T.$$

If $\mathcal{P}^{(1)}, \mathcal{P}^{(2)}$ or $\mathcal{Q}^{(1)}, \mathcal{Q}^{(2)}$ are asymptotically separable, then $\mathcal{F}^{(1)}$ and $\mathcal{F}^{(2)}$ are asymptotically separable too.

Proof. The proposition follows immediately from Lemma 1. □

The following example of asymptotic separability can be shown using Proposition 2 and Remark 1.

Example 4. The 2D-array given by $f^{(1)}(i,j) = \cos(2\pi\omega_1 i)\ln(j+1) + \ln(i+1)\cos(2\pi\omega_2 j)$ is asymptotically separable from a constant 2D-array $f^{(2)}(i,j) = \text{const}$.

Example 4 demonstrates that separability in the 2D case is more varied than in the 1D case. For instance, nothing but periodic 1D-sequences are separable from a constant sequence.

The next example is an analogue of Example 3.

Example 5. Two 2D sine-wave arrays given by

$$f^{(l)}(i,j) = \sum_{k=1}^{m} c_k^{(l)} \cos(2\pi\omega_{1k}^{(l)} i + \varphi_{1k}^{(l)}) \cos(2\pi\omega_{2k}^{(l)} j + \varphi_{2k}^{(l)}),\ l = 1,2,$$

with different frequencies are asymptotically separable by 2D-SSA.

However, the problem of lack of strong separability in presence of weak separability appears more frequently in the 2D case. The wider is the range of eigenvalues of the HbH matrix corresponding to a 2D-array, the more likely is mixing of components produced by the 2D-array and other constituents. This becomes a problem at Grouping step. For example, if two 1D-sequences have eigenvalues from the range $[\lambda_2, \lambda_1]$, then the range of eigenvalues of their product, by Proposition 1, is wider: $[\lambda_2^2, \lambda_1^2]$.

3.4 Checking the separability: Weighted correlations

Following the 1D case, we introduce a necessary condition of separability, which can be applied in practice.

Definition 9. *A* weighted inner product *of 2D-arrays* $\mathrm{F}^{(1)}$ *and* $\mathrm{F}^{(2)}$ *is defined as follows:*

$$\left\langle \mathrm{F}^{(1)}, \mathrm{F}^{(2)} \right\rangle_w \stackrel{\text{def}}{=} \sum_{i=0}^{N_x-1} \sum_{j=0}^{N_y-1} f^{(1)}(i,j) \cdot f^{(2)}(i,j) \cdot w_x(i) \cdot w_y(j),$$

where

$$w_x(i) = \min(i+1, L_x, K_x, N_x - i) \quad \textit{and} \quad w_y(j) = \min(j+1, L_y, K_y, N_y - j).$$

In fact, the functions $w_x(i)$ and $w_y(j)$ define the number of entries on secondary diagonals of Hankel $L_x \times K_x$ and $L_y \times K_y$ matrices respectively. More precisely,

$$w_x(i) = \#\{(k,l): 1 \leqslant k \leqslant K_x, 1 \leqslant l \leqslant L_x,\ k+l = i+1\},$$
$$w_y(j) = \#\{(k,l): 1 \leqslant k \leqslant K_y, 1 \leqslant l \leqslant L_y,\ k+l = j+1\}.$$

Hence, for a Hankel-block-Hankel matrix $\mathbf{W}$ generated by F, the product $w_x(i)w_y(j)$ is equal to the number of entries in $\mathbf{W}$ corresponding to the entry (i,j) of the 2D-array F. The same holds for the number of entries in a 2D-trajectory matrix $\mathbf{X}$. This observation implies the following proposition.

Proposition 3.

$$\left\langle \mathrm{F}^{(1)}, \mathrm{F}^{(2)} \right\rangle_w = \left\langle \mathbf{X}^{(1)}, \mathbf{X}^{(2)} \right\rangle_{\mathcal{M}} = \left\langle \mathbf{W}^{(1)}, \mathbf{W}^{(2)} \right\rangle_{\mathcal{M}}.$$

With the help of the weighted inner product, we can formulate a necessary condition for separability.

Proposition 4. *If* $\mathrm{F}^{(1)}$ *and* $\mathrm{F}^{(2)}$ *are separable, then* $\langle \mathrm{F}^{(1)}, \mathrm{F}^{(2)} \rangle_w = 0$.

Finally, we introduce weighted correlations to measure approximate separability and the matrix of weighted correlations to provide an additional information useful for grouping.

Definition 10. *A* weighted correlation (w-correlation) ρ_w *between two 2D-arrays* $\mathrm{F}^{(1)}$ *and* $\mathrm{F}^{(2)}$ *is defined as*

$$\rho_w(\mathrm{F}^{(1)}, \mathrm{F}^{(2)}) = \frac{\langle \mathrm{F}^{(1)}, \mathrm{F}^{(2)} \rangle_w}{\|\mathrm{F}^{(1)}\|_w \|\mathrm{F}^{(2)}\|_w}.$$

Consider the 2D-array F and apply 2D-SSA with parameters (L_x, L_y). If we choose the maximal grouping (10), namely $m = d$ and $I_k = \{k\}, 1 \leqslant k \leqslant d$, then each $\widetilde{\mathrm{F}}_{I_k}$ is called the kth *elementary reconstructed component* and the *matrix of weighted correlations* $\mathrm{R} = (r_{ij})_{i,j=1}^{d}$ is given by

$$r_{ij} = |\rho_w(\widetilde{\mathrm{F}}_{I_i}, \widetilde{\mathrm{F}}_{I_j})|. \tag{30}$$

For an example of application see §5.

4 2D-SSA ranks of 2D-arrays. Examples of calculation

4.1 Basic properties

Let us first introduce a definition of the *2D-SSA rank*.

Definition 11. *The* (L_x, L_y)-rank *(*2D-SSA rank *for window sizes* (L_x, L_y)*) of the 2D-array* F *is defined to be*

$$\operatorname{rank}_{L_x,L_y}(F) \stackrel{\text{def}}{=} \dim \mathfrak{L}^{(L_x,L_y)} = \dim \mathfrak{L}^{(K_x,K_y)} = \operatorname{rank} \mathbf{W}.$$

It is immediate that the (L_x, L_y)-rank is equal to the number of components in the SVD (9) of the Hankel-block-Hankel matrix generated by F. There is another way to express the rank through the 2D-trajectory matrix (17).

Lemma 2. *If for fixed window sizes* (L_x, L_y) *there exists representation*

$$\mathbf{X} = \sum_{i=1}^{m} A_i \otimes B_i, \quad B_i \in \mathcal{M}_{L_x,L_y}, \quad A_i \in \mathcal{M}_{K_x,K_y}, \tag{31}$$

then $\operatorname{rank}_{L_x,L_y} F$ *does not exceed* m. *Furthermore, if each system* $\{A_i\}_{i=1}^m$, $\{B_i\}_{i=1}^m$ *is linearly independent, then* $\operatorname{rank}_{L_x,L_y}(F) = m$.

Proof. The proof is evident, since equality (31) can be rewritten as

$$\mathbf{W} = \sum_{i=1}^{m} \operatorname{vec} B_i (\operatorname{vec} A_i)^T$$

by (6). □

By Theorem 1, the 2D-SSA rank of a product of 1D-sequences 2D-SSA rank is equal to the product of the ranks:

$$\operatorname{rank}_{L_x,L_y}(PQ^T) = \operatorname{rank}_{L_x}(P) \operatorname{rank}_{L_y}(Q), \tag{32}$$

where $\operatorname{rank}_L(\cdot)$ stands for $\operatorname{rank}_{L,1}(\cdot)$.

For a sum of products of 1D-sequences $F = \sum_{i=1}^{n} P^{(i)}(Q^{(i)})^T$, the 2D-SSA rank is not generally equal to the sum of products of ranks due to possible linear dependence of vectors. In order to calculate 2D-SSA ranks for this kind of 2D-arrays, the following lemma may be useful.

Lemma 3. *If for fixed window sizes* (L_x, L_y) *there exist linearly independent systems* $\{A_j\}_{j=1}^n$ *and* $\{B_i\}_{i=1}^m$ *such that*

$$\mathbf{X} = \sum_{i,j=1}^{m,n} c_{ij} A_j \otimes B_i, \; B_i \in \mathcal{M}_{L_x,L_y}, \; A_j \in \mathcal{M}_{K_x,K_y}, \tag{33}$$

then $\operatorname{rank}_{L_x,L_y}(F) = \operatorname{rank} \mathbf{C}$, *where* $\mathbf{C} = (c_{ij})_{i,j=1}^{m,n}$.

Proof. Let us rewrite the condition (3) in the same way as in the proof of Lemma 2:

$$\mathbf{W} = \sum_{i,j=1}^{m,n} c_{ij} \operatorname{vec} B_i (\operatorname{vec} A_j)^T.$$

If we set $\mathbf{A} = [\operatorname{vec} A_1 : \ldots : \operatorname{vec} A_n]$ and $\mathbf{B} = [\operatorname{vec} B_1 : \ldots : \operatorname{vec} B_m]$, then $\mathbf{W} = \mathbf{BCA}^T$. Since $\mathbf{A}$ and $\mathbf{B}$ have linearly independent columns, the ranks of $\mathbf{W}$ and $\mathbf{C}$ coincide. □

4.2 Ranks of time series

In the 1D case, class of series having constant rank within a range of window length is called *time series of finite rank* [7]. This class mostly consist of sums of products of polynomials, exponents and cosines:

$$f_n = \sum_{k=1}^{d'} P_{m_k}^{(k)}(n)\, \rho_k^n \cos(2\pi\omega_k n + \varphi_k) + \sum_{k=d'+1}^{d} P_{m_k}^{(k)}(n)\, \rho_k^n. \tag{34}$$

Here $0 < \omega_k < 0.5$, $\rho_k \neq 0$, and $P_l^{(k)}$ are polynomials of degree l. The time series (34) form the class of time series *governed by linear recurrent formulae* (see [3, 7]).

It happens that SSA ranks of time series like (34) can be explicitly calculated.

Proposition 5. *Let a time series* $F_N = (f_0, \ldots, f_{N-1})$ *be defined in (34) with* $(\omega_k, \rho_k) \neq (\omega_l, \rho_l)$ *for* $1 \leqslant k, l \leqslant d'$ *and* $\rho_k \neq \rho_l$ *for* $d' < k, l \leqslant d$. *Then* $\operatorname{rank}_L(F_N)$ *is equal to*

$$r = 2\sum_{k=1}^{d'} (m_k + 1) + \sum_{k=d'+1}^{d} (m_k + 1) \tag{35}$$

if $L \geqslant r$ *and* $K \geqslant r$.

Proof. Equality (34) can be rewritten as a sum of complex exponents:

$$f_n = \sum_{k=1}^{d'} P_{m_k}^{(k)}(n)\, (\alpha_k(\lambda_k)^n + \beta_k(\lambda_k')^n) + \sum_{k=d'+1}^{d} P_{m_k}^{(k)}(n)\, \rho_k^n,$$

where $\lambda_k = \rho_k e^{2\pi i\omega_k}$, $\lambda_k' = \rho_k e^{-2\pi i\omega_k}$ and $\alpha_k, \beta_k \neq 0$. The latter equality yields a canonical representation (see [1, §8]) of the Hankel matrix $\mathbf{W}$ with rank r. Under the stated conditions on L and K, $\operatorname{rank} \mathbf{W} = r$ by [1, Theorem 8.1]. □

4.3 Calculation of 2D-SSA ranks

Proposition 5 together with (32) gives possibility to calculate 2D-SSA ranks for 2D-arrays that are products of 1D-sequences. However, the general 2D case is much more complicated. In this section, we introduce results concerning 2D-SSA ranks for 2D exponential, polynomial and sine-wave arrays.

In the examples below, one can observe the effect that the 2D-SSA rank of a 2D-array given by $f(i,j) = p_{i+j}$ is equal to the SSA rank of the sequence (p_i). It is not surprising, since 2D-SSA is in general invariant to rotation (and to other linear maps) of arguments of a 2D-function $f(i,j)$.

4.3.1 Exponent.

The result on rank of a sum of 2D exponents is quite simple.

Proposition 6. *For an exponential 2D-array* $F = \left(f(i,j)\right)_{i,j=0}^{N_x-1,N_y-1}$ *defined by*

$$f(i,j) = \sum_{n=1}^{m} c_n \rho_n^i \mu_n^j, \qquad \rho_n, \mu_n \neq 0, \tag{36}$$

$\mathrm{rank}_{L_x,L_y}(F) = m$ *if* $L_x, L_y, K_x, K_y \geqslant m$ *and* $(\rho_l, \mu_l) \neq (\rho_k, \mu_k)$ *for* $l \neq k$.

Proof. The proof is based on Lemma 2. Let us express entries of the matrix $\mathbf{X}$ using equality (16):

$$(F_{k,l})_{i,j} = f(i+k-2, j+l-2) = \sum_{n=1}^{m} c_n \rho_n^{(i-1)} \mu_n^{(j-1)} \rho_n^{(k-1)} \mu_n^{(l-1)}. \tag{37}$$

It is easy to check that equality (37) defines decomposition

$$\mathbf{X} = \sum_{n=1}^{m} A_n \otimes B_n, \quad \text{where}$$

$$A_n = (\rho_n^0, \ldots, \rho_n^{(K_x-1)})^T (\mu_n^0, \ldots, \mu_n^{(K_y-1)}),$$

$$B_n = (\rho_n^0, \ldots, \rho_n^{(L_x-1)})^T (\mu_n^0, \ldots, \mu_n^{(L_y-1)}).$$

Obviously, each system $\{A_i\}_{i=1}^m$, $\{B_i\}_{i=1}^m$ is linearly independent. Applying Lemma 2 finishes the proof. □

4.3.2 Polynomials.

Let P_m be a polynomial of degree m:

$$P_m(i,j) = \sum_{s=0}^{m} \sum_{t=0}^{m-s} g_{st} i^s j^t$$

and at least one of leading coefficients $g_{s,m-s}$ for $s = 0, \ldots, m$ is non-zero. Consider the 2D-array F of sizes $N_x, N_y > 2m+1$ with $f(i,j) = P_m(i,j)$.

Proposition 7. *If* $L_x, L_y, K_x, K_y \geqslant m+1$, *then*

$$\operatorname{rank}_{L_x,L_y}(F) = \operatorname{rank}_{m+1,m+1}(G'),$$

where

$$G' = \begin{pmatrix} G'' & \mathbf{0} \\ \mathbf{0} & \mathbf{0}_{m\times m} \end{pmatrix}, \quad G'' = \begin{pmatrix} g'_{00} & \cdots & g'_{0m} \\ \vdots & \cdot^{\cdot^{\cdot}} & \\ g'_{m0} & & \mathbf{0} \end{pmatrix}, \quad g'_{st} = g_{st}s!\,t!.$$

In addition, the following inequality holds:

$$m+1 \leqslant \operatorname{rank}_{L_x,L_y}(F) \leqslant \begin{cases} (m/2+1)^2, & \text{for even } m, \\ ((m+1)/2+1)\,(m+1)/2, & \text{for odd } m. \end{cases} \tag{38}$$

Proof. The first part of the proposition is proved in the same way as Proposition 6 except for using Lemma 3 instead of Lemma 2. Let us apply Taylor formula

$$\begin{aligned}(F_{k,l})_{i,j} &= P_m(i+k-2, j+l-2) \\ &= \sum_{s=0}^{m}\sum_{t=0}^{m}(i-1)^s(j-1)^t\frac{1}{s!\,t!}\left(\frac{\partial^{s+t}P_m}{\partial i^s \partial j^t}\right)(k-1,l-1) \\ = \sum_{s=0}^{m}\sum_{t=0}^{m}&\frac{(i-1)^s}{s!}\frac{(j-1)^t}{t!}\sum_{u=0}^{m-s}\sum_{\nu=0}^{m-t} g_{u+s,\nu+t}(u+s)!(\nu+t)!\frac{(k-1)^u}{u!}\frac{(l-1)^\nu}{\nu!}.\end{aligned} \tag{39}$$

If we set $g'_{st} = 0$ for $s+t > m+1$, then we can rewrite (39) as

$$\mathbf{X} = \sum_{s,t,u,\nu=0}^{m} g'_{u+s,\nu+t} A_{u+(m+1)\nu} \otimes B_{s+(m+1)t}, \quad \text{where} \tag{40}$$

$$A_{u+(m+1)\nu} = \frac{1}{u!\ \nu!}(0^u, \ldots, (K_x-1)^u)^T(0^\nu, \ldots, (K_y-1)^\nu) \text{ for } 0 \leqslant u, \nu \leqslant m$$

$$B_{s+(m+1)t} = \frac{1}{s!\ t!}(0^s, \ldots, (L_x-1)^s)^T(0^t, \ldots, (L_y-1)^t) \text{ for } 0 \leqslant s, t \leqslant m.$$

Let $\mathbf{W}^{(g)}$ be the Hankel-block-Hankel matrix generated by G' with window sizes $(m+1, m+1)$. Then (40) can be rewritten as

$$\mathbf{X} = \sum_{i,j=0}^{(m+1)^2-1} (\mathbf{W}^{(g)})_{ji} A_i \otimes B_j.$$

The systems $\{A_i\}_{i=0}^{(m+1)^2-1}$ and $\{B_j\}_{j=0}^{(m+1)^2-1}$ are linearly independent due to restrictions on L_x, L_y. By Lemma 3, the first part of the proposition is proved.

The bounds in (38) can be proved using the fact that

$$\operatorname*{rank}_{m+1,m+1}(G') = \dim \mathcal{L}^{(m+1,m+1)}(G') = \dim \operatorname{span}\left(\{G'_{k,l}\}_{k,l=1}^{m+1,m+1}\right),$$

where $G'_{k,l}$ is the $(m+1) \times (m+1)$ submatrix of G' beginning from the entry (k, l). Define by $\mathcal{T}_n$ the space of $(m+1) \times (m+1)$ matrices with zero entries below the nth secondary diagonal:

$$\mathcal{T}_n \stackrel{\text{def}}{=} \{A = (a_{ij})_{i,j=0}^{m,m} \in \mathcal{M}_{m+1,m+1} : a_{ij} = 0 \quad \text{for} \quad i+j > n\}.$$

Then $G'_{k,l}$ belongs to $\mathcal{T}_n$ for $n \geqslant m - (k+l) + 2$ and does not, in general, for smaller n. Let us introduce

$$\mathcal{C}_n \stackrel{\text{def}}{=} \text{span}\left(\{G'_{k,l}\}_{k+l=m-n+2}\right) \subseteq \mathcal{T}_n,$$
$$\mathcal{S}_n \stackrel{\text{def}}{=} \text{span}(\mathcal{C}_0, \ldots, \mathcal{C}_n) = \text{span}(\mathcal{S}_{n-1}, \mathcal{C}_n) \subseteq \mathcal{T}_n.$$

Then $\mathcal{L}^{(m+1,m+1)}(G') = \mathcal{S}_m$.

By the theorem conditions, there exists i such that $g_{i,m-i} \neq 0$. Hence, there exist $C_0, \ldots, C_m \in \mathcal{M}_{m+1,m+1}$ such that $C_n \in \mathcal{C}_n \subseteq \mathcal{T}_n, C_n \notin \mathcal{T}_{n-1}$. Therefore, the system $\{C_0, \ldots, C_m\}$ is linearly independent and the lower bound is proved.

To prove the upper bound, note that

$$\dim \mathcal{S}_n \leqslant \min(\dim \mathcal{S}_{n-1} + \dim \mathcal{C}_n, \dim \mathcal{T}_n).$$

Since $\dim \mathcal{C}_n \leqslant m + 1 - n$ and $\dim \mathcal{T}_n = \sum_{k=1}^{n+1} k$, one can show that

$$\dim \mathcal{S}_m \leqslant \sum_{n=0}^{m} \min(n+1, m-n+1) = \begin{cases} (m/2+1)^2, & m \text{ even}, \\ ((m+1)/2+1)\,(m+1)/2, & m \text{ odd}. \end{cases} \qquad \square$$

Let us demonstrate two examples that meet the bounds in inequality (38) exactly: the 2D-SSA rank of the 2D array given by $f(k, l) = (k + l)^2$ $(m = 2)$ equal to 3, while the 2D-SSA rank for $f(k, l) = kl$ is equal to 4.

4.3.3 Sine-wave 2D-arrays. Consider a sum of sine-wave functions

$$h_d(k, l) = \sum_{m=1}^{d} A_m(k, l), \tag{41}$$

$$A_m(k, l) = \begin{pmatrix} \cos(2\pi\omega_m^{(X)} k) \\ \sin(2\pi\omega_m^{(X)} k) \end{pmatrix}^T \begin{pmatrix} a_m & b_m \\ c_m & d_m \end{pmatrix} \begin{pmatrix} \cos(2\pi\omega_m^{(Y)} l) \\ \sin(2\pi\omega_m^{(Y)} l) \end{pmatrix}, \tag{42}$$

where $1 \leqslant k \leqslant N_x$, $1 \leqslant l \leqslant N_y$, at least one coefficient in each group $\{a_m, b_m, c_m, d_m\}$ is non-zero and the frequencies meet the following conditions:

$$(\omega_n^{(X)}, \omega_n^{(Y)}) \neq (\omega_m^{(X)}, \omega_m^{(Y)}), \quad \text{for } n \neq m, \ \omega_m^{(X)}, \omega_m^{(Y)} \in (0, 1/2). \tag{43}$$

Proposition 8. *For window sizes* (L_x, L_y) *such that* $L_x, L_y, K_x, K_y \geqslant 4d$ *the 2D-SSA rank of* $F = (h_d(k,l))_{k,l=0}^{N_x-1,N_y-1}$ *is equal to*

$$\operatorname{rank}_{L_x,L_y}(F) = \sum_{m=1}^{d} \nu_m, \quad \text{where} \quad \nu_m = 2 \text{ or } 4;$$

and numbers ν_m *can be expressed as*

$$\nu_m = 2\operatorname{rank}\begin{pmatrix} a_m & b_m & c_m & d_m \\ d_m & -c_m & -b_m & a_m \end{pmatrix}. \tag{44}$$

Proof. Summands A_m of (42) can be rewritten as a sum of complex exponents:

$$\begin{aligned} 4A_m(k,l) &= (a_m - d_m - i(c_m + b_m))\, e^{\ 2\pi i\omega_m^{(X)}k} e^{\ 2\pi i\omega_m^{(Y)}l} \\ &+ (a_m - d_m + i(c_m + b_m))\, e^{-2\pi i\omega_m^{(X)}k} e^{-2\pi i\omega_m^{(Y)}l} \\ &+ (a_m + d_m + i(c_m - b_m))\, e^{-2\pi i\omega_m^{(X)}k} e^{\ 2\pi i\omega_m^{(Y)}l} \\ &+ (a_m + d_m - i(c_m - b_m))\, e^{\ 2\pi i\omega_m^{(X)}k} e^{-2\pi i\omega_m^{(Y)}l}. \end{aligned}$$

Note that the coefficients of the first pair of complex exponents become zero at once if $a_m = d_m$ and $b_m = -c_m$. The second pair of complex exponents vanishes if $a_m = -d_m$ and $b_m = c_m$. Therefore, the number of non-zero coefficients of the complex exponents corresponding to each summand $A_m(k,l)$ is equal to ν_m defined in (44). Then the 2D-array can be represented as a sum of products:

$$h_d(k,l) = \sum_{n=1}^{r} x_n {y_n}^k {z_n}^l, \quad r = \sum_{m=1}^{d} \nu_m, \tag{45}$$

where all the coefficients $x_n \in \mathbb{C}$ are non-zero, while y_n and z_n have the form $y_n = e^{2\pi i\omega_n^{(X)}}$, $z_n = e^{2\pi i\omega_n^{(Y)}}$, and pairs (y_n, z_n) are distinct due to conditions (43), namely $(y_n, z_n) \neq (y_m, z_m)$ for $n \neq m$.

Due to [5], the rank of the Hankel-block-Hankel matrix **W** generated by the 2D-array (45) is equal to r at least for $L_x, L_y \geqslant 4d$. □

Note that the condition $L_x, L_y \geqslant 4d$ is just sufficient for the result of Proposition 8. The same result is valid for a larger range of L_x, L_y; this range depends on the input 2D array, see [5] for the case of complex exponents.

Let us apply the proposition to two examples. Let $f(k,l) = \cos(2\pi\omega^{(X)}k + 2\pi\omega^{(Y)}l)$. Then the 2D-SSA rank equals 2. If $f(k,l) = \cos(2\pi\omega^{(X)}k) \cdot \cos(2\pi\omega^{(Y)}l)$, then the 2D-SSA rank equals 4.

5 Example of analysis

Consider a real-life digital image of Mars (275×278) obtained by web-camera[2] (see Fig. 2). As one can see, the image is corrupted by a kind of periodic noise,

[2] Source: Pierre Thierry

probably sinusoidal due to possible electromagnetic nature of noise. Let us try to extract this noise by 2D-SSA. It is more suitable to use the 2D-trajectory matrix notation. After choosing window sizes (25, 25) we obtain expansion (18). As we will show, these window sizes are enough for separation of periodic noise.

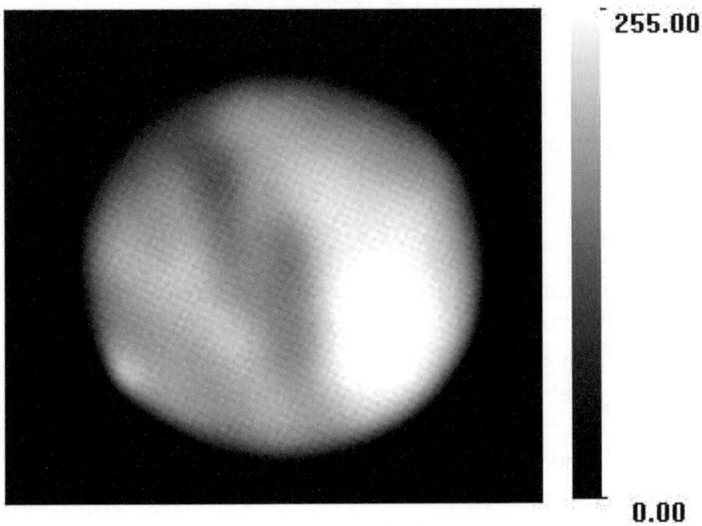

Fig. 2. 2D-array: Mars

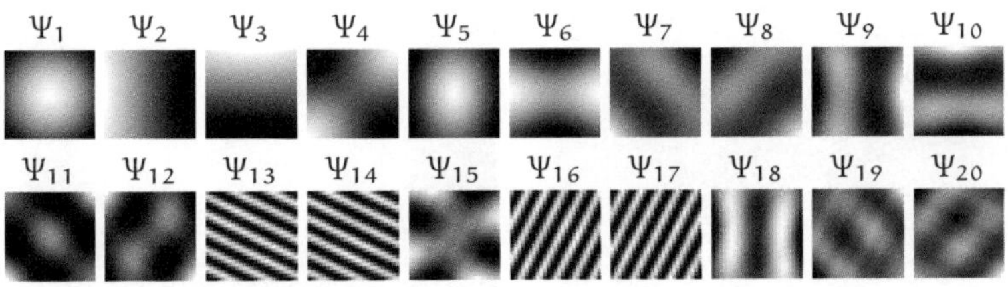

Fig. 3. Eigenarrays

Let us look at the eigenarrays (Fig. 3). The eigenarrays from the eigentriples with indices $\mathcal{N} = \{13, 14, 16, 17\}$ have periodic structure similar to the noise. The factor arrays have the same periodicity too. This observation entitles us to believe that these eigentriples constitute the periodic noise. In addition, 4 is a likely rank for sine-wave 2D-arrays.

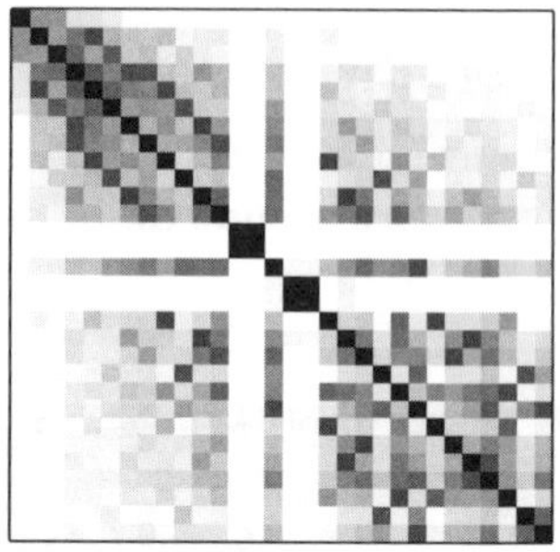

Fig. 4. Weighted correlations for the leading 30 components

Let us validate our conjecture examining the plot of weighted correlations matrix (see Fig. 4). The plot depicts w-correlations r_{ij} (30) between elementary reconstructed components (the left-top corner represents the entry r_{11}). Values are plotted in grayscale, white stands for 0 and black does for 1.

The plot contains two blocks uncorrelated to the rest. This means that the sum of elementary reconstructed components corresponding to indices from $\mathcal{N}$ is separable from the rest. Reconstruction of a 2D-array by the set $\mathcal{N}$ gives us the periodic noise, while the residual produces a filtered image.

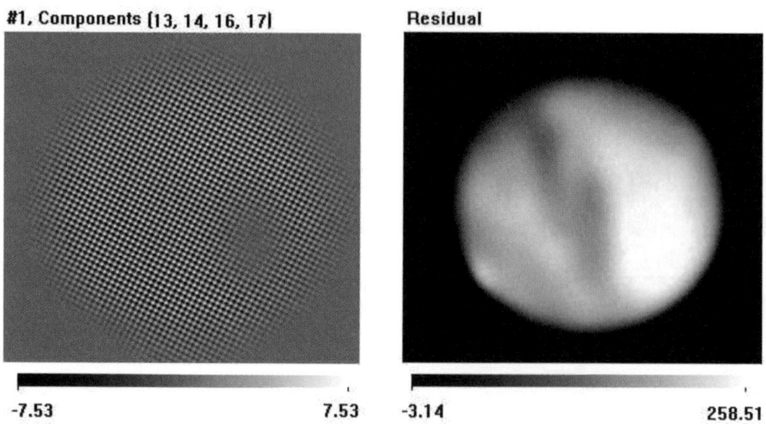

Fig. 5. Reconstructed noise and residual (filtered image)

As the matter of fact, the noise is not pure periodic and is in a sense modulated. This happens due to clipping of the signal values range to $[0, 255]$.

References

1. G. Heinig and K. Rost *Algebraic methods for Toeplitz-like matrices and operators*, Akademie Verlag, Berlin, 1984.
2. C.F. Van Loan and N.P. Pitsianis *Approximation with Kronecker products* in M.S.Moonen and G. H. Golub, eds., *Linear Algebra for Large Scale and Real Time Applications*, Kluwer Publications, pp. 293–314, 1993.
3. V.M. Buchstaber *Time series analysis and grassmannians* in S. Gindikin, ed., *Applied Problems of Radon Transform*, AMS Transaction – Series 2, Vol. 162, Providence (RI), pp. 1–17, 1994.
4. J. Elsner and A. Tsonis *Singular Spectrum Analysis. A New Tool in Time Series Analysis*, Plenum Press, New York, 1996.
5. H. Hua Yang and Y. Hua *On Rank of Block Hankel Matrix for 2-D Frequency Detection and Estimation*, IEEE Transactions on Signal Processing, Vol. 44, Issue 4, pp. 1046–1048 1996.
6. D. Danilov and A. Zhigljavsky, eds., *Principal Components of Time Series: the "Caterpillar" method*, St.Petersburg State University, St.Petersburg, 1997 (in Russian).
7. N. Golyandina, V. Nekrutkin, and A. Zhigljavsky *Analysis of Time Series Structure: SSA and Related Techniques*, Chapman & Hall/CRC, Boca Raton, 2001.
8. J.R. Magnus and H. Neudecker *Matrix Differential Calculus with Applications to Statistics and Econometrics*, John Wiley & Sons, 2004.
9. A.J. Laub *Matrix Analysis for Scientists and Engineers*, SIAM, 2004.
10. D. Stepanov and N.Golyandina *SSA-based approaches to analysis and forecast of multidimensional time series*, Proceedings of the 5th St.Petersburg Workshop on Simulation, St.Petersburg State University, St.Petersburg, pp. 293–298, 2005.

Application of Radon Transform for Fast Solution of Boundary Value Problems for Elliptic PDE in Domains With Complicated Geometry

Alexandre I. Grebennikov

Facultad de Ciencias Fisico Matematicas, Benemerita Universidad Autonoma de Puebla, Av. San Claudio y Rio Verde, Col. San Manuel, Ciudad Universitaria, Puebla, Puebla, 72570, Mexico
agrebe@fcfm.buap.mx

Abstract. A new approach for solution of the boundary value problems for wide class of elliptic partial differential equations of mathematical physics is proposed. This class includes the Laplace, Poisson, and Helmholtz equations. The approach discovered by author is based on the Local Ray Principle and leads to new General Ray (GR) method, which presents the solution of the Dirichlet boundary problems by explicit analytical formulas that include the direct and inverse Radon transform. GR-method is realized by fast algorithms and MATLAB software, whose quality is demonstrated by numerical experiments.

Keywords: Dirichlet problem for the Laplace equation, direct and inverse Radon transform.

1 Introduction

The traditional scheme of solving inverse problems of mathematical physics requires, as a rule, solution of a sequence of direct problems [1]. That is why development of new fast methods for solution of direct problems is very important for solving inverse problems [2, p. 311].

There are two main approaches for solving boundary value problems for partial differential equations in analytical form: the Fourier decomposition and the Green function method [2]. The Fourier decomposition is used, as a rule, only in theoretical investigations. The Green function method is the explicit one, but it is difficult to construct the Green function for the complex geometry of the considered domain Ω. The known numerical algorithms are based on the Finite Differences method, Finite Elements (Finite Volume) method and the Boundary Integral Equation method. Numerical approaches lead to solution of systems of linear algebraic equations [3] that require a lot of computer time and memory.

A new approach for the solution of boundary value problems on the base of the General Ray Principle (GRP) was proposed by the author in [4], [5] for the stationary waves field. GRP leads to explicit analytical formulas (GR-method) and fast algorithms, developed and illustrated by numerical experiments in [5]–[8] for solution of direct and coefficient inverse problems for the equations of mathematical physics. But there were some difficulties with the strict theoretical justification of that version of GR-method.

Here we extend the proposed approach to construction of another version of GR-method based on application of the direct Radon transform [9] to the PDE [10]–[12]. This version of GR-method is justified theoretically, formulated in algorithmic form, implemented as a program package in MATLAB system and illustrated by numerical experiments.

2 General Ray Principle

The General Ray Principle (GRP) was proposed in [4], [5]. It gives no traditional mathematical model for considered physical field and corresponding boundary problems. GRP consists in the following main assumptions:

1. the physical field can be simulated mathematically by the superposition of plane vectors (general rays) that form field $\mathbf{V}(l)$ for some fixed straight line l; each vector of field $\mathbf{V}(l)$ is parallel to the direction along this line l, and the superposition corresponds to all possible lines l that intersect domain Ω;
2. the field $\mathbf{V}(l)$ is characterized by some potential function $u(x, y)$;
3. we know some characteristics such as values of function $u(x, y)$ and/or flow of the vector $\mathbf{V}(l)$ in any boundary point $P_0 = (x^0, y^0)$ of the domain.

Application of the GRP to the problem under investigation means to construct an analogue of given PDE in the form of family of ODEs describing the distribution of the function $u(x, y)$ along the "General Rays", which are presented by a straight line l with some parameterization. We use the traditional Radon parameterization with a parameter t: $x = p\cos\varphi - t\sin\varphi$, $y = p\sin\varphi + t\cos\varphi$. Here $|p|$ is a length of the perpendicular from the origin to the line l, $\varphi \in [0, 2\pi]$ is the angle between the axis X and this perpendicular. Using this parameterization, we considered in [4], [5] the variant of GRP that reduces the Laplace equation to the assemblage (depending on p, φ) of ordinary differential equations with respect to variable t. This family of ODEs was used as the local analogue of the PDE. There we constructed corresponding version of the General Ray method for the convex domain Ω. It consists in the following steps:

1. solution of boundary value problems for the obtained assemblage of ODEs in explicit analytical or approximate form, using well known standard formulas and numerical methods;
2. calculation of the integral average for this solution along the line l;

3. transformation of these solutions by the inverse Radon transform producing the required superposition.

The numerical justification of this version of GR-method was given for the case of domain Ω being the unit circle [5]. For some more complicated domains the quality of the method was illustrated by numerical examples.

The reduction of the considered PDE to the family of ODEs with respect to the variable t makes it possible to satisfy directly boundary conditions, to construct the efficient and fast numerical algorithms. At the same time, there are some difficulties with implementation of this method for the complicated geometry of the domain Ω, as well as with its theoretical justification even for the simple cases.

3 Formulation and theoretical justification of p-version of GR-method

Let us consider the Dirichlet boundary problem for the Poisson equation:

$$\triangle u(x,y) = \psi(x,y), \quad (x,y) \in \Omega; \tag{1}$$

$$u(x,y) = f(x,y), \quad (x,y) \in \Gamma \tag{2}$$

for the function $u(x,y)$ that has two continuous derivatives with respect to both variables inside the plane domain Ω bounded by a continuous curve Γ. Here $\psi(x,y)$, $(x,y) \in \Omega$ and $f(x,y)$, $(x,y) \in \Gamma$, are given functions.

In [10]–[12], investigations are presented on the possibility of reduction of solution of PDE to the family of ODEs using the direct Radon transform [9]. This reduction leads to ODE with respect to variable p and can be interpreted in the frame of the introduced General Ray Principle. But at first glance, using the variable p makes it impossible to satisfy directly the boundary conditions expressed in (x,y) variables. Possibly by this reason the mentioned and other related investigations were concentrated only at theoretical aspect of construction of some basis of general solutions of PDE. Unfortunately, this approach was not used for construction of numerical methods and algorithms for solution of boundary value problems, except for some simple examples [10]. The important new element, introduced here into this scheme, consists in satisfying the boundary conditions by their reduction to homogeneous ones. The p-version of the GR-method can be formulated as the sequence of the following steps:

1. reduce the boundary value problem to homogeneous one;
2. represent the distribution of the potential function along the general ray (a straight line l) by its direct Radon transform $u_\varphi(p)$;
3. construct the family of ODEs in the variable p with respect the function $u_\varphi(p)$;

4. solve the constructed ODEs with zero boundary conditions;
5. calculate the inverse Radon transform of the obtained solution;
6. revert to the initial boundary conditions. We present below the implementation of this scheme.

We suppose that the boundary Γ can be represented in polar coordinates (r, α) by some one-valued positive function that we denote $r_0(\alpha)$, $\alpha \in [0, 2\pi]$. It is always possible for the simple connected star-shaped domain Ω with the centre at the origin. Let us write the boundary function as

$$\bar{f}(\alpha) = f(r_0(\alpha)) \cos\alpha, r_0(\alpha) \sin\alpha). \tag{3}$$

Supposing that functions r_0 and $\bar{f}(\alpha)$ have the second derivative we introduce the functions

$$f_0(\alpha) = \frac{\bar{f}(\alpha)}{r_0^2(\alpha)}, \qquad (x, y) \in \Omega \tag{4}$$

$$\psi_0(x, y) = \psi(x, y) - 4f_0(\alpha) - f_0''(\alpha) \tag{5}$$

$$u_0(x, y) = u(x, y) - r^2 f_0(\alpha). \tag{6}$$

To proceed with the first step of the scheme, we can write the boundary-value problem with respect to the function $u_0(x, y)$ as the following two equations:

$$\Delta u_0(x, y) = \psi_0(x, y), \quad (x, y) \in \Omega; \tag{7}$$

$$u_0(x, y) = 0, \quad (x, y) \in \Gamma. \tag{8}$$

To make the second and the third steps we need the direct Radon transform [7]:

$$R[u](p, \varphi) = \int_{-\infty}^{+\infty} u(p\cos\varphi - t\sin\varphi, p\sin\varphi + t\cos\varphi)dt.$$

After application of the Radon transform to the equation (7) and using formula (2) at the pp. 3 of [8] we obtain the family of ODEs with respect to the variable p:

$$\frac{d^2 u_\varphi(p)}{dp^2} = R[\psi_0](p, \varphi), \quad (p, \varphi) \in \widehat{\Omega} \tag{9}$$

where $\widehat{\Omega}$ is the domain of possible values of parameters p, φ. As a rule, $\varphi \in [0, 2\pi]$, while modulus of the parameter p is equal to the radius in the polar coordinates and varies in the limits determined by the boundary curve Γ. In the considered case, for some fixed φ the parameter p is in the limits $-r_0(\varphi - \pi) < p < r_0(\varphi)$.

Unfortunately, boundary condition (8) cannot be modified directly by Radon transform to the corresponding boundary conditions for every equation of the family (9). For the fourth step we propose to use the following boundary conditions for every fixed $\varphi \in [0, 2\pi]$:

$$u_\varphi(-r_0(\varphi - \pi)) = 0; \quad u_\varphi(r_0(\varphi)) = 0. \tag{10}$$

Denote by $\widehat{u}_\varphi(p)$ the solution of the problem (9)-(10) that can be univocally determined as function of variable p for every $\varphi \in [0, 2\pi], p \in (-r_0(\varphi - \pi), r_0(\varphi))$, and outside of this interval we extend $\widehat{u}_\varphi(p) \equiv 0$ for all φ with continuity in p.

Let us denote the inverse Radon transform as an operator R^{-1}, which for any function $z(p, \varphi)$ can be represented by formula (9):

$$R^{-1}[z] = \frac{1}{2\pi^2} \int_{-\pi}^{\pi} \int_{-\infty}^{\infty} \frac{z'_p(x \cos\varphi + y \sin\varphi, \varphi)}{(x \cos\varphi + y \sin\varphi) - t} dt d\varphi.$$

The justification of the fifth step of the scheme is contained in the following theorem.

Theorem 1. *The following formula for the solution of boundary value problems (7)-(8) is true:*

$$\bar{u}_0(x, y) = R^{-1}[\widehat{u}_\varphi(p)], \quad (x, y) \in \Omega. \tag{11}$$

Proof. Substituting function defined by (11) into left-hand side of equation (7) and using [8, Lemma 2.1, p. 3] we obtain the following relations:

$$\triangle \bar{u}_0(x, y) = R^{-1}[\frac{d^2 \widehat{u}_\varphi(p)}{dp^2}] = R^{-1}[R[\psi_0](p, \varphi)] = \psi_0(x, y) \tag{12}$$

which mean that the equation (7) is satisfied (see also [8], p. 40). From the condition $\widehat{u}_\varphi(p) \equiv 0$, $p \notin (-r_0(\varphi - \pi), r_0(\varphi))$, $\varphi \in [0, \pi]$ and Theorem 2.6 (the support theorem) from [8, p.10] it follows that $\bar{u}_0(x, y) \equiv 0$ for $(x, y) \notin \Omega$ and, due its continuity, satisfies the boundary conditions (8). This finishes the proof. The sixth step of GR-method is presented in detail in the following theorem.

Theorem 2. *The solution $\bar{u}(x, y)$ of boundary-value problems (1), (2) is presented by the following formulas*

$$\bar{u}(x, y) = R^{-1}[(\hat{\psi}_2(p, \varphi) - \frac{(p + r_0(\varphi - \pi))}{(r_0(\varphi) + r_0(\varphi - \pi))} \hat{\psi}_2(r_0(\varphi), \varphi))] + r^2 f_0(\alpha) \tag{13}$$

$$\hat{\psi}_2(p, \varphi) = \int_{-r_0(\varphi-\pi)}^{p} \int_{-r_0(\varphi-\pi)}^{p} \widehat{\psi}_0(p, \varphi) dp,$$

$$\widehat{\psi}_0(p, \varphi) = R[\psi_0(x, y)]. \tag{14}$$

Justification of this theorem obviously follows from the explicit formula for the solution of equation (9) with conditions (10).

The direct and inverse Radon transforms in explicit formulas (13), (14) can be implemented numerically by fast Fourier discrete transformation (FFDT) which ensures the efficiency of the proposed method.

4 Results of numerical experiments

We have constructed the fast algorithmic and program implementation of GR-method for considered problem in MATLAB system. We used the uniform discretization of variables $p \in [-1, 1]$, $\varphi \in [0, \pi]$, as well as the discretization of variables x, y, with n nodes. We made tests of mathematically simulated model examples with known exact functions $u(x, y)$, $f(x, y)$, $\psi(x, y)$. Graphic illustrations of numerical examples of solution by p-version of GR-method are presented in Fig. 1(a)-1(d). From Fig. 1(a), 1(d) we can see that the method gives a good approximation also for a non-differentiable curve Γ.

5 Conclusion

New version of GR-method is constructed. It is based on the application of the Radon transform directly to the Poisson equation. This version of GR-method for arbitrary simply connected star-shaped domains is justified theoretically, formulated in algorithmic form, implemented as a program package in MATLAB system and illustrated by numerical experiments. Proposed version can be applied for the solution of boundary value problems for other PDEs with constant coefficients. In perspective, it seems interesting to develop this approach for the solution of direct and inverse problems involving the equations of mathematical physics with variable coefficients.

Acknowledgments

Author acknowledges VIEP BUAP for the support in the frame of the Project No 04/EXES/07 and also SEP and CONACYT for support in the frame of the Project No CB 2006-01/0057479.

References

1. A.N. Tikhonov, V.Y. Arsenin, *Solutions of Ill-Posed Problems*, V.H. Winston & Sons, Washington, D.C., 1977.
2. S. L. Sobolev, *Partial differential equations mathematical physics*, Pergamon Press, 1964.
3. A.A. Samarskii, *The theory of difference schemes*, Marcel Dekker, Inc., New York, 2001.
4. A. I. Grebennikov, *Fast algorithm for solution of Dirichlet problem for Laplace equation*, WSEAS Transaction on Computers Journal, 2(4), 1039–1043 (2003).
5. A. I. Grebennikov, *The study of the approximation quality of GR-method for solution of the Dirichlet problem for Laplace equation*, WSEAS Transaction on Mathematics Journal, 2(4), 312–317 (2003).

6. A. I. Grebennikov, *Spline Approximation Method and Its Applications*, MAX Press, Russia, 2004.
7. A. I. Grebennikov, *A novel approach for the solution of direct and inverse problems of some equations of mathematical physics*, Proceedings of the 5-th International Conference on Inverse Problems in Engineering: Theory and Practice, (ed. D. Lesnic), Vol. II, Leeds University Press, Leeds, UK, Chapter G04, 1–10. (2005).
8. A. Grebennikov, *Linear regularization algorithms for computer tomography*, Inverse Problems in Science and Engineering, Vol. 14, No. 1, January, 53–64 (2006).
9. J. Radon, *Über die Bestimmung von Funktionen durch ihre Integralwerte längs gewisser Mannigfaltigkeiten*, 75 years of Radon transform (Vienna, 1992), Conf. Proc. Lecture Notes Math. Phys., IV, 324–339 (1994).
10. Helgason Sigurdur, *The Radon Transform*, Birkhauser, Boston-Basel-Berlin, 1999.
11. M. Gelfand and S. J. Shapiro, *Homogeneous functions and their applications*, Uspekhi Mat. Nauk, 10, 3–70 (1955).
12. V. A. Borovikov, *Fundamental solutions of linear partial differential equations with constant coefficients*, Trudy Moscov. Mat. Obshch., 8, 877–890 (1959).

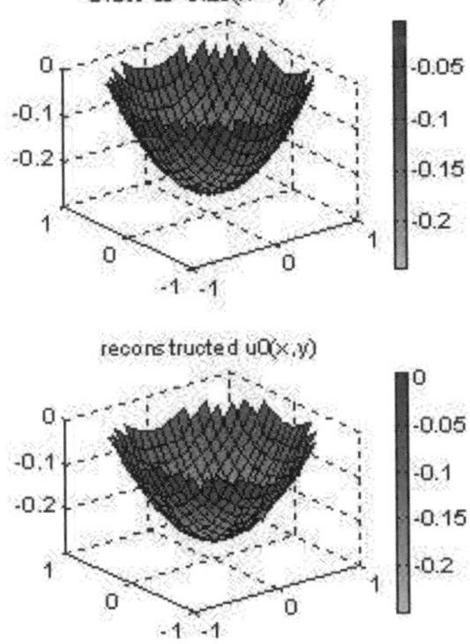

(a) Solution of the Poisson equation in the unit circle with the homogeneous Dirichlet condition.

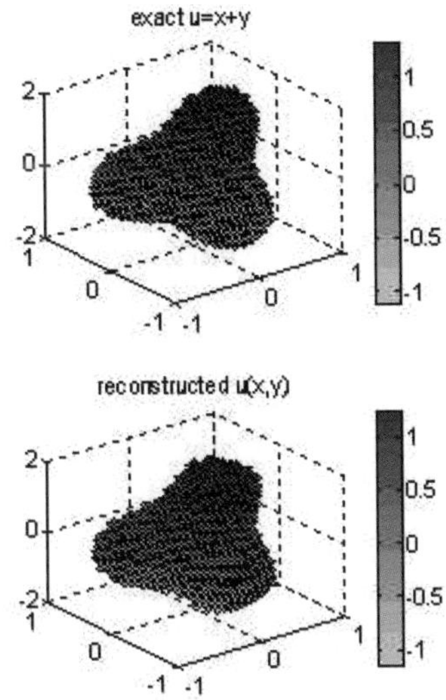

(b)

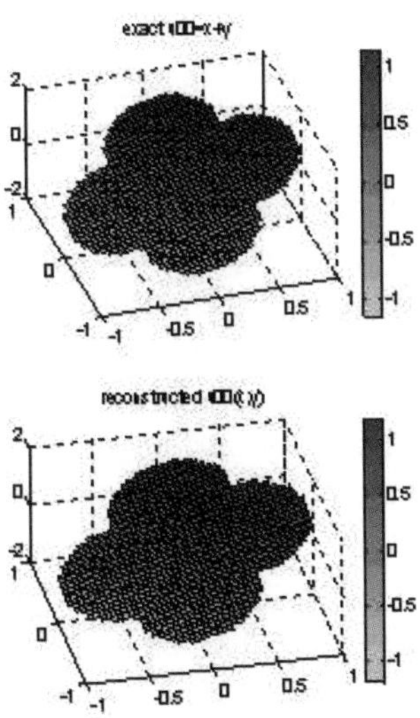

(c)

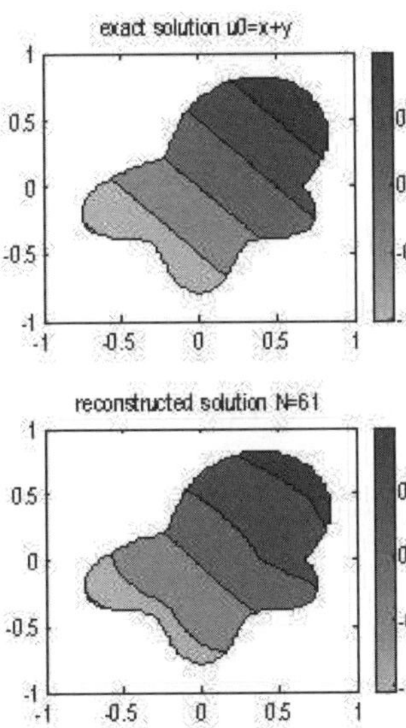

(d)

Fig. 1.

Application of a Multigrid Method to Solving Diffusion-Type Equations*

M. E. Ladonkina*, O. Yu. Milyukova**, and V. F. Tishkin***

Institute for Mathematical Modeling, RAS, Moscow, Russia
* ladm@imamod.ru, ** miliukova@imamod.ru, *** tishkin@imamod.ru

Abstract. A new efficient multigrid algorithm is proposed for solving parabolic equations. It is similar to implicit schemes by stability and accuracy, but the computational complexity is substantially reduced at each time step. Stability and accuracy of the proposed two-grid algorithm are analyzed theoretically for one- and two-dimensional heat diffusion equations. Good accuracy is demonstrated on model problems for one- and two-dimensional heat diffusion equations, including those with thermal conductivity defined as a discontinuous function of coordinates.

Keywords: parabolic equations, multigrid methods, stability, accuracy.

1 Introduction

Numerical simulation of many problems in mathematical physics must take into account diffusion processes modeled by parabolic equations. Explicit schemes lead to severe CFL restrictions on the time step [1], [2]. Implicit schemes are free from stability restrictions, but difficult to use because if high computational complexity of the corresponding linear algebraic equations. Application of classical multigrid methods [3] may also be costy and not much better than explicit schemes. Therefore, new algorithms should be developed for parabolic equations.

In this paper, we present a new efficient multigrid algorithm. We analyze the stability and accuracy of the two-grid algorithm applied to model problems for one- and two-dimensional heat diffusion equations with constant and variable coefficients. The proposed algorithm is similar to an implicit scheme in regard to stability and accuracy and substantially reduces the computational complexity at each time step.

* This work was supported by the RFBR (Grant N 08-01-00435).

2 Description of the algorithm

As an example, we consider an initial-boundary value problem for one- and two-dimensional heat diffusion equations,

$$\rho c_v \frac{\partial T}{\partial t} = \operatorname{div}(k \operatorname{grad} T) + f, \qquad x \in G, \tag{1}$$
$$T(x,t) = g(x,t), \qquad x \in \gamma,$$
$$T(x,0) = T_0(x),$$

where C_v is heat at a unit volume, ρ is density, k is thermal conductivity, T is temperature at point x at time t, f is the heat source density, γ is the computational domain boundary, and $g(x,t)$ and $T_0(x)$ are given functions. To approximate problem (1) in the computational domain $G = \{0 < x < l_1,\ \ 0 < y < l_2,\ \ 0 < t \leqslant T\}$, we use the fully implicit finite-difference scheme

$$\begin{aligned}
(\rho c_v)_{ij} \frac{u_{ij}^{n+1} - u_{ij}^{n}}{\tau} &= k_{i+0.5,j} \frac{u_{i+1,j}^{n+1} - u_{ij}^{n+1}}{h_x^2} - k_{i-0.5,j} \frac{u_{i,j}^{n+1} - u_{i-1,j}^{n+1}}{h_x^2} \\
&\quad + k_{i,j+0.5} \frac{u_{i,j+1}^{n+1} - u_{ij}^{n+1}}{h_y^2} - k_{i,j-0.5} \frac{u_{i,j}^{n+1} - u_{i,j-1}^{n+1}}{h_y^2} + \Phi_{ij}, \qquad (2) \\
&\qquad 0 < i < N_1,\ 0 < j < N_2, \\
u_{0j}^{n+1} = u_1(y_j, t_{n+1}), &\quad u_{N_1,j}^{n+1} = u_2(y_j, t_{n+1}), \quad 0 \leqslant j \leqslant N_2, \\
u_{i,0}^{n+1} = u_3(x_i, t_{n+1}), &\quad u_{i,N_2}^{n+1} = u_4(x_i, t_{n+1}), \quad 0 \leqslant i \leqslant N_1, \\
u_{ij}^{0} &= T_0(x_i, y_j), \quad 0 \leqslant i \leqslant N_1, \quad 0 \leqslant j \leqslant N_2,
\end{aligned}$$

where h_x and h_y are constant mesh sizes in the x and y directions and τ is a time step. Finite-difference scheme (2) is a system of linear algebraic equations in the unknown values of the solution at the (n+1)th time level:

$$A_h u_h = f_h. \tag{3}$$

The proposed algorithm for calculating the grid function at the next time level consists of the following steps.

Step 1. One or several smoothing iterations of equation (2) or (3) are performed using the formula

$$\begin{aligned}
u_{ij}^{s+1} &= \sigma \Bigg[\frac{(\rho c_v)_{ij} \check{u}_{ij}}{\tau} + \frac{k_{i+0.5,j} u_{i+1,j}^{s} + k_{i-0.5,j} u_{i-1,j}^{s}}{h_x^2} \\
&\qquad + \frac{k_{i,j+0.5} u_{i,j+1}^{s} + k_{i,j-0.5} u_{i,j-1}^{s}}{h_y^2} + \Phi_{ij} \Bigg] \\
&\times \left[\frac{(\rho c_v)_{ij}}{\tau} + \frac{k_{i+0.5,j} + k_{i-0.5,j}}{h_x^2} + \frac{k_{i,j+0.5} + k_{i,j-0.5}}{h_y^2} \right]^{-1} + (1-\sigma) u_{ij}^{s}, \qquad (4)
\end{aligned}$$

where $i = 1, 2, ..., N_1-1, j = 1, 2, ..., N_2-1$, σ is a weight coefficient ($0 < \sigma \leqslant 1$), and $u_{ij}^0 = u_{ij}^n$. In formula (4), index $n+1$ is omitted and $\check{u}_{ij} = u_{ij}^n$. The resulting grid function is denoted by u_{ij}^{sm}. Then, the residual is calculated as

$$r_h = A_h u_h^{sm} - f_h.$$

Step 2. The residual is restricted to the coarse grid:

$$R_{lp} = r_{2i_1,2j_1}, \qquad l = i_1 = 1, ..., N_1/2 - 1, \quad p = j_1 = 1, ..., N_2/2 - 1.$$

Step 3. A coarse grid correction equation is solved. For the two-dimensional problem analyzed here, it has the form

$$\begin{gathered}(\rho c_v)_{lp} \frac{\Delta_{lp}}{\tau} - \acute{k}_{l+0.5,p} \frac{\Delta_{l+1,p} - \Delta_{lp}}{H_x^2} + \acute{k}_{l-0.5,p} \frac{\Delta_{l,p} - \Delta_{l-1,p}}{H_x^2} \\ -\acute{k}_{l,p+0.5} \frac{\Delta_{l,p+1} - \Delta_{lp}}{H_y^2} + \acute{k}_{l,p-0.5} \frac{\Delta_{l,p} - \Delta_{l,p-1}}{H_y^2} = R_{lp}, \\ \Delta_{l0} = \Delta_{l,N_2/2} = \Delta_{0p} = \Delta_{N_1/2,p} = 0, l = 1, 2, ..., N_1/2 - 1, p = 1, 2, ..., N_2/2 - 1,\end{gathered} \tag{5}$$

where $H_x = 2h_x, H_y = 2h_y$.

Step 4. The coarse grid correction Δ_{lp} is interpolated to the fine grid by performing a 4-point face-centered and a 16-point cell-centered interpolation:

$$\delta_{ij} = \begin{cases} \Delta_{lp}, & i = 2l, j = 2p, \\ \frac{9}{16}(\Delta_{lp} + \Delta_{l+1,p}) - \frac{1}{16}(\Delta_{l-1,p} + \Delta_{l+2,p}), & i = 2l+1, j = 2p, \\ \frac{9}{16}(\Delta_{lp} + \Delta_{l,p+1}) - \frac{1}{16}(\Delta_{l,p-1} + \Delta_{l,p+2}), & i = 2l, j = 2p+1, \\ \frac{81}{256}(\Delta_{lp} + \Delta_{l+1,p} + \Delta_{l,p+1} + \Delta_{l+1,p+1}) & \\ -\frac{9}{256}(\Delta_{l-1,p} + \Delta_{l-1,p+1} + \Delta_{l,p+2} + \Delta_{l+1,p+2} & \\ +\Delta_{l+2,p+1} + \Delta_{l+2,p} + \Delta_{l+1,p-1} + \Delta_{l,p-1}) & i = 2l+1, \\ +\frac{1}{256}(\Delta_{l-1,p-1} + \Delta_{l+2,p+2} + \Delta_{l-1,p+2} + \Delta_{l+2,p-1}), & j = 2p+1 \end{cases} \tag{6}$$

where $i = 1, 2, ..., N_1 - 1$, $j = 1, 2, ..., N_2 - 1$. Note that $\delta_{0j} = \delta_{N_1,j} = \delta_{i,0} = \delta_{i,N_2} = 0$.

Step 5. Finally, the grid function is calculated at the next time level as

$$u_{ij} = u_{ij}^{sm} - \delta_{ij}. \tag{7}$$

Thus, a single iteration of the two-grid cycle is performed. Even though the system of linear equations remains incompletely solved, the algorithm is similar to an implicit scheme in terms of stability and accuracy. This is demonstrated below both theoretically and numerically for several model problems. Moreover, when the number of fine grid points is sufficiently large, the computational cost is lower in the proposed algorithm as compared to the implicit scheme used on the fine grid, because the solution of coarse grid correction equation (5) has a much lower computational complexity as compared to the solution of implicit scheme (2).

3 Theoretical stability analysis

We use Fourier analysis [4], [5] to examine stability of the two-grid algorithm with respect to initial conditions. As a model example, we consider the Dirichlet problem for the one-dimensional heat diffusion equation with unit coefficients on the interval [0, 1].

Suppose that N is an even number and a single smoothing iteration is performed. The implicit scheme used on the fine grid is

$$\frac{u_i^{im} - \breve{u}_i^{im}}{\tau} = \frac{u_{i+1}^{im} - 2u_i^{im} + u_{i-1}^{im}}{h_x^2} + \Phi_i, \qquad 0 < i < N, \tag{8}$$

$$u_0^{im} = u_{N_1}^{im} = 0,$$

$$u_i^0 = T_0(x_i), \quad 0 \leqslant i \leqslant N_1,$$

where u_i^{im} is the solution of the implicit scheme for the heat diffusion equation at the next time level, $h = 1/N$, and $(T_0)_i$ is a given grid function.

We represent the solution at the nth level as a Fourier series,

$$\breve{u}_i = \sum_{k=1}^{N-1} a_k \sin k\pi x_i \sqrt{2}.$$

The Fourier series expansion of the solution at the (n+1)th time level obtained in [6] is

$$u_i = \sum_{k=1, k\neq N/2}^{N-1} \{[q_{sm}^k - 0.5(1 + \acute{q}_k)Q_{cor}^k q_{res}^k]a_k + 0.5(1 + \acute{q}_k)Q_{cor}^k q_{res}^{N-k} a_{N-k}\}$$
$$\times \sin k\pi x_i \sqrt{2} + q_{sm}^{N/2} a_{N/2} \sin 0.5N\pi x_i \sqrt{2}, \tag{9}$$

where

$$q_{sm}^k = 1 + \frac{\sigma R}{R+1}(q_k - 1), \quad q_{res}^k = \frac{q_{sm}^k[1 + R(1 - q_k)] - 1}{\tau},$$
$$Q_{cor}^k = \frac{\tau}{1 + 0.5R(1 - q_k^2)}, \tag{10}$$
$$q_k = \cos k\pi/N, \qquad \acute{q}_k = q_k[1 + 0.5(1 - q_k^2)], \qquad R = 2\tau/h^2.$$

In the one-dimensional problem, the interpolation at Step 4 is performed as follows:

$$\delta_i = \begin{cases} \Delta_l, & i = 2l, \\ \frac{9}{16}(\Delta_l + \Delta_{l+1}) - \frac{1}{16}(\Delta_{l-1} + \Delta_{l+2}), & i = 2l+1, \end{cases}$$

where $i = 1, 2, ..., N-1$.

Now, we show that the algorithm is absolutely stable in a certain norm with respect to initial conditions when $\sigma = 0.5$. We define the linear subspace H^k as the span of the Fourier modes $\sqrt{2}\sin k\pi x_i$ and $\sqrt{2}\sin(N-k)\pi x_i$, where $k = 1, 2, ...N/2-1$. By virtue of representation (9) combined with the equalities $Q^k_{cor} = Q^{N-k}_{cor}$ and $\acute{q}_k = -\acute{q}_{N-k}$, the vector

$$\mathbf{x}^k = a_k \sin k\pi x_i \sqrt{2} + a_{N-k}\sin(N-k)\pi x_i\sqrt{2} \in H^k$$

is transformed into vector

$$\mathbf{y}^k = A_k \mathbf{x}^k,$$

where

$$A_k = \begin{pmatrix} q^k_{sm} - 0.5(1+\acute{q}_k)Q^k_{cor}q^k_{res} & 0.5(1+\acute{q}_k)Q^k_{cor}q^{N-k}_{res} \\ 0.5(1-\acute{q}_k)Q^k_{cor}q^k_{res} & q^{N-k}_{sm} - 0.5(1-\acute{q}_k)Q^k_{cor}q^{N-k}_{res} \end{pmatrix},$$

$1 \leqslant k \leqslant N/2-1$.

It was shown in [6] that the eigenvalues of A_k satisfy the inequalities

$$\lambda^k_1 \neq \lambda^k_2, \qquad |\lambda^k_1| \leqslant 1, \qquad |\lambda^k_2| \leqslant 1. \tag{11}$$

We define the norm $\|\check{u}\|_1$ on the space of grid functions as

$$\|\check{u}\|_1^2 = \sum_{k=1}^{N/2-1} (\alpha^k_1)^2 + (\alpha^k_2)^2 + a^2_{N/2},$$

where α^k_1, α^k_2 are the components of $\check{u}$ in the basis $\mathbf{e}^k_1$, $\mathbf{e}^k_2$, $\sqrt{2}\sin 0.5N\pi x_i$ $(k = 1, 2, ..., N/2-1)$; $\mathbf{e}^k_1$ and $\mathbf{e}^k_2$ are the eigenvalues associated with eigenvalues λ^k_1 and λ^k_2, respectively.

Combining (11) with the inequality $|q^k_{sm}| \leqslant 1$, we have

$$\|u\|_1 \leqslant \|\check{u}\|_1.$$

This proves the absolute stability in the norm $\|\|_1$ of the algorithm with respect to initial conditions. We note here that the norms $\|\|_1$ and $\|\|_{L_2}$ are equivalent [6].

It was shown in [6] that the algorithm is stable with respect to the right-hand side. Thus, it is proved that the algorithm is absolutely stable with respect to initial conditions and right-hand side. For the one-dimensional model problem, it holds that

$$\|u\|_1 \leqslant \|u^0\|_1 + \tau Q_1 \sum_{j=0}^{n} \|\Phi^j\|_2,$$

where $Q_1 = \mathrm{const}$ is independent of h, τ. The norm $\| \ \|_2$ is defined by analogy with $\| \ \|_1$.

4 Solution error

As a model example, we consider an initial-boundary value problem for the two-dimensional heat diffusion equation with unit coefficients on the unit square, subject to zero boundary conditions:

$$\begin{aligned} \frac{\partial u}{\partial t} = \frac{\partial^2 u}{\partial x^2} + \frac{\partial^2 u}{\partial y^2}, \qquad & 0 < x < 1, 0 < y < 1, \qquad 0 < t \leqslant T, \\ u(x,0,t) = 0, \qquad u(x,1,t) = 0, \qquad & 0 \leqslant x \leqslant 1, \qquad 0 \leqslant t \leqslant T, \\ u(0,y,t) = 0, \qquad u(1,y,t) = 0, \qquad & 0 \leqslant y \leqslant 1, \qquad 0 \leqslant t \leqslant T, \\ u(x,y,0) = T_0(x,y), \qquad & 0 \leqslant x \leqslant 1, \qquad 0 \leqslant y \leqslant 1. \end{aligned} \tag{12}$$

We assume here that $T_0(x,y)$ is an infinitely differentiable function.

The implicit scheme used on the fine grid is

$$\begin{aligned} \frac{u_{ij}^{im} - \tilde{u}_{ij}^{im}}{\tau} = \frac{u_{i+1,j}^{im} - 2u_{ij}^{im} + u_{i-1,j}^{im}}{h^2} + \frac{u_{i,j+1}^{im} - 2u_{ij}^{im} + u_{i,j-1}^{im}}{h^2}, \\ 0 < i < N, 0 < j < N, \\ u_{0,j}^{im} = u_{N,j}^{im} = 0, \qquad 0 < j < N, \\ u_{i,0}^{im} = u_{i,N}^{im} = 0, \qquad 0 < i < N, \\ u_{i,j}^{0} = T_0(x_i, y_j), \quad 0 \leqslant i \leqslant N, 0 \leqslant j \leqslant N, \end{aligned} \tag{13}$$

where $h = 1/N$, the grid function $(T_0)_{ij}$ approximates $T_0(x,y)$, and N is an even number. Suppose that a single smoothing iteration is performed and $\sigma = 0.5$. We represent the solution at the nth time level as a Fourier series expansion:

$$u_{ij}^n = \sum_{k=1}^{N-1} \sum_{m=1}^{N-1} a_{km} 2 \sin k\pi x_i \sin m\pi y_j, \qquad 0 < i < N, 0 < j < N. \tag{14}$$

We calculate the Fourier series expansion of the solution at the next time level. Following [6], we demonstrate each step of the algorithm. Substituting expansion (14) into the right-hand side of (4) and setting $C_\nu\rho \equiv 0$, $k_{ij} \equiv 1$, $s = 0$, and $h_x = h_y = h$, we perform the smoothing step to obtain

$$u_{ij}^{sm} = \sum_{k=1}^{N-1}\sum_{m=1}^{N-1} q_{sm}^{km} a_{km} 2\sin k\pi x_i \sin m\pi y_j, \tag{15}$$

where

$$q_{sm}^{km} = 1 + \frac{0.5R}{2R+1}(q_k + q_m - 2), \tag{16}$$

q_k, $\acute{q}_k$ are defined in (10), and $R = 2\tau/h^2$. Replacing u_{ij}^{im} with u_{ij}^{sm} given by (15) and $\breve{u}_{ij}^{im}$ with u_{ij}^{n} defined by (14) in (13), we find a Fourier series expansion for the residual on the fine grid:

$$r_{ij} = \sum_{k=1}^{N-1}\sum_{m=1}^{N-1} q_{res}^{km} a_{km} 2\sin k\pi x_i \sin m\pi y_j, \qquad 0 < i < N, 0 < j < N,$$

where

$$q_{res}^{km} = \frac{q_{sm}^{k}[1 + R(2 - q_k - q_m)] - 1}{\tau}. \tag{17}$$

Performing Step 2 (restricting the residual to the coarse grid) and using the identities $\sin(N-k)\pi x_{2i} = -\sin k\pi x_{2i}$ and $\sin(0.5\pi N x_{2i}) = 0$, we obtain

$$R_{lp} = \sum_{k=1}^{\frac{N}{2}-1}\sum_{m=1}^{\frac{N}{2}-1} (q_{res}^{km} a_{km} - q_{res}^{k,N-m} a_{k,N-m} - q_{res}^{N-k,m} a_{N-k,m} \tag{18}$$
$$+ q_{res}^{N-k,N-m} a_{N-k,N-m}) \times 2\sin k\pi x_l \sin m\pi y_p,$$

where $x_l = x_{2i}$ and $y_p = y_{2j}$ ($l = i = 1, 2, ..., N/2-1$, $p = j = 1, 2, ..., N/2-1$).

We represent the solution Δ_{lp} of the coarse grid correction equation

$$\frac{\Delta_{lp}}{\tau} - \frac{\Delta_{l+1,p} - 2\Delta_{lp} + \Delta_{l-1,p}}{H^2} - \frac{\Delta_{l,p+1} - 2\Delta_{lp} + \Delta_{l,p-1}}{H^2} = R_{lp}, \tag{19}$$
$$\Delta_{l0} = \Delta_{l,N/2} = \Delta_{0p} = \Delta_{N/2,p} = 0$$

($l = 1, 2, ..., N/2-1, p = 1, 2, ..., N/2-1$, $H = 2h$) as Fourier series,

$$\Delta_{lp} = \sum_{k=1}^{N/2-1}\sum_{m=1}^{N/2-1} \tilde{a}_{km} 2\sin k\pi x_l \sin m\pi y_p.$$

Substituting (18) and (4) into (19), we obtain

$$\Delta_{lp} = \sum_{k=1}^{N/2-1}\sum_{m=1}^{N/2-1} Q_{cor}^{km}(q_{res}^{km} a_{km} - q_{res}^{k,N-m} a_{k,N-m} - q_{res}^{N-k,m} a_{N-k,m}$$
$$+ q_{res}^{N-k,N-m} a_{N-k,N-m}) 2\sin k\pi x_l \sin m\pi y_p,$$

where

$$Q_{cor}^{km} = \frac{\tau}{1 + 0.5R(2 - q_k^2 - q_m^2)}. \tag{20}$$

We interpolate Δ_{lp} to the fine grid in two substeps. First, interpolation is performed to the grid $\{(ih, pH), i = 1, ..., N-1, p = 1, ..., N/2-1\}$ as follows:

$$\tilde{\Delta}_{ip} = \begin{cases} \Delta_{lp}, & i = 2l, \\ \frac{9}{16}(\Delta_{lp} + \Delta_{l+1,p}) - \frac{1}{16}(\Delta_{l-1,p} + \Delta_{l+2,p}), & i = 2l+1. \end{cases} \tag{21}$$

Then, $\tilde{\Delta}_{ip}$ is interpolated to the fine grid by formulas analogous to (21). It can be shown that this procedure is equivalent to interpolation by (6). Following [6] in each substep, we find the Fourier series expansion of the grid function δ_{ij}:

$$\delta_{ij} = \sum_{k \neq N/2} \sum_{m \neq N/2} 0.25(1 + \acute{q}_k)(1 + \acute{q}_m) Q_{cor}^{km} (q_{res}^{km} a_{km} - q_{res}^{k,N-m} a_{k,N-m}$$
$$- q_{res}^{N-k,m} a_{N-k,m} + q_{res}^{N-k,N-m} a_{N-k,N-m}) 2 \sin k\pi x_l \sin m\pi y_p.$$

Finally, formula (7) at Step 5 yields

$$u_{ij} = \sum_{k \neq \frac{N}{2}} \sum_{m \neq \frac{N}{2}} (b_{km}^1 a_{km} + b_{km}^2 a_{k,N-m} + b_{km}^3 a_{N-k,m} - b_{km}^4 a_{N-k,N-m})$$
$$\times 2 \sin k\pi x_i \sin m\pi y_j + \sum_{m=1}^{N-1} q_{sm}^{N/2,m} a_{N/2,m} 2 \sin 0.5N\pi x_i \sin m\pi y_j$$
$$+ \sum_{k \neq \frac{N}{2}} q_{sm}^{k,N/2} a_{k,N/2} 2 \sin k\pi x_i \sin 0.5N\pi y_j, \tag{22}$$

where

$$b_{km}^1 = q_{sm}^{km} - (1 + \acute{q}_k)(1 + \acute{q}_m) Q_{cor}^{km} q_{res}^k / 4,$$

$$b_{km}^2 = (1 + \acute{q}_k)(1 + \acute{q}_m) Q_{cor}^{km} q_{res}^{k,N-m} / 4,$$

$$b_{km}^3 = (1 + \acute{q}_k)(1 + \acute{q}_m) Q_{cor}^{km} q_{res}^{N-k,m} / 4,$$

$$b_{km}^4 = (1 + \acute{q}_k)(1 + \acute{q}_m) Q_{cor}^{km} q_{res}^{N-k,N-m} / 4,$$

q_{sm}^{km}, q_{res}^{km}, Q_{cor}^{km} are defined by (16), (17), and (20), respectively, and q_k, $\acute{q}_k$ are defined in (10), $R = 2\tau/h^2$.

As a result, we have Fourier series expansion (22) of the solution at the next time level obtained by the proposed algorithm.

To analyze the accuracy of the solution, we start with estimating the truncation error of implicit scheme (13) on this solution. In (13), we substitute

u_{ij} given by (22) for u_{ij}^{im} and replace $\tilde{u}_{ij}^{im}$ with u_{ij}^{n} represented by (14). The resulting residual is

$$\varphi_{ij} = \sum_{k\neq N/2}\sum_{m\neq N/2}(r^1_{km}a_{km} + r^2_{k,m}a_{k,N-m} + r^3_{k,m}a_{N-k,m} - r^4_{k,m}a_{N-k,N-m})2\sin k\pi x_i \sin m\pi y_j$$
$$+ \sum_{m=1}^{N-1} r^5_m a_{N/2,m} 2\sin 0.5N\pi x_i \sin m\pi y_j + \sum_{k\neq N/2} r^6_k a_{k,N/2} 2\sin k\pi x_i \sin 0.5N\pi y_j,$$

where

$$r^1_{km} = \frac{b^1_{km}-1}{\tau} + \frac{2b^1_{km}(2-q_k-q_m)}{h^2}, \qquad r^{2,3,4}_{km} = b^{2,3,4}_{km}\left[\frac{1}{\tau} + \frac{2(2-q_k-q_m)}{h^2}\right],$$
$$r^5_m = \frac{q^{N/2,m}_{sm}-1}{\tau} + \frac{2q^{N/2,m}_{sm}(2-q_m)}{h^2}, \qquad r^6_k = \frac{q^{k,N/2}_{sm}-1}{\tau} + \frac{2q^{k,N/2}_{sm}(2-q_k)}{h^2}. \tag{23}$$

Applying the triangle inequality and the Parseval identity, we obtain

$$\|\varphi\|_{L_2} \leqslant \|\varphi^1\|_{L_2} + \|\varphi^2\|_{L_2} + \|\varphi^3\|_{L_2} + \|\varphi^4\|_{L_2}, \tag{24}$$

where the terms on the right-hand side are defined as

$$\begin{aligned}
\|\varphi^1\|^2_{L_2} &= \textstyle\sum_{k\neq N/2}\sum_{m\neq N/2}(r^1_{km})^2(a_{km})^2,\\
\|\varphi^2\|^2_{L_2} &= \textstyle\sum_{k\neq N/2}\sum_{m\neq N/2}(r^2_{k,N-m})^2(a_{km})^2,\\
\|\varphi^3\|^2_{L_2} &= \textstyle\sum_{k\neq N/2}\sum_{m\neq N/2}(r^3_{N-k,m})^2(a_{km})^2,\\
\|\varphi^4\|^2_{L_2} &= \textstyle\sum_{k\neq N/2}\sum_{m\neq N/2}(r^4_{N-k,N-m})^2(a_{km})^2\\
&\quad + \textstyle\sum_{m=1}^{N-1}(r^5_m)^2(a_{N/2,m})^2 + \sum_{k\neq N/2}(r^6_k)^2(a_{k,N/2})^2.
\end{aligned} \tag{25}$$

Suppose that

$$\tau = h^\beta, \text{ where } 0 < \beta < 2.$$

We assume that u^n_{ij} has 2p bounded finite-difference derivatives with respect to both coordinates.

To obtain an upper bound for the first term in (24), the (k, m) index domain Ω is partitioned into four subdomains, $\Omega = \Omega_1 \cup \Omega_2 \cup \Omega_3 \cup \Omega_4$ (see Fig. 1).

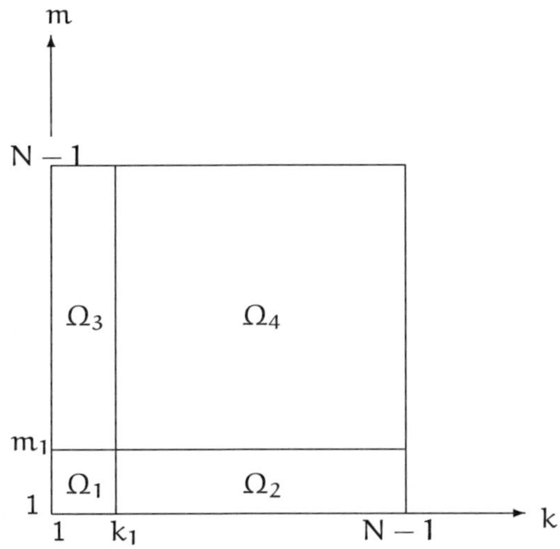

Figure 1. Decomposition of the (k, m) index domain Ω into subdomains: $k_1 = m_1 = [N^{\beta\delta}]$, $0 < \delta < 1/7$, $[N^{\beta\delta}]$ is the integer part of $N^{\beta\delta}$.

We find upper bounds for $|r^1_{km}|$ and $|a_{km}|$ in each subdomain.

In Ω_1, it holds that $k\pi h << 1$, $m\pi h << 1$, and $m^2\pi^2\tau << 1$. Using the Taylor series expansions of $\cos(k\pi h)$ and $f(x) = (1+x)^{-1}$ and estimates analogous to those given in [6] for the one-dimensional problem, we obtain

$$|r^1_{km}| \leqslant c_1 h^2 h^{\beta(1-6\delta)}, \qquad \text{where } c_1 = \pi^6/3 = \text{const}. \tag{26}$$

In Ω_2, Ω_3, Ω_4, we combine the inequalities $0 < 1 - q_k < 2$, $0 < 0.5(1 + \acute{q}_k) < 1$, and $0 < (1 + \acute{q}_k) < 9(1 + q_k)/8$ to show that $|b^1_{km}| < 17/16$ for b^1_{km} defined in (4). This result is used to obtain an upper bound for r^1_{km} defined in (23):

$$|r^1_{km}| \leqslant c_2/h^2, \qquad \text{where } c_2 = 4\frac{3}{16} = \text{const}. \tag{27}$$

Following [6], we find that

$$|a_{km}| \leqslant g_1 = \text{const in } \Omega_1, \tag{28}$$

$$|a_{km}| \leqslant \frac{g_2}{k^{2p}} \text{ in } \Omega_2, \qquad \text{where } g_2 = \text{const}, \tag{29}$$

$$|a_{km}| \leqslant \frac{g_3}{m^{2p}} \text{ in } \Omega_3, \qquad \text{where } g_3 = \text{const}, \tag{30}$$

and both (29) and (30) hold in Ω_4.

We represent $\|\varphi^1\|^2_{L_2}$ defined in (25)) as the sum of four terms corresponding to the subdomains Ω_1, Ω_2, Ω_3, and Ω_4:

$$\|\varphi^1\|^2_{L_2} = A_{11} + A_{12} + A_{13} + A_{14}, \tag{31}$$

where

$$A_{11} = \sum_{(k,m)\in\Omega_1} (r^1_{k,m})^2 (a_{km})^2, \tag{32}$$

$$A_{12} = \sum_{(k,m)\in\Omega_2, k\neq N/2} (r^1_{k,m})^2 (a_{km})^2,$$

$$A_{13} = \sum_{(k,m)\in\Omega_3, m\neq N/2} (r^1_{k,m})^2 (a_{km})^2, \tag{33}$$

$$A_{14} = \sum_{(k,m)\in\Omega_4, k\neq N/2, m\neq N/2} (r^1_{k,m})^2 (a_{km})^2.$$

Using (26)–(30) and taking into account the number of terms in each sum, we find that

$$A_{11} \leqslant \acute{A}_{11} = c_{11}h^4h^{2\beta(1-7\delta)}, \quad A_{12} \leqslant \acute{A}_{12} = c_{12}h^4h^{2p2\beta\delta-\beta\delta-9},$$
$$A_{13} \leqslant \acute{A}_{13} = c_{13}h^4h^{2p2\beta\delta-\beta\delta-9}, \quad A_{14} \leqslant \acute{A}_{14} = c_{14}h^4h^{2p2\beta\delta-10}, \tag{34}$$

where $c_{1,s} = \mathrm{const}, s = 1,2,3,4$. If $2p$ is sufficiently large that $2p2\beta\delta - 6 > 2\beta(1-3.5\delta)$, then $\acute{A}_{14} = o(\acute{A}_{11})$, $\acute{A}_{12} = o(\acute{A}_{11})$, and $\acute{A}_{13} = o(\acute{A}_{11})$, and it follows from (34) that

$$\|\varphi^1\|_{L_2} \leqslant \bar{c}_1 h^2 h^{\beta(1-7\delta)}, \qquad \text{where } \bar{c}_1 = \mathrm{const}. \tag{35}$$

Analogously, it can be shown that if $2p$ is sufficiently large, then

$$\|\varphi^2\|_{L_2} \leqslant \bar{c}_2 h^2 h^{\beta(1-7\delta)}, \quad \|\varphi^3\|_{L_2} \leqslant \bar{c}_3 h^2 h^{\beta(1-7\delta)}, \quad \|\varphi^4\|_{L_2} \leqslant \bar{c}_4 h^2 h^{\beta(1-7\delta)}, \tag{36}$$

where $\bar{c}_s = \mathrm{const}, s = 2,3,4$. Combining (35) and (36) with (24), we obtain

$$\|\varphi\|_{L_2} \leqslant c h^2 h^{\beta(1-7\delta)}, \qquad \text{where } c = \mathrm{const}. \tag{37}$$

Inequality (37) is an upper bound for the truncation error of the implicit scheme on the fine grid on the solution obtained by the proposed algorithm.

Since the implicit scheme for the heat diffusion equation is stable in the L_2 norm [1], it holds for sufficiently large $2p$ that

$$\|u_{ij} - u^{im}_{ij}\|_{L_2} = O(h^2 h^{\beta(1-7\delta)}) = O(h^2 \tau^{(1-7\delta)}) = o(\tau + h^2), \tag{38}$$

where u^{im}_{ij} and u_{ij} are, respectively, the solutions at the next time level obtained by using the implicit scheme on the fine grid and the proposed algorithm.

Thus, we have the following lemma.

Lemma 1. *If the solution u^n_{ij} to the model problem at the current time level has any number of bounded even-order finite-difference derivatives with respect to both coordinates and $\tau = h^\beta$, where $0 < \beta < 2$, then $\|u_{ij} - u^{im}_{ij}\|_{L_2} \to 0$ as $h \to 0$, $\tau \to 0$, and estimate (38) holds for any real number δ such that $0 < \delta < 1/7$.*

Estimate (38) quantifies the discrepancy between the solutions of the problem obtained by using the algorithm and the implicit scheme on the fine grid.

Applying the triangle inequality $\|u_{ij}^T - u_{ij}\|_{L_2} \leqslant \|u_{ij}^T - u_{ij}^{im}\|_{L_2} + \|u_{ij} - u_{ij}^{im}\|_{L_2}$, where u_{ij}^T is the restriction of the exact solution to the coarse grid and using the estimate (see [1])

$$\|u_{ij}^T - u_{ij}^{im}\|_{L_2} = O(\tau + h^2) \tag{39}$$

we obtain

$$\|u_{ij}^T - u_{ij}\|_{L_2} = O(\tau + h^2). \tag{40}$$

Estimate (40) quantifies the accuracy of the algorithm as applied to the model problem. Comparing (38) with (39), we see that $\|u_{ij} - u_{ij}^{im}\|_{L_2}$ goes to zero faster than does $\|u_{ij}^T - u_{ij}^{im}\|_{L_2}$ as $h \to 0$, $\tau \to 0$.

In summary, we have shown that the two-grid algorithm can be used to obtain a smooth solution of the two-dimensional problem considered here. The solution order of accuracy is similar to that of the fully implicit scheme on the fine grid.

5 Results of numerical experiments

The algorithm is tested by computing nine model problems.

Problem 1: One-dimensional heat diffusion equation with unit coefficients under boundary conditions of the first kind,

$$\begin{aligned} &\frac{\partial u}{\partial t} = \frac{\partial^2 u}{\partial x^2}, \qquad 0 < x < 1, \qquad 0 < t \leqslant T, \\ &u(0,t) = 0, \qquad u(1,t) = 0, \qquad 0 \leqslant t \leqslant T, \\ &u(x,0) = \sin(\pi x), \qquad 0 \leqslant x \leqslant 1, \end{aligned}$$

where T=0.0679. The exact solution is $u_T(x,t) = \exp^{-\pi^2 t}\sin(\pi x)$. A single smoothing iteration is performed ($\sigma = 0.5$). The correction equations are solved by a tridiagonal method [1], [7]. Table 1 lists the maximum relative error for several values of $K = \tau/h^2$. The results presented here demonstrate the stability and accuracy of the algorithm.

Problem 2: Two-dimensional heat diffusion equation with unit coefficients under boundary conditions of the first kind,

$$\begin{aligned} &\frac{\partial u}{\partial t} = \frac{\partial^2 u}{\partial x^2} + \frac{\partial^2 u}{\partial y^2}, \qquad 0 < x < 1, 0 < y < 1, \qquad 0 < t \leqslant T, \\ &u(x,0,t) = 0, \qquad u(x,1,t) = 0, \qquad 0 \leqslant x \leqslant 1, \qquad 0 \leqslant t \leqslant T, \\ &u(0,y,t) = 0, \qquad u(1,y,t) = 0, \qquad 0 \leqslant y \leqslant 1, \qquad 0 \leqslant t \leqslant T, \\ &u(x,y,0) = \sin(\pi x)\sin(\pi y), \qquad 0 \leqslant x \leqslant 1, \qquad 0 \leqslant y \leqslant 1, \end{aligned}$$

Table 1. $\max_{x_i,t} |(u(x_i,t) - u_T(x_i,t))/u_T(x_i,t)|$ in Problem 1.

K	N=100	N=500	N=1000
1	0.039	0.00159	0.0005
10	0.34	0.0139	0.0036
100	3.25	0.136	0.034
1000		1.3	0.34

where $T = 0.199$. The exact solution is $u_{ij}^T = \exp^{-2\pi^2 t}\sin(\pi x_i)\sin(\pi y_j)$. A single smoothing iteration is performed, with $\sigma = 0.5$. The correction equations are solved by the Gauss-Seidel method [1], [7]. The computations are performed for $h_x = h_y = 1/N$. In Table 2, the maximum absolute and relative errors are listed in the first and second columns corresponding to each N, respectively.

Table 2. $\max_{i,j,t} |u_{ij} - u_{ij}^T|$ and $\max_{i,j,t} |(u_{ij} - u_{ij}^T)/u_{ij}^T|$ in Problem 2.

K	N=100		N=200		N=500	
1	$8.405 \cdot 10^{-5}$	$4.356 \cdot 10^{-3}$	$2.097 \cdot 10^{-5}$	$1.087 \cdot 10^{-3}$		
5	$3.864 \cdot 10^{-4}$	$2.003 \cdot 10^{-2}$	$9.626 \cdot 10^{-5}$	$4.988 \cdot 10^{-3}$		
10	$7.665 \cdot 10^{-4}$	$3.973 \cdot 10^{-2}$	$1.905 \cdot 10^{-4}$	$9.872 \cdot 10^{-3}$	$3.043 \cdot 10^{-4}$	$1.577 \cdot 10^{-3}$
20	$1.534 \cdot 10^{-3}$	$7.949 \cdot 10^{-2}$	$3.794 \cdot 10^{-4}$	$1.966 \cdot 10^{-2}$		
100	$7.976 \cdot 10^{-3}$	0.4134	$1.912 \cdot 10^{-3}$	$9.907 \cdot 10^{-2}$	$3.019 \cdot 10^{-4}$	$1.564 \cdot 10^{-2}$
1000	$9.393 \cdot 10^{-2}$	4.868	$2.112 \cdot 10^{-2}$	1.0948	$3.081 \cdot 10^{-3}$	0.1596

These results demonstrate the stability and accuracy of the algorithm.

Table 3 lists the maximum absolute and relative errors of solutions obtained by using the proposed algorithm and the implicit scheme. It is obvious that the difference between corresponding values is insignificant.

Problems 3-9: Initial-boundary value problems for the heat diffusion equation,

$$\begin{aligned}
&\frac{\partial u}{\partial t} = \frac{\partial}{\partial x}k\frac{\partial u}{\partial x} + \frac{\partial}{\partial y}k\frac{\partial u}{\partial y}, && 0 < x < 1, && 0 < y < 1, && 0 < t \leqslant T,\\
&u(x,0,t) = 1 + x, \quad u(x,1,t) = 2 + x, && && 0 \leqslant x \leqslant 1, && 0 \leqslant t \leqslant T,\\
&u(0,y,t) = 1 + y, \quad u(1,y,t) = 2 + y, && && 0 \leqslant y \leqslant 1, && 0 \leqslant t \leqslant T,\\
&u(x,y,0) = 1 + x + y, && && 0 \leqslant x \leqslant 1, && 0 \leqslant y \leqslant 1,
\end{aligned}$$

where $T = 0.199$ and the function $k(x,y)$ is specified below for each particular problem. Problems 3-9 are solved by the two-grid algorithm described above.

Table 3. $\max_{i,j,t}|u_{ij} - u^T_{ij}|$, $\max_{i,j,t}|(u_{ij} - u^T_{ij})/u^T_{ij}|$, $\max_{i,j,t}|u^{im}_{ij} - u^T_{ij}|$, and $\max_{i,j,t}|(u^{im}_{ij} - u^T_{ij})/u^T_{ij}|$ in Problem 2.

K	N=100		N=100, implicit schema	
1	$8.405 \cdot 10^{-5}$	$4.356 \cdot 10^{-3}$	$8.389 \cdot 10^{-5}$	$4.348 \cdot 10^{-3}$
5	$3.864 \cdot 10^{-4}$	$2.003 \cdot 10^{-2}$	$3.885 \cdot 10^{-4}$	$1.999 \cdot 10^{-2}$
10	$7.665 \cdot 10^{-4}$	$3.973 \cdot 10^{-2}$	$7.653 \cdot 10^{-4}$	$3.966 \cdot 10^{-2}$
20	$1.534 \cdot 10^{-3}$	$7.949 \cdot 10^{-2}$	$1.531 \cdot 10^{-3}$	$7.936 \cdot 10^{-2}$
100	$7.976 \cdot 10^{-3}$	0.4134	$7.947 \cdot 10^{-3}$	0.4124
1000	$9.393 \cdot 10^{-2}$	4.868	$9.378 \cdot 10^{-2}$	4.861

In the difference equations (2) and (5) thermal conductivity is approximated by the half-sum of its values at the neighboring grid points. The coarse grid correction equation is solved by the MICCG(0) (modified incomplete Cholesky conjugate gradient) method [8]. The solution obtained by the two-grid algorithm is compared with that of the implicit scheme used on the fine grid.

In Problems 3-7, thermal conductivity is discontinuous as a function of coordinates. In Problem 3,

$$k = \begin{cases} 100, & \text{if } x \in (0.25, 0.75) \text{ and } y \in (0.25, 0.75), \\ 1, & \text{otherwise.} \end{cases}$$

In problem 4,

$$k = \begin{cases} 100, & \text{if } x \in [0.25, 0.75] \text{ and } y \in [0.25, 0.75], \\ 1, & \text{otherwise.} \end{cases}$$

When thermal conductivity is approximated on the coarse grid by the formula $\acute{k}_{lp} = k(x_l, y_p)$, where $x_l = lH$ and $y_p = pH$, then the algorithm using a single smoothing iteration is unstable if the point of discontinuity of thermal conductivity is an odd fine-grid point. To preserve stability, more than one smoothing iteration is required.

At the coarse-grid points, thermal conductivity is approximated as follows:

$$\acute{k}_{lp} = \frac{k_{ij}}{4} + \frac{k_{i+1,j} + k_{i-1,j} + k_{i,j+1} + k_{i,j-1}}{8} + \\ + \frac{k_{i+1,j+1} + k_{i-1,j-1} + k_{i-1,j+1} + k_{i+1,j-1}}{16}, \tag{41}$$

where $i = 2l, j = 2p$.

Table 4 lists the values of $\max_{i,j,t}|(u^{im}_{ij} - u_{ij})/u^{im}_{ij}|$ in Problem 4 evaluated over the interval $0 < t < 0.199$ ($n \geqslant 2$) and $\max_{i,j}|(u^{im}_{ij} - u_{ij})/u^{im}_{ij}|$ at $t = 0.199$ in the first and second columns corresponding to each N, respectively (s is the number of smoothing iterations).

Table 4. $\max_{i,j,t} |(u_{ij}^{im} - u_{ij})/u_{ij}^{im}|$ and $\max_{i,j} |(u_{ij}^{im} - u_{ij})/u_{ij}^{im}|$ at $t = 0.199$ in Problem 4, with k approximated by (41).

	N=100			N=200			N=500		
K	s=1	s=2	s=1	s=1	s=2	s=1	s=1	s=2	s=1
10	.061	.038	$.143 \cdot 10^{-9}$	.092	.060	$.164 \cdot 10^{-12}$			
100	.048	.019	$.189 \cdot 10^{-2}$	.077	.033	$.128 \cdot 10^{-3}$	.045	.026	$.247 \cdot 10^{-12}$
1000			.0492	.035	.025	$.397 \cdot 10^{-2}$	.037	.013	$.184 \cdot 10^{-14}$

It is clear from Table 4 that the algorithm is stable as applied to Problem 4. However, more than one smoothing iteration may be required to achieve good accuracy. Similar results are obtained in Problem 3.

In Problem 5,

$$k = \begin{cases} 100, & \text{if } (x-0.5)^2 + (y-0.5)^2 < 1/16, \\ 1, & \text{otherwise.} \end{cases}$$

In Problem 6,

$$k = \begin{cases} 100(1 + 0.3 \sin 10\pi x \sin 10\pi y), & \text{if } (x-0.5)^2 + (y-0.5)^2 < 1/16, \\ 1, & \text{otherwise.} \end{cases}$$

In Problem 7,

$$k = \begin{cases} 100(1 + 0.3 \sin 10\pi x), & \text{if } (x-0.5)^2 + (y-0.5)^2 < 1/16, \\ 1, & \text{otherwise.} \end{cases}$$

We used formula (41) to approximate thermal conductivity on the coarse grid. In each row of Tables 5–7, the first two numbers corresponding to a particular N are the values of $\max_{i,j,t} |(u_{ij}^{im} - u_{ij})/u_{ij}^{im}|$ at $0 < t < 0.199$ ($n \geqslant 2$), and the third one is the value of $\max_{i,j} |(u_{ij}^{im} - u_{ij})/u_{ij}^{im}|$ at $t = 0.199$ in Problems 5-7, respectively.

Table 5. $\max_{i,j,t} |(u_{ij}^{im} - u_{ij})/u_{ij}^{im}|$ and $\max_{i,j} |(u_{ij}^{im} - u_{ij})/u_{ij}^{im}|$ at $t = 0.199$ in Problem 5.

	N=100			N=200			N=500		
K	s=1	s=2	s=1	s=1	s=2	s=1	s=1	s=2	s=1
10	.045	.031	$.96 \cdot 10^{-9}$	.030	.021	$.245 \cdot 10^{-10}$			
50	.026	.014	$.62 \cdot 10^{-5}$	.021	.011	$.169 \cdot 10^{-8}$			
100	.02	.007	$.265 \cdot 10^{-3}$	.016	.007	$.129 \cdot 10^{-6}$	.028	.015	$.333 \cdot 10^{-10}$

Table 6. $\max_{i,j,t}|(u_{ij}^{im}-u_{ij})/u_{ij}^{im}|$ and $\max_{i,j}|(u_{ij}^{im}-u_{ij})/u_{ij}^{im}|$ at $t = 0.199$ in Problem 6.

	N=100			N=200			N=500		
K	s=1	s=2	s=1	s=1	s=2	s=1	s=1	s=2	s=1
10	.047	.032	$.108 \cdot 10^{-8}$	.030	.022	$.259 \cdot 10^{-10}$			
50	.027	.014	$.137 \cdot 10^{-4}$	.022	.011	$.203 \cdot 10^{-8}$			
100	.021	.009	$.313 \cdot 10^{-3}$	.017	.007	$.170 \cdot 10^{-6}$	.029	.016	$.353 \cdot 10^{-10}$

Table 7. $\max_{i,j,t}|(u_{ij}^{im}-u_{ij})/u_{ij}^{im}|$ and $\max_{i,j}|(u_{ij}^{im}-u_{ij})/u_{ij}^{im}|$ at $t = 0.199$ in Problem 7.

	N=100			N=200			N=500		
K	s=1	s=2	s=1	s=1	s=2	s=1	s=1	s=2	s=1
10	.044	.030	$.495 \cdot 10^{-6}$	.030	.021	$.122 \cdot 10^{-4}$			
50	.031	.015	$.137 \cdot 10^{-4}$	.024	.016	$.304 \cdot 10^{-4}$			
100	.03	.013	$.41 \cdot 10^{-2}$	.016	.007	$.126 \cdot 10^{-4}$	.033	.016	$.133 \cdot 10^{-4}$

These results demonstrate that the proposed algorithm provides good accuracy as applied to an initial-boundary value problem for the heat diffusion equation.

To examine the dependence of accuracy on the magnitude of the jump in thermal conductivity, we compare the results for problems 3-7 presented above with the results obtained for a relatively small jump in k and with those for thermal conductivity defined as a continuous function of coordinates. In Problem 8,

$$k = \begin{cases} 1 + 0.3\sin 10\pi x, & \text{if } (x-0.5)^2 + (y-0.5)^2 < 1/16, \\ 1, & \text{otherwise.} \end{cases}$$

In Problem 9,

$$k = 1 + 0.3\sin 10\pi x.$$

Tables 8 and 9 list the values of $\max_{i,j,t}|(u_{ij}^{im}-u_{ij})/u_{ij}^{im}|$ at $0 < t < 0.199$, and $\max_{i,j}|(u_{ij}^{im}-u_{ij})/u_{ij}^{im}|$ at $t = 0.199$ in the first and second columns corresponding to each value of N in Problems 8 and 9, respectively. These results are obtained by using approximation (41) for thermal conductivity on the coarse grid.

It is clear from comparison between Tables 7, 8 and 9 that higher accuracy is achieved when k is continuous or has a small jump, as compared to the case of a large jump in thermal conductivity.

Table 8. $\max_{i,j,t} |(u_{ij}^{im} - u_{ij})/u_{ij}^{im}|$ and $\max_{i,j} |(u_{ij}^{im} - u_{ij})/u_{ij}^{im}|$ at $t = 0.199$ in Problem 8 (s=1).

K	N=100		N=200		N=500	
10	.000313	$.213 \cdot 10^{-7}$	.000197	$.207 \cdot 10^{-7}$		
100	.000457	$.979 \cdot 10^{-6}$	.000276	$.951 \cdot 10^{-7}$	$.898 \cdot 10^{-4}$	$.102 \cdot 10^{-7}$

Table 9. $\max_{i,j,t} |(u_{ij}^{im} - u_{ij})/u_{ij}^{im}|$ and $\max_{i,j} |(u_{ij}^{im} - u_{ij})/u_{ij}^{im}|$ at $t = 0.199$ in Problem 9 (s=1).

K	N=100		N=200		N=500	
10	.000243	$.416 \cdot 10^{-7}$	$.614 \cdot 10^{-4}$	$.207 \cdot 10^{-7}$		
100	.00029	$.898 \cdot 10^{-6}$	$.846 \cdot 10^{-4}$	$.378 \cdot 10^{-7}$	$.24 \cdot 10^{-4}$	$.188 \cdot 10^{-8}$

6 Conclusion

A new efficient algorithm is developed for solving diffusion-type equations. By applying the algorithm to several model problems, it is shown both theoretically and numerically that the algorithm is similar to an implicit scheme in terms of stability and accuracy. The new algorithm substantially reduces the the computational complexity at each time level, as compared to implicit schemes.

References

1. A.A. SAMARSKY, *Difference scheme theory*, Nauka, 1989 (in Russian).
2. N.S. BAHVALOV, N.P. ZHIDKOV AND G.M. KOBELKOV, *Numerical methods*, Nauka, 1987 (in Russian).
3. R.P. FEDORENKO, *A relaxation method for solving difference elliptic equations*, Zh. Vychisl. Mat. Mat. Fiz., Vol.1 (1961), N 5, pp. 922-927 (in Russian).
4. S.K. GODUNOV, V.S.RYABENKIY, *A relaxation method for solving difference elliptic equations*, Zh. Vychisl. Mat. Mat. Fiz., Vol.1 (1961), N 5, pp. 922-927 (in Russian).
5. R. RIHTMAER, K.MORTON, *Difference methods for solving of boundary value problem*, Mir, 1972 (in Russian).
6. M.E. LADONKINA, O.YU. MILYUKOVA, V.F. TISHKIN, *A numerical algorithm for diffusion-type equations based on the multigrid methods*, Mat. Model., Vol.19, (2007), N 4, pp. 71-89, (in Russian).
7. A.A. SAMARSKY, YE.S. NIKOLAYEV, *Methods for solving finite-difference equations*, Nauka, (1978).
8. I. GUSTAFSSON, *A Class of First Order Factorization Methods*, BIT, V.18 (1978), pp.142-156.
9. M.E. LADONKINA, O.YU. MILYUKOVA, V.F. TISHKIN, *Application of the multigrid method for calculation diffusion processes*, CD-Proceedings of

West-East Speed Flow Field Conference 19-22, November 2007, Moscow, Russia (http://wehsff.imamod.ru/pages/s7.htm).

Monotone Matrices and Finite Volume Schemes for Diffusion Problems Preserving Non-Negativity of Solution

I. V. Kapyrin

Institute of Numerical Mathematics, Russian Academy of Sciences,
ul. Gubkina 8, Moscow, 119333 Russia
ivan.kapyrin@gmail.com

Abstract. A new finite volume scheme for 3D diffusion problems with heterogeneous full diffusion tensor is considered. The discretization uses nonlinear two-point flux approximation on unstructured tetrahedral grids. Monotonicity of the linearized operator allows us to guarantee non-negativity of the discrete solution.

Introduction

The simulation of substance transport in porous media [1] necessitates the discretization of the diffusion operator. In such problems, the diffusion tensor is strongly inhomogeneous and anisotropic and the geometry of the computational domain requires the use of unstructured condensing meshes. Under these conditions, the solutions produced by some modern numerical schemes [2] exhibit unphysical oscillations and negative values. Negative solution values may lead to incorrectly computed chemical interactions between the substances. As a result, the scheme becomes nonconservative.

In the present paper a finite volume (FV) method for numerical solution of three-dimensional diffusion problems with anisotropic full diffusion tensor on tetrahedral grids is being considered. The method was introduced in [3] for problems with homogeneous Dirichlet boundary conditions, here we extend it to the case of nonhomogeneous conditions of Dirichlet and Neumann types. For the formulation of the schemes we use a special nonlinear diffusive flux approximation, introduced for two-dimensional diffusion problems by C.Le Potier in [4] and modified in [5]. The resulting schemes are conservative and monotone in the sense of ensuring the nonnegativity of solution for respective sources and boudary conditions (see [6], Section 2.4). The proof of the latter feature of the method is based on the monotonicity property of the linearized operator matrix.

1 Nonlinear finite volume method

Let Ω be a convex polyhedral domain in $\mathbb{R}^3$ with boundary $\partial\Omega$. Consider the stationary diffusion equation with two types of boundary conditions in the mixed statement:

$$\nabla \cdot \mathbf{r} = f, \tag{1a}$$

$$\mathbf{r} = -\mathcal{D}\nabla C \text{ in } \Omega, \tag{1b}$$

$$C|_{\Gamma_D} = g^D(x), \quad \mathbf{r} \cdot \mathbf{n}|_{\Gamma_N} = g^N(x). \tag{1c}$$

Here, C is the concentration of the substance, $\mathbf{r}$ is the diffusion flux, f is the source function, and $\mathcal{D}$ is a symmetric positive definite diffusion tensor of dimension 3×3 that is piecewise constant in Ω. The boundary $\partial\Omega$ consists of two parts Γ_D and Γ_N. On Γ_D the concentration is specified by a continuous function $g^D(x)$. On Γ_N the continuous function $g^N(x)$ prescribes the diffusive flux through the boundary. In the following we assume that Γ_N is the union of nonintersecting planar fragments.

In the computational domain Ω, we construct a conformal tetrahedral mesh ε_h, such that the diffusion tensor is constant on each of its elements T. Let N_T be the number of tetrahedra $T \in \varepsilon_h$, N_P be the number of vertices, N_e be the total number of faces, and N_B be the number of external faces in ε_h.

The mass conservation law (1a) can be integrated with respect to $T \in \varepsilon_h$ by using Green's identity:

$$\int_{\partial T} \mathbf{r} \cdot \mathbf{n} \, ds = \int_T f \, dx \quad \forall T \in \varepsilon_h, \tag{2}$$

where $\mathbf{n}$ denotes the unit outward normal to ∂T. Let $\mathbf{n}_e$ be an outward normal to the face e of T whose length is numerically equal to the surface area of the corresponding face; i.e., $|\mathbf{n}_e| = |e|$. Relation (2) can be rewritten as:

$$\sum_{e \in \partial T} \mathbf{r}_e \cdot \mathbf{n}_e = \int_T f \, dx \quad \forall T \in \varepsilon_h, \tag{3}$$

where $\mathbf{r}_e$ is the mean diffusion flux density through the face e:

$$\mathbf{r}_e = \frac{1}{|e|} \int_e \mathbf{r} \, ds$$

The diffusion flux $\mathbf{r}_e \cdot \mathbf{n}_e$ through e can be approximated as follows. For each $T \in \varepsilon_h$ and each external face e, we introduce their degrees of freedom. The set of support points of these degrees of freedom is defined as $\mathbb{B} = \{X_j\}_{j=1}^{N_T+N_B}$. For each tetrahedron T, $\mathbb{B}$ includes some point X_T inside T (its coordinates will be specified later). Let the tetrahedron T have a face e belonging to $\partial\Omega$ and $\mathbf{n}_e$

be the outward normal to e. Then if $e \in \Gamma_D$ we add its center of mass X_e to $\mathbb{B}$, otherwise, if $e \in \Gamma_N$ we add to $\mathbb{B}$ the projection X_e of the internal point X_T along the vector $\mathcal{D}\mathbf{n}_e$ (the choice of X_T will guarantee that X_e lies inside the face e). Since Ω is convex, for any internal vertex O_i of ε_h, there are four points $X_{i,j}$ $(j = 1, 2, 3, 4)$ from $\mathbb{B}$ such that O_i lies inside the tetrahedron formed by them (the nearest points are picked). Therefore, there are nonnegative coefficients $\lambda_{i,j}$ satisfying the conditions

$$\sum_{j=1}^{4} \lambda_{i,j} \cdot \overrightarrow{O_i X_{i,j}} = 0, \quad \sum_{j=1}^{4} \lambda_{i,j} = 1.$$

The coefficients $\lambda_{i,j} \geqslant 0$ are used for linear interpolation of the concentration at interior nodes of the initial mesh from its values at points of $\mathbb{B}$:

$$C_{O_i} = \sum_{j=1}^{4} \lambda_{i,j} C_{X_{i,j}}. \tag{4}$$

A similar formula can be written for the concentrations at points $O_i \in \Gamma_N$ using the values at three vertexes of a triangle in Γ_N, which contains O_i. For the points $O_i \in \Gamma_D$ the interpolation is not needed because the respective concentration values are known from the Dirichlet boundary conditions.

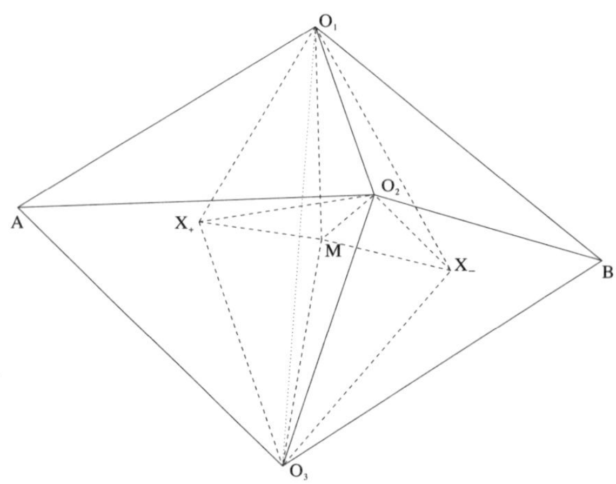

Fig. 1. Geometric constructions for the nonlinear finite-volume method.

Consider two neighboring tetrahedra $T_+ = AO_1O_2O_3$ and $T_- = BO_1O_2O_3$ in the initial mesh ε_h (see figure 1), X_+,X_- are the corresponding elements in $\mathbb{B}$, $\mathcal{D}_+$ and $\mathcal{D}_-$ are diffusion tensors, and V^+ and V^- – are their volumes. Let

M be the center of mass of the common face e, $e = O_1O_2O_3$. We introduce the following notation (here and below, i, j and k are assumed to be different; i.e., $\{i, j, k\} = \{1, 2, 3\}, \{2, 1, 3\}, \{3, 1, 2\}$):

- T_i^+ and (T_i^-) are the tetrahedra $X_+MO_jO_k$ and $X_-MO_jO_k$ respectively, and V_i^+ and V_i^- are their respective volumes.
- $\mathbf{n}_e$ is the normal to the common face $O_1O_2O_3$, that is external with respect to T_+.
- $\mathbf{n}_{ei}^+$ and $\mathbf{n}_{ei}^-$ are the normals to the face MO_jO_k, that are external with respect to T_i^+ and T_i^-, respectively.
- $\mathbf{n}_{ij}^+$ and $\mathbf{n}_{ij}^-$ are the normals to the respective faces MX_+O_k and MX_-O_k, that are external with respect to T_i^+ and T_i^-, respectively.
- $\mathbf{n}_i^+$ and $\mathbf{n}_i^-$ are the normals to the respective faces $X_+O_jO_k$ and $X_-O_jO_k$, are external with respect to T_i^+ and T_i^-, respectively.
- The lengths of all the above normals are numerically equal to the surface areas of the corresponding faces.

Each pair of tetrahedra T_i^+ and T_i^- is associated with an auxiliary variable $C_{M,i}$, that is the substance concentration at the point M. The diffusion flux $\mathbf{r}_i^*$ (here and below, the star denotes either a plus or a minus) on each tetrahedron T_i^* is defined by using Green's identity $\int\limits_{T_i^*} \nabla C \, dx = \int\limits_{\partial T_i^*} C\mathbf{n} \, ds$, integrating it to second-order accuracy, and taking into account $\mathbf{n}_i^* + \mathbf{n}_{ei}^* + \mathbf{n}_{ij}^* + \mathbf{n}_{ik}^* = 0$:

$$V_i^* \mathcal{D}_*^{-1} \mathbf{r}_i^* = \frac{1}{3} \left(\mathbf{n}_i^* C_{M,i} + \mathbf{n}_{ei}^* C_{X_*} + \mathbf{n}_{ij}^* C_{O_j} + \mathbf{n}_{ik}^* C_{O_k} \right). \tag{5}$$

The introduced degrees of freedom $C_{M,i}$ are eliminated using the assumption of flux continuity through e: $\mathbf{r}_i^+ \cdot \mathbf{n}_e = \mathbf{r}_i^- \cdot \mathbf{n}_e$. As a result, the flux in (5) is defined in terms of the concentrations C_{X_+}, C_{X_-} at the points X_+ and X_- and in terms of C_{O_j} and C_{O_k}, for which we use linear interpolation (4). The total diffusion flux $\mathbf{r}_e \cdot \mathbf{n}_e$ through e is represented as a linear combination of three fluxes $\mathbf{r}_i^+ \cdot \mathbf{n}_e$:

$$\mathbf{r}_e \cdot \mathbf{n}_e = \mu_1^e \mathbf{r}_1^+ \cdot \mathbf{n}_e + \mu_2^e \mathbf{r}_2^+ \cdot \mathbf{n}_e + \mu_3^e \mathbf{r}_3^+ \cdot \mathbf{n}_e. \tag{6}$$

To determine the coefficients μ_i^e, $i = 1, 2, 3$, we set the following conditions on diffusion flux (6) through e.

- If the values $\mathbf{r}_i^+ \cdot \mathbf{n}_e / |\mathbf{n}_e|$ approximate the diffusion flux density, then $\mathbf{r}_e \cdot \mathbf{n}_e / |\mathbf{n}_e|$ is also its approximation:

$$\sum_{j=1}^{3} \mu_j^e = 1. \tag{7}$$

- The approximation stencil for the flux is two-point and nonlinear:

$$\mathbf{r}_e \cdot \mathbf{n}_e = K_+(C_{O_1}, C_{O_2}, C_{O_3}) C_{X_+} - K_-(C_{O_1}, C_{O_2}, C_{O_3}) C_{X_-}. \tag{8}$$

This condition is ensured by the equation

$$(a_{12}C_{O_2}+a_{13}C_{O_3})\mu_1^e+(a_{21}C_{O_1}+a_{23}C_{O_3})\mu_2^e+(a_{31}C_{O_1}+a_{32}C_{O_2})\mu_3^e=0, \tag{9}$$

where

$$a_{ij}=\frac{(\mathcal{D}_+\mathbf{n}_j^+,\mathbf{n}_a)(\mathcal{D}_-\mathbf{n}_i^-,\mathbf{n}_a)-(\mathcal{D}_-\mathbf{n}_j^-,\mathbf{n}_a)(\mathcal{D}_+\mathbf{n}_i^+,\mathbf{n}_a)}{(\mathcal{D}_+\mathbf{n}_i^+,\mathbf{n}_a)V_i^- -(\mathcal{D}_-\mathbf{n}_i^-,\mathbf{n}_a)V_i^+}.$$

Equations (7) and (9) define a family of solutions with parameter p^e:

$$\mu_1^e(p^e)=\mu_1^e(0)+p^e[C_{O_1}(a_{31}-a_{21})+C_{O_2}a_{32}-C_{O_3}a_{23}], \tag{10a}$$
$$\mu_2^e(p^e)=\mu_2^e(0)+p^e[C_{O_2}(a_{12}-a_{32})+C_{O_3}a_{13}-C_{O_1}a_{31}], \tag{10b}$$
$$\mu_3^e(p^e)=\mu_3^e(0)+p^e[C_{O_3}(a_{23}-a_{13})+C_{O_1}a_{21}-C_{O_2}a_{12}]. \tag{10c}$$

Here, $\mu_1^e(0),\mu_2^e(0)$ and $\mu_3^e(0)$ comprise a particular solution to system (7),(9):

$$\mu_i^e(0)=\frac{[(\mathcal{D}_-\mathbf{n}_i^-,\mathbf{n}_e)V_i^+-(\mathcal{D}_+\mathbf{n}_i^+,\mathbf{n}_e)V_i^-]C_{O_i}}{\sum\limits_{j=1}^{3}[(\mathcal{D}_-\mathbf{n}_j^-,\mathbf{n}_e)V_j^+-(\mathcal{D}_+\mathbf{n}_j^+,\mathbf{n}_e)V_j^-]C_{O_j}}. \tag{11}$$

Remark 1. Coefficients (11) are identical to those in the two-dimensional nonlinear finite-volume method with the volumes replaced by areas. In the two-dimensional case, μ_1^e and μ_2^e are unique and precisely determined by conditions (7) and (8) on two-point approximations of the diffusion flux.

In case when $O_1O_2O_3\in\Gamma_N$, we have the following diffusive flux approximation

$$\mathbf{r}_e\cdot\mathbf{n}_e=\int_e g^N(x)\,ds. \tag{12}$$

If the face $O_1O_2O_3$ belongs to Γ_D, Green's identity on the tetrahedron $X_+O_1O_2O_3$ with volume V^+ yields the equation

$$V^+\mathcal{D}^{-1}\mathbf{r}=\frac{1}{3}(C_{X_+}\mathbf{n}_e+C_{O_1}\mathbf{n}_1^+ +C_{O_2}\mathbf{n}_2^+ +C_{O_3}\mathbf{n}_3^+), \tag{13}$$

where C_{O_i}, $i\in\{1,2,3\}$ are known from the boundary conditions. For the external face $e\in\Gamma_D$, we can write

$$\mathbf{r}_e\cdot\mathbf{n}_e=K_{B_+}C_{X_+}+K_{B_-}, \tag{14}$$

where $K_{B_+}=\dfrac{(\mathcal{D}\mathbf{n}_e,\mathbf{n}_e)}{3V^+}$, and

$$K_{B_-}=\frac{(\mathcal{D}\mathbf{n}_1^+,\mathbf{n}_e)C_{O_1}+(\mathcal{D}\mathbf{n}_2^+,\mathbf{n}_e)C_{O_2}+(\mathcal{D}\mathbf{n}_3^+,\mathbf{n}_e)C_{O_3}}{3V^+}. \tag{15}$$

Thus, the diffusion flux $\mathbf{r}_e \cdot \mathbf{n}_e$ is defined by formulas (6), (10) and (5) for internal mesh faces and by formulas (12), (14) for external mesh faces.

Let C_T be the concentration at the point X_T corresponding to tetrahedron T having the face $e \in \Gamma_N$. We eliminate the concentration C_e at the point X_e on the face e using the approximation of diffusive flux through e:

$$\frac{C_e - C_T}{l} = -g^N(X_e),$$

where $l = \frac{\|X_e - X_T\|}{\|\mathcal{D}\mathbf{n}\|}$ and $\mathbf{n}$ is the unit normal vector to the face e. It is to be mentioned here that with nonnegative C_{T_i}, $i = 1, .., N_T$ and a nonpositive function $g^N(x)$ the nonnegativity of C_{O_i} in (4) is guaranteed after the elimination of C_e for all faces $e \in \Gamma_N$.

The formulation of the method is completed by substituting the flux expressions into mass conservation law (3). Discretization of (3) produces a nonlinear system of equations

$$A(C_X)C_X = F, \tag{16}$$

where C_X is the N_T-vector of unknown concentrations at the points X_T of the set $\mathbb{B}$. The matrix $A(C_X)$ can be represented as the union of submatrices

$$A(C_X) = \sum_{e \in \partial \varepsilon_h} \mathcal{N}_e A_e(C_X) \mathcal{N}_e^T, \tag{17}$$

$\mathcal{N}_e$ being the respective assembling matrices, consisting of zeros and ones. Here $A_e(C_X)$ is a 2×2 matrix of the form

$$A_e(C_X) = \begin{pmatrix} K_{e_+} & -K_{e_-} \\ -K_{e_-} & K_{e_+} \end{pmatrix} \tag{18}$$

for any internal face e and a 1×1 matrix of the form $A_e(C_X) = K_{B_+}$ for any $e \in \Gamma_D$. For the component F_T of the right-hand-side vector F corresponding to tetrahedron T the following relation holds:

$$F_T = \int_T f dx - \sum_{e \in \partial T \cap \Gamma_D} K_{B_-} - \sum_{e \in \partial T \cap \Gamma_N} \int_e g^N ds. \tag{19}$$

System (16) is solved using the Picard iteration

$$A(C_X^k)C_X^{k+1} = F \tag{20}$$

with some initial approximation C_X^0. To construct monotone schemes, we define the location of a point $X_T \in \mathbb{B}$ corresponding to an arbitrary tetrahedron $T = ABCD$ in the initial mesh ε_h with faces a, b, c and d opposite to A, B, C, D and D, respectively. Let $\mathbf{R}_A, \mathbf{R}_B, \mathbf{R}_C$ and $\mathbf{R}_D$ be the position vectors of the corresponding vertices of T. The vectors $\mathbf{n}_a, \mathbf{n}_b, \mathbf{n}_c$ and $\mathbf{n}_d$ are outward normals to the faces. Their lengths are numerically equal to the surface areas of the corresponding faces. Define

$$\mathbf{R}_{X_T} = \frac{\mathbf{R}_A\|\mathbf{n}_a\|_{\mathcal{D}} + \mathbf{R}_B\|\mathbf{n}_b\|_{\mathcal{D}} + \mathbf{R}_C\|\mathbf{n}_c\|_{\mathcal{D}} + \mathbf{R}_D\|\mathbf{n}_d\|_{\mathcal{D}}}{\|\mathbf{n}_a\|_{\mathcal{D}} + \|\mathbf{n}_b\|_{\mathcal{D}} + \|\mathbf{n}_c\|_{\mathcal{D}} + \|\mathbf{n}_d\|_{\mathcal{D}}}, \tag{21}$$

where $\|\mathbf{n}_\beta\|_{\mathcal{D}} = \sqrt{(\mathcal{D}\mathbf{n}_\beta, \mathbf{n}_\beta)}$ and $\beta \in \{a, b, c, d\}$. Note that, for an isotropic tensor, expression (21) gives the coordinates of the center of the sphere inscribed in T.

2 Monotonicity of the method

Hereafter we formulate the monotonicity property that is the main feature of the proposed FV method.

Theorem 1. *Let the right-hand side in system (16) of the nonlinear finite-volume method be nonnegative (i.e., $F_i \geqslant 0$); the boundary conditions satisfy $g^D(x) \geqslant 0$ on Γ_D and $g^N(x) \leqslant 0$ on Γ_N. Let (16) be the corresponding nonlinear system of FV discretization for (1); the support points of the degrees of freedom on the tetrahedra be given by formula (21); the initial approximation be $(C_X^0)_i \geqslant 0$; and, for any internal face e, the nonnegative values μ_i^e, $i \in \{1,2,3\}$ be chosen from solutions (10a)-(10c) on every Picard iteration (20). Then all the iterative approximations to C_X are nonnegative:*

$$(C_X^k)_i \geqslant 0, \quad i = 1, \dots, N_T, \quad \forall k \geqslant 0.$$

Proof. We rely on the following definition of a monotone matrix: *The matrix A is called a monotone matrix if the condition $Ax \geqslant 0$ implies that the vector x is positive.* Assume that the matrix $A(C_X)$ is monotone for any nonnegative vector C_X, and the right-hand-side F is nonnegative. Then the solution C_X^{k+1} to system (20) is also a nonnegative vector. Taking into account $(C_X^0)_i \geqslant 0$, we find by induction that $(C_X^k)_i \geqslant 0$, $\forall k \geqslant 0$, $\forall i = 1, \dots, N_T$.

Let us prove that the matrix $A(C_X)$ is monotone for any nonnegative vector C_X, and the right-hand-side F is nonnegative. Consider the coefficients $K_+(C_{O_1}, C_{O_2}, C_{O_3})$, $K_-(C_{O_1}, C_{O_2}, C_{O_3})$, K_{B+} and K_{B-} in expressions (8) and (14) for the diffusion flux through a face. The coefficient K_{B+} is positive because $\mathcal{D}$ is positive definite. Plugging (5) (after eliminating $C_{M,i}$) into (6) gives formulas for K_+ and K_-:

$$K_+ = \sum_{i=1}^{3} \mu_i^e \cdot \frac{(\mathcal{D}_+\mathbf{n}_e, \mathbf{n}_e)}{3V^+} \cdot \frac{(\mathcal{D}_-\mathbf{n}_i^-, \mathbf{n}_e)V_i^+}{(\mathcal{D}_-\mathbf{n}_i^-, \mathbf{n}_e)V_i^+ - (\mathcal{D}_+\mathbf{n}_i^+, \mathbf{n}_e)V_i^-},$$

$$K_- = -\sum_{i=1}^{3} \mu_i^e \cdot \frac{(D_-\mathbf{n}_e, \mathbf{n}_e)}{3V^-} \cdot \frac{(D_+\mathbf{n}_i^+, \mathbf{n}_e)V_i^-}{(\mathcal{D}_-\mathbf{n}_i^-, \mathbf{n}_e)V_i^+ - (\mathcal{D}_+\mathbf{n}_i^+, \mathbf{n}_e)V_i^-}.$$

For K_+ and K_- to be positive and for K_{B-} to be nonpositive, it is sufficient to show that

$$(\mathcal{D}_- \mathbf{n}_i^-, \mathbf{n}_e) > 0, \quad (\mathcal{D}_+ \mathbf{n}_i^+, \mathbf{n}_e) < 0. \tag{22}$$

Consider the tetrahedron $ABCD \in \mathcal{E}_h$ with faces a, b, c and d opposite to the vertices A, B, C and D, respectively, and with normals $\mathbf{n}_a, \mathbf{n}_b, \mathbf{n}_c$ and $\mathbf{n}_d$ to these faces (the lengths of the normals are numerically equal to the surface areas of the corresponding faces). The point X_T inside the tetrahedron is defined by formula (21). Let $\mathbf{n}_{ab}$ be defined as the normal (external with respect to X_TBCD) to the plane X_TCD, $\mathbf{n}_{bc}$ be defined as the normal (external with respect to X_TACD) to the plane X_TAD, and so on for $\mathbf{n}_{\beta\gamma}$, where $\beta, \gamma \in \{a, b, c, d\}$, $\beta \neq \gamma$. Since the length of a normal is not important for the proof of (22), $\mathbf{n}_{ab}$ can be calculated as

$$\mathbf{n}_{ab} = \frac{1}{2}(\|\mathbf{n}_a\|_{\mathcal{D}} + \|\mathbf{n}_b\|_{\mathcal{D}} + \|\mathbf{n}_c\|_{\mathcal{D}} + \|\mathbf{n}_d\|_{\mathcal{D}})(\overrightarrow{CX_T} \times \overrightarrow{DX_T}). \tag{23}$$

For the vectors $\overrightarrow{CX_T}$ and $\overrightarrow{DX_T}$, we have the expressions

$$\overrightarrow{CX_T} = \frac{\overrightarrow{CA}\|\mathbf{n}_a\|_{\mathcal{D}} + \overrightarrow{CB}\|\mathbf{n}_b\|_{\mathcal{D}} + \overrightarrow{CD}\|\mathbf{n}_d\|_{\mathcal{D}}}{\|\mathbf{n}_a\|_{\mathcal{D}} + \|\mathbf{n}_b\|_{\mathcal{D}} + \|\mathbf{n}_c\|_{\mathcal{D}} + \|\mathbf{n}_d\|_{\mathcal{D}}},$$
$$\overrightarrow{DX_T} = \frac{\overrightarrow{DA}\|\mathbf{n}_a\|_{\mathcal{D}} + \overrightarrow{DB}\|\mathbf{n}_b\|_{\mathcal{D}} + \overrightarrow{DC}\|\mathbf{n}_c\|_{\mathcal{D}}}{\|\mathbf{n}_a\|_{\mathcal{D}} + \|\mathbf{n}_b\|_{\mathcal{D}} + \|\mathbf{n}_c\|_{\mathcal{D}} + \|\mathbf{n}_d\|_{\mathcal{D}}}.$$

Substituting them into vector product (23) gives

$$\mathbf{n}_{ab} = \mathbf{n}_b\|\mathbf{n}_a\|_{\mathcal{D}} - \mathbf{n}_a\|\mathbf{n}_b\|_{\mathcal{D}}.$$

Let us show that $(\mathcal{D}\mathbf{n}_a, \mathbf{n}_{ab}) < 0$ and $(\mathcal{D}\mathbf{n}_b, \mathbf{n}_{ab}) > 0$ by using the Cauchy–Schwarz inequality

$$\begin{aligned}(\mathcal{D}\mathbf{n}_a, \mathbf{n}_{ab}) &= (\mathcal{D}\mathbf{n}_a, \mathbf{n}_b)\|\mathbf{n}_a\|_{\mathcal{D}} - (\mathcal{D}\mathbf{n}_a, \mathbf{n}_a)\|\mathbf{n}_b\|_{\mathcal{D}} \\ &= \|\mathbf{n}_a\|_{\mathcal{D}}\left[(\mathbf{n}_a, \mathbf{n}_b)_{\mathcal{D}} - \|\mathbf{n}_a\|_{\mathcal{D}}\|\mathbf{n}_b\|_{\mathcal{D}}\right] < 0. \end{aligned} \tag{24}$$

Here, $(\cdot, \cdot)_{\mathcal{D}}$ is the scalar product in the metric defined by the tensor $\mathcal{D}$. Similarly, we can prove $(\mathcal{D}\mathbf{n}_b, \mathbf{n}_{ab}) > 0$ and inequalities of the form $(\mathcal{D}\mathbf{n}_\beta, \mathbf{n}_{\beta\gamma}) < 0$ and $(\mathcal{D}\mathbf{n}_\gamma, \mathbf{n}_{\beta\gamma}) > 0$, $\beta \neq \gamma$, where $\beta, \gamma \in \{a, b, c, d\}$. In (22), $\mathbf{n}_i^-$ and $\mathbf{n}_i^+$ are replaced by the corresponding vectors $\mathbf{n}_{\beta\gamma}$ and $\mathbf{n}_e$ is replaced by $\mathbf{n}_\beta$ or $\mathbf{n}_\gamma$. Then, using (24), we prove (22). Therefore, K_+ and K_- are positive and K_{B-} is nonpositive. Thus, the matrix $A(C_X)$ has the following properties.

- All the diagonal elements of $A(C_X)$ are positive.
- All the off-diagonal elements of $A(C_X)$ are nonpositive.
- The matrix is column diagonally dominant; this diagonal dominance is strict for columns corresponding to elements that have faces on the boundary of the computational domain with Dirichlet conditions.

Therefore, $A^T(C_X)$ is an M-matrix and all the elements of $(A^T(C_X))^{-1}$ are nonnegative. Since the transposition and inversion of matrices are commuting operations, we have $(A^T(C_X))^{-1} = (A^{-1}(C_X))^T$. Therefore, all the elements of $A^{-1}(C_X)$ are nonnegative and $A(C_X)$ is monotone.

The nonnegativity of right-hand-side F represented by the formula (19) is provided by the conditions of the theorem and the nonpositivity of coefficients K_{B-}. □

Remark 2. The validity of (22) implies that $\mu_i^e \geqslant 0,\ i \in \{1,2,3\}$ required in the assumption of the theorem can always be chosen by setting $p^e = 0\ \ \forall e$ in (10a)-(10c). The range of p^e for which μ_i^e are positive is an interval; it may degenerate into the point $p^e = 0$ when two of the three C_{O_i} are zero. If $C_{O_i} = 0\ \ \forall i \in \{1,2,3\}$, then solution (10a)-(10c) is always positive and does not depend on p^e.

Remark 3. The point X_T given by (21) is a solution to the system of six equations determining the equality of the angles in the $\mathcal{D}$-metric between the vectors $\mathbf{n}_\beta, \mathbf{n}_{\beta\gamma}$ and $\mathbf{n}_\gamma, -\mathbf{n}_{\beta,\gamma}$, where $\beta, \gamma \in \{a,b,c,d\}$ and $\beta \neq \gamma$.

Corollary 1. *Consider the nonstationary diffusion equation*

$$\frac{\partial C}{\partial t} - \nabla \cdot \mathcal{D}\nabla C = f \tag{25}$$

with a nonnegative right-hand side, a nonnegative initial condition, and a nonnegative Dirichlet boundary condition. The nonlinear FV method is used to construct the implicit scheme

$$\left(\frac{V}{\Delta t} + A(C_X^{n+1})\right) C_X^{n+1} = \frac{V}{\Delta t} C_X^n + F^{n+1},$$

where V is a diagonal matrix of elements' volumes and F involves the right-hand side and the boundary conditions. At every time step, the system is solved by the Picard method

$$\left(\frac{V}{\Delta t} + A(C_X^{n+1,k})\right) C_X^{n+1,k+1} = \frac{V}{\Delta t} C_X^n + F^{n+1}, \quad k = 1,2\ldots, \quad C_X^{n+1,0} = C_X^n.$$

If $\mu_i^e\ \ \forall e,\ i \in \{1,2,3\}$ are positive, then $(C_X^{n+1,k})_j \geqslant 0,\ j = 1,\ldots,N_T,\quad k = 1,2\ldots.$

Corollary 2. *In the explicit scheme for the discretization of* (25)

$$\frac{V}{\Delta t} C_X^{n+1} = \left(\frac{V}{\Delta t} - A(C_X^n)\right) C_X^n + F^{n+1},$$

the solution C_X^{n+1} can be made nonnegative by choosing a sufficiently small Δt ensuring that the diagonal elements of $V/\Delta t - A(C^n)$ are nonnegative *(its off-diagonal elements are obviously nonnegative). Moreover,* $\Delta t \sim h^2$ *(where* h *is the size of a quasi-uniform mesh), which is similar to the stability condition for explicit schemes.*

Although the convergence of the discrete solution to the solution of differential problem (1a)-(1c) is not proved, test computations have revealed that the nonlinear finite-volume method with coefficients (11) has quadratic convergence with respect to the concentration and linear convergence with respect to diffusion fluxes. At the same time the convergence of Picard iterations is not guaranteed and this problem may become a key question in the further development of this method.

Acknowledgements

The author is grateful to Yu. V. Vassilevski, C. Le Potier, D. A. Svyatski, and K. N. Lipnikov for fruitful discussions of the problem and the ideas used in the development of the method. This work was supported in part by the Russian Foundation for Basic Research (project no. 04-07-90336), by the program "Computational and Information Issues of the Solution to Large-Scale Problems" of the Department of Mathematical Sciences of the Russian Academy of Sciences, and by a grant from the Foundation for the Support of National Science for best graduate students of the Russian Academy of Sciences.

References

1. A. Bourgeat, M. Kern, S. Schumacher and J. Talandier. The COUPLEX test cases: Nuclear waste disposal simulation. *Computational Geosciences*, 2004, 8, pp.83-98.
2. G. Bernard-Michel, C. Le Potier, A. Beccantini, S. Gounand and M. Chraibi. The Andra Couplex 1 test case: Comparisons between finite element, mixed hybrid finite element and finite volume discretizations. *Computational Geosciences*, 2004, 8, pp.83-98.
3. I. V. Kapyrin. A family of monotone methods for the numerical solution of three-dimensional diffusion problems on unstructured tetrahedral meshes.*Doklady Mathematics*, 2007, Vol.76, No.2, pp.734-738.
4. C. Le Potier. Schema volumes finis monotone pour des operateurs de diffusion fortement anisotropes sur des maillages de triangle non structures. *C. R. Acad. Sci. Paris*, 2005, Ser. I 341, pp.787-792.
5. K. Lipnikov, M. Shashkov, D. Svyatski and Yu. Vassilevski. Monotone finite volume schemes for diffusion equations on unstructured triangular and shape-regular polygonal meshes. *Journal of Computational Physics*, 2007, Vol.227, No.1, pp.492-512.
6. A. A. Samarskii and P. N. Vabishchevich. Numerical Methods for Solving Convection–Diffusion Problems Editorial URSS, Moscow, 1999, 248p. [in Russian].

Sparse Approximation of FEM Matrix for Sheet Current Integro-Differential Equation*

Mikhail Khapaev[1] and Mikhail Yu. Kupriyanov[2]

[1] Dept. of Computer Science, Moscow State University, 119992 Moscow, Russia
vmhap@cs.msu.su

[2] Nuclear Physics Institute, Moscow State University, 119992 Moscow, Russia
mkupr@pn.sinp.msu.ru

Abstract. We consider two-dimensional integro-differential equation for currents in thin superconducting films. The integral operator of this equation is hypersingular operator with kernel decaying as $1/R^3$. For numerical solution Galerkin Finite Element Method (FEM) on triangular mesh with linear elements is used. It results in dense FEM matrix of large dimension. As the kernel is quickly decaying then off-diagonal elements of FEM matrix are small. We investigate simple sparsification approach based on dropping small entries of FEM matrix. The conclusion is that it allows to reduce to some extent memory requirements. Nevertheless for problems with large number of mesh points more complicated techniques as one of hierarchical matrices algorithms should be considered.

Keywords: superconductivity, FEM, sparse matrix.

1 Introduction

In this paper we consider the problem of numerical solution of boundary value problem for integro-differential equation for sheet current in thin superconducting films. The simplest form of this equation for a single conductor is

$$-\lambda_\perp \Delta \psi(r) + \frac{1}{4\pi} \iint_S \left(\nabla \psi(r'), \nabla' \frac{1}{|r - r'|} \right) ds + H_z(r) = 0, \qquad (1)$$

where $\lambda_\perp$ is constant parameter, S is 2D bounded domain on plane (x, y), $r = (x, y)$. $\psi(r)$ is unknown function. It is stream function potential representation for 2D sheet current. $H_z(r)$ is the right hand side and has the sense of z component of external magnetic field. The boundary condition for (1) is

$$\psi(r) = F(r), \quad r \in \partial S. \qquad (2)$$

Here function $F(r)$ is completely defined by inlet and outlet currents over conductor boundary ∂S and currents circulating around holes in S. In the paper we

* The paper is supported by ISTC project 3174.

evaluate the problem in more general form accounting several single-connected conductors with holes and finite thickness of films.

Our interest to problem (1), (2) is motivated by computations of inductances and current fields in microelectronic superconductor structures [1,2].

Traditionally problems for surface, sheet or volume currents are equally solved using PEEC (Partial Element Equivalent Circuit) technique [3,4]. This approach brings to equation with weakly singular kernel. In our case it is

$$\lambda_{\perp}\mathbf{J}(r) + \frac{1}{4\pi}\iint\limits_{S} \frac{\mathbf{J}(r')}{|r - r'|}\, ds = -\nabla\chi(r), \tag{3}$$

$$\nabla \cdot \mathbf{J}(r) = 0, \quad \Delta\chi = 0. \tag{4}$$

In (3) $\mathbf{J}(r)$ is unknown current, $\chi(r)$ is one more unknown function (phase). (1) can be obtained from (3) using differentiation. Equation (3) needs boundary conditions for function $\chi(r)$ and current $\mathbf{J}(r)$. Equations similar to (3) are well known for normal conductors. Approaches similar to PEEC for (3) for superconductors are also known [6,7]. For normal conductor function $\chi(r)$ has sense of voltage potential. Recently fast multipoles technique based program FASTHENRY [5] for (3) was adopted for superconductors [8].

The main problem in numerical solution of (1) or (3) is dense matrix of large size. It is necessary to fill this matrix fast and then store it or it's approximation. It is also necessary to have a fast and reliable method for solution of system of linear equations with this matrix. In other case simulation of many practical problems can be unfeasible.

We prefer to solve equation (1) instead of (3) because (1) accounts important physical features of the problem and because of numerical efficiency considerations:

- Many superconductivity problems are based solely on currents and magnetic field. In these cases it is difficult to define boundary conditions for $\chi(r)$.
- Holes in S is a problem for (3) and is an easy task for (1). Given currents circulating around holes are accounted in boundary conditions in function $F(r)$ (1). Non-decaying currents circulating around holes are typical for problems in superconductivity.
- FEM for (1) has better numerical approximation then PEEC and thus can give smaller system of linear equations.
- FEM off-diagonal matrix elements for (1) quickly tends to zero with the distance between finite elements.

In this paper we outline the evaluation of boundary value problem for integro-differential equations for sheet currents in thin superconducting films. Properties of operators are discussed and finite element method is formulated. We study decaying of matrix elements and formulate simple strategy for dropping small elements of the matrix. Then direct sparse solver is used for factorisation and solution. Two numerical examples are considered.

The sparsification technique we developed allows to extend the set of problems that can be efficiently solved. It is also shown that even for quickly decaying kernels more complicated methods of solving large dense FEM (Galerkin) systems of equations like [9, 10] should be used.

2 Equations evaluation

2.1 Preliminaries

In this paper we study the currents in conducting layers separated by layers of dielectric. Let t_m be the thickness of conducting layers and d_k be the thickness of dielectric layers, k, m — the numbers of the layers. Conducting layers can contain few single-connected conductors of arbitrary shape. Let the number of conductors in all layers be N_c and the total number of holes in all conductors will be N_h. Each conductor can have current terminals where inlet or outlet currents are given.

For large class of microwave and digital circuits it can be assumed [11, 6] $d_k \ll l, \quad t_m \ll l$, where l is the typical lateral size of circuit in plane (x, y).

Each conductor occupy space domain $V_m = S_m \times [h_m^0, h_m^1]$, $m = 1, \ldots, N_c$. Two-dimensional domain S_m is the projection of the conductor on the plane (x, y). We call the boundary of the conductor ∂S_m the boundary of the projection S_m. Let $\partial S_{h,k}$ be the boundary of the hole with number k, $\partial S_{ext,m}$ — external boundary of m-th conductor. We assume that all current terminals are on the external boundary of the conductors.

The magnetic field is excited by external magnetic field, currents circulating around holes and currents through chains of terminals on the conductors.

For further convenience, let P, P_0 stands for points in 3D space, r, r_0 — for points on plane. Also, consider differential operators $\partial_x = \partial/\partial x$, $\partial_y = \partial/\partial y$, $\nabla_{xy} = (\partial_x, \partial_y)$.

2.2 London equations for conductors of finite thickness

The basic equations for further consideration are static London equations [1]. Let $\mathbf{j}$ be current density and $\mathbf{H}$ — total magnetic field including self-field of $\mathbf{j}$ and external magnetic field, λ — so called London penetration depth [1]. Then basic equations are:

$$\lambda^2 \nabla \times \mathbf{j} + \mathbf{H} = 0, \tag{5}$$

$$\nabla \times \mathbf{H} = \mathbf{j}. \tag{6}$$

Typically λ and film thickness are of same order. As film is assumed thin $\mathbf{j} \approx \mathbf{j}(x, y)$ and problem reduces to z- component of (5) [12]

$$\lambda^2(\partial_x j_y(P_0) - \partial_y j_x(P_0)) + H_z(P_0) = 0. \tag{7}$$

Consider the sheet current density $\mathbf{J}_m(r)$:

$$\mathbf{J}_m(r) = \int_{h_m^0}^{h_m^1} \mathbf{j}(P)dz, \quad r \in S_m. \tag{8}$$

Self magnetic field in (7) is calculated by means of average current density $\mathbf{J}_n(r)/t_n$ and Biot-Savart formula:

$$\mathbf{H}(P_0) = \frac{1}{4\pi} \sum_{n=1}^{N_c} \int_{V_n} \frac{1}{t_n} \mathbf{J}_n(r) \times \nabla_P \frac{1}{|P - P_0|} dv_P. \tag{9}$$

Consider London penetration depth for films

$$\lambda_m^s = \lambda_m^2 / t_m. \tag{10}$$

Averaging (7) over the thickness of conductors we obtain the following equations for the sheet currents in conductors

$$\lambda_m^s(\partial_x J_{m,y}(r_0) - \partial_y J_{m,x}(r_0)) +$$
$$\frac{1}{4\pi} \sum_{n=1}^{N_c} \iint_{S_n} (\mathbf{J}_n(r) \times \nabla_{xy} G_{mn}(r, r_0))_z \, ds_r + H_z(r_0) = 0, \tag{11}$$

where $r_0 \in S_m$, $m = 1, \ldots, N_c$, $H_z(r)$ is z component of external magnetic field and

$$G_{mn}(r, r_0) = \frac{1}{t_m t_n} \int_{h_m^0}^{h_m^1} dz_0 \int_{h_n^0}^{h_n^1} \frac{1}{|P - P_0|} dz. \tag{12}$$

The equations (11) must be completed by the charge conservation low $\nabla \cdot \mathbf{J}_m = 0, \quad m = 1, \ldots, N_c$. Our goal is to take into account small but finite thickness of conductors. Therefore we substitute the both of one-dimensional integrals in (12) by quadrature formulas of rectangles or trapezoids.

For $h_n \in [h_n^0, h_n^1]$ formula of rectangles gives us the following kernels:

$$G_{mn}(r, r_0) = 1 \Big/ \sqrt{|r - r_0|^2 + (h_m - h_n)^2}. \tag{13}$$

Obviously (13) is infinitely thin current sheet approximation where sheets have heights h_n. For single conductor we have equation (1).

If trapezoid formula is used then for $m = n$ the kernel has form

$$G_{mm}(r, r_0) = \frac{1}{2} \left(\frac{1}{|r - r_0|} + \frac{1}{\sqrt{|r - r_0|^2 + t_m^2}} \right). \tag{14}$$

This approach show good precision and numerical stability for problems with finite thickness of conducting and dielectric layers [12].

2.3 Stream function

For the sheet current well-known stream function representation is used. Stream function (T - function) $\psi_m(r)$ is defined for each single-connected conductor on the base of charge conservation law. In our case

$$J_{m,x}(r) = \partial_y \psi_m(r), \quad J_{m,y}(r) = -\partial_x \psi_m(r). \tag{15}$$

Or, if $\Psi = (0, 0, \psi(r))$ then $\mathbf{J}_m = \nabla \times \Psi$.

$\psi_m(r)$ has the sense and dimension of full current. Let $\Gamma \in S$ be any open curve in S_m with the origin r_0 and end r_1. Then full current through this curve is $I(r_0, r_1) = \psi_m(r_1) - \psi_m(r_0)$. The full current does not depend on specific curve joining P_0 and P_1.

Let us introduce the necessary agreements concerning functions $\psi_m(r)$. We assume the normal current distribution through terminals is homogeneous and $\psi_m(r) = 0$ on a non-terminal part ∂S_m^0 of boundary S_m. We define the total current circulating around a hole in S_m as the total current through any curve joining ∂S_m^0 and the boundary of the hole. The total current does not depend on the choice of the joining curve.

Substituting (15) into (11), we obtain the set of equations for functions $\psi_m(r)$, $m = 1, \ldots, N_c$:

$$-\lambda_m^s \Delta\psi_m(r_0) + \frac{1}{4\pi} \sum_{n=1}^{N_c} \iint_{S_n} (\nabla\psi_n(r), \nabla_{xy} G_{mn}(r, r_0))\, ds_r = 0. \tag{16}$$

Let $I_{h,k}$ be full currents circulating around the holes, $k = 1, \ldots, N_h$, then the boundary conditions for (16) are:

$$\psi_m(r) = I_{h,k}, \ r \in \partial S_{h,k}, \ k = 1, \ldots, N_k; \tag{17}$$

$$\psi_m(r) = F_m(r), \ r \in \partial S_{ext,m}, \ m = 1, \ldots N_c. \tag{18}$$

Function $F_m(r)$ is defined by the properties of $\psi(r)$ and terminal current distribution. We assume $F_m(r)$ is linear in the limits of terminals. It is constant on the non-terminal boundaries.

Equations (16) together with boundary conditions (17), (18) completely define the current distribution in the circuit and allow to calculate inductance matrix [12].

It is also possible to calculate approximately full energy ($H_z(r) = 0$) for a solution of (16), (17), (18):

$$E = \frac{1}{2} \sum_{n=1}^{N_c} \iint_{S_n} \lambda_s^n (\nabla\psi_n)^2\, ds_n +$$

$$\frac{1}{8\pi} \sum_{n=1}^{N_c} \sum_{m=1}^{N_c} \iint_{S_n} ds_n \iint_{S_m} (\nabla\psi_n, \nabla\psi_m)\, G_{mn} ds_m. \tag{19}$$

If external magnetic field $H_z(r) = 0$ then matrix of inductances can be calculated using this expression. Let us combine all inlet and circulating currents into a vector $\mathbf{I}$. As the problem is linear then $2E = (L\mathbf{I}, \mathbf{I})$. L is the matrix of inductances. This matrix is symmetric. For kernel $1/R$ it is possible to prove positive definetness of quadratic form (19) and L.

2.4 Integral operator properties

Consider principal value integral operator (1)

$$L\psi(r) = \frac{1}{4\pi} \iint\limits_S \left(\nabla\psi(r'), \nabla' \frac{1}{|r - r'|} \right) ds'. \tag{20}$$

Integral operator (20) is well known [13]. If $\psi(r) = 0$ on ∂S then using integration by parts it can be reduced to hypersingular integral

$$L\psi(r) = -\frac{1}{4\pi} \iint\limits_S \frac{\psi(r')}{|r - r'|^3} ds'. \tag{21}$$

It is shown in [13] that in proper function spaces $L > 0$. In our case we have operator $-\Delta + L$. This operator is positive self-adjoint if $\psi(r) = 0$ on ∂S. Exact formulations for hypersingular part in proper Sobolev spaces can be found in [13], [14]. Thus problem (1), (2) is very similar to boundary problem for Laplace equation.

The problem (16), (17), (18) is more complicated. Nevertheless the singular part of integral operator in (16) is similar to (20).

3 Finite Element Method

3.1 Formulation

For simplicity we consider the case of one conductor. The extension on multiconductor case is straightforward. For short let $H_z(r) = 0$.

Our version of Galerkin FEM is based on triangular meshing and linear elements [15].

The bilinear form $a(u, v)$ for "weak" formulation of the problem (16) is:

$$a(u, v) = \lambda^s \iint_S (\nabla u(r), \nabla v(r))\, ds +$$
$$\frac{1}{4\pi} \iint_S ds' \iint_S (\nabla u(r), \nabla v(r')) G(r, r')\, ds. \tag{22}$$

The principal value integral in (16) was integrated by parts.

For the triangulation of S let I be the set of indexes of internal points of the mesh and J the indexes of all nodes including boundary nodes. The unknown function $\psi(r)$ is approximated by linear finite elements. This approximation takes into account the prescribed boundary values of $\psi(r)$:

$$\psi(r) \approx \psi^h(r) = \sum_{j \in J} \psi_j^h u_j^h(r). \tag{23}$$

Here ψ_j^h are approximate values of $\psi(r)$ in the nodes of mesh r_j, $u_j^h(r)$ are basic functions of finite element interpolation. For linear finite elements that we use $u_j^h(r) \neq 0$ if $r \in S_j$. S_j are polygons combined from several triangles with common central vertice r_j. All $u_j^h(r)$ are facet pyramidal functions. Setting in (22) $u(r) = \psi^h(r)$, $v(r) = u_i^h(r)$ one can derive the following system of linear equations:

$$\sum_{j \in J} a(u_i^h, u_j^h) \cdot \psi_j^h = 0 \quad i \in I, \tag{24}$$

Then (24) can be rewritten as a system of linear equations for ψ_i^h, $i \in I$ with symmetric dense matrix where the non-zero right part is formed by terms with prescribed boundary values $\psi(r)$ (17), (18).

Diagonal elements of FEM matrix are positive. If $S_i \bigcap S_j \neq \varnothing$ then elements can be of any sign. If $S_i \bigcap S_j = \varnothing$ the elements are negative (21).

3.2 Implementation

The program implementation of finite element method meets some difficulties. The expression (22) contains quadruple integrals. Calculation of these integrals over triangles is the most CPU time consuming part of the algorithm.

Consider expression for matrix elements of FEM equations:

$$a(u_i^h, u_j^h) = \lambda^s \iint_{S_i \cap S_j} (\nabla u_i^h, \nabla u_j^h)\, ds$$
$$+\frac{1}{4\pi} \iint_{S_i} ds \iint_{S_j} (\nabla u_j^h(r'), \nabla u_i^h(r)) G(r, r')\, ds'. \tag{25}$$

As S_i, S_j are unions of several triangles having as vertices nodes i and j then expression (25) is the sum of double integrals over triangles. For fast and accurate calculation of integrals over triangles we use the following procedure. For non-singular integrals we use Gauss quadratures for triangles [16]. These quadratures are of high order if triangles are close to each other and are simple three or one nodes formulas for far spaced triangles. Special problem is singular integrals over coinciding or having common vertices triangles T_i and T_j:

$$I_{ij} = \iint_{T_i} ds \iint_{T_j} \frac{1}{|r - r'|}\, ds'. \tag{26}$$

For calculation of these integrals we use formulas [17]. These formulas take into account relation

$$\frac{1}{|r-r'|} = -\nabla \cdot \nabla' |r-r'|, \quad \nabla = \left(\frac{\partial}{\partial x}, \frac{\partial}{\partial y}\right).$$

It allows to reduce area integrals to contour integrals. As result integrals over coinciding triangles are taken analytically [17]. Other near-singular integrals are reduced to simple regular contour integrals and can be evaluated numerically using 1D Gauss quadratures.

3.3 Sparsification procedure

The FEM matrix is the sum of sparse matrix resulted from Laplace operator and dense part resulted from integral operator. If number of nodes in triangulation of S is moderate then finite element equations can be solved using Cholesky factorization. The problem arise when number of nodes is several thousands or tens thousands of nodes. It is not possible to store this matrix in dense format or to factor it. One of possible solutions is to evaluate a sparsification procedure.

Consider matrix elements for hypersingular integral operator (25). For non-intersecting finite elements supports $S_i \bigcap S_j = \varnothing$ we have

$$b_{ij} = \iint_{S_i} ds \iint_{S_j} \frac{(\nabla u_j^h(r'), \nabla u_i^h(r))}{|r-r'|} ds' = -\iint_{S_i} ds \iint_{S_j} \frac{u_j^h(r') u_i^h(r)}{|r-r'|^3} ds'. \tag{27}$$

The value of b_{ij} in calculations can be very small. Let us assume that triangular mesh is approximately snapped on rectangular square grid. Then we can assume that diameter of any triangle can be estimated by certain value h. We call this value mesh step. Then we can assume that $m_{ij}h = |r_i - r_j|$, m_{ij} is a number and r_i and r_j are center nodes for finite element supports S_i and S_j. m_{ij} has sense of approximate number of mesh cells between finite elements centers. From (27) follows simple estimation:

$$|b_{ij}| \leqslant M \frac{h}{m_{ij}^3}. \tag{28}$$

Thus matrix elements (25) are quickly decaying when distance between finite elements grow up. Equation with weak singularity (3) has only first order of decaying.

Our sparsification procedure is very simple. We introduce small tolerance value t. All off-diagonal matrix elements a_{ij} (25) below $|a_{ij}| < t \cdot a_{ii}$ are dropped thus making matrix sparse. We also modify diagonal terms a_{ii} and a_{jj}:

$$a_{ii} = a_{ii} + a_{ij}, \quad a_{jj} = a_{jj} + a_{ij}. \tag{29}$$

As result we obtain sparse symmetric matrix. Positive definetness of matrix can be preserved if dropping tolerance is taken small enough.

To solve equations with sparse matrix we use CSparse package direct Cholesky solver [18].

4 Results of calculations

4.1 Spiral inductor

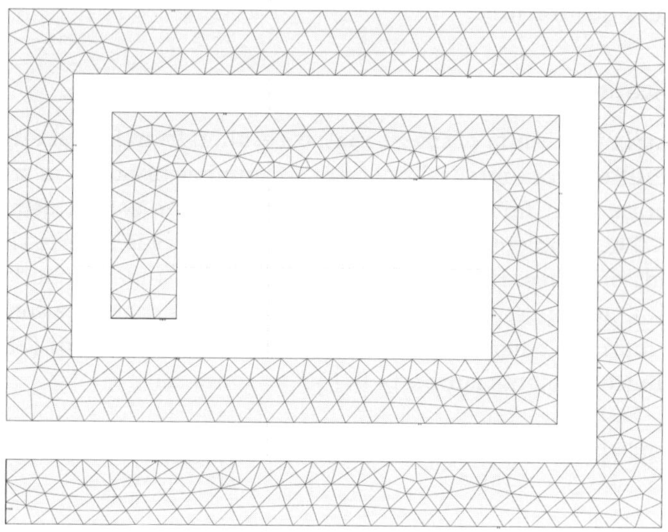

Fig. 1. Spiral inductor: domain S and triangular mesh. Two shortest sides are terminals for inlet and outlet currents.

First example is spiral inductor presented on Fig. 1. Non-dimensional sizes of this figure are 2.5×2 and strip was 0.25 width. Parameter $\lambda_\perp = 0.1$. Mesh size on Fig. 1 is 0.25 but some triangles created by mesh generator are little larger. We calculated matrix filling and inductance error to compare with full dense matrix case. The results for mesh size $h = 0.025$ are in the table 1, where it was $L = 7.8$ for full dense matrix. We also fixed dropping tolerance as 0.01 and calculated filling in dependence of mesh size. The results are in table 2. It is seen that filling is decreasing if mesh is densing. This fact gives us the opportunity to solve large problems. Calculations show that we can significantly reduce the memory requirements die to moderate decreasing of accuracy.

4.2 Flux flow oscillator

Flux flow oscillator (FFO) is one of superconductivity high frequency microelectronic devices [1, 2]. FFO contain long Josephson junction as non-linear element

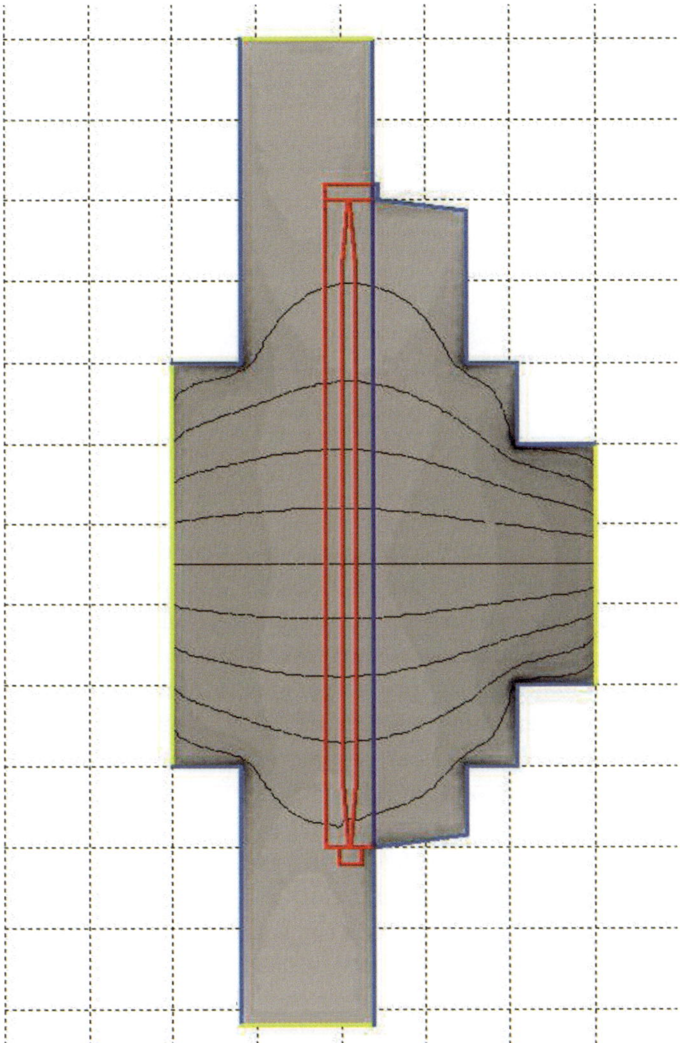

Fig. 2. Flux Flow Oscillator: domain S and current flow lines for two pairs of terminals. Josephson junction inside microstrip is shown. Grid size 50μm.

Table 1. Sparse matrix filling and relative inductance error, dependence on dropping tolerance t.

t	Filling %	Δ L %
10^{-2}	0.36	7.5
10^{-3}	1.72	3.5
10^{-4}	5.1	1.3

Table 2. Sparse matrix filling for tolerance $t = 0.001$ and different h.

h	Filling %
0.1	11
0.05	4.3
0.025	1.72

inside a microstrip and has thin film input-output conductors ended by current terminals. The top view on typical FFO device [2] is presented on Fig. 2. The problem can be splitted on two consecutive subproblems [19]. First problem is linear problem for calculation of currents in single sheet conductor as it is shown on Fig. 2. Domain S is the projection of two-layer microstrip structure on plane (x, y). This problem solely reduces to the solution of equation (1). Second problem is simulation of non-linear Josephson junction. It isn't concidered in this paper. Josephson junction simulation needs boundary conditions. Boundary conditions are taken from the solution of first problem. Thus we need to consider problem of calculation of current field using evaluated FEM technique.

For accurate reproducing of small details of device we need rather fine triangulation. For dense matrix calculations the dimension of FEM equations can't be larger then several thousands. In this case triangulation of S is not fine enough to approximate small details near junction ends even for refined mesh. Thus we evaluated and implemented sparsification procedure described above.

First we compared solutions with dense and sparsified with tolerance $2 \cdot 10^{-5}$ matrices for mesh step $5\mu m$. In both cases matrix dimension was 3266. Sparse matrix had filling 22%. Minimal row filling was 240 and maximal row filling was 1266 non-zero elements. There was no reduction of CPU time because all matrix elements were calculated in both cases. Maximal error for inductances was 10%. Then we reduced mesh step in twice and repeated calculations. Filling was 7.25% and there was minimal 301 and maximal 1517 non-zero elements in the row. There was some difference of order 10% in inductances values with dense matrix solution on coarse mesh. This difference can be explained by better accuracy on fine mesh. Thus we can use more dense mesh and calculate current field with better resolution. Nevertheless accurate tunining of tolerance with accuracy control are necessary.

Thus calculations for this test case demonstrate efficiency of algorithm on large practical problem. Our calculations also show the neccesity of tuning mesh size and tolernace parameter.

5 Conclusions

Sparsification technique for FEM matrix of integro-differential equation was evaluated. Systematic calculations for verifying the accuracy and possible gain in memory preservation were performed. The conclusion is that the size of problems amenable for practical solutions was extended in several times. Other conclusion is that further improvements needs more complicated algorithms as skeleton matrix approximations [9] and perhaps implementation of library of algorithms [10]. Then simple dropping of small entries in matrix can be used as construction of preconditioner for iterative solution.

Acknowledgments

We wish to thank Profs. V. P. Koshelets and E. V. Zakharov for comments and inspiring discussions and A. Sobolev for help with FFO problem simulations.

References

1. T. Van Duser and C. W. Turner, *Principles of Superconductive Devices and Circuits.* North Holland: Elsevier, 1981.
2. V. Koshelets and J. Mygind, *Flux flow oscillators for superconducting integrated submm wave receivers*, Studies of High Temperature Superconductors, vol. 39, pp. 213–244, March 2001. ed. A. V. Narlikar, NOVA Science Publishers, New York.
3. P. A. Brennan, N. Raver and A. E. Ruehli, *Three-dimensional inductance computations with partial element equivalent circuits*, IBM J. Res. Develop., vol. 23, N. 6, pp. 661–668, November 1979.
4. G. Antonini, A. E. Ruehli, *Fast Multipole and Multifunction PEEC Methods*, IEEE Trans. on Mobile Computing, vol. 2, N. 4, pp. 288–298, October–December 2003.
5. M. Kamon, M. J. Tsuk and J. K. White, *FastHenry: a multipole-accelerated 3-D inductance extraction program*, IEEE Trans. Microwave Theory and Techn., vol. 42, pp. 1750–1758, September 1994.
6. B. Guan, M. J. Wengler, P. Rott and M. J. Feldman, *Inductance estimation for complicated superconducting thin film structures with a finite segment method*, IEEE Trans. Appl. Superconductivity, 1997, vol. 7, pp. 2776-2779, June 1997.
7. G. Hildebrandt and F. H. Uhlmann, *Inductance calculation for integrated superconducting structures by minimizing free energy*, IEEE Trans. Appl. Superconductivity, 1995, vol.5, pp. 2766–2769, June 1995.

8. http://www.wrcad.com/
9. S. A. Goreinov, E. E. Tyrtyshnikov, and N. L. Zamarashkin, *A theory of pseudo-skeleton approximations*, Linear Algebra Appl., 261: 1–21 (1997)
10. http://www.hlib.org/
11. K. Gaj, Q. P. Herr, V. Adler, A. Krasniewski, E. G. Friedman, M. J. Feldman, *Tools for the Computer-Aided Design of Multigigahertz Superconducting Digital Circuits*, IEEE Trans. Appl. Superconductivity, 1999, vol. 9, pp. 18–38, March 1999.
12. M. Khapaev *Inductance Extraction of Multilayer Finite Thickness Superconductor Circuits*, IEEE Trans. Microwave Theory and Techn., 2001 v. 49, pp. 217–220.
13. V. J. Ervin, E. P. Stephan *A Boundary Element Galerkin Method for a Hypersingular Integral Equation on Open Surfaces*, Mathematical Methods in the Applied Sciences, 1990, v. 13, pp. 281–289.
14. M. Costabel, *Principles of Boundary Element Methods*, *Computer Physics Reports*, 6(1987), pp. 243–274.
15. G. Strang and G. J. Fix, *An Analysis of the Finite Element Method.* Prentice-Hall, Inc., 1973.
16. O. C. Zienkiewicz, K. Morgan, *Finite Elements and Approximation.* John Wiley & Sons, New York, 1983.
17. P. Arcioni, M. Bressan and L. Perregrini, *On the Evaluation of the Double Surface Integrals Arising in the Application of the Boundary Integral Method to 3-D Problems*, IEEE Trans. Microwave Theory and Techn., vol. 45, pp. 436–439, March 1997.
18. T. A. Davis, *Direct Methods for Sparse Linear Systems*, SIAM, Philadelphia, 2006.
19. K. K. Likharev, *Dynamics of Josephson Junctions and Circuits*, Gordon and Breach, New York (1986).

The Method of Magnetic Field Computation in Presence of an Ideal Conductive Multiconnected Surface by using the Integro-Differential Equation of the First Kind

Tatiana Kochubey* and V. I. Astakhov**

Southern Scientific Centre of Russian Academy of Sciences, Rostov-on-Don, Russia
*tako84@yandex.ru, **v.astakhov@mail.ru

Abstract. The problem of the stationary magnetic field modelling in the presence of an ideal-conductive surface is reduced to the integro-differential equation of the first kind. The analysis of the equation is carried out by the variational method. The novel software package has been created for its solving. The examples of its usage are represented.

Keywords: magnetostatic field, integro-differential equation of the first kind, variational method.

1 Introduction

Many engineering problems require the mathematical modelling and the computation of stationary and quasi-stationary magnetic fields in the presence of the conductive layers (cases, plates). Such layers play a part of protective screens, frames and load-bearing elements of electro technical and electrical survey devices. At present time membranous and printing technologies are developed therefore the interest to such problems has grown.

In particular cases the problems of stationary magnetic field computation in presence of a conductive surface can be solved analytically. If the surface has the complicated geometrical form and has no boundary then the most known methods for problem solving are the methods of potential theory. These methods are the most economical in computer realization. Using them, we obtain integral equations of the second kind for densities of secondary sources [1, 2].

The problems of the magnetic field modelling in the presence multiconnected surface with boundary are very difficult. Known integral equations of the second kind are not defined on the unclosed surface or have very complicated kernels. This is the main difficulty. These equations are not used by this reason. The usage of the finite elements method for such problems leads to the nonstable systems of linear algebraic equations of very big dimension.

In this paper the problem of the magnetic field modelling in presence of multiconnected surface with ideal conductivity is reduced to the integro-differential equation of the first kind. We prove that the problem is well defined in the special pair of functional spaces. These spaces are established by the physics properties of the problem. We have also established the effective method for reduction of the equation to the SLAE.

2 Problem formulation

Here we consider the problem of the magnetic field modelling in presence of the ideal conductive piecewise-smooth Lipschitz surface Γ' with the boundary $\delta\Gamma$ (see fig. 1). We suppose that the surface has the finite number $N+1$ of holes $\Gamma_0'', \Gamma_1'', \Gamma_2'', \ldots, \Gamma_N''$ and denote their union as Γ''.

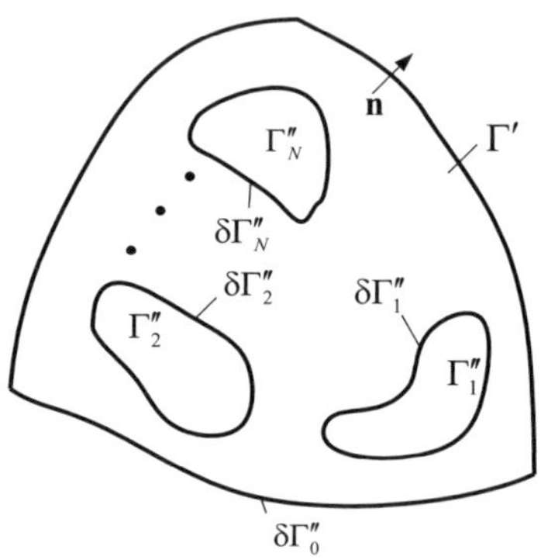

Fig. 1.

Let the non-perturbed magnetic field (or equivalent for it in terms of [3]) is given by sources that are located outside Γ' in a bounded domain of space. This field has finite energy, i.e.,

$$\int_\Omega \left|\mathbf{B}^0\right|^2 \mathrm{d}\Omega < \infty.$$

Suppose the external medium is homogeneous with a finite positive magnetic conductivity. The property of ideal conductivity is written as

$$\mathbf{Bn} = 0 \quad \text{on } \Gamma', \tag{1}$$

where $\mathbf{B}$ is the induction of the resulting magnetic field; $\mathbf{n}$ is the exterior normal to the surface Γ'. Here it is also taken into account that on passing through Γ' the limiting values of the normal component in (1) coincide.

Besides, it is necessary to note the presence or the absence of so-called "frozen" fluxes in the internal holes of the surface. Let

$$\int_{\Gamma''_k} \mathbf{B} d\Gamma = \Phi_k, \quad k = \overline{1, N},$$

where Φ_k are given.

We have the following asymptotic for the field $\mathbf{B}$:

$$|\mathbf{B}(M)| \xrightarrow{M \to \infty} 0\left(1/r^3\right).$$

Due to linearity properties of the medium, it is possible to decompose the induction $\mathbf{B}$ into two components: $\mathbf{B} = \mathbf{B}^0 + \mathbf{B}^*$, where $\mathbf{B}^0$ is the induction of the non-perturbed magnetic field; $\mathbf{B}^*$ is the induction of the magnetic reaction field. The last one is induced by micro currents on the surface.

The field of reaction is potential outside $\Gamma' \cup \Gamma''$, therefore its induction can be represented in the form

$$\mathbf{B}^* = -\operatorname{grad} \varphi^* \text{ outside } \Gamma' \cup \Gamma'',$$

where φ^* is a scalar potential.

Taking into account the solenoidality of $\mathbf{B}$ and $\mathbf{B}^*$, the computation of the magnetic field in the terms of the potential φ^* is reduced to the scalar boundary problem. This problem has the following form

$$\begin{aligned} \Delta\varphi^* &= 0 \text{ outside } \Gamma' \cup \Gamma'', \\ \varphi^*_+ - \varphi^*_- &= \tau \text{ on } \Gamma' \cup \Gamma'', \\ \frac{\partial \varphi^*}{\partial n_+} = \frac{\partial \varphi^*}{\partial n_-} &= \frac{\partial \varphi^*}{\partial n} \text{ on } \Gamma' \cup \Gamma'', \\ \frac{\partial \varphi^*}{\partial n} &= B_n^0 \text{ on } \Gamma', \\ \int_{\Gamma''_k} \frac{\partial \varphi^*}{\partial n} d\Gamma &= \int_{\Gamma''_k} B_n^0 d\Gamma - \Phi_k, \; k = \overline{1, N}, \\ \varphi^*(M) &\xrightarrow{M \to \infty} 0(1/r^2), \end{aligned} \tag{2}$$

where symbols "+" and "−" in subscript denote the limiting values along positive and negative directions of the normal $\mathbf{n}$, respectively; τ is the jump of the limiting values of potential. Since the magnetic field is solenoidal, we have

$$\int_{\Gamma''} \left(\frac{\partial \varphi^*}{\partial n} - B_n^0\right) d\Gamma = 0.$$

This fact is also taken into account.

It is obvious that the conditions of problem (2) are held if

$$\varphi^*(M) = \frac{1}{4\pi} \int_{\Gamma} \tau(Q) \frac{\partial}{\partial n_Q} \frac{1}{r_{QM}} d\Gamma_Q, \; M \notin \Gamma.$$

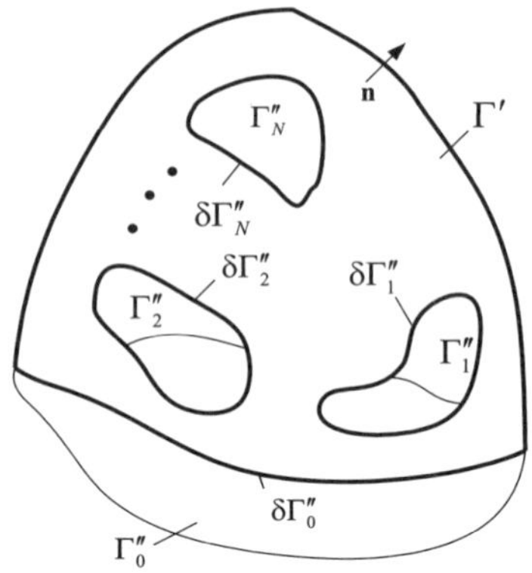

Fig. 2.

Then problem (2) is reduced to the equation

$$\frac{1}{2\pi}\frac{\partial}{\partial n_M}\int_{\Gamma}\tau(Q)\frac{\partial}{\partial n_Q}\frac{1}{r_{QM}}d\Gamma_Q = 2B_n^0(M), \quad M \to \Gamma' \tag{3}$$

with the conditions

$$\int_{\Gamma_k''}\frac{\partial}{\partial n_M}\frac{1}{2\pi}\int_{\Gamma}\tau(Q)\frac{\partial}{\partial n_Q}\frac{1}{r_{QM}}d\Gamma_Q\Bigg|_{M\to\Gamma_k''}d\Gamma_M = 2\int\limits_{\Gamma_k''}B_n^0 d\Gamma - 2\Phi_k, \quad k=\overline{1,N}, \tag{4}$$

$$\tau(M) = c_k, \; M \in \Gamma_k'' \cup \delta\Gamma_k'', \; c_k = \text{const}, \; k = \overline{0,N}. \tag{5}$$

The considering surface Γ is shown in fig. 2. Here $\delta\Gamma_k''$ is the adjacent boundary for Γ' and Γ_k''; $\delta\Gamma_0''$ is the external boundary of Γ'.

3 Analysis of the equation

Now we analyse the equation (3) using the variational method [4], i.e., we reduce the equation to an equivalent variational problem for some functional. Using the Riss theorem [4], we establish the variational problem as well defined. The variational method is useful, because effective way of the variational problem solving is well known. It is the construction of the minimizing Ritz sequence. This way formally coincides with using Bubnov-Galerkin method for the equation.

We must extend equation (3) to the closed surface Γ for a further analysis. From condition (5) and annihilating properties of the integro-differential operator from (3) it follows that τ will be defined with accuracy to within the arbitrary constant. In the sequel, this property can be used suitably. Thus conditions (4) are transformed to the following form

$$\frac{1}{2}\int_{\Gamma_k''} K_{\tau\sigma}\tau d\Gamma = -\int_{\Gamma_k''} B_n^0 d\Gamma + \Phi_k,\ k = \overline{1,N}, \tag{6}$$

where

$$K_{\tau\sigma}\tau(M) = -\frac{1}{2\pi}\frac{\partial}{\partial n_M}\int_\Gamma \tau(Q)\frac{\partial}{\partial n_Q}\frac{1}{r_{QM}}d\Gamma_Q,\ M \to \Gamma.$$

Further, we introduce the operator P

$$P\xi = \begin{cases} \xi & \text{on } \Gamma'; \\ \frac{1}{mes(\Gamma_k')}\int_{\Gamma_k''}\xi d\Gamma & \text{on } \Gamma_k'',\ \ k = \overline{0,N}. \end{cases}$$

After that equation (3) and conditions (6) are reduced to unified operator equation:

$$A_\tau\tau = f_\tau', \tag{7}$$

where

$$A_\tau = \frac{1}{2}PK_{\tau\sigma},$$

$$f_\tau'(M) = \begin{cases} -B_n^0(M), & M \in \Gamma'; \\ -\frac{1}{mes(\Gamma_k''))}\left(\int_{\Gamma_k''} B_n^0 d\Gamma - \Phi_k\right), & M \in \Gamma_k'',\ \ k = \overline{0,N}. \end{cases} \tag{8}$$

Moreover, the following condition is carried out

$$\int_\Gamma f_\tau' d\Gamma = 0. \tag{9}$$

We choose $\widetilde{L}_2(\Gamma)$ as initial space for the analysis of the equation. It is the Hilbert space of real square-integrable functions with constant values on Γ_k'', $k = \overline{0,N}$. In this space the inner product and the norm are defined by

$$\langle a_1, a_2\rangle_{\widetilde{L}_2} = \int_\Gamma a_1 a_2 d\Gamma, \qquad \|a\|_{\widetilde{L}_2} = \langle a, a\rangle_{\widetilde{L}_2}^{1/2}.$$

Lemma 1. *The operator A_τ is linear, self-adjoint and positive in $\widetilde{L}_2(\Gamma)$.*

Proof. The linearity of the operator is obvious. We prove its self-adjointment and positivity below. Write

$$\langle A_\tau\tau_1, \tau_2\rangle_{\widetilde{L}_2} = \frac{1}{2}\langle PK_{\tau\sigma}\tau_1, \tau_2\rangle_{\widetilde{L}_2} = \frac{1}{2}\langle K_{\tau\sigma}\tau_1, P\tau_2\rangle_{\widetilde{L}_2} = \frac{1}{2}\langle K_{\tau\sigma}\tau_1, \tau_2\rangle_{\widetilde{L}_2} =$$
$$\int_{\Omega\setminus\Gamma} \operatorname{grad}\varphi_{\tau 1}\operatorname{grad}\varphi_{\tau 2} d\Omega = \frac{1}{2}\langle \tau_1, K_{\tau\sigma}\tau_2\rangle_{\widetilde{L}_2} = \langle \tau_1, A_\tau\tau_2\rangle_{\widetilde{L}_2}\ \forall \tau_1, \tau_2 \in \widetilde{L}_2(\Gamma),$$

where $\varphi_{\tau 1}$, $\varphi_{\tau 2}$ are potentials of double layers with densities τ_1, τ_2, respectively. Also, the isomorphism between the space of surface potentials and the space of their densities [5] is used.

It now follows that

$$\langle A_\tau \tau, \tau \rangle_{\widetilde{L}_2} = \int_{\Omega \setminus \Gamma} |\text{grad}\, \varphi_\tau|^2 \, d\Omega \geqslant 0.$$

Here, the equality to zero is possible if and only if φ_τ is equal to constant outside Γ'. Therefore we have

$$\tau = c_\tau = \text{const on } \Gamma.$$

It follows from the theorem on the jump [4] of limiting values for double layer potential. Thus, it is enough for the positivity of the operator A_τ, if we add a subsidiary condition for τ. For example, the condition is similar to (9). In addition, we can restrict the initial space. □

The calibration, which is obtained in the lemma, allows to choose the subspace $\widetilde{L}_2^0(\Gamma)$ in $\widetilde{L}_2(\Gamma)$. It consists of the elements of $\widetilde{L}_2(\Gamma)$ with zero mean value on Γ. The received subspace can be used as the initial space for the variational principle. Then, using the properties of the operator A_τ in $\widetilde{L}_2^0(\Gamma)$ from the lemma, we get the following theorem.

Theorem 1. *The equation (3) and the variational problem*

$$F(\tau) = \langle A_\tau \tau, \tau \rangle_{\widetilde{L}_2^0} - 2\langle f'_\tau, \tau \rangle_{\widetilde{L}_2^0} \to \min \tag{10}$$

are equivalent in the energetic space of the operator A_τ in $\widetilde{L}_2^0(\Gamma)$.

The energetic space of the operator A_τ in $\widetilde{L}_2^0(\Gamma)$ is $H_\tau^0(\Gamma)$. It is the Hilbert space of double layer potential densities with constant values on Γ''_k, $k = \overline{0, N}$ and zero mean value on Γ. In this space the inner product and the norm are defined by

$$\langle \tau_1, \tau_2 \rangle_{H_\tau^0} = \langle A_\tau \tau_1, \tau_2 \rangle_{\widetilde{L}_2^0}, \qquad \|\tau\|_{H_\tau^0} = \langle \tau, \tau \rangle_{H_\tau^0}^{1/2}.$$

Besides, we recall that $P\xi = \xi$ for $\forall \xi \in \widetilde{L}_2^0(\Gamma)$.

Let us show that the functional $G(\tau) = \langle f'_\tau, \tau \rangle_{\widetilde{L}_2^0}$ is bounded in $H_\tau^0(\Gamma)$ for the given formulation problem. Then, according to [4] the variational problem has a unique solution.

Theorem 2. *The functional $G(\xi) = \langle f'_\tau, \xi \rangle_{\widetilde{L}_2^0}$ is bounded in $H_\tau^0(\Gamma)$. At the same time we have the following estimate*

$$|G(\xi)| \leqslant \left(\int_\Omega \left| \widetilde{\mathbf{B}}^0 \right|^2 d\Omega \right)^{1/2} \|\xi\|_{H_\tau^0}$$

where $\widetilde{\mathbf{B}}^0 = \mathbf{B}^0 + \mathbf{B}^\Phi$, $\mathbf{B}^\Phi$ *is the induction of any magnetic field with finite energy. In this case the fluxes of such a field, which pass through internal holes, coincide with* Φ_k, $k = \overline{1, N}$. *The distribution of such field can be assigned by different ways. For example, it can be assigned in a form of a magnetic tubes bunch. Each tube passes through external and one internal holes, and carries corresponding flux.*

Proof. Consider

$$|G(\xi)| = \left|\langle f'_\tau, \xi\rangle_{\widetilde{L}_2^0}\right| =$$

$$\left|-\int_{\Gamma'} B_n^0 \xi d\Gamma + \sum_{k=0}^{N} \frac{1}{\text{mes}\,(\Gamma_k'')} \int_{\Gamma''} \xi_k \left(-\int_{\Gamma_k''} B_n^0 d\Gamma + \Phi_k\right) d\Gamma\right| =$$

$$\left|-\int_{\Gamma'} B_n^0 \xi d\Gamma + \sum_{k=0}^{N} \xi_k \left(-\int_{\Gamma_k''} B_n^0 d\Gamma + \Phi_k\right)\right| =$$

$$\left|-\int_{\Gamma} B_n^0 \xi d\Gamma + \sum_{k=0}^{N} \xi_k \Phi_k\right| = \left|\int_{\Gamma} \widetilde{B}_n^0 \xi d\Gamma\right| = \left|\int_{\Omega\setminus\Gamma} \text{grad}\,\varphi_\xi \widetilde{\mathbf{B}}^0 d\Omega\right| \leqslant$$

$$\left(\int_{\Omega\setminus\Gamma} |\text{grad}\,\varphi_\xi|^2 d\Omega \int_{\Omega} \left|\widetilde{\mathbf{B}}^0\right|^2 d\Omega\right)^{1/2} = \|\xi\|_{H_\tau^0} \left(\int_{\Omega} \left|\widetilde{\mathbf{B}}^0\right|^2 d\Omega\right)^{1/2} \quad \forall \xi \in H_\tau^0(\Gamma)$$

where expression (8) for f'_τ, the Ostrogradskii-Gauss theorem [6] and the Cauchy-Buniakovsky inequality [6] were used. □

From the theorem, we get

$$\|G\|_{H_\tau^0} \leqslant \left(\int_{\Omega} \left|\widetilde{\mathbf{B}}^0\right|^2 d\Omega\right)^{1/2}.$$

Using the Riss theorem we conclude that variational problem (9) and equation (7) are both solved uniquely in $H_\tau^0(\Gamma)$. Moreover,

$$\|G\|_{H_\tau^0} = \|\tau\|_{H_\tau^0} \leqslant \left(\int_{\Omega} \left|\widetilde{\mathbf{B}}^0\right|^2 d\Omega\right)^{1/2}.$$

Thus, the error of the magnetic reaction field energy of the conducting surface Γ' will be less than the total error of the initial fields energy. We have the last-named error in case if the initial fields are assigned inaccurate or are approximated. The solution of equation (7) is stable in this sense.

4 Numerical solving of equation

The numerical solving of the variational problem is carried out by constructing the minimizing Ritz sequence $\tau^{(1)}, \tau^{(2)}, \ldots, \tau^{(n)}$. $\tau^{(n)}(M) = \sum_{k=1}^{n} c_k^{(n)} \tau_k(M)$ is the approximate solution that we have by using the basis $\{\tau_k\}_{k=1}^{n}$. The coefficients $\left\{c_k^{(n)}\right\}_{k=1}^{n}$ are defined by solving the following SLAE

$$\begin{bmatrix} \alpha_{11} & \alpha_{12} & \dots & \alpha_{1n} \\ \alpha_{21} & \alpha_{22} & \dots & \alpha_{2n} \\ \dots & \dots & \dots & \dots \\ \alpha_{n1} & \alpha_{n2} & \dots & \alpha_{nn} \end{bmatrix} \begin{bmatrix} c_1^{(n)} \\ c_2^{(n)} \\ \dots \\ c_n^{(n)} \end{bmatrix} = \begin{bmatrix} \beta_1 \\ \beta_2 \\ \dots \\ \beta_n \end{bmatrix}. \tag{11}$$

The SLAE has the real symmetric matrix. The elements of the system are defined by the following formulas

$$\alpha_{ij} = \frac{1}{4\pi} \int_{\Gamma_i} \int_{\Gamma_j} \frac{\operatorname{grad} \tau_j(Q)}{r_{QM}} d\Gamma_Q \operatorname{grad} \tau_i(M) d\Gamma_M, \quad i,j = \overline{1,n},$$

$$\beta_i = -\int_{\Gamma_i} \left(\mathbf{A}^{\Phi} - \mathbf{A}^{0}\right) \sigma_i d\Gamma, \quad i = \overline{1,n},$$

where $\sigma_i = [\mathbf{n} \operatorname{grad} \tau_i]$; $\mathbf{A}^0$ is the vector potential of the field $\mathbf{B}^0$ ($\mathbf{B}^0 = \operatorname{rot}\mathbf{A}^0$);

$$\mathbf{A}^{\Phi}(M) = \frac{1}{4\pi} \sum_{k=1}^{N} \Phi_k \operatorname{rot} \oint_{L_k} \frac{\mathbf{dl}_Q}{r_{QM}}, \; M \in \Gamma.$$

Here $L_1, L_2, \dots, L_N$ is the system of closed contours. This contours pass through Γ_0'' and corresponding internal holes of surface Γ; Φ_k is a given magnetic flux passing through the hole k.

The basis functions τ_i, $i = \overline{1,n}$ can be taken from a class of continuous and piecewise continuous differentiable on Γ functions. For example, it can be piecewise-polynomial functions which are shown in fig. 3. In this case the elements of the main matrix of SLAE (11) are calculated analytically. Moreover the matrix of the SLAE is a positively defined one according to [7]. From [8] it follows that Gauss-Seidel method is one of the most effective methods for solving such systems.

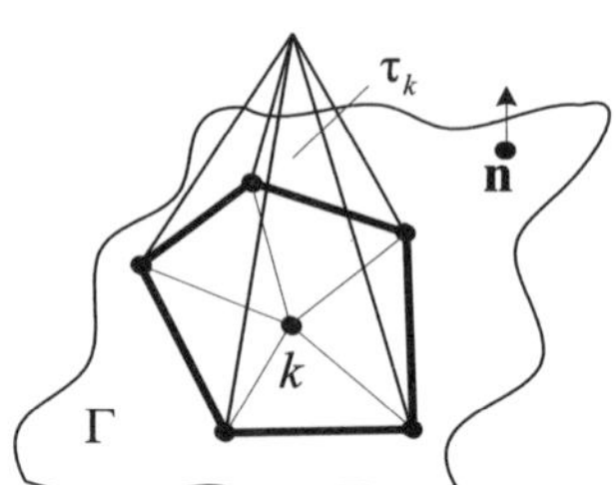

Fig. 3.

In order to solve the problem the novel software package has been created in the high-level programming language Microsoft Visual C# 2005. Using this package, we can assign both a non-perturbed field as its sources and directly as the formula. The results of its usage are shown in fig. 4–6.

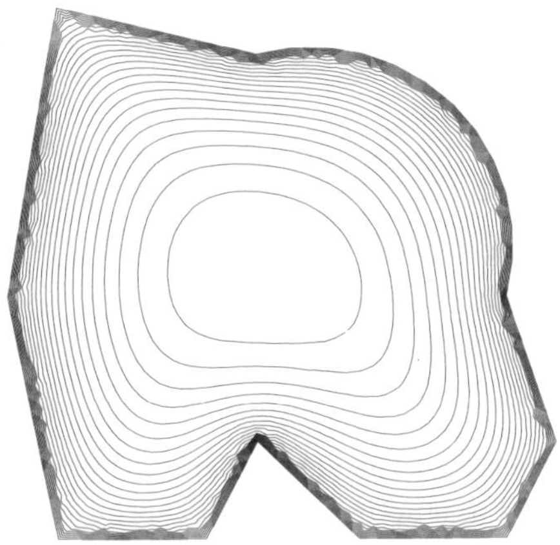

Fig. 4.

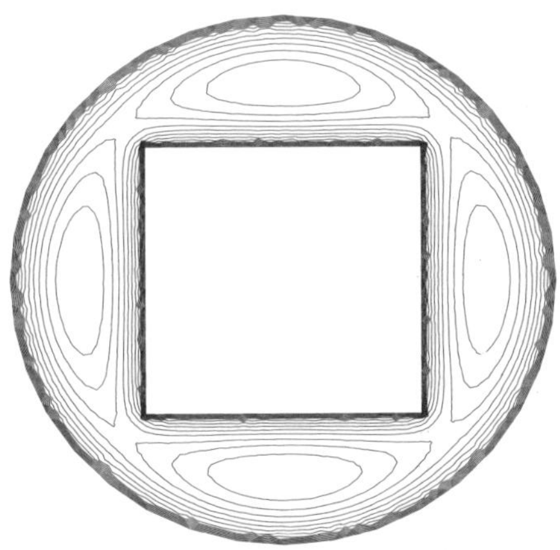

Fig. 5.

The eddy currents distribution on the simply connected ideal-conductive plate in the homogeneous cross magnetic field is represented in fig. 4. The flow function of the currents is the computed density τ. The distribution of the currents on the multiconnected plate in this field is shown in fig. 5. The

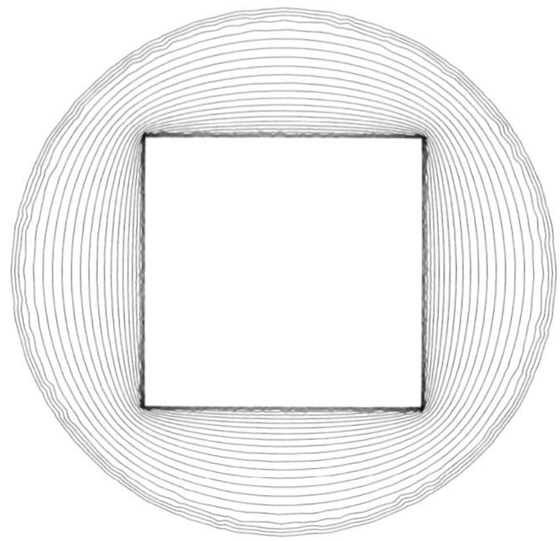

Fig. 6.

distribution of the currents on the same plate is shown in fig. 6 in the case if there is no non-perturbed magnetic field, and there is the magnetic flux passing through the hole.

References

1. O. V. TOZONI, *Method of Secondary Sources in Electrotechnics*, Energia, Moscow, 1975 [in Russian].
2. V. L. CHECHURIN, K. S. DEMIRCHAN, *Electric Magnetic Fields Computation*, Vysshaia Shkola, Moscow, 1986 [in Russian].
3. V. I. ASTAKHOV, *About an admissibility of idealization of borders of polarizable bodies and some energetic identities for stationary magnetic and electrostatic fields*, Izv. vuzov. Elektromekhanika, 1 (2000), pp. 3–14 [in Russian].
4. S. G. MIKHLIN, *Linear Partial Differential Equations*, Vysshaya Shkola, Moscow, 1977 [in Russian].
5. V. I. ASTAKHOV, *Surface potentials and operators of potential theory in Dirichlet spaces*, Izv. vuzov. Elektromekhanika, 2 (2000), pp. 3–18 [in Russian].
6. N. E. KOCHIN, *Vector Calculus and Foundations of Tensor Calculus*, Akad. Nauk SSSR, Moscow, 1961 [in Russian].
7. PH. CIARLET, *Finite Element Method for Elliptic Problems*, Society for Industrial and Applied Mathematics, Philadelphia, PA, USA, 2002.
8. J. ORTEGA, W. POOLE, *Numerical Methods for Differential Equations*, Pitman Publishing Inc., 1981.

Spectral Model Order Reduction Preserving Passivity for Large Multiport RCLM Networks

Yu. M. Nechepurenko[1,*], A. S. Potyagalova[2], and I. A. Karaseva[3,**]

[1] Institute of Numerical Mathematics, Russian Academy of Sciences, ul. Gubkina 8, Moscow, 119333 Russia
yumn@inm.ras.ru

[2] Cadence Design Systems LLC, ul. B. Ordynka 44, Bldg.4, Moscow, 119017 Russia
alena@cadence.com

[3] Moscow Institute of Physics and Technology, 141700, 9, Institutskii per., Dolgoprudny, Moscow Region, Russia
irina_kar@bk.ru

Abstract. For RCLM networks we present a novel algebraic spectral model order reduction algorithm equipped with efficient tools for preserving the passivity. For RC networks our approach is similar to the well-known spectral reduction technique PACT (pole analysis via congruence transformations). The accuracy and reduction ratio of resulting reduced-order models are demonstrated with several industrial examples.

1 Introduction

Complexity reduction for generating compact models of interconnect networks have been an intensive research area in the past decade due to increasing signal integrity effects and rising couplings modeled with parasitic capacitors and inductors. Since the original systems were passive one of the main requirements to the reduced systems was the preservation of passivity. Many algebraic model order reduction methods preserving passivity were proposed based on implicit moment matching, congruence transformations and truncated balance realizations [1, 4, 5, 7].

The spectral model order reduction method for RCLM networks, which we propose in this paper, can be construed as a generalization of the well-known model order reduction approach proposed in [4] for RC networks. Up to the present any such a generalization preserving passivity has seemed impossible

* This work was supported by the Russian Foundation for Basic Research (project №07-01-00658) and Russian Academy of Sciences (project "Optimization of numerical algorithms for solving the problems of mathematical physics").

** This work was supported by the Russian Foundation for Basic Research (project №07-01-00658).

(see, for instance, [5]) and other model order reduction methods were applied for the reduction of RCL and RCLM networks.

2 Background

Linear control systems modelling multiport RCLM networks [1] can be presented in the form

$$E\frac{dx}{dt} + Ax = Bu, \quad y = B^T x, \tag{1}$$

where $u \in \mathbf{R}^{n_1}$ is the control or input, $y \in \mathbf{R}^{n_1}$ is the observation or output and $x \in \mathbf{R}^n$ is the state vector; $B \in \mathbf{R}^{n\times n_1}$ and $A,\ E \in \mathbf{R}^{n\times n}$ have the following block-structures

$$A = \begin{bmatrix} A_{11} & A_{12} & A_{13} & -I_{n_1} \\ A_{12}^T & A_{22} & A_{23} & 0 \\ -A_{13}^T & -A_{23}^T & A_{33} & 0 \\ I_{n_1} & 0 & 0 & 0 \end{bmatrix}, \quad E = \begin{bmatrix} E_{11} & E_{12} & 0 & 0 \\ E_{12}^T & E_{22} & 0 & 0 \\ 0 & 0 & E_{33} & 0 \\ 0 & 0 & 0 & 0 \end{bmatrix}, \quad B = \begin{bmatrix} 0 \\ 0 \\ 0 \\ I_{n_1} \end{bmatrix}, \tag{2}$$

with $A_{ij},\ E_{ij} \in \mathbf{R}^{n_i\times n_j}$ where $n_1+n_2+n_3+n_1 = n$. Here and after I_m denotes the identity matrix of order m.

According to the above partition $x = (x_1^T, x_2^T, x_3^T, x_4^T)^T$ with $x_i \in \mathbf{R}^{n_i}$ for $i = 1, 2, 3$ and $x_4 \in \mathbf{R}^{n_1}$. Note that components of $x_1 = u$, x_2, x_3 and $y = x_4$ are respectively port voltages, inner node voltages, inductor currents and currents injected into the ports. For RC networks $n_3 = 0$, i.e. x has no the subcolumn x_3 and all the matrices have no blocks with the indices equal to 3.

The considered system is passive [1], i.e. it does not generate energy. Moreover it satisfies the following

Passivity criterion 1. *The system (1) is passive if*

$$A + A^T \geqslant 0, \ E = E^T \geqslant 0. \tag{3}$$

We report in this paper a balanced dichotomy approach which we proposed and justified in [3] for the model order reduction of (1), (2) to linear systems of the same form, input vector and dimension of the output vector but significantly smaller dimension of the state vector. The reduction error is estimated with the transfer functions for a given set of frequencies.

3 Proposed techniques

Congruence transformations proposed in [3] and based on known properties of (1), (2) reduce this system to an equivalent one of the same form with

$$A = \begin{bmatrix} A_{11} & V & -I_{n_1} \\ JV^T & F & 0 \\ I_{n_1} & 0 & 0 \end{bmatrix}, \quad E = \begin{bmatrix} E_{11} & 0 & 0 \\ 0 & I_m & 0 \\ 0 & 0 & 0 \end{bmatrix}, \quad B = \begin{bmatrix} 0 \\ 0 \\ I_{n_1} \end{bmatrix}, \tag{4}$$

where $m = n_2 + n_3$ and

$$F = \begin{bmatrix} A_{22} & A_{23} \\ -A_{23}^T & A_{33} \end{bmatrix}, \; V = \begin{bmatrix} A_{12} \, A_{13} \end{bmatrix}, \; J = \begin{bmatrix} I_{n_2} & 0 \\ 0 & -I_{n_3} \end{bmatrix}.$$

At that, the blocks of new matrices A and E are (generally speaking) modified while n_1 and n_3 stay the same and n_2 decreases. But the new matrices A and E are still satisfying Passivity criterion 1 and the finite spectrum of the pencil $A + sE$ is preserved.

3.1 Block diagonalization

The aim of the procedure, which we will call *the block diagonalization*, consists of transforming the system (1), (4) to an equivalent one of the same form (1) with the same matrices E and B but with the matrix A of the form

$$A = \begin{bmatrix} A_{11} & V_1 \ldots V_p & -I_{n_1} \\ J_1 V_1^T & F_1 & \\ \vdots & \ddots & \\ J_p V_p^T & \quad F_p & \\ I_{n_1} & & 0 \end{bmatrix} \tag{5}$$

where

$$F_j = \begin{bmatrix} A_{22}^j & A_{23}^j \\ -A_{23}^{jT} & A_{33}^j \end{bmatrix}, \; V_j = \begin{bmatrix} A_{12}^j \, A_{13}^j \end{bmatrix}, \; J_j = \begin{bmatrix} I_{n_2^j} & 0 \\ 0 & -I_{n_3^j} \end{bmatrix}, \tag{6}$$

blocks A_{kk}^j are symmetric nonnegative definite matrices and each block with indices kl is of size $n_k^j \times n_l^j$ with

$$n_1^j = n_1, \quad \sum_{j=1}^{p} n_k^j = n_k, \; k = 2, 3.$$

At that, if $n_k^j = 0$ then there are no blocks with index k in matrices F_j, V_j, J_j. For instance, $n_2^j = 0$ implies

$$F_j = A_{33}^j, \; V_j = A_{13}^j, \; J_j = -I_{n_3^j}.$$

We propose to reduce A to (5) with some similarity transformations increasing p step by step. In the first step we multiply the second block row by Y^{-1} from the left and the second block column by Y from the right where Y is a matrix of order $m = n_2 + n_3$ such that

$$Y^{-1} = \begin{bmatrix} J_1 & 0 \\ 0 & J_2 \end{bmatrix} Y^T J, \quad Y^{-1} F Y = \begin{bmatrix} F_1 & 0 \\ 0 & F_2 \end{bmatrix}, \quad J_j = \begin{bmatrix} I_{n_2^j} & 0 \\ 0 & -I_{n_3^j} \end{bmatrix}, \tag{7}$$

and spectra of F_1 and F_1 are some separate mutually disjoint self-conjugate subsets of the spectrum of F. At that $n_2^1 + n_2^2 = n_2$, $n_3^1 + n_3^2 = n_3$. As a result we obtain (5), (6) with $p = 2$ and $[V_1, V_2] = VY$.

Then we apply the above transformation, named *the balanced dichotomy* in [3], to the second (or third) block row and column of the new matrix A to obtain (5), (6) with $p = 3$. And so on. Since each step is the similarity transformation of A but not generally speaking the congruence transformation, the matrix A in (5) with $p \geqslant 2$ may not satisfy the inequality $A + A^T \geqslant 0$. The following criterion enables an efficient verification of the inequality after each balanced dichotomy.

Passivity criterion 2. *The system (5), (6) satisfies (3), and hence is passive, if*

$$A_{22}^j > 0,\ A_{33}^j \geqslant 0,\ j = 1, \ldots, p; \quad \Delta = \sum_{j=0}^{p} \Delta_j \geqslant 0,\ E_{11} \geqslant 0,$$

where

$$\Delta_0 = A_{11}, \quad \Delta_j = -A_{12}^j (A_{22}^j)^{-1} A_{12}^{jT}.$$

To apply this criterion we need to store for each pair of the $(j+1)$-th block row and column the corresponding matrix Δ_j and subtract the matrix from Δ and add to Δ the corresponding contributions of new block rows and columns when the block row and column are splitted with the balanced dichotomy. So the usage of Passivity criterion 2 allows us to avoid eigenvalue computations for matrices of order n. Note that the criterion can be extend to the case $A_{22}^j \geqslant 0$ [3].

Furthermore, since

$$Y^{-1} = (Y^T Y)^{-1} Y^T,$$

the quality $\nu = \mathrm{cond}_2(Y^T Y)$ can be used as a distance between the similarity and congruence transformations with Y. A balanced dichotomy procedure was proposed in [3, 6] for computing Y that satisfies (7). This procedure is based on the real Schur decomposition [2] and includes an a priori estimation of ν for some possible choices of Y. So it takes possible to choose Y with ν not larger than a given tolerance. This procedure needs $\mathcal{O}(m^3)$ arithmetic operations. If at each split with the procedure sizes of two new blocks are approximately equal then the total computational cost amounts as $\mathcal{O}(m^3)$ as well.

3.2 Reduction

Let the block diagonalization have been fulfilled and (5), (6) satisfying Passivity criterion 2 have been obtained. Then the transfer function of the system can be represented as

$$G(s) = \sum_{j=0}^{p} G_j(s) \tag{8}$$

where $G_0(s) = A_{11} + sE_{11}$ and for each $j \geqslant 1$

$$G_j(s) = -V_j(F_j + sI_{m_j})^{-1} J_j V_j^T \ (m_j = n_2^j + n_3^j)$$

and satisfies the following equalities:

$$G_j(s) = G_{j0} + sG_{j1}(s) = G_{j0} + sG_{j10} + s^2 G_{j11}(s)$$

with

$$G_{j0} = -V_j F_j^{-1} J_j V_j^T, \ G_{j10} = V_j F_j^{-1} J_j F_j^{-T} V_j^T,$$

$$G_{j1}(s) = V_j(F_j + sI_{m_j})^{-1} J_j F_j^{-T} V_j^T, \ G_{j11}(s) = -V_j F_j^{-1}(F_j + sI_{m_j})^{-1} J_j F_j^{-T} V_j^T.$$

The proposed reduction is curried out by removing some terms in (8). The error is checked directly as $\|G(i\omega) - G^{new}(i\omega)\|_2 / \|G(i\omega)\|_2$ for a given set of frequencies ω. Consider the following types of removing:

a) the term $G_j(s)$ is removed completely,

b) the zeroth moment G_{j0} is preserved,

c) the first two moments G_{j0} and G_{j10} are preserved.

The corresponding reduced system is obtained from the initial one by removing the $j+1$-th block rows and columns in A and E and the $j+1$-th block row in B. When the zeroth moment is preserved we modify in addition A_{11}: $A_{11}^{new} = A_{11} + G_{j0}$. When the first two moments are preserved the matrix E_{11} is modified as well: $E_{11}^{new} = E_{11} + G_{j10}$. The reduced system is still satisfying Passivity criterion 2 if $\Delta^{new} \geqslant 0$ where $\Delta^{new} = \Delta - \Delta_j$ in the case of complete removing of $G_j(s)$ and $\Delta^{new} = \Delta - \Delta_j + G_{j0}$ in the case of preserving the zeroth moment. When the first moment is preserved we need to verify the inequality $E_{11}^{new} \geqslant 0$ as well.

When the reduction has been fulfilled we continue the block diagonalization to decrease the number of nonzero entries in A. At this stage we care of computational stability only.

4 Experimental results

The proposed method was implemented in MATLAB and tested with 4 industrial interconnect RCLM networks. The order of corresponding matrices varied from 41 to 142. Three first networks were RCL, the fourth one was RCLM. Amounts of resistors (R), capacitors (C), self-inductances (L) and mutual inductances (M) and the orders of corresponding initial and reduced matrices are demonstrated in Table 1.

Table 1. The orders of initial and reduced matrices

test	R	C	L	M	initial order	reduced order
1	32	15	13	–	41	20
2	141	48	46	–	142	12
3	95	48	46	–	142	15
4	500	1070	41	610	141	9

Table 2. Matrix fill-ins

test	nnz	reduced nnz	reduction ratio
1	142	120	0.85
2	517	56	0.1
3	420	57	0.14
4	4677	54	0.01

Table 2 shows an efficiency of the proposed reduction method in terms of matrix fill-ins (the number of non-zero entries of matrices A, E and B).

Note that the proposed method was designed to reduce small and medium size multiport RCLM networks with high efficiency.

References

1. M. Celik, L. Pileggi, and A. Odabasioglu, *IC Interconnect analysis.* Kluwer Academic Publishers, 2002.
2. G. H. Golub and C. F. Van Loan, *Matrix computations.* London: The John Hopkins University Press, 1991.
3. I. A. Karaseva, Yu. M. Nechepurenko, and A. S. Potyagalova, Spectral model order reduction for control systems modelling passive integration circuits, *Comp. Maths. Math. Phys.* (to appear in 2008).
4. K. J. Kerns and A. T. Yang, Stable and efficient reduction of large multiport RC networks by pole analysis via congruence transforms, *IEEE Trans. Computer-Aided Desing,* vol. 16, pp. 734-744, 1997.
5. K. J. Kerns and A. T. Yang, Preservation of passivity during RLC network reduction via congruence transforms, *IEEE Trans. Computer-Aided Desing,* vol. 17, pp. 582-590, 1998.
6. Yu. M. Nechepurenko, A new spectral analysis technology based on the Schur decomposition, *Russ. J. Numer. Anal. Math. Modelling,* vol. 14, pp. 265–274, 1999.
7. J. R. Phillips, L. Daniel, and L. M. Silveira, Guaranteed passive balancing transformations for moder order reduction, *IEEE Trans. Computer-Aided Desing,* vol. 22, pp. 1027-1041, 2003.
8. G. Stewart and J. Sun, *Matrix perturbation theory.* San Diego, California: Academic Press, 1990.

New Smoothers in Multigrid Methods for Strongly Nonsymmetric Linear Systems*

Galina V. Muratova** and Evgenia M. Andreeva***

Computing Center, Southern Federal University, Rostov-on-Don, Russia
**muratova@sfedu.ru, ***andreeva@sfedu.ru

Abstract. New smoothers resulted from a special class of triangular skew-symmetric splitting iteration methods for the multigrid methods were used to solve the systems of linear equations with strongly non-symmetric coefficient matrices, which may be produced by the central-difference approximation of a convection-diffusion equation with dominant convection.

Keywords: multigrid method, smoothing procedure, triangular skew-symmetric splitting iteration methods.

1 Introduction

Multi-grid methods (MGMs) are fast iterative solvers based on the multilevel or the multi-scale approach. A typical application of the multigrid method is to numerical solution of partial differential equations in two or higher dimension. Alternatively, the MGM can be applied in combination with any of the common discretization techniques; in such case, it is among the fastest solution techniques known today.

Multigrid method does not mean only a fixed multigrid algorithm applying to all boundary value problems in the general setting, and it may usually refer to an algorithmic framework of the multigrid techniques. The efficiency of the MGM heavily depends on the adjustment of the involved components, e.g., the basic linear solver or the smoothing procedure.

We suggest new smoothers resulted from a special class of triangular skew-symmetric splitting iteration methods for the multigrid methods used to solve the systems of linear equations with strongly nonsymmetric coefficient matrices, which may be produced by the central-difference approximation of the stationary convection-diffusion equation with dominant convection.

Mathematical models that involve a combination of convective and diffusive processes are among the most widespread in all the sciences. Research of these processes is especially important and difficult when convection is dominant [1].

* This work was supported by the Russian Foundation for Basic Research (RFBR) under the Grants №06-01-39002, 06-01-00038.

At the same time convection - diffusion equations are used as tests in researching iterative methods for solving systems of strongly non- symmetric linear equations. Applying upwind differences we obtain an M- matrix [2], and using central differences, we get a positive real or dissipative matrix [3]. We have used central- difference approximation of convective terms. In this case the resulting system of linear algebraic equations is strongly nonsymmetric one. Special class of triangular skew-symmetric iteration methods are intended to for this type systems [4]. We have used the triangular iterative methods (TIMs) from this class as smoothers in multigrid method.

2 Model problem

We consider the model problem of the steady-state convection-diffusion process in domain Ω

$$\begin{cases} -\frac{1}{Pe}\triangle u + \frac{1}{2}\left[v_1 u_x + v_2 u_y + (v_1 u)_x + (v_2 u)_y\right] = F\,, \\ v_i = v_i(x,y),\; i = 1,2, \;\; u = u(x,y), \;\; F = F(x,y)\,, \\ (x,y) \in \Omega = [0,\, 1] \times [0,\, 1]\,, \\ u\,|_{\partial\Omega} = 0, \end{cases} \tag{1}$$

where F is selected so that the solution of (1) is defined as

$$\tilde{u}(x,y) = e^{xy} \sin \pi x \sin \pi y\,.$$

The initial form of the convection - diffusion equation is rather important for such problems. There exist three forms of the convective operator , which are equivalent to a differential level of incompressible environments, but result in various forms of the difference equations distinguished on the properties.

FDM with central differences was used for discretization of (1). We obtain linear algebraic equation system (2)with strongly nonsymmetric matrix.

$$Au = f \tag{2}$$

Present the matrix A as

$$A = A_0 + A_1 \tag{3}$$

where

$$A_0 = \tfrac{1}{2}(A + A^*) > 0, \qquad A_1 = \tfrac{1}{2}(A - A^*) = -A_1^*, \tag{4}$$

A_0 is symmetric part, A_1 is skew-symmetric part of initial matrix A. In this case A is strongly nonsymmetric matrix, i.e. the following inequality is held in some matrix norm

$$\|A_0\| \ll \|A_1\| \tag{5}$$

Let's note that in this case matrix A is a real positive one, that is $A_0 = A_0^* > 0$. The following decomposition of matrix A_1 is used

$$A_1 = K_l + K_u, \tag{6}$$

where K_l and K_u are the strictly lower and upper triangular parts of matrix A_1. We use multigrid method with specially created smoothers for solving system (2).

3 Smoothing procedure in multigrid method

To solve problem (2) we suggest using MGM , where the following triangular iterative method (TIM) will be used as a smoother of MGM

$$B(y_{n+1} - y_n)/\tau + Ay_n = f, \qquad n = 0, 1, 2, ... \tag{7}$$

with operator

$$B = I + 2\tau K_l \qquad \text{or} \qquad B = I + 2\tau K_u \tag{8}$$

where $\tau > 0$ is a scalar parameter.

As the matrix A we can present B as

$$B = B_0 + B_1 \tag{9}$$

where

$$B_0 = \tfrac{1}{2}(B + B^*) > 0, \qquad B_1 = \tfrac{1}{2}(B - B^*) = -B_1^*, \tag{10}$$

B_0 is a symmetric part, B_1 is a skew-symmetric part of matrix B.

This method was suggested in [3]. The splitting used as a smoother is related to the ones proposed in [5], [6], [4].

Suggested way of operator B construction defines the class of triangular skew-symmetric iterative methods (TIM). Any method from this class behaviors in the same way as Gauss-Seidel one: it quickly reduces the high-, but not low-frequency components of error frequencies. This is the necessary property of the smoother of MGM, that's why we have used these methods as the smoothers.

We also consider two methods from the class of triangular skew-symmetric methods — TIM1 and TIM2.

For TIM1 the operator B is under construction as follows:

$$B = \alpha I + 2K_\ell \quad \text{or} \quad B = \alpha I + 2K_u. \tag{11}$$

For TIM2:

$$B = \alpha_i I + 2K_\ell \quad \text{or} \quad B = \alpha_i I + 2K_u. \tag{12}$$

.

Parameters of the offered methods $\alpha_i, \alpha > 0$ get out under formulas:

$$\alpha = \|M\|$$

$$\alpha_i = \sum_{j=0}^{n} |m_{ij}|, i = \overline{0, n}$$

where $M = \{m_{ij}\}_0^n$ is a symmetric matrix which is constructed in the following way $M = A_0 + K_u - K_l$, n is a dimension of a matrix A.

The proof of the convergence of new modification of Multigrid Method is given in [7].

4 Numerical results

We consider the problem (1) to research properties of MGM modifications with suggested smoothers. We research four model problems with different velocity fields, presented in table 1.

Table 1. Velocity coefficients for test problems

Problem N	v_1	v_2
1	1	-1
2	$1 - 2x$	$2y - 1$
3	$x + y$	$x - y$
4	$\sin 2\pi x$	$-2\pi y \cos 2\pi x$

The different Peclet numbers were considered: $Pe = 1000, 10000, 100000$. The central-difference approximation was used on a grid 33×33, 512×512.

The problem (1) was solved with multigrid method where three kinds of smoothes were used: TIM, TIM1 and TIM2. The number of smoothing iterations in MGM is 15. This iteration number is rather large but it is optimal for this MGM modification. We have obtained this value having carried out the Fourier analysis of MGM modification and besides TIM iteration is cheap in sense of CPU time. In table 2 the results of comparison of the suggested MGM modifications with triangular skew-symmetric methods and Gauss-Seidel method as the smoothers on a grid 33×33 are presented.The symbol D means that for a given problem iterations of a method do not converge.

In table 3 the same results of comparison of the suggested MGM modifications on a grid 512×512 are presented. MGM with Gauss-Seidel method as the smoother on a grid 512×512 does not converge.

5 Conclusions

a) the suggested multigrid method modification with triangular iterative smoothers proved to be effective for solution of the systems of linear equations with strongly nonsymmetric coefficient matrices.
b) the multigrid method with the smoothers TIM1 and TIM2 is more effective for the problems, than MGM with TIM the smoother. Under consideration the most effective method for convection-diffusion problem with dominant convection is MGM with smoother TIM2.
c) the coefficient of skew-symmetry κ= Pe*h*$|\nu|/2$ has the greatest influence on the behaviour of the method (not neither the size of grid nor the coefficients of equation in particular).

References

1. H.Elman, D.Silvester, A.Wathen, *Finite Elements and Fast Iterative Solvers: with Applications in Incompressible Fluid Dynamics*, 2005, Oxford University Press, 400 p.
2. A.A. Samarskii, *On Monotone Difference Schemes for Elliptic and Parabolic Equations in the Case of a Non-self-adjoint Elliptic Operator*, Z. Vychisl. Mat. i Mat. Fiz. 5 1965, P. 548–551.
3. L. Krukier, *Implicit Difference Schemes and an Iterative Method for Their Solution for One Class of Quasilinear Systems of Equations*, Izvestija Vuzov, Mathematics, **7** 1979, P.41–52.
4. Z.-Z. Bai, L.A. Krukier and T.S. Martynova, *Two-step iterative methods for solving the stationary convection-diffusion equation with a small parameter at the highest derivative on a uniform grid*, Comput. Math. Math. Phys., 46(2006), P.282–293.
5. O. Axelsson, Z.-Z. Bai and S.-X. Qiu, *A class of nested iteration schemes for linear systems with a coefficient matrix with a dominant positive definite symmetric part*, Numer. Algorithms, 35(2004), P.351–372.
6. Z.-Z. Bai, G.H. Golub and M.K. Ng, *Hermitian and skew-Hermitian splitting methods for non-Hermitian positive definite linear systems*, SIAM J. Matrix Anal. Appl., 24(2003), P.603–626.
7. G. Muratova, L. Krukier, *Multigrid method for the iterative solution of strongly nonselfadjoint problems with dissipative matrix*, Proceedings of the Conference on AMLI'96, Nijmegen, 1996, v.2, 169–178

Table 2. MGM iteration number and CPU-time on the grid 32×32

Pe	MGM (Seidel)	MGM (TIM)	MGM (TIM1)	MGM (TIM2)
Problem 1: $v_1(\mathbf{x}) = 1 \quad v_2(\mathbf{x}) = -1$				
10	13 0:00:31	35 0:00:94	30 0:00:93	30 0:00:109
100	63 0:00:188	7 0:00:16	5 0:00:15	5 0:00:15
1000	D	13 0:00:31	9 0:00:47	9 0:00:31
10000	D	78 0:00:250	58 0:00:203	58 0:00:188
Problem 2: $v_1(\mathbf{x}) = 1 - 2x_1 \quad v_2(\mathbf{x}) = 2x_2 - 1$				
10	22 0:00:62	72 0:00:188	53 0:00:172	50 0:00:171
100	18 0:00:47	24 0:00:63	19 0:00:63	14 0:00:47
1000	D	16 0:00:47	12 0:00:31	6 0:00:15
10000	D	59 0:00:187	51 0:00:171	32 0:00:109
Problem 3: $v_1(\mathbf{x}) = x_1 + x_2 \quad v_2(\mathbf{x}) = x_1 - x_2$				
10	16 0:00:47	43 0:00:125	35 0:00:110	34 0:00:110
100	23 0:00:62	9 0:00:31	7 0:00:15	5 0:00:15
1000	D	17 0:00:47	12 0:00:31	8 0:00:31
10000	D	74 0:00:219	55 0:00:187	36 0:00:125
Problem 4: $v_1(\mathbf{x}) = \sin 2\pi x_1 \quad v_2(\mathbf{x}) = -2\pi x_2 \cos 2\pi x_1$				
10	17 0:00:47	39 0:00:109	32 0:00:110	27 0:00:94
100	D	16 0:00:47	12 0:00:47	7 0:00:31
1000	D	29 0:00:94	22 0:00:78	10 0:00:31
10000	D	193 0:00:625	159 0:00:562	57 0:00:187

Table 3. MGM Number of iterations and CPU time on the grid 512×512

Pe	MGM (TIM)	MGM (TIM1)	MGM (TIM2)	κ= Pe*h*|v|/2
Problem 1: $v_1(\mathbf{x}) = 1 \quad v_2(\mathbf{x}) = -1$				
1000	82 1:7:110	56 0:50:31	56 0:47:609	0,976562
10000	62 0:51:265	36 0:34:984	36 0:30:703	9,765625
100000	119 1:38:109	110 1:38:844	110 1:33:781	97,65625
Problem 2: $v_1(\mathbf{x}) = 1 - 2x_1 \quad v_2(\mathbf{x}) = 2x_2 - 1$				
1000	408 5:12:953	251 3:48:234	153 2:10:407	0,976562
10000	345 4:23:187	149 2:11:78	49 0:41:828	9,765625
100000	221 2:54:422	189 2:46:250	67 0:57:157	97,65625
Problem 3: $v_1(\mathbf{x}) = x_1 + x_2 \quad v_2(\mathbf{x}) = x_1 - x_2$				
1000	153 1:59:953	98 1:26:94	67 0:57:141	1,953125
10000	138 1:48:422	72 1:3:320	17 0:14:532	19,53125
100000	177 2:18:984	86 1:15:531	59 0:50:438	195,3125
Problem 4: $v_1(\mathbf{x}) = \sin 2\pi x_1 \quad v_2(\mathbf{x}) = -2\pi x_2 \cos 2\pi x_1$				
1000	242 3:15:734	158 2:24:625	83 1:13:703	6,135923
10000	240 3:13:765	157 2:23:844	56 0:49:750	61,35923
100000	349 4:41:938	333 4:40:0	79 1:10:328	613,5923

Operator Equations for Eddy Currents on Singular Carriers

Jan Naumenko

Southern Scientific Centre of Russian Academy of Sciences, Rostov-on-Don, Russia
jan@novoch.ru

Abstract. The problem of eddy currents computation on the conducting surface is considered. It is reduced to the integral equations. The existence, uniqueness and stability of their solutions is proved. The numerical method for the mentioned problem is described. The example of computing is given.

Keywords: integral equation, potential of single layer, magnetic field.

1 Introduction

A wide range of practical tasks requires numerical computation of magnetic fields in the medium with conducting bodies. Nowadays the theory of these problems is well developed, whereas another important issue - computation of magnetic fields in the presence of multiconnected crack (conducting surface) is paid less attention and provides an area for further study. The main objective of this article is to develop a theory and a numerical method for these actual problems solution.

2 Main equations

Let's consider the space $L_2(S;\mathbb{R}^2)$ or $L_2(S;\mathbb{C}^2)$ that consists of two-component real or complex square-integrable on S vector functions accordingly. To simplify we write $L_2(S)$ in case it does not lead to misunderstanding. We suppose that the multiconnected Riemannian surface S and its boundary satisfy the Lipschitz's conditions [6] (see fig. 1). The space $L_2(S)$ can be decomposed to the sum[1, 2]:

$$L_2(S) = L^{(P)} \oplus \mathfrak{L}.$$

Here $L^{(P)}$ consists of potential fields generalized by the Weyl [1]. It can be represented as the closure by the norm of $L_2(S)$ of the gradients of smooth functions [3]. It is obvious that for all $\mathbf{b} \in \mathfrak{L}$ is

$$\iint_S \mathbf{b}\,\mathrm{grad}\,\varphi\, dS = 0$$

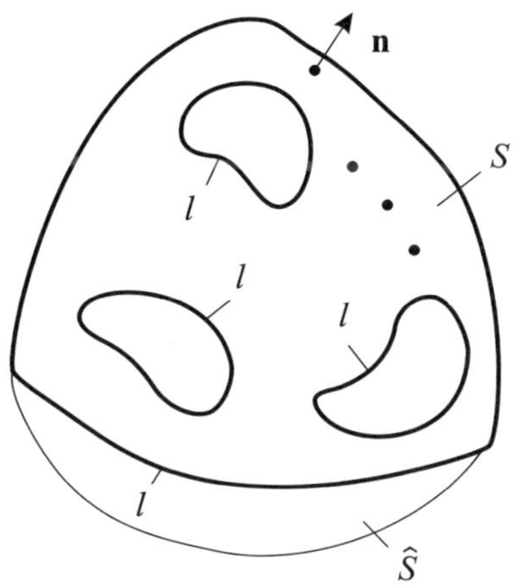

Fig. 1.

where φ is smooth function (the operators grad and rot are understood here and below in the sense of Hamilton-Beltrami).

The space $\mathfrak{L}$ can be also decomposed to the pair

$$\mathfrak{L} = \mathfrak{G} \oplus \mathrm{L}^{(S)}.$$

Here $\mathfrak{G}$ is the finite dimensional space that consists of generalized by Weyl harmonic fields. Its dimension depends directly on the connectness of the S. For example if S is simple connected then $\mathfrak{G} = \varnothing$. In other words, the space $\mathfrak{G}$ consists of cycles. The components of the elements of $\mathfrak{G}$ are smooth and the property

$$\operatorname{div} \mathbf{g} = 0, \operatorname{rot} \mathbf{g} = 0,$$
$$\mathbf{g}\nu = 0$$

is correct for all $\mathbf{g} \in \mathfrak{G}$ [2]. Here ν is the tangential to S normal to the S border.

The important property [5] of the $\mathrm{L}^{(S)}$ elements is that their cycles are equal to zero. It means in case of smooth components of $\mathrm{b} \in \mathrm{L}^{(S)}$ that

$$\oint_C \mathbf{b} d\mathbf{l} = 0$$

for any smooth closed curve $\mathrm{C} \subset \mathrm{S}$.

We use the orthoprojector $\mathrm{P} = \mathrm{P}^{\mathrm{L}}\mathrm{P}^{\mathrm{S}}$ below where P^{S} vanishes normal to S field component (a normal to Lipschitz the manifold exists almost everywhere [6]) and P^{L} is orthoprojector $\mathrm{L}_2(\mathrm{S}) \to \mathfrak{L}$.

The computing of quasi-stationary magnetic fields in presence of the conducting surface (the conducting body with degenerated third dimension) can be reduced to the following operator equations for eddy (Foucault) currents density:

$$K\sigma = \mathbf{f}_1, \tag{1}$$

$$\delta = \lambda K\delta + \mathbf{f}_2, \tag{2}$$

$$\delta = \lambda \frac{\partial}{\partial t} K\delta + \mathbf{f}_3(\mathbf{t}). \tag{3}$$

Here $K = P\Gamma$, $\Gamma\xi = \frac{1}{4\pi}\iint\limits_S \frac{\xi}{r_{NM}} dS_N$, r is the distance between points M and N, λ is some parameter, $\mathrm{Im}\lambda \neq 0$ in equation (2) and $\mathrm{Im}\lambda = 0$ in equation (3), $\mathbf{f}_1, \mathbf{f}_2, \mathbf{f}_3(\mathbf{t}) \in \mathfrak{L}$. It is necessary to add an initial condition to equation (3). We suppose that $\delta(0) = \delta^0 \in \mathfrak{L}$.

The equation (1) describes eddy currents in case of infinite conductivity of S, the equation (1) describes eddy currents in case of finite conductivity of S in the time-harmonic mode, the equation (3) describes eddy currents the case of finite conductivity of S in the transitive mode. The operator K is analyzed in work [5] and there the following theorem was proved.

Theorem 1. *The operator* K *is linear, self-adjoint, positive and compact in the space* $\mathfrak{L}$.

By using the properties of K we can introduce the energy space $\mathfrak{L}_K$. It is the closure of the space $\mathfrak{L}$ by the following norm:

$$(\sigma_1, \sigma_2)_{\mathfrak{L}_K} = (K\,\sigma_1, \sigma_2)_{\mathfrak{L}}, \|\sigma\|_{\mathfrak{L}_K} = (\sigma, \sigma)^{1/2}_{\mathfrak{L}_K}. \tag{4}$$

Because the operator P is self-adjoint (P is orthoprojector) it is possible to omit P in (4) and write (4) in following form:

$$(\sigma_1, \sigma_2)_{\mathfrak{L}_K} = (\Gamma\sigma_1, \sigma_2)_{\mathfrak{L}}. \tag{5}$$

The representation (5) is usable in numerical implementations because the direct numerical realization of P is the serious problem.

As it is shown in [5] the following theorems are correct.

Theorem 2. *The equation* (1) *has unique and stable solution in* $\mathfrak{L}_K$ *if* $\mathbf{A}^0 \in B_2^{1/2}(S)$.

Here B is the Besov space [6].

The proof of this theorem leads to the following inequalities:

$$\|\mathbf{u}\|_{\mathfrak{L}_K} \leqslant \vartheta\,\|\mathbf{f}_1\|_{B_2^{1/2}(S)}.$$

The constant ϑ depends only on the geometry of S. This inequality ais usable fore the aposteriory estimation of the computational error of the numerical solution.

Theorem 3. *The equation* (2) *has unique and stable solution in* $\mathfrak{L}$ *if* $\mathbf{f}_2 \in \mathfrak{L}$. *The inequality*

$$\|\delta\|_{\mathfrak{L}} \leqslant \frac{\|\mathbf{f}_2\|_{\mathfrak{L}}}{1+|\lambda|/|\lambda_1|} \tag{6}$$

is correct. Here λ_1 *is the first characteristic number of operator* K.

Theorem 4. *The equation* (3) *has unique and stable solution if* $\|\mathbf{f}_3\|_{\mathfrak{L}} \in C^1[0,\infty]$.

Proof. Because the operator K is self-adjoint and positive it has a full in $\mathfrak{L}$ system $\{\omega_i\}_{i=1}^{\infty}$ of eigenfunctions with the system of characteristic numbers $\{\lambda_i\}_{i=1}^{\infty}$. We can write formally:

$$\delta = \sum_{i=1}^{\infty} c_k(t)\,\omega_i.$$

By using this representation in (3) we get:

$$\sum_{i=1}^{\infty} c_i\omega_i = -\lambda\frac{\partial}{\partial t}K\sum_{i=1}^{\infty} c_i\omega_i + \mathbf{f}_3.$$

Let's transform:

$$c_i = -\lambda\frac{c_i'}{\lambda_i} + (\mathbf{f}_3,\omega_i)_{\mathfrak{L}},$$

$$c_i' = -\frac{\lambda_i}{\lambda}c_i - (\mathbf{f}_3,\omega_i)_{\mathfrak{L}},$$

and then

$$c_i(t) = e^{-\frac{\lambda_i}{\lambda}t}\left[c_i^0 + \int_0^t e^{\frac{\lambda_i}{\lambda}\tau} f_i(\tau)\,d\tau\right].$$

Here $c_i^0 = \left(\delta^0,\omega_i\right)_{\mathfrak{L}}$ and $f_i(t) = (\mathbf{f}_3,\omega_i)_{\mathfrak{L}}$. Finally, we have:

$$\delta(x,t) = \sum_{i=1}^{\infty}(\delta^0,\omega_k)e^{-\frac{\lambda_i}{\lambda}t}\omega_i(x) + \sum_{i=1}^{\infty}\omega_i(x)\int_0^t e^{-\frac{\lambda_i}{\lambda}(t-\tau)} f_i(\tau)\,d\tau. \tag{7}$$

Let's analyze the properties of the series (7).

$$\left\{\int_0^t e^{-\frac{\lambda_i}{\lambda}(t-\tau)} f_i(\tau)\,d\tau\right\} \leqslant \int_0^t e^{-\frac{2\lambda_i}{\lambda}(t-\tau)}\,d\tau\cdot\int_0^t f_i^2(\tau)\,d\tau$$

$$= \frac{1-\frac{2\lambda_i}{\lambda}t}{2\frac{\lambda_i}{\lambda}}\cdot\int_0^t f_i^2(\tau)\,d\tau < \frac{\lambda}{2\lambda_i}\cdot\int_0^t f_i^2(\tau)\,d\tau \leqslant \frac{\lambda}{2\lambda_1}\int_0^t f_i^2(\tau)\,d\tau.$$

Because the series $\sum\limits_{i=1}^{\infty}(\delta^0,\omega_i)^2$ converges and $\sum\limits_{i=1}^{\infty} f_i^2 = \|\mathbf{f}_3\|_{\mathfrak{L}}$, the series (7) converges almost everywhere and has continuous sum in time domain. Consequently from the theorem of Dini the series (7) converges regular almost everywhere.

The series

$$\frac{\partial}{\partial t}\delta = \sum_{i=1}^{\infty}\left\{-\frac{\lambda_i}{\lambda}(\delta^0,\omega_i)\,e^{-\frac{\lambda_i}{\lambda}t} + f_i(t) - \frac{\lambda_i}{\lambda}\int_0^t e^{-\frac{\lambda_i}{\lambda}(t-\tau)} f_i(\tau)d\tau\right\}\cdot\omega_i. \quad (8)$$

converges regular almost everywhere in every time domain $[\bar{t}, T]$, $0 < \bar{t} < T < \infty$. We have estimation for squares of coefficients of (8):

$$3\left(\frac{\lambda_i}{\lambda}e^{-\frac{\lambda_i}{\lambda}t}\right)^2(\delta^0,\omega_i)^2 + 3f_i^2(0)\cdot e^{-2\frac{\lambda_i}{\lambda}t} + 3\left(\int_0^t e^{-\frac{\lambda_i}{\lambda}(t-\tau)} f_i'(\tau)d\tau\right)^2$$

$$\leqslant \frac{3}{(\bar{t}\cdot e)^2}(\delta^0,\omega_i)^2 + 3f_i^2(0) + \frac{3\lambda}{2\lambda_i}\int_0^T [f_i'(\tau)]^2\,d\tau.$$

Here the Cauchy inequality is used.

The representation (7) of the equation (3) solution leads to the inequality:

$$\|\Delta\delta\|_{\mathfrak{L}} \leqslant \|\Delta\delta^0\|_{\mathfrak{L}}\,e^{-\frac{\lambda_1}{\lambda}t}.$$

Here $\Delta\delta^0$ is the deviation of initial condition, $\Delta\delta$ is the solution error.

3 Numerical algorithm

Now we will discuss the main object of the article: the numerical implementations of the equations (1 - 3). The basic ideas of the numerical algorithms are very similar because the main part of all three equations is the operator K.

3.1 Interpolation schemes for the spaces $\mathfrak{L}$ and $\mathfrak{L}_k$

We use the interpolation approach introduced in [4].

Let's consider the approximation system $\{\mathbf{b}_i\}_{i=1}^n$ which consists of linearly-independent solenoidal piecewise-constant fields like FEM. See an example on fig. 2. We designate $S_i = \operatorname{supp}\mathbf{b}_i$.

As it is proved in [4] such system is complete in $L^{(S)}$.

The completeness of the interpolation system in $L^{(S)}$ is not enough in case of multi-connected surface S. We describe a set Δ_Σ of triangles which lay near

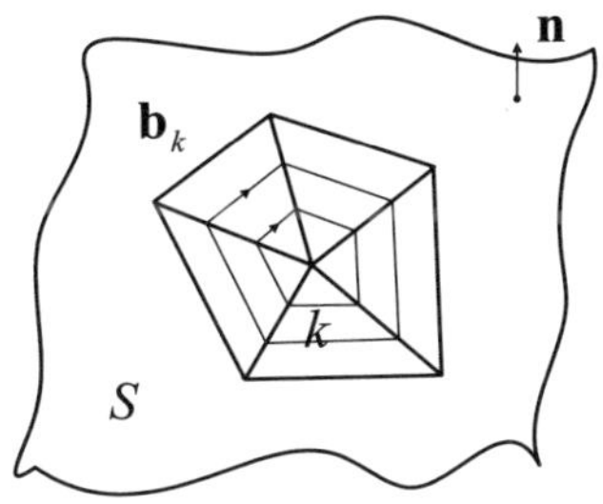

Fig. 2.

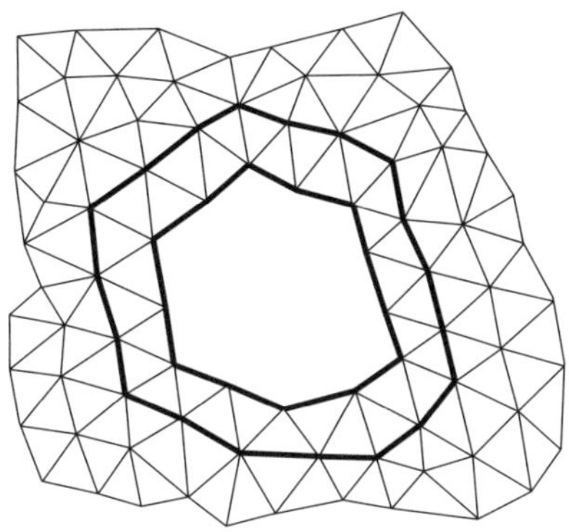

Fig. 3.

the border of hole. See an example on fig. 3. We set on Δ_Σ solenoidal piecewise-constant field similarly to one mentioned above. This field is in space $\mathfrak{L}$. Moreover, such field includes nonzero component from the space $\mathfrak{G}$. We obtain the system complete in $\mathfrak{L}$, having put in conformity with each of holes a similar additional element.

It is important that the constructed system is complete in $\mathfrak{L}_k$ as $\mathfrak{L}$ is dense set in $\mathfrak{L}_k$.

As it is shown shown in [5] the approximation error goes to zero when the maximal diameter of the elements of triangulation goes to zero and all of the angles in triangles are less than $\pi/2$.

The usage of Riesz method leads to the following numerical problems for the equations (1 - 3)

$$\sum_{k=1}^{m} (\Gamma \mathbf{b}_k, \mathbf{b}_i)_{\mathfrak{L}}\, C_k^{(m)} = (\mathbf{f}_1, \mathbf{b}_i)_{\mathfrak{L}}\,, i = 1, 2...m, \tag{9}$$

$$\sum_{k=1}^{m} [(\mathbf{b}_k, \mathbf{b}_i)_{\mathfrak{L}} - \lambda\, (\Gamma \mathbf{b}_k, \mathbf{b}_i)_{\mathfrak{L}}]\, C_k^{(m)} = (\mathbf{f}_2, \mathbf{b}_i)_{\mathfrak{L}}\,, i = 1, 2...m, \tag{10}$$

$$\lambda \sum_{k=1}^{m} (\Gamma \mathbf{b}_k, \mathbf{b}_i)_{\mathfrak{L}} \frac{d}{dt} C_k^{(m)}(t) = -\sum_{k=1}^{m} (\mathbf{b}_k, \mathbf{b}_i)_{\mathfrak{L}} C_k^{(m)}(t) - (\mathbf{f}_3, \mathbf{b}_i)_{\mathfrak{L}}, \ i = 1, 2...m \tag{11}$$

The main part of all these systems is the matrix A:

$$a_{i,j} = (\Gamma \mathbf{b}_i, \mathbf{b}_j)_{\mathfrak{L}}.$$

The most important property of the described approximation system is that the numerical dimension of the problems (9 - 11) is equal to scalar formulation (one unknown value for each node of triangulation).

3.2 Matrix A construction

It is simple to prove [5] that the matrix is symmetric and positive-definite.

The most difficult part of the numerical algorithm is the construction of the matrix A. The calculation of the elements of the matrix A can be reduced to the integral:

$$(\Gamma \mathbf{b}_k, \mathbf{b}_i)_{\mathfrak{L}} = \frac{1}{4\pi} \iint_{S_i} \iint_{S_k} \frac{\mathbf{b}_k(M)\, \mathbf{b}_i(N)}{r_{NM}} dS_N dS_M \tag{12}$$

This four-fould integral has high computational complexity. Moreover it has singularities when $S_i \cap S_k \neq \varnothing$.

The calculation of (12) can be reduced to the integral:

$$\iint_{\Delta_m} \iint_{\Delta_n} \frac{1}{r_{NM}} dS_N dS_M.$$

This integral is calculated analytically in work [4] in particular case when the triangles Δ_n and Δ_n lies in the same plane. On the base of the idea from [4] we found the analytical representation for the integral (12) in general case but it is too cumbersome to show it here. The numerical representation of this formula consists of approximately one thousand lines of program code.

However, even the analytical representation of the (12) is not a panacea. Complete construction of the A by using of the (12) is time-expensive. Moreover full storage in the computer memory of the A may be impossible in high presicion computations because of high dimension of A.

This problem can be solved in different ways. The most general approach is developed in the works of Tyrtyshnikov [7]. We use here an alternative approach that is specialized for the form of kernel of integral equations and for approximations used in this work.

Without loss of generality we describe the case when S is plane. All the results are correct in general case but formulas are more cumbersome. Let's suppose that the distance r_{NM} between the geometrical centers of S_i and S_k is much more than $h = maxdiam(h_i)$, i.e.

$$\frac{h}{r_{NM}} = \eta << 1.$$

We expand the function $\frac{1}{r_{NM}}$ into Taylor series near the geometrical centers of finite elements supports:

$$\frac{1}{r_{NM}} = \sum_{i=0}^{\infty}\sum_{j=0}^{\infty}\sum_{k=0}^{\infty}\sum_{m=0}^{\infty} P_{i,j}(M - M_0) P_{k,m}(N - N_0)$$
$$\times \frac{\partial^{i+j+k+m}}{\partial^i x_M \partial^j y_M \partial^k x_N \partial^m y_N} \frac{1}{r_{NM}} (M_0, N_0).$$

Here $P_{i,j}$, $P_{k,m}$ are some polinomials of the $i + j$ and $k + m$ orders accordingly. Let's use this representation in (12)

$$\iint_{S_i}\iint_{S_k} \frac{\mathbf{b}_k(M)\,\mathbf{b}_i(N)}{r_{NM}} dS_N dS_M$$
$$= \sum_{i=0}^{\infty}\sum_{j=0}^{\infty}\sum_{k=0}^{\infty}\sum_{m=0}^{\infty} \frac{\partial^{i+j+k+m}}{\partial^i x_M \partial^j y_M \partial^k x_N \partial^m y_N} \frac{1}{r_{NM}} (M_0, N_0)$$
$$\times \iint_{S_i}\iint_{S_k} \mathbf{b}_k(M)\,\mathbf{b}_i(N)\,P_{i,j}(M - M_0)\,P_{k,m}(N - N_0)\, dS_N dS_M.$$

It is important that all components of the solenoidal fields has zero median value. Some integrals in the sum vanish. The last representation can be transformed to the following form:

$$\iint_{S_i}\iint_{S_k} \frac{\mathbf{b}_k(M)\,\mathbf{b}_i(N)}{r_{NM}} dS_N dS_M$$
$$= \sum_{i=1}^{\infty}\sum_{j=1}^{\infty}\sum_{k=1}^{\infty}\sum_{m=1}^{\infty} \frac{\partial^{i+j+k+m}}{\partial^i x_M \partial^j y_M \partial^k x_N \partial^m y_N} \frac{1}{r_{NM}} (M_0, N_0)$$
$$\times \iint_{S_k} \mathbf{b}_k(M)\,P_{i,j}(M - M_0)\, dS_M \iint_{S_i} \mathbf{b}_i(N)\,P_{k,m}(N - N_0)\, dS_N. \qquad (13)$$

Here all integrals components $i + j = 0$ or $k + m = 0$ are equal to zero. We use the following designations for the integrals:

$$\mathbf{M}^i_{k,m} = \iint_{S_i} \mathbf{b}_i P_{k,m} dS.$$

It is obvious that the integrals $\mathbf{M}^i_{k,m}$ can be calculated for each finite element once, stored in array and then used when it is necessary. We will refer to these integrals as "moments".

This consideration has an additional consequence. There is an opinion in literature that the usage of the equations (1 - 2) is not so effective as the usage of equal scalar equations with supersingular kernel. This equations can be formulated on the base (1 - 2) by using of fluid function. There are two reasons for this transformation. The first reason is that the numerical dimension of the scalar model is lower than of the vectorial model. As it was mentioned above this dimensions are equal. And the second one is that the asymptotic of the integral (12) is $O\left(\frac{1}{r_{NM}}\right)$ but it is $O\left(\frac{1}{r^3_{NM}}\right)$ for the supersingular equations. It is clear from the representation (13) that the asymptotic of integral (12) is $O\left(\frac{1}{r^3_{NM}}\right)$ too if the correct approximation system is used. Moreover, it is obvious that this asymptotic will be correct for all finite element systems that is in $\mathfrak{L}$ or $\mathfrak{L}_k$.

Let's return to the our approximation system. By using the moments we can introduce the approximative formulas for elements of matrix A of different orders. For example:

$$I_1(i,j) = \mathbf{M}^i_{1,0}\mathbf{M}^i_{1,0}\frac{\partial^2}{\partial x_i \partial x_j}\frac{1}{r} + \mathbf{M}^i_{1,0}\mathbf{M}^i_{0,1}\frac{\partial^2}{\partial x_i \partial y_j}\frac{1}{r}$$

$$+\mathbf{M}^i_{0,1}\mathbf{M}^i_{1,0}\frac{\partial^2}{\partial y_i \partial x_j}\frac{1}{r} + \mathbf{M}^i_{0,1}\mathbf{M}^i_{0,1}\frac{\partial^2}{\partial y_i \partial y_j}\frac{1}{r}.$$

The construction of the array of moments may be carried out by using of the analytical integration. We use alternative approach that is based on the Radon's cubature formula for triangles [8]. This formula is equal to analytical integration for the polynomials of the order $n \leqslant 4$. This approach allows to construct the unified procedure for calculation of the moments. The method can be used for approximative formulas from I_1 to I_4.

The formulas I_k when $k > 4$ is not very usable in practical computations. We use the approximation I_3 in practice.

The estimation of the computational error of the I_k is cumbersome. We don't show it here. Practically the approximation I_3 does not provide additional computational error in case $\eta < 0.05$.

The approximative formula I_3 has low computational complexity. It is possible do not storage all matrix A in the computer memory in case of usage of iterative methods. We store in the memory only the part of the matrix that is calculated by the full formula (12). The elements for which the formula I_3 is applicable are calculated "on the fly". Practically it means that a fixed number of stored elements per row. In the first row of A this number is ≈ 200. In practical calculations it means that in computer memory stored $\approx 1\%$ of all elements of matrix A.

3.3 The numerical computation of the solutions

The standard algorithms for the systems with positive matrices are allowed for the system of linear algebraic equations (9 - 10). We use stabilized conjugate gradient method [9]. The free member column is used as initial value.

The system of ordinary differential equations (11) is not resolved for the high-order derivatives. The inversion of the matrix has too much computational complexity. We use implicit RungeKutta method of of the fourth order [10]. This method leads to necessity of the calculation of solution of the system of the linear algebraic equations like

$$Ax = b$$

a few time during integration of the system (11). We use stabilized method of conjugated gradients. It is effective to use the solutions of the system of the linear algebraic equations obtained on previous step as the initial values of the iterative processes on current step of integration.

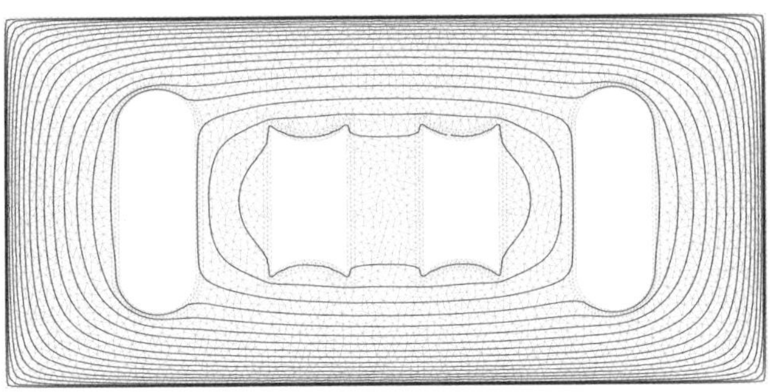

Fig. 4.

The example of numerical solution of the equation (1) is shown on fig. 4. S is plane and magnetic field is homogeneous. Usually integral equations of the first kind are very unstable but the equation (1) has good numerical properties.

References

1. Weyl H., *The method of orthogonal projection in potential theory*, Duke Math. J. (1940), vol. 7, pp. 411–444.
2. K. O. Friedrichs, *Differential forms on Riemannian manifolds*, Comm. on Pure and Appl. Math. (1955), vol. VIII. pp. 551–590.
3. M. Birman and M. Solomyak, L^2*-theory of the Maxwell operator in arbitrary domains* Russ. Math. Surv. 42 (1987), pp. 75-96.

4. J. Naumenko and V. Astakhov, *Mathematical modeling of magnetic field in presence of ideal conducting surface with boundary*, Electromechanics #5 (2003), pp. 11–16.
5. J. Naumenko, *Local attribute of correctness and numerical computing of integral equation of first kind for density of eddy currents*, Bulletin of the Southern Science Centre of RAS #4, (2005), pp. 3–8.
6. O. Besov, V. Il'in and S. Nikolsky *Integral representations of functions and embedding theorems*, Nauka, Moscow, 1996.
7. S. Goreinov and E. Tyrtyshnikov, *The maximal-volume concept in approximation by low-rank matrices*, Contemporary Mathematics, (2001), vol. 208, pp. 47–51.
8. J. Radon, *Zur mechanischen Kubatur*, Monatsh. für Math. 52, #4, (1948), pp. 286–300.
9. Y. Saad, *Iterative methods for sparse linear systems, Second Edition*, SIAM, 2003.
10. J. Ortega, W. Poole, *Numerical Methods for Differential Equations*, Pitman Publishing Inc., 1981.

Matrix Approach to Modelling of Polarized Radiation Transfer in Heterogeneous Systems

T. A. Sushkevich, S. A. Strelkov, and S. V. Maksakova

Keldysh Institute for Applied Mathematics, Russian Academy of Sciences,
4, Miusskaia Ploschad, Moscow, 125047 Russia
tamaras@keldysh.ru

Abstract. The general vector boundary-value problem for the integro-differential kinetic Boltzmann equation, which describes the polarized radiation transfer in a heterogeneous plane layer with a horizontally non-homogeneous and anisotropically reflective boundary, cannot be solved by the finite-difference methods. A mathematical model is proposed and justified that gives an asymptotically exact solution of the boundary-value problem in the class of the functions of slow growth. The new model is constructed by the influence function method and is efficient for algorithms using parallel computing.

1 Introduction

The international cooperation in the aerospace global monitoring of the Earth as well as international centers of the space operative information concerning the natural disasters and hazards phenomena are being created now. USA, Japan and other nations are setting up the most powerful high-performance multi-processor computer systems, which require the elaboration of new mathematical tools for solving the direct and inverse problems of the radiation transfer theory, for the interests of international global project of studying the evolution of the Earth and the climate using the hyper-bases of data obtained from space.

The radiation field of the Earth, i.e. the short-wave, the solar and the long-wave thermal radiation of the "atmosphere–Earth surface" system, is an indispensable component of the Earth ecosystem that determines many processes in the living organisms (man, fauna, microbes, viruses, bacteria, etc.). From the other side, the radiation characteristics measured from the ground-based and aerospace instruments are carriers of the information on the environmental condition and miscellaneous outbreaks containing the pollutants (consequences of anthropogenic hazards, military actions, volcano eruptions, etc.).

The new global applied problems and the new technological computational resources set up the new problems for the applied mathematics, i.e. the new physical and mathematical models, the new or improved mathematical tools and the new optimal parallel computing algorithms are needed.

An original universal mathematical tool is proposed for the radiation transfer modelling taking into account the multiple scattering, absorption and polarization in the multi-layer, non-permanent heterogeneous natural and artificial systems with the essentially different radiation modes in the separate areas of the system. The approach is based on the construction of the generalized solutions of the boundary-value problems for the radiation transfer equation in the form of the linear vector-valued functionals, the kernels of which are given by vectors of the influence functions for the separate layers of the system. The influence functions of the layers are calculated using the analytical formulas or the numerical methods as the solutions of the boundary-value problems of the first kind for the integro-differential kinetic Boltzmann equation or their modifications.

2 Mathematical statement of the polarized radiation transfer problem

The vector of the Stokes parameters Φ and its first component I, called the radiation intensity (or the radiance), being considered as the most complete characteristics of the quasi-monochromatic electromagnetic field, is in fact a complex functional of the parameters of the medium and the boundaries as well as of the sources. In SP-representation (Stokes-Poincaré) the components of any column vector $\Phi = (I, Q, U, V)^T$ have the normalization by intensity:

$$Q = Ip\cos 2\chi \cos 2\beta, \quad U = Ip\sin 2\chi \cos 2\beta, \quad V = Ip\sin 2\beta,$$

where χ is the azimuth of the polarization plane, β is the ellipticity degree, $0 \leqslant p \leqslant 1$ is the polarization degree. These components possess the following important properties:

$$I \geqslant 0, \quad I^2 \geqslant Q^2 + U^2 + V^2, \quad |Q| + |U| + |V| \leqslant \sqrt{3}I;$$

in the functional space $L(x)$, $x \in X$, X is the phase space:

$$\|\Phi\|_{L^{(4)}} = \sum_{m=1}^{M} \|\Phi_m\|_L \leqslant (1 + \sqrt{3})\|I\|_L,$$

$$\|I\|_L = \int_X |I(x)|dx = \int_X I(x)dx.$$

Let us search the Stokes vector of an incoherent multiply scattered light beam in a non-linear system approach as a solution of general vector boundary-value problem of the polarized radiation transfer theory in a plane layer with reflective boundary. The non-linear property is due to the relevant dependence of the solution on the characteristics of the reflection in the boundary condition and due to the multiple up and down reflection of the radiation on the boundary.

The approach elaborated by the authors and named the influence functions (IF) and the space-frequency characteristics (SFC) method [1] turns out to be most efficient and natural. The idea of this approach is in the representation of the boundary-value problem solution by a functional (a generalized solution) called the transfer operator. The angular and space distributions of the Stokes parameters inside and outside the transfer system can be calculated using this functional, from one side, and the explicit relations of these parameters with the characteristics of the transfer system (the radiation sources, the reflective and transmitting boundaries) are found, from the other side, that enables to develop the new approaches to solution of the inverse problems.

The influence functions of the boundary-value problems of the transfer theory or the space-frequency characteristics, i.e. the Fourier transformations from horizontal coordinates of the influence functions, serve as the functional kernels. In other words, the solution of the first and general boundary-value problems is found by a fundamental solution that is determined by the Fourier transformation method. The influence functions are the characteristics of the radiation transfer system, which are invariant with respect to the particular angular and space structures of the sources and the coefficients of the boundaries reflectivity and transmittance.

Considered is the problem of the polarized radiation transfer in a plane layer, unlimited in the horizontal direction $(-\infty < x, y < \infty,\ r_\perp = (x, y))$ and finite in the vertical direction $(0 \leqslant z \leqslant H)$ of the three-dimensional Cartesian space: $r = (x, y, z)$ is the radius vector. The radiation transfer system described as "the horizontally homogeneous plane medium – the reflective and transmitting boundary on the level $z = H$", is considered as non-multiplicative (without multiplication).

The set of all directions of the light beam propagation s is described in spherical coordinate system, i.e. each s corresponds to two quantities: $s = (\vartheta, \varphi)$, where $\vartheta \in [0, \pi]$ is the zenith angle, $\varphi \in [0, 2\pi]$ is the azimuth, and all directions build up the unit sphere $\Omega = \Omega^+ \cup \Omega^-$; here Ω^+ Ω^- are the hemispheres for the directions with $\vartheta \in [0, \pi/2)$ and $\vartheta \in (\pi/2, \pi]$ respectively.

Let us introduce the following notation:

$$s = (\mu, \varphi)\,, \quad \mu = \cos\vartheta\,, \quad \mu \in [-1, 1]\,, \quad \varphi \in [0, 2\pi]\,, \quad s \in \Omega = \Omega^+ \cup \Omega^-\,;$$

$$\Omega^+ = \left\{s^+ = (\mu^+, \varphi) \;:\; \mu^+ \in (0, 1]\,,\ \varphi \in [0, 2\pi]\right\}\,;$$

$$\Omega^- = \left\{s^- = (\mu^-, \varphi) \;:\; \mu^- \in [-1, 0)\,,\ \varphi \in [0, 2\pi]\right\}\,.$$

The phase areas are introduced to write the boundary conditions (t — "top", b — "bottom"):

$$t = \{z, r_\perp, s :\ z = 0, s = s^+ \in \Omega^+\},\quad b = \{z, r_\perp, s :\ z = H, s = s^- \in \Omega^-\}.$$

The Stokes vectors generate a real vector space with the scalar product $\Phi \mathbf{E} = \{\Phi_m E_m\}$, $m = 1, ..., M$, $M \leqslant 4$. The Stokes vector can be found as a solution of the general vector boundary-value problem of the transfer theory (with $\hat{R} \not\equiv 0$)

$$\hat{K}\Phi = \mathbf{F}, \quad \Phi|_t = \mathbf{F}^0, \quad \Phi|_b = \varepsilon \hat{R}\Phi + \mathbf{F}^H \tag{1}$$

with the linear operators: the transfer operator

$$\hat{D} \equiv (s, grad) + \sigma(z) = \hat{D}_z + \left(s_\perp, \frac{\partial}{\partial r_\perp}\right), \quad \hat{D}_z \equiv \mu \frac{\partial}{\partial z} + \sigma(z) ;$$

the collision integral

$$\hat{S}\Phi \equiv \sigma_s(z) \int_\Omega \hat{P}(z, s, s')\Phi(z, r_\perp, s')\, ds', \quad ds' = d\mu'\, d\varphi' ;$$

the uniformly bounded reflection operator

$$[\hat{R}\Phi](H, r_\perp, s) \equiv \int_{\Omega^+} \hat{q}(r_\perp, s, s^+)\Phi(H, r_\perp, s^+)\, ds^+ ; \tag{2}$$

the integro-differential operator $\hat{K} \equiv \hat{D} - \hat{S}$; the one-dimensional operator $\hat{K}_z \equiv \hat{D}_z - \hat{S}$; $\hat{P}(z, s, s')$ is the phase matrix of the scattering; $\sigma(z)$ and $\sigma_s(z)$ are the vertical profiles of the extinction and scattering coefficients; $\hat{q}(r_\perp, s, s^+)$ is the phase matrix of reflection; the parameter $0 \leqslant \varepsilon \leqslant 1$ fixes the act of the radiation interaction with the underlying surface; $\mathbf{F}(z, s)$, $\mathbf{F}^0(r_\perp, s)$, $\mathbf{F}^H(r_\perp, s)$ are the sources of the insolation. If at least one of the functions $\mathbf{F}^0$, $\mathbf{F}^H$, $\hat{q}$ depends on $r_\perp$, the solution of the problem (1)–(2) belongs to the 5D-phase space $(x, y, z, \vartheta, \varphi)$, or to the 3D-phase space (z, ϑ, φ), if there is no dependence on $r_\perp$. The phase matrix of the scattering

$$\hat{P}(z, s, s') = \hat{L}(\alpha)\hat{\gamma}(z, \vartheta_s)\hat{L}(\alpha')$$

is determined through the rotation matrix $\hat{L}(\alpha)$ and the scattering matrix $\hat{\gamma}(z, \vartheta_s)$ which is the function of the scattering angle ϑ_s between the directions of the incident s' and scattered s light beams.

Let us formulate a new mathematical model of the polarized radiation transfer that is asymptotically coincident with the problem (1)–(2) in the space of the vector-valued generalized functions of slow growth $\mathcal{S}'$ in the horizontal coordinate $r_\perp \in R^2$. As it was shown by V.S. Vladimirov, the influence function of the stationary differential operator of the transfer equation is a finite function of slow growth in the space R^3. The angular variables of the Stokes vector can be considered as the element of Hilbert space $L_2^{(4)}(\Omega)$, if one uses the expansion in vector spherical functions, or as the element of the normalized vector space $C^{(4)}(\Omega)$. G.A. Mikhailov has shown that the Neumann series corresponding to multiplicities of scattering converges in spaces L_∞ and L.

Let us consider the following four classes of the problems from a united methodological point of view: horizontally homogeneous and non-homogeneous in space coordinates, isotropic and anisotropic boundary-value conditions on angular variables (2) in the space of the vector-valued real linear continuous functionals on the phase set $x = (r_\perp, s) \in X = \Omega \times R^2$. We propose a new model, asymptotically exact and efficiently realizable by parallel computing, instead of the initial model, i.e. the general boundary-value problem (1)–(2), that cannot be solved by the finite-difference methods if the boundary condition is horizontally inhomogeneous (2).

The boundary-value problem (1)–(2) is linear and its solution can be found in the additive form

$$\Phi = \Phi_0 + \Phi_R .$$

The background radiation Φ_0 is determined as a solution of the first vector boundary-value problem of the transfer theory with the "vacuum" conditions (FVBP with $\hat{R} \equiv 0$)

$$\hat{K}\Phi_0 = \mathbf{F}, \quad \Phi_0|_t = \mathbf{F}^0, \quad \Phi_0|_b = \mathbf{F}^H \tag{3}$$

for the layer with the transparent or absolutely black (without reflection) boundaries and with insolation sources $\mathbf{F}$, $\mathbf{F}^0$, $\mathbf{F}^H$.

The problem for the inside illumination Φ_R due to the influence of the reflective boundary is the general vector boundary-value problem of the transfer theory (GVBP with $\hat{R} \neq 0$, $\mathbf{E} \neq 0$)

$$\hat{K}\Phi_R = 0, \quad \Phi_R|_t = 0, \quad \Phi_R|_b = \varepsilon\hat{R}\Phi_R + \varepsilon\mathbf{E}, \tag{4}$$

where the source $\mathbf{E}(r_\perp, s) \equiv \hat{R}\Phi_0$ is the radiance of the boundary produced by the background radiation.

3 The influence function of the vector boundary-value problem of the transfer theory

Let us consider the first vector boundary-value problem of the transfer theory

$$\hat{K}\Phi = 0, \quad \Phi|_t = 0, \quad \Phi|_b = \mathbf{f}(s^H; r_\perp, s). \tag{5}$$

The parameter $s^H \in \Omega^-$ may be absent. The different possible polarization states of a plane transverse electric wave in general case are represented by the vector $\Phi(s^H; z, r_\perp, s)$ composed of four real variables Φ_m, $m = 1, ..., M$, $M = 4$, which are the expansion coefficients of the vector Φ in terms of the unit vectors $\mathbf{i}_m$ of some coordinate system

$$\Phi = \mathbf{i}_1\Phi_1 + \mathbf{i}_2\Phi_2 + \mathbf{i}_3\Phi_3 + \mathbf{i}_4\Phi_4 ,$$

which depends on the method of the polarized radiation description.

The polarization states of the insolation source $\mathbf{f} = \{f_n(s^H; r_\perp, s)\}$, $n = 1, ..., N$, $N \leqslant 4$, and the radiation Φ may be different. In general case, when the Stokes vector of the source $\mathbf{f}$ contains the incoincident anisotropic horizontally non-homogeneous components $f_n(s^H; r_\perp, s)$, the solution of the problem (5) can be represented in the form of the following sum

$$\Phi(r, s) = \sum_{n=1}^{N} \Phi_n(r, s) ,$$

whose terms are the solutions of the following simultaneous problems

$$\hat{K}\Phi_n = 0, \quad \Phi_n|_t = 0, \quad \Phi_n|_b = \mathbf{t}_n f_n \tag{6}$$

with the vectors $\mathbf{t}_n = \{\delta_{mn}\}$, $m = 1, \ldots, M$, $n = 1, \ldots, N$, where δ_{mn} is the Kronecker's delta.

The solution of the problem (6) for the fixed value of n is obtained in the form of the vector-valued linear functional

$$\Phi_n = (\Theta_n, f_n) = \int_{\Omega^-} ds^- \int_{-\infty}^{\infty} \Theta_n(s^-; z, r_\perp - r'_\perp, s) f_n(s^H; r'_\perp, s^-)\, dr'_\perp .$$

The vector influence functions (VIF) $\Theta_n = \{\Theta_{mn}\}$, $n = 1, \ldots, N$, whose components are the Stokes parameters $\Theta_{mn}(s^-; z, r_\perp, s)$, $m = 1, ..., M$,, are found as a set of solutions of the first vector boundary-value problems

$$\hat{K}\Theta_n = 0, \quad \Theta_n|_t = 0, \quad \Theta_n|_b = \mathbf{t}_n f_\delta, \quad f_\delta(s^-; r_\perp, s) = \delta(r_\perp)\delta(s - s^-),$$

with the parameter $s^- \in \Omega^-$. The parameters of the Stokes vectors $\Phi_n = \{\Phi_{mn}(z, r_\perp, s)\}$ are calculated as scalar linear functionals

$$\Phi_{mn} = [\hat{\mathcal{P}}(\mathbf{f})]_{mn} \equiv (\Theta_{mn}, f_n)$$

$$\equiv \int_{\Omega^-} ds^- \int_{-\infty}^{\infty} \Theta_{mn}(s^-; z, r_\perp - r'_\perp, s) f_n(s^H; r'_\perp, s^-)\, dr'_\perp . \tag{7}$$

Introduce the influence functions tensor (TIF), defined by N Stokes vectors Θ_n, and present it in the form of the matrix

$$\hat{\Pi} = \begin{bmatrix} \Theta_{11} & \ldots & \Theta_{1n} & \ldots & \Theta_{1N} \\ \ldots & \ldots & \ldots & \ldots & \ldots \\ \Theta_{m1} & \ldots & \Theta_{mn} & \ldots & \Theta_{mN} \\ \ldots & \ldots & \ldots & \ldots & \ldots \\ \Theta_{M1} & \ldots & \Theta_{Mn} & \ldots & \Theta_{MN} \end{bmatrix} . \tag{8}$$

The first index $m = 1, ..., M$, $M \leqslant 4$, of the Θ_{mn} component of the tensor $\hat{\Pi}$ refers to the ordinal number of the parameter in the Stokes vector Θ_n, while the second index $n = 1, ..., N,\ N \leqslant 4$, corresponds to the index of the source vector $\mathbf{t}_n$ in the set of the problems (6) describing the calculation model of the vector influence functions Θ_n, and, hence, of the components of the tensor $\hat{\Pi}$ (8).

Introduce the linear vector functional of the vector $\mathbf{f}$

$$\Phi = \hat{\mathcal{P}}(\mathbf{f}) \equiv (\hat{\Pi}, \mathbf{f}) = \{\Phi_m\}, \quad m = 1, \ldots, M, \quad M \leqslant 4. \tag{9}$$

Components of the solution of the problem (5), determined via the scalar product (9):

$$\Phi_m = [\hat{\mathcal{P}}(\mathbf{f})]_m \equiv \sum_{n=1}^{N} (\Theta_{mn}, f_n),$$

are linear combinations of the linear scalar functionals (7).

If the source $f_n(r_\perp)$ is angular isotropic and horizontally non-homogeneous, the solution of the problem (6) is found via the vector linear functionals:

$$\Phi_n(z, r_\perp, s) = [\hat{\mathcal{P}}_r(\mathbf{f})]_n \equiv (\Theta_{rn}, f_n) = \int_{-\infty}^{\infty} \Theta_{rn}(z, r_\perp - r'_\perp, s) f_n(r'_\perp)\, dr'_\perp,$$

whose kernels are the vector influence functions

$$\Theta_{rn}(z, r_\perp, s) = \int_{\Omega^-} \Theta_n(s^-; z, r_\perp, s)\, ds^- \tag{10}$$

satisfying the problems

$$\hat{K}\Theta_{rn} = 0, \quad \Theta_{rn}\,|_{\,t} = 0, \quad \Theta_{rn}\,|_{\,b} = \mathbf{t}_n \delta(r_\perp). \tag{11}$$

In the case of angular anisotropic and horizontally homogeneous source $f_n(s^H; s)$ the solution of the problem (6) is found in the form of linear vector functional

$$\Phi_n(s^H; z, s) = [\hat{\mathcal{P}}_z(\mathbf{f})]_n \equiv (\Theta_{zn}, f_n) = \int_{\Omega^-} \Theta_{zn}(s'; z, s) f_n(s^H; s')\, ds'$$

whose kernel is the vector influence function

$$\Theta_{zn}(s^-; z, s) = \int_{-\infty}^{\infty} \Theta_n(s^-; z, r_\perp, s)\, dr_\perp, \tag{12}$$

which is the solution of the one-dimensional problem

$$\hat{K}_z\Theta_{zn} = 0, \quad \Theta_{zn}\,|_{\,t} = 0, \quad \Theta_{zn}\,|_{\,b} = \mathbf{t}_n \delta(s - s^-); \quad s^- \in \Omega^-. \tag{13}$$

In the case of angular isotropic and horizontally homogeneous source the solution of the problem (6)

$$\Phi_n(z, s) = [\hat{\mathcal{P}}_c(\mathbf{f})]_n \equiv f_n \mathbf{W}_n(z, s), \quad f_n = \text{const},$$

is calculated via the vector influence function

$$\mathbf{W}_n(z,s) = \int_{\Omega^-} ds^- \int_{-\infty}^{\infty} \Theta_n(s^-;z,r_\perp,s)\,dr_\perp$$

$$= \int_{-\infty}^{\infty} \Theta_{rn}(z,r_\perp,s)\,dr_\perp = \int_{\Omega^-} \Theta_{zn}(s^-;z,s)\,ds^-, \tag{14}$$

which is also called the vector transmittance function with the multiple scattering contribution and which is determined as the solution of the one-dimensional vector problem

$$\hat{K}_z\mathbf{W}_n = 0, \quad \mathbf{W}_n|_t = 0, \quad \mathbf{W}_n|_b = \mathbf{t}_n. \tag{15}$$

The relations (10), (12), (14) can be used as the error tolerance criteria for the calculation of $\Theta_n, \Theta_{rn}, \Theta_{zn}$ via the solutions of more simple problems (11), (13), (15). The tensor $\hat{\Pi}$ defined by the components of the vectors $\Theta_n(s^-;z,r_\perp,s)$ virtually describes the polarized radiation field in a layer with non-reflective boundaries. This field is due to the processes of multiple scattering of stationary elliptically polarized narrow beam with the direction $s^- \in \Omega^-$, whose source is located on the boundary $z = H$ in the center of the system of horizontal coordinates x ,y.

The tensor $\hat{\Pi}$ determined by the components of the vectors $\Theta_{zn}(s^-;z,s)$ describes the polarized radiation field formed in a layer, whose border $z = H$ undergoes from outside the elliptically polarized wide flux coming in the direction $s^- \in \Omega^-$.

The vector influence functions form a complete set of the base models of the influence functions for the boundary-value problems (4) and (5) of the polarized radiation transfer theory for a plane layer.

4 The vector transfer operator

Let us use the formulated above models of the vector influence functions and the representation of the solution of the first boundary-value problem (5) in the form of the vector linear functional (9) whose kernel is given by the tensor (8), to construct the solution of the general boundary-value problem (4). If the source in the problem (4) is defined through the single reflection of the background radiation, then the degree of the parameter ε corresponds to the degree of the dependence of solution of the problem (1)–(2) on the characteristics of the reflection operator.

Introduce a parametric set of the regular perturbations

$$\Phi_R = \sum_{k=1}^{\infty} \varepsilon^k \Phi_k,$$

whose terms satisfy the recurrent system of the first vector boundary-value problems

$$k=1: \qquad \hat{K}\Phi_1 = 0, \qquad \Phi_1\,|_{\,t} = 0, \qquad \Phi_1\,|_{\,b} = \mathbf{E}; \tag{16}$$

$$k \geqslant 2: \qquad \hat{K}\Phi_k = 0, \qquad \Phi_k\,|_{\,t} = 0, \qquad \Phi_k\,|_{\,b} = \hat{R}\Phi_{k-1}. \tag{17}$$

Introduce a matrix operation describing the interaction of the polarized radiation with the boundary via the tensor (8):

$$[\hat{G}\mathbf{f}](s^-; H, r_\perp, s) \equiv \hat{R}(\hat{\Pi}, \mathbf{f}) = \int_{\Omega^+} \hat{q}(r_\perp, s, s^+)(\hat{\Pi}, \mathbf{f})\, ds^+.$$

It can be shown that the solutions of the problems (16), (17) may be represented via the tensor (8):

$$\Phi_1 = (\hat{\Pi}, \mathbf{E}); \quad \Phi_k = (\hat{\Pi}, \hat{R}\Phi_{k-1}) = (\hat{\Pi}, \hat{G}^{k-1}\mathbf{E})$$

and the asymptotically exact solution of the problem (4) may be explicitly expressed (we assume $\varepsilon = 1$)) in the form of the vector transfer operator (VTO):

$$\Phi_R = (\hat{\Pi}, \mathbf{Y}), \tag{18}$$

where the vector

$$\mathbf{Y} \equiv \hat{Y}\mathbf{E} \equiv \sum_{k=0}^{\infty} \hat{G}^k \mathbf{E} = \sum_{k=0}^{\infty} \hat{R}\Phi_k \tag{19}$$

is the sum of Neumann series corresponding to the multiplicities of the radiation reflection from boundary, — a “scenario” of an optical image, or a radiance of the reflective boundary.

For the “scenario” $\mathbf{Y}(x)$, $x = (r_\perp, s^-) \in X^-$, $X^- = \Omega^- \times R^2$ we get an estimate

$$\|\mathbf{Y}\| \leqslant \sum_{k=0}^{\infty} \|\hat{R}\Phi_k\| \leqslant \|\hat{R}\Phi_0\| \sum_{k=0}^{\infty} (q_* c_0)^k = \frac{\|\mathbf{E}\|}{1 - q_* c_0} \leqslant \frac{q_* \|\Phi_0\|}{1 - q_* c_0},$$

where c_0 is the spherical albedo of the system, i.e. the Neumann series (19) converges as the geometric series. The convergence is estimated in the vector space of the linear continuous functionals via the following norms and the relations:

$$\|\Phi_k\| = \max_m \text{vrai} \sup_{z, r_\perp, s} |\Phi_{mk}| \leqslant q_*^{k-1} c_0^k \|\mathbf{E}\|,$$

$$\|\hat{R}\Phi_k\| \leqslant \|\hat{R}(\mathbf{1})\| \|\Phi_k\| \leqslant q_*^k c_0^k \|\mathbf{E}\|,$$

$$\|\hat{R}(\mathbf{1})\| \leqslant \max_m \text{vrai} \sup_{r_\perp, s^-} \sum_{n=1}^{N} \int_{\Omega^+} |\, q_{mn}(r_\perp, s^-, s^+)|\, ds^+ = q_* \leqslant 1,$$

$$\|\hat{\mathcal{P}}(\mathbf{f})\| \leqslant \|\hat{\mathcal{P}}(\mathbf{1})\| \, \|\mathbf{f}\|, \quad \|\hat{\mathcal{P}}(\mathbf{1})\| \leqslant \max_{m} \text{vrai} \sup_{z,s} |W_{m1}| = c_0 < 1 \, .$$

The norm of the functional is defined via the solution of the problem (15)

$$[\hat{\mathcal{P}}(\mathbf{1})]_m = \sum_{n=1}^{N} W_{mn}(z,s) = W_{m1}(z,s).$$

It can be shown that the "scenario" satisfies the Fredholm equation of the second kind

$$\mathbf{Y} = \hat{R}(\hat{\Pi}, \mathbf{Y}) + \mathbf{E},$$

that is called the equation of "ground photography".

By virtue of the new mathematical model (18)–(19), instead of the calculation of series corresponding to reflection multiplicities in the complete phase volume of the solution of the problem (4), it is sufficient to calculate the finite Neumann series (19) only for the "scenario" on the boundary $z = H$ in the phase volume X^-, and further search the angular and space distributions of the Stokes parameters, i.e. the solutions of the problem (4) using the functional (18).

For the problems with the phase space dimension 5 or 4, i.e. for the problems with a horizontal non-homogeneity, the vector influence functions are calculated by Fourier transformation method as an inverse transform of the space-frequency characteristics (Sushkevich, 2005) or by the Monte-Carlo method. The solution of the problems with the dimension of the phase space 3, i.e. one-dimensional in space, is classical and calculated by means of widely used analytical or numerical methods.

5 The mathematical model of polarized radiation transfer in two-media system

Let us construct the mathematical model of the polarized radiation transfer in a plane layer with two media and internal reflective and transmittive boundary. The solution of the general vector boundary-value problem for the kinetic equation is reduced to the calculation of the matrix transfer operator (MTO). Tensors of the influence functions of both media serve as the kernels of the MTO.

We construct an asymptotically exact solution of the general boundary-value problem for the vector integro-differential equation of the polarized radiation transfer

$$\begin{cases} \hat{K}\Phi = \mathbf{F}, \quad \Phi|_t = \mathbf{F}^0, \quad \Phi|_b = \hat{R}\Phi + \mathbf{F}^H, \\ \Phi|_{d1} = \varepsilon(\hat{R}_1\Phi + \hat{T}_{21}\Phi) + \mathbf{F}^1, \quad \Phi|_{d2} = \varepsilon(\hat{R}_2\Phi + \hat{T}_{12}\Phi) + \mathbf{F}^2 \, . \end{cases} \tag{20}$$

using the influence functions method. The boundary between both media able to transmit and to reflect the radiation is on the level $z = h$ inside the layer. The phase areas are introduced to write the inner boundary conditions:

$$d1 = \{z, r_\perp, s : z = h, s \in \Omega^-\}, \quad d2 = \{z, r_\perp, s : z = h, s \in \Omega^+\}.$$

The radiation transmission through the boundary between the media is described by the uniformly bounded reflection $\hat{R}_1$, $\hat{R}_2$ and transmission $\hat{T}_{12}$, $\hat{T}_{21}$ operators, where the index 1 refers to the layer with $z \in [0, h]$ and the index 2 to the layer with $z \in [h, H]$:

$$
\begin{aligned}
[\hat{R}_1\Phi](h, r_\perp, s) &= \int_{\Omega^+} \hat{q}_1(r_\perp, s, s^+)\,\Phi(h, r_\perp, s^+)\,ds^+, \quad s \in \Omega^-;\\
[\hat{R}_2\Phi](h, r_\perp, s) &= \int_{\Omega^-} \hat{q}_2(r_\perp, s, s^-)\,\Phi(h, r_\perp, s^-)\,ds^-, \quad s \in \Omega^+;\\
[\hat{T}_{12}\Phi](h, r_\perp, s) &= \int_{\Omega^+} \hat{t}_{12}(r_\perp, s, s^+)\,\Phi(h, r_\perp, s^+)\,ds^+, \quad s \in \Omega^+;\\
[\hat{T}_{21}\Phi](h, r_\perp, s) &= \int_{\Omega^-} \hat{t}_{21}(r_\perp, s, s^-)\,\Phi(h, r_\perp, s^-)\,ds^-, \quad s \in \Omega^-.
\end{aligned}
$$

The parameter $0 \leqslant \varepsilon \leqslant 1$ fixes the act of the radiation interaction on the internal boundary $z = h$; $\hat{q}_1(r_\perp, s, s^+)$, $\hat{q}_2(r_\perp, s, s^-)$ are the phase reflection matrices, $\hat{t}_{12}(r_\perp, s, s^+)$, $\hat{t}_{21}(r_\perp, s, s^-)$ are the phase transmittance matrices for the internal boundary.

The boundary-value problem (20) is linear and its solution can be found in the additive form

$$\Phi = \Phi_0 + \Phi_c.$$

The background radiation Φ_0 is found as a solution of the first vector boundary-value problem of the transfer theory with the "vacuum" conditions

$$
\begin{cases}
\hat{K}\Phi_0 = \mathbf{F}, \quad \Phi_0|_t = \mathbf{F}^0, \quad \Phi_0|_b = \mathbf{F}^H,\\
\Phi_0|_{d1} = \mathbf{F}_{d1}, \quad \Phi_0|_{d2} = \mathbf{F}_{d2}\,.
\end{cases}
\tag{21}
$$

The problem (21) for the layer $z \in [0, H]$ is split into two independent FVBPs:
for layer with $z \in [0, h]$

$$\hat{K}\Phi_0^1 = \mathbf{F}_1, \quad \Phi_0^1|_t = \mathbf{F}^0, \quad \Phi_0^1|_{d1} = \mathbf{F}_{d1}$$

and for layer with $z \in [h, H]$

$$\hat{K}\Phi_0^2 = \mathbf{F}_2, \quad \Phi_0^2|_b = \mathbf{F}^H, \quad \Phi_0^2|_{d2} = \mathbf{F}_{d2},$$

where $\mathbf{F}_1 = \mathbf{F}$ for first medium $\mathbf{F}_2 = \mathbf{F}$ for second medium.

The contribution Φ_c due to the radiation exchange between the two media on the border $z = h$ and due to the influence of the reflective boundary at $z = H$, is determined as the solution of GVBP

$$\begin{cases} \hat{K}\Phi_c = 0, \quad \Phi_c|_t = 0, \quad \Phi_c|_b = \hat{R}\Phi_c + \mathbf{E}^H, \\ \Phi_c|_{d1} = \varepsilon(\hat{R}_1\Phi_c + \hat{T}_{21}\Phi_c + \mathbf{E}^1), \\ \Phi_c|_{d2} = \varepsilon(\hat{R}_2\Phi_c + \hat{T}_{12}\Phi_c + \mathbf{E}^2) \end{cases} \tag{22}$$

with the given incoming fluxes at the boundary

$$\mathbf{E}^H(r_\perp, s) \equiv \hat{R}\Phi_0, \quad \mathbf{E}^1(r_\perp, s) \equiv \hat{R}_1\Phi_0 + \hat{T}_{21}\Phi_0, \quad \mathbf{E}^2(r_\perp, s) \equiv \hat{R}_2\Phi_0 + \hat{T}_{12}\Phi_0,$$

which are created by the background radiation.

Without loss of generality of the obtained results, we restrict our consideration to GVBP

$$\begin{cases} \hat{K}\Phi_d = 0, \quad \Phi_d|_t = 0, \quad \Phi_d|_b = 0, \\ \Phi_d|_{d1} = \varepsilon(\hat{R}_1\Phi_d + \hat{T}_{21}\Phi_d + \mathbf{E}^1), \\ \Phi_d|_{d2} = \varepsilon(\hat{R}_2\Phi_d + \hat{T}_{12}\Phi_d + \mathbf{E}^2), \end{cases} \tag{23}$$

that results from the GVBP (22) under the non-reflective and non-radiative boundary condition at the level $z = H$ ($\hat{R} \equiv 0$, $\mathbf{F}^H \equiv 0$) and describes the influence of radiation exchange of given two media through the internal border $z = h$ on the formation of the total radiation field of the system

$$\Phi = \Phi_0 + \Phi_d .$$

Let us search the GVBP solution (23) in the form of a parametric series of regular perturbations for the two vectors of Stokes parameters

$$\Phi_d^1 = \sum_{k=1}^{\infty} \varepsilon^k \Phi_k^1, \quad \Phi_d^2 = \sum_{k=1}^{\infty} \varepsilon^k \Phi_k^2, \tag{24}$$

where Φ_d^1 is responsible for the radiation field in the layer with $z \in [0, h]$, while Φ_d^2 corresponds to the layer with $z \in [h, H]$.

The terms of the series (24) satisfy the recurrent system of the first vector boundary-value problems that is split into the problems for the first medium with $z \in [0, h]$:

$$k = 1: \quad \hat{K}\Phi_1^1 = 0, \quad \Phi_1^1|_t = 0, \quad \Phi_1^1|_{d1} = \mathbf{E}^1; \tag{25}$$

$$k \geqslant 2: \quad \hat{K}\Phi_k^1 = 0, \quad \Phi_k^1|_t = 0, \quad \Phi_k^1|_{d1} = \hat{R}_1\Phi_{k-1}^1 + \hat{T}_{21}\Phi_{k-1}^2 \tag{26}$$

and for second medium with $z \in [h, H]$:

$$k = 1: \quad \hat{K}\Phi_1^2 = 0, \quad \Phi_1^2|_b = 0, \quad \Phi_1^2|_{d2} = \mathbf{E}^2; \tag{27}$$

$$k \geqslant 2: \quad \hat{K}\Phi_k^2 = 0, \quad \Phi_k^2|_b = 0, \quad \Phi_k^2|_{d2} = \hat{R}_2\Phi_{k-1}^2 + \hat{T}_{12}\Phi_{k-1}^1. \tag{28}$$

Each problem in (25), (26) is FVBP of the form

$$\hat{K}\Phi^1 = 0, \quad \Phi^1\big|_t = 0, \quad \Phi^1\big|_{d1} = \mathbf{f}^1(s^1; r_\perp, s), \tag{29}$$

while each problem in (27), (28) is FVBP of the form

$$\hat{K}\Phi^2 = 0, \quad \Phi^2\big|_b = 0, \quad \Phi^2\big|_{d2} = \mathbf{f}^2(s^2; r_\perp, s). \tag{30}$$

The parameters $s^1 \in \Omega^-$ and $s^2 \in \Omega^+$ may be absent.

6 The vector influence functions for two-media system

Let us use the results obtained above for the one-medium system of the polarized radiation transfer, to construct the model of the polarized radiation transfer in the two-media system with the internal reflective and transmittive boundary. The solution of the linear FVBP (29) may be represented in the additive form

$$\Phi^1(s^1; r, s) = \sum_{n=1}^{N_1} \Phi_n^1(s^1; r, s),$$

whose terms are determined by the solution of the FVBP set

$$\hat{K}\Phi_n^1 = 0, \quad \Phi_n^1\big|_t = 0, \quad \Phi_n^1\big|_{d1} = \mathbf{t}_n f_n^1\,. \tag{31}$$

The FVBP solution (31) for the fixed n is obtained in the form of the vector linear functional:

$$\Phi_n^1 = (\Theta_n^1, f_n^1) = \int_{\Omega^-} ds^- \int_{-\infty}^{\infty} \Theta_n^1(s^-; z, r_\perp - r'_\perp, s)\, f_n^1(s^1; r'_\perp, s^-)\, dr'_\perp.$$

The vector influence functions $\Theta_n^1 = \{\Theta_{mn}^1\}$, $n = 1, ..., N_1$, whose components are the Stokes parameters $\Theta_{mn}^1(s^-; z, r_\perp, s)$, $m = 1, ..., M_1$, and may be found as a solution of the FVBP set

$$\hat{K}\Theta_n^1 = 0, \quad \Theta_n^1\big|_t = 0, \quad \Theta_n^1\big|_{d1} = \mathbf{t}_n\, f_\delta^1 \tag{32}$$

with parameter $s^- \in \Omega^-$ and the source $f_\delta^1(s^-; r_\perp, s) = \delta(r_\perp)\delta(s - s^-)$. The parameters of the Stokes vector $\Phi_n^1 = \{\Phi_{mn}^1(z, r_\perp, s)\}$ are calculated as scalar functionals:

$$\Phi_{mn}^1(s^1; z, r_\perp, s) = (\Theta_{mn}^1, f_n^1)$$

$$= \int_{\Omega^-} ds^- \int_{-\infty}^{\infty} \Theta^1_{mn}(s^-; z, r_\perp - r'_\perp, s)\, f^1_n(s^1; r'_\perp, s^-)\, dr'_\perp . \quad (33)$$

Introduce the influence functions tensor defined by N_1 Stokes vectors Θ^1_n:

$$\hat{\Pi}^1 = \begin{bmatrix} \Theta^1_{11} & \dots & \Theta^1_{1n} & \dots & \Theta^1_{1N_1} \\ \dots & \dots & \dots & \dots & \dots \\ \Theta^1_{m1} & \dots & \Theta^1_{mn} & \dots & \Theta^1_{mN_1} \\ \dots & \dots & \dots & \dots & \dots \\ \Theta^1_{M_1 1} & \dots & \Theta^1_{M_1 n} & \dots & \Theta^1_{M_1 N_1} \end{bmatrix} . \quad (34)$$

The solution of FVBP (29) is found in the form of the linear vector functional of the vector $\mathbf{f}^1$

$$\Phi^1 = (\hat{\Pi}^1, \mathbf{f}^1) = \{\Phi^1_m\}, \quad m = 1, ..., M_1, \quad M_1 \leqslant 4 , \quad (35)$$

whose kernel is given by the influence functions tensor (34). The parameters of the Stokes vector (35) are linear combinations of the linear scalar functions (33) :

$$\Phi^1_m = \sum_{n=1}^{N_1} \Phi^1_{mn} = \sum_{n=1}^{N_1} (\Theta^1_{mn}, f^1_n) .$$

The solution of the linear FVBP (30) can be represented in the additive form

$$\Phi^2(s^2; r, s) = \sum_{n=1}^{N_2} \Phi^2_n(s^2; r, s),$$

whose terms are the solutions of the FVBP set

$$\hat{K}\Phi^2_n = 0, \quad \Phi^2_n \big|_b = 0, \quad \Phi^2_n \big|_{d2} = \mathbf{t}_n f^2_n . \quad (36)$$

The solution of FVBP (36) for the fixed n is obtained in the form of the vector linear functional:

$$\Phi^2_n = (\Theta^2_n, f^2_n) = \int_{\Omega^+} ds^+ \int_{-\infty}^{\infty} \Theta^2_n(s^+; z, r_\perp - r'_\perp, s)\, f^2_n(s^2; r'_\perp, s^+)\, dr'_\perp .$$

The vector influence functions $\Theta^2_n = \{\Theta^2_{mn}\}$, $n = 1, ..., N_2$, whose components are the Stokes parameters $\Theta^2_{mn}(s^+; z, r_\perp, s)$, $m = 1, ..., M_2$, and may be found as the solution of FVBP set

$$\hat{K}\Theta^2_n = 0, \quad \Theta^2_n \big|_b = 0, \quad \Theta^2_n \big|_{d2} = \mathbf{t}_n f^2_\delta \quad (37)$$

with the parameter $s^+ \in \Omega^+$ and the source $f^2_\delta(s^+; r_\perp, s) = \delta(r_\perp)\delta(s - s^+)$. The parameters of the Stokes vector $\Phi^2_n = \{\Phi^2_{mn}(z, r_\perp, s)\}$ are calculated as the scalar functionals:

$$\Phi^2_{mn}(s^2; z, r_\perp, s) = (\Theta^2_{mn}, f^2_n)$$

$$= \int_{\Omega^+} ds^+ \int_{-\infty}^{\infty} \Theta^2_{mn}(s^+; z, r_\perp - r'_\perp, s)\, f^2_n(s^2; r'_\perp, s^+)\, dr'_\perp \,. \tag{38}$$

Introduce the influence functions tensor defined by N_2 the Stokes vectors Θ^2_n:

$$\hat{\Pi}^2 = \begin{bmatrix} \Theta^2_{11} & \dots & \Theta^2_{1n} & \dots & \Theta^2_{1N_2} \\ \dots & \dots & \dots & \dots & \dots \\ \Theta^2_{m1} & \dots & \Theta^2_{mn} & \dots & \Theta^2_{mN_2} \\ \dots & \dots & \dots & \dots & \dots \\ \Theta^2_{M_21} & \dots & \Theta^2_{M_2n} & \dots & \Theta^2_{M_2N_2} \end{bmatrix} \tag{39}$$

and the linear vector functional of the vector $\mathbf{f}^2$ in the form

$$\Phi^2 = (\hat{\Pi}^2, \mathbf{f}^2) = \{\Phi^2_m\}, \quad m = 1, ..., M_2, \quad M_2 \leqslant 4, \tag{40}$$

where the parameters of the Stokes vector determined as the solution FVBP (30)

$$\Phi^2_m = \sum_{n=1}^{N_2} \Phi^2_{mn} = \sum_{n=1}^{N_2} (\Theta^2_{mn}, f^2_n)$$

are linear combinations of the linear scalar functionals (38).

7 Matrix transfer operator for two-media system

Let us use the formulated above models of the vector influence functions and the representations of the FVBP solutions (29) and (30) in the form of vector linear functionals (35) and (40), whose kernels are the influence functions tensors (34) and (39), to construct the solution to GVBP (23). If the source in GVBP (23) is determined through the single interaction of the background radiation Φ_0 with the border $z = h$, then the degree of the parameter ε corresponds to the degree of the dependence of the solution of the problem (23) from the characteristics of the operators of the reflection $\hat{q}_1, \hat{q}_2$ and the transmission $\hat{t}_{12}, \hat{t}_{21}$.

Introduce the algebraic vectors in the form of columns:

$$\Phi_d = \begin{bmatrix} \Phi^1_d \\ \Phi^2_d \end{bmatrix}, \quad \Phi_k = \begin{bmatrix} \Phi^1_k \\ \Phi^2_k \end{bmatrix}, \quad \mathbf{E} = \begin{bmatrix} \mathbf{E}^1 \\ \mathbf{E}^2 \end{bmatrix}, \quad \mathbf{f} = \begin{bmatrix} \mathbf{f}^1 \\ \mathbf{f}^2 \end{bmatrix}, \quad \mathbf{Z} = \begin{bmatrix} \mathbf{Z}^1 \\ \mathbf{Z}^2 \end{bmatrix},$$

$$\Theta = \begin{bmatrix} \Theta^1 \\ \Theta^2 \end{bmatrix}, \quad \hat{\Pi} = \begin{bmatrix} \hat{\Pi}^1 \\ \hat{\Pi}^2 \end{bmatrix}, \quad (\hat{\Pi}, \mathbf{f}) = \begin{bmatrix} (\hat{\Pi}^1, \mathbf{f}^1) \\ (\hat{\Pi}^2, \mathbf{f}^2) \end{bmatrix}$$

and determine a matrix operation corresponding to the single radiation pass through the boundary of two media taking into account the multiple scattering, absorption and polarization in both media through the influence functions tensor:

$$\hat{\mathcal{P}}_{rt}\mathbf{f} \;\equiv\; \hat{P}_{rt}(\hat{\Pi},\mathbf{f}) = \begin{bmatrix} \hat{R}_1(\hat{\Pi}^1,\mathbf{f}^1) + \hat{T}_{21}(\hat{\Pi}^2,\mathbf{f}^2) \\ \\ \hat{R}_2(\hat{\Pi}^2,\mathbf{f}^2) + \hat{T}_{12}(\hat{\Pi}^1,\mathbf{f}^1) \end{bmatrix}, \tag{41}$$

where $\hat{P}_{rt}$ denotes the matrix composed from the reflection and transmission operators on the internal border of the two media:

$$\hat{P}_{rt} \;\equiv\; \begin{bmatrix} \hat{R}_1 & \hat{T}_{21} \\ \\ \hat{T}_{12} & \hat{R}_2 \end{bmatrix}.$$

The boundary-value problems (25) and (27) for a linear approximation may be solved using the vector linear functionals (35) (40):

$$\Phi_1 = \begin{bmatrix} \Phi_1^1 \\ \\ \Phi_1^2 \end{bmatrix} = \begin{bmatrix} (\hat{\Pi}^1,\mathbf{E}^1) \\ \\ (\hat{\Pi}^2,\mathbf{E}^2) \end{bmatrix} = (\hat{\Pi},\mathbf{E}).$$

Let us describe the first approximations, i.e. the solutions of FVBPs (26) and (28), in operator form using the definition (41):

$$\mathbf{F}_1 \equiv \hat{P}_{rt}\Phi_1 = \hat{P}_{rt}(\hat{\Pi},\mathbf{E}) = \hat{\mathcal{P}}_{rt}\mathbf{E};$$

$$\Phi_2 = (\hat{\Pi},\mathbf{F}_1) = (\hat{\Pi},\hat{P}_{rt}\Phi_1) = (\hat{\Pi},\hat{\mathcal{P}}_{rt}\mathbf{E});$$

$$\mathbf{F}_2 \equiv \hat{P}_{rt}\Phi_2 = \hat{P}_{rt}(\hat{\Pi},\mathbf{F}_1) = \hat{\mathcal{P}}_{rt}\mathbf{F}_1 = \hat{\mathcal{P}}^2_{rt}\mathbf{E};$$

$$\Phi_3 = (\hat{\Pi},\mathbf{F}_2) = (\hat{\Pi},\hat{P}_{rt}\Phi_2) = (\hat{\Pi},\hat{\mathcal{P}}^2_{rt}\mathbf{E}).$$

Using induction one can show that the two consecutive k-approximations are linked by the recurrent relation

$$\Phi_k = (\hat{\Pi},\hat{P}_{rt}\Phi_{k-1})$$

while for $k \geqslant 1$ ($\mathbf{F}_0 \equiv \mathbf{E}$) the algebraic vector of the source is

$$\mathbf{F}_k = \hat{P}_{rt}\Phi_k = \hat{\mathcal{P}}_{rt}\mathbf{F}_{k-1} = \hat{\mathcal{P}}^k_{rt}\mathbf{E},$$

and the algebraic vector of k-approximation to solution of FVBPs (26) and (28) is

$$\Phi_k = (\hat{\Pi},\mathbf{F}_{k-1}) = (\hat{\Pi},\hat{\mathcal{P}}^{k-1}_{rt}\mathbf{E}).$$

Eventually we obtain the asymptotically exact solution to GVBP (23) :

$$\Phi_d = (\hat{\Pi},\hat{Z}\mathbf{E}) = (\hat{\Pi},\mathbf{Z}). \tag{42}$$

The "scenario" on the media boundary

$$\mathbf{Z} = \hat{Z}\mathbf{E} \equiv \sum_{k=0}^{\infty} \hat{\mathcal{P}}^k_{rt}\mathbf{E} \tag{43}$$

is given by the Neumann series corresponding to the multiplicities of the radiation transmission through the boundary taking into account the influence of the multiple scattering, absorption and polarization in both media using the influence functions tensors.

The representation of the solution to GVBP (23) in the form of the vector (non-linear) functional (42), which gives an explicit relation between the registered radiation and the "scenario" (43) on both sides of the media boundary, is called the matrix transfer operator (MTO) for the two-media transfer system. In its turn, the "scenario" is explicitly described through the reflection and transmission characteristics of the boundary under its given illumination. The Neumann series (43) determines the "scenario" of the optical image on both sides of the internal boundary of the media. This image is formed as a result of the multiple scattering and absorption of radiation in both media as well as the multiple passing of the radiation through the boundary taking into account the radiation polarization and depolarization mechanisms both within the layer and on the boundary. Actually, the universal representation of the MTO (42) is valid for all cases of the spatial and angular dependence of the media boundary and sources characteristics, which were considered above for the one-medium system of radiation transfer.

8 Mathematical model of polarized radiation transfer in multi-media heterogeneous system

Consider the radiation transfer system consisting of M layers with the boundaries $h_m, \quad m = 1 \div M+1$:

$$z \in [0, H], \quad H = \bigcup_{m=1}^{m=M} [h_m, h_{m+1}], \quad h_1 = 0\,, \quad h_m < h_{m+1}, \quad h_{M+1} = H\,.$$

The phase areas are introduced to write the boundary conditions:

$$d\downarrow, m = \{z, s : z = h_m, s \in \Omega^{\downarrow}\}; \quad d\uparrow, m = \{z, s : z = h_m, s \in \Omega^{\uparrow}\}; \quad \Omega = \Omega^{\downarrow} \cup \Omega^{\uparrow}\,;$$

$$\mu^{\downarrow} = \cos\vartheta^{\downarrow}\,, \quad \vartheta^{\downarrow} \in [0, \pi/2)\;; \quad \Omega^{\downarrow} = \left\{s^{\downarrow} = (\mu^{\downarrow}, \varphi)\;:\; \mu^{\downarrow} \in (0, 1]\,,\; \varphi \in [0, 2\pi]\right\}\;;$$

$$\mu^{\uparrow} = \cos\vartheta^{\uparrow}\,, \quad \vartheta^{\uparrow} \in (\pi/2, \pi]\;; \quad \Omega^{\uparrow} = \left\{s^{\uparrow} = (\mu^{\uparrow}, \varphi)\;:\; \mu^{\uparrow} \in [-1, 0)\,,\; \varphi \in [0, 2\pi]\right\}\,.$$

The total Stokes vector of radiation $\Phi_\lambda(r, s)$, where the index λ is the wavelength (omitted below), is found as a solution to the general boundary-value problem of the transfer theory for multi-layer heterogeneous system

$$\hat{K}\Phi = \mathbf{F}^{in}, \quad \Phi|_{t\downarrow} = \mathbf{F}^{\downarrow}_{t}\,, \quad \Phi|_{b\uparrow} = \hat{R}^{\uparrow}_{b}\Phi + \mathbf{F}^{\uparrow}_{b} \tag{44}$$

with the boundary conditions on the inner boundaries of the layers for $m = 2 \div M$:

$$\Phi|_{d\uparrow,m} = \varepsilon(\hat{R}^{\uparrow}_{m}\Phi + \hat{T}^{\uparrow}_{m}\Phi) + \mathbf{F}^{\uparrow}_{m-1}\,, \quad \Phi|_{d\downarrow,m} = \varepsilon(\hat{R}^{\downarrow}_{m}\Phi + \hat{T}^{\downarrow}_{m}\Phi) + \mathbf{F}^{\downarrow}_{m} \quad (45)$$

and for the outer boundaries of the system

$$\mathbf{F}^{\downarrow}_{1} = \mathbf{F}^{\downarrow}{}_{t}\,; \quad \mathbf{F}^{\uparrow}_{M} = \mathbf{F}^{\uparrow}{}_{b}\,; \quad d\downarrow,1 = t\downarrow\,; \quad d\uparrow,M+1 = b\uparrow\,. \quad (46)$$

We search the solution in the form of a regular perturbation series

$$\Phi = \sum_{n=0}^{\infty} \varepsilon^{n}\Phi^{(n)}\,.$$

We introduce the algebraic vectors of dimension $2M$:
the complete solution

$$\Phi = \{\Phi^{\downarrow}_{1}, \Phi^{\uparrow}_{1}, \Phi^{\downarrow}_{2}, \Phi^{\uparrow}_{2}, \ldots, \Phi^{\downarrow}_{m}, \Phi^{\uparrow}_{m}, \ldots, \Phi^{\downarrow}_{M}, \Phi^{\uparrow}_{M}\}\,;$$

n-approximation to the sources

$$\mathbf{F}^{(n)} = \{\mathbf{F}^{\downarrow(n)}_{1}, \mathbf{F}^{\uparrow(n)}_{1}, \mathbf{F}^{\downarrow(n)}_{2}, \mathbf{F}^{\uparrow(n)}_{2}, \ldots, \mathbf{F}^{\downarrow(n)}_{m}, \mathbf{F}^{\uparrow(n)}_{m}, \ldots, \mathbf{F}^{\downarrow(n)}_{M}, \mathbf{F}^{\uparrow(n)}_{M}\}\,;$$

n-approximation to the solution

$$\Phi^{(n)} = \{\Phi^{\downarrow(n)}_{1}, \Phi^{\uparrow(n)}_{1}, \Phi^{\downarrow(n)}_{2}, \Phi^{\uparrow(n)}_{2}, \ldots, \Phi^{\downarrow(n)}_{m}, \Phi^{\uparrow(n)}_{m}, \ldots, \Phi^{\downarrow(n)}_{M}, \Phi^{\uparrow(n)}_{M}\}\,;$$

initial approximation to the sources

$$\mathbf{E} = \{\mathbf{E}^{\downarrow}_{1}, \mathbf{E}^{\uparrow}_{1}, \mathbf{E}^{\downarrow}_{2}, \mathbf{E}^{\uparrow}_{2}, \ldots, \mathbf{E}^{\downarrow}_{m}, \mathbf{E}^{\uparrow}_{m}, \ldots, \mathbf{E}^{\downarrow}_{M}, \mathbf{E}^{\uparrow}_{M}\}\,;$$

the "scenario" at the boundaries

$$\mathbf{Z} = \{\mathbf{Z}^{\downarrow}_{1}, \mathbf{Z}^{\uparrow}_{1}, \mathbf{Z}^{\downarrow}_{2}, \mathbf{Z}^{\uparrow}_{2}, \ldots, \mathbf{Z}^{\downarrow}_{m}, \mathbf{Z}^{\uparrow}_{m}, \ldots, \mathbf{Z}^{\downarrow}_{M}, \mathbf{Z}^{\uparrow}_{M}\}\,;$$

vector influence functions of the layers

$$\Theta = \{\Theta^{\downarrow}_{1}, \Theta^{\uparrow}_{1}, \Theta^{\downarrow}_{2}, \Theta^{\uparrow}_{2}, \ldots, \Theta^{\downarrow}_{m}, \Theta^{\uparrow}_{m}, \ldots, \Theta^{\downarrow}_{M}, \Theta^{\uparrow}_{M}\}\,;$$

influence functions tensors of the layers

$$\hat{\Pi} = \{\hat{\Pi}^{\downarrow}_{1}, \hat{\Pi}^{\uparrow}_{1}, \hat{\Pi}^{\downarrow}_{2}, \hat{\Pi}^{\uparrow}_{2}, \ldots, \hat{\Pi}^{\downarrow}_{m}, \hat{\Pi}^{\uparrow}_{m}, \ldots, \hat{\Pi}^{\downarrow}_{M}, \hat{\Pi}^{\uparrow}_{M}\}\,.$$

The reflection operators $\hat{R}^{\uparrow}_{m}$, $\hat{R}^{\downarrow}_{m}$ and the transmission operators $\hat{T}^{\uparrow}_{m}$, $\hat{T}^{\downarrow}_{m}$ are defined by the phase functions $\hat{\gamma}^{\uparrow}_{m}$ and $\hat{\gamma}^{\downarrow}_{m}$ of the media scattering in accordance with the following rules:

$$[\hat{R}^{\uparrow}_{m}(\hat{\gamma}^{\uparrow}_{m-1})\mathbf{f}^{\downarrow}_{m-1}](h_{m}, s^{-}_{m-1})$$

$$= \int_{\Omega^+} \hat{\gamma}^{\uparrow}_{m-1}(h_m, s^+_{m-1}, s^-_{m-1})\, \mathbf{f}^{\downarrow}_{m-1}(h_m, s^+_{m-1})\, ds^+_{m-1} \; ;$$

$$[\hat{T}^{\downarrow}_m(\hat{\gamma}^{\downarrow}_m)\mathbf{f}^{\downarrow}_{m-1}](h_m, s^+_m) = \int_{\Omega^+} \hat{\gamma}^{\downarrow}_m(h_m, s^+_{m-1}, s^+_m)\, \mathbf{f}^{\downarrow}_{m-1}(h_m, s^+_{m-1})\, ds^+_{m-1} \; ;$$

$$[\hat{T}^{\uparrow}_m(\hat{\gamma}^{\uparrow}_{m-1})\mathbf{f}^{\uparrow}_m](h_m, s^-_{m-1}) = \int_{\Omega^-} \hat{\gamma}^{\uparrow}_{m-1}(h_m, s^-_m, s^-_{m-1})\, \mathbf{f}^{\uparrow}_m(h_m, s^-_m)\, ds^-_m \; ;$$

$$[\hat{R}^{\downarrow}_m(\hat{\gamma}^{\downarrow}_m)\mathbf{f}^{\uparrow}_m](h_m, s^+_m) = \int_{\Omega^-} \hat{\gamma}^{\downarrow}_m(h_m, s^-_m, s^+_m)\, \mathbf{f}^{\uparrow}_m(h_m, s^-_m)\, ds^-_m \; .$$

We split the original problem (44)–(46) into 2M problems with their own boundary conditions. The initial approximation is the radiation from the sources without the radiation exchange between the layers ($\mathbf{F}^{\downarrow}_1 = \mathbf{F}^{\downarrow}_t$; $\mathbf{F}^{\uparrow}_M = \mathbf{F}^{\uparrow}_b$.) for $m = 1 \div M$:

$$\hat{K}\Phi^{\downarrow(0)}_m = \mathbf{F}^{\downarrow in}_m \,, \quad \Phi^{\downarrow(0)}_m\Big|_{d\downarrow,m} = \mathbf{F}^{\downarrow}_m \,, \quad \Phi^{\downarrow(0)}_m\Big|_{d\uparrow,m+1} = 0 \; ;$$

$$\hat{K}\Phi^{\uparrow(0)}_m = \mathbf{F}^{\uparrow in}_m \,, \quad \Phi^{\uparrow(0)}_m\Big|_{d\downarrow,m} = 0 \,, \quad \Phi^{\uparrow(0)}_m\Big|_{d\uparrow,m+1} = \mathbf{F}^{\uparrow}_m \, .$$

The n-approximations with $n \geqslant 1$ are described by the system of 2M equations for layers $m = 1 \div M$:

$$\hat{K}\Phi^{\downarrow(n)}_m = 0 \,, \quad \Phi^{\downarrow(n)}_m\Big|_{d\downarrow,m} = \mathbf{F}^{\downarrow(n-1)}_m \,, \quad \Phi^{\downarrow(n)}_m\Big|_{d\uparrow,m+1} = 0 \; ;$$

$$\hat{K}\Phi^{\uparrow(n)}_m = 0 \,, \quad \Phi^{\uparrow(n)}_m\Big|_{d\downarrow,m} = 0 \,, \quad \Phi^{\uparrow(n)}_m\Big|_{d\uparrow,m+1} = \mathbf{F}^{\uparrow(n-1)}_m$$

with the sources at the inner boundaries h_m with $m = 2 \div M$:

$$\mathbf{F}^{\downarrow(n)}_m = \hat{T}^{\downarrow}_m \Phi^{\downarrow(n)}_{m-1} + \hat{T}^{\downarrow}_m \Phi^{\uparrow(n)}_{m-1} + \hat{R}^{\downarrow}_m \Phi^{\downarrow(n)}_m + \hat{R}^{\downarrow}_m \Phi^{\uparrow(n)}_m \; ;$$

$$\mathbf{F}^{\uparrow(n)}_m = \hat{R}^{\uparrow}_{m+1} \Phi^{\downarrow(n)}_m + \hat{R}^{\uparrow}_{m+1} \Phi^{\uparrow(n)}_m + \hat{T}^{\uparrow}_{m+1} \Phi^{\downarrow(n)}_{m+1} + \hat{T}^{\uparrow}_{m+1} \Phi^{\uparrow(n)}_{m+1}$$

and at the outer boundaries h_m with $m = 1$ and $m = M$:

$$\mathbf{F}^{\downarrow(n)}_1 = 0 \; ; \quad \mathbf{F}^{\uparrow(n)}_M = \hat{R}^{\uparrow}{}_b \Phi^{\downarrow(n)}_M + \hat{R}^{\uparrow}{}_b \Phi^{\uparrow(n)}_M \, .$$

The solutions are found in the form of the vector linear functionals for each layer with $m = 1 \div M$:

$$\Phi^{\downarrow(n)}_m = \left(\hat{\Pi}^{\downarrow}_m, \mathbf{F}^{\downarrow(n-1)}_m\right) \; ; \quad \Phi^{\uparrow(n)}_m = \left(\hat{\Pi}^{\uparrow}_m, \mathbf{F}^{\uparrow(n-1)}_m\right) \, .$$

The kernels of the functionals are the influence functions tensors $\hat{\Pi}^{\downarrow}_m = \{\Theta^{\downarrow}_m\}$, $\hat{\Pi}^{\uparrow}_m = \{\Theta^{\uparrow}_m\}$ of layers and their elements are determined from the boundary-value problems $m = 1 \div M$

$$\hat{K}\Theta^{\downarrow}_m = 0, \quad \Theta^{\downarrow}_m\big|_{d\downarrow,m} = \mathbf{f}^{\downarrow}_{\delta,m}, \quad \Theta^{\downarrow}_m\big|_{d\uparrow,m+1} = 0 \; ;$$

$$\hat{K}\Theta^{\uparrow}_{m} = 0, \quad \Theta^{\uparrow}_{m}\Big|_{d\downarrow,m} = 0, \quad \Theta^{\uparrow}_{m}\Big|_{d\uparrow,m+1} = \mathbf{f}^{\uparrow}_{\delta,m} .$$

In the vector form, the n-approximation to the solution is

$$\Phi^{(n)} = \left(\hat{\Pi}, \mathbf{F}^{(n-1)}\right) .$$

The source in the $(n-1)$-approximation is

$$\mathbf{F}^{(n-1)} = \hat{P}\Phi^{(n-1)} .$$

The matrix $\hat{P}$ is banded with the characteristics of reflectivity and transmittivity of the boundaries:
$m = 1 , \quad i = 1$ is the first line: $\quad P_{1,j} = 0 , \quad j = 1 \div 2M$;
$m = 1 \div M-1 , \quad i = 2m , \quad i = 2 \div 2M$ are even lines:

$$P_{2m,j} = \begin{cases} 0 , & j = 1 \div 2(m-1) \\ \hat{R}^{\uparrow}_{m+1} , & j = 2(m-1)+1 , \quad 2(m-1)+2 \\ \hat{T}^{\uparrow}_{m+1} , & j = 2(m-1)+3 , \quad 2(m-1)+4 \\ 0 , & j = 2(m-1)+5 \div 2M ; \end{cases}$$

$m = 2 \div M , \quad i = 2m-1 , \quad i = 3 \div 2M-1$ are odd lines:

$$P_{2m-1,j} = \begin{cases} 0 , & j = 1 \div 2(m-1) \\ \hat{T}^{\downarrow}_{m} , & j = 2(m-1)+1 , \quad 2(m-1)+2 \\ \hat{R}^{\downarrow}_{m} , & j = 2(m-1)+3 , \quad 2(m-1)+4 \\ 0 , & j = 2(m-1)+5 \div 2M ; \end{cases}$$

$m = M , \quad i = 2M$ is the last line:

$$P_{2M,j} = \begin{cases} 0 , & j = 1 \div 2(M-1) \\ \hat{R}^{\uparrow}_{M} , & j = 2M-1 , \quad 2M . \end{cases}$$

The matrix-vector operation describes only the act of radiation interaction at the boundaries and takes into account the multiple scattering, absorption and polarization in the layers through their influence functions tensors:

$$\hat{G}\mathbf{F} = \hat{P}(\hat{\Pi}, \mathbf{F}) =$$

$$\begin{bmatrix} 0 \\ \hat{R}_2^{\uparrow}(\hat{\Pi}_1^{\downarrow}, \mathbf{F}_1^{\downarrow}) + \hat{R}_2^{\uparrow}(\hat{\Pi}_1^{\uparrow}, \mathbf{F}_1^{\uparrow}) + \hat{T}_2^{\uparrow}(\hat{\Pi}_2^{\downarrow}, \mathbf{F}_2^{\downarrow}) + \hat{T}_2^{\uparrow}(\hat{\Pi}_2^{\uparrow}, \mathbf{F}_2^{\uparrow}) \\ \vdots \\ \hat{T}_m^{\downarrow}(\hat{\Pi}_{m-1}^{\downarrow}, \mathbf{F}_{m-1}^{\downarrow}) + \hat{T}_m^{\downarrow}(\hat{\Pi}_{m-1}^{\uparrow}, \mathbf{F}_{m-1}^{\uparrow}) + \hat{R}_m^{\downarrow}(\hat{\Pi}_m^{\downarrow}, \mathbf{F}_m^{\downarrow}) + \hat{R}_m^{\downarrow}(\hat{\Pi}_m^{\uparrow}, \mathbf{F}_m^{\uparrow}) \\ \hat{R}_{m+1}^{\uparrow}(\hat{\Pi}_m^{\downarrow}, \mathbf{F}_m^{\downarrow}) + \hat{R}_{m+1}^{\uparrow}(\hat{\Pi}_m^{\uparrow}, \mathbf{F}_m^{\uparrow}) + \hat{T}_{m+1}^{\uparrow}(\hat{\Pi}_{m+1}^{\downarrow}, \mathbf{F}_{m+1}^{\downarrow}) + \hat{T}_{m+1}^{\uparrow}(\hat{\Pi}_{m+1}^{\uparrow}, \mathbf{F}_{m+1}^{\uparrow}) \\ \vdots \\ \hat{T}_M^{\downarrow}(\hat{\Pi}_{M-1}^{\downarrow}, \mathbf{F}_{M-1}^{\downarrow}) + \hat{T}_M^{\downarrow}(\hat{\Pi}_{M-1}^{\uparrow}, \mathbf{F}_{M-1}^{\uparrow}) + \hat{R}_M^{\downarrow}(\hat{\Pi}_M^{\downarrow}, \mathbf{F}_M^{\downarrow}) + \hat{R}_M^{\downarrow}(\hat{\Pi}_M^{\uparrow}, \mathbf{F}_M^{\uparrow}) \\ \hat{R}^{\uparrow}{}_b(\hat{\Pi}_M^{\downarrow}, \mathbf{F}_M^{\downarrow}) + \hat{R}^{\uparrow}{}_b(\hat{\Pi}_M^{\uparrow}, \mathbf{F}_M^{\uparrow}) \end{bmatrix} .$$

Two successive n-approximations are connected by the following recurrent relation,

$$\Phi^{(n)} = \left(\hat{\Pi}, \hat{P}\Phi^{(n-1)}\right) = \left(\hat{\Pi}, \hat{G}^{n-1}\mathbf{E}\right)$$

where $\mathbf{E}$ is an initial approximation.

Asymptotically exact solution is obtained in the form of the matrix-vector linear functional, i.e. the matrix transfer operator:

$$\Phi = \left(\hat{\Theta}, \mathbf{Z}\right) .$$

"Scenario" is given by the vector $\mathbf{Z}$ of the radiance distributions on the boundaries

$$\mathbf{Z} \equiv \hat{Z}\mathbf{E} \equiv \sum_{n=0}^{\infty} \hat{G}^n\mathbf{E} = \mathbf{E} + \sum_{n=1}^{\infty} \hat{P}\Phi^{(n)} = \mathbf{E} + \sum_{n=1}^{\infty} \mathbf{F}^{(n)}$$

and is the sum of Neumann series corresponding to multiplicities of the radiation transfer through the boundaries taking into account the multiple scattering impact using the influence functions tensors of each layer. The calculation is implemented in the following stages:

1. Calculation of the vector influence functions with parametric dependence for each layer is implemented using parallel computing. The functions and written into archives of solutions. The computational method selected in each layer depends on the radiation mode of this layer. There are two sources of parallelism: with respect to the layers ("domain decomposition" of the system) and with respect to parameters of the influence functions.

2. Calculation of the "scenario" vector on the boundaries of the layers through the matrix-vector procedure.

3. Calculation of the angular and spatial distributions of the radiation inside the system and on its boundaries using the matrix transfer operator.

Acknowledgements

These studies are supported by the Russian Fund for Basic Research (Projects №06-01-00666, 08-01-0024) and Russian Academy of Sciences (Project No. OMN-3(4)).

References

1. SUSHKEVICH T.A., *Mathematical models of radiation transfer*, Moscow, BINOM, Laboratory of Knowledge Publishers, 2005 (in Russian).

The Method of Regularization of Tikhonov Based on Augmented Systems

A. I. Zhdanov* and T. G. Parchaikina

Samara State Aerospace University, Moskovskoe sh. 34, Samara, 443086 Russia
*zhdanov@ssau.ru

Abstract. The method of the solving of ill-posed problems turned into the solving of arbitrary systems of linear algebraic equations is considered. This method is based on the reduction of an arbitrary (in general, inconsistent) linear system to an equivalent consistent augmented system with a symmetric matrix. The problem of choosing of regularization parameter is considered.

Keywords: regularization solutions, augmented systems, regularization parameter.

1 Formulation of the problem

Solving an approximate systems of linear algebraic equations (SLAEs) is fundemental problem of numerical analysis.

Let the exact SLAE

$$Au = f \tag{1}$$

be specified by the (a priori unknown) initial data $d = \{A, f\}$, where $A = (a_{ij}) \in \mathbb{R}^{m \times n}$, $f = (f_1, \dots, f_m)^T \in \mathbb{R}^m$ and $u = (u_1, \dots, u_n)^T \in \mathbb{R}^n$.

In the general case, the "solution" to exact SLAE (1) is understood as its normal pseudosolution

$$u_* = A^\dagger f, \tag{2}$$

where $A^\dagger$ is the *pseudoinverse* or the Moore-Penrose generalized of A.

Then, the *inconsistency measure* of the exact SLAE (1) is defined by

$$\mu = \inf_{u \in \mathbb{R}^n} \|Au - f\| = \|Au_* - f\| \geqslant 0.$$

Throughout the report, the vector norms in $\mathbb{R}^n$ and $\mathbb{R}^m$ are the Euclidean (quadratic) norms; i.e., $\|r\| = \|r\|_2$, where $r = f - Au$ is the residual.

The information on system (1) is given by the approximate data $\tilde{d} = \{\tilde{A}, \tilde{f}\}$ (i.e., by an individual approximate SLAE $\tilde{A}u = \tilde{f}$) such that

$$\|A - \tilde{A}\| \leqslant h, \quad \|f - \tilde{f}\| \leqslant \delta,$$

where the scalars $h \geqslant 0$ and $\delta \geqslant 0$ specify the errors in the assignment of the approximate data $\tilde{d}$ and $\|A\|$ is the spectral norm of A.

If rank $A < \min(n, m)$, then solving system (1) (as specified in (2)) on the basis of approximate data $\tilde{d}$ with $h > 0$ is an ill-posed problem in the sense of Hadamard, because the approximate normal pseudosolution

$$\tilde{u}_* = \tilde{A}^\dagger \tilde{f}$$

is unstable for infinitesimal perturbations in the initial data.

To determine stable solutions to system (1) from approximate data $\tilde{d}$, various regularization method are used. The regularization method of A.N. Tikhonov is among the most universal ones. It is well known that, in this method, the regularized solution $\tilde{u}_\alpha$ is determined as a (unique) solution to the *Euler equation*

$$(\tilde{A}^T \tilde{A} + \alpha I_n)u = \tilde{A}^T \tilde{f}, \quad \alpha > 0,$$

where I_n is the identity matrix of order n and α is the regularization parameter.

2 Method of augmented systems

In this report, a approach is proposed for systems with $\mu > 0$. This approach is based on the reduction of the original system to an equivalent consistent augmented system. Moreover, the approach makes it possible to obtain efficient numerical algorithms for solving the problem under discussion.

The normal system of equations is equivalent to the augmented system

$$\begin{aligned} \tilde{r} + \tilde{A}u &= \tilde{f}, \\ \tilde{A}^T \tilde{r} &= 0 \end{aligned} \Leftrightarrow \begin{pmatrix} I_m & \tilde{A} \\ \tilde{A}^T & 0 \end{pmatrix} \begin{pmatrix} \tilde{r} \\ u \end{pmatrix} = \begin{pmatrix} \tilde{f} \\ 0 \end{pmatrix} \Leftrightarrow \tilde{G}z = \tilde{b}, \tag{3}$$

where $z = (\tilde{r}^T, u^T)^T \in \mathbb{R}^{m+n}$.

Hence, the regularized solution $\tilde{z}_\alpha$ to system (3) is determined as a (*unique*) solution to the *Euler equation*

$$(\tilde{G}^2 + \alpha I_{m+n})z = \tilde{G}\tilde{b}. \tag{4}$$

Thus, using equivalent augmented systems, we can directly extend the basic results obtained for consistent systems ($\mu = 0$) to the class of inconsistent systems ($\mu > 0$).

Using the symmetry of $\tilde{G}$, we can reduce the condition number of the regularized (4). To this end, we apply the method of an imaginary shift of the spectrum (Faddeeva's method).

According of this method, we replace Eq. (4) by the equation

$$(\tilde{G} + i\sqrt{\alpha} I)z = \tilde{b}, \tag{5}$$

where $i = \sqrt{-1}$ is the imaginary unit. An immediate implication of (5) is that $x = \operatorname{Re} z$ is a solution to the equation (4).

To investigate the condition number of our problem, we examine the spectrum of the matrix

$$\begin{pmatrix} I_m & A \\ A^T & 0 \end{pmatrix}.$$

Let $\sigma_1 \geqslant \sigma_2 \geqslant \ldots \geqslant \sigma_\tau > \sigma_{\tau+1} = \ldots = \sigma_n = 0$ be the singular values of A; i.e. $\sigma_k = \sqrt{\lambda_k(A^T A)}$, where $\lambda_k(A^T A)$ are the eigenvalues of $A^T A$, $k = 1, 2, \ldots, n$; $\tau = \operatorname{rank}(A)$.

Proposition 1. *The eigenvalues of the augmented matrix* $G \in \mathbb{R}^{(m+n)\times(m+n)}$ *are given by the formula*

$$\lambda = \begin{cases} \frac{1}{2} \pm \sqrt{\frac{1}{4} + \sigma_k^2},\ k = 1, 2, \ldots, \tau; \\ \qquad 1,\ \textit{of multiplicity } m - \tau; \\ \qquad 0,\ \textit{of multiplicity } n - \tau. \end{cases}$$

Proof. If $Gz = \lambda z$, where $z = (v^T, u^T)^T \neq 0$, then $v + Au = \lambda v$ and $A^T v = \lambda u$. Eliminating v from these equations, we obtain

$$A^T A u = (\lambda^2 - \lambda) u.$$

Hence, if $u \neq 0$, then u is an eigenvector of $A^T A$ and $\lambda^2 - \lambda$ is the corresponding eigenvalue, i.e., $\lambda^2 - \lambda = \sigma_k^2$.

If $u = 0$, then

$$A^T v = 0, \quad v = \lambda v, \quad v \neq 0,$$

hence, 1 is eigenvalue of G. If $\operatorname{rank} A = \tau \leqslant n$, then the multiplicity of 1 is equal to $(m - \tau)$ and 0 is an eigenvalue of multiplicity $(n - \tau)$. □

Let us examine the condition of regularized system (4) (i.e., the condition of the Euler equation).

Let $\tilde{G}$ be a matrix of full rank, i.e., $\tilde{\tau} = \operatorname{rank}(\tilde{A}) = n$. Then, the minimal singular value of $\tilde{A}$ is positive: $\tilde{\sigma}_n = \tilde{\sigma}_{\min}(\tilde{A}) > 0$ and

$$\operatorname{cond}_2(\tilde{G}^2 + \alpha I_{m+n}) < \operatorname{cond}_2(\tilde{G}^2) = \left(\frac{1}{2} + \sqrt{\frac{1}{4} + 2\left(\frac{\tilde{\sigma}_1}{\tilde{\sigma}_n}\right)}\right)^2 \leqslant (\sqrt{2}\operatorname{cond}_2(\tilde{A}) + 1)^2,$$

where $\operatorname{cond}_2(\tilde{A}) = \frac{\tilde{\sigma}_1}{\tilde{\sigma}_n}$ is the spectral condition number of $\tilde{A}$.

Let $\tilde{G}$ is not a matrix of full rank, $\tilde{\tau} = \operatorname{rank}(\tilde{A}) < n$. Then

$$\lambda_{\max}(\tilde{G}^2 + \alpha I_{m+n}) = \left(\frac{1}{2} + \sqrt{\frac{1}{4} + \tilde{\sigma}_1^2}\right)^2 + \alpha \leqslant 1 + \tilde{\sigma}_1 + \tilde{\sigma}_1^2 + \alpha.$$

It is obvious, that if $\tilde{\tau} < n$, then $\lambda_{\min}(\tilde{G}) = 0$. Hence,

$$\lambda_{\min}(\tilde{G}^2 + \alpha I_{m+n}) = \alpha.$$

Thus,

$$\text{cond}_2(\tilde{G}^2 + \alpha I_{m+n}) \leqslant 1 + \frac{1 + \tilde{\sigma}_1 + \tilde{\sigma}_1^2}{\alpha}.$$

The singular values $\sigma_k(\tilde{G} + i\sqrt{\alpha} I_{m+n})$ of matrix $\tilde{G} + i\sqrt{\alpha} I_{m+n}$ are given by

$$\sigma_k(\tilde{G} + i\sqrt{\alpha} I_{m+n}) = \sqrt{\lambda_k(\tilde{G}^2 + \alpha I_{m+n})} = \sqrt{\tilde{\lambda}_k^2 + \alpha},$$

where $\tilde{\lambda}_k$ are the eigenvalues of $\tilde{G}$. Hence,

$$\text{cond}_2(\tilde{G}^2 + \alpha I_{m+n}) = \text{cond}_2^2(\tilde{G} + i\sqrt{\alpha} I_{m+n}).$$

Condition number of system (5) is estimate by inequality

$$\text{cond}_2(\tilde{G} + i\sqrt{\alpha} I_{m+n}) < \text{cond}_2(\tilde{G}) \leqslant \sqrt{2}\text{cond}_2(\tilde{A}) + 1,$$

when $\tilde{\tau} = n$ and by inequality

$$\text{cond}_2(\tilde{G} + i\sqrt{\alpha} I_{m+n}) \leqslant \sqrt{1 + \frac{\tilde{\sigma}_1^2 + \tilde{\sigma}_1 + 1}{\alpha}},$$

when $\tilde{\tau} < n$.

The problem of choosing of regularization parameter is payed much attention in the theory methods of approximate solution to ill-posed problems.

Regularization parameter α is choosing according to input error and the less the error is, the less regularization parameter is taken, so $\alpha = \alpha(h, \delta)$.

Several methods of choosing of regularization parameter were considered in this report. First is heuristic method. According to this method, when $\alpha = h^{2/3}$, found regularized solution approximates normal pseudosolution to system to accuracy of $O(h^{2/3})$.

Second is method of choosing of regularization parameter $\alpha = h$ by V. A. Morozov, S. F. Gilyazov [3]. This method provides the error $O(h+\delta)$ for regularized solutions.

The third is the following: for some α find solution z_α from (4) and calculate closure $r_\alpha = \tilde{G} z_\alpha - \tilde{b}$. Then compare r_α with approximate data δ and $h\|z_\alpha\|$. Make accounts for different α. Optimal is value when $\|r_\alpha\| \approx \delta + h\|z_\alpha\|$. This approach provides the error $O(h + \delta)$ for regularized solutions.

There are some special problems, which can not be find solutions to according to this recommendations. They are the problems of $\sigma_{\min} = o(0,001)$, where $\sigma_{\min}$ is minimal not null singular value of matrix A. Minimization of functions

$$\varphi(\alpha) = \|\tilde{z}_{\alpha_{k+1}} - \tilde{z}_{\alpha_k}\|, \quad \psi(\alpha) = \|\tilde{G}\tilde{z}_{\alpha_k} - \tilde{b}\|$$

was used to choose regularization parameter for these problems. Consistency $\alpha_k = \alpha_0 q^k$, $q > 0$, $\alpha_0 > 0$ was used and $k = 1, \ldots, K$, when $\|\tilde{z}_{\alpha_{k+1}} - \tilde{z}_{\alpha_k}\| = O(h)$ and $\|\tilde{G}\tilde{z}_{\alpha_k} - \tilde{b}\| = O(h)$. Such choice of regularization parameter provides error $O(h)$ for regularized solutions.

3 Numerical examples

Now we turn to the results of investigation of this method by solving some test problems. We show that accuracy of approximate solution complies with the theoretical results.

1. Let

$$A = \begin{pmatrix} 2 & -1 & 0 \\ -1 & 1 & 1 \\ 0 & 1 & 2 \end{pmatrix}, \quad f = \begin{pmatrix} 18 \\ 27 \\ -18 \end{pmatrix}.$$

The normal pseudosolution to (exact) system is

$$u_* = A^{\dagger} f = \begin{pmatrix} 1 \\ -1 \\ -1 \end{pmatrix}.$$

Let perturbation a_{13} such as $|a_{13} - \tilde{a}_{13}| = 0,0001$. Then, we obtain the approximate system of equations

$$\begin{pmatrix} 2 & -1 & 0 \\ -1 & 1 & 1 \\ 0,0001 & 1 & 2 \end{pmatrix} \begin{pmatrix} u_1 \\ u_2 \\ u_3 \end{pmatrix} = \begin{pmatrix} 18 \\ 27 \\ -18 \end{pmatrix}. \tag{6}$$

The system (6) is nonsingular, because $\det \tilde{A} = -0.0001 \neq 0$. We find its solution by any classical method (in machine arithmetic):

$$\tilde{u}_* \approx \begin{pmatrix} -900000 \\ -1800018 \\ 900045 \end{pmatrix}.$$

It is obvious that the errors in the approximate data of system (6) are determined by the quantities:

$$\|A - \tilde{A}\| \leqslant h = 0,0001.$$

The regularized solutions to system (6) were determined from SLAE (4). According to the recommendations, we choose the value $\alpha = h = 0.0001$ for the regularization parameter. Solving system (6) by any classical method, we obtain the regularized solution:

$$\tilde{u}_\alpha \approx \begin{pmatrix} 0,9995 \\ -1,0005 \\ -0,9999 \end{pmatrix}.$$

Comparing the regularized solution with the exact normal pseudosolution u_*, we have

$$\|\tilde{u}_\alpha - u_*\| < 8 \cdot 10^{-5},$$

which complies with the theoretical results obtained in this paper.

Now we turn to the results of investigation of this method for several h:

Table 1.

h	10^{-3}	10^{-5}	10^{-6}	10^{-7}	10^{-8}	10^{-9}
$\|\tilde{u}_\alpha - u_*\|$	$6 \cdot 10^{-2}$	$6 \cdot 10^{-4}$	$6 \cdot 10^{-5}$	$6 \cdot 10^{-6}$	$6 \cdot 10^{-7}$	$6 \cdot 10^{-8}$
$\|\tilde{u}_* - u_*\|$	$3 \cdot 10^{4}$	$1,7 \cdot 10^{6}$	$1,7 \cdot 10^{7}$	$1,7 \cdot 10^{8}$	$1,7 \cdot 10^{9}$	$1,5 \cdot 10^{8}$

2. Let

$$A = \begin{pmatrix} 3 & -7 & 1 \\ 3 & -7,01 & 1 \\ 6 & -14,02 & 2 \end{pmatrix}, \quad f = \begin{pmatrix} 1 \\ 1 \\ 2 \end{pmatrix}.$$

The normal pseudosolution to (exact) system is

$$u_* = A^\dagger f = \begin{pmatrix} 0,3 \\ 0 \\ 0,1 \end{pmatrix}.$$

Let perturbation of matrix A such as

$$\|A - \tilde{A}\| = h = 0,00001.$$

Then, we obtain the approximate system of equations

$$\begin{pmatrix} 2,9999943 & -6,9999999 & 1,00000249 \\ 3,0000038 & -7,0099992 & 1,0000051 \\ 5,9999949 & -14,0199952 & 1,9999957 \end{pmatrix} \begin{pmatrix} u_1 \\ u_2 \\ u_3 \end{pmatrix} = \begin{pmatrix} 0,999994 \\ 1,000008 \\ 2,0000024 \end{pmatrix}. \tag{7}$$

The system (7) is nonsingular, because $\det \tilde{A} = 0,0000003 \neq 0$. We find its solution by any classical method:

$$\tilde{u}_* \approx \begin{pmatrix} 0,5716 \\ 0,0049 \\ -0,7124 \end{pmatrix}.$$

The regularized solutions to system (7) were determined from SLAE (4). According to the recommendations, we choose the value α by minimization of functions $\|\tilde{z}_{\alpha_{k+1}} - \tilde{z}_{\alpha_k}\|$ and $\|\tilde{G}\tilde{z}_{\alpha_k} - \tilde{b}\|$, because $\sigma_{\min} = 0,0038$. Consistency

Table 2.

α	$5 \cdot 10^{-6}$	$8 \cdot 10^{-9}$	$2 \cdot 10^{-12}$	$5 \cdot 10^{-13}$	$4 \cdot 10^{-15}$
$\|\tilde{u}_{\alpha_{k+1}} - \tilde{u}_{\alpha_k}\|$	$2 \cdot 10^{-5}$	$5 \cdot 10^{-3}$	$1 \cdot 10^{-2}$	$2 \cdot 10^{-3}$	$3 \cdot 10^{-1}$
$\|\tilde{G}\tilde{z}_{\alpha_k} - \tilde{b}\|$	$4 \cdot 10^{-6}$	$4 \cdot 10^{-6}$	$2 \cdot 10^{-7}$	$1 \cdot 10^{-8}$	$7 \cdot 10^{-9}$
$\|\tilde{u}_\alpha - u_*\|$	$28 \cdot 10^{-2}$	$28 \cdot 10^{-2}$	$4 \cdot 10^{-3}$	$1 \cdot 10^{-3}$	$2 \cdot 10^{-3}$

$\alpha_k = 2 \cdot 0,1^k$ was used and $k = 3, 4, \ldots, 16$. Now we turn to the results of investigation of this method

We choose the value $\alpha = 5 \cdot 10^{-13}$ for the regularization parameter. Solving system (7) by any classical method, we obtain the regularized solution:

$$\tilde{u}_\alpha \approx \begin{pmatrix} 0,29856 \\ -0,00075 \\ 0,09905 \end{pmatrix}$$

and

$$\|\tilde{u}_\alpha - u_*\| < 2 \cdot 10^{-3},$$

which complies with the theoretical results obtained in this paper.

The method of regularization of Tikhonov based on augmented systems makes it possible to reduce an arbitrary inconsistent SLAE to an equivalent consistent augmented system. This approach makes it possible to solve arbitrary ill-posed SLAEs with approximate data, including systems not having a full rank and inconsistent systems. The coefficient matrix of the augmented SLAE is symmetric. This property makes it possible to use the method of an imaginary shift of the spectrum (Faddeevas method) for calculating regularized solutions. In this way, a significant reduction of the condition number of the problem is possible.

References

1. ZHDANOV A.I., *Regularization of unstable finite-dimensional linear problems based on augmented systems*, Comp. Math. and Math. Physics, 45 (2005), pp. 1845–1853.
2. BJÖRK Å *Numerical stability of methods for solving augmented system*, Contemporary Math., 204 (1997), pp. 51–60.
3. MOROZOV V. A., *Algorithmic foundations of methods for solving ill-posed problems*, Vychisl. Metody Program., 45 (2003), pp. 130–141.

Author Index